Alpha and Gamma Motor Systems

Alpha and Gamma Motor Systems

Edited by

Anthony Taylor
Sherrington School of Physiology
UMDS, St. Thomas' Hospital
London, England

Margaret H. Gladden
Institute of Biomedical and Life Sciences
University of Glasgow
Glasgow, Scotland

and

Rade Durbaba
Sherrington School of Physiology
UMDS, St. Thomas' Hospital
London, England

Springer Science+Business Media, LLC

Library of Congress Cataloging-in-Publication Data

On file

Proceedings of a symposium on Alpha and Gamma Motor Systems, held July 11–14, 1994, in London, England

ISBN 978-1-4613-5793-3 ISBN 978-1-4615-1935-5 (eBook)
DOI 10.1007/978-1-4615-1935-5

Originally published by Plenum Press,New York in 1995
Softcover reprint of the hardcover 1st edition 1995

10 9 8 7 6 5 4 3 2 1

FOREWORD

The Sherrington School of Physiology at St. Thomas' Hospital provided a natural venue for four days of enthusiastic debate on Sherrington's Final Common Path - the alpha motoneurone and related matters; Sherrington himself held a lecturership at St. Thomas' just over a century ago. The occasion was a happy one. Most participants already knew each other, the topics were familiar, the discussion was vigorous and critical but without personal rancor. The program had set out to encourage debate with 'critiques' both of the oral papers and the posters; their inclusion in the present volume helps to show where we are standing on rock rather than sand. In spite of a venerable history some surprisingly simple matters remain controversial, such as the information content of the signals from the Golgi tendon organs.

To those working on alpha and gamma motoneurones and their control this volume provides an essential up-dating. Classical problems continue to be attacked on a broad front; advance is steady and continuous, a swelling tide rather than a sudden view of the summit. But in some sectors the difficulties are so great that to the outsider little might seem to change, the same battles continuing; but even here, the terms of debate change and concensus develops. Other branches of biology claim breakthrough upon breakthrough from the routine application of the new technologies, so we have to be the first to ask whether our classical approach to science can still be justified. The answer is a confident yes, provided we point to our strengths, embrace new technologies, develop new theories and emphasize our contribution to the wider scene.

There are plenty of encouraging examples here. The variety of approach is sobering; moreover, so much of the equipment is personalized rather than simply 'bought off the shelf'. Histology is here, albeit with the newest machinery and on the finest scale. Computer modelling is in, from the pacemakers of the muscle spindle to the need for 'fuzzy logic' in motor control. With the new dyes, calcium imaging allows the neural activity underlying 'walking' to be followed in the isolated chick spinal cord and selected groups of neurones to be destroyed. In spite of their cost, experiments on animals continue to be essential, and continue to yield new insights. Observations on human subjects offer many advantages. Neurological patients are studied for their deficits, whether in their potassium channels leading to death of the motoneurones or for their more integrative actions. There is something here for everyone, and much for all to learn from those working in related areas. The rich colours in the tapestry confirm the health of our subject.

The crucial importance of our work is that we are studying the interface between the brain and the outside world; the brain cannot act or communicate without muscle and its alpha motoneurones. The mere existence of the gamma motoneurones shows the importance of sensory feedback, and emphasizes that we must seek principles of neural regulation and control as well as studying the motoneurones for themseleves. One of our strengths is that we are studying a low level of motor synthesis, from the interaction of synaptic currents to the operation of elementary neural circuits. A reasonably full understanding of this level of organization should be within our grasp with current technology.

Such progress on the foundations is probably an essential prelude to achieving understanding of the workings of higher motor structures, such as the basal ganglia; the current impressive knowledge of their chemistry and microcircuitry shows little sign of answering wider questions of function. The spinal cord reads the code written by the higher motor centers, and sends its own coded messages to them. The codes are probably dictated by the workings of the spinal cord, since it came first in evolution; the higher centers had to be 'designed' to fit in. The coding can probably be most readily cracked by working on spinal mechanisms.

But too many seem to believe that the foundations are completed and can be forgotten, that all attention can be concentrated on the elegant structures that now supervize and regulate everything done by the spinal cord, even that 'slice' preparations will provide all that is worth knowing. This volume shows what nonsense this is for the understanding of motor control; so much needs to be done on the lower levels in the whole animal or man, with attack required on all sides, deploying all available techniques. This volume provides a snap-shot of the current state of affairs, and describes a series of mini-symposia with their papers, posters and critiques.

Peter B.C. Matthews Oxford

PREFACE

This book is composed of papers presented at a symposium of the same title organized by the editors at the Sherrington School of Physiology, U.M.D.S at the St. Thomas' Hospital Campus from July 11–14, 1994. The plan of the meeting owed much to ideas tried successfully at the meetings on similar themes at St. Thomas' in 1980, Glasgow, 1984 and at Rheinfelden in 1988. The draft manuscripts of groups of oral presentations were made available to chosen experts before the meeting so that a prepared critique could be presented at the end of each session. The critics subsequently formalized these into the manuscripts printed here. Because of the great demand for opportunities to present new research findings, the invited talks were interspersed with poster sessions and the presenters given the opportunity to submit short papers for the book. The value of the posters was also enhanced by persuading chosen participants to review them orally after each session. Unfortunately, space constraints have not permitted publication of these reviews. The final form of the book otherwise reflects that of the meeting. Groups of full papers are followed by critiques and then by the short papers based on the most closely related posters.

The editors are most grateful to the many contributors who ensured the success of the meeting and of this book by attending, mostly at their own expense, to present their latest work. We were anxious to try to reward them by striving for a rapid publication, so that the work presented is still fresh. If our editorial efforts have been a little more energetic than is often the case for such symposium books, they have been taken in very good part by the authors. We are grateful to all concerned for the good spirit in which all the interchanges have been conducted. Our thanks are also due to Miss Joanna Lawrence of Plenum Press for her guidance in the production of the camera-ready text.

Medical research demands a very particular dedication from individual workers and nowhere more so than in the neuroscience of motor control. Part of the effort which they make is to satisfy their scientific curiosity, but there is always in the background the desire to do something for the relief of suffering. There is still much to be done. To give some idea of the scale of the problem, in Glasgow alone, with a population of 900,000, there are reported to be 18,000 people with long-term motor disabilities. Although more investment in health care and social services could undoubtedly bring some relief, for the underlying problems there are no quick and easy solutions. Only patient research which leads to understanding of normal function and how it is changed in disease can lead to rational and economical therapies. We must have the confidence and strength of purpose to keep reminding those who control resources that this is the way forward. It is therefore with some satisfaction that we can offer this volume as the outcome of so much dedicated research by workers in many countries as a contribution to progress in facing the realities of human neurological disease.

Anthony Taylor	London
Margaret H. Gladden	Glasgow
Rade Durbaba	London

ACKNOWLEDGMENTS

The organizers gratefully acknowledge the following companies and organizations for their financial support which made the symposium possible:

The Wellcome Trust

The Guarantors of Brain

International Spinal Research Trust

The Physiological Society

Plenum Press

Magstim Ltd

MacLab AD Instruments

Cambridge Electronic Design Ltd

The cover design is based on a photomicrograph kindly supplied by Robert Fyffe.

CONTENTS

PART 1
MOTONEURONE INPUTS

PART 2
MOTOR UNIT RECRUITMENT

PART 4
INNERVATION PATTERNS

PART 5
SENSORY RECEPTOR PROPERTIES

PART 6
ANALYSIS AND MODELLING

PART 7
CENTRAL CONTROL

PART 8
PHARMACOLOGY OF CENTRAL CONTROL

PART 9
CLINICAL IMPLICATIONS

PART 10
NATURAL MOTOR PATTERNS - 1

PART 11
NATURAL MOTOR PATTERNS - 2

PART 1

MOTONEURONE INPUTS

SYNAPTIC DIFFERENTIATION ON TYPE IDENTIFIED MOTONEURONES

Lorne M. Mendell

Department of Neurobiology and Behaviour
SUNY at Stony Brook
Stony Brook, NY 11794, USA.

The connection between Ia afferents and motoneurones is characterised by considerable divergence from individual afferents to the homonymous and heteronymous motoneurone pools and a consequent convergence from afferents onto individual motoneurones (reviewed in Henneman and Mendell, 1981). The profuse connectivity has made this a very convenient model system to study synaptic function in the central nervous system. The ability to activate the afferents either individually (single fibre EPSPs) or in combination with all the other afferents (composite EPSPs), along with the relative simplicity implied by a monosynaptic connection, has facilitated these studies.

A recurring theme in the analysis of this system is that Ia-evoked EPSPs differ systematically across the motoneurone pool. In early work using low frequency stimulation, it was found that single fibre EPSPs in low rheobase, long AHP, high resistance motoneurones are relatively large in amplitude on the average whereas EPSPs in low resistance, large rheobase, short AHP motoneurones are relatively small (Fleshman, Munson & Sypert, 1981). These findings were interpreted as an inverse correlation between EPSP amplitude in a motoneurone and its position in the recruitment rank order, i.e., easily recruited motoneurones display large EPSPs and vice versa. This functional correlation was taken as evidence of the importance of EPSP amplitude differences in determining motoneurone function. These differences in EPSP amplitude were attributed to variation in motoneurone size; synaptic current was considered to be uniform across the motoneurones in the pool (reviewed in Mendell, Collins & Koerber, 1990).

When Ia afferent synapses on motoneurones in the cat were activated with more physiologically realistic input patterns, a more complex pattern of synaptic differentiation on motoneurones of the pool was observed (Collins, Honig & Mendell, 1984). Instead of single shocks delivered at relatively low frequencies, the afferents were activated with short, high frequency bursts more characteristic of spindle discharge during natural movements such as stepping (Loeb & Duysens, 1979). Large EPSPs in motoneurones with properties indicative of being easily recruited tended to exhibit a decrease in amplitude during the burst whereas small EPSPs in motoneurones high on the recruitment rank order scale tended to exhibit an increase in amplitude.

These conclusions were established initially using single fibre stimulation (Collins et al., 1984). In subsequent experiments where EPSPs were recorded either from the same motoneurone in response to stimulation of different single afferent fibres or in more than one motoneurone in response to stimulation of the same single afferent fibre, it was determined that the motoneurone rather than the afferent fibre is the element whose identity is correlated with the type of response during high frequency stimulation (Koerber & Mendell, 1991). This was further confirmed by noting the similarity in the behaviour of composite EPSPs during maximal group Ia high frequency stimulation to the behaviour in response to single fibre stimulation. These findings suggested that the synapses made by all Ia afferents on a given motoneurone behave in a similar manner during high frequency stimulation (Koerber & Mendell, 1991). Further insight into this synaptic differentiation can be obtained from studies of the response to high frequency stimulation in the neonatal rat. In the immediate postnatal period these connections exhibit substantial depression during high frequency stimulation (Lev Tov & Pinco, 1992), and this gradually changes towards values seen in the adult as the animal matures, i.e., less depression (Seebach & Mendell, 1994). However, even at 15 days after birth the synapses do not function at their adult capability. EPSP amplitude rundown during high frequency stimulation is not the result of failure for the afferent fibres to conduct impulses at these frequencies (Lev Tov & Pinco, 1992). It is interesting that these changes in synaptic properties occur at a time when motor units are changing in their characteristics. When the rat is born, all muscle units contract slowly and speed up in the initial postnatal period, with the units destined to be Type S undergoing a slowing of twitch contraction later during the initial 5 week maturation period (Close, 1964). Thus during the immediate 15 day postnatal period where EPSPs are becoming less prone to synaptic depression, the motor units are speeding up. This is what one would expect qualitatively according to the differences between Type F and Type S in the adult (as measured in the cat). It will be interesting to see whether synaptic changes occur on Type S motoneurones as they undergo their final maturational phase of slowdown in contraction time.

The mechanisms accounting for these changes in synaptic properties as the animal matures are not known. They are consistent with a retrograde specification, i.e., from muscle unit to motoneurone to synapse (see Mendell, Collins & Munson, 1994 and companion paper, this volume by Munson). At present, however, we are unable to decide what might be specified by such a retrograde mechanism, whether it might be in local networks underlying primary afferent depolarisation as recently shown by Rudomin and colleagues (Eguibar, Quevedo, Jimenez & Rudomin, 1993) or in the intrinsic properties of the synaptic boutons (size, active zone size, etc.).

The fact that the behaviour during high frequency stimulation differs according to properties of the motoneurone could mean either that receptor properties on individual motoneurones differ systematically and/or that the presynaptic terminals are different. On the postsynaptic side the rate of desensitisation of excitatory amino acid receptors might be correlated with EPSP amplitude such that small EPSPs would exhibit less desensitisation. The residual calcium accumulating in the Ia terminals during high frequency stimulation would result in more synaptic facilitation in these motoneurones during high frequency stimulation than in motoneurones whose receptors exhibited more desensitisation. At present, however, the evidence for desensitisation of receptors at these stimulus frequencies is not compelling (see Mendell et al., 1994). On the presynaptic side a balance between depression due to transmitter depletion and facilitation due to the action of residual Ca^{2+} would predict that large and small EPSPs differ in the amount of transmitter released (Collins et al., 1984). For large EPSPs, effects which entailed depletion (large release of transmitter) would overwhelm those due to residual Ca^{2+} and lead to synaptic depression. For small EPSPs the facilitation due to residual Ca^{2+} would more than counterbalance any depression since there would be relatively small transmitter depletion. Manipulation of transmitter release has been achieved in some cases by changing extracellular Ca^{2+}/Mg^{2+} levels (Lev Tov & Pinco, 1992; Seebach & Mendell, 1994) or by application of $GABA_B$ agonists and antagonists to reduce or increase Ca^{2+} entry into the terminals (Peshori, Collins & Mendell, 1993). In all cases, reduction of EPSP amplitude measured at low rates of Ia fibre stimulation was accompanied by a tendency for the EPSPs to be depressed less (or to change to facilitation) during high frequency stimulation. Increases in EPSP amplitude drove the changes in amplitude occurring during high frequency stimulation towards greater levels of depression.

These findings indicating that manipulation of presynaptic release can influence behaviour during high frequency stimulation are consistent with the idea that there may normally be intrinsic differences in transmitter release among terminals on motoneurones generating small and large EPSPs. This would supplement the widely held view that differences in EPSP amplitude can be accounted for largely by differences in the passive electrical properties of the motoneurone, i.e., input resistance and input capacitance (reviewed in Mendell et al., 1990). Additional presynaptic factors may also play a role, perhaps accounting to some degree for the fact that individual EPSPs vary in amplitude by a much greater ratio (100/1 or more) than differences in input resistance among motoneurones (10/1). In addition, the number (Burke, Fleshman & Segev, 1988) or efficacy of the individual boutons (Pierce & Mendell, 1993) may also account for these differences in EPSP amplitude (see below).

In a recent set of experiments (Mendell, Taylor, Johnson & Munson, 1995) these issues were re-examined in the cat using the electrical typing scheme introduced by Zengel, Reid, Sypert & Munson (1985) to characterise motoneurone function. The same stimulation sequence was delivered to the Ia fibres (bursts of 32 shocks at 167 Hz every 2 s, with the responses being averaged in register: see Figure 1). It was found that Type S motoneurones, defined by afterhyperpolarisation (AHP) duration > 30 ms (1/2 fall time), exhibited composite EPSPs that consistently diminished (modulation < 0; Figure 2) during high

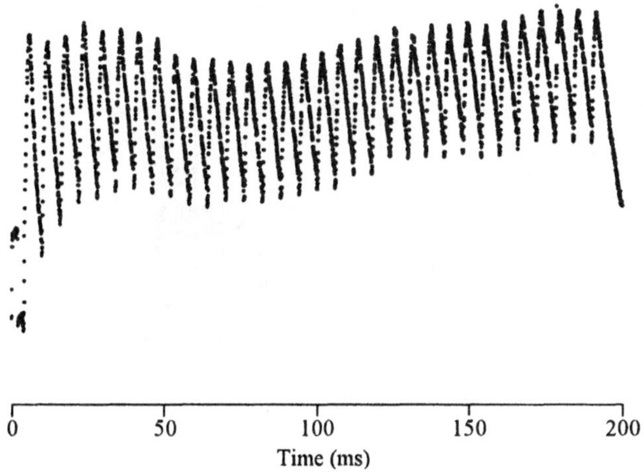

Figure 1. Response of LGS motoneurone (AHP = 56 ms) to stimulation of MG nerve with 16 bursts of 32 shocks at 167 Hz presented at interburst interval of 2s. Responses averaged in register (1,2,...,32) over the 16 burst presentations. Modulation was computed as $100*[(EPSP_{30+31}/2*EPSP_1) - 1]\%$. In this case Modulation was negative since EPSP amplitude decreased during the burst; in other cases it was positive (see text). Calibration pulse prior to stimulation = 1 mV. Data from Mendell et al. (1995).

frequency stimulation. Type F motoneurones (AHP < 30 ms) were split between those exhibiting negative modulation and those exhibiting facilitation of EPSP amplitude (Figure 2), with a weak tendency for EPSPs in Type FF motoneurones (rheobase/ input resistance > 18) to show more positive modulation than those in Type FR motoneurones (rheobase/ input resistance < 18). Thus the synapses on Type F and Type S motoneurones appear to differ substantially in their responses to high frequency stimulation with somewhat smaller differences between those on Type FF and Type FR motoneurones.

In these same experiments EPSPs of a given amplitude tended to exhibit more depression if generated in a Type S motoneurone than in a Type F motoneurone (Figure 2). This raises some interesting questions concerning the differentiation of synapses on Type F and Type S motoneurones. Over the entire range of amplitude values Type F motoneurones had smaller values of input resistance than Type S motoneurones (Figure 3), as anticipated from previous results (Zengel et al., 1985). Therefore, to achieve a given EPSP amplitude, one would anticipate more quanta being released from synapses on Type F motoneurones than from those on Type S motoneurones assuming equal numbers of converging afferents. This means either that on the average there are more release sites for transmitter per afferent on Type F motoneurones or that the probability of release on Type F motoneurones is higher. At present it is difficult to choose definitively between these alternatives. The anatomy at the light microscopic level indicates similar numbers of boutons on the different motoneurones of the pool (Burke et al., 1988), but the more detailed evidence from ultrastructural observations demonstrate that these boutons are extremely diverse in their physical size and in the extent of their synaptic machinery such as vesicle number, active zone dimensions, etc. (Pierce & Mendell, 1993). It is interesting that the ultrastructural evidence indicates a clear tendency for the largest diameter dendrites (putatively from Type F motoneurones) to have synapses with the smallest active zones on the average. Active zone size is generally considered to correlate with probability of release (reviewed in Pierce & Mendell, 1993) which would suggest that the Type F motoneurones have synapses with low, not high, probability of release. Thus in choosing from the alternatives listed above, it seems more likely that there are more release sites from Ia fibres on the large dendrites of Type F motoneurones, and that these have a lower probability of release making them less

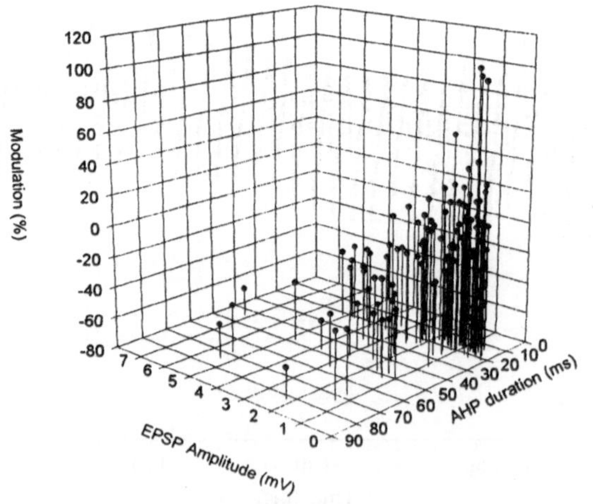

Figure 2. 3-D plot of modulation of EPSP amplitude as a function of motoneurone AHP and EPSP amplitude measured at 0.5 Hz. Heteronymous composite EPSPs: LGS to MG and MG to LGS. Note that large EPSPs in Type S motoneurones (AHP > 30 ms) tend to have negative modulation whereas those in Type F motoneurones tend to have positive modulation. Note also that large EPSPs tend to exhibit more negative values of amplitude modulation. Data from Mendell et al. (in press).

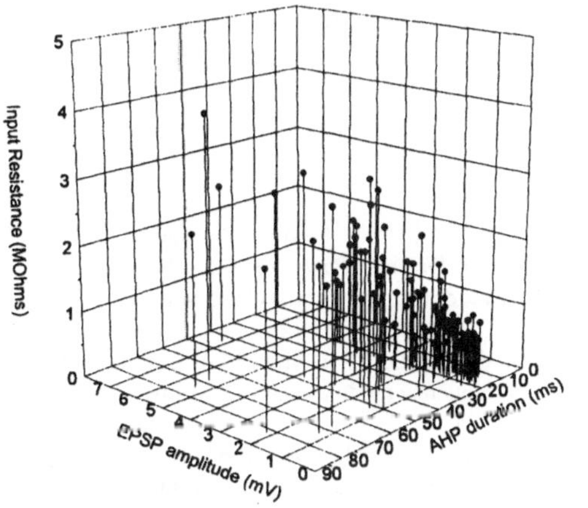

Figure 3. 3-D plot demonstrates that Type S (AHP > 30 ms) and Type F (AHP < 30 ms) motoneurones with equal values of EPSP amplitude (e.g. 2 mV) recorded at 0.5 Hz stimulation have different values of input resistance (ResTypeS > ResTypeF). Further discussion in text. Data from Mendell et al. (in press).

prone to synaptic depression.

The functional importance of these synaptic properties may relate to the fact that individual group Ia fibres diverge to contact all the motoneurones in the homonymous pool as well as the majority of heteronymous ones (Henneman & Mendell, 1981). The necessity that some motoneurones be systematically recruited before others by stretch (i.e., the Size Principle of Henneman, Somjen & Carpenter, 1965) constrains these motoneurones to develop large EPSPs since the evidence is that motoneurones do not differ systematically in the voltage threshold for the initiation of action potentials (Pinter, Curtis & Hosko, 1983). However, since spindles typically discharge in high frequency bursts, reaching discharge rates of as much as 500 Hz (Prochazka, Hulliger, Trend, Llewellyn & Durmuller, 1989), typically there will be considerable amounts of temporal summation after recruitment which would exaggerate the differences in level of depolarisation between motoneurones which develop small or large EPSPs in response to each of their inputs. It is in this context that the function of these differences in synaptic behaviour can be viewed: motoneurones generating the largest EPSPs during low frequency stimulation will be recruited first, but their membrane will not be depolarised to the extent anticipated on the basis of linear summation during the high frequency component of the afferent discharge. In contrast motoneurones generating small EPSPs will be recruited last but temporal summation will tend to be exaggerated. This may maintain these neurones so as to be appropriately susceptible to inhibitory inputs during rhythmic movements such as stepping or paw shaking.

In conclusion, it appears that the differentiation of motoneurones described in detail over the past 25 years are matched by differentiation of synapses upon them. This has developmental, biophysical and functional aspects that remain to be fully elucidated. Furthermore, as seen in the companion paper (Munson, this volume), these synapses are subject to alterations in response to manipulation of the peripheral target of the afferents and motoneurones, supporting the possibility for retrograde regulation of their properties.

ACKNOWLEDGEMENTS

Some of this work was performed in collaboration with Dr. John Munson at the University of Florida. The author's research was supported principally by NIH: RO1- NS

16696 (Javits Neuroscience Award). Additional support was provided by PO1 NS 14899 and RO1 NS 32264.

REFERENCES

BURKE, R.E., FLESHMAN, J.W., & SEGEV, I. (1988) Factors that control the efficacy of group Ia synapses in alpha-motoneurons. *J. Physiol. (Paris)* **83**, 133-140.

CLOSE, R. (1964) Dynamic properties of fast and slow skeletal muscles of the rat during development. *J. Physiol.* **173**, 74-95.

COLLINS, W.F., III, DAVIS, B.M. & MENDELL, L.M. (1986) Amplitude modulation of EPSPs in motoneurons in response to a frequency modulated trains in single Ia afferent fibers. *J. Neurosci.* **6**, 1463-1468.

COLLINS, W.F., III, HONIG, M.G. & MENDELL, L.M. (1984) Heterogeneity of group Ia synapses on homonymous motoneurons as revealed by high frequency stimulation of Ia afferent fibers. *J. Neurophysiol.* **52**, 980-993.

EGUIBAR, J.R., QUEVEDO, J.N., JIMENEZ, I. & RUDOMIN, P. (1993) Differential control exerted by the motor cortex on the synaptic effectiveness of two intraspinal branches of the same afferent fiber. *Neurosci. Abstr.* **19**, 588.6.

FLESHMAN, J.W., MUNSON, J.B. & SYPERT, G.W. (1981) Homonymous projection of individual group Ia fibers to physiologically characterized medial gastrocnemius motoneurons in the cat. *J. Neurophysiol.* **46**, 1339-1348.

HENNEMAN, E. & MENDELL, L.M. (1981) Functional organization of the motoneuron pool and its inputs. In *Handbook of Physiology. The Nervous System. Motor Control.* eds. BROOKHART, J.M. & MOUNTCASTLE, V.B. pp 423-507. American Physiological Society, Bethesda.

HENNEMAN, E., SOMJEN, G.G. & CARPENTER, D.O. (1965) Functional significance of cell size in spinal motoneurons. *J. Neurophysiol.* **28**, 560-580.

KOERBER, H.R. & MENDELL, L.M. (1991a) Modulation of functional synaptic transmission during high frequency stimulation: role of postsynaptic target. *J. Neurophysiol.* **65**, 590-597.

KOERBER, H.R. & MENDELL, L.M. (1991b) Modulation of synaptic transmission at Ia afferent connections on motoneurons during high frequency afferent stimulation: dependence on motor task. *J. Neurophysiol.* **65**, 1313-1320.

LEV-TOV, A. & PINCO, M. (1992) In vitro studies of prolonged synaptic depression in the neonatal rat spinal cord. *J. Physiol.* **447**, 149-169.

MARTIN, A.R. (1977) Junctional transmission. II. Presynaptic mechanisms. In *Handbook of Physiology. The Nervous System. Sec. 1. Vol. 1.* ed. KANDEL, E.R. pp. 329-356. American Physiological Society, Bethesda.

MENDELL, L. M., COLLINS, W.F. III & KOERBER, H.R. (1990) How are synapses distributed on spinal motoneurons to permit orderly recruitment? In *The Segmental Motor System* eds. BINDER, M.D. & MENDELL, L.M. pp. 328-340. Oxford University Press. New York.

MENDELL, L.M, COLLINS, W.F., III & MUNSON, J.B. (1994) Retrograde determination of motoneuron properties and their synaptic input. *J. Neurobiol.* **25**, 707-721.

MENDELL, L.M., TAYLOR, J.S., JOHNSON, R.D. & MUNSON, J.B. (1995) Rescue of motoneuron and muscle afferent function in cats by regeneration into skin II. The Ia-motoneuron synapse. *J. Neurophysiol.* **73**, 662-673.

PESHORI, K.R., COLLINS, W.F., III & MENDELL, L.M. (1993) Effects of GABA-B manipulation on EPSP amplitude and frequency-dependent modulation at rat Ia-motoneuron synapses-evidence for tonic GABA-B activity. *Neurosci. Abstr.* **19**. 224.12

PIERCE, J.P & MENDELL, L.M. (1993) Quantitative ultrastructure of Ia boutons in the ventral horn: scaling and positional relationships. *J. Neurosci.* **13**, 4748-4763.

PINTER, M.J., CURTIS, R.L. & HOSKO, M.J. (1983) Voltage threshold and excitability among variously sized cat hindlimb motoneurons. *J. Neurophysiol.* **50**, 644-657.

PROCHAZKA, A., HULLIGER, M., TREND, P., LLEWELLYN, M. & DURMULLER, N. (1989) Muscle afferent contribution to paw shakes in normal cats. *J. Neurophysiol.* **61**, 550- 562.

SEEBACH, B.S. & MENDELL, L.M. (1994) Postnatal maturation of frequency-dependent behavior of rat Ia-motoneuron synapses. *Neurosci. Abst.* **20**. 330.3.

ZENGEL, J.E., REID, S.A., SYPERT, G.W. & MUNSON, J.B. (1985) Membrane electrical properties and prediction of motor-unit type of medial gastrocnemius motoneurons in the cat. *J. Neurophysiol.* **53**, 1323-1344.

RETROGRADE EFFECTS OF TARGET ON MOTONEURONES AND MUSCLE AFFERENTS

John B. Munson

Department of Neuroscience, College of Medicine
University of Florida Brain Institute
Gainesville, Florida 32610-0244, USA

INTRODUCTION

Alpha motoneurones and muscle spindle sensory neurones are closely linked functionally. Here we present evidence from cat that they share also a common dependency on target innervation: their functional properties are altered when they are deprived of target tissue, and these properties are in large measure rescued when the neurones regenerate into either hairy skin or into muscle.

Muscle Afferents

When muscle afferents are chronically axotomised, their conduction velocity reduces to half or less of normal (e.g. Collins, Mendell & Munson, 1986). At first the target-deprived afferents retain their characteristic slowly-adapting mechanosensitivity (Johnson & Munson, 1991) but this is progressively lost over the succeeding months in the absence of target innervation (Munson & Johnson, 1994). The ability of the muscle afferents to generate a cord dorsum potential is also reduced or lost (Johnson, Taylor, Mendell & Munson, 1995) as is the ability to generate EPSPs in spinal motoneurones (Goldring, Kuno, Nunez & Snider, 1980; Johnson et al., 1995).

Muscle afferents permitted to reinnervate muscle (estimated to take about 5 weeks) exhibit nearly normal properties. Conduction velocity remains somewhat reduced overall but most afferents are sensitive to mechanical stimulation of the muscle (Collins et al., 1986). Cord dorsum potentials and EPSPs generated by muscle-reinnervated muscle afferents are of normal configuration but may be reduced in amplitude (Johnson et al., 1995). However, in these cases it was not known to what extent the regenerated muscle afferents had reinnervated their native vs. a foreign receptor (e.g. muscle spindle, tendon organ) or persist simply as intramuscular free nerve endings (but see Collins et al., 1986). Depending on the delay between axotomy and target reinnervation, the observed "rescue" may represent either retention of or restoration of function.

Interestingly, regeneration of muscle afferents into hairy skin of the cat via the caudal cutaneous sural nerve may be equally effective in rescuing functional properties of muscle afferents. In such afferents, conduction velocity is virtually normal, almost all skin-innervated afferents exhibit normal slowly-adapting mechanosensitivity, and electrical stimulation produces cord dorsum potentials and EPSPs of normal configuration and of somewhat reduced amplitude (Johnson et al., 1995). We estimate that these afferents reach skin about 10 weeks after axotomy, a delay at which axotomised afferents retain mostly normal properties (excepting conduction velocity). Thus innervation of skin, as of muscle, both restores function (conduction velocity) and prevents further loss of function by muscle afferents. Experiments in progress will test the abilities of skin and muscle to restore such functions once lost following prolonged target deprivation.

Motoneurones

Alpha motoneurones are similarly dependent upon target innervation for their normal function: chronically axotomised motoneurones exhibit reduced conduction velocity, rheobase and after-hyperpolarisation (AHP) half-decay time, increased input resistance, and EPSPs of reduced amplitude (e.g. Foehring, Sypert & Munson, 1987). These losses are prevented if the motoneurones regenerate into their own or a similar muscle (e.g. Foehring et al., 1987); they appear to be rescued as well if the motoneurones regenerate into hairy skin of the cat via the caudal cutaneous sural nerve (Nishimura, Johnson & Munson, 1991).

RECENT OBSERVATIONS

High-frequency Stimulation

We have recently subjected these skin-regenerated muscle afferents and motoneurones to more rigorous tests in order to determine whether skin and muscle as target are truly equally effective in rescuing motor nerve neuronal function (Mendell, Taylor, Johnson & Munson, 1995). The medial gastrocnemius (MG) nerve was sectioned in the popliteal fossa and joined to the sural nerve, thus permitting both the MG afferents and MG motoneurones to regenerate via the sural into the hairy skin of the ankle. As controls, we examined (i) normal unoperated cats, and cats with the MG nerve (ii) chronically axotomised or (iii) self-regenerated. Terminal acute experiments were conducted after 5 weeks to 30 months. Using intracellular techniques, we examined (i) the properties of the normal or operated MG motoneurones and of the unoperated lateral gastrocnemius/soleus (LGS) motoneurones, (ii) the EPSPs generated by the normal or operated MG muscle afferents in normal LGS motoneurones, and (iii) the EPSPs generated by normal LGS afferents in the normal or operated MG motoneurones.

A technical innovation in these experiments was the use of bursts of high-frequency stimulation to elicit EPSPs; i.e. trains of 32 stimuli at 167 Hz, summated in register (Collins, Honig & Mendell, 1984; Figure 1 of Mendell, this volume). Such bursts are typical of muscle afferent activity occurring during normal locomotion (Loeb & Duysens, 1979). In normal cats the increase or decrease in EPSP amplitude that occurs during the burst (i.e., positive or negative modulation) is related closely to the AHP of the recorded motoneurone: those with AHP half-decay >30ms ("slow" motoneurones: Zengel, Reid, Sypert & Munson, 1985) virtually all exhibit negative modulation; those with AHP <30ms ("fast" motoneurones) may exhibit positive or negative modulation (see Figure 2 of Mendell, this volume). Furthermore, large EPSPs, especially those in putative type-S motoneurones, typically exhibit negative modulation while small EPSPs typically exhibit positive modulation (see Figure 2 of Mendell, this volume).

In operated cats, 5 - 15 weeks after axotomy of the MG nerve, we found exclusively negative modulation of EPSPs during the burst, both of those generated by the axotomised MG afferents in the normal LGS motoneurones, and of those generated by the normal LGS afferents in the axotomised MG motoneurones (Figure 1). Following cross-regeneration of the MG nerve into skin (14 - 30 months) there was considerable rescue of the ability of both operated afferents and operated motoneurones to generate high-frequency EPSPs. In both cases the plot of the data has shifted to the right - away from the axotomy values and toward normal values (Figure 1), indicating an ability of skin to rescue these functions of neurones which normally innervate muscle. Because profound negative modulation was exhibited in MG nerves axotomised for 5 or 10 weeks, whereas the regenerating MG nerve would not reach skin for at least 10 weeks, it is clear that skin has enabled some restoration of muscle nerve function, as well as prevention of additional loss.

We have made similar studies of MG motor nerves regenerated into the MG muscle, 6 months after surgery. Modulation of EPSPs during the burst was returned to near-normal ranges of positive and negative values, both for the operated afferents and the operated motoneurones (Figure 1). Thus 6 months reinnervation of muscle was somewhat more effective in rescuing muscle afferent and motoneuronal function than was 30 months cross-innervation of skin, although skin clearly provided a considerable rescue function.

Cellular Mechanisms of Altered Amplitude Modulation

As discussed elsewhere (Mendell, this volume) the mechanisms responsible for amplitude modulation during high-rate stimulation are at present not known. Possibilities include transmitter depletion, receptor desensitisation, and Ca^{2+} accumulation. A number of correlations exist during the burst. Motoneurones with low rheobase, high resistance and

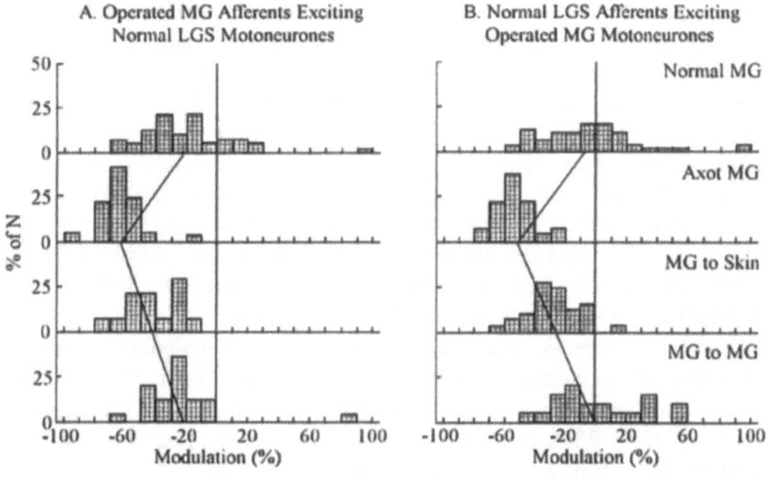

Figure 1. Histograms of modulation of EPSPs during trains of 167 Hz stimulation of muscle afferents in unoperated and in operated animals. A. Stimulated MG afferents are normal, axotomised (5 - 15 wk), cross-regenerated into skin (14 - 30 months) or self-regenerated (6 months; Ns = 43, 37, 28, 25, respectively); recipient LGS motoneurones are normal. B. Stimulated LGS afferents are normal; recipient MG motoneurones are normal, axotomised, cross- or self-regenerated (times as in A; Ns = 54, 27, 33, 20, respectively). Note that axotomy of either afferents or motoneurones results in exclusively negative modulation; self-regeneration largely restores normal ranges of modulation; cross-regeneration into skin partially restores normal modulation. Diagonal lines connect median values.

long AHP have large EPSPs which generally decline, whilst motoneurones with high rheobase, low resistance and short AHP have small EPSPs which generally increase in amplitude.

Throughout the present experiments on altered muscle nerves, the more negative values of amplitude modulation were suggestive of what is normally observed in low rheobase, high resistance motoneurones. In fact, the rheobase of axotomised motoneurones had decreased and the input resistance had increased; thus qualitatively the agreement between synaptic transmission and motoneurone properties during high-frequency stimulation was maintained. However, the EPSPs evoked in axotomised motoneurones in response to low frequency (0.5 Hz) stimulation were similar in amplitude to those in intact cats. Since negative modulation in intact preparations is normally associated with the largest EPSPs, the changes in modulation of EPSP amplitude were not explained by changes in low-frequency EPSP amplitude.

In cases where the afferents had been axotomised the motoneurones exhibited no change in rheobase, input resistance or AHP; thus there was a mismatch between the properties of the motoneurones and the properties of their EPSPs measured at high frequency. Furthermore EPSP amplitude measured at low frequency became considerably smaller and so the finding of more negative modulation after axotomy could not be attributed to a change in EPSP amplitude; i.e., it was not the result of more transmitter depletion offsetting the residual facilitation. A different change, related directly to the damage to the afferent must have been responsible, perhaps an inability for residual Ca^{2+} to accumulate in the terminals. Such an explanation would explain the uniform findings in these experiments that positive modulation was affected disproportionately in comparison with negative modulation.

When MG motor axons innervate skin, they form associations that permit recovery from the effects of axotomy (Nishimura et al., 1991). Modulation of EPSP amplitude during high-frequency stimulation recovered only partially despite the virtually complete recovery of motoneurone cellular properties (14 - 30 months survival times). Reinnervation of the original muscle resulted in virtually complete recovery of synaptic properties despite the fact that motoneurone properties had not recovered completely (6 months survival time). Thus in both cases the recovery of synaptic properties was dissociated from the recovery of motoneurone properties. As with axotomy, the role of the periphery in affecting motoneurone function does not appear to occur via changes in motoneurone cell properties; i.e. the synaptic properties and cellular properties are not tightly linked.

On the afferent side, skin reinnervation yielded partial rescue from the effects of axotomy; muscle reinnervation was even more effective in promoting the recovery of central synaptic properties. In none of these cases was there any consistent alteration in motoneurone properties; i.e. the changes were apparently the result of different degrees of recovery of the afferent terminals from the effects of axotomy.

Overall these experiments indicate that contact with the periphery plays an important role in the function of the central connections made by these fibres. However the effect of the periphery is not exerted in a simple manner through changes in the properties of the motoneurones and the EPSPs that are normally closely correlated with the behaviour of these connections under conditions of high-frequency activation.

POSSIBLE ROLE OF NEUROTROPHINS

Current research on the nurturing role of the neurotrophins (reviewed in Mendell, 1994) suggests that they contribute to these rescue functions. Because muscle afferent function may be rescued by regeneration into either muscle or skin, one might predict that these muscle afferents express receptors for neurotrophins found in both muscle and skin. Consistent with this, the *trk*C receptor predominates for muscle afferents and NT-3 (the neurotrophin of choice for the *trk*C receptor) is present in both dermis and muscle.

Similarly, motoneurones might express receptors for neurotrophins found in both tissues. Obligingly, adult motoneurones express both *trk*B and *trk*C receptors, whose preferred neurotrophins (BDNF and NT-3, respectively) are expressed in both skin and muscle.

Compelling as these relationships are, they cannot be telling the complete story. While regeneration into skin provides impressive rescue of muscle afferent and motoneurone function, reinnervation of muscle is more effective (Figure 1). This suggests that there may be either an inadequate supply of an appropriate skin-derived neurotrophin, or that the neurotrophin from the skin may not be entirely-appropriate.

ACKNOWLEGEMENTS

Some of this research was carried out in collaboration with Dr. L.M. Mendell. Research support to the author derived from RO1 NS-15913 and PO1 NS-27511.

REFERENCES

COLLINS, W.F., III, HONIG, M.G. & MENDELL, L.M. (1984) Heterogeneity of group Ia synapses on homonymous α-motoneurons as revealed by high-frequency stimulation of Ia afferent fibers. *J. Neurophysiol.* **52**, 980-993.

COLLINS, W.F., III, MENDELL, L.M. & MUNSON, J.B. (1986) On the specificity of sensory reinnervation of cat skeletal muscle. *J. Physiol.* **375**, 587-609.

FOEHRING, R.C., SYPERT, G.W. & MUNSON, J.B. (1987) Motor-unit properties following cross-innervation of cat lateral gastrocnemius and soleus muscles with medial gastrocnemius nerve. II. Influence of muscle on motoneurons. *J. Neurophysiol.* **57**, 1227-1245.

GOLDRING, J.M., KUNO, M., NUNEZ, R. & SNIDER, W.D. (1980) Reaction of synapses on motoneurones to section and restoration of peripheral sensory connections in the cat. *J. Physiol.* **309**, 185-198.

JOHNSON, R.D. & MUNSON, J.B. (1991) Regenerating sprouts of axotomized cat muscle afferents express characteristic firing patterns to mechanical stimulation. *J. Neurophysiol.* **66**, 2155-2158.

JOHNSON, R.D., TAYLOR, J.S., MENDELL, L.M. & MUNSON, J.B. (1995) Rescue of motoneuron and muscle afferent functions in cats by regeneration into skin. I. Properties of afferents. *J. Neurophysiol.* **73**, 651-661.

LOEB, G.E. & DUYSENS, J. (1979) Activity patterns in individual hindlimb primary and secondary muscle spindle afferents during normal movements in unrestrained cats. *J. Neurophysiol.* **42**, 420-440.

MENDELL, L.M. (1994) Neurotrophic factors and the specification of neural function. *The Neuroscientist* **1** (in press).

MENDELL, L.M. (1995) Synaptic differentiation on type-identified motoneurones. This Volume.

MENDELL, L.M., TAYLOR, J.S., JOHNSON, R.D. & MUNSON, J.B. (1995) Rescue of motoneuron and muscle afferent functions in cats by regeneration into skin. II. The Ia-motoneuron synapse. *J. Neurophysiol.* **73**, 662-673.

MUNSON, J.B. & JOHNSON, R.D. (1994) Inherent firing patterns of muscle afferent axon terminals are gradually lost in chronic neuromas. *Soc. Neurosci. Abs.* **20**, 1382.

NISHIMURA, H., JOHNSON, R.D. & MUNSON, J.B. (1991) Rescue of motoneurons from the axotomized state by regeneration into a sensory nerve in cats. *J. Neurophysiol.* **66**, 1462-1470.

ZENGEL, J.E., REID, S.A., SYPERT, G.W. & MUNSON, J.B. (1985) Membrane electrical properties and prediction of motor-unit type of medial gastrocnemius motoneurons in the cat. *J. Neurophysiol.* **53**, 1323-1344.

EFFECTIVE SYNAPTIC CURRENTS GENERATED IN CAT SPINAL MOTONEURONES BY ACTIVATING DESCENDING AND PERIPHERAL AFFERENT FIBRES

Marc D. Binder and Randall K. Powers

Department of Physiology & Biophysics
University of Washington School of Medicine
Seattle, Washington 98195, USA

INTRODUCTION

Understanding how synaptic inputs from segmental and descending systems shape motor output from the spinal cord requires detailed descriptions of the relative magnitudes of the synaptic currents produced by the different systems and their pattern of distribution within a motoneurone pool (Burke, 1981; Heckman & Binder, 1990). Knowledge of how distinct synaptic inputs interact when they are activated concurrently and quantitative expressions for how synaptic currents are transduced into spike trains by motoneurones are equally important.

Until recently, the relative magnitude and distribution of a specific synaptic input to the constituents of a motoneurone pool has only been inferred from the analysis of the amplitude and time course of synaptic potentials measured at the motoneurones' resting potentials (reviewed in Burke, 1981; Munson, 1990). However, the amplitudes of synaptic potentials are strongly dependent on the intrinsic properties of the cells they are recorded in and, therefore, cannot provide a direct measure of the magnitude of a synaptic input (Heckman & Binder, 1988; 1990). Moreover, synaptic potentials are also dependent on motoneurone membrane potential. This is particularly true for inhibitory inputs and for mixtures of excitatory and inhibitory inputs for which the reversal potential can be at or near the resting potential (Lindsay & Binder, 1991; Stuart & Redman, 1990; Powers, Robinson, Konodi & Binder, 1993).

Another limitation of synaptic potential measurements as an indicator of synaptic efficacy is the absence of a general, quantitative expression for the relationship between the amplitude of a synaptic potential measured in a cell at rest and the effect it has on the firing rate of the same cell. The relation between single fibre EPSP amplitude and firing rate modulation (Cope, Fetz & Matsumura, 1987) has been used to estimate the effects produced when many afferents are active, based on the assumption that the effects of individual afferent fibres are independent of one another (Fetz, Cheney, Mewes & Palmer,

1989). Unfortunately, this assumption is probably not valid, because the effect of a given presynaptic impulse on postsynaptic discharge probability depends upon its timing with respect to previous presynaptic and postsynaptic spikes (Brillinger, Bryant & Segundo, 1976).

As an alternative, we have recently proposed that measurement of synaptic current reaching the soma provides a more useful means of quantifying synaptic input magnitude than do measures of synaptic potentials. We have called this measure the "effective synaptic current" or I_N (Heckman & Binder, 1988; see also Redman, 1976). Direct measurements of effective synaptic current avoid the confounding effects of input resistance, incorporate the effect of changes in synaptic driving force, and may lead to a simple, quantitative relationship between synaptic input magnitude and synaptically-evoked changes in motoneurone discharge rate (Heckman & Binder, 1990; 1993; Powers, Robinson, Konodi & Binder, 1992).

MEASUREMENT OF EFFECTIVE SYNAPTIC CURRENT

Effective synaptic currents are measured in motoneurones by combining injected current with repetitive activation of an identified source of synaptic input as shown in Figure 1. Part A of this figure shows intracellular recordings from a cat medial gastrocnemius (MG) motoneurone in response to a series of depolarising and hyperpolarising injected current pulses and synaptic current produced by electrical stimulation of the common peroneal nerve (10 x group I threshold) at 200 Hz. The experimental records are divided into three 500 ms periods: (1) injected current alone; (2) injected and synaptic currents together; and (3) synaptic current alone. The steady-state response of the neurone to injected current is designated as V_i, that to injected+synaptic current as V_{i+s}, and the difference between these two values, as ΔV_s. Plots of V_i and V_{i+s} versus the magnitude of injected current (I; Figure 1B) yield three measures: (1) I_N, the magnitude of effective synaptic current; (2) $R_{N\ SYN}$, the effective steady-state input resistance during synaptic activation; and (3) $R_{N\ SS}$, the steady-state input resistance in the absence of synaptic stimulation. The effective synaptic current at rest ($I_{N\ rest}$) is defined as the current required to clamp the membrane potential at the resting potential during the activation of the synaptic input (Figure 1B, filled circles; Heckman & Binder, 1988), and is calculated by multiplying the x-intercept of the best-fit linear regression line of V_{i+s} versus I by -1 (i.e., the magnitude I_N is assumed to be equal in magnitude and opposite in sign to the injected current required to clamp the membrane potential at the resting value; Lindsay & Binder, 1991; Powers et al., 1992; 1993).

The slope of the linear relation between V_{i+s} and I provides a measure of the input resistance during synaptic activation ($R_{N\ SYN}$), whereas that of V_i versus I yields the steady-state input resistance ($R_{N\ SS}$) in the absence of synaptic activation. A significant difference between the slopes indicates that the synaptic input altered the conductance of the motoneurone measured at the soma (Heckman & Binder, 1991; Lindsay & Binder, 1991; Powers et al., 1993).

The relation between ΔV_s (the amplitude of the steady-state synaptic potential) and V_i (the membrane potential relative to rest just prior to the onset of synaptic activation) is used to determine the reversal potential and the voltage-dependence of the effective synaptic current (Lindsay & Binder, 1991; Powers et al., 1992; 1993). A plot of ΔV_s versus V_i reveals the dependence of the steady-state synaptic potential on somatic membrane potential, and the x-intercept of the linear fit approximates the somatic reversal potential for the effective synaptic current (Figure 1C).

When a significant linear correlation is found between ΔV_s and V_i, this relation can be used to predict ΔV_s at the threshold for repetitive firing ($V_i = V_{thr}$). The effective synaptic current flowing at threshold for repetitive discharge ($I_{N\ thr}$) can be estimated by dividing this extrapolated value of ΔV_s by $R_{N\ SYN}$ (Powers et al., 1992). Alternatively, $I_{N\ thr}$ can be determined by calculating how the change in synaptic driving force during repetitive discharge would increase or decrease the effective synaptic current measured at rest ($I_{N\ rest}$; Powers et al., 1992; 1993): $I_{N\ thr} = I_{N\ rest} * (V_{thr} - V_{rev})/ (V_{rest} - V_{rev})$, where V_{rev} is the reversal potential for the net effective synaptic current and V_{rest} is the resting potential (Powers et al., 1992; 1993). Five input systems have been studied and the results are summarised in Table 1 and reviewed below.

DISTRIBUTION OF EFFECTIVE SYNAPTIC CURRENT FROM IDENTIFIED INPUT SYSTEMS

Monosynaptic Ia Afferent Input

The covariance between Ia EPSPs and motoneurone input resistance has generally been assumed to result from an approximately constant synaptic current applied to cells of

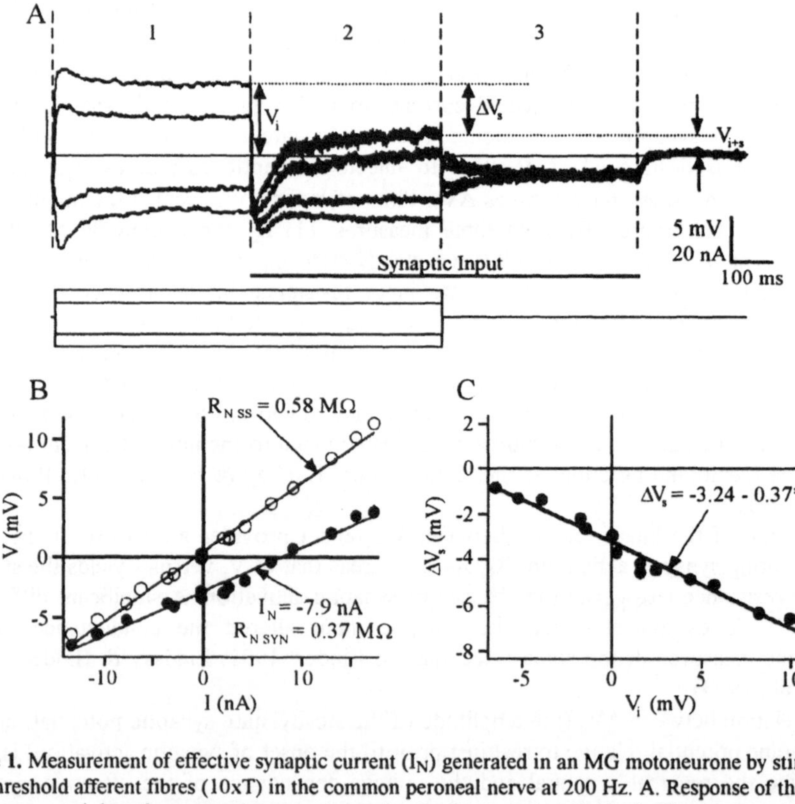

Figure 1. Measurement of effective synaptic current (I_N) generated in an MG motoneurone by stimulating high-threshold afferent fibres (10xT) in the common peroneal nerve at 200 Hz. A. Response of the motoneurone to injected currents (lower traces) and synaptic currents (solid bar). The mean resting potential (measured prior to current injection) has been subtracted from each trace. B. Voltage responses versus injected current (I). Solid lines indicate best linear fit to the data points (V_i: open circles; V_{i+s}: filled circles). The effective synaptic current (I_N) is taken to be equal in magnitude and opposite in sign to the current at which $V_{i+s} = 0$ (estimated from the zero intercept of the fit to V_{i+s} vs. I). C. Dependence of steady-state synaptic potential (ΔV_S) on somatic membrane potential (V_i). See text for further details.

varying R_N values, in keeping with the original size principle (reviewed in Heckman & Binder, 1990). Although the amplitudes of steady-state Ia EPSPs are highly correlated with R_N, measurements of the underlying effective synaptic currents indicate that this covariance results from systematic variance in both I_N and R_N (Heckman & Binder, 1988). Within the cat MG motoneurone pool, effective synaptic currents (I_N) generated by homonymous Ia afferents display a wide range of values that are systematically related to R_N and to other motoneurone properties: I_N is about twice as large on average in motoneurones with high R_N values as in those with low R_N values (Heckman & Binder, 1988). Combining Ia afferent fibres from the synergist lateral gastrocnemous and soleus muscles approximately doubles the magnitude of I_N (mean value MG Ia afferents = 2nA; mean value triceps surae Ia afferents = 4.2nA; Westcott, Powers, Robinson, Konodi & Binder, 1993). These surprising results clearly illustrate the difficulties in assessing the magnitude of synaptic inputs from analysis of synaptic potential amplitudes.

The monosynaptic input from Ia afferents has no appreciable effect on the input resistance of the motoneurone measured at the soma (Heckman & Binder, 1988). Moreover, since the reversal potential for the synapses lies betweeen 0 and +20mV (Finkel & Redman, 1983), the I_N from Ia afferent fibres is probably reduced by 15% or less as the motoneurone reaches its threshold for repetitive firing.

Reciprocal Ia Inhibition

Comparisons of reciprocal Ia IPSPs with homonymous Ia EPSPs generated in cat hindlimb motoneurones have revealed the existence of a strong correlation between these two synaptic input systems (Burke, Rymer & Walsh, 1976). However, an analysis of the data of Burke et al. (1976) by Stein & Bertoldi (1981) showed that there is a steeper gradient of synaptic strength across the motoneurone pool for Ia excitation than for Ia inhibition. Thus, one would expect that differences in the effective synaptic currents underlying Ia inhibition within a motoneurone pool would be considerably less than those found for excitatory Ia inputs. This appears to be the case, since amplitudes of the inhibitory effective synaptic currents generated in antagonist motoneurones by activation of MG Ia afferent fibres extended over a five-fold range (-0.5 to -2.7 nA), but I_N was correlated neither with intrinsic motoneurone properties nor with presumed motor unit type (Heckman & Binder, 1991). The average value of the steady-state Ia-inhibitory synaptic current in presumed type F motoneurones was -1.60 ± 0.64 nA, whereas that for presumed type S motoneurones was -1.65 ± 0.71 nA.

Since many of the synaptic boutons of Ia-inhibitory interneurones lie on the somata of motoneurones (Fyffe, personnal communication), they can generate a substantial change in

Table 1. Distribution of effect synaptic current for five different input systems. [1]Heckman & Binder (1988); [2]Lindsay & Binder (1991); [3]Heckman & Binder (1991); [4]Westcott, Powers, Robinson, Konodi & Binder (1993); [5]Powers, Robinson, Konodi & Binder (1993).

SYNAPTIC INPUT	EFFECTIVE SYNAPTIC CURRENT (nA) AT V_{rest}		
	RANGE	MEAN	
Ia Afferent Fibres[1]	+0.6 to +4.3	+2.0	Type S > Type F
Renshaw Interneurones[2]	0.0 to -1.2	-0.4	Type S = Type F
Ia-Inhibitory Interneurones[3]	-0.5 to -2.7	-1.6	Type S = Type F
Deiter's Nucleus (LVST)[4]	-4.6 to +10.7	+2.5	Type F > Type S
Rubrospinal Neurones[5]	-7.4 to +15.4	-2.7 +4.4	Type S - Type F +

the motoneurone input resistance. Comparison of the input resistance during steady-state activation of Ia-inhibitory interneurones ($R_{N\ SYN}$) with resting input resistance ($R_{N\ SS}$) indicates net synaptic conductances of up to 800 nS, with an average value of 175 nS. As was the case for the effective synaptic currents from Ia-inhibitory interneurones, there was no systematic relationship between the steady-state synaptic conductance and the electrical properties of the motoneurone.

Recurrent inhibition

It has been reported previously that the amplitudes of recurrent IPSPs are correlated with a number of motoneurone intrinsic properties (e.g., Friedman, Sypert, Munson & Fleshman, 1981; Hultborn, Katz & Mackel, 1988). However, the effective synaptic currents underlying recurrent inhibition (RC I_N) in triceps surae motoneurones appear to be uniformly distributed within the pool, i.e., the magnitude of RC I_N is entirely independent of motoneurone input resistance and rheobase (Lindsay & Binder, 1991). The mean RC I_N measured at rest was only -0.4nA, considerably smaller than comparable values derived for other input systems (cf. Table 1). Since the Renshaw input was activated in this study by stimulating the synergist motor axons at 100 Hz, which produces steady-state IPSPs that are about 50% maximum value (Lindsay, Heckman & Binder, unpublished data), the total RC I_N at rest is still likely to be < -1nA on average.

Motoneurones are more likely to receive recurrent inhibition while they are firing repetitively than when they are quiescent, and several hypotheses of the role of recurrent inhibition in motor control emphasize the possible effects of recurrent inhibition on firing frequency (e.g. Hultborn, Lindstrom & Wigstrom, 1979). The distribution of effective synaptic currents from Renshaw cells at threshold was no different from that measured at resting potential. However, the mean RC I_N at threshold was -1.25nA, three times the mean value at resting potential. These results suggest that the effect of recurrent inhibition on the firing frequencies of motoneurones will be uniform and quite modest. Using an average value for the slope of the motoneurone firing frequency-injected current relationship (f/I curve) of 1.5 impulses/s/nA (Kernell, 1979), the average change in firing frequency produced by maximal recurrent inhibition (-2.5nA) should be less than 4 impulses/s. Since the slope of the f/I curve does not appear to covary with other motor unit properties (Kernell, 1979), the change in firing frequency should not vary systematically within a motoneurone pool. The maximum effective synaptic current calculated at threshold (-4.6nA) would only decrease firing frequency by about 7 impulses/s, a value similar to that observed by in the experiments of Granit & Renkin (1961) in which whole ventral roots were stimulated to activate the Renshaw cells.

Rubrospinal Input

The rubrospinal system is one of several oligosynaptic pathways in the cat that generate qualitatively different synaptic potentials within hindlimb motoneurone pools: low-threshold motoneurones receive predominantly inhibitory input, whereas high-threshold motoneurones receive predominantly excitatory input (e.g., Burke, Jankowska & ten Bruggencate, 1970). It has been proposed that this pattern of synaptic input may act to disrupt the normal hierarchy of recruitment thresholds as well as to alter the gain of the input-output function of a motoneurone pool (Burke et al., 1970; Burke, 1981; Heckman & Binder, 1990; 1993; Powers et al., 1993).

We have recently measured the effective synaptic currents produced in cat triceps surae motoneurones by stimulating the hindlimb projection area of the contralateral magnocellular red nucleus (Powers et al., 1993). At motoneurone resting potential, the distribution of effective synaptic currents from the red nucleus was qualitatively similar to the distribution of synaptic potentials: 86% of presumed type F motoneurones received a net depolarising effective synaptic current from the red nucleus stimulation, whereas only 38% of presumed type S units did so. However, at threshold the distribution was markedly altered. Inhibition continued to predominate in the type S cells, but among the type F cells, half received net excitatory effective synaptic currents and half received net inhibitory effective synaptic currents. This suggests the existence of a more complex pattern of red nucleus input to the pool than that deduced from the analysis of PSPs (Burke et al., 1970). Other surprising features of these data were the enormous range of effective synaptic currents observed in different cells and the fact that they were often three times larger than comparable inputs from segmental pathways (Table 1). Within the cat triceps motoneurone pool, activation of red nucleus synaptic input reduced motoneurone input resistance by 40%, on average (Powers et al., 1993.) The effect on input resistance was most pronounced in those motoneurones that received hyperpolarising effective synaptic currents. The data on effective synaptic currents indicate that the red nucleus input may provide a powerful source of synaptic drive to some high threshold motoneurones, while concurrently inhibiting low threshold cells. Thus, this input system can potentially alter the gain of the input-output function of the motoneurone pool, change the hierarchy of recruitment thresholds within the pool, and mediate rate limiting of discharge in low-threshold motoneurones (Heckman & Binder, 1990, 1993; Powers et al., 1993).

Lateral Vestibulospinal (Deiter's Nucleus) Input

Stimulation of Deiter's nucleus (DN) produces mono- and disynaptic EPSPs in cat triceps motoneurones (Grillner, Hongo & Lund, 1970). Although all monosysnaptic EPSPs were quite small (< 2.2mV), they did display a wide range of amplitudes that were not correlated with the duration of the after hyperpolarisation. Similarly, the amplitudes of monosynaptic EPSPs produced by stimulation of DN axons within the ipsilateral ventral funiculus were not related to motor unit type (Burke et al., 1976).

Based on the dependence of PSP amplitude on R_N, the PSP data suggest that the underlying effective synaptic currents are inversely related to R_N. Recent direct measurements of the effective synaptic currents produced by DN stimulation in triceps surae motoneurones of the cat demonstrated that this is indeed the case (Westcott et al. 1993). The effective synaptic currents caused by DN stimulation were primarily small and depolarising in both medial gastrocnemius (MG) and lateral gastrocnemius-soleus (LGS) motoneurones (mean = 2.5nA). The DN input tended to be larger in presumed type F motoneurones. The amplitudes of the steady-state DN synaptic potentials found in this study were similar to those previously reported for transient PSPs (Grillner et al., 1970; Burke et al., 1976). The amplitude of the DN PSPs showed no correlation with rheobase, resting membrane potential, or input resistance. The DN synaptic input caused no significant change in motoneurone input resistance.

EFFECTS OF SYNAPTIC INPUTS ON MOTONEURONE DISCHARGE

The functional significance of a synaptic input system ideally should be assessed in terms of its effects on both motoneurone recruitment and firing rate modulation. In general, previous analyses of the input-output properties of motoneurone pools have concentrated on motoneurone recruitment alone. However, since nearly 75% of motor unit force modulation is generally achieved through changes in discharge rate (Kernell, 1983), understanding how synaptic inputs alter the discharge rates of activated motoneurones is of critical importance.

In cat hindlimb motoneurones, the steady-state relation between injected current magnitude and discharge frequency (f-I relation) can be described by one or two line segments (primary and secondary range) with average slopes of about 1.5 and 3.0 impulses/s/nA, respectively (Kernell, 1983). It has been demonstrated that synaptic and injected currents are usually equivalent with respect to their effects on repetitive firing within the primary range and that during motoneurone discharge, input currents normally sum algebraically (e.g. Schwindt & Calvin, 1973a, b). Even synaptic inputs leading to large changes in motoneurone input resistance simply shift the f-I relation along the current axis, without producing a change in slope (Schwindt & Calvin, 1973a, b), indicating that synaptic inputs add a constant amount of depolarising or hyperpolarising current regardless of the background firing rate (see, however, Kernell, 1966 and Shapavolov, 1972).

Based on these results, Schwindt & Calvin (1973b) postulated that one could infer the effective, steady-state current delivered by any synaptic input from the slope of a motoneurone's f-I relation and the change in discharge produced by activating the synaptic input. This postulate leads to a simple, quantitative expression for steady-state motoneurone behavior, namely that the change in motoneurone discharge (ΔF) is equal to the product of the net effective synaptic current (I_N) and the slope of the frequency-current curve (f/I) in the primary range ($\Delta F = I_N * f/I$).

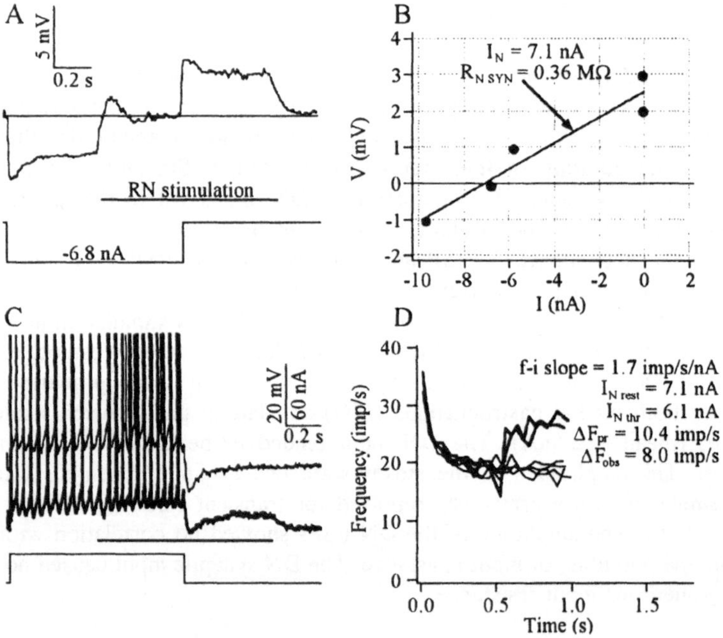

Figure 2. Effective synaptic current and firing rate modulation produced in a motoneurone by activation of the contralateral red nucleus. A. Intracellular recordings from an MG motoneurone in response to hyperpolarising injected current and excitatory synaptic current produced by supramaximal 200 Hz stimulation within the contralateral red nucleus (stimulation period noted by line between voltage and current traces; cf. Figure 1). B. Plotting the membrane potential during the epoch of simultaneous injected and synaptic activation as a function of the injected current provides a means of estimating the effective synaptic current (I_N; cf. Figure 1). C. Responses of the same motoneurone to injected current alone (upper voltage trace) and the injected current plus the synaptic current (lower voltage trace) in alternate trials. D. Instantaneous frequency vs. time for 4 control trials (injected current alone; thin lines) and 2 trials with superimposed synaptic current (thick lines). See text for further details.

We have tested the validity of this expression by measuring each of these three quantities (ΔF, I_N and f/I) in the same motoneurones (Powers et al., 1992). Following the measurement of I_N (Figure 1 and Figure 2A & B), the f-I slope (f/I) is measured by injecting a series of 1s current pulses of different amplitude and plotting the relation between f and I. The effects of the synaptic input on the motoneurone's discharge rate (ΔF) can be determined by applying suprathreshold 1s depolarising current pulses of fixed amplitude, and superimposing the synaptic stimulation (Figure 2C & D). The change in firing rate (ΔF) produced by the synaptic input is calculated from the difference between the average firing rate over the last 300ms of current injection in test and control (injected current alone) trials (Powers et al., 1992). When possible the trials were repeated at one or more additional levels of injected current. For the cell shown in Figure 2, the synaptic current produced by red nucleus stimulation produced an average firing rate increase of 8 impulses/s. This is quite close to the predicted change in firing rate (10 impulses/s) based on the product of effective synaptic current (6 nA at threshold for repetitive firing) and the slope of the f-I curve in the primary range of firing (1.7 impulses/s/nA).

Similar results were obtained for motoneurones that received inhibitory input from the red nucleus (Powers et al., 1992) and completely analogous results have been obtained for synaptic input from Ia afferents, Ia-inhibitory interneurones, Renshaw interneurones, interneurones activated by low-threshold cutaneous afferents from the sural nerve, and flexor reflex afferent (FRA) input from the common peroneal nerve (cf. Figure 1). For each of the input systems, the observed changes in motoneurone discharge frequency are usually closely approximated by the product of the effective synaptic current and the slope of the f-I relation in the primary range. Thus, these results validate the measurement of effective synaptic current as a quantitative index of synaptic efficacy. This expression simplifies attempts to assess the operation of neural circuits quantitatively since the effective synaptic current already subsumes all of the factors governing current delivery from the dendrites to the soma, and therefore does not require any detailed information about the electrotonic architecture of the postsynaptic cell (Segev, Fleshman & Burke, 1990) or information about the precise location of the presynaptic boutons (Rall, Burke, Smith, Nelson & Frank, 1967; Stuart & Redman, 1990).

ACKNOWLEDGEMENTS

We thank our colleagues Drs. C.J. Heckman, Mark Konodi, Amy D. Lindsay, Farrel R. Robinson, Andrea Sawczuk, and Sarah L. Westcott and for their valuable contributions to the experiments described here. This work was supported by grant NS-26840 from the National Institutes of Health.

REFERENCES

BRILLINGER, D.R., BRYANT, H.L. & SEGUNDO, J.P. (1976) Identification of synaptic interactions. *Biol. Cybern.* **22**, 213-228.

BURKE, R.E. (1981) Motor units: anatomy, physiology, and functional organization. In *Handbook of Physiology, The Nervous System, Motor Control* ed. BROOKS, V.B. pp 345-422. American Physiological Society. Bethesda.

BURKE, R.E., JANKOWSKA, E. & TEN BRUGGENCATE, G. (1970) A comparison of peripheral and rubrospinal input to slow and fast twitch motor units of triceps surae. *J. Physiol.* **207**, 709-732.

BURKE, R.E., RYMER, W.Z. & WALSH, J.V. (1976) Relative strength of synaptic input from short-latency pathways to motor units of defined type in cat medial gastrocnemius. *J. Neurophysiol.* **39**, 447-458.

COPE, T.C., FETZ, E.E. & MATSUMURA, M. (1987) Cross-correlation assessment of synaptic strength of single Ia fibre connections with triceps surae motoneurones in cats. *J. Physiol.* **390**, 161-188.

FETZ, E.E., CHENEY, P.D., MEWES, K. & PALMER, S. (1989) Control of forelimb muscle activity by populations of corticomotoneuronal and rubromotoneuronal cells. *Prog. Brain Res.* **80**, 437- 449.

FINKEL, A.S. & REDMAN, S.J. (1983) The synaptic current evoked in cat spinal motoneurones by impulses in single group Ia axons. *J. Physiol.* **342**, 615-32.

FRIEDMAN, W.A., SYPERT, G.W., MUNSON, J.B. & FLESHMAN, J.W. (1981) Recurrent inhibition in type-identified motoneurons. *J. Neurophysiol.* **46**, 1349-1359.

GRANIT, R. & RENKIN, B. (1961) Net depolarization and discharge rate of motoneurones, as measured by recurrent inhibition. *J. Physiol.* **158**, 461-475.

GRILLNER, S., HONGO, T. & LUND, S. (1970) The vestibulospinal tract. Effects on alpha-motoneurones in the lumbosacral spinal cord in the cat. *Exp. Brain Res.* **10**, 94-120.

HECKMAN, C.J. & BINDER, M.D. (1988) Analysis of effective synaptic currents generated by homonymous Ia afferent fibers in motoneurons of the cat. *J. Neurophysiol.* **60**, 1946-1966.

HECKMAN, C.J. & BINDER, M.D. (1990) Neural mechanisms underlying the orderly recruitment of motoneurons. In *The Segmental Motor System.* eds. BINDER, M.D. & MENDELL, L.M. pp 182-204. Oxford University Press. New York.

HECKMAN, C.J. & BINDER, M.D. (1991) Analysis of Ia-inhibitory synaptic input to cat spinal motoneurons evoked by vibration of antagonist muscles. *J. Neurophysiol.* **66**, 1888-1893.

HECKMAN, C.J. & BINDER, M.D. (1993) Computer simulations of motoneuron firing rate modulation. *J. Neurophysiol.* **69**, 1005-1008.

HULTBORN, H., KATZ, R. & MACKEL, R. (1988) Distribution of recurrent inhibition within a motor nucleus. II. Amount of recurrent inhibition in motoneurons to fast and slow units. *Acta Physiol. Scand.* **134**, 363-374.

HULTBORN, H., LINDSTROM, S. & WIGSTORM, H. (1979) On the function of recurrent inhibition in the spinal cord. *Exp. Brain Res.* **37**, 399-403.

KERNELL, D. (1966) The repetitive discharge of motoneurones. In *Muscular Afferents and Motor Control. Nobel Symp. I.* ed. GRANIT, R. pp 351-362. Almqvist and Wiksell, Stockholm.

KERNELL, D. (1979) Rhythmic properties of motoneurones innervating muscle fibres of different speed in m. gastrocnemius medialis of the cat. *Brain Res.* **160**, 159-162.

KERNELL, D. (1983) Functional properties of spinal motoneurons and gradation of muscle force. *Adv. Neurol.* **39**, 213-226.

LINDSAY, A.D. & BINDER, M.D. (1991) Distribution of effective synaptic currents underlying recurrent inhibition in cat triceps surae motoneurons. *J. Neurophysiol.* **65**, 168-177.

MUNSON, J.B. (1990) Synaptic inputs to type-identified motor units. In *The Segmental Motor System.* eds. BINDER, M.D. & MENDELL, L.M. pp 291-307. Oxford University Press, New York.

POWERS, R.K., ROBINSON, F.R., KONODI, M.A. & BINDER, M.D. (1992) Effective synaptic current can be estimated from measurements of neuronal discharge. *J. Neurophysiol.* **68**, 964-968.

POWERS, R.K., ROBINSON, F.R., KONODI, M.A. & BINDER, M.D. (1993) Distribution of rubrospinal synaptic input to cat triceps surae motoneurons. *J. Neurophysiol.* **70**, 1460-1468.

RALL, W., BURKE, R.E., SMITH, T.G., NELSON, P.G. & FRANK, K. (1967) Dendritic location of synapses and possible mechanisms for the monosynaptic EPSP in motoneurons. *J. Neurophysiol.* **30**, 1169-1193.

REDMAN, S. (1976) A quantitative approach to the integrative function of dendrites. In *International Review of Physiology: Neurophysiology.* ed. PORTER, R. pp. 1-36. University Park Press, Baltimore.

SCHWINDT, P.C. & CALVIN, W.H. (1973a) Equivalence of synaptic and injected current in determining the membrane potential trajectory during motoneuron rhythmic firing. *Brain Res.* **59**, 389-394.

SCHWINDT, P.C. & CALVIN, W.H. (1973b) Nature of conductances underlying rhythmic firing in cat spinal motoneurons. *J. Neurophysiol.* **36**, 955-973.

SEGEV, I., FLESHMAN, J.W.J. & BURKE, R.E. (1990) Computer simulation of group Ia EPSPs using morphologically realistic models of cat alpha-motoneurons. *J. Neurophysiol.* **64**, 648-660.

SHAPOVALOV, A.I. (1972) Extrapyramidal monosynaptic and disynaptic control of mammalian alpha-motoneurons. *Brain Res.* **40**, 105-115.

STEIN, R.B. & BERTOLDI, R. (1981) *The Size Principle: A Synthesis of Neurophysiological Data.* Karger, Basel.

STUART, G.J. & REDMAN, S.J. (1990) Voltage dependence of Ia reciprocal inhibitory currents in cat spinal motoneurones. *J. Physiol.* **420**, 111-125.

WESTCOTT, S.L., POWERS, R.K., ROBINSON, F.R., KONODI, M.A. & BINDER, M.D. (1993) Comparison of vestibulospinal synaptic input and Ia afferent synaptic input in cat triceps surae motoneurons. *The Physiologist* **36**, A20.

AFFERENT, PROPRIOSPINAL AND DESCENDING CONTROL OF LUMBAR MOTONEURONES IN THE NEONATAL RAT

Aharony Lev-Tov

Department of Anatomy
The Hebrew University Medical School
P.O.Box 12272, Jerusalem 91120, Israel

INTRODUCTION

The motor unit is considered as the unitary output element of the motor system. The motoneurone which comprises the neural part of the motor unit, receives synaptic input from various sources. These include afferent input from muscle, tendon, and skin receptors, descending input from supraspinal centres, and ascending and descending input from propriospinal projections (Baldissera, Hultborn & Illert, 1981). The connectivity of these pathways and the efficacy of their synapses during the embryonic and post-embryonic development of the central nervous system, determine the complexity of the behavioral Repertoire exhibited by the embryo and the newborn animal.

Recent studies have shown that the synaptic input to lumbar motoneurones from afferent, propriospinal and most supraspinal (except corticospinal) pathways is formed during embryonic development (Kudo, Furakawa & Okado, 1993). It has also been shown that rat embryos and neonatal rats are capable of generating complex motor functions (Beckoff & Lau, 1980; Kudo, Ozaki & Yamada, 1991; Greer, Smith & Feldman, 1992), which might require the involvement of these pathways. It is therefore interesting to assess whether the early-established connectivity of synaptic input to lumbar motoneurones is functional, and whether the physiological properties of these synaptic contacts reflect a mature functional specificity. Electrophysiological studies of reticulospinal and propriospinal pathways in the adult (see Baldissera et al., 1981; Armstrong, 1988) and recent ultrastructural studies of axonally transported tracers (Holsteg & Kuypers, 1987) have shown, that these pathways, have in addition to their intermediate grey terminations, extensive direct projections to lamina IX motoneurones. Based on our recent studies (Lev-Tov & Pinco, 1992; Pinco & Lev-Tov, 1993a & b; Floeter & Lev-Tov, 1993; Pinco & Lev-Tov, 1994) the present work describes the synaptic pharmacology and the physiological properties of monosynaptic projections of afferent, supraspinal and propriospinal pathways to lumbar motoneurones in the neonatal rat. All experiments were performed on the isolated spinal cord or brain stem-spinal cord preparations of 1-6 day old rats, using sharp electrode intracellular recordings from lumbar motoneurones, suction electrode recordings from ventral roots, and extracellular stimulation of dorsal and ventral roots, of white matter funiculi, and of descending tracts in the brain stem.

RECEPTORS INVOLVED IN MONOSYNAPTIC EXCITATION OF LUMBAR MOTONEURONES

Dorsal root afferent and medial longitudinal fasciculus (MLF) projections to lumbar motoneurones in the neonatal rat include mono- and polysynaptic excitatory, and polysynaptic inhibitory pathways (for details see Ziskind-Konhaim, 1990; Lev-Tov & Pinco, 1992; Pinco & Lev-Tov, 1993a; Floeter & Lev-Tov, 1993). The projections of ventrolateral funiculus (VLF) axons (which carry mainly propriospinal and medullary reticulospinal fibres) to lumbar motoneurones include, in addition to the pathways mentioned above, a presumed monosynaptic inhibitory pathway (Pinco & Lev-Tov, 1994; see Figure 1). Short latency EPSPs of a presumed monosynaptic nature could be resolved from the composite PSPs produced by stimulation of dorsal root afferent (Lev-Tov & Pinco, 1992; Pinco & Lev-Tov, 1993a), MLF (Floeter & Lev-Tov, 1993) and VLF (Pinco & Lev-Tov, 1994) projections to lumbar motoneurones. This is achieved by addition of mephenesin (which has been shown specifically to reduce polysynaptic activity in the spinal cord) and the $GABA_A$ and glycine receptor antagonists, bicuculline and strychnine. These short latency EPSPs were shortened by addition of the NMDA receptor blocker 2-amino-5-phosphonovaleric acid (APV), and revealed a 6-cyano-7-nitroquinoxaline (CNQX; the non-NMDA receptor antagonist) resistant component which could have been then blocked by APV.

It has therefore been suggested that the EPSPs produced by activation of the three different sources (dorsal root afferents, MLF and VLF) of monosynaptic excitatory input to lumbar motoneurones are mediated primarily by non-NMDA receptors, with a significant contribution of an NMDA receptor-mediated component that is activated at normal concentrations (1mM) of Mg^{2+} and at normal resting potentials.

DYNAMIC PROPERTIES OF SYNAPTIC INPUT CONVERGING TO LUMBAR MOTONEURONES

Presynaptic modulation of synaptic transmission is one of the factors by which the

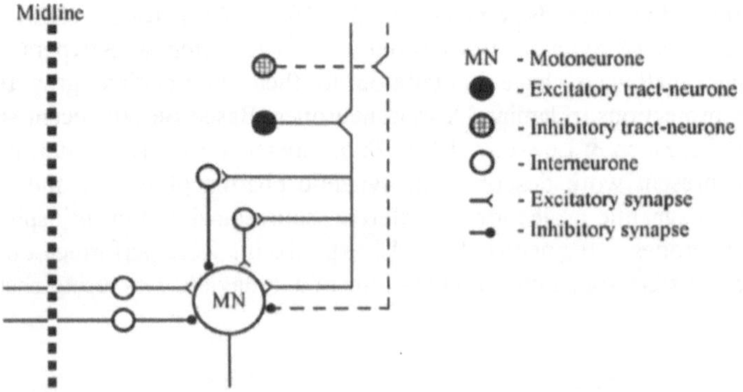

Figure 1. Projections of propriospinal axons travelling in the VLF to lumbar motoneurones in the neonatal rat. The suggested projections include ipsilateral excitatory and inhibitory monosynaptic pathways, and ipsi- and contralateral excitatory and inhibitory polysynaptic pathways (Lev-Tov & Pinco, 1994).

input to output ratio of central and peripheral synapses is governed. Such presynaptic modulation can be achieved by autoreceptors located on the presynaptic elements, by inhibitory axo-axonic synapses (presynaptic inhibition) and by the frequency of presynaptic activation. Tetanisation of Ia afferents in the adult cat spinal cord induces facilitation and potentiation of Ia EPSPs in this way, which are followed by synaptic depression and post-tetanic potentiation (Curtis & Eccles, 1960; Kuno, 1964; Lev-Tov, Pinter & Burke, 1983; Lev-Tov, Meyers & Burke, 1988). Studies of the neuromuscular junction have suggested that frequency modulation of synaptic potentials is generally determined by a dynamic interplay between processes that act to increase the release of transmitter (facilitation and potentiation) and those that act to decrease it (synaptic depression; see Barrett & Magleby, 1976, for review). The balance between these frequency modulated processes can vary for synaptic input from different sources and among synaptic contacts formed by the same source on different target neurones (see Mendell, 1984).

Dorsal root afferent projections

Our studies of the monosynaptic projections of dorsal root afferents to lumbar motoneurones in the neonatal rat spinal cord have shown that the EPSPs developed a frequency dependent depression even over a range of low stimulation frequencies (0.0166-1 Hz; Lev-Tov & Pinco, 1992). Tetanisation of these afferents resulted in a more severe depression with no sign of frequency potentiation (Pinco & Lev-Tov, 1993b). The depression was not accompanied by changes in the passive properties of the motoneurones, and had similar effects on the non-NMDA and the NMDA receptor-mediated components of the dorsal root afferent EPSPs (Lev-Tov & Pinco, 1992; Pinco & Lev-Tov, 1993a). Superfusion of the preparations with low-calcium high-magnesium Krebs saline, or addition of low concentrations of the $GABA_B$ receptor agonist L (-) baclofen reduced the amplitude of the EPSPs by a substantial amount. Under these conditions, the depression induced by low frequency stimulation was virtually abolished and EPSP facilitation and potentiation were revealed by double-pulse and high frequency stimulation (Pinco & Lev-Tov, 1993b).

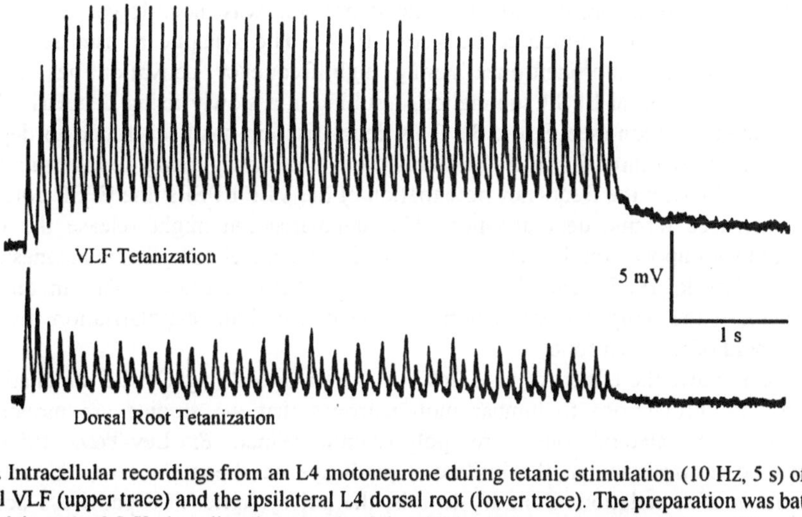

Figure 2. Intracellular recordings from an L4 motoneurone during tetanic stimulation (10 Hz, 5 s) of the ipsilateral VLF (upper trace) and the ipsilateral L4 dorsal root (lower trace). The preparation was bathed in normal calcium (1 mM) Krebs saline that contained (in addition to its usual constituents, see Pinco & Lev-Tov, 1994) 1mM mephenesin, 5mM strychnine and 10 mM bicuculline. Note that the substantial potentiation of VLF EPSPs was accompanied by slight (3-4 mV) depolarisation and that the dorsal root afferent EPSPs exhibited marked tetanic depression during the stimulus train.

Based on these results, it has been suggested that the presynaptic properties of immature afferent projections to motoneurones in the neonatal rat spinal cord dictate prolonged EPSP depression during low and high frequency stimulation (Lev-Tov & Pinco, 1992; Pinco & Lev-Tov, 1993b) and that superimposed facilitation and potentiation are normally masked, and can be expressed only when the basic level of transmitter release is substantially reduced (Pinco & Lev-Tov, 1993b).

MLF projections

The monosynaptic projections of reticulospinal MLF axons to lumbar motoneurones have been shown to exhibit twin-pulse facilitation at short interpulse intervals (like the adult Ia afferent input) and to develop depression with a similar (although somewhat shorter) time course to that observed for the afferent input in the neonate during low-frequency repetitive stimulation (Floeter & Lev-Tov, 1993). These findings have also been ascribed to possible immature properties of the medial reticulospinal pathway.

VLF projections

It might be argued that the predominant prolonged synaptic depression described for afferent and reticulospinal MLF projections to lumbar motoneurones (see above) is a consequence of the *in vitro* conditions under which the experiments were performed. This possibility has been excluded recently by our studies of the projections of propriospinal and medullary reticulospinal axons travelling in the ventrolateral funiculus (VLF) to lumbar motoneurones (Pinco & Lev-Tov, 1994). These studies have shown that the short latency excitatory VLF projections to lumbar motoneurones in the neonatal rat were characterised by the absence of depression during low frequency stimulation, and by substantial frequency potentiation during tetanic stimulus trains under normal in vitro conditions (see Figure 2). The frequency potentiation of the VLF evoked EPSPs was further increased as the bathing solution in the experiments was switched to low-calcium high-magnesium Krebs saline. The presence of a slight synaptic depression of VLF EPSPs could be detected toward the end of tetanic trains, because the EPSPs were less potentiated than in the beginning of the train (see Figure 2). As expected, the extent of this depression was much lower in low-calcium high-magnesium solutions (Pinco & Lev-Tov, 1994). The synaptic transmission of VLF projections to lumbar motoneurones during high frequency activation has also been characterised by temporal EPSP summation, resulting in a slowly rising depolarisation due to the tetanic stimulation (see Figure 3). This depolarisation has been found to depend on the frequency and the intensity of the stimulus train. The slowly rising depolarisation might act synergistically with the frequency potentiation of the VLF EPSPs: the EPSP potentiation leads to increased tetanic depolarisation. This depolarisation might release the voltage-dependent block imposed on the NMDA receptor/ionic channel complex by magnesium ions (Ascher & Nowak, 1987), and thereby prolong the EPSP duration. This in turn might further increase the temporal summation and the concomitant depolarisation toward the firing threshold of motoneurones.

Figure 3 shows the effects of ipsi- (left) vs. contra-lateral (right) VLF stimulation. The ipsilateral VLF projections to lumbar motoneurones that we studied are monosynaptic whereas the contralateral ones are polysynaptic (Pinco & Lev-Tov, 1994). The monosynaptic transmission in the ipsilateral pathways has been shown to be reliable at different rates of stimulation. Contrary to this, the polysynaptic VLF pathways developed intermittent transmission at high stimulation rates (as is the case with polysynaptic pathways in many other systems). The decay of the depolarisation induced by polysynaptic activation (right record) corresponded to the appearance of polysynaptic transmission failures. These findings suggest that the slowly rising depolarisation reflects temporal PSP summation rather than some alternative mechanism (such as extracellular accumulation of ions or specific conductance changes).

FUNCTIONAL SPECIALISATION OF SYNAPTIC INPUT CONVERGING ON NEONATAL RAT MOTONEURONES

In summary, the work reviewed in this chapter shows that the excitatory monosynaptic afferent and MLF projections to lumbar motoneurones are characterised by prolonged synaptic depression, and that the monosynaptic projections of the VLF to the same motoneurones are characterised by strong frequency potentiation and a slowly rising background depolarisation. Generally speaking, the reticulospinal and propriospinal pathways travelling in the VLF are involved in some of the fundamental and early-established behaviours of the embryo and the neonate. These include (a) supraspinal control of reflex pathways (reviewed in Baldissera et al. 1981), (b) control of the spinal central pattern generator for locomotion (reviewed in Armstrong 1988; Jordan, 1991), (c) multisegmental coordinated actions of the spinal cord (Bekoff & Trainer 1979, Bekoff & Lau, 1980), and (d) synchronisation of motoneurone activity and transmission of rhythmic synaptic drive to caudal motoneurones, as suggested by recent studies (Ho & O'Donovan, 1993). Our results indicate that projections of VLF (mainly of the propriospinal component) on lumbar motoneurones in the developing spinal cord are functionally specialised, and that a reliable synaptic transmission across these synapses may well be a basic requirement for an efficient execution of the motor control functions described above.

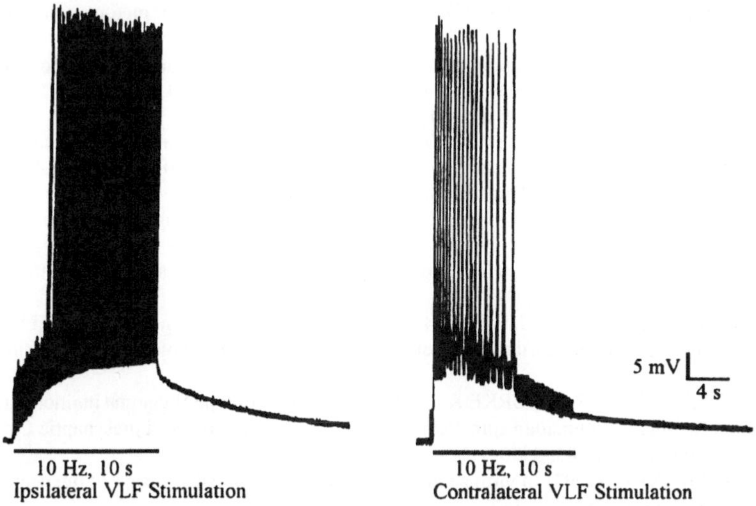

Figure 3. Temporal summation of VLF PSPs. Slow sweep display of intracellular recordings from an L4 motoneurone during and following tetanic trains (10Hz, 10s) that were applied to the ipsi- and contralateral VLF at the L6 level. Tetanic potentiation accompanied by temporal summation of the ipsilaterally elicited short-latency PSPs, leads to motoneurone firing (left). The depolarisation induced by temporal summation of the contralaterally elicited polysynaptic PSPs at the beginning of the train (right), was blocked with the occurrence of transmission failures toward the end of the train. The preparation was bathed in normal Krebs saline. The depolarisation seen here is more substantial than that seen in Figure 2 due to the higher intensity of stimulation used here. This enabled the recruitment of slower monosynaptic projections and thereby increased the probability of temporal EPSP summation.

ACKNOWLEDGMENTS

This work was supported by grant No. 91-0005311 From the US-Israel Binational Science Foundation (BSF), Jerusalem, Israel.

REFERENCES

ASCHER P. & NOWAK, L. (1987) Electrophysiological studies on NMDA receptors. *TINS* **10**, 284-288.

ARMSTRONG, D.M. (1988) The supraspinal control of mammalian locomotion. *J. Physiol.* **405**, 1-37.

BALDISSERA, F., HULTBORN, H., & ILLERT, M. (1981) Integration in spinal neuronal systems. In *Handbook of Physiology, The Nervous System, Vol II, part 1* ed. BROOKS, V.B. pp 509-595. American Physiological Society, Bethesda.

BARRETT, E. F. & MAGLEBY, K. L. (1976). Physiology of cholinergic transmission. In *Biology of Cholinergic Function.* eds. GOLDBERG, A.M. & HANIN, E.. pp 29-100. Raven Press, New York.

BEKOFF, A. & LAU, B. (1980) Interlimb coordination in 20-day-old rat fetuses. *J. Exp. Zool.* **214**, 173-175.

BEKOFF, A. & TRAINER, W. (1979) The development of interlimb coordination during swimming in postnatal rats. *J. Exp. Biol.* **83**, 1-11.

CURTIS, D.R. & ECCLES, J.C. (1960) Synaptic action during and after repetitive stimulation. *J. Physiol.* **150**, 374-398.

FLOETER, M.K. & LEV-TOV, A. (1993) Excitation of lumbar motoneurons by the medial longitudinal fasciculus in the in vitro brainstem spinal cord preparation of the neonatal rat. *J. Neurophysiol.* **70**, 2241-2250.

GREER, J.J., SMITH, J.C. & FELDMAN, J.L. (1992) Respiratory and locomotor patterns generated in the fetal rat brainstem-spinal cord in vitro. *J. Neurophysiol.* **67**, 996-999.

HO, S. & O'DONOVAN, M.J. (1993) Regionalization and intersegmental coordination of rhythm generating net works in the spinal cord of the chick embryo. *J. Neurosci.* **13**, 1354-1371.

HOLSTEG, J.C. & KUYPERS, H.G.J.M. (1987) Brainstem projections to spinal motoneurons: an update. *Neuroscience* **23**, 809-821.

JORDAN, L.M. (1991) Brainstem and spinal cord mechanisms for the initiation of locomotion. In *Neurological Basis of Human Locomotion.* eds. SHIMAMURA, M., GRILLNER, S. & EDGERTON, V.R. pp.3-20. Japanese Science Society, Tokyo.

KUDO, N. OZAKI, S. & YAMADA, T. (1991) Ontogeny of rhythmic activity in the spinal cord of the rat. In *Neurological Basis of Human Locomotion.* eds. SHIMAMURA, M., GRILLNER, S. & EDGERTON, V.R. pp.127-136. Japanese Science Society, Toyko.

KUDO, N. FURAKAWA, F. & OKADO, N. (1993) Development of descending fibers to the rat embryonic spinal cord. *Neurosci. Res.* **16**, 131-141.

KUNO, M. (1964) Mechanisms of facilitation and depression of the excitatory synaptic potentials in spinal motoneurones. *J. Physiol.* **175**, 100-112.

LEV-TOV, A., PINTER, M.J. & BURKE, R.E. (1983) Post-tetanic potentiation of group Ia EPSPs: possible mechanisms for differential distribution among medial gastrocnemius motoneurons. *J. Neurophysiol.* **50**, 379-398.

LEV-TOV, A., MEYERS D.E.R. & BURKE R.E. (1988) Activation of type B gamma aminobutiric acid receptors in the intact mammalian spinal cord mimics the effects of reduced presynaptic Ca^{++} influx. *Proc. Nat. Acad. Sci. USA* **85**, 5530-5534.

LEV-TOV, A. & PINCO M. (1992) In vitro studies of prolonged synaptic depression in the neonatal rat spinal cord. *J. Physiol.* **447**, 149-169.

MENDELL, L.M. (1984) Modifiability of spinal synapses. *Physiol. Rev.* 64, 260-324.

PINCO, M. & LEV-TOV, A. (1993a) Synaptic excitation of α-motoneurons by dorsal root afferents in the neonatal rat spinal cord. *J. Neurophysiol.* **70**, 406-417.

PINCO, M. & LEV-TOV, A. (1993b) Modulation of monosynaptic excitation in the neonatal rat spinal cord. *J. Neurophysiol.* **70**, 1151-1158.

PINCO, M. & LEV-TOV, A. (1994) Synaptic transmission between ventrolateral funiculus axons and lumbar motoneurons in the isolated spinal cord of the neonatal rat. *J. Neurophysiol.* (in press).

ZISKIND-CONHAIM, L. (1990) NMDA receptors mediate poly- and monosynaptic potentials in motoneurons of rat embryos. *J. Neurosci.* **10**, 125-135.

MODULATION OF TRANSMISSION IN REFLEX PATHWAYS OF TRIGEMINAL MOTONEURONES

Kwabena Appenteng, John C. Curtis,
Paul Grimwood, Ming-Yuan Min
and Hsiu-Wen Yang

Department of Physiology
University of Leeds
Leeds LS2 9NQ, UK

MONOSYNAPTIC EPSPS ELICITED BY TRIGEMINAL SPINDLE AFFERENTS AND LAST-ORDER INTERNEURONES

Evidence obtained by intracellular spike-triggered averaging has led to the belief that the monosynaptic excitatory connections of jaw-elevator muscle spindle afferents with jaw-elevator motoneurones may be weak. In the first such report, EPSPs were found of mean amplitude 18μV (range = 3.1 - 60μV: pentobarbitone anaesthetised cats: Appenteng, O'Donovan, Somjen, Stephens & Taylor, 1978). A subsequent study in the guinea-pig (ketamine anaesthesia) which gave a mean averaged EPSP amplitude of 21.5μV, appeared to support this (Nozaki, Iriki & Nakamura, 1985). However, a more recent study in the rat (pentobarbitone anaesthetisia) gave a higher mean amplitude of 64μV (range = 7 - 289μV) and a connectivity of 52% (Grimwood, Appenteng & Curtis, 1992). Even with this revised estimate the steady depolarisation which might be caused by all the muscle spindles in masseter would only amount to some 2.2mV (see Harrison & Taylor, 1981; Munson, Fleshman & Sypert, 1982). This is based on 200 afferents (Gottlieb, Taylor & Bosley, 1984) firing at 100 impulses/s.

A study of the excitatory connections of premotor interneurones located in the rostral pole of the trigeminal nucleus oralis gave figures of 17μV for the mean averaged EPSP amplitude (range = 7 - 48μV) and a connectivity of 50% (Grimwood et al., 1992). There appear to be about 150 of these interneurones (Appenteng & Girdlestone, 1987),and assuming that all of them connect with each elevator motoneurone, and that they fire at a steady rate of 100 impulses/s, then they would produce a steady depolarisation of only 405μV (Grimwood et al., 1992). Thus, both types of excitatory input to elevator motoneurones so far studied appear to be weak.

Such estimates of steady depolarisation by excitatory synapses depend on the reliability of estimates of average EPSP amplitude and functional connectivity. If a significant number of presynaptic spikes fail to elicit detectable postsynaptic events (i.e. failures of transmission), then the potential strength and incidence of connections may be underestimated. Underestimation of both mean EPSP amplitude and functional connectivity would result from tonic pre- or postsynaptic inhibition. Quantal analysis has provided the one generally accepted means of estimating both the magnitude and probability of occurrence of the underlying components of an EPSP. However, this method does make stringent demands with regard to the noise which can be associated with such recordings (Jack, Redman & Wong, 1981; Redman, 1990). The low signal to noise ratio in the *in-vivo* recordings of Grimwood et al. (1992) prevented use of this method and so an alternative method, based on the visual analysis of sweeps, was developed for estimating the incidence of failures.

This method depends on examining each sweep for an EPSP within a time window of ±0.3ms relative to the start of the averaged EPSP (Figure 1: Grimwood et al., 1992; Curtis & Appenteng, 1993), and the use of a correction for randomly occuring EPSPs. Figure 2 shows the relationship between corrected incidence of EPSPs and averaged EPSP amplitude for a sample of spindle afferents. Two points follow from this relationship. First, if there were no failures of transmission, spindle afferents should give averaged EPSPs of at least 150μV. Thus, the potential strength of connections may have been underestimated in all studies so far performed. Secondly, variations in the incidence of failures of transmission may be a means of modulating transmission at synapses on trigeminal motoneurones.

Some support for these conclusions has come from more recent *in-vitro* studies using the whole cell patch method to obtain intracellular recordings from motoneurones. Histograms of the amplitude of spontaneous EPSPs are multimodal, with regularly spaced peaks being evident when the data are plotted at a number of different bin widths (Figure

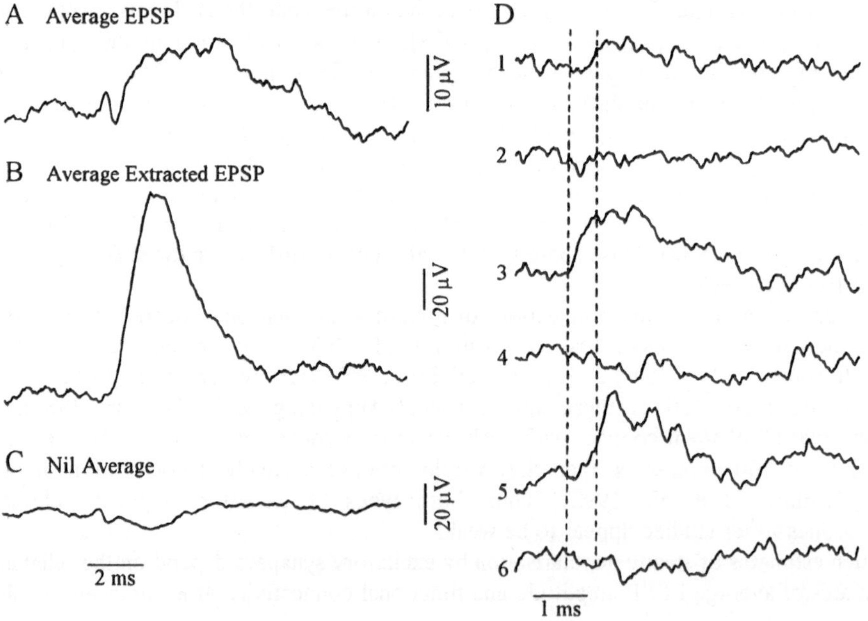

Figure 1. EPSP elicited by a temporalis spindle afferent intracellularly from a masseter motoneurone. EPSPs were obtained by averaging from (A) all sweeps (n = 367), (B) those sweeps containing an EPSP (n = 48; see D) and (C) those sweeps without an EPSP (n = 319). (D) Fluctuations in the EPSPs. The dashed lines enclose a window of ±0.3 ms relative to the start of the averaged EPSP (shown in A). Traces "D1", "D3" and "D5" show examples of EPSPs which commenced within this window and so were assumed to be triggered by the afferent under study. "D2", "D4" and "D6" show examples of failures of transmission.

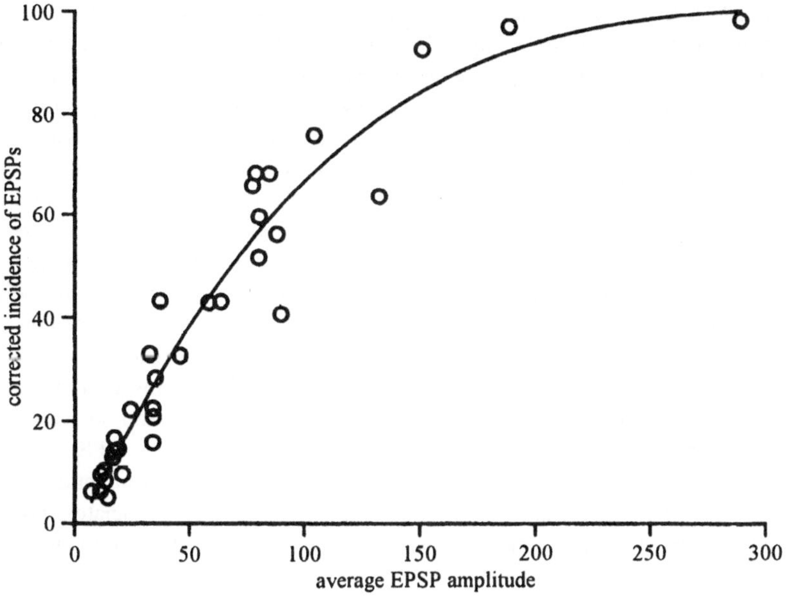

Figure 2. Plot of the corrected incidence of EPSPs against amplitude of averaged EPSP obtained by inclusion of all sweeps (r = 0.97).

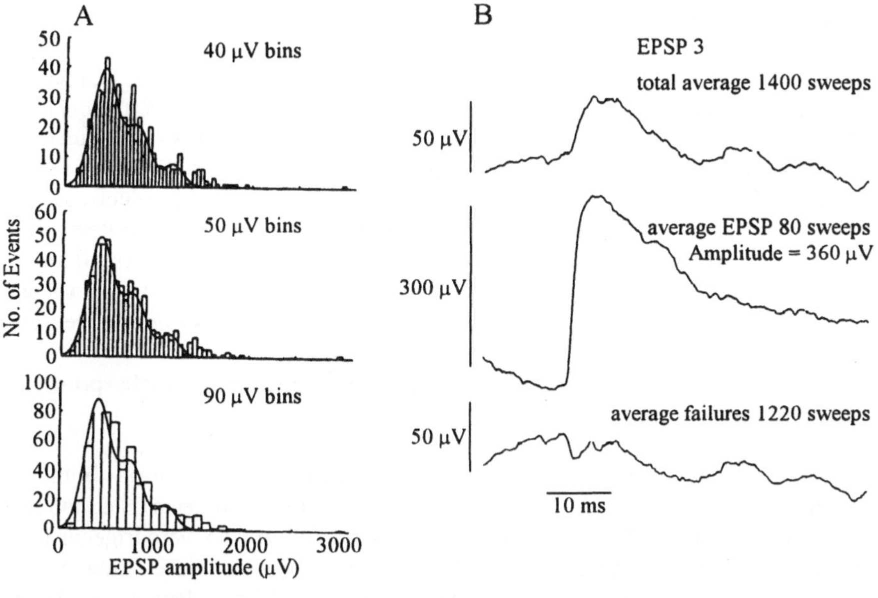

Figure 3. Analysis of spontaneous (A) and triggered (B) EPSPs recorded in a motoneurone *in-vitro* using the whole-cell patch method. A. Histogram plots of distribution of amplitudes of single EPSPs. The plots show the same data plotted at different bin widths. The continuous line shows the fit obtained by using the sum of 3 Gaussians. The means (± s.d.) of the 3 Gaussians were 405±128, 760±136 and 1154±104μV (fit range = 101.6 -1358 μV; n = 553). The mean quantal amplitude was 385±26.27μV. B. Averaged EPSP elicited by a single interneurone in the same motoneurone. Top trace was obtained by inclusion of all sweeps, middle trace by using only sweeps judged to contain EPSPs and the lower trace sweeps judged to contain no EPSPs.

3A). The position of the peaks has been determined by fitting the unbinned data with the sum of three Gaussian curves using maximum likelihood statistics (Jonas, Major & Sakmann, 1993). The mean (± s.d.) separation between the peaks (i.e. quantal amplitude) in the example shown in Figure 3A was 385±26.27μV, but values of mean separation of 210 - 481μV have been obtained for other cells studied. Regularly spaced peaks were not observed in the amplitude histograms of the background noise in the cells. Taken together, these observations are compatable with excitatory transmission at synapses on trigeminal motoneurones being quantal in nature, with the spacing between the peaks representing the quantal amplitude. An implication of this result is that the quantal amplitude in trigeminal motoneurones may be uniform, as in spinal motoneurones (see Redman, 1990). Figure 3B shows the averaged EPSP (amplitude 50μV) recorded in the same cell when trigerring from a single interneurone. Visual inspection revealed that while 13% (Figure 3B) of the sweeps comprising the average contained an EPSP occuring within ±0.5ms of the start of the averaged EPSP, the majority of sweeps (87%: Figure 3C) did not, and so were judged to represent failures of transmission. The amplitude of the averaged EPSP obtained by excluding failures was 360μV (Figure 3B), a value close to the mean quantal amplitude of 385μV. This suggests that transmission at these synapses was associated with a high incidence of failures, and that release in this case was from a single functional site (Curtis, Appenteng & Min, unpublished observation).

MODULATION OF TRANSMISSION AT SYNAPSES OF SPINDLE AFFERENTS ON MOTONEURONES

Frequency-dependent modulation of compound EPSP

Frequency-dependent potentiation of transmission has been described previously at synapses of spindle afferents on spinal motoneurones (Curtis & Eccles, 1960; Honig, Collins & Mendell, 1983) but not on trigeminal motoneurones. Figure 4A shows the potentiation of the compound EPSP elicited in masseter synergist motoneurones following paired pulse stimulation of the masseter nerve. At intervals of 5-19ms paired pulses produced statistically significant potentiation of transmission, and as shown in Figure 4B, the potentiation was maintained during repetitive stimulation (Grimwood et al, in press). Thus, for afferent firing frequencies up to 200 imp/s, changes in the firing frequency may result in changes of approximately 30% in amplitude of the averaged EPSP elicited in elevator motoneurones.

Frequency-independent modulation of averaged EPSP elicited by single spindle afferents and interneurones

The evidence for frequency-independent modulation comes from both *in-vivo* (pentobarbitone anaesthetised rat) and *in-vitro* (tissue slice) experiments using intracellular spike-triggered averaging. The presynaptic neurones studied *in-vivo* were trigeminal muscle spindle afferents and interneurones, while only interneurones were studied *in-vitro*. Interneurones studied in each case were in the rostral pole of the nucleus oralis (Appenteng & Girdlestone, 1987).

Figure 5 shows an example from the *in-vivo* data of the modulation in transmission at the connection of a muscle spindle afferent with a motoneurone. Averages were computed from consecutive blocks of 100 sweeps in order to assess the variability in amplitude of the averaged EPSP with time. Its amplitude was initially 118μV and rose to 194μV before decreasing (A). The 60% change was not accompanied by any change in the afferent firing rate (C) and the change in membrane potential of the motoneurone during the recording period was not sufficient to account for the increase (B). Out of 80 averaged EPSPs obtained *in-vivo*, only 8 were recorded under conditions in which the membrane potential of the motoneurone remained approximately constant (changed by <10mV) and the firing rate of the presynaptic neurone remained constant. In 3 of the 8 cases, averaged EPSPs showed modulations in amplitude, between 40 and 80μV (80% to 90%).

Four out of 9 averaged EPSPs from interneurones changed in amplitude by approximately 100μV during the period of recording. In each case the changes occurred in the absence of significant changes in either the membrane potential or the firing rate of the presynaptic neurone (Figure 6).

SITE OF THE EPSP MODULATION

The 7 cases in which the EPSPs were significantly modulated (3 *in-vivo*, 4 *in-vitro*) were analysed visually to estimate the probability of release. This method is most useful when the probability is low, but is less useful in cases where the probability is high enough to result in release from at least one functional site with each presynaptic impulse. Under the latter conditions, any variations in release probability from other functional sites would not be detected as a change in incidence. Four of the averaged EPSPs (3/4 *in-vitro* and 1/3 *in-vivo*) showed statistically significant linear relationships between the corrected incidence of EPSPs and averaged EPSP amplitude in the block. This implies that the modulations involved an alteration in the probability of release and so could be ascribed, at least in part, to a presynaptic locus.

POSSIBLE MECHANISMS MEDIATING THE PRESYNAPTIC MODULATION

The *in-vitro* observations show that the modulation is mediated by mechanisms, or pathways, within or close to the motor nucleus. There is at present no specific evidence as to what these may involve but, in addition to ruling out changes in either the motoneurone membrane potential or firing rate of the presynaptic neurone, we can also rule out an involvement of the N-methyl-D-aspartate (NMDA) receptor as the modulations can be seen in tissue-slices continuously bathed in the NMDA receptor antagonist APV (e.g Figure 6). This suggests that activation of the NMDA sub-type of the glutamate receptor is not required for either the induction or maintenance of the changes seen at synapses on

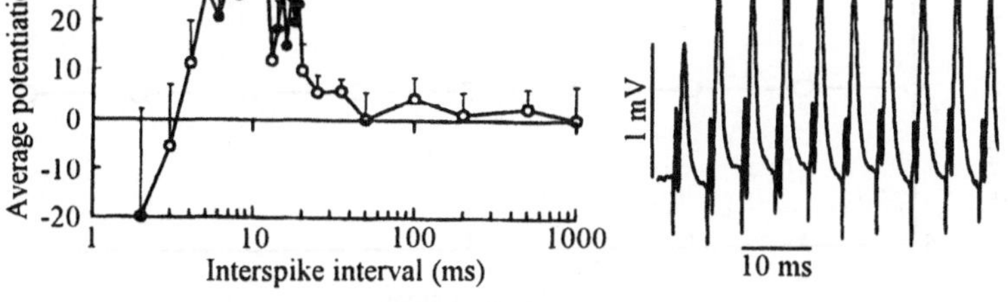

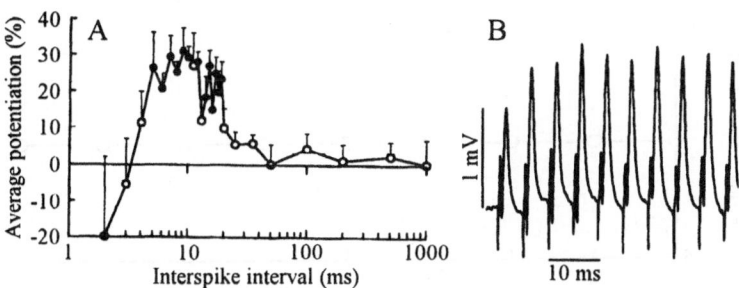

Figure 4. Frequency potentiation of compound monosynaptic EPSP. A. Variation of mean (n=10 sweeps) potentiation with interval between first and second pulse in a pair. Potentiation calculated from average amplitude of $(EPSP_2-EPSP_1)/EPSP_1$. Each point shows the mean (± S.E.) of data from at least 7 motoneurones. Filled circles show points with statistically (t-test) significant changes in amplitude of $EPSP_2$. B. Average (n=10 sweeps) change in amplitude of monosynaptic EPSP during repetitive stimulation of masseter nerve at 100 Hz.

trigeminal motoneurones. Evidence that glutamate is the transmitter mediating the EPSPs, at least those seen *in-vitro*, has come from recent work involving the recording of spontaneous EPSPs recorded in slices bathed in tetrodotoxin (i.e. miniture EPSPs; mEPSPs) so as to block action potential generation. Under these conditions mEPSP activity can be completely abolished by bath application of the glutamate receptor antagonist CNQX, indicating that the activity is mediated by glutamate acting via AMPA/kainate receptors (Yang, Appenteng, Curtis, Min & Saha, 1994).

Our current research is concerned with identifying the receptor sub-types present which can modulate transmission at synapses on motoneurones. The approach adopted has been to assess the effects of various agonists and antagonists on mEPSP activity. One preliminary result is that bath application of $GABA_B$ receptor antagonist (2-

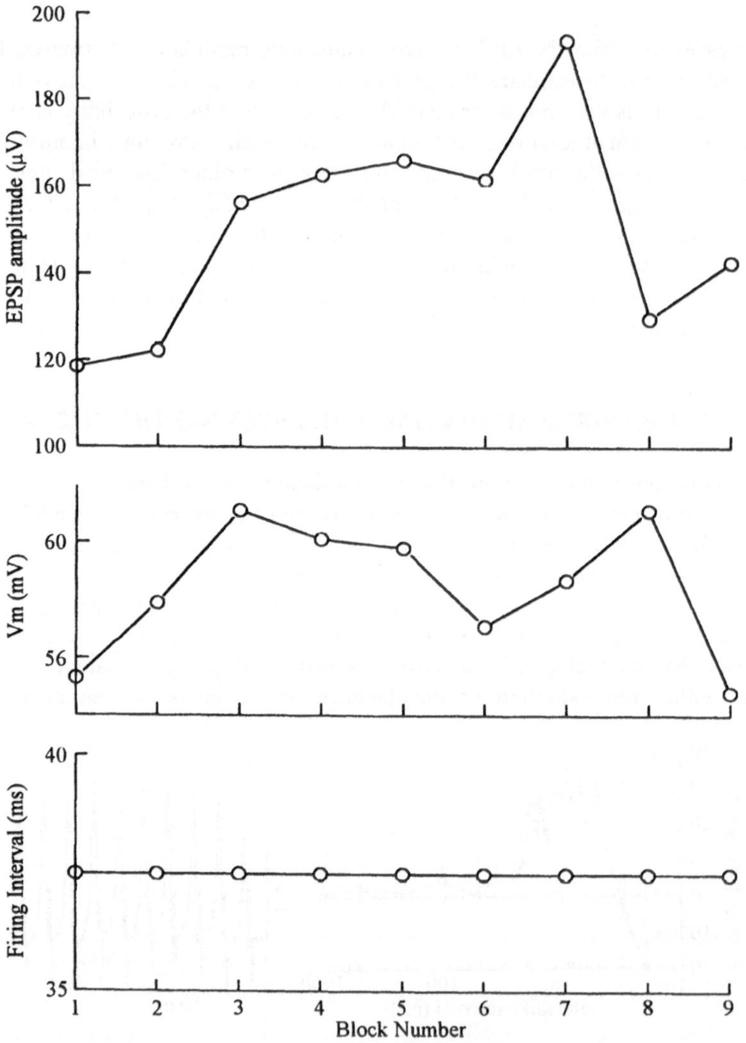

Figure 5. Frequency-independent modulation of monosynaptic EPSPs elicited by a spindle afferent in an elevator motoneurone. Top panel shows variations in amplitude of averaged EPSPs obtained by averaging sequential blocks of 100 sweeps. Middle trace shows the mean V_m of the motoneurone and the lower trace the mean firing frequency of the afferent during the blocks of sweeps. The changes in EPSP amplitude were not correlated with either the motoneurone V_m or the afferent firing frequency.

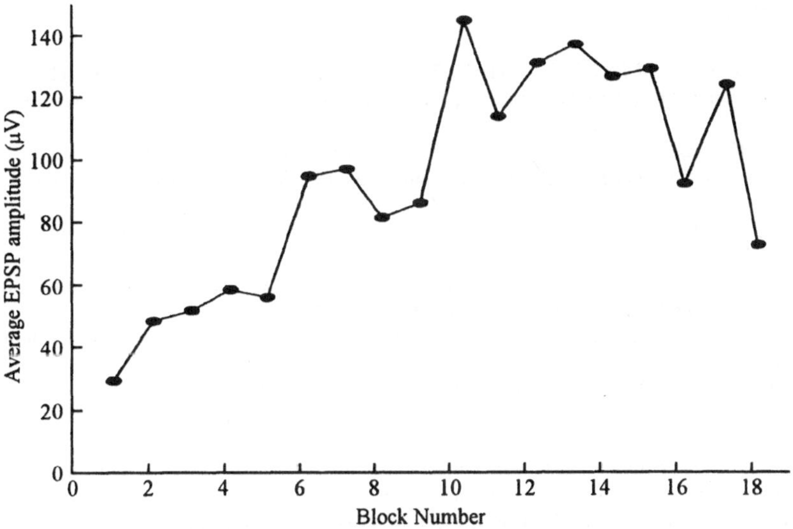

Figure 6. Variations in amplitude of averaged EPSPs obtained *in-vitro* from a trigeminal motoneurone. Note the progressive three-fold increase in the averaged EPSP amplitude. Firing rate of the interneurone was irregular but varied between 5-10 imp/s throughout the recording period. Membrane potential of the motoneurone was 53mV during the first block, decreased to 50mV by the start of block 10, and then decreased to 45mV between block 10 and the last block in the run.

hydroxysaclofen; 50 or 100 μM) can increase both frequency and amplitude of mEPSPs (Figure 7). The increase in mEPSP frequency is assumed to represent a presynaptic change while the increase in the mEPSP amplitude may represent an additional postsynaptic change. These results are compatible with the suggestion that transmission at motoneurone synapses may normally be subject to a tonic inhibition and that alterations in the level of this inhibition may underly the frequency-independent modulations in transmission seen both *in-vivo* and *in-vitro*. The reason why transmission should be subject to such control may be that the jaw-closer muscles are far more powerful than the jaw-openers. Thus the two sets of muscles operating at the temporomandibular joint are unequal antagonists and we suggest that effective antagonism of the actions of the jaw-closing muscles may require inhibition of excitatory transmission to the closing motoneurones. Thus the tonic inhibition of excitatory transmission to elevator motoneurones may be part of a conservative strategy to guard against their uncontrolled activity.

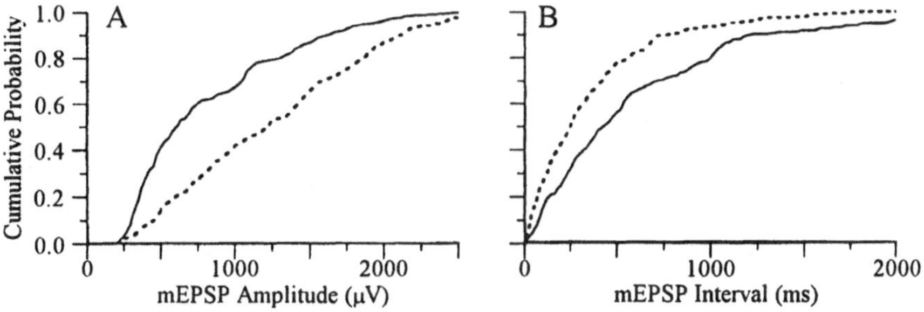

Figure 7. Plots of cummulative probability of mEPSP amplitudes (A) and the interval between mEPSPs (B). Continuous lines show data obtained in control conditions (i. e. TTX alone) and dashed lines, data obtained in the presence of 2-hydroxysaclofen (100 μM).

REFERENCES

APPENTENG, K. & GIRDLESTONE, D. (1987) Transneuronal transport of wheat germ agglutinin-conjugated horseradish peroxidase into trigeminal interneurones of the rat. *J. Comp. Neurol.* **258**, 387-396.

APPENTENG, K., O'DONOVAN, M.J., SOMJEN, G., STEPHENS, J.A. & TAYLOR, A. (1978) The projection of jaw elevator muscle spindle afferents to the fifth nerve motor nucleus. *J. Physiol.* **279**, 409-423.

CURTIS, D.R. & ECCLES, J.C. (1960) Synaptic action during and after repetitive stimulation. *J. Physiol.* **150**, 374-398.

CURTIS, J.C. & APPENTENG, K. (1993) The electrical geometry, electrical properties and synaptic connections onto rat V motoneurones *in-vitro*. *J. Physiol.* **465**, 85-119.

GOTTLIEB, S., TAYLOR, A. & BOSLEY, M.A. (1984) The distribution of afferent neurones in the mesencephalic nucleus of the fifth nerve of the cat. *J. Comp. Neurol.* **228**, 273-283.

GRIMWOOD, P.D., APPENTENG, K. & CURTIS, J.C. (1992) Monosynaptic EPSPs elicited by single interneurones and spindle afferents in trigeminal motoneurones of anaesthetised rats. *J. Physiol.* **455**, 641-662.

HARRISON, P.J. & TAYLOR, A. (1981) Individual excitatory postsynaptic potentials due to muscle spindle Ia afferents in cat triceps surae motoneurones. *J. Physiol.* **312**, 455-470.

HONIG, M.G., COLLINS, W.F. III & MENDELL, L.M. (1983) α-Motoneuron EPSPs exhibit different frequency sensitivities to single Ia-afferent fiber stimulation. *J. Neurophysiol.* **49**, 886-901.

JACK, J.J.B., REDMAN, S.J. & WONG, K. (1981). The components of synaptic potentials evoked in cat spinal motoneurones by impulses in single group Ia afferents. *J. Physiol.* **321**, 65-96.

JONAS, P., MAJOR, G. & SAKMANN, B. (1993) Quantal analysis of unitary EPSCs at the mossy fibre synapse on CA3 pyramidal cells of rat hippocampus. *J. Physiol.* **472**, 615-663.

MUNSON, J.B., FLESHMAN, J.W. & SYPERT, G.W. (1982) Properties of single-fiber spindle group II EPSPs in triceps surae motoneurons. *J. Neurophysiol.* **44**, 713-725.

NOZAKI, S., IRIKI, A. & NAKAMURA, Y. (1985) Trigeminal mesencephalic neurons innervating functionally identified muscle spindles involved in the monosynaptic stretch reflex of the lateral pterygoid muscle of the guinea pig. *J. Comp. Neurol.* **236**, 106-120.

REDMAN, S.J. (1990) Quantal analysis of synaptic potentials in neurons of the central nervous system. *Physiol. Revs.* **70**, 165-198.

YANG, H.-W., APPENTENG, K., CURTIS, J.C., MIN, M.-Y. & SAHA, S. (1994) Glutamate mediated synaptic interactions in the rat trigeminal motor nucleus. *J. Physiol.* **479**, 138P.

CRITIQUE OF PAPERS ON MOTONEURONE INPUTS

Michael J. O'Donovan

Section on Developmental Neurobiology
Laboratory of Neural Control
NINDS, NIH, Bethesda
MD 20892, USA

All the papers considered in this critique are concerned with the synaptic connections of various input systems with motoneurones. The papers by Mendell, Munson and Lev-Tov deal with the properties and regulation of synaptic projections to lumbar motoneurones in the cat or the neonatal rat. The remaining two papers by Appenteng, Curtis, Grimwood, Min & Yang, and Binder & Powers focus on methodological issues concerning the measurement of synaptic strength, but also consider applications of these methods.

Mendell's paper is concerned with the factors that regulate the behaviour of motoneuronal EPSPs during high frequency stimulation of muscle afferents. He reviews earlier work demonstrating that all of the muscle afferent synapses on a particular motoneurone exhibit similar behaviour when subjected to a high frequency train. In addition, individual afferent EPSPs recorded in the same motoneurone tended to show the same type of frequency-dependent modulation suggesting that this property was somehow dictated by the motoneurone (Koerber & Mendell, 1991). This work is extended in Mendell's chapter by examining the frequency-dependence of the composite EPSP evoked in type-identified motoneurones. The results show that the behaviour of the composite EPSP during high frequency stimulation is correlated with the duration of the after-hyperpolarization (AHP) of the motoneurone in which the EPSP was recorded. Depression of EPSP amplitude (negative modulation) was characteristic of EPSPs recorded in motoneurones with a long AHP whereas facilitation (positive modulation) was seen in motoneurones with the shortest AHP. A correlation with motoneurone type was not as clear. EPSPs produced in type S motoneurones (AHP > 30ms) generally exhibited negative modulation while those produced in Type F motoneurones could either facilitate or depress. The high frequency behaviour was not simply related to the initial size of the EPSP because the amplitude modulation differed between similarly sized EPSPs, according to the duration of the motoneuronal AHP.

The most likely site for this effect is at the presynaptic terminal but this may be difficult to prove in experiments on adult cats. To establish a presynaptic mechanism will probably require studies in vitro where it is possible to perform ionic manipulations to test presynaptic function and to obtain whole cell recordings of high enough resolution to allow the application of quantal analysis. Such experiments should establish also whether or not receptor desensitization plays a role in modulation of EPSP amplitude during high frequency stimulation of muscle afferents.

Despite these uncertainties Mendell's results raise the intriguing possibility that at least some properties of the muscle afferent synapses on motoneurones may be retrogradely determined by the motoneurone. At the present time the nature of this putative retrograde signal is unknown. One possibility is that it might be similar to the signal invoked to explain the retrograde modification of transmitter release that appears to underlie some forms of LTP in the hippocampus. In the hippocampus nitric oxide has been implicated as a possible retrograde messenger. One difficulty of this hypothesis in the spinal cord is that adult motoneurones do not appear to express the enzyme nitric oxide synthase unless they are damaged (Wu, Liuzzi, Schinco, Depto, Li, Mong, Dawson & Snyder, 1994).

Further evidence for a retrograde specification of muscle afferent properties was presented in the paper by Munson. Munson and his colleagues had shown previously that regeneration of muscle afferents into hairy skin was effective in reversing several of the changes in afferent function that accompany axotomy performed in adult cats. These include: a decline in conduction velocity, the loss of cord dorsum potentials and a reduction in the amplitude of motoneuronal EPSPs. Similarly, some of the changes in properties that accompany motoneurone axotomy can be corrected if motoneurones are induced to regenerate into hairy skin. In his chapter Munson explores further the degree of functional recovery that follows regeneration of motoneurones and muscle afferents into the skin. He and his colleagues examined the degree of modulation of muscle afferent-evoked EPSPs during high frequency stimulation under two different experimental conditions, using the stimulation protocols described in the chapter by Mendell. In the first, *muscle afferents* innervating the medial gastrocnemius muscle (MG) were cut and left axotomized or allowed to regenerate into the hairy skin or, alternatively, to reinnervate the medial gastrocnemius muscle. The effects of these manipulations were addressed by stimulating the afferents and recording the composite EPSPs evoked in normal motoneurones innervating lateral gastrocnemius and soleus muscles (LGS). In the second condition, the MG *motor axons* were cut and either remained axotomized, regenerated into the hairy skin or reinnervated the MG muscle. EPSPs were recorded from the manipulated MG motoneurones following high frequency stimulation of normal LGS muscle afferents. The results were similar for both types of procedure. Afferents or motoneurones reinnervating the MG muscle exhibited the normal range of modulation with the majority of cells exhibiting EPSP depression during high frequency stimulation. By contrast, exclusively negative modulation was seen when either axotomized muscle afferents were stimulated and the EPSPs recorded in normal LGS motoneurones or when normal LGS afferents were activated and the responses recorded in axotomized MG motoneurones. Motoneurones and muscle afferents regenerating into the skin were associated with EPSPs that displayed a smaller degree of negative modulation than after axotomy, suggesting some level of functional recovery.

In Mendell's paper it was argued that the type of modulation of muscle afferent motoneurone EPSPs was dictated by the motoneurone - possibly by some retrograde influence derived from the motoneurones. The finding that axotomized or skin-regenerated MG muscle afferents exhibit negative modulation even though they are projecting to normal and undamaged motoneurones suggests the high frequency behaviour at muscle afferent synapses depends on the characteristics of the afferent as well as the motoneurone in which the EPSP is recorded. One possibility is that the axotomized or skin-regenerated muscle afferents have a reduced capacity to respond to a retrograde motoneuronal signal.

Munson's experiments also reveal a dissociation between the amplitude of the composite EPSP and its behaviour during high frequency stimulation. In normal cats the large EPSPs tend to exhibit negative modulation while the small ones show positive modulation. Following afferent axotomy, all of the EPSPs were reduced in amplitude (at low frequency stimulation) and yet all exhibited negative modulation during high frequency stimulation. In addition, normal afferents projecting to axotomized motoneurones all exhibited negative modulation even though the amplitude of the EPSPs evoked by normal LGS muscle afferents was within the normal range. The reasons for these differences in the behaviour of the experimental and normal synapses are unknown but presumably reflect the fact that transmitter release and its dynamics are regulated by multiple factors.

Munson's results also strongly suggest that some signal derived from the muscle is important in determining the frequency-dependent properties of muscle afferent synapses on motoneurones. Such a retrograde specification may be part of a cascade from muscle to motoneurone to afferents. However, his results also suggest that orthograde signals from muscle to muscle afferent may play a role in determining the high frequency behaviour of motoneuronal EPSPs. This is because composite EPSPs recorded in normal motoneurones, but evoked by activation of axotomized or skin-regenerated axons, exhibited altered high frequency function.

The paper by Lev-Tov examined the properties of three afferent systems projecting to lumbar motoneurones of the neonatal rat spinal cord maintained *in vitro*. The three systems were muscle afferents, descending axons running in the medial longitudinal fasciculus (MLF) and axons travelling in the ventrolateral funiculus (VLF). All of the input pathways have an excitatory monosynaptic component that is mediated primarily by non-NMDA receptors but with a significant NMDA component.

Lev-Tov also showed that synaptic responses generated by high frequency stimulation differed for each input system. The MLF and muscle afferent pathways exhibited a frequency-dependent depression of the amplitude of EPSPs recorded in lumbar motoneurones. This depression was particularly severe for EPSPs evoked by muscle afferent stimulation which declined at stimulus frequencies as low as 0.1 Hz. Stimulation of the VLF, by contrast, resulted in potentiation of the EPSPs although some depression could be observed towards the end of the stimulus train. These results are interesting for several reasons. First, they show that the frequency-dependence of the afferent input is not simply a consequence of the *in-vitro* conditions since all of the inputs were tested under the same conditions and in some cases in the same motoneurone. Second, they suggest that maturity of the afferent system rather than the motoneurone, is important in determining the high frequency properties of developing synapses. The VLF connections, which are among the first to form in the developing cord, exhibited robust transmission a few days after birth when the electrophysiological experiments were performed. By contrast, transmission through the muscle afferent pathway was much weaker presumably because muscle afferent EPSPs, which appear just before birth, are still quite immature at the time of the recordings.

We don't yet know why immature synapses display synaptic depression. Lev-Tov provides evidence that the long-term depression of muscle afferent EPSPs probably results from a reduced level of transmitter output but what aspect of the transmitter machinery is responsible for this diminished output is not known. It will be of considerable interest to establish whether or not retrograde influences from motoneurones to synapses regulate the maturation of synaptic properties during development. If so, such a mechanism must be coordinated with the differing rates of maturation of the various input pathways.

The remaining two papers, by Appenteng et al., and by Binder & Powers, focus on the measurement of synaptic efficacy and the application of these methods to several different input systems to motoneurones.

Appenteng and his colleagues examined the projections from muscle spindle afferents and last order excitatory interneurones to motoneurones innervating the jaw elevator muscles using the technique of spike triggered averaging. Based on estimates of the average amplitude of single fibre EPSPs they determined that the depolarizing drive available to jaw elevator motoneurones from activation of all of the interneurones and afferents firing at 100 Hz was only about 3 mV. This figure seemed to be surprisingly low for the summed action of two excitatory input systems to these motoneurones. They proposed that the strength and connectivity of the connections might have been underestimated if a significant number of transmission failures (from presynaptic spike to EPSP) occurred during the accumulation of the spike-triggered average. Therefore, they estimated the number of failures in individual sweeps of the averaged single fibre EPSPs and removed those sweeps in which failures occurred. In addition, they eliminated sweeps which contained randomly occurring EPSPs. The "corrected" averages generated from this set of data were considerably larger than the originals. Of course, the visual method of estimating the failures presents some difficulties since the designation of a failure is based on a qualitative judgement rather than a quantitative set of criteria. In addition, as pointed out by the authors, this method of estimating the potential maximal amplitude of the EPSP (by correcting for failures) only produces a quantitative estimate of the EPSP if there is a single release site. With multiple release sites the corrected EPSP amplitude obtained by eliminating trials in which failures occurred is bound to underestimate the possible size of the EPSP. Nonetheless, these results demonstrate that a significant degree of modulation of the EPSP amplitude is possible. The authors then address the factors that might regulate the amplitude of the EPSP. They show that the amplitude of the EPSP can be increased by high frequency stimulation but their more interesting observations concern spontaneous and maintained changes in the amplitude of EPSPs generated by interneuronal and muscle afferent synapses on motoneurones. By averaging the EPSPs in a 100 trial blocks they establish the existence of substantial variability in the amplitude of the EPSP with time. They illustrate an example where the muscle afferent-evoked EPSP recorded in a jaw elevator motoneurone increases over 10 epochs (100 trials each) from about 30μV to over 140μV. Using the visual method for estimating the incidence of transmission the authors argue that the changes in EPSP amplitude can be accompanied by variations in the number of failures in each epoch, implying alterations in the probability of transmitter release. From these data it seems reasonable to conclude that the amount of transmitter released per impulse can fluctuate over the course of the average recording period. These results are instructive. First they show that caution must be exercised when using the spike triggered averaging method when the data are non-stationary and secondly they reveal an intriguing and unexpected fluctuation in the efficacy of connectivity between pre- and postsynaptic neurones in the pathways controlling jaw elevator muscles. The mechanisms responsible for these fluctuations remain to be established.

The final paper I will discuss, by Binder & Powers, describes two independent methods for measuring the effective synaptic current injected into the soma from synaptic inputs. The virtue of this approach is that it provides a direct and quantitative measure of the efficacy of synaptic input. Indeed, once the effective current is known together with the input resistance of the cell during synaptic activation it is easy to calculate the effect of the synaptic current on the firing behaviour of the cell. No knowledge of the electrotonic architecture of the neurone or the distribution of the synapses is required.

The first method employs a series of depolarizing and hyperpolarizing current pulses injected into a motoneurone. Halfway through the injected current pulse an input system is stimulated at high frequency to obtain a steady level of synaptic input. The duration of the synaptic current lasts somewhat longer than the current injected through the intracellular electrode. The essence of the method is to determine the current that has to be injected to null the synaptic potential. In practice this is accomplished by plotting several values of injected current against the sum of the synaptic potential and the voltage response to the injected current (V_{i+s}) and determining the value of the current for which V_{i+s} is zero. The effective synaptic current is defined as being equal and opposite in sign to this current.

One problem with this technique is the possibility of activating conductances in the cell by the injected current. However, the linearity of the plots of V_{i+s} and V_i against the injected current argues against a significant degree of contamination by intrinsic currents. Another concern is the ability to produce a true steady-state synaptic input particularly if the PSPs exhibit significant potentiation or synaptic depression. Related to this is the likelihood that high frequency stimulation may lead to recruitment or derecruitment of polysynaptic pathways that may complicate comparison with the distribution of EPSPs generated by low frequency stimulation.

The authors describe a second and independent method for obtaining the effective synaptic current for a steady-state synaptic input. In this method the firing rate of the cell is measured for a series of injected currents. From such data a linear plot can be constructed of the frequency of firing against the injected current. If a steady state synaptic input is superimposed on a portion of the injected current then the effective synaptic input can be calculated from the slope of the f/I plot and the change in firing rate produced by activation of the synaptic input.

The authors detail measurements of the effective synaptic current for several input systems to motoneurones including: Ia muscle afferent projections, Renshaw inhibition, Ia inhibitory interneurones, cutaneous inputs and flexor reflex afferent inputs. In reviewing their results, the authors emphasize differences in the distribution of the effective synaptic currents within a motoneurone pool when compared to the distribution of EPSPs activated at low frequency. For instance, they find that the effective synaptic current generated by activation of Ia synapses in MG motoneurones is about twice as high in motoneurones with a high input impedance than in motoneurones with a low input impedance, contradicting the assumption that synaptic currents from muscle afferent activation are uniformly distributed across the motoneurone pool. The mechanism responsible for this difference in the effective synaptic input onto different motoneurones is unknown. It could result from a differences in the number and properties of afferent terminals on motoneurones, or alternatively because of variations in the electrotonic architecture of small and large motoneurones. Another possibility is that the probability of transmitter release might be correlated with the motoneurone innervated, perhaps by the sort of retrograde influence postulated by Mendell and his colleagues.

REFERENCES

KOERBER, H. R. & MENDELL, L.M. (1991) Modulation of functional synaptic transmission during high frequency stimulation: role of postsynaptic target. *J. Neurophysiol.* **65**, 590-597.

WU, W., LIUZZI, F.J., SCHINCO, F.P., DEPTO, A.S., LI, Y., MONG, J.A., DAWSON, T.M. & SNYDER, S.H. (1994) Neuronal nitric oxide synthase is induced in spinal neurons by traumatic injury. *Neuroscience.* **61**, 719-726.

QUANTITATIVE ANALYSIS OF MOTONEURONE FIRING RATE MODULATION IN RESPONSE TO SIMULATED REPETITIVE SYNAPTIC INPUT

Randall K. Powers and Marc D. Binder

University of Washington
School of Medicine
Seattle, WA 98195, USA

The integrative function of the nervous system is dependent on the input-output properties of its individual neural elements. Although there are a number of empirical and analytical models of neuronal input-output behaviour, none of them have general predictive value. We have compared the ability of three simple models to predict the input-output behaviour of mammalian motoneurones. The first two models are derived from empirical observations and the third is based on a basic threshold-crossing neurone model. The first empirical model predicts that a synaptically-evoked change in motoneurone discharge rate (ΔF) is equal to the product of the net effective synaptic current (I_N) and the slope of the motoneurone's steady-state frequency-current relation (f/I): $\Delta F = I_N * f/I$ (Powers, Robinson, Konodi & Binder, 1992; see Figure 1A). The second empirical prediction is based on the characteristics of the post-synaptic potential (PSP) rather than the net current reaching the soma. Cope, Fetz & Matsumura (1987) found a linear relation between the amplitude of the PSP produced by a single Ia afferent fibre and the number of extra counts (above baseline) in the correlogram peak. Fetz and colleagues (Fetz, Cheney, Mewes & Palmer, 1989) have proposed that as a first approximation, the total change in motoneurone firing rate produced by activating a group of presynaptic fibres at a given rate is simply the product of the total arrival rate of PSPs (number of fibres times their average discharge frequency: PSPs/sec) and the average effect of a single-fibre PSP (spikes/mV of PSP): ΔF = (spikes/mV of PSP)* PSPs/sec (see Figure 1C). The third prediction is based on a threshold-crossing model with a fixed spike threshold and stereotyped interspike membrane potential trajectories during repetitive discharge (Fetz & Gustafsson, 1983). According to this model, the average change in firing rate is related to the mean depolarisation produced by the repetitive synaptic input at rest (e_{mean}) divided by the slope of the membrane potential trajectory during repetitive firing (dv/dt): $\Delta F = f_o^2 * e_{mean}/(dv/dt)$, where f_o is the mean motoneurone firing frequency in the absence of synaptic stimulation (see Figure 1D).

We tested these three predictions in cat and rat alpha motoneurones by eliciting background discharge with depolarising current steps, and on alternate trials superimposing brief (0.5 - 5 ms) current pulses of different amplitudes and frequencies to mimic repetitive synaptic input. This basic protocol, together with measurement of the motoneurone's frequency-current relation and previously published crosscorrelation results (Cope et al., 1987), provides all of the data needed to test the three predictions of firing rate modulation. ΔF was calculated as the difference in the mean steady-state firing rate averaged over trials

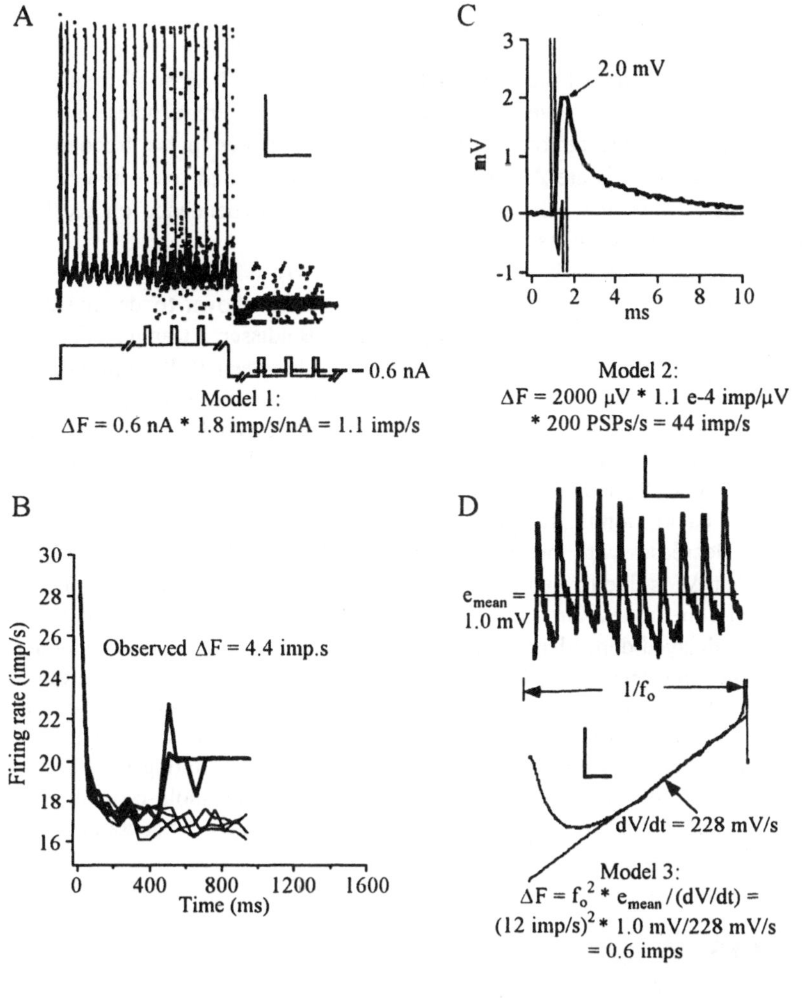

Figure 1. Predicted and observed firing rate modulation of a cat α motoneurone in response to repetitive current pulses. A. Response to a 1 sec suprathreshold current step alone (thin lines) and with superimposed 200 Hz pulses (dots). Lower trace is a schematic illustration of the injected current. The net current added by the pulses is 0.6nA. Model 1 predicts ΔF to be the product of the net current added by the pulses and the slope of the motoneurone's f-I relation. B. Instantaneous firing rate vs. time for the trials illustrated in A and similar trials. The observed firing rate modulation is taken as the mean difference between firing rate in trials with and without superimposed current pulses. C. Peak depolarisation produced at rest by the current pulse. (Solid trace is depolarisation after correction for bridge imbalance and removal of capacitative artifacts). Model 2 predicts ΔF to be th product of the peak depolarisation, the relation between peak depolarisation and cross-correlogram peak area (from Cope et al., 1987), and pulse arrival rate. D. Average depolarisation produced by the pulse train alone (upper trace) and average interspike membrane trajectory (lower trace) during repetitive discharge elicited by the 1 sec current step alone. Model 3 predicts ΔF to be the product of the square of the background discharge rate and the mean depolarisation produced by the pulses, divided by the slope of the membrane trajectory (Fetz & Gustafsson, 1983). Calibration bars are 20mV and 250ms in A, 0.5 mV and 10ms in upper trace in D, 5mV and 10ms in lower trace in D.

with superimposed current pulses and the rate averaged over bracketing trials with current steps alone (see Figure 1B). We measured I_N as the mean current added by the pulses and we measured the peak and mean amplitudes of the voltage deflections produced by the pulses. The f_0 and the dv/dt were obtained from the responses to the current steps without superimposed pulses.

The relation between current pulse frequency and ΔF was curvilinear, generally exhibiting a rapid rise in ΔF at low pulse rates, and flattening out for pulse rates exceeding 100 Hz. In contrast, both of the empirical models predicted linear relations between pulse frequency and ΔF. The current-based prediction underestimated ΔF at all frequencies, while the voltage-based prediction overestimated ΔF at all pulse frequencies. The threshold-crossing model predicted sigmoidal-shaped relations between pulse frequency and ΔF, which tended to underestimate ΔF at low pulse frequencies and overestimate ΔF at high pulse frequencies. Moreover, ΔF did not show a strong dependence on the background firing rate of the motoneurone, as predicted by this model.

The underestimate of ΔF based on the product of I_N and the steady-state f/I slope is not suprising, since motoneurone discharge rate has been shown to depend both on the magnitude and the rate of change of current (e.g, Baldissera, Campadelli & Piccinelli, 1982). The overestimate of ΔF based on the relation between PSP amplitude and cross-correlogram peak area results from two factors. First, the transient increase in discharge probabiliity indicated by the correlogram peak was followed by a period of decreased discharge probability. Prediction of firing rate modulation based on peak area alone will overestimate the net effect on discharge, which is better represented by the average cusum value over the background interspike interval. Second, the spike-triggering efficacy of current pulses decreases with increasing pulse rate. The inability of the threshold-crossing model to predict ΔF is due to two major differences between the features of the model and those of real motoneurones. First, the spike threshold of real motoneurones is not fixed, but in fact shows a delayed dependence on somatic membrane potential. This dependence leads to a decrease in spike threshold when the membrane is hyperpolarised during the interspike interval, resulting in an increase in the spike-triggering efficacy of current pulses. Second, due to variations in effective neuronal conductance during repetitive discharge, the mean depolarisation produced by a train of current pulses is different during repetitive discharge than that obtained at the resting potential. A modified threshold-crossing model that included a variable spike threshold and variations in membrane conductance more closely approximated the observed firing rate behaviour.

REFERENCES

BALDISSERA, F., CAMPADELLI, P. & PICCINELLI, L. (1982) Neural encoding of input transients investigated by intracellular injection of ramp currents in cat α-motoneurones. *J. Physiol.* **328**, 73-86.

COPE, T. C., FETZ, E. E. & MATSUMURA, M. (1987) Cross-correlation assessment of synaptic strength of single Ia fibre connections with triceps surae motoneurones in cats. *J. Physiol.* **390**, 161-188.

FETZ, E. E., CHENEY, P. D., MEWES, K. & PALMER, S.(1989) Control of forelimb muscle activity by populations of corticomotoneuronal and rubromotoneuronal cells. *Prog. Brain Res.* **80**, 437- 449.

FETZ, E. E., & GUSTAFSSON, B. (1983) Relation between shapes of post-synaptic potentials and changes in firing probability of cat motoneurones. *J. Physiol.* **341**, 387-410.

POWERS, R. K., ROBINSON, F. R., KONODI, M. A. & BINDER, M. D. (1992) Effective synaptic current can be estimated from measurements of neuronal discharge. *J. Neurophysiol.* **68**, 964-968.

THE EFFECT OF TEMPERATURE ON CHEMICAL SYNAPTIC TRANSMISSION TO SPINAL MOTONEURONES IN THE FROG

H.P. Clamann and A.E. Dityatev

Department of Physiology
University of Berne
3012 Berne, Switzerland

INTRODUCTION

It has long been known that action potentials may fail to be conducted past branches in axons in the central (reviewed by Lüscher & Clamann, 1992) or peripheral nervous systems (Westerfield, Joyner & Moore, 1978). Conduction past branch points depends critically on their geometry and on the local conductance. The conductance is in turn influenced by ionic concentrations and by temperature. Experiments (Westerfield et al., 1978, Lüscher & Clamann, 1992) and simulations (see Lüscher & Clamann, 1992 for review) have shown that propagation past such branch points failed above a critical ratio of post-branch to pre-branch diameters. This ratio is very sensitive to temperature, so that conduction is more reliable at lower temperatures.

The present experiments examine the effects of temperature and paired stimuli to test the hypothesis that branch point failure prevents conduction of impulses into synaptic terminals. While branch point failure can explain some of our results, the synaptic response is too complex to be completely explained by this hypothesis. Of particular interest are differences shown by synaptic responses evoked by stimulation of two different inputs, namely afferents in dorsal roots (DR) and fibres of the ventro-lateral columns (VLC).

METHODS

Experiments were performed on hemisected spinal cords dissected from frogs (Rana ridibunda) and placed into a small plexiglass chamber in normal frog Ringer's solution. The chamber was provided with a Peltier element to control its temperature between 0° and 25°C. A dorsal and a ventral root, usually root 9 or 10, were held with suction electrodes for stimulation or recording. Motoneurones were impaled with glass pipette microelectrodes filled with 3 M KCl. Only cells with membrane potentials more negative than -60 mV were used. Reflex responses in the form of compound action potentials were recorded from a ventral root. EPSPs or action potentials were recorded intracellularly from motoneurones. Reflex size was measured from the amplitude of the reflex volley, taken from baseline to its first peak, or from the area under that waveform measured from its onset to that peak. These two measures correlated well.

RESULTS

A reflex response with a small monosynaptic component was obtained by single shocks applied to a DR. Unlike similarly evoked responses in mammals, the reflex response invariably had a dominant second component which was polysynaptic, judging from its latency. Simultaneous intracellular recordings from a motoneurone in some experiments showed that the motoneurone would discharge two or more times, or show a spike followed by an EPSP. This second response was synchronised with the second phase of the reflex volley.The two components tended to overlap, and it was not possible to separate them for quantitative analysis. The size of the reflex declined with increasing temperature whether measured as an amplitude or as the area under the rising phase. Action potentials recorded intracellularly in several experiments showed changes in amplitudes and areas with temperature. When a correction based on these shape changes was applied to the reflex volleys, the changes attributable to temperature were strongly reduced: in a typical experiment, an amplitude ratio of a cold to a warm reflex response of 210% was reduced to a ratio of 110% by this correction.

When pairs of shocks were applied 62 ms apart, the second response was strongly reduced at low temperatures and slightly potentiated at high temperatures (Figure 1). At 16°C in most experiments, the second response was the same size as the first. Like the first response, the second showed a pattern of a monosynaptic, followed by or overlapping with a polysynaptic component. This temperature effect could not be attributed to shape changes in the individual action potentials making up the reflex volleys.

Reflex responses induced by VLC stimulation showed several differences from reflex

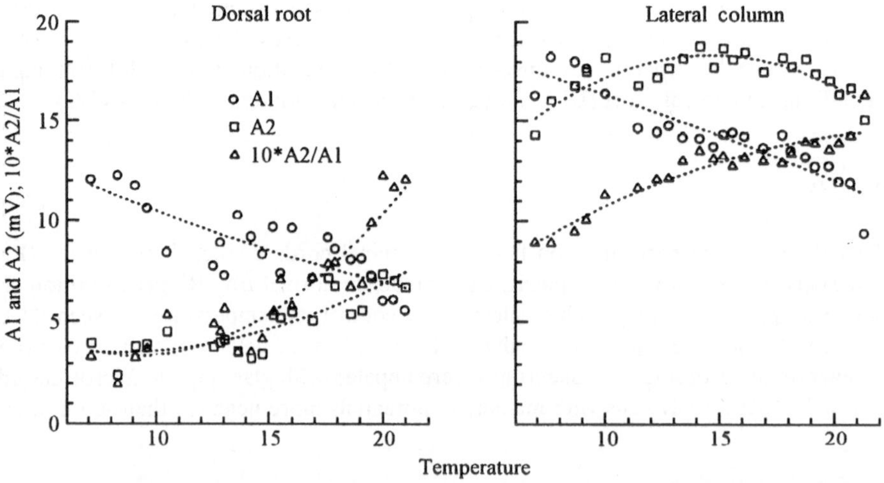

Figure 1. Effect of temperature on the amplitude of the first (A1) and second (A2) responses to pairs of stimuli 62 ms apart. Open triangles show a measure of the degree of change, A2/A1. Although an increase in this ratio with temperature occurs independently of input source, the shape of the A2/A1 - temperature relationship is qualitatively different in the two cases.

evoked by DR stimulation. The reflex volleys were shorter lasting and had a clearer monosynaptic component than did those induced by DR stimulation. For example, the time to peak was 5-12 ms, as opposed to 7-25 ms for responses to DR stimuli. Potentiation of a second response was 2-3 times as great as that seen in DR stimulation. In contrast, the degree of potentiation increased with temperature only to about 15°C after which it declined slightly (Figure 1).

DISCUSSION

Although reflex amplitude and area varied with temperature as predicted (Brooks, Koizumi & Malcolm, 1955), much of that size change could be accounted for by changes in the shapes of individual action potentials making up the reflex volley, and in changes in their degree of synchrony (Clamann, Gillies, Skinner & Henneman, 1974). When corrections were made for such factors, the remaining size change was around 10%.

By contrast, a conditioning stimulus had an effect on the response to a later test stimulus which was also temperature dependent. We conclude that this temperature sensitivity has a pre-synaptic origin, since stimulation of different inputs to motoneurones produced quantitatively and qualitatively different effects (Figure 1).

Branch point failure need not be an all-or-none phenomenon. There is evidence from experiments (Mallart, 1984) and from simulation studies (Lüscher & Clamann, 1992) that action potentials do not regularly invade terminal boutons, but influence them by passive spread of depolarisation which may vary in amplitude. A synapse may not only be turned on or off by failure of action potential propagation; it may be depolarised to a greater or lesser degree depending on its past history and on its geometry (Westerfield et al., 1978). Mallart (1984) has suggested that the distribution of ion channels is non-uniform near the ends of axons, which adds additional complications to our ability to predict the effect of an action potential. It is very likely that changes in the degree of depolarisation of pre-synaptic elements contributes to the variability of transmitter release.

ACKNOWLEDGEMENTS

Supported by a grant from the Swiss National Science Foundation.

REFERENCES

BROOKS, C. McC., KOIZUMI, K.& MALCOLM, J.L. (1955) Effects of changes in temperature on reactions of spinal cord. *J. Neurophysiol.* **18**, 205-216.

CLAMANN, H.P., GILLIES, J.D., SKINNER, R.D. & HENNEMAN, E. (1974) Quantitative measures of output of a motoneuron pool during monosynaptic reflexes. *J. Neurophysiol.* **37**, 1328-1337.

LÜSCHER, H.-R. & CLAMANN, H.P. (1992) Relation between structure and function in information transfer in spinal monosynaptic reflex. *Physiol. Rev.* **72**, 71-99.

MALLART, A. (1984) Presynaptic currents in frog motor endings. *Pflügers Arch.* **400**, 8-13.

WESTERFIELD, M., JOYNER, R.W. & MOORE, J.W. (1978) Temperature-sensitive conduction failure at axon branch points. *J. Neurophysiol.* **41**, 1-8.

THE ROLES OF CENTRAL CHOLINERGIC AND ELECTRICAL SYNAPSES MADE BY SPINAL MOTONEURONES IN *XENOPUS* EMBRYOS

Ray Perrins

School of Biological Sciences
University of Bristol
Bristol, BS8 1UG, U.K.

The hatchling tadpole of *Xenopus laevis* has been used successfully as a simple model system in which to study the spinal neural circuits that control locomotion (Roberts, 1990). It is therefore an ideal simple system in which to study possible functions for central synapses made by spinal motoneurones. Such synapses are known to exist, being made on premotor interneurones (for example Renshaw cells) as well as on other motoneurones, in a variety of vertebrates from fish to adult mammals. However, very little is known of their role during locomotion, which is probably in part due to the complexity of the systems in which they have been found (e.g. Noga, Shefchyk, Jamal & Jordan, 1987). We have recently shown that motoneurones are integral parts of the simple *Xenopus* circuit which produces the drive during swimming, rather than just being output devices to the muscles, as was previously supposed. By making simultaneous intracellular recordings from pairs of spinal motoneurones we have directly demonstrated the presence of cholinergic synapses between motoneurones, the activation of which gave rise to fast EPSPs which were blocked by nicotinic antagonists (Perrins & Roberts, 1995; Figure 1A). A more local electrical coupling between motoneurones was also sometimes observed (Perrins & Roberts, 1995). Both these types of interaction were only found between motoneurones on the same side of the spinal cord. For these pairs about 50% were chemically coupled and 10% electrically coupled. Since motoneurones on the same side of the spinal cord spike roughly in phase, excitation from these two types of connection would be expected to occur on-cycle during swimming. We therefore investigated the composition of the fast on-cycle excitation which underlies spiking activity in spinal neurones, which had previously been thought to be due entirely to kainate/AMPA receptor activation by an excitatory amino acid. By the local application of the nicotinic antagonists dihydro-β-erythroidine and d-tubocurarine (both 10 μM), we showed that 20% of the on-cycle excitation received by motoneurones during swimming is provided by cholinergic EPSPs. The local application of 100 μM Cd^{2+}, which blocks all chemical neurotransmission, demonstrated that 50% of the fast on-cycle excitation was due to electrotonic coupling with other, spiking neurones. The anatomical

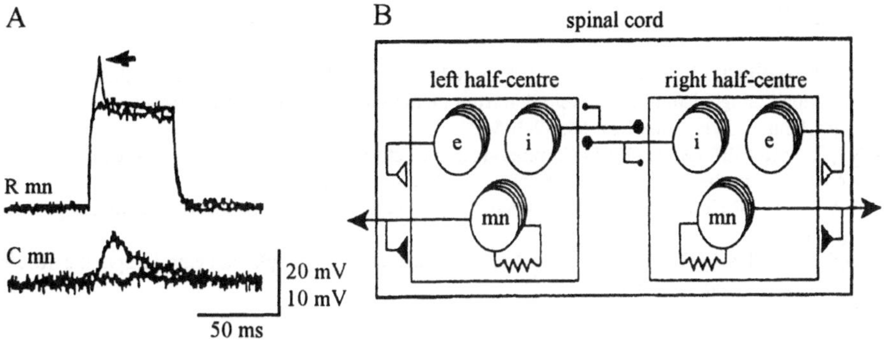

Figure 1. Central motoneurone feedback in the *Xenopus* spinal cord. A. Cholinergic EPSPs between spinal motoneurones. A suprathreshold current pulse in a rostral motoneurone (R mn) results in an action potential (arrow) and a fast EPSP in the caudal motoneurone (C mn). A trace from a subthreshold current pulse which elicits no EPSP is superimposed. This EPSP was blocked by 10mM mecamylamine, a specific nicotinic antagonist in this system. B. Revised rhythm generating network for swimming in *Xenopus* embryos. The spinal cord contains left and right "half-centres". Inhibitory interneurones (i) inhibit the opposite half-centre strongly and their own half-centre weakly. Excitatory interneurones (e) excite neurones within their own half-centre (including other excitatory interneurones) with an excitatory amino acid (open triangles). The motoneurones (mn) have axons leaving the spinal cord to innervate the swimming muscles. They also make connections within the spinal cord, feeding back to their own half-centre with cholinergic excitation (closed triangle) and to each other with both cholinergic and electrical (resistor sign) connections.

simplicity of the *Xenopus* spinal cord (which contains just eight morphological classes of neurone; Roberts & Clarke, 1982), along with the results of the paired recording study mean that the most likely source for these two types of excitation is other motoneurones. The local application of CNQX showed that the remaining 30% of the fast excitation was due to kainate/AMPA receptor activation.

Local application of drugs to premotor interneurones of the central pattern generator for swimming showed that they also receive on-cycle cholinergic excitation, again presumably from motoneurones, and that this excitation increases the reliability of interneurones firing during swimming. When cholinergic excitation to the central pattern generator is blocked by the application of nicotinic antagonists over the whole spinal cord, the frequency of fictive swimming is reduced and the total episode duration is shortened. Therefore the increased reliability of spiking in premotor interneurones through this cholinergic positive feedback excitation appears to help to maintain spinal rhythm generation. As previously, the most likely source of this cholinergic excitation is the motoneurones. The remainder of the excitation in interneurones was mediated by kainate/AMPA receptors, and there appeared to be no electrotonic input.

In conclusion, we propose that motoneurones (1) strongly excite each other by cholinergic and electrical synapses, possibly increasing their reliability and the local synchrony of motoneurone firing and (2) weakly excite the central pattern generator for swimming via the cholinergic synapses only, helping to maintain swimming after a brief stimulus. These new connections are summarised in Figure 1B. If the motoneurones do contribute excitation in this fashion, they must be considered as integral members of the central pattern generator, a role which has yet to be shown in vertebrates.

ACKNOWLEDGEMENTS

Supported by the Wellcome Trust.

REFERENCES

NOGA, B. R., SHEFCHYK, S. J., JAMAL, J. & JORDAN, L. M. (1987). The role of Renshaw cells in locomotion: antagonism of their excitation from motor axon collaterals with intravenous mecamylamine. *Exp. Brain Res.* **66**, 99-105.

PERRINS, R. & ROBERTS, A. (1995) Cholinergic and electrical synapses between synergistic spinal motoneurones in the *Xenopus laevis* embryo. *J. Physiol.* (in press).

ROBERTS, A. (1990) How does a nervous system produce behaviour? A case study in neurobiology. *Sci. Prog.* **74**, 31-51.

ROBERTS, A. & CLARKE, J. D. W. (1982) The neuroanatomy of an amphibian embryo spinal cord. *Phil. Trans. R. Soc. Lond. B* **269**, 195-212.

SYNAPTIC LOSS FROM AXOTOMISED α- AND γ-MOTONEURONES

I.P. Johnson[1], Y.S. Simaika[1]
and T.A. Sears[2]

[1]Department of Anatomy
Queen Mary & Westfield College
Mile End Road, London
[2]Department of Physiology, UMDS
St. Thomas' Hospital, London

INTRODUCTION

Loss of synaptic terminals from the cell bodies of α-motoneurones within about a week of axotomy is well-documented (Blinzinger & Kreutzberg, 1968; Chen, 1978). Such loss may be due to the loss of peripheral target contact, since it also occurs after the interruption of axonal transport in peripheral nerves and when neuromuscular transmission is blocked (Cull, 1975; Sumner, 1979). Previously (Johnson, Pullen & Sears, 1990), we reported that the restoration of peripheral target contact is associated with the return towards normal of synaptic terminals contacting cat thoracic α-motoneurones. No such restoration occurred for motoneurones prevented from reinnervating muscle and we suggested that this difference reflected the influence of the peripheral target on the synapses contacting α-motoneurones. There is no information on the long-term effects of axotomy on the synapses of γ-motoneurones, although axotomy-induced changes in their central connectivity are also likely to affect the degree to which recovery of movement control occurs. In this study, therefore, we have compared the long-term effects of reversible- or permanent- axotomy on the synaptic terminals of α- and γ-motoneurones to determine the extent to which axotomy-induced changes are modified by the restoration of peripheral target contact.

METHODS

Intercostal nerves in anaesthetised (45 mg/kg sodium pentobarbitone, I.P.) adult cats ($n = 2$ per time-point) were either (i) crushed or (ii) transected, ligated proximally and a portion of the distal stump removed. After 7 days or 63 days, horseradish peroxidase (HRP) was applied to the newly-lesioned proximal nerves to label retrogradely the cell bodies of α-

and γ-motoneurones. Control cats (n = 3) received injections of HRP into the intercostal muscles. One day after the application of HRP, all experimental- and control- cats were anaesthetised then perfused with fixative. 70 μm sections of spinal cord were processed to demonstrate peroxidase activity using 3,3' diaminobenzidine and osmicated sections then flat-embedded in Araldite for electron microscopy. The cell bodies of large (> 40μm diameter) and small (< 30μm diameter) motoneurones were considered to be α- and γ-motoneurones, since the respective diameters of these motoneuronal types after electrophysiological identification in the cat spinal cord are reported to be 25.5 - 39.5 μm and 40.0 - 75.5 μm, respectively (see Johnson & Sears, 1989 for references). Twenty γ- and 10 α- motoneurones were quantified from control animals. Ten γ- and 10 α- motoneurones were quantified from each post-axotomy time-point.

RESULTS

In controls, synaptic frequency (number/100 μm plasma membrane) and cover (length of synaptic terminal apposition/100 μm plasma membrane) for the cell bodies of γ-motoneurones were only 56% and 48%, respectively, of α-motoneurone values (Table 1.). Synaptic terminals with predominantly round- (S-type) or oval-(F-type) synaptic vesicles were seen on both motoneuronal types, but the large C-type terminal was seen only on α-motoneurones.

Eight days following axotomy, synaptic frequency and cover for α-motoneurones had fallen to 78% and 49%, respectively, of control α-motoneurones, while for γ-motoneurones, synaptic frequency and cover had fallen further to 38% and 26%, respectively, of control γ-motoneurones. This change for γ-motoneurones affected particularly those synaptic terminals with flattened synaptic vesicles (Table 1).

Table 1. Different morphological types of synaptic terminals contacting the cell bodies of α- and γ-motoneurones after permanent- or reversible-axotomy. F = Synaptic frequency, C = % coverage. Mean ± S. E. M. *, $p < 0.05$ vs. control (t-test).

			CONTROL	8d CUT	64d CUT	64d CRUSH
S-type	α	F	6.29±0.66	4.97±0.58*	3.39±0.63*	5.96±1.06
		C	11.28±1.51	5.82±0.61*	4.20±0.90	7.19±1.66
	γ	F	3.68±0.71	2.67±0.67	6.57±1.63*	5.40±1.24
		C	6.71±1.31	3.14±0.82	10.85±3.25	12.51±3.47
F-type	α	F	8.88±1.14	7.30±0.44	3.47±0.87	7.67±1.40
		C	17.94±2.61	9.46±1.47*	4.64±1.32*	9.73±2.29*
	γ	F	5.33±1.22	0.78±0.19*	2.09±0.77	3.30±1.35
		C	9.89±2.16	1.21±0.31*	3.35±1.19*	6.30±2.86
C-type	α	F	0.99±0.38	0.39±0.14	0.65±0.23	1.05±0.57
		C	5.43±2.17	1.74±0.87	2.22±0.90	3.36±1.73
	γ	F	0.00±0.00	0.00±0.00	0.00±0.00	0.00±0.00
		C	0.00±0.00	0.00±0.00	0.00±0.00	0.00±0.00
Total	α	F	16.16±1.53	12.66±1.33	7.51±1.32*	14.68±2.31
		C	34.65±3.37	17.02±1.54*	11.06±2.26*	20.28±3.91
	γ	F	9.01±1.63	3.45±0.76*	8.66±2.32	8.70±2.32
		C	16.60±2.90	4.35±0.87*	14.21±4.28	18.81±4.96*

Sixty-four days following axotomy, partial restoration to control values for synaptic frequency and cover of α-motoneurones was seen after nerve crush (frequency and cover 91% and 59% of α controls), but not after nerve transection and ligation (frequency and cover 47% and 32% of α controls). Recovery of synaptic frequency and cover of γ-motoneurones was found following both nerve crush (frequency and cover 97% and 113% of γ controls) and nerve transection with ligation (frequency and cover 96% and 86% of γ controls). This restoration for γ-motoneurones was associated with an increase in the ratio of S- to F-type synapses (Table 1.).

Compared to α-motoneurones, γ-motoneurones have fewer synapses normally and lose proportionately more synapses 8 days after axotomy. Whether this reflects intrinsic differences in the retrograde response of these two motoneuronal types, differences in the nature of their afferent inputs, or reflects their disconnection from peripheral targets which are structurally and functionally distinct is unknown. Whatever the mechanism, these findings urge caution when extrapolating the well-known features of the retrograde response of α-motoneurones to the less well-studied response of γ-motoneurones. Differences in the long-term synaptic response of α- and γ-motoneurones serve to emphasise this point: partial restoration of synapses occurred for α-motoneurones only when conditions were suitable for axonal regeneration, while synaptic restoration for γ-motoneurones occurred even when conditions were unsuitable for axonal regeneration. These results indicate that recovery of the central connectivity of axotomised γ-motoneurones in the cat thoracic spinal cord may not be dependent on target-derived influences.

ACKNOWLEDGEMENTS

Supported by the MRC.

REFERENCES

BLINZINGER, K. & KREUTZBERG, G.W. (1968) Displacement of synaptic terminals from regenerating motoneurons by microglial cells. *Z. Zell. Miksok. Anat.* **85**, 145-157.

CHEN, D.H. (1978) Qualitative and quantitative study of synaptic displacement in chromatolysed spinal motoneurons of the cat. *J. Comp. Neurol.* **177**, 635-664.

CULL, R.E. (1975) Role of axonal transport in maintaining central synaptic connections *Exp. Brain Res.* **24**, 97-101.

JOHNSON, I.P., PULLEN, A.H. & SEARS, T.A. (1990) Synaptic terminal changes on cat thoracic motoneurones following either nerve crush or nerve transection without reinnervation. *Eur. J. Neurosci.* Suppl **3**, 2068.

JOHNSON, I.P. & SEARS, T.A. (1989) Ultrastructure of axotomised alpha and gamma motoneurones in the cat thoracic spinal cord. *Neuropath. Appl. Neurobiol.* **15**, 149-163.

SUMNER, B.E.H. (1979) Responses of boutons and glia in the hypoglossal nucleus to injection of alpha- or beta-bungarotoxin into the tongue. *Expl. Brain Res.* **36**, 387-392.

SYNAPTIC REMODELLING DURING THE POSTNATAL DIFFERENTIATION OF PERONEAL MOTONEURONES IN KITTENS

M. Simon, J. Destombes, G. Horcholle-Bossavit
and D. Thiesson

URA CNRS 1448
Université René Descartes
45 rue des Saints-Pères
75270 Paris Cedex 06, France

In the adult cat, ultrastructural features of alpha (α) motoneurones (MNs) differ from those of gamma (γ) MNs and it was shown in peroneal motor nuclei that α and γ MNs of similar size could be distinguished without ambiguity (Destombes, Horcholle-Bossavit, Thiesson & Jami, 1992): α MNs have a vacuolated nucleolus and a somatic membrane with a rich equipment of various axonic terminals. They are contacted by five types of synaptic boutons, namely F, S, M, P, C types according to the terminology of Conradi (1969). γ MNs have a compact nucleolus and a somatic membrane with few synaptic boutons, they only receive F and S terminals.

Peroneal MNs of kittens have now been examined by electron microscopy to study their maturation, the differentiation of the two types of MNs and the evolution of their synaptic equipment in the postnatal period.

In eight kittens, peroneus brevis MNs were identified by retrograde transport of horseradish peroxidase (HRP), injected in the muscle. The distribution of the mean soma diameter is unimodal in 3-8 day-old kittens and bimodal in 18-22 day-old kittens (Horcholle-Bossavit, Jami, Thiesson & Zytnicki, 1990). Therefore kittens were assembled in two groups: four kittens in a one-week age group and four in a 3-week age group. After fixation, the L7 and S1 segments of the spinal cord were cut and serial 70-80 µm sections were processed for HRP. Sections were embedded in epon 812 and blocks containing the selected MNs were cut for electron microscopy in ultrathin sections at 4-7µm intervals in order to analyse 2-4 profiles of each MN. Profiles were photographed at a magnification of x5,000. Photomontages covering the entire profiles were examined at a final magnification of x10,000. The contours of the cell body and the origins of primary dendrites were delimited at the points of interruption of the somatic profile convexity. Two types of terminal profiles were considered: 1) synaptic boutons containing vesicles and characterised by pre and/or post-synaptic densities indicating the active zones 2) axonic terminals with a size >0.5μm, containing clusters of vesicles and without marked membrane specialisation. The synaptic frequencies (Fs), defined as the mean number (N) of boutons apposed on 100μm of membrane length (Fs = N/100μm) were calculated for somatic and proximal dendritic compartments. Apposition lengths and lengths of active zones were measured in samples of terminals examined at high magnification (x 20,000) where pre- and postsynaptic membrane thickenings, synaptic vesicles closely apposed to the presynaptic membrane and synaptic cleft were visible. All measurements obtained in the two age groups were compared to similar samples examined and measured in material from the adult cat.

Samples of 21 and 18 MNs were analysed for the 1- and 3-week stages respectively. At both stages, motoneurone immaturity appeared in nuclear membrane invaginations which were deeper for γ than for α MNs, changes in the aspect of the nucleoli in α MNs, numerous somatic and dendritic spines and small size of terminals. On ultrastructural criteria (aspect of nucleolus, terminal types, synaptic frequency), most of the 1- and 3-week MNs could be identified as γ or α although 4 γ MNs were contacted by small "C-like" boutons. However, 4 MNs of various size showed mixed characteristics. Such unidentifiable MNs indicate that at birth, differentiation of motoneurone type is not completed.

The graphs of Figure 1 show a significant reduction of the synaptic frequency on the somatic compartment of both γ and α MNs between 1 and 3 weeks. This reduction was mainly of the F-type synaptic frequency which decreased by nearly one third on both α MNs and γ MNs (see also Conradi & Ronnevi, 1975 and Arvidsson, Svedlund, Lagerback & Cullheim, 1987). In contrast, the dendritic frequencies did not show significant changes in the peroneal MNs during postnatal development. The total synaptic frequencies on the proximal dendritic compartment of α MNs (36.8 and 37.8 at one and three weeks respectively) were always higher than on the somatic compartment (32.9 and 25.9 at one and three weeks respectively). Similarly, the total synaptic frequencies on the proximal dendritic compartment of γ MNs were 21.9 and 16.9 at one and three weeks respectively whereas on the somatic compartment, the values were 9.4 and 6.0 at one and three weeks respectively.

On α MNs, the increase in size of F-type boutons occurred continuously between one week (Mean apposition length±S.D: 1.50±0.6μm), three weeks (1.82±0.64μm) and the adult stage (2.21±0.9μm). For the S-type boutons, the mean apposition length was about the same in one week and three week stages (1.57±0.70μm), the increase occurred beyond three weeks since the mean apposition length in the adult was 2.21±0.9μm. In γ MNs, the increase of mean apposition length mainly occurred between one week (1.38±0.51μm and

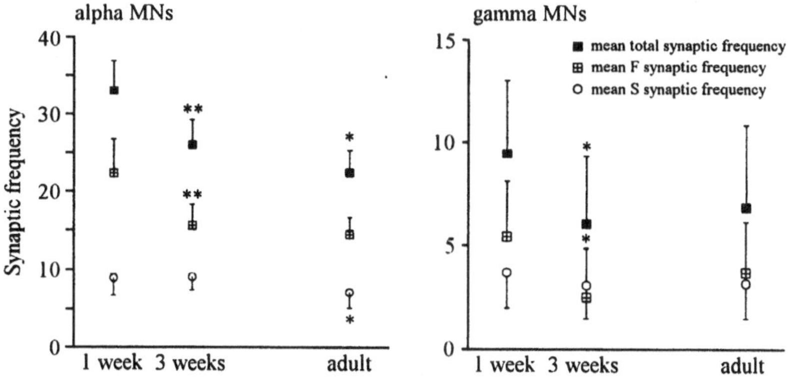

Figure 1. Mean synaptic frequencies on α and γ MNs cell bodies at different ages. Vertical bars represent the standard deviation of the values. Significant differences with values of the preceding stage are indicated: **: $p < 0.01$ and *: $p < 0.05$ (Mann-Whitney test)

1.30±0.3µm respectively for F and S boutons) and three weeks (1.69±0.69µm and 1.56±0.71µm respectively for F and S boutons). Significant changes were also observed in the lengths of the active zones which were larger in kittens than in adult cats. The mean length of active zones for F-type boutons were 0.62±0.25µm and 0.49±0.17µm for α and γ MNs respectively in three week kittens while in the adult, the values were 0.44± 0.21µm and 0.32±0.14µm. Altogether, our results indicate that important synaptic remodelling occurs during the postnatal differentiation of MNs and the changes have different temporal sequences for α and γ MNs.

REFERENCES

ARVIDSSON, U., SVEDLUND, J., LAGERBACK, P-A. & CULLHEIM, S. (1987) An ultrastructural study of the synaptology of γ-motoneurones during the postnatal development of the cat. *Dev. Brain Res.* **37**, 303-312.

CONRADI, S. (1969) Ultrastructure and distribution of neuronal and glial elements on the motorneurone surface of the spinal cord of the adult cat. *Acta physiol. scand.* **332**, pp 5-48.

CONRADI, S. & RONNEVI, L-O. (1975) Spontaneous elimination of synapses on cat spinal motoneurons after birth: do half of the synaspes on the cell body disappear? *Brain Res.* **92**, 505-510.

DESTOMBES, J. HORCHOLLE-BOSSAVIT, G., THIESSON, D. & JAMI, L. (1992) Alpha and gamma motoneurones in the peroneal nuclei of the cat spinal cord: an ultrastructural study. *J. Comp. Neurol.* **317**, 79-90.

HORCHOLLE-BOSSAVIT, G., JAMI, L., THIESSON, D. & ZYTNICKI, D. (1990) Postnatal development of peroneal motoneurons in the kitten. *Dev. Brain Res.* **54**, 205-215.

PHYSIOLOGY AND MORPHOLOGY OF THE NEURONAL CIRCUITS OF JAW REFLEXES IN CATS

A. Yoshida[1], Y. C. Bae[1], Y. Shigenaga[1]
and B. J. Sessle[2]

[1]Department of Oral Anatomy
Osaka University Faculty of Dentistry
Osaka, Japan
[2]Department of Oral Physiology
Faculty of Dentistry, University of Toronto
Toronto, Canada

INTRODUCTION

Although it is well known that periodontal afferent excitation evokes jaw-closing and jaw-opening reflexes, details of the neuronal circuits related to these reflexes are incomplete. We have therefore carried out a series of morphological and physiological investigations of 43 periodontal mesencephalic (Vmes) afferents. At the same time we have gathered data on 102 jaw-closing and jaw-opening motoneurones, 13 jaw-closing muscle spindle afferents and 14 premotor interneurones in trigeminal subnucleus oralis (Vor).

METHODS

Cells and axons in the vicinity of the trigeminal motor nucleus were penetrated with glass micropipettes loaded with horseradish peroxidase. After recording their physiological responses, they were filled iontophoretically and serial sections prepared for histological reconstruction. Full details are provided in the references quoted.

RESULTS

While jaw-opening motoneurones showed little variability in their dendritic configuration, masseter motoneurones could be divided into three groups on the basis of their morphology. One of these groups, which had a simpler dendritic configuration and longer latency of evoked antidromic spikes than the others, were considered to be γ-

motoneurones (Yoshida, Tsuru, Mitsuhiro, Otani & Shigenaga, 1987; Shigenaga, Yoshida, Tsuru, Mitsuhiro, Otani & Cao, 1988). Antidromic spike potentials of the masseter and jaw-opening motoneurones were evoked at a constant latency of 0.4 - 1.4 ms (mean 0.86 ± 0.27 ms, n = 70) and of 0.4 - 1.5 ms (0.83 ± 0.27 ms, n = 32), respectively. When the electrical shock intensity to the masseter nerve was reduced, EPSPs were revealed in the masseter motoneurones with latencies of 1.64 ± 0.19 ms (n = 34).

Stimulation of periodontal (or pulp) afferents produced long-latency EPSPs (3.45 ± 0.94 ms, n = 10) as well as short-latency EPSPs (1.42 ± 0.43 ms, n = 10) of small amplitude that preceded IPSPs (latency 3.12 ± 0.90 ms, n = 45) in the masseter motoneurones. Patterns of postsynaptic potentials evoked were classified into 4 types: hyperpolarisation (n = 40), depolarisation-hyperpolarisation (n = 10), hyperpolarisation - depolarisation (n = 5), and depolarisation with spike potentials (n = 5). On the other hand, in the jaw-opening motoneurones, EPSPs with spike potentials (latency 3.18 ± 1.02 ms, n = 19) or IPSPs (n = 5) were induced (Shigenaga et al., 1988).

On the basis of their axonal trajectory, jaw-closing muscle spindle afferents could be classified as type I and type II, with different axon diameters (4.7 ± 0.3 and 4.0 ± 0.6 μm respectively, $P < 0.05$; based on measurements of the "united" or U fibre component of the afferent, see Figure 1). Type I and type II could conceivably represent primary and secondary spindle afferents respectively. Type I afferents sent their collaterals mainly into the jaw-closing motoneurone pool of the trigeminal motor nucleus (i.e. Vmo.dl), as well as into the intertrigeminal region (Vint) and juxtatrigeminal region (Vjux). On the other hand, type II afferents terminated mainly in the supratrigeminal nucleus (Vsup), as well as Vmo.dl, Vint and Vjux (Shigenaga, Mitsuhiro, Shirana & Tsuru, 1990).

Periodontal Vmes afferents had either slowly adapting (SA) or fast adapting (FA) properties and both types projected to Vsup. All periodontal Vmes afferents with a peripheral (P) fibre passing through the trigeminal motor nerve tract (M type) terminated in Vmo.dl. All SA afferents also projected to Vjux, and all FA afferents to Vint and Vor (Shigenaga, Doe, Suemune, Mitsuhiro, Tsuru, Otani, Shirana, Hosoi, Yoshida & Kagawa,

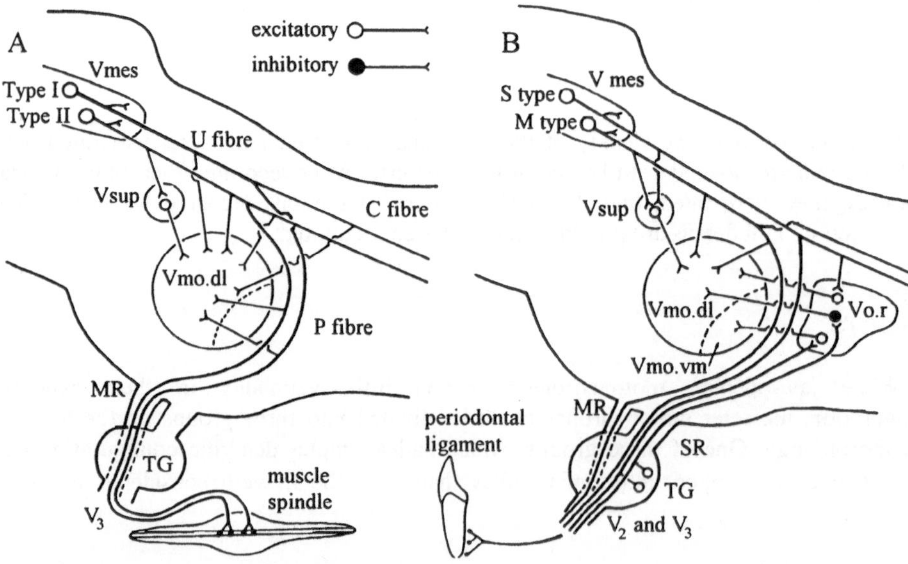

Figure 1. Neuronal circuits of jaw reflexes evoked by muscle spindle (A) and periodontal afferent (B) excitation. MR, motor root; SR, sensory root; V2, maxillary nerve; V3, mandibular nerve.

1989). All Vmes neurones examined had axon collaterals that were given off from the U, P and central (C) fibres, and, thus, they were unipolar rather than pseudounipolar. Vor neurones receiving input from periodontal tissues projected to either Vmo.dl or the jaw-opening motoneurone pool (Vmo.vm; Yoshida et al., 1987; Yoshida, Yasuda, Dostrovsky, Bae, Takemura, Shigenaga & Sessle, 1994).

DISCUSSION

Neuronal circuits related to the jaw-closing and jaw-opening reflexes can be represented by the diagrams in Figure 1. The findings support the view that the jaw-closing reflex elicited by spindle afferent excitation is monosynaptic. Our studies have also documented direct projections of type I and II spindle afferents to the Vmo.dl. Furthermore our data suggest that disynaptic pathways may be involved in the jaw-closing reflex in which some Vsup neurones serve as excitatory interneurones intercalated between type II spindle afferents and jaw-closing α-motoneurones (Figure 1A). This was supported by the finding that electrical stimulation of the masseter nerve produced EPSPs but never elicited short-latency IPSPs in the jaw-closing motoneurones.

In addition, our studies, suggest circuits (Figure 1B) whereby M and S (P fibre passing through sensory root or tract) types of Vmes periodontal afferents excite jaw-closing α-motoneurones disynaptically through interneurones located in Vsup or Vor, while the M type may also excite jaw-closing α-motoneurones monosynaptically. This was supported by the electrophysiological findings.

It is a general feature that inferior alveolar or lingual nerve stimulation in cats elicits predominantly IPSPs in jaw-closing motoneurones and EPSPs in jaw-opening motoneurones. According to our studies, stimulation of periodontal (or pulp) afferents produced IPSPs in masseter motoneurones and EPSPs in jaw-opening motoneurones. Together with our morphological data, these findings suggest that activation of periodontal afferents originating in the trigeminal ganglion (TG) are involved in the disynaptic jaw-opening reflex through inhibitory and excitatory interneurones located in Vor. These inhibitory and excitatory interneurones act directly on jaw-closing and jaw-opening α-motoneurones, respectively (Figure 1B).

In summary, these findings provide details of the neural circuitry underlying jaw reflexes, and indicate that Vmes afferents contribute to the jaw-closing reflex and TG afferents to the jaw-opening reflex.

REFERENCES

SHIGENAGA, Y., DOE, K., SUEMUNE, S., MITSUHIRO, Y., TSURU, K., OTANI, K., SHIRANA, Y., HOSOI, M., YOSHIDA, A. & KAGAWA, K. (1989) Physiological and morphological characteristics of periodontal mesencephalic trigeminal neurons in the cat - intra-axonal staining with HRP. *Brain Res.* **505**, 91-110.

SHIGENAGA, Y., MITSUHIRO, Y., SHIRANA, Y. & TSURU, H. (1990) Two types of jaw-muscle spindle afferents in the cat as demonstrated by intra-axonal staining with HRP. *Brain Res.* **514**, 219-237.

SHIGENAGA, Y., YOSHIDA, A., TSURU, K., MITSUHIRO, Y., OTANI, K. & CAO, C.Q. (1988) Physiological and morphological characteristics of cat masticatory motoneurons - intracellular injection of HRP. *Brain Res.* **461**, 238-256.

YOSHIDA, A., TSURU, K., MITSUHIRO, Y., OTANI, K. & SHIGENAGA, Y. (1987) Morphology of masticatory motoneurons stained intracellularly with horseradish peroxidase. *Brain Res.* **416**, 393-401.

YOSHIDA, A., YASUDA, K., DOSTROVSKY, J.O., BAE, Y.C., TAKEMURA, M., SHIGENAGA, Y. & SESSLE, B.J. (1994) Two major types of premotoneurons in the feline trigeminal nucleus oralis as demonstrated by intracellular staining with HRP. *J. Comp. Neurol.* **347**, 495-514.

CENTRAL TERMINATIONS OF INTRA-AXONALLY LABELLED JAW-ELEVATOR MUSCLE SPINDLE AFFERENTS IN THE RAT

Revers Donga[1] and Dean Dessem[2]

[1]Sherrington School of Physiology, UMDS
St. Thomas' Hospital, London SE1 7EH, UK
[2]Department of Physiology
Dental School, University of Maryland
Baltimore, Maryland 21201, USA

INTRODUCTION

Primary and secondary muscle spindle afferents are usually separated on the basis of conduction velocity (CV). The reliability of this in hindlimb studies is supported by the correlations between CV and dynamic sensitivity (Matthews, 1972) as well as by their central terminations (Mendell & Henneman, 1971; Kirkwood & Sears, 1974; Stauffer, Watt, Taylor, Reinking & Stuart, 1976; Brown & Fyffe, 1978). However, the use of CV in the classification of jaw muscle afferents has proved unsatisfactory since conduction distances are short and difficult to measure. Also, afferent fibre diameters have not been found to be bimodally distributed (Morimoto, Inoue & Kawamura, 1982). This suggests that other ways of distinguishing between primary and secondary spindle afferents in these muscles must be used. In the present study we correlate passive responses of jaw-elevator muscle spindle afferents with the morphology of their central projections.

METHODS

In anaesthetised and paralysed rats, recordings were made with micropipettes filled with horseradish peroxidase (HRP) from axons of stretch sensitive afferents in the region dorsal to the trigeminal motor nucleus (NVmot). After making recordings of their responses to ramp and hold and sinusoidal jaw movements, axons were impaled. In those with stable resting membrane potentials ≤ -40 mV, depolarising current pulses (40 ms, 100 Hz, 0.5 - 5 nA) were applied through the micropipette. Current injection was stopped if the membrane potentials were ≥ -30 mV. Animals were killed after 3hrs with anaesthetic overdose and perfused with saline, fixed with 1.25% glutaraldehyde and 1% paraformaldehyde in phosphate buffer and rinsed with 10% sucrose solution. Sagittal brain sections were cut at 100 μm by vibrotome and the tissue reacted for demostration of peroxidase activity using the method of Metz, Schneider & Fyffe (1989). Axonal morphology was reconstructed using a Eutectics software-based 3-dimensional computer reconstruction system. The locations of swellings on fine collaterals taken to be synaptic boutons were examined at x1000. Bouton morphology was further analysed using a Zeiss videoplan manual image analysis computer.

RESULTS

A total of 8 jaw-muscle spindle afferent axons were labelled with HRP in the tract of the mesencephalic nucleus of the fifth nerve. They responded to stretches in one of two ways. One group (n = 2) had high dynamic indices (≥ 60 imp/s) and high peak frequencies during ramp and hold stretches. By analogy with hindlimb muscle spindle afferents, these features are suggestive of primary behaviour. The other group (n = 6) had characteristics reminiscent of secondary behaviour, owing to their low dynamic indices (≤ 20 imp/s) and low peak frequencies during ramp and hold stretches.

A typical reconstruction of a primary-like spindle afferent is illustrated in Figure 1A.

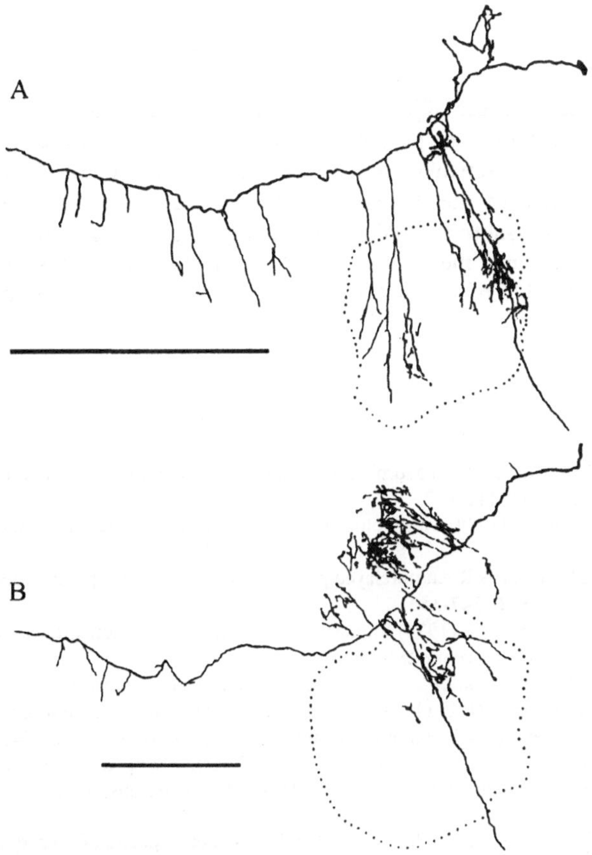

Figure 1. Sagittal reconstructions of a primary-like (A) and secondary-like (B) jaw muscle spindle afferents. The NVmot, as defined by Nissl staining, is shown by the dotted line. Calibration bars = 1 mm.

The incoming axons passed through the NVmot where some collaterals were given off before bifurcating into the rostrally directed tract of the mesencephalic nucleus and the caudally directed tract of Probst. This morphology is similar to that described in previous studies without distinction of primary and secondary types (Dessem & Taylor, 1989). In all afferents, bifurcations occured dorsal to the NVmot. All labelled afferents were found to have pseudounipolar somata within the rostro-caudal extent of the MesV. As can be seen from Figure 1A, the heaviest density of collaterals of suspected primary-like afferents was confined to the trigeminal motor nucleus (NVmot). In contrast, suspected secondary-like afferents (Figure 1B) had their heaviest density of collaterals in the region dorsal to the NVmot, comprising the supratrigeminal area. This group of afferents also had terminations in the NVmot, although very sparingly. Afferents occasionally appeared to terminate on cells of the trigeminal mesencephalic nucleus. The densities of collaterals of both categories of afferents were comparable, although their distribution within and outside the NVmot suggested significant differences.

A noteworthy finding was that one afferent of each category, had axon collaterals that branched from the tract of the MesV and coursed dorsally into the superior cerebellar peduncle. Although these collaterals were extremely fine, they were nevertheless clear. This finding is significant as the first anatomical demonstration of physiologically characterised first-order spindle afferents projecting directly to the cerebellum (Donga & Dessem, 1993).

CONCLUSIONS

The main projection areas of jaw-elevator muscle spindle afferents, as indicated by the presence of axon collaterals, en passant and terminal boutons, were in the NVmot, the region dorsal to it and the reticular formation caudal to the NVmot. Primary-like afferents projected most strongly to the NVmot, whilst secondary-like afferents projected most strongly to the region dorsal to the NVmot, including the supratrigeminal region. Both types sent collaterals directly into the superior cerebellar peduncle, suggesting that these first-order afferents are capable of transmitting unprocessed muscle spindle information directly to the cerebellum.

REFERENCES

BROWN, A.G. & FYFFE, R.W. (1978) Morphology of group Ia afferent fibre collaterals in the spinal cord of the cat. *J. Physiol.* **274**, 111-127.

DESSEM, D. & TAYLOR, A. (1989) Morphology of jaw-muscle spindle afferents in the rat. *J. Comp. Neurol.* **282**, 389-403.

DONGA, R. & DESSEM, D. (1993) An unrelayed projection of jaw-muscle spindle afferents to the cerebellum. *Brain Res.* **626**, 347-350.

KIRKWOOD, P.A. & SEARS, T.A. (1974) Monosynaptic excitation of motoneurones from secondary endings of muscle spindles. *Nature* **252**, 243-244.

MATTHEWS, P.B.C. (1972) *Mammalian Muscle Receptors and their Central Actions*. Arnold, London.

MENDELL, L.M. & HENNEMAN, E. (1971) Terminals of single Ia fibres: Location, density and distribution within a pool of 300 homonymous motoneurones. *J. Neurophysiol.* **34**, 171-187.

METZ, C.B., SCHNEIDER, S.P. & FYFFE, R.E.W. (1989) Selective suppression of endogenous peroxidase activity: application for enhancing appearance of HRP-labeled neurons in vitro. *J. Neurosci. Meth.* **26**, 181-188.

MORIMOTO, T., INOUE, H. & KAWAMURA, Y. (1982) Diameter spectra of sensory and motor fibres in nerves to jaw-closing and jaw-opening muscles in the cat. *Jap. J. Physiol.* **32**, 171-182.

STAUFFER, E.K., WATT, D.G., TAYLOR, A., REINKING, R.M. & STUART, D.G. (1976) Analysis of muscle receptor connections by spike triggered averaging. 2. Spindle group II afferents. *J. Neurophysiol.* **39**, 1392-1402.

PART 2

MOTOR UNIT RECRUITMENT

THE "SIZE PRINCIPLE" OF HUMAN α-MOTONEURONES

Friedemann Awiszus[1] and Helmut Feistner[2]

Klinik für Orthopädie[1] and
Klinik für Neurophysiologie[2]
Otto-von-Guericke-Universität
Magdeburg, Germany

INTRODUCTION

The recruitment order of mammalian α–motoneurones in a motoneurone pool obeys the so-called size principle (Henneman, Somjen & Carpenter, 1965), i.e. small motoneurones are recruited before larger ones. The intrinsic basis for this behaviour is given by the fact that an excitatory input system (for example the Ia fibres) evokes large EPSPs in small and small EPSPs in large α–motoneurones. It has been proposed that this is a universal feature of all excitatory input systems to α–motoneurones (Lüscher & Clamann, 1992; but see Heckman & Binder, 1993).

Almost all experimental work on the relationship between EPSP amplitudes and motoneuronal size was performed in cats. As the size principle was also proposed to be valid for human motoneurones (see Freund, 1983 for review), one would expect that the excitatory inputs promote orderly recruitment of motor units by the same principle as in cats. We tested this proposal in several experiments on different muscles of healthy volunteers. Excitatory input to the investigated α–motoneurones was obtained through Ia fibre stimulation in the periphery and transcranial magnetic stimulation of the motor cortex. EPSP size was estimated by a cross correlation of stimuli with motor unit discharges recorded through an intramuscular needle electrode (Ashby & Zilm, 1982).

MATERIALS AND METHODS

All experiments were performed on healthy volunteers who had given informed consent. Additionally, all experimental procedures were approved by the Local Ethical Commitee.

Either electrical stimuli to a peripheral nerve or transcranial magnetic stimuli were applied. Electrical stimuli had a duration of 0.5ms which is optimal for the stimulation of Ia afferents (Panizza, Nilsson & Hallett, 1989). Their strength was adjusted to 90% of the motor threshold. Transcranial magnetic stimuli were delivered by a Novametrix Magstim 200 stimulator using either a circular coil at the vertex for hand muscle stimulation or a specific figure of eight coil at the vertex for leg muscles. Magnetic stimulus strength was adjusted to be just below the stimulator output necessary for a visible contraction of the target muscle which was slightly voluntarily activated.

A needle electrode was inserted into the muscle of interest and positioned close to a motor unit made active by a slight voluntary contraction. The EMG signal was amplified and fed into a Schmitt trigger converting the action potentials of the unit under investigation into digital pulses. A laboratory computer recorded the incoming spike train. Additionally, the computer triggered the stimulator in such a way, that a flat pre-stimulus spike density was achieved (Awiszus, 1993a).

The stimulus correlated spike train data from 100 stimuli were recorded from each unit. The response data were analysed by procedures described in detail elsewhere (Awiszus, 1992; 1993b) which avoid the binning of spike times. The peri-stimulus spike density, usually estimated with a peri-stimulus time histogram (PSTH), was estimated with a procedure that is based on *r*ate estimation of an *i*nhomogenous *P*oisson *p*rocess (RIPP density estimate; Awiszus, 1992). In order to arrive at an RIPP estimate the unbinned spike times relative to the stimulus were sorted into ascending order. Then the local density of ten consecutive spike times was calculated and this local estimate was assigned to the midpoint of the peri-stimulus time interval in which the ten spikes occurred. Moving the ten spike window through out the sorted spike times yields a series of local density estimates throughout the peri-stimulus time. By connecting consecutive local estimates one arrives at the RIPP density estimate. The window width of ten was found to give valid spike density estimates for the human spike train data obtained in this laboratory, although the exact value for the window width is not crucial for this method as long as it is larger than three (Awiszus, 1992).

It has been shown experimentally, at least for large EPSPs in mammalian motoneurones, that the EPSP rising phase recorded intracellularly is very similar to the time integral of the spike density peak evoked by the same EPSP during extracellular spike train recording (Fetz & Gustafsson, 1983). Moreover, it has been shown previously that the

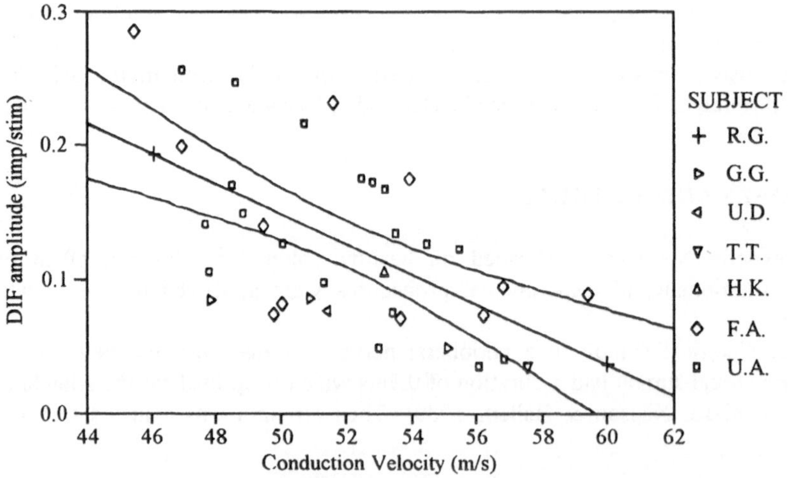

Figure 1. Correlation between the single motor unit conduction velocity and the DIF amplitude, which was obtained from a cross correlation of 100 electrical stimuli to the tibial nerve with ongoing motor unit discharges. DIF amplitude is taken to be a measure of Ia EPSP amplitude. Results from 38 soleus units of seven subjects are presented. Also shown is the linear regression line with 95% confidence intervals.

difference of the cumulative spike distribution functions observed with and without stimulation gives a direct estimate of the time integral of the spike density (Awiszus, 1993b). The empirical spike distribution functions of the sorted peri-stimulus spike times, i.e. the staircase functions jumping upward by a fixed amount each time a spike is encountered while scanning unbinned spike times from left to right, served as an estimate for the required cumulative distribution functions. The resulting function obtained after a proper scaling measures the number of displaced impulses per stimulus within the peri-stimulus time interval of interest and was thus termed the *d*isplaced *i*mpulse *f*unction (DIF; Awiszus, 1993b). Due to the fact that this function also represents a direct estimate of the time integral of the peri-stimulus spike density, the DIF rising phase may be assumed to be quite similar to the rising phase of the underlying EPSPs (Fetz & Gustafsson, 1983).

For each unit, the motor unit-triggered average potential from an additional nonspecific electrode (Lemon, Mantel & Rea, 1990) was obtained which in turn could be used to obtain an estimate of the axonal conduction velocity of the α–motoneurone under investigation (Awiszus & Feistner, 1993).

RESULTS

Correlation between the Ia EPSP and motoneurone size

Thirty eight soleus motor units from seven subjects were exposed to electrical stimuli of the tibial nerve at the popliteal fossa. For each unit a clear short-latency peak was found in the peri-stimulus spike density representing the single unit H-reflex. The size of the underlying Ia EPSP was estimated from the corresponding DIF amplitude. The relationship between the Ia EPSP estimate and the motor axon conduction velocity is given in Figure 1. Conduction velocity is taken to correlate with motoneurone size.

It can be seen that there is a significant correlation between the Ia EPSP amplitude estimate and the motoneuronal size parameter. The lower the conduction velocity (i.e. the smaller the motoneurone), the larger is the Ia EPSP (indicated by the larger DIF amplitude).

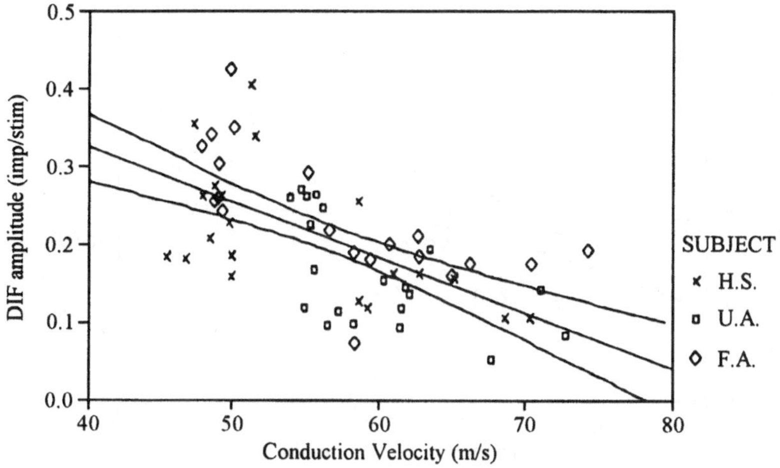

Figure 2. Correlation between the single motor axon conduction velocity and the DIF amplitude evoked by transcranial magnetic stimulation, which is taken as a measure of EPSP amplitude. Results from 60 first dorsal interosseus units of three subjects are presented. Also shown is the linear regression line with 95% confidence intervals.

Thus, in the human soleus muscle the homonymous Ia input promotes the orderly recruitment of motor units.

Correlation between the EPSP evoked by transcranial magnetic stimulation and motoneurone size

In sixty units from the first dorsal interosseus muscle from three subjects, the size of the EPSP evoked by transcranial magnetic stimulation was estimated. The majority of units (90%) responded to the series of 100 magnetic stimuli with a single short-latency spike density peak. Only six units (10%) responded with two peaks which might be attributed to multiple corticospinal volleys evoked by the stimulus. Therefore, for the majority of units this near-threshold stimulus evoked a simple EPSP similar to the Ia EPSP in soleus motor units. The correlation of the transcranial magnetically induced EPSP size with the axonal conduction velocity is given in Figure 2.

As in the case of the Ia EPSP, a significant correlation exists between motoneuronal size and the size of the EPSP evoked by the transcranial magnetic stimulus. Small motoneurones receive larger monosynaptic EPSPs through the corticospinal tract. Consequently, the excitation of the direct corticomotoneuronal pathway also enhances the orderly recruitment of motor units in the first dorsal interosseus muscle.

Correlation between the EPSP evoked by transcranial magnetic stimulation and the Ia EPSP

The results given above indicate that there should be a direct correlation between Ia EPSP amplitude and the amplitude of the EPSP evoked by transcranial magnetic stimulation in a single α–motoneurone. We tested this hypothesis directly in units of the tibialis anterior muscle for which both types of EPSPs are easily evoked (Priori, Bertolasi, Dressler, Rothwell, Day, Thompson & Marsden, 1993). For each unit cross-correlations of motor unit discharges with 100 electrical stimuli to the peroneal nerve at the head of the fibula and

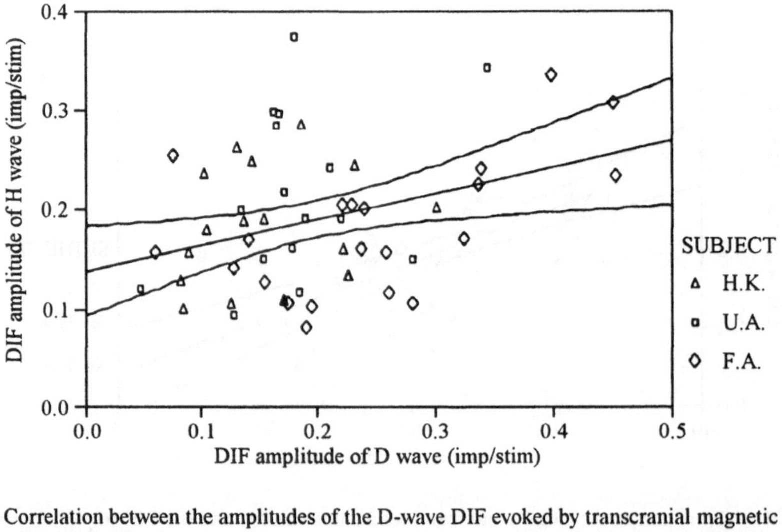

Figure 3. Correlation between the amplitudes of the D-wave DIF evoked by transcranial magnetic stimulation and the Ia DIF. Results from 54 tibialis anterior units of three subjects are presented. Also shown is the linear regression line with 95% confidence intervals. It is inferred that the DIF amplitude estimates the EPSP amplitudes.

with 100 transcranial magnetic stimuli were estimated. The DIF responses obtained for tibialis anterior with Ia stimulation were very similar to those of our soleus units. However, DIF responses of tibialis anterior motor units to transcranial magnetic stimulation were more complex than those of the first dorsal interosseus units. Most tibialis anterior motor units (79%) responded with several spike density peaks, indicating that the corticomotoneuronal volley is likely to have several subcomponents. Following the interpretation of Priori et al. (1993) for the tibialis anterior muscle, we interpreted the first spike density peak as a D-wave effect (i.e. as a consequence of direct stimulation of the corticomotoneuronal cells) and subsequent peaks as I-wave effects (i.e. as consequences of indirect trans-synaptic activation of the corticomotoneuronal cells). For 54 units from 3 subjects both a D-wave and an I-wave effect were seen after transcranial magnetic stimulation as well as a single-unit H-reflex after peripheral electrical stimulation. The correlation between the DIF amplitude of the D-wave and the DIF amplitude representing the Ia EPSP for these units is given in Figure 3.

As can be seen there is a significant correlation between the estimates of the D-wave and Ia EPSP amplitudes, although the correlation is weaker than those shown in Figures 1 and 2. Correlation coefficients calculated between the estimates of the amplitude of the D-wave and I-wave EPSPs and of the I-wave and Ia EPSPs did not differ significantly from zero. Neither was there any correlation found between the sum of D- and I-wave EPSPs and the Ia EPSP amplitude.

CONCLUSION

The results support the assumption that the homonymous Ia input and the direct corticomotoneuronal input evoke large EPSPs in small and small EPSPs in large human α–motoneurones. Consequently, these input systems promote the orderly recruitment of motor units.

On the other hand, the EPSP induced by *indirect* excitation of the corticomotoneuronal cells does not appear to promote the size principle. One should keep in mind, however, that the I-wave EPSP appears shortly after the D-wave EPSP and correlations between D- and I-wave EPSP amplitude could have been masked by nonlinear summations at the membrane of the spinal motoneurone. Additionally, the I-wave EPSP amplitude may be much more task dependent than the D-wave EPSP (Flament, Goldsmith, Buckley & Lemon, 1993). Consequently, some additional scatter is to be expected in the amplitude of the I-wave EPSP which could be responsible for the observed results.

Nevertheless, the major excitatory input systems to α–motoneurones testable in humans appear to promote the orderly recruitment of motor units in several different muscles. No conclusion, however, is possible about the extent to which this is also true for other input systems as for example for the rubrospinal input.

REFERENCES

ASHBY, P. & ZILM, D. (1982) Relationship between EPSP shape and cross-correlation profile explored by computer simulation for studies on human motoneurons. *Exp. Brain Res.* **47**, 33-40.

AWISZUS, F. (1992) The RIPP density estimate: an alternative method for the estimation of peri-stimulus spike density. *J. Neurosci. Meths.* **45**, 55-62.

AWISZUS, F. (1993a) Sensitivity of different stimulus-timing strategies for the detection of small excitations in noisy spike train data. *Biol. Cyber.* **68**, 553-558.

AWISZUS, F. (1993b) Quantification and statistical verification of neuronal stimulus responses from noisy spike train data. *Biol. Cyber.* **68**, 267-274.

AWISZUS, F. & FEISTNER, H. (1993) The relationship between estimates of Ia-EPSP amplitude and conduction velocity in human soleus motoneurons. *Exp. Brain Res.* **95**, 365-370.

FETZ, E.E. & GUSTAFSSON, B. (1983) Relation between shapes of post-synaptic potential and changes in firing probability of cat motoneurones. *J. Physiol.* **341**, 387-410.

FLAMENT, D., GOLDSMITH, P., BUCKLEY, C. J. & LEMON, R.N. (1993) Task dependence of responses in first dorsal interosseus muscle to magnetic brain stimulation in man. *J. Physiol.* **464**, 361-378.

FREUND, H.-J. (1983) Motor unit and muscle activity in voluntary motor control. *Physiol. Revs.* **63**, 387-436.

HECKMAN, C.J. & BINDER, M.D. (1993) Computer simulations of the effects of different synaptic input systems on motor unit recruitment. *J. Neurophysiol.* **70**, 1827-1840.

HENNEMAN, E., SOMJEN, G. & CARPENTER, D.O. (1965) Functional significance of cell size in spinal motoneurons. *J. Neurophysiol.* **28**, 560-580.

LEMON, R.N., MANTEL, G.W.H. & REA, P.A. (1990) Recording and identification of single motor units in the free-to-move primate hand. *Exp. Brain Res.* **81**, 95-106.

LÜSCHER, H.R. & CLAMANN, H.P. (1992) Relation between structure and function in information transfer in spinal monosynaptic reflex. *Physiol. Revs.* **72**, 71-99.

PANIZZA, M., NILSSON, J. & HALLETT, M. (1989) Optimal stimulus duration for the H reflex. *Muscle Nerve* **12**, 576-579.

PRIORI, A., BERTOLASI, L., DRESSLER, D., ROTHWELL, J C., DAY, B.L., THOMPSON, P.D. & MARSDEN, C.D. (1993) Transcranial electric and magnetic stimulation of the leg area of the human motor cortex: single motor unit and surface EMG responses in the tibialis anterior muscle. *Electroenceph. Clin. Neurophysiol.* **89**, 131-137.

ARE THERE IMPORTANT EXCEPTIONS TO THE SIZE PRINCIPLE OF α-MOTONEURONE RECRUITMENT?

Timothy C. Cope[1] and Brian D. Clark[2]

[1]Department of Physiology
Emery University
Atlanta
GA 30322, USA
[2]Department of Biomedical Sciences
Pennsylvania College of Podiatric Medicine
8th and Race Sts.
Philadelphia
PA 19107, USA

INTRODUCTION

The size principle states that motoneurones are recruited in a sequence that follows a size rule; small motoneurones are recruited before large ones. By extension, exceptions to the principle would be identified when that size order is reversed for motoneurones within a functional group. At first glance, and ignoring for the moment the exact meaning of motoneurone size, these notions appear unambiguous. Some uncertainty arises, however, as to the nature of the experimental observations that would be necessary and sufficient to demonstrate exceptions to the size principle. In addition, it is unclear whether any such exceptions should be regarded as important to our understanding of neuromotor operation or function.

In this brief review, recent experimental evidence for the suggestion of non-size ordered recruitment is evaluated. Whereas some observations, notably those involving electrical stimulation of cutaneous afferents succeed in demonstrating true exceptions to the size principle, others fail to do so because of the confounding possibility that separate pools of motoneurones may have been studied together. Although the studies using electrical stimulation might lead to important knowledge about synaptic circuitry, the artificial, simultaneous activation of multiple afferent fibers devalues their relevance to normal neuromotor function. We conclude that there exists little evidence for functionally important exceptions to the size principle. This conclusion leads to the proposal that size-ordered recruitment could serve as a baseline assumption for studies of the assignment of motor units into different functional groups.

SIZE PRINCIPLE

A pattern of physiological variation emerges from the sequence in which motor units are recruited. This pattern has been observed in assorted skeletal muscles engaged in diverse movements and postures, and it has been amply documented for the few physiological measures that are experimentally obtainable from motoneurones or motor units during recruitment trials (Henneman & Mendell, 1981; Calancie & Bawa, 1990; Cope & Clark, 1991). For the axons of motoneurones these measures are the velocity of conduction or amplitude of current associated with action potentials, and for the groups of muscle fibers supplied by motoneurones, i.e. muscle units, these measures are tension or the current generated by the compound action potential. By these parameters, the sequential recruitment of motor units can be seen to progress in a regular pattern from units with the smallest values for conduction velocity, tension and action potential current to units with larger values. Furthermore, these measures are directly related to the physical size of a motoneurone and its motor unit (Henneman & Mendell, 1981). These observations give reason enough to use the term "size principle" as a descriptor of this pattern, and it is commonly understood that rank ordering of the recruitment of motor units (small before large in relation to these measures) is what is meant by "size-ordered recruitment". Because of this consensus understanding, together with the credit the term imparts to its author, Professor Elwood Henneman, the term size principle remains valid and appropriate, even though its original formulation may not adequately account for the underlying mechanisms of orderly recruitment (see Sypert & Munson, 1981; Burke, 1981).

POSSIBLE EXCEPTIONS TO THE SIZE PRINCIPLE

Random

In nearly all recruitment studies, some fraction of the sample of motor units is recruited in reverse size order (Henneman & Mendell, 1981, but see Zajac, 1990). This irregularity was evident in our own studies of the reflex recruitment of motor units in the medial gastrocnemius muscle of decerebrate cats (Cope & Clark, 1991; Clark, Dacko & Cope, 1993). In some cases, motor units sampled as pairs were recruited in reverse size order, e.g. the unit with faster conduction velocity was recruited in advance of the one with slower conduction velocity, or the unit producing more isometric tension was recruited before the one generating less tension. These reversals were unequivocal; even though the differences between pairs were not extreme (the largest difference was 14 m/s for conduction velocity and 11 g for maximum isometric tension), they exceeded measurement uncertainty for 10-15% of our sample. Moreover, the reversed recruitment did not change in repeated trials.

It is worth considering whether these exceptions belie a recruitment order that is actually perfect. In our recruitment studies we examined five properties of motor units: axonal conduction velocity and muscle unit twitch and tetanic tension, contraction time, and fatigability. Although recruitment sequence of the pairs in our sample was significantly correlated with all of these parameters, it was not perfectly correlated with any of them (Cope & Clark, 1991). For some pairs of motor units, recruitment order was reversed with respect to all of these measures. It seems unlikely that there exists some additional, unmeasured descriptor of size that runs counter to several others in these exceptional cases so as to make perfect recruitment predictions. Still, apparent exceptions to the size principle might be nullified by some combination of parameters, and in this regard it is of interest that motor units characterised by type according to multiple unit properties are recruited without exception in the order S before FR before FF (Zajac, 1990). While this scheme achieves perfect prediction in categorising units into three broad groups, it is incomplete because it fails to account for the observed orderly recruitment of units within each type. Considering all of the above, we propose that no comprehensive recruitment scheme is perfect; there is a small but real random error in size-ordered recruitment.

Systematic

Two kinds of studies suggest the occurrence of systematic rather than random exceptions to the size principle. In one, units are recruited by electrical stimulation of peripheral sensory afferents. A good example is found in a report by Garnett & Stephens (1981). In their study, human subjects recruited motor units of the first dorsal interosseous muscle in slowly graded abduction of the index finger. Under this condition, the size principle prevailed. However, with the application of continuous electrical stimulation to the skin at the base of the index finger, recruitment order reversed after a time lag of several seconds. Order returned in most cases upon removal of the electrical stimulus. These observations, and similar ones made in anaesthetised cats (Mizote, 1982), seem to suggest that the size principle can be overridden.

Electrical stimulation of muscle afferents also seems to produce exceptions to the size principle. Davies, Wiegner & Young (1993) drew this inference from comparisons of units recruited in the human soleus muscle during voluntary contractions with those recruited in H-reflexes elicited by electrical stimulation of the posterior tibial nerve. Electrical stimulation recruited units in advance of the volitionally recruited ones. This observation, as well as indirect evidence of faster axonal conduction velocities for units recruited first in H-reflexes again suggests that the size principle can be superceded.

Well-documented effects of electrical stimulation in modulating the ongoing firing of motoneurones and in generating synaptic potentials give added support to the possibility of systematic switching of recruitment organisation. Examples include studies of the effects of electrical stimulation of skin afferents on volitionally controlled repetitive firing of motor units in the first dorsal interosseous (Garnett & Stephens, 1980; Masakado, Kamen & De Luca, 1991) and tibialis anterior muscles (Nielsen & Kagamihara, 1993) in human subjects. Both studies demonstrated tendencies toward inhibition of firing in early recruited units and facilitation in later recruited ones. These findings are consonant with direct measurements of synaptic potentials produced in motoneurones by electrical stimulation of selected afferent pathways in experimental animals. In their review, Stuart and Enoka (1983) counted as evidence for exceptions to orderly recruitment those cases in which afferent stimulation generates inhibitory potentials in small motoneurones and excitatory potentials in larger motoneurones. Results of computer modelling generally support this presumption by showing that, for example, electrical activation of the red nucleus simulated an increased incidence of reversals in size-ordered recruitment by reason of its facilitatory effects on large motoneurones and inhibitory effects on small ones (Heckman & Binder, 1993).

The report by Kanda, Burke & Walmsley (1977) has been notably influential through its suggestion that natural as well as electrical stimulation of skin afferents can produce exceptions to the size principle. Two observations made in cats motivated this investigation. One was the finding that, in contrast to usual operation, stimulation of the cutaneous sural nerve preferentially excited the predominantly fast-twitch medial gastrocnemius muscle and inhibited the slow-twitch soleus muscle. Extension of this unusual relative behavior of whole muscles to predictions about the activities of single units would mean that, contrary to the size principle, fast-twitch units were recruited before slow twitch ones. The other reason for suspecting size-principle exceptions derived from the tendency for sural nerve stimulation to produce inhibitory potentials in slow-twitch and excitatory potentials in fast-twitch motoneurones supplying the medial gastrocnemius muscle. To test for preferential recruitment, Kanda et al. (1977) used decerebrate cats to examine sural nerve effects on medial gastrocnemius motor axons that were brought to fire in tonic vibration reflexes. In each of two cases illustrated, the cutaneous afferents, whether activated by electrical stimulation or by skin pinch, decelerated the firing of units with low conduction velocity and accelerated or recruited units with higher conduction velocity. Thus it seemed as though synaptic input from cutaneous sources might produce exceptions to the size principle.

A second line of evidence for apparent exceptions to the size principle surfaces in studies of human subjects performing voluntary changes in the direction of movement or force (see Calancie & Bawa, 1990). An interesting recent case is presented by Nardone, Romano & Schieppati (1989). In that study, subjects engaged triceps surae muscles in voluntary contractions to produce two directions of ankle rotation; either plantar flexion by concentric contraction of the triceps surae or dorsiflexion by eccentric contraction in which the muscles slowly yielded to an external force. Apparent exceptions to the size principle occurred during eccentric contractions; units judged to be fast-twitch were preferentially recruited while other units that had exhibited low thresholds during concentric contractions were not activated. This result has been promoted as strong evidence of a major departure from fixed recruitment order (Burke, 1991).

WHICH EXAMPLES ARE TRULY EXCEPTIONS TO THE SIZE PRINCIPLE?

In discussing recruitment patterns up to this point, we have cited apparent "exceptions" to the size principle that may not in fact meet the criteria for "true" exceptions. Our caution about defining exceptions as true may be appreciated from the following illustrations. Suppose that one observes upon sampling the activity of two motor units that the larger of the two, e.g. the one with faster conduction velocity or greater tension, is recruited while the other is not. Does this constitute an exception to the size principle? If the two units belonged to different muscles, then it seems most reasonable to answer in the negative and to conclude instead that the smaller motor unit belonged to a muscle that was not recruited in that motor task. What if, however, both units belonged to the same muscle or to the same portion of a muscle? In these cases would inactivity of the smaller unit in the presence of activity of the larger one constitute a genuine exception to the size principle? Unfortunately, this question has no clear answer. Once again it is possible that the smaller unit belonged to a group that was not recruited in the motor task under examination, and therefore the unit's inactivity does not necessarily signal a size exception. On the other hand, the smaller unit may have been excluded from the recruited units strictly because of its size, in which case we may be willing to consider its relative inactivity as exceptional. These two potential explanations for inactivity of small units are not distinguishable based upon the measures of unit spike activity typically obtained in recruitment studies.

Exceptions to the size principle might become experimentally distinguishable if comparisons were restricted to units belonging to the same "motor pool". Although the term is inconsistently used, we assume some agreement in defining a motor pool as a collection of synergistic motor units, i.e. units that would act together in some circumstance given sufficient excitatory input. Once motor units are assigned to a single pool, cases in which they are recruited out of order by size or in which large motor units are selectively recruited should be counted as exceptions to the size principle. The determination of which motor units belong to a pool is difficult, however; anatomical boundaries between and within muscles do not necessarily segregate motor pools (Burke, 1991). While information about synaptic input would assist in recognising the motoneurones that belong to a given pool, such information is not available in typical recruitment studies. Thus the current construct of "motor pool" does little more than shift the investigator's quandary from identifying unusual patterns of recruitment to classifying motor units into pools.

Given the difficulty with defining motor pool membership, exceptions to the size principle might be judged with greater confidence by considering only the recruitment order among motor units that are activated together. Using this criterion, cases of random reversals in size ordered recruitment would stand as exceptions to the size principle. Other instances among the cases described above that would qualify as exceptions are the findings that electrical stimulation of the skin or electrical activation of the H-reflex recruit units in reverse size order.

By contrast, the exclusion of small units during eccentric contractions as described by Nardone et al. would not constitute a true exception to the size principle according to our criterion because these units were not engaged in activity. The observed activation pattern may be merely an expression of a process that engages different groups of motor units in different motor tasks, just as elbow flexion and supination involve selective activation of different subsets of motor units in the biceps brachii muscle (Denier van der Gon, Tax, Gielen & Erkelens, 1991).

Evaluation of the results reported by Kanda et al. (see above) requires special consideration. A remarkable result of that study was that cutaneous stimulation selectively inhibited ongoing discharge of some motor axons having the slowest conduction velocities. Under very similar experimental conditions, Clark et al. (1993) also found evidence of decelerating firing rates in some trials of skin pinch for a few (2/7) units that had slow conduction velocities and other characteristics of type S units. By extending the argument made above to derecruitment patterns, we suggest that these data express the selective deactivation of motor unit groups and not necessarily an exception to the size principle. Moreover, our direct examinations of units recruited by skin pinch failed to reveal any reversals of recruitment order in 37 pairs of units, seven of which include one type S and one type F unit. Thus, support is not robust for the popular notion that skin pinch reverses size-ordered recruitment in decerebrate cats.

WHY MIGHT EXCEPTIONS TO THE SIZE PRINCIPLE BE IMPORTANT?

Exceptions to the size principle might be considered important for at least three reasons. First, they pertain to our understanding of the neural mechanisms that control movement. The paucity of exceptions observed among motoneurones recruited by sundry classes of presynaptic input in decerebrate cats brought Henneman and co-workers (see Henneman & Mendell, 1981) to the forceful conclusion that postsynaptic properties of motoneurones predominate in the recruitment process. In debates that are now long-standing, others point to exceptions in order to bolster arguments for the primacy of synaptic inputs in determining recruitment order (see Burke, 1981). Both pre- and postsynaptic factors are included in the recent formulation of "recruitment gain" by Kernell & Hultborn (1990). According to this idea, synaptic effects play a key role in regulating the relative excitabilities of motoneurones. Whether changes in recruitment gain could actually reverse recruitment order rests on the demonstration of exceptions to the size principle.

Second, exceptions to the size principle as we define them might reflect functional advantages of other recruitment schemes. Large motor units tend to have relatively fast contraction times, so it is reasonable to suspect that their early recruitment may facilitate the performance of rapid movements. In addition, the early recruitment of forceful units may facilitate precision grip (Kanda & Desmedt, 1983) or enhance power generation upon foot contact during locomotion (Kernell & Hultborn, 1990). Note that while these benefits would accrue from exceptions to the size principle, they might also be expected from the rapid yet orderly recruitment of large units after smaller ones.

Finally, the frequency of exceptions in comparison to the normal state could be valuable in assessing injury or disease of the neuromuscular system. Cope & Clark (1993) examined recruitment order for cat muscles that were reinnervated by their own cut or crushed nerves. This study showed a nearly normal restoration of the size principle for units recruited by stimulation of an uninjured cutaneous nerve. Apart from demonstrating recovery of the match between the relative excitabilities and physiological properties of motor units, the infrequency of exceptions lessens concern that recruitment disorder contributes to motor dysfunction after self-reinnervation.

WHICH EXCEPTIONS ARE IMPORTANT ACCORDING TO OUR CRITERIA?

Considering the criteria described above, we believe that exceptions produced by electrical activation of skin afferents are important from two standpoints. First, they verify the capacity of presynaptic input to outweigh the effects of postsynaptic properties in determining motoneurone excitability. Also with respect to spinal mechanisms, these findings map the synaptic effects of cutaneous afferents within a motor nucleus, i.e. among the motoneurones supplying single muscles. Second, these and indeed any reproducible and systematic exceptions can be used to probe neuromuscular organisation in a way that could be valuable in assessing disease, injury or the effectiveness of various rehabilitation therapies. We are dubious, however, about attributing functional significance to those exceptions to the size principle that are produced by electrical stimulation. Electrical stimulation sends synchronised volleys of action potentials that are probably rarely received by spinal circuits in natural settings. Perhaps this explains why natural tactile stimulation of the index finger failed to reproduce the exceptions to the size principle that were observed with electrical stimulation (Kanda & Desmedt, 1983).

The apparent exceptions observed in H-reflexes can also be appreciated for demonstrating that particular presynaptic pathways can override the size principle. It seems odd, however, that the Ia afferents excited in the H-reflex could produce exceptions to the size principle - the well established synaptology of this sensory system leads to the prediction of size ordered recruitment (see Heckman & Binder, 1993). The unexpected result of Davies et al. (1993) may be explained by the involvement of a mixture of heteronymous Ia afferents. It is likely that a sizeable fraction of the Ia excitatory drive in their study originated from heteronymous sources, because H-reflexes were evoked in the soleus by percutaneous stimulation applied to the whole tibial nerve. This aspect of the study bears consideration because Clamann, Ngai, Kulkulka & Goldberg (1983) report large changes in motoneurone recruitability produced by switching stimulation of group I afferents between different dorsal rootlets. This finding is reasonably interpreted as an expression of non-uniformity in the distribution of afferents that arrive from heteronymous sources and that are activated in artificial combinations. The results of Davies et al. (1993) likewise suggest irregularity in the projection of heteronymous afferents to soleus motoneurones. We hesitate to conclude, however, that this irregularity is functionally important.

CONCLUSION/PROPOSAL

Writing this brief report has proven a useful exercise. First, it has given us the opportunity to explain why we believe that there are virtually no functionally important exceptions to the size principle of alpha motoneurone recruitment. In arriving at this conclusion we concur with Henneman & Mendell (1981) and with Calancie & Bawa (1990) who together give much fuller accounts of the pertinent literature prior to 1990. Second, it has brought us to realise a possible utility of the size principle - it might serve to uncover schemes used by the central nervous system to organise the activities of motor units belonging to different muscles and muscle compartments. In other words, the size principle might assist in defining "functional groups" of motor units. By "functional group" we mean a collection of motor units that act together, or that would act together given sufficient excitation, in a particular motor act. The modifier "functional" is used to emphasize that the motor unit composition of groups may vary in a way that is dependent upon motor task. The experimental strategy that we propose for defining functional groups is based on the contention that the size principle is inviolate, and it would be implemented as follows. Pairs of motor units that consistently obey the size principle during a given motor action would be considered to belong to the same functional group regardless of their muscular origin. Conversely, units manifesting reverse order, i.e. exceptions, would be considered to belong to different functional groups. In this way we might discover whether active motor units are organised as a single group or as separate groups in a given task. We recognise that there is some degree of error inherent in this approach and that no group assignment will be made with complete confidence. Our work shows that some exceptions will be random and not indicative of different group membership, but these should amount to roughly 10% of the sample when ranking units by conduction velocity. In addition, randomly chosen units from unrelated functional groups will show some degree of size ordering. Even so, we believe that this tack is likely to yield statistically verifiable patterns in the grouping of motor units engaged in a task. Knowledge of these functional groups could greatly assist studies of spinal motor systems. First, it might yield additional understanding of known synaptic connections in the spinal cord and produce predictions about as yet unstudied connections. Second, features that distinguish motor units in different functional groups, e.g. torque vector, could provide insight into the strategies used by the neuromotor system to control movement.

ACKNOWLEDGMENTS

We thank Drs. T. Richard Nichols and Alan J. Sokoloff for their comments on the manuscript and acknowledge the support of NIH Grant NS 21023.

REFERENCES

BURKE, R.E. (1991) Selective recruitment of motor units. In *Dahlem Workshop Reports. Motor Control: Concepts and Issues.* eds HUMPHREY, D.R. AND FREUND, H.-J. pp. 5-21, John Wiley & Sons, Chichester.

BURKE, R.E. (1981) Motor units: anatomy, physiology and functional organization. In *Handbook of Physiology. The Nervous System. Motor Control.* eds. BROOKHART, J.M. AND MOUNTCASTLE, V.B. pp. 345-422, American Physiological Society, Bethesda.

CALANCIE, B. & BAWA, P. (1990) Motor unit recruitment in humans. In *The Segmental Motor System.* eds. BINDER, M.D. & MENDELL, L.M. pp. 75-95, Oxford University Press, New York.

CLAMANN, H.P., NGAI, A.C., KULKULKA, C.G. & GOLDBERG, S.J. (1983) Motor pool organization in monosynaptic reflexes: responses in three different muscles. *J. Neurophysiol.* **50**, 725-742.

CLARK, B.D., DACKO, S.M. & COPE, T.C. (1993) Cutaneous stimulation fails to alter motor unit recruitment in the decerebrate cat. *J. Neurophysiol.* **70**, 1433-1439.

COPE, T.C. & CLARK, B.D. (1991) Motor unit recruitment in the decerebrate cat: several unit properties are equally good predictors of order. *J. Neurophysiol.* **66**, 1127-1138.

COPE, T.C. & CLARK, B.D. (1993) Motor-unit recruitment in self reinnervated muscle. *J. Neurophysiol.* **70**, 1787-1796.

DAVIES, L., WIEGNER, A.W. & YOUNG, R.R. (1993) Variation in firing order of human soleus motoneurons during voluntary and reflex activation. *Brain Res.* **602**, 104-110.

DENIER VAN DER GON, J.J., TAX, T., GIELEN, S. & ERKELENS, C. (1991) Synergism in the control of force and movement of the forearm. *Revs. Physiol. Biochem. Pharmacol.* **118**, 97-124.

GARNETT, R. & STEPHENS, J.A. (1980) The reflex responses of single motor units in human first dorsal interosseous muscle following cutaneous afferent stimulation. *J. Physiol.* **303**, 351-364.

GARNETT, R. & STEPHENS, J.A. (1981) Changes in the recruitment threshold of motor units in human first dorsal interosseous muscle produced by skin stimulation. *J. Physiol.* **311**, 463-473.

HECKMAN, C.J. & BINDER, M.D. (1993) Computer simulations of the effects of different synaptic input systems on motor unit recruitment. *J. Neurophysiol.* **70**, 1827-1840.

HENNEMAN, E. & MENDELL, L.M. (1981) Functional organization of motoneuron pool and its inputs. In *Handbook of Physiology. The Nervous System. Motor Control.* eds. BROOKHART, J.M. & MOUNTCASTLE, V.B. pp. 423-507, American Physiological Society, Bethesda.

KANDA, K., BURKE, R.E. & WALMSLEY, B. (1977) Differential control of fast and slow twitch motor units in the decerebrate cat. *Exp. Brain Res.* **29**, 57-74.

KANDA, K. & DESMEDT, J.E. (1983) Cutaneous facilitation of large motor units and motor control of human fingers in precision grip. In *Motor Control Mechanisms in Health and Disease.* ed. DESMEDT, J.E. pp. 253-261, Raven Press, New York.

KERNELL, D. & HULTBORN, H. (1990) Synaptic effects on recruitment gain: a mechanism of importance for the input- output relation of motoneuron pools? *Brain Res.* **507**, 176-179.

MASAKADO, Y., KAMEN, G. & DE LUCA, C.J. (1991) Effects of percutaneous stimulation on motor unit firing behavior in man. *Exp. Brain Res.* **86**, 426-432.

MIZOTE, M. (1982) The effect of digital nerve stimulation on recruitment order of motor units in the first deep lumbrical muscle of the cat. *Brain Res.* **248**, 245-255.

NARDONE, A., ROMANO, C. & SCHIEPPATI, M. (1989) Selective recruitment of high-threshold human motor units during voluntary isotonic lengthening of active muscles. *J. Physiol.* **409**, 451-471.

NIELSEN, J. & KAGAMIHARA, Y. (1993) Differential projection of the sural nerve to early and late recruited human tibialis anterior motor units: change of recruitment gain. *Acta physiol. scand.* **147**, 385-401.

STUART, D.G. & ENOKA, R.M. (1983) Motoneurons, motor units, and the size principle. In *The Clinical Neurosciences. Neurobiology.* eds. ROSENBERG, R.N. & WILLIS, W.D. pp. 471-517, Churchill Livingstone, New York.

SYPERT, G.W. & MUNSON, J.B. (1981) Basis of segmental motor control: motoneuron size or motor unit type? *Neurosurg.* **8**, 608-621.

ZAJAC, F.E. (1990) Coupling of recruitment order to the force produced by motor units: the "size principle hypothesis" revisited. In *The Segmental Motor System.* eds. BINDER, M.D. & MENDELL, L.M. pp. 96-111, Oxford University Press, New York.

PLASTICITY OF MOTONEURONE RECRUITMENT ORDER IN THE FLEXION REFLEX FOLLOWING SPINAL CORD TRANSECTION

Russell G. Durkovic

Department of Physiology
SUNY Health Science Center
Syracuse, NY 13210, U.S.A.

INTRODUCTION

The observation that motoneurones are generally activated in order (small motoneurones are recruited before large) is known as 'the size principle' of motoneurone recruitment (Henneman, 1977). While a number of exceptions to this principle have been noted, there is abundant evidence that it does hold for a variety of inputs in a wide range of animals (Henneman & Mendell, 1981). Interestingly, for the cutaneous evoked flexion reflex the first report of orderly recruitment (Olson, Carpenter & Henneman, 1968) received little additional experimental study or verification. In one of the few other studies dealing with recruitment order during the flexion reflex it was reported that motor unit recruitment order can be variable, despite application of standardised stimuli in spinal man (Grimby & Hannerz, 1970). In the present experiments the results of activation of tibialis anterior motor nerve fibres by stimulation of saphenous and superficial peroneal cutaneous nerves in acute and three month chronic spinal cats support the orderly recruitment results of Olson et al. (1968). However, significant deviations from the standard order of recruitment were found for these flexion reflexes in two week chronic spinal animals.

METHODS

Adult female cats with T_{10} spinal transections made three months, two weeks or immediately before decerebration were used in these studies. For the chronic transections, twelve cats (six two week chronic, six three month chronic) were deeply anaesthetised with sodium pentobarbitone and using aseptic techniques a complete T_{10} spinal cord transection was performed using thermal cautery. Cats were kept in excellent health under supervision of a licensed veterinarian. Experimental procedures followed NIH guidelines for the care and use of experimental animals. At two weeks or three months after spinal cord transection, experiments were conducted as described below for acute transection animals.

Six acute animals were given atropine (0.3 mg in 1 ml saline s.c.), anaesthetised with ether and tracheotomised. Anaesthesia was continued with a mixture of oxygen, halothane and nitrous oxide. A T_{10} spinal transection and anaemic decerebration were performed, anaesthesia was discontinued, and the animals were respired with room air.

The motor nerve to the tibialis anterior (TA) muscle was isolated, and a branch of the nerve was cut and placed in a saline solution containing trypsin, chymotrypsin and hyaluronidase (5 mg/ml each) for 2-4 minutes in order to break up connective tissue. The nerve branch was then teased into finer divisions, and an individual filament was placed on a pair of silver recording electrodes. In order to determine recruitment orders, thresholds of action potentials were measured for pairs of motor nerve axons activated reflexly by single electrical stimuli (0.2 ms square waves) to saphenous or superficial peroneal cutaneous nerves. Occasionally more than two units were recruited from a single filament. When this occurred each unit was ranked according to its stimulus threshold and action potential amplitude relative to the other units, and each comparison was considered an individual pair of units. When recruitment thresholds of two units were indistinguishable, "ties" were recorded. An "orderly" recruitment was designated when the smaller of two units had the lower threshold for activation. A "reversed" recruitment order was recorded when the larger of two units had a lower threshold for activation by the cutaneous nerve stimuli. For statistical purposes the tests on recruitment order were carried out on the combined data from saphenous and superficial peroneal nerves, since statistical tests on each group indicated that no significant differences existed for the results from the two nerves.

RESULTS

In acute spinal transected animals there was a well-defined relationship between the motoneurones recruited at lowest stimulus intensity and the sizes of the motoneurones as

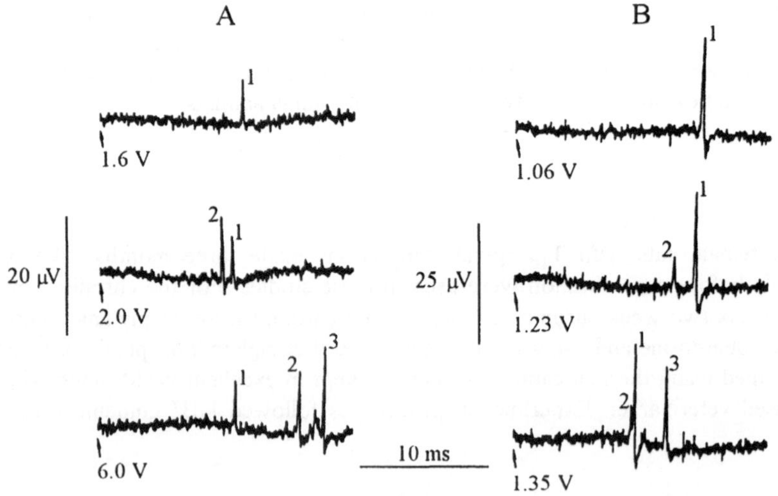

Figure 1. Reflex activation of tibialis anterior nerve action potentials at varying cutaneous nerve stimulus intensities. A. Saphenous nerve stimuli activate larger units as stimulus intensities increase (acute spinal cat). Three pairs of "orderly recruited" units were obtained from these records. B. Superficial peroneal nerve stimuli in a two week chronic spinal animal. One "orderly" and two pairs of "reversed" recruited units were obtained from these records.

Table 1. Relationships between action potential size and threshold of activation for pairs of tibialis anterior units. Number of unit pairs in each category are shown for the three groups of animals. Orderly: the smaller of the two spikes was activated at the lower stimulus intensity. Reversed: the larger of the two spikes was activated at the lower stimulus intensity. Tied: small and large spikes had the same stimulus threshold.

	ACUTE			2 WEEK CHRONIC			3 MONTH CHRONIC		
	Orderly	Reversed	Tied	Orderly	Reversed	Tied	Orderly	Reversed	Tied
Saphenous nerve	30	4	10	14	27	8	31	7	7
Superficial Peroneal nerve	31	5	12	23	28	13	40	11	7
TOTAL	61	9	22	37	55	21	71	18	14

judged by the relative amplitudes of the action potentials recorded (Figure 1A). Stimulation of either cutaneous nerve usually evoked activity at lowest thresholds in the axon with the smaller action potential (Table 1). Excluding pairs of units with equal thresholds, in 87% of the unit pairs examined, those units with the smallest action potentials had the lowest threshold to cutaneous nerve stimuli. Eleven pairs of units were found that could be tested for recruitment by both cutaneous nerves, and all eleven were recruited in the same order for each cutaneous nerve. To estimate the probability that the observed distributions of "orderly" and "reversed" unit pairs occurrred at random the sums of the recruitment order results for the two cutaneous nerves were used in a χ^2 analysis. "Tied" units were assigned half scores and assigned equally to "orderly" and "reversed" categories. For acute spinal transected animals the probability that the observed distribution of recruitment orders occurred at random was less than 0.001.

The outcome for two week chronic spinal cats was different from that of acute spinal animals. At lowest thresholds, inputs from both saphenous and superficial peroneal cutaneous nerves more often activated units with larger action potentials. As stimulus intensity increased, smaller units were often recruited (Table 1 and Figure 1B). Excluding pairs of units with equal thresholds, in only 40% of unit pairs were the smallest sized spikes recruited at lower cutaneous nerve thresholds. In this group 25 pairs of units could be tested for recruitment order by both cutaneous nerves and of these, 22 exhibited the same order. A χ^2 analysis on the sums of "orderly" and "reversed" pairs for saphenous and superficial peroneal nerves in the two week chronic group ("ties" treated as for acute spinal animals) suggested a tendency toward recruitment order reversal (χ^2 = 2.867, $0.05<P<0.1$; χ^2 = 3.841 required for a 0.05 level of significance) for this group. Excluding ties, χ^2 for this group was 3.522.

The three month chronic spinal group exhibited recruitment orders similar to those observed for the acute spinal animals. That is, those units with the smallest action potentials were generally recruited at lower thresholds than units with larger spikes (Table 1). For 80% of the unit pairs in this group the smaller unit was recruited at the lower threshold (not including "tied" cases). For the same unit pairs tested for recruitment order with both cutaneous nerve inputs, 14 of the 15 pairs had the same order. The χ^2 analysis indicated a probability less than 0.001 that the summed data for the two nerves were randomly distributed (Table 1).

Individual animal's percentages of orderly recruited pairs ("tied" pairs equally split) for each group of animals were tested using an analysis of variance, and significance was found ($F = 5.568$; df = 2, 15; $P < 0.05$). Scheffé's test (Edwards, 1968) for multiple comparisons ($\alpha = 0.05$) indicated that the two week chronic group was significantly different from both the acute and three month chronic groups with respect to the percentage of orderly recruitment. The difference between the acute and three month chronic group was not significant.

DISCUSSION

In acute and three month chronic spinal animals a strong correlation between excitation threshold and axon spike height has been demonstrated. Direct correlations have been shown to exist among a number of motoneurone size variables: motoneurone size, axon diameter and axon conduction velocity (Cullheim, 1978); axon spike amplitude and axon diameter (Boyd & Kalu, 1979); motor unit tetanic tension and axonal conduction velocity (Jami & Petit, 1975; cf., Goslow, Cameron & Stuart, 1977). Therefore these results for the cutaneously evoked flexion reflex clearly support the size principle of Henneman (1977). In the present study the orderly recruitment results held equally well for both saphenous and superficial peroneal cutaneous nerve inputs. In contrast, recruitment order in the two week chronic spinal animals showed a tendency for recruitment in the reverse order, from large to small, for both saphenous and superficial peroneal nerve inputs.

There are several caveats regarding the interpretation of the present data. First, the apparent alteration in recruitment order in chronic spinal animals relies on the assumption that axon sizes, and therefore spike amplitudes, have not changed relative to their sizes at the time of transection. Secondly, it is assumed that the action potential recordings represent motoneurone spikes. However, the recordings could include muscle afferent nerve action potentials activated (antidromically) via the dorsal root reflex (Schmidt & Willis, 1963).

If one assumes that the present recordings represent motoneurone spikes (see below) another question concerns the type of motoneurone involved (alpha and/or gamma), because conduction velocity measurements could not be made. Gamma motoneurones often exhibit spontaneous activity, and their action potentials are very small relative to alpha motoneurones (Kanda, Burke & Walmsley, 1977). These features describe activity that was often recorded in the present experiments but was not used for analysis (tiny, spontaneous spikes in Figure 1).

The concerns expressed above are minimised by the parallel results obtained from recruitment order studies of pairs of TA unit EMG spikes in animals with spinal transections of the same duration as in the present study (Durkovic & Misulis, 1994). Such results which are free of the potential interpretive problems mentioned above, strongly suggest that the results of the present study are valid for recruitment orders of alpha motoneurones, since TA muscle fibre size alterations in the chronic spinal animals were minor and could not be a factor in explaining the recruitment order alterations (Durkovic & Misulis, 1994).

In another motor unit study from this laboratory, expression of orderly recruitment was shown to occur in the process of classically conditioned long-lasting potentiation of flexion reflexes in the acute spinal cat (Misulis & Durkovic, 1982). Given the present results, it would not be surprising if conditioning of reflex potentiation is altered significantly in two week chronic compared to acute spinal animals.

Given that the major factors thought to be responsible for orderly recruitment are: (i) organisation of presynaptic inputs to the motoneurone pool and (ii) the intrinsic properties of the motoneurones (Heckman & Binder, 1990), processes potentially influencing these factors include degeneration followed by axonal sprouting to former tract synapses (Murray & Goldberger, 1974) and denervation supersensitivity (Tremblay, Bédard, Maheux & Di Paolo, 1985). While the present study provides no clues regarding underlying mechanisms, whatever leads to the tendency for recruitment order reversal in two week chronic spinal animals, it is compensated for either by its loss over time or additional plastic changes that force the standard pattern of recruitment to be reinstated. One possibility is that intrinsic properties of motoneurones are altered in two week chronics as a consequence of the loss of descending monosynaptic inputs that include the extensive bulbospinal serotonergic inputs (Pearson, Alvarez, Dewey, Harrington & Fyffe, 1993) and the excitatory inputs from the medial longitudinal fasciculis (Floeter, Sholomenko, Gossard & Burke, 1993). Sprouting of spinal inputs to motoneurones and/or interneurones may then have resulted in a return to normal motoneurone/recruitment order properties in the three month chronics.

It should be emphasized that neural plasticity following spinal transection is not unusual, since significant changes in other spinal neural characteristics observed soon after transection have also been demonstrated to subside weeks to months later. Examples include increases in the magnitude of recurrent inhibition (Goldfarb & Sharpless, 1971), decreases in synapse numbers on motoneurones (Pullen & Sears, 1978; Bernstein & Gelderd, 1973), increases in Ia EPSP magnitude and Ia projection frequency to extensor motoneurones (Nelson & Mendell, 1979) and serotonin immunoreactivity losses (Goldberger, 1986). Nevertheless, it is surprising that such a widespread and fundamental property as orderly recruitment exhibits the plasticity demonstrated to occur following spinal transection.

ACKNOWLEDGEMENTS

This research was supported by NSF grant 9220206. The author expresses his gratitude to Professor J. B. Preston for his comments on the early version of this manuscript.

REFERENCES

BERNSTEIN, J.J. & GELDERD, J.B. (1973). Synaptic reorganization following regeneration of goldfish spinal cord. *Exp. Neurol.* **41**, 402-410.

BOYD, I.A., & KALU, K.U. (1979). Scaling factor relating conduction velocity and diameter for myelinated afferent nerve fibers in the cat hind limb. *J. Physiol.* **289**, 277-297.

CULLHEIM, S. (1978). Relations between cell body size, axon diameter and axon conduction velocity of cat alpha-motoneurons stained with horseradish peroxidase. *Neurosci. Lett.* **8**, 17-20.

DURKOVIC, R.G., & MISULIS, K.E. (1994). Flexion reflex and muscle plasticity following spinal cord transection. (submitted).

EDWARDS, A.L. (1968) *Experimental Design in Psychological Research*. pp. 150-153. Holt, Rinehart and Winston Inc., New York.

FLOETER, M.K., SHOLOMENKO, G.N., GOSSARD, J.P. & BURKE, R.E. (1993). Disynaptic excitation from the medial longitudinal fasciculus to lumbosacral motoneurons: modulation by repetitive activation, descending pathways, and locomotion. *Exp. Brain Res.* **92**, 407-419.

GOLDBERGER, M.E. (1986). Autonomous spinal motor function and the infant lesion effect. In *Development and Plasticity of the Mammalian Spinal Cord.* edS. GOLDBERGER, M.E., GORIO, A. & MURRAY, M., pp. 363-380. Springer Verlag, New York.

GOLDFARB, J. & SHARPLESS, S.K. (1971). Effects of nicotine and recurrent inhibition on monosynaptic reflexes in acute and chronic spinal cats. *Neuropharmacology* **10**, 413-423.

GOSLOW, G.E., Jr., CAMERON, W.E. & STUART, D.G. (1977). The fast twitch motor units of cat ankle flexors. 2. Speed-force relations and recruitment order. *Brain Res.* **134**, 47-57.

GRIMBY, L. & HANNERZ, J. (1970). Differences in recruitment order of motor units in phasic and tonic flexion reflex in 'spinal man'. *J. Neurol. Neurosurg. Psychiat.* **33**, 562-570.

HECKMAN, C.J. & BINDER, M.D. (1990). Neural mechanisms underlying the orderly recruitment of motoneurons. In *The Segmental Motor System*. edS. BINDER, M.D. & MENDELL, L.M., pp. 181-204. Oxford University Press, New York.

HENNEMAN, E. (1977). Functional organization of motoneuron pools: The size principle. *Proc. Int. Union Physiol. Sci.* **12**, 50.

HENNEMAN, E. & MENDELL, L.M. (1981). Functional organization of motoneuron pool and its inputs. In *Handbook of Physiology, The Nervous System Vol. II.* ed. BROOKS, V.B., pp. 423-507. American Physiological Society, Bethesda.

JAMI, L. & PETIT, J. (1975). Correlation between axonal conduction velocity and tetanic tension of motor units in four muscles of the cat hind limb. *Brain Res.* **96**, 114-118.

KANDA, K., BURKE, R.E. & WALMSLEY, B. (1977). Differential control of fast and slow twitch motor units in the decerebrate cat. *Exp. Brain Res.* **29**, 57-74.

MISULIS, K.E., & DURKOVIC, R.G. (1982). Classically conditioned alterations in single motor unit activity in the spinal cat. *Behav. Brain Res.* **5**, 311-317.

MURRAY, M. & GOLDBERGER, M.E. (1974). Restitution of function and collateral sprouting in the cat spinal cord: the partially hemisected animal. *J. Comp. Neurol.* **158**, 19-36.

NELSON, S.G. & MENDELL, L.M. (1979) Enhancement in Ia-motoneuron synaptic transmission caudal to chronic spinal cord transection. *J. Neurophysiol.* **42**, 642-654.

OLSON, C.B., CARPENTER, D.O & HENNEMAN, E. (1968). Orderly recruitment of muscle action potentials. *Arch. Neurol.* **19**, 591-597.

PEARSON, J.C., ALVAREZ, F.J., DEWEY, D.E., HARRINGTON, D. & FYFFE, R.E.W. (1993). Serotonergic innervation of motoneurons in the cat's lumbar spinal cord. *Soc. Neurosci. Abs.* **19**, 983.

PULLEN, A.H. & SEARS, T.A. (1978). Modification of 'C' synapses following partial central deafferentation of thoracic motoneurons. *Brain Res.* **145**, 141-146.

SCHMIDT, R.F. & WILLIS, W.D. (1963) Depolarization of central terminals of afferent fibers in the cervical spinal cord of the cat. *J. Neurophysiol.* **26**, 44-60.

TREMBLAY, L.E., BÉDARD, P.J., MAHEUX, R. & DI PAOLO, T. (1985). Denervation supersensitivity to 5-hydroxytryptamine in the rat spinal cord is not due to the absence of 5-hydroxytryptamine. *Brain Res.* **330**, 174-177.

HYSTERESIS IN RECURRENT INHIBITION AND PROPRIOCEPTIVE FEEDBACK: DO THEY COMPENSATE FOR HYSTERESIS OF MOTOR UNITS?

U. Windhorst

The University of Calgary, Faculty of Medicine,
Departments of Clinical Neurosciences
and Medical Physiology
Calgary, Alberta T2N 4N1, Canada

INTRODUCTION

The problems to be solved by the CNS of higher animals in controlling its motor periphery are daunting. Thus, although there may be some indications that, at cortical level, movement trajectories are represented in kinematic terms and, at peripheral level, motoneurones activate muscles so as to produce forces or torques, little is known about the implementation of the required intermediate transformations. Mutatis mutandis, the same applies to rhythmic movements generated by central pattern generators.

Many of the problems the CNS faces in organising posture and movement arise from the multiple interactions between different peripheral elements, and the nonlinearities and time-dependent properties of the skeleto-muscular apparatus, which have rather seldom been addressed (e.g. Houk & Rymer, 1981). Muscles in particular are notorious for dependencies of force production on length, direction and velocity of length change, thixotropy, hysteresis, potentiation and depression (Partridge & Benton, 1981). For space limits, the following discussion will concentrate on hysteresis because it should play a significant role in organising locomotor movements. Hysteresis was originally described as a phenomenon of magnetism, but in engineering it refers to a system's ambiguous static input-output characteristic arising from its having two limbs depending on the direction in which the input variable changes. This is a history-dependent effect because which limb of the characteristic is followed at any time depends on the direction from which the working point was approached. In a wider sense, the term hysteresis may be used to describe history-dependent dynamic nonlinearities.

When a muscle or motor unit is activated repetitively with a cyclically, e.g. sinusoidally, varying activation rate, isometric muscle force is higher on the descending than the ascending phase for the same stimulus rate, the dynamic input-output relation usually forming a non-elliptical curve. Even assuming that a motoneurone receives a sinusoidal input signal in the form of synaptic current and translates it linearly into a sinusoidal discharge pattern (Baldissera, Campanelli & Piccinelli, 1984; see below), this input would nevertheless be distorted into a strongly non-sinusoidal force signal (under isometric conditions), or length signal (under isotonic conditions: Partridge, 1966). How the CNS deals so efficiently with such phenomena, i.e., what mechanisms it uses to compensate for the signal distortion, is to a large extent unknown. There are several possibilities which are not mutually exclusive: (i) *internal models* mimicking muscle behaviour and enabling the CNS to compensate for this behaviour in an open-loop fashion; (ii) *afferent closed-loop feedback* to estimate these properties; (iii) "*neural networks*". In the following, the first two possibilities will be dealt with.

RECURRENT INHIBITION: AN INTERNAL MODEL?

Internal models are well known from eye-movement control. At spinal level, there is no hard evidence for their existence. Nonetheless, it has been proposed that Renshaw cells mediating spinal recurrent inhibition provide an internal model of the muscle force component generated neurally (Loeb & Levine, 1990; Windhorst, 1993).

Since Renshaw cells are predominantly excited by motor axon collaterals, it appears superficially plausible that the discharge of sets of them reflects the activity in a group of α-motoneurones and thus provides an internal representation of motor output. It does not appear at all obvious why they should represent muscle activation or the related component of force production. As pointed out above, muscle has some peculiar properties such as activation history-dependent force production (facilitation, depression, post-tetanic potentiation etc.), and these should then be reflected in correspondingly peculiar properties of Renshaw cells or the motor axon-Renshaw cell subsystem. Moreover, these cells receive inputs in addition to that from motoneurones. The arguments for the provocative hypothesis that Renshaw cell activity acts as estimator of muscle activation are as yet indirect and speculative. They rest on intriguing comparisons (from different experimental data sets) between muscle force and Renshaw cell output as functions of particular motor axon activation patterns. The following points should be considered:

1. From anatomical and indirect physiological evidence, it can be inferred that, on average, the strength of excitation of Renshaw cells by motorneurones increases in the same order as the twitch and tetanic forces of their units, namely S<FR<FF. Hence, it may be predicted that with orderly *recruitment* of motor units, Renshaw cell excitatory input from motor axons should increase nonlinearly in about the same way as cumulative tetanic muscle force would increase (Hultborn, Lipski, Mackel & Wigström, 1988; Windhorst, 1993).

2. Under static conditions the dependence of muscle force on activation frequency shows an S-shaped form (Partridge & Benton, 1981). The location of the steepest portion of such curves on the frequency axis determines around which mean activation frequency frequency-modulation is most effective in changing force. The force/frequency curve of slow motor units is shifted to the left as compared to that of fast motor units. On theoretical grounds, it may be assumed that a similar relation holds for the input/output curves of Renshaw cells which receive inputs of different strength from slow and fast motor units.

HYSTERESIS IN RECURRENT INHIBITION

From the preceding section, one might suspect that recurrent inhibition could show signal-transmission properties that in part parallel those of muscle. For brevity, the following discussion will be restricted to rhythmic rate modulation of motor output.

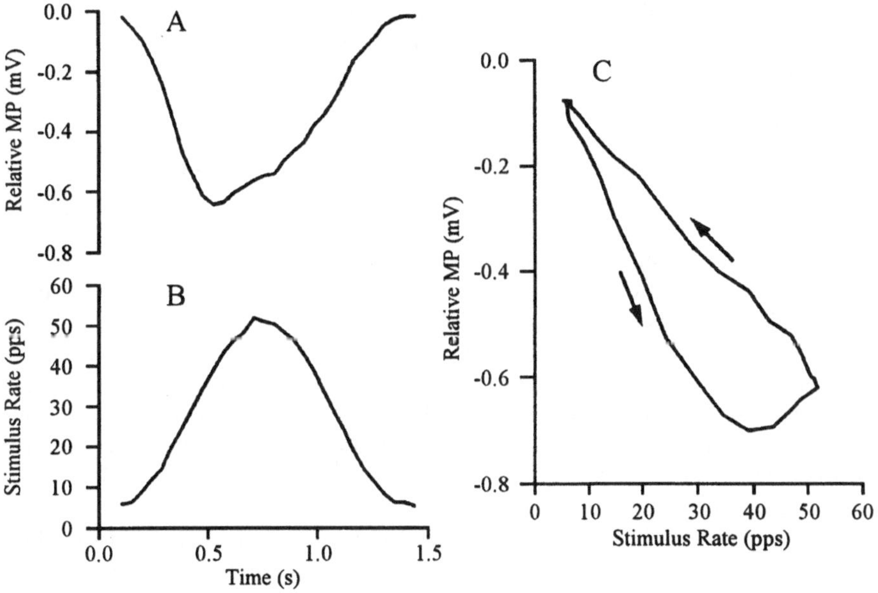

Figure 1. A. Cycle-averaged motoneurone membrane potential (MP) changes caused by motor axon stimulation. The rate was varied at 0.75 Hz modulation frequency between 5 and 53 pps (as shown in B). C. Dynamic relationship between frequency and MP change. Arrows indicate the direction of time. Intracellular recordings were obtained in a decerebrate cat from a medial gastrocnemius motoneurone with axon conduction velocity of ca. 71 m/s, afterhyperpolarisation duration of ca. 120 ms and a mean membrane potential of ca. -60 mV. Recurrent inhibition was elicited by supramaximal stimulation of the lateral gastrocnemius-soleus nerve. Relative membrane potential was obtained by removing the dc value.

In parallel to skeletal muscle, one would expect that recurrent inhibition should show hysteresis, based on medium-term processes of facilitation and/or depression. This is the case. A typical example is shown in Figure 1. In a decerebrate cat with dorsal roots L_5 to S_2 cut, a medial gastrocnemius motoneurone was recorded intracellularly, and the lateral gastrocnemius-soleus nerve was repetitively stimulated at rates varying sinusoidally from ca. 5 pulses/s (pps) to 53 pps at a modulation frequency of 0.75 Hz (plot B). Such rates may occur during cat locomotion (Hoffer, Sugano, Loeb, Marks, O'Donovan & Pratt, 1987). The cycle-averaged membrane potential change (plot A) displayed an initial rapid hyperpolarisation which peaked during the ascending phase of the stimulus cycle and then declined more slowly during its remainder. Plotting, in graph C, relative hyperpolarisation vs. stimulus rate yielded a typical hysteresis curve, the arrows indicating the direction of time. Such hysteresis is also seen at the level of Renshaw cells whose discharge rate is higher on the ascending than the descending limb of a stimulus cycle. The form of this hysteresis curve bears some similarity to that of skeletal muscle (Partridge, 1966). Could this hysteresis in recurrent inhibition compensate for muscle hysteresis such as to eliminate or at least diminish the distortion of signal transmission from motoneurone input to muscle force?

To answer this question, the effect of hysteresis in recurrent inhibition on motoneurone discharge pattern would have to be known, which is difficult to establish experimentally. But a thought experiment may help. When, in deeply anaesthetised cats, a motoneurone is injected with sinusoidally varying, suprathreshold current (Baldissera et al., 1984), the resulting firing rate is sinusoidal as well, particularly at low modulation frequencies. The firing rate is related to input current by Bode plots corresponding to a linear first-order lead system. From the data of Baldissera et al. (1984) the phase advance at 0.5 Hz can be estimated to average about 4.5 and at 1.0 Hz about 8 degrees, yielding a clockwise ellipse (see Figure 2). When many motoneurones are excited rhythmically, e.g. during locomotion, recurrent inhibition may be strong enough to add a non-sinusoidal (distorted) hyperpolarising current to the sinusoidal depolarising current. If these two currents simply added linearly, a distorted current and, hence, output firing rate would result. This of course could not produce the recurrent inhibition assumed here, but one of different time course which in turn would have led to a different resultant current. What in effect would happen due to this non-linear negative feedback system is that the input to the motoneurone encoder and, hence, the output firing rate would be temporally distorted, such that the upstroke would be less steep than the downstroke and the peak would be delayed, as compared to sinusoidal driving input current. Since even sinusoidal activation patterns entail distorted muscle force profiles, the described distortion of motoneurone output rate due to nonlinear recurrent inhibition would not amend the situation but worsen it. As considered here, nonlinear recurrent inhibition would therefore not linearise signal transmission from motoneurone input to muscle output. What would be required to do so would be recurrent inhibition with a hysteresis which has a reverse time course.

Is there a rescue? There may be, because Renshaw cells are subject to modulating influences by-passing motoneurones from descending and afferent sources (Baldissera, Hultborn & Illert, 1981; Windhorst, 1988) as well as possibly from the central locomotor pattern generator, all of which could act so as to alter the form of hysteresis in recurrent inhibition. At present, there is hardly any evidence that they do. A small hint is given by data from Pratt & Jordan (1987). They published firing patterns and 'locomotor drive potentials' of quadriceps motoneurones and firing patterns of related Renshaw cells during fictive locomotion. Hysteresis plots constructed from their data in fact have a time course opposite to that shown in Figure 1.

PROPRIOCEPTIVE FEEDBACK

Afferent feedback from muscle and other receptors could also act on motoneurones so as to alter appropriately their discharge patterns in response to sinusoidal inputs. The receptors believed to be most closely related to monitoring muscle force are Golgi tendon organs which will be dealt with first.

Feedback from Golgi tendon organs

When an isometrically suspended muscle, e.g. the soleus, is activated with a sinusoidally varying pattern of repetitive pulses, Golgi tendon organ discharge shows a phase advance over muscle force (Anderson, 1974), representing it in a slightly distorted way (Hulliger, Weytjens & Windhorst, unpublished data; cf. Figure 2). However, this is not enough to compensate for the effects of muscle hysteresis in the overall signal line from motor axons to Ib afferents. Plotting Golgi tendon organ discharge (ordinate) vs. stimulus rate (abscissa) still yields a hysteresis curve akin to that of muscle itself, with an ascending limb below the descending one. To some extent, this hysteresis curve has a form similar to that of Renshaw cells, but with the opposite time course. In fact, then, if in the classical pathway of autogenetic inhibition Ib discharge were linearly translated into membrane hyperpolarisation of homonymous and synergistic motoneurones, the hyperpolarisation vs. rate plots would be similar to those for the recurrent inhibitory pathway. The decisive difference would be the direction of time. The two curves would be traversed in opposite directions. That is, in the autogenetic pathway, hyperpolarisation would first slowly increase during increasing stimulus rate, remain nearly stable for the initial part of decreasing stimulus rate and then more rapidly return to its starting value. This time course would be appropriate to alter motoneuronal membrane current time course so as to compensate for signal distortion through skeletal muscle.

There are a number of problems, however. First, little if anything is known about dynamic signal transmission properties of the spinal autogenetic reflex pathway. It could well add its own peculiar dynamics, which remains to be tested. Second, there is increasing evidence that during locomotion, signals from extensor Ib afferents during the stance phase are routed through an alternative excitatory pathway (Pearson & Collins, 1993). This would upset the above argument. Again, here as in the autogenetic pathway, the dynamics of intercalated spinal pathways are unknown and urgently need to be determined. Third, as with recurrent inhibition, all pathways from Ib afferents to motoneurones are subject to modulating influences, for instance presynaptic inhibition, which changes rhythmically through a locomotor cycle (Dueñas & Rudomin, 1988) and affects signal transmission through any Ib pathway in an unpredictable manner.

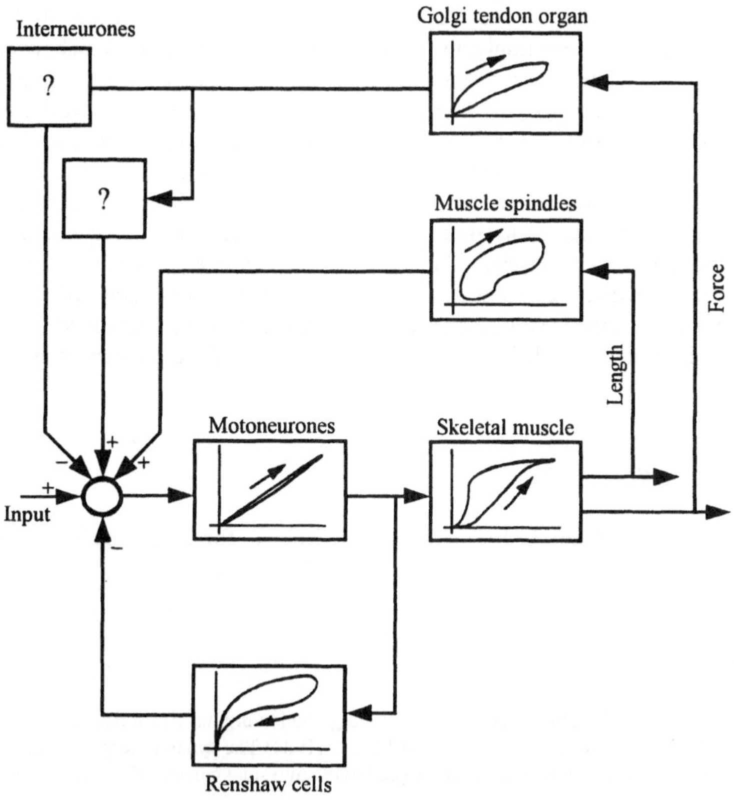

Figure 2. Block diagram of various homonymous and synergistic feedback systems involved in shaping motoneurone discharge. The blocks contain dynamic relationships between respective time-varying inputs (abscissae) and outputs (ordinates), the arrows indicating the direction of time. The input for motoneurones is supposed to be synaptic current and the output discharge rate. The input to Renshaw cells is here supposed to be motoneurone discharge rate, and the output synaptic inhibitory current. Muscle also has rate as input, but is supposed to have two outputs, length and force, for spindles and Golgi tendon organs, respectively. Interneurones mediating effects from Ib afferents have unknown dynamic characteristics, as indicated by the question marks. Both inhibitory and excitatory pathways are known and operate under different conditions.

Feedback from muscle spindles

Muscle spindle afferents show hysteresis as well, even under isometric conditions, where they respond to internal muscle fibre length changes (Windhorst, 1988), but they have not been well studied under these conditions. In acute preparations where muscle length is repeatedly changed to different values and back (Kostyukov & Cherkassky, 1992) and during locomotion (Prochazka, Trend, Hulliger & Vincent, 1989), discharge is higher during length increases than decreases, the firing rate-length relation being roughly the same under both conditions (Figure 2). Could this help to compensate for force distortion? Again, as in the case of Golgi tendon organs, this is difficult to tell, the major reasons here being different fusimotor patterns in different muscles and presynaptic inhibition.

CONCLUDING REMARKS

For simplicity and brevity, this paper has dealt with a circumscribed problem discussed for limited conditions. For instance, only rate modulation was considered. What about motor unit recruiment? Interestingly, in regard to recurrent inhibition, Renshaw cells appear to show the same dynamics to changes in input rate as to changes in the number of inputs from motor axon collaterals (Ross, Cleveland, & Kuschmierz, 1982). It would certainly be of importance to study this aspect more extensively with respect to the questions touched upon here. But it is not easy to design and implement a suitable experiment.

In summary, then, we must admit that we do not yet know much about how the CNS copes with the peculiarities of its musculo-skeletal periphery, including nonlinearities and activation dependencies of muscles. To find this out, however, remains an important task because without knowing the extent to which CNS processes are concerned with properties of its executive apparatus, it will be difficult to delineate processes involved in dealing with the body's environment. Modelling in the widest sense will be an important tool in this endeavour.

ACKNOWLEDGEMENTS

I am grateful to G. Boorman, L. Eldridge and D. Kirmayer for assistance in experiments and data analysis.

REFERENCES

ANDERSON, J.H. (1974) Dynamic characteristics of Golgi tendon organs. *Brain Res.* **67**, 531-537

BALDISSERA, F., CAMPANELLI, P. & PICCINELLI, L. (1984) The dynamic response of α-motoneurones investigated by intracellular injection of sinusoidal current. *Exp. Brain Res.* **54**, 275-282.

BALDISSERA, F., HULTBORN, H. & ILLERT, M. (1981) Integration in spinal neuronal systems. In *The Nervous System.* ed. BROOKS, V.B., pp. 509-595. American Physiological Society, Bethesda.

DUEÑAS, S.H. & RUDOMIN, P. (1988) Excitability changes of ankle extensor group Ia and Ib fibers during fictive locomotion in the cat. *Exp. Brain Res.* **70**, 15-25.

HOFFER, J.A., SUGANO, N., LOEB, G.E., MARKS, W.B., O'DONOVAN, M.J. & PRATT, C.A. (1987) Cat hindlimb motoneurons during locomotion. II. Normal activity patterns. *J. Neurophysiol.* **57**, 530-553

HOUK, J.C. & RYMER, W.Z. (1981) Neural control of muscle length and tension. In *The Nervous System.* ed. BROOKS, V.B., pp. 257-323. American Physiological Society, Bethesda.

HULTBORN, H., LIPSKI, J., MACKEL, R. & WIGSTRÖM, H. (1988) Distribution of recurrent inhibition within a motor nucleus. I. Contribution from slow and fast motor units to the excitation of Renshaw cells. *Acta physiol. scand.* **134**, 347-361.

KOSTYUKOV, A.I. & CHERKASSKY, V.L. (1992) Movement-dependent after-effects in the firing of the spindle endings from the de-efferented muscles of the cat hindlimb. *Neuroscience* **46**, 989-999.

LOEB, G.E. & Levine, W.S. (1990) Linking musculoskeletal mechanics to sensorimotor neurophysiology. In *Multiple Muscle Systems: Biomechanics and Movement Organization.* eds. WINTERS, J.M. & WOO, S.L.-Y., pp. 165-181. Springer-Verlag, New York.

PARTRIDGE, L.D. (1966) Signal-handling characteristics of load-moving skeletal muscle. *Am. J. Physiol.* **210**, 1178-1191.

PARTRIDGE, L.D. & BENTON, L.A. (1981) Muscle, the motor. In *The Nervous System.* ed. BROOKS, V.B., pp. 43-106. American Physiological Society, Bethesda.

PEARSON, K.G. & COLLINS D.F. (1993) Reversal of the influence of group Ib afferents from plantaris on activity in medial gastrocnemius muscle during locomotor activity. *J. Neurophysiol.* **70**, 1009-1017.

PRATT, C.A. & JORDAN, L.M. (1987) Ia inhibitory interneurons and Renshaw cells as contributors to the spinal mechanisms of fictive locomotion. *J. Neurophysiol.* **57**, 56-71.

PROCHAZKA, A., TREND, P., HULLIGER, M. & VINCENT, S. (1989) Ensemble proprioceptive activity in the cat step cycle: towards a representative look-up chart. *Prog. Brain Res.* **80**, 61-74.

ROSS, H.-G., CLEVELAND, S. & KUSCHMIERZ, A. (1982) Dynamic properties of Renshaw cells: equivalence of responses to step changes in recruitment and discharge frequency of motor axons. *Pflügers Archiv* **394**, 239-242.

WINDHORST, U. (1988) How brain-like is the spinal cord? Interacting cell assemblies in the nervous system. Springer-Verlag, Berlin

WINDHORST, U. (1993) A new concept of the role of proprioceptive and recurrent inhibitory feedback in motor control. In *Robots and biological systems: towards a new bionics?* ed. DARIO, P., SANDINI, G. & AEBISCHER, P., pp. 295-318. Springer, Berlin.

CONTRIBUTION OF THE β MOTOR SYSTEM TO A POSITIVE FEEDBACK OF CONTRACTION

N. Kouchtir, J.-F. Perrier,
D. Zytnicki and L. Jami

CNRS URA 1448, Université René Descartes
45 Rue des Saints-Pères, 75270 Paris Cx 06,
France

INTRODUCTION

Among cat leg muscles, peroneus brevis (PB) and peroneus tertius (PT) are known to receive a rich supply of skeleto-fusimotor or β innervation, that is, motor axons distributing terminal branches to extra- and intrafusal muscle fibres (Bessou, Emonet-Dénand & Laporte, 1965). In both PB and PT, the motor unit complements include one third of β units and 50-75% of spindles receive at least one β axon in addition to regular fusimotor (or γ) innervation (Jami, Murthy & Petit, 1982; Emonet-Dénand, Petit & Laporte, 1992). The effects of β motoneurones on spindle sensitivity are similar to those of γ motoneurones, allowing distinction of dynamic and static subpopulations. But in contrast to γ motoneurones, the β subpopulations have clearly segregated conduction velocities, with most dynamic β axons under 80m/s and static β axons between 80 and 105m/s. In keeping with this segregation, dynamic and static β motor units, respectively, have extrafusal components of the slow and fast types. Whether dynamic or static, actions on spindles are detectable when single β axons are stimulated at frequencies of 10-40/s, that is, within the presumed physiological range of motoneurone discharge rates (Emonet-Dénand & Laporte, 1983; Jami, Petit & Scott, 1985).

As the discharge of β motoneurones entails spindle activation concomitant with extrafusal contraction, the skeleto-fusimotor innervation of PB and PT muscles provides a hard-wired basis for spindle excitation during contraction, hence for a potential positive feedback of contraction on homonymous motoneurones, independent of α–γ coactivation. However, the actual operation of such feedback had to be ascertained because 1) it was not known whether β-activation at physiological frequencies would be efficient enough to produce spindle discharges in the midst of extensive extrafusal contractions; 2) muscle contraction unloads some spindles and, in the absence of γ activity, pauses occur in their discharges, so that it was not known whether β-activation could produce enough Ia afferent input to compensate for these pauses and actually excite homonymous motoneurones; and 3) there was a possibility that excitatory effects of input from β-activated spindles would be offset by inhibitory effects of input from contraction-activated tendon organs.

The present experiments were designed to investigate this feedback in anesthetised cats. Peroneal motoneurones were recorded intracellularly during sustained submaximal isometric contractions of PB muscle elicited by stimulation of cut ventral root filaments containing several motor axons. The stimulated axons were not identified individually, but they were likely to include at least a few β axons since one out of three PB motor units is β-innervated.

METHODS

The experiments were carried out on adult cats anesthetised with pentobarbitone sodium (initial dose of 45mg/kg i.p., supplemented i.v. whenever necessary). The PB muscle was dissected without disturbing its blood supply and its tendon was attached to a low compliance force transducer. The limb was rigidly fixed and the muscle length was set close to the physiological maximum. Except for PB, all the rest of the hindlimb was denervated. In order to locate motoneurones for intracellular recording the intact nerve to PB and the central end of the cut nerve to peroneus longus (PL) were dissected and mounted on stimulating electrodes. These two nerves were usually stimulated together because PT nerve is short and often difficult to isolate over a length sufficient to rule out diffusion of stimulation to PB nerve. The muscle and nerves prepared for stimulation were covered by a pool of mineral oil kept at 38°C. The lumbo-sacral spinal cord segments were exposed and a filament of the L7 ventral root containing motor axons for PB was cut to be placed on stimulating electrodes. Stimulation frequencies were kept between 5 and 40/s, that is within the rates of activation of hindlimb muscle motor units observed in freely moving cats (Hoffer, Sugano, Loeb, Marks, O'Donovan & Pratt, 1987). Conventional glass micropipettes were used for intracellular recordings. Simultaneous records of motoneurone membrane potential and muscle force were amplified and fed via DC-coupled channels into a Nicolet 4094-A digital oscilloscope. From 4 to 100 responses were averaged, stored on a floppy disk and subsequently displayed on a HP 7550A digital plotter.

RESULTS

The effects induced in peroneal motoneurones by PB contractions were strikingly different from those previously observed in triceps surae and plantaris motoneurones on contraction of gastrocnemius medialis muscle, as reported by Zytnicki, Lafleur, Horchaolle-Bossavit, Lamy & Jami (1990). There, contraction-induced inhibition, elicited by Ib (and group II) afferents, occurred in more than 85% of motoneurones and although this inhibition quickly declined during a maintained contraction, excitatory potentials were weak (Figure 1A, left) or absent. In contrast, a PB tetanus evoked pure excitation in PB-PT motoneurones. Successive excitatory potentials, appearing in phase with the oscillations of tetanic force, were superimposed on a slow component of excitation and persisted throughout the period of contraction (Figure 1A, right). Similar effects, with amplitudes of excitatory potentials in a range of 0.2-6.5mV, were observed in 43 of 53 peroneal motoneurones, including a small sample of four PL motoneurones. Of the ten motoneurones in which excitation was absent, five did not respond at all to PB contractions and five others showed membrane hyperpolarisations that did not represent actual inhibitory potentials since they were not reversed by intracellular injection of hyperpolarising current.

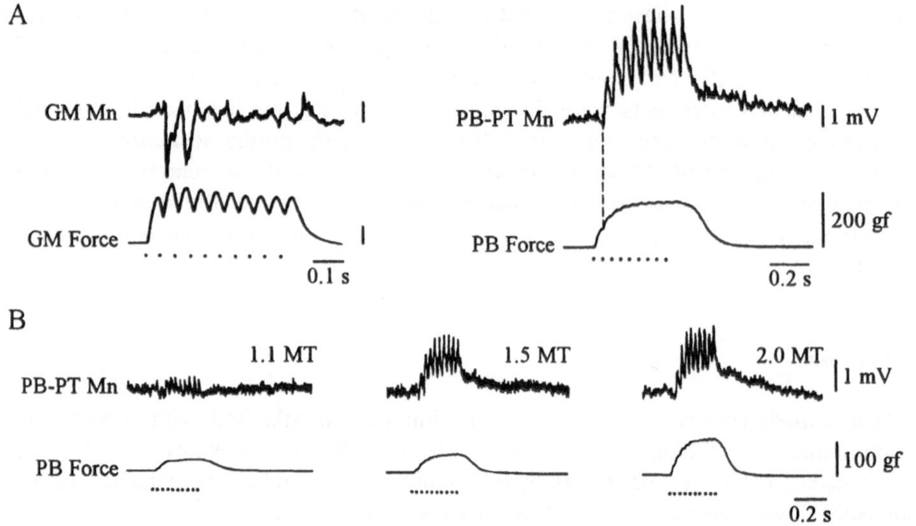

Figure 1. Contraction-induced excitation in peroneal motoneurones. A. Comparison of the effects on homonymous motoneurones of the contraction of gastrocnemius medialis (left) and peroneus brevis (right) muscles. In each panel, the upper trace is membrane potential and the lower trace is muscle force. Averages of ten responses in each trace. The vertical dashed line indicates the onset of excitatory potentials in a peroneal motoneurone. B. Increase of excitatory potential amplitudes in a peroneal motoneurone on increase of ventral root filament stimulation strength (MT, motor threshold). Same arrangement as in A.

Contraction-induced Ib inhibition was never observed in peroneal motoneurones although the present experimental conditions were similar to those of the previous study (Zytnicki et al., 1990). As it is known that PB has an important complement of tendon organs (the ratio of motor units to tendon organs in this muscle is 1.9) that are readily activated by contractions (see Jami, 1992), there still was a possibility that the oscillations of excitatory potentials observed in peroneal motoneurones (Figure 1A, right) represented Ib inhibition indenting the membrane depolarisation. If it were the case, however, one would have expected a reversal of these indentations on injection of hyperpolarising current in the motoneurone, but whenever attempted, such reversal was never obtained. In triceps surae and plantaris motoneurones, the contraction-induced Ib inhibition was found to decline during sustained tetani and the cause of the decline could be ascribed to presynaptic inhibition of Ib afferents (Lafleur, Zytnicki, Horcholle-Bossavit & Jami, 1992). Possibly, presynaptic inhibition was even stronger in the Ib pathways to peroneal motoneurones, totally preventing any expression of contraction-induced Ib inhibition in these motoneurones.

The origin of the contraction-induced excitation of peroneal motoneurones could be ascribed, at least partly, to spindle afferent input because the amplitude of excitatory potentials increased when the strength of ventral root filament stimulation was increased to recruit γ axons whose actions increased spindle discharges. The threshold strength for engagement of γ axons was determined at the end of each experiment by recording electroneurograms from the cut PB nerve and was occasionally (see e.g. Figure 2B) found to be under the traditional limit of two times the motor threshold (MT), that is, the stimulation strength producing the smallest contraction detectable by the force transducer.

However, excitatory potentials appeared in peroneal motoneurones even when the ventral root filament was stimulated at strengths under γ threshold, as shown by the example in Figure 1B. In this experiment, the electroneurograms recorded at the end of the experiment indicated that a few γ axons were recruited when the stimulation strength was increased from 1.5 to 2MT, and their actions on spindles might account for the small increase in excitatory potential amplitudes visible on the records. Further increases in stimulation strength, recruiting more γ axons, produced still larger excitatory potentials, occasionally leading to action potential discharge (not shown). But the first two records in Figure 1B, obtained with stimulation strengths of 1.1 and 1.5MT respectively, show excitatory potentials that cannot be ascribed to γ-activated afferents and might be due to inputs arising from β-innervated spindles.

This diagnosis could not be ascertained before effects from early spindle discharges or from purely mechanical activation of spindles by extrafusal contraction had been ruled out. Early dicharges of spindle primary endings, occurring near the onset of a muscle contraction were described by Hunt & Kuffler (1951) and ascribed to intramuscular mechanical perturbations or ephaptic excitation of Ia afferent fibres within the electrical field of muscle action potentials. Early spindle discharges would be expected to produce motoneurone excitation with a very short delay after the onset of contraction, but most of the contraction-induced excitatory potentials appeared with latencies of 50-70 ms after the onset of contraction, that is after the first oscillation of force (see the vertical dashed line in Figure 1A, right). Moreover these latencies did not change when increase of stimulation strength elicited increase in force output.

Evidence that the input which elicited contraction-induced excitatory potentials had another origin than mechanically activated spindles was difficult to obtain. Completely convincing evidence for motoneuronal excitation by input from β-activated spindles would require the demonstration that excitation persists after all extrafusal contractions have been eliminated, either by curare (Bessou et al., 1965) or by fatigue (Emonet-Dénand & Laporte, 1974). This was difficult for two reasons: 1. It was difficult to maintain satisfactory intracellular recordings from motoneurones during repeated, long-lasting periods of ventral root filament stimulation. However, in one case, a reduction of 93% of the initial extrafusal force was acheived, but nevertheless the motoneurone excitation persisted unchanged. 2. Total suppression of extrafusal contraction could not be reached, probably because PB muscle contains a particularly high proportion of fatigue resistant motor units (Emonet-Dénand, Hunt, Petit & Pollin, 1988).

For these reasons we had recourse to indirect evidence as employed by Jami et al. (1982) to identify static β axons. They made use of the fact that intrafusal chain fibres (i.e., the fibres receiving static β innervation) have higher fusion frequencies than extrafusal fibres so that 1) a primary afferent discharge can still reflect the oscillations of β-innervated chain fibres when a static β axon is stimulated at frequencies above extrafusal fusion rates and 2) when the stimulation frequency is abruptly increased, for instance from 100 to 200/s, a distinct step occurs in the discharge frequency of the primary ending without counterpart in the force profile, because these frequencies are far above the fusion frequency for extrafusal contraction. An example displaying these features in contraction-induced excitation is illustrated by Figure 2. The record shows an average of 100 responses obtained after more than 1hour of repeated stimulations of a cut ventral root filament. A stage was reached where extrafusal contraction developed only 25% of the initial force during 100/s stimulation and was unable to sustain its output on abrupt increase of stimulation frequency from 100 to 200/s.

The response to contraction of the motoneurone in Figure 2 showed a large biphasic initial event that has been cut from the record because it disappeared on further extrafusal fatigue (and was therefore considered to reflect changes in spindle discharges due to contraction-induced mechanical perturbations). After this initial event, small oscillations of membrane potential at 100/s are apparent in the record (inset A in Figure 2), which occur at the stimulation frequency applied to the ventral root filament. At this frequency the profile of extrafusal force was smooth. These oscillations of excitatory potentials were therefore likely to represent responses to Ia afferent discharges "driven" at the stimulation frequency during unfused intrafusal contractions, which is known to occur during static β activation (see Jami et al. 1985). Although very small in amplitude, the oscillations became apparent on the record because the large number of averaged sweeps improved the signal-to-noise ratio. The latency between each stimulus and the next oscillation of excitatory potential was 2-3ms (Figure 2 inset A), that is, too short to accomodate the successive events between ventral root stimulation and motoneurone excitation : 1) conduction in β motor axon(s) from ventral root to the muscle takes about 2ms; 2) spindle activation within the muscle takes 5-12ms (see Jami et al., 1985); 3) conduction from the muscle to the spinal cord in Ia afferent(s) takes about 2ms; and 4) a monosynaptic central delay takes a minimum of 0.5ms. In total, the minimum latency should be in a range of 9.5-16.5ms. The 12-13ms latency between a particular stimulus and the oscillation of excitatory potential occurring after the next stimulus would fall within this range.

Evidence that *intrafusal* action might be responsible for the driving input were firstly suggested by the oscillations of excitatory potentials "driven" by the stimulation frequency in the absence of oscillation of the force record. Second, there was a small but clear increase in motoneurone depolarisation (upward arrow in Figure 2) when the frequency of ventral

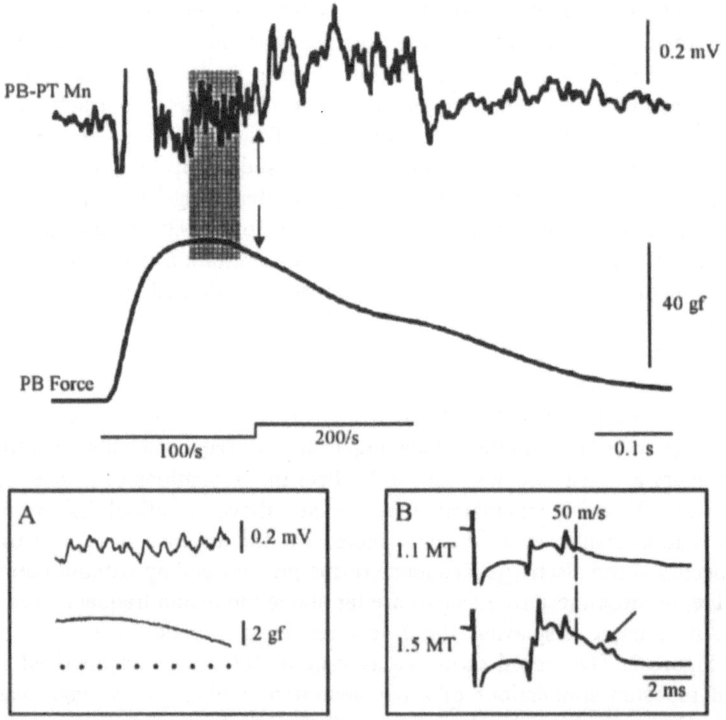

Figure 2. Excitatory potentials recorded intracellularly from a peroneal motoneurone during fatiguing PB contractions. 100 responses were averaged to obtain the membrane potential (top) and force (bottom) traces. Stimulation frequency started at 100/s and was increased to 200/s after 0.2s. Downward arrow points to decline in force output and upward arrow to increase in excitatory potential amplitude after the step in stimulation frequency. A large diphasic event appearing at the onset of the motoneurone response has been cut (see text). The shaded area is expanded in inset A. Inset B shows electroneurograms recorded from the cut PB-PT nerve at the end of the experiment, showing that 1.1MT stimulation did not excite any axon with a conduction velocity under 50m/s (dashed vertical line), whereas stimulation at 1.5MT did (arrow).

root filament stimulation was suddenly increased from 100 to 200/s. This occurred while extrafusal force was decreasing (downward arrow in Figure 2) and could not be due to any mechanical effect of extrafusal contraction. The step in excitatory potential could only reflect an increase in excitatory afferent input due to an increase in spindle excitation of intrafusal origin.

Third, the response shown in Figure 2 was obtained with stimulation of the ventral root filament at 1.1MT and the electroneurogram recorded from PB nerve at the end of the experiment showed that no γ axon was recruited for this strength (inset B in Figure 2). It follows that both the 100/s oscillations and the step increase in excitatory potential could only result from effects of β-activated spindle afferents. Instances for which this conclusion could be reached were not common. In only two motoneurones did the contraction-induced excitation satisfy the criteria allowing identificaton of β effects. But, however rare, these observations demonstrate for the first time an actual influence of the feedback from β-activated spindles on homonymous motoneurones.

DISCUSSION

The high incidence of β innervation in PB (Emonet-Dénand et al., 1992) was likely to produce spindle activation concurrent with contraction. The actual demonstration that β axons stimulated in ventral root filaments contributed to the contraction-induced excitation of motoneurones was not easy but the results of several previous investigations (Jami et al., 1982; 1985; Murthy, 1983; Emonet-Dénand et al., 1992) support the diagnostic value of the tests used in the present study, pointing to the conclusion that β innervation accounted, at least partly, for the contraction-induced excitation of peroneal motoneurones.

The effects of β innervation are likely to become functionally significant when contraction-induced Ib inhibition is absent as seems to be the case in peroneal motoneurones, and these effects may be useful when contraction occurs without γ-coactivation. A positive feedback of contraction to motoneurones might have a particular significance for the peroneal muscles as they are involved in complex biomechanical functions including ankle flexion, foot eversion and toe abduction (Lawrence, Nichols & English, 1993). In various motor tasks, peroneal muscles can work as either synergists or antagonists of the major flexor and extensor muscle groups in the leg. For instance, PB is activated together with soleus in locomotion (Loeb, 1993) but also contributes to ankle flexion while working as an antagonist of tibialis anterior in foot eversion (Lawrence et al., 1993). The low muscle compliance prevailing under the present experimental conditions helped to reveal the effects of β-innervation. Possibly, if PB muscle were allowed to shorten during contraction, less input would be generated in β-activated spindles. However, for anatomical reasons, PB muscle is appreciably stretched in most of the usual positions of the ankle and is therefore unlikely to display a very high compliance under natural conditions. Positive feedback might be helpful to resist antagonistic influences from either tibialis anterior or soleus. However, cutaneous inputs from the foot were found to modulate this positive feedback deeply (unpublished observations), and further investigations in this direction will be necessary to understand fully the function of contraction-induced excitation in peroneal motoneurones.

REFERENCES

BESSOU, P., EMONET-DENAND, F. & LAPORTE, Y. (1965). Motor fibres innervating extrafusal and intrafusal muscle fibres in the cat. *J. Physiol.* **180**, 649-672.

EMONET-DENAND, F., HUNT, C.C., PETIT, J. & POLLIN, B. (1988). Proportion of fatigue-resistant motor units in hindlimb muscles of cat and their relation to axonal conduction velocity. *J. Physiol.* **400**, 135-158.

EMONET-DENAND, F. & LAPORTE, Y. (1974). Blocage neuromusculaire sélectif des jonctions extrafusales des axones squelettofusimoteurs produit par leur stimulation répétitive à fréquence élevée. *Comp. Rend. de l'Acad. Sci. Paris* **D279**, 2083-2085.

EMONET-DENAND, F. & LAPORTE, Y. (1983). Observations on the effects on spindle primary endings of the stimulation at low frequency of dynamic β-axons. *Brain Res.* **258**, 101-104.

EMONET-DENAND, F., PETIT, J. & LAPORTE, Y. (1992). Comparison of skeletofusimotor innervation in cat peroneus brevis and peroneus tertius muscles. *J. Physiol.* **458**, 519-525.

HOFFER, J.A., SUGANO, N., LOEB, J.E., MARKS, W.B., O'DONOVAN, M.J. & PRATT, C.A. (1987). Cat hindlimb motoneurones during locomotion. II. Normal activity patterns. *J. Neurophysiol.* **57**, 530-553.

HUNT, C.C. & KUFFLER, S.W. (1951). Stretch receptor discharges during muscle contraction. *J. Physiol.* **113**, 298-315.

JAMI, L. (1992) Golgi tendon organs in mammalian skeletal muscle : functional properties and central actions. *Physiol. Revs.* **72**, 623-666.

JAMI, L., MURTHY, K.S.K. & PETIT, J. (1982). A quantitative study of skeletofusimotor innervation in the cat peroneus tertius muscle. *J. Physiol.* **325**, 125-144.

JAMI, L., PETIT, J. & SCOTT, J.J.A. (1985). Activation of cat muscle spindles by static skeletofusimotor axons. *J. Physiol.* **369**, 323-335.

LAFLEUR, J., ZYTNICKI, D., HORCHOLLE-BOSSAVIT, G. & JAMI, L. (1992). Depolarization of Ib afferent axons in the cat spinal cord during homonymous muscle contraction. *J. Physiol.* **445**, 345-354.

LAWRENCE, J.H., NICHOLS, R. & ENGLISH, A.W. (1993). Cat hindlimb muscles exert substantial torques outside the sagittal plane. *J. Neurophysiol.* **69**, 282-285.

LOEB, G.E. (1993). The distal hindlimb musculature of the cat : interanimal variability of locomotor activity and cutaneous reflexes. *Exp. Brain Res.* **96**, 125-140.

MURTHY, K.S.K. (1983). Physiological identification of static β-axons in primate muscle. *Exp. Brain Res.* **52**, 6-8.

ZYTNICKI, D., LAFLEUR, J., HORCHOLLE-BOSSAVIT, G., LAMY, F. & JAMI, L. (1990). Reduction of Ib autogenetic inhibition in motoneurons during contractions of an ankle extensor muscle in the cat. *J. Neurophysiol.* **64**, 1380-1389.

CRITIQUE OF PAPERS BY AWISZUS & FEISTNER; COPE & CLARK; DURKOVIC; WINDHORST; KOUCHTIR, PERRIER, ZYTNICKI & JAMI

Hans Hultborn

Department of Medical Physiology
University of Copenhagen, Copenhagen,
Denmark

SOME GENERAL REMARKS ON "RECRUITMENT ORDER" AND "RECRUITMENT GAIN"

The discussion on recruitment order has been much dominated by Henneman's 'size principle'. With our present understanding this hypothesis implies that the major factors determining the recruitment order are localised to the motoneurones themselves. That may be the sum of several mechanisms (such as membrane resistivity, threshold levels etc.) which are *correlated* to motoneuronal size. In addition to these properties of the motoneurones themselves, it is obvious that the distribution of the synaptic input in the motoneuronal pool must also play a role. The predominant projection of Ia excitation to slow motor units - rather than to fast units - surely contributes to reinforce and conserve the normal recruitment order.

In a discussion on this topic Kernell & Hultborn (1990) pointed out that the distribution of intrinsic properties and of synaptic excitation or inhibition among individual motoneurones in a pool will determine not only the recruitment *order* during a gradually increasing excitatory drive, but also the ease of recruitment gradation, i.e. the recruitment *gain*. At any time, the ability of a motoneurone to generate an action potential will depend both on its intrinsic excitability and the sum of its synaptic inputs. Kernell & Hultborn (1990) pointed out that synaptic systems with an uneven distribution among the motoneurones in the pool may expand or compress the range of functional thresholds among the motoneurones and thereby change the ease with which an additional, and more equally distributed input (e.g. the 'command system'), will activate the pool. More precisely, the term "recruitment gain" was used to indicate the relation between the synaptic input (the drive) to a motoneuron pool and the output, i.e. the number of recruited neurones. It was demonstrated that with a background synaptic bias with a skewed distribution to low or high threshold units (as compared to the "drive") would indeed change this gain.

In an extreme case it would be possible to imagine that activation of an afferent system or descending pathway with a strong excitation of motoneurones to fast motor units, while inhibiting those of slow units could "override" the effects given by the distribution of intrinsic properties and actually result in a reversed recruitment order. From the discussion above it should be obvious that long before a manifest change in recruitment *order*, significant changes in recruitment *gain* would have taken place.

INDEPENDENT MEASUREMENT OF THE DISTRIBUTION OF SYNAPTIC INPUTS TO THE MOTONEURONAL POOL

From the lengthy introductory remarks above, it now seems necessary to establish the relative distribution of synaptic excitation and inhibition to different parts of the motoneuronal pool. This is actually the purpose of the study by Awiszus & Feistner. They are using their particular bin-free version of the traditional PSTH paradigm on motoneurones which are firing at low rate by voluntary activation. They confirm that the Ia EPSP amplitudes are larger in slow conducting soleus motoneurones than in fast ones. Similarly they demonstrate that the initial peak of the corticospinal excitation (following magnetic brain stimulation) of first dorsal interosseous motor units is larger in slow conducting units than in fast conducting ones. This seems to introduce a very promising approach, but I feel that it would be of great importance to compare different inputs to *the same motoneuron pool*, even though different pools may offer particular advantages for certain inputs. The other problem is to standardise the stimulus strength so that easy and valid comparisons can be made between different experimental sessions and between different subjects. Obviously the strength of the stimulus cannot be so strong that too many motor units are fired, which would compromise selective recording of the investigated motor unit.

Alternatives to the present approach would be to study in humans the conditioning effects on H-reflexes of *single motor units at rest* with different "critical firing levels" - in other words to use the paradigm introduced by Henneman, Clamann, Gillies & Skinner (1974). This has now been done in a beautiful study by Shindo, Yanagawa, Morita & Hashimoto (1994). The obvious restriction is that only units in the low threshold part can be investigated and that the method is limited to muscles which show H-reflexes at rest.

CHANGES IN RECRUITMENT ORDER

The contribution by Cope & Clark questions if there really are significant exceptions to the 'size principle'. Although they agree with the experimental results of *apparent exceptions*, in a number of cited studies they call into question whether these results really should be interpreted as *true exceptions*. It all depends on the definition of the pool of motoneurones in which the reversals are studied. While most original studies in which reversals were studied simply considered all motoneurones innervating a single muscle to be members of the same pool (with some obvious difficulties in the case of anatomically complex multifunctional muscles) Cope & Clark take a more restrictive point of view and argue for a more *functional* definition and they write "*...we assume some agreement in defining a motor pool as a collection of synergistic motor units, i.e. units that would act together in some circumstance given sufficient excitatory input.*" For several years I have used the different recruitment pattern in triceps surae during concentric and eccentric contractions (Nardone, Romano & Schieppati, 1989) as one of the most beautiful and functionally interpretable examples of 'reversed' recruitment order. That type of evidence is questioned by Cope & Clark, as the slow units were not recruited at all during the eccentric contraction - perhaps they did not belong to the motor pool during this task, or perhaps they did, but the contraction strength in that particular experiment was not great enough to demonstrate that they were "recruitable". Although I follow this reasoning, I feel that such a "functional" definition would introduce even further problems in the field. Up to this point I have considered a motor pool to be strictly related to the muscle it innervated, and that recruitment of different parts of the pool may be an obvious consequence of the muscle being used in different tasks. Of course I would reserve the term "*recruitment reversal*" to the case when the systematic differences refers to the aspect of recruitment of *slow versus fast units*.

Differences in recruitment following experimental spinal lesions are discussed by Durkovic. His finding of an orderly recruitment in the spinal flexion reflex among recorded pairs of units in acute spinal animals agrees with the original findings from Henneman's group. The dominance of reversed (and tied) couples two weeks after the transection is indeed interesting. In his chapter he refers to similar findings in spinal humans (Grimby & Hannerz, 1970). Referring to my introductory remarks it may be that the "recruitment gain" is increased following chronic spinal lesions and that the normal order is lost in a more sudden and "explosive" recruitment. In this relation I would like to draw attention to the work by Tang & Rymer (1981). They demonstrated that spastic patients require a larger integrated EMG to obtain a given force than healthy subjects (or on the healthy control side). That might also be explained by an increased recruitment gain in the following way: if the threshold spacings between the motoneurones are reduced (increased recruitment gain) the *rate increase* of the previously recruited units is not sufficient to attain an optimal force output from those units before new units are recruited. Thereby, an increased recruitment gain could contribute to the "uneconomical" EMG/force ratio in these spastic patients.

PROPRIOCEPTIVE FEEDBACK AND MOTOR OUTPUT

Windhorst has much experience in dealing with the dynamic aspects of recurrent inhibition, as well as the proprioceptive feedback from Golgi tendon organs and muscle spindles. It is well known that the output units producing force and movement - the motor units - have highly nonlinear and time-dependent properties. How does the central nervous system deal with these nonlinearities? Are the control systems mentioned above contributing to compensate for these problems? Although I agree that it is important - indeed mandatory - to study the dynamic properties of recurrent inhibition and the two proprioceptive systems to understand their role in motor control, I am less sure that it is fruitful to link these observations to the nonlinearities of the skeleto-muscular apparatus. Although it may be easier for an engineer to deal with a simple linear system, we do not know enough about how the brain controls the movement to assume that the three control systems mentioned are needed to "solve" the particular problem of non-linearity and hysteresis of the skeleto-muscle system. As pointed out in the paper it is even very unlikely, as they themselves propose an hysteresis that - at least partly - would enhance the hysteresis given by the motor units themselves. In addition, all the control systems mentioned can be effectively controlled from the brain and their effects are distributed to several motor pools controlling muscles even at some distance from the muscle of origin.

The last report by Kouchtir, Perrier, Zytnicki & Jami addresses the particular problem of whether β-innervation of muscle spindles may contribute a positive feedback during contraction. In the present work they now give direct and very strong evidence for a synaptic excitation of interneurones evoked by β-innervation. They have used the peroneus brevis muscle in line with their earlier anatomical and physiological work. The present observations obviously will underestimate the normal contribution, as the activation of the spindles was left solely to the β-motoneurones without the normal cooperation with γ-motoneurones. It would be of great importance to extend the studies to other muscle groups to learn to what extent the knowledge from peroneus brevis can be extrapolated to other major muscle groups. Once more this raises the question of whether a strong β-innervation could explain the seemingly much stronger "fusimotor"-linkage in humans than seen from studies in the cat.

REFERENCES

GRIMBY, L. & HANNERZ, J. (1970) Differences in recruitment order of motor units in phasic and tonic flexion reflex in 'spinal man'. *J. Neurol. Neurosurg. Psychiat.* **33**, 562-570.

HENNEMAN, E., CLAMANN, H.P., GILLIES, J.D. & SKINNER, R.D. (1974) Rank order of motoneurons within a pool: law of combination. *J. Neurophysiol.* **37**, 1338-1349.

KERNELL, D. & HULTBORN, H. (1990) Synaptic effects on recruitment gain: a mechanism of importance for the input-output relations of motoneurone pools. *Brain Res.* **507**,176-179.

NARDONE, A., ROMANO, C. & SCHIEPPATI, M. (1989) Selective recruitment of high-threshold human motor units during voluntary isotonic lengthening of active muscles. *J. Physiol.* **409**, 451-471.

SHINDO, M., YANAGAWA, S., MORITA, H. & HASHIMOTO, T. (1994) Conditioning effect in single human motoneurones. A new method using the unitary H reflex. *J. Physiol.* (in press).

TANG, A. & RYMER, W.Z. (1981) Abnormal force-EMG relations in paretic limbs of hemiparetic human subjects. *J. Neurol. Neurosurg. Psychiat.* **44**, 690-698.

TIME COURSE OF EXCITABILITY CHANGES OF HUMAN α-MOTONEURONES FOLLOWING SINGLE SPIKES AND DOUBLETS

K. E. Jones[1], B. Calancie[2], A. Hall[2] and P. Bawa[1]

[1]School of Kinesiology, Simon Fraser University
Burnaby, B.C., Canada V5A 1S6
[2]The Miami Project to Cure Paralysis
and Department of Neurological Surgery,
University of Miami School of Medicine
1600 NW 10th Avenue, Miami FL 33136, USA

For rhythmically firing α-motoneurones of the cat, biophysical properties such as membrane voltage, membrane conductance, and spike threshold have been measured and shown to vary systematically along different phases of the interspike interval (ISI) (Calvin & Schwindt, 1972; Schwindt & Calvin, 1973; Calvin, 1974). In man one can estimate the collective effect of all such parameters and noise on motoneurone excitability by measuring the response probability to a phasic input volley. The resulting profile of the response probabilities at various phases of the ISI will be termed the "response trajectory" in contrast to the underlying afterhyperpolarisation (AHP) membrane voltage trajectory, which is one of the factors that influences the response.

Response trajectories were obtained for low threshold, rhythmically firing motoneurones innervating the fast flexor carpi radialis (FCR) and slow soleus muscles. A homonymous Ia volley was repeatedly (0.2/s) evoked at a specific delay, d, following a single spike of a motor unit spike train. Peri-stimulus time histograms were constructed for approximately 40-50 stimuli at each post-spike delay. Response probability, $P = C/N$, was computed for each delay (C is the number of counts in the response peak of the PSTH, N is the number of stimuli). The onset time of the PSTH provided an approximation of the dead time, t, during which the excitability of the motoneurone could not be tested. The effective times at which the Ia volley arrived at the motoneurone is given by $T = t + d$. For each response trajectory, both the stimulus strength and the firing rate were kept constant. Figure 1A shows 4 response trajectories for FCR and 4 for soleus when the subjects were asked to fire the motor units at slow rhythmic firing rates. The thick lines are the means for each set indicating the duration and shape of the mean response trajectory of a slow or fast motoneurone. The slow motoneurones have a response trajectory of a long duration

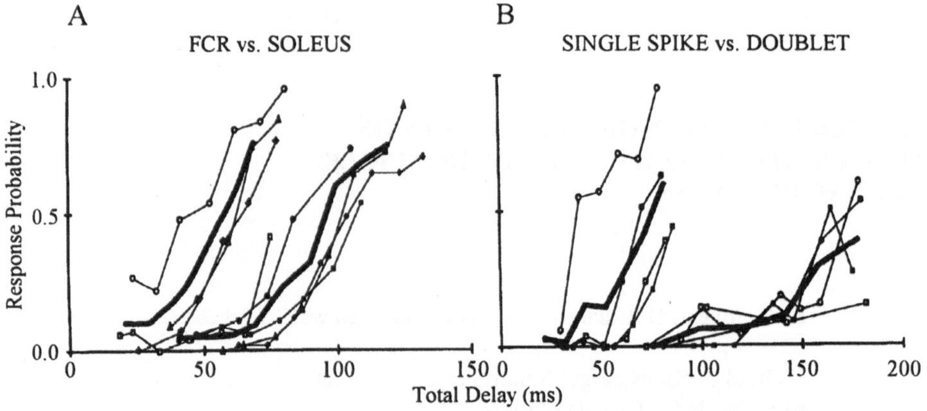

Figure 1. Response trajectories from motoneurones of two different muscles and during two different firing patterns in the same muscle. A: Response trajectories are shown for FCR (open symbols) and soleus (closed symbols) motoneurones. The interspike interval (ISI) range for the FCR motoneurones was 109 - 129 ms with a mean (± sd) of 120 ± 9 ms. For the soleus motoneurones, ISI ranged from 147 to 159 ms with a mean of 155 ± 8 ms. The thick solid lines represent the mean response trajectories from the 4 motoneurones illustrated for each muscle. B: Response trajectories are shown for 4 FCR motoneurones in both normal single spike rhythmic firing and doublet firing. The ISI range during rhythmic single spike firing was 82 - 108 ms with a mean of 96 ± 10 ms. Again, the thick solid lines are the mean response trajectories of the 4 motoneurones in each firing mode. The total delay represents the time of arrival of the Ia EPSP w.r.t. the motoneurone spike. It is approximately equal to the spike-to-stimulus delay plus the dead time as estimated from the onset of the PSTH.

characteristic of their longer ISIs. The response trajectories also depend on the intensity of the stimulus. Stronger stimuli produce shallower response trajectories of shorter duration.

A motoneurone can voluntarily be made to fire in a normal (single spike) rhythmic pattern and a doublet firing pattern (Bawa & Calancie, 1983). Response trajectories were compared for single spike rhythmic firing and doublet firing. Repetitive firing of doublets for long periods was not possible, therefore, interdoublet intervals were not controlled; any doublet available was used as a trigger. Figure 1B shows response trajectories associated with single spikes versus doublets for the same motoneurone. The decrease in response probability following a doublet is dramatic. The doublet response trajectories at these stimulus intensities approach zero at times when it is possible to see a response probability of 1 during normal single spike rhythmic firing with the same stimulus intensity. It has previously been shown that generally the depth of the AHP following a doublet is twice that following a single spike and the duration of the AHP is prolonged. Since the response trajectories following a doublet are not just twice the depth of those following a single spike, it shows that other factors such as membrane conductance, have a profound effect on responses.

The response trajectories show collectively how the underlying biophysical processes affect the input-output relations of rhythmically firing human motoneurones. If any of these parameters should be altered, perhaps in pathological cases, the alteration may reveal itself in the response trajectories.

REFERENCES

BAWA, P. & CALANCIE, B. (1983) Repetitive doublets in human flexor carpi radialis muscle. *J. Physiol.* **339**, 123-132.

CALVIN, W.H. & SCHWINDT, P.C. (1972) Steps in production of motoneuron spikes during rhythmic firing. *J. Neurophysiol.* **35**, 297-310.
CALVIN, W.H. (1974) Three modes of repetitive firing and the role of threshold time course between spikes. *Brain Res.* **59**, 341-346.
SCHWINDT, P.C. & CALVIN, W.H. (1973) Nature of conductances underlying rhythmic firing in cat spinal motoneurons. *J. Neurophysiol.* **36**, 955-973.

MOTOR UNIT SYNCHRONISATION DURING THE PRECISION GRIP

E.J. Hüsler, G.C. Maissen, M.A. Maier
and M.-C. Hepp-Reymond

Brain Research Institute, University of Zürich
8029 Zürich, Switzerland

INTRODUCTION

Control of force in the hand requires the synergistic co-operation of many intrinsic and extrinsic finger muscles (Maier, Hepp-Reymond & Meyer, 1991; Maier, 1992). An area of interest lies in the coordination and neuronal circuitry of the muscles involved, both at the global and at the motor unit (MU) level. Recent studies of MU synchronisation, found within and between muscles (Bremner, Baker & Stephens, 1991; Nordstrom, Fuglevand & Enoka, 1992; Schmied, Ivarsson & Fetz, 1993), have resulted in a better understanding of motoneuronal connectivity. The purpose of this investigation was to determine whether intra- and intermuscular (IAM, IRM) synchronisation is present between MUs in activated finger muscles in the precision grip and how it might be related to muscle synergies.

METHODS

In a visuo-motor step-tracking task, five normal subjects maintained an isometric force on a transducer held between the tip of thumb and index finger, matching three consecutive target forces (1, 2, 3 N). Up to four MUs from maximally five different muscles recorded simultaneously were discriminated from the global intramuscular EMG signals (Haas & Meyer, 1989). Cross-correlation analysis calculation of IAM and IRM MU pairs was performed at each force level separately, with at least 300 spikes/channel, yielding 3 test cases/MU pair. Synchronisation peaks and their statistical significance were determined using the cumulative sum derivative (CUSUM; Davey, Ellaway & Stein, 1986). The peaks were quantified with 5 synchronisation indices: relative peak amplitude: k (Sears & Stagg, 1976); relative mean peak amplitude: k'; peak area normalised with total number of trigger plus response spikes: b (Bremner et al., 1991); peak area normalised to total recording time: common input strength CIS (Nordstrom et al., 1992); mean percent increase above baseline: mpi.

RESULTS

Motor unit synchronisation was tested in 11 IAM and 23 IRM combinations, resulting in 43 and 114 MU pairs respectively. The percentage of significant cross-correlogram peaks for IAM and IRM pairs was 60% and 47% respectively (52/86 and 95/201). The mean phase shift between two synchronised MUs was 13±30ms for IAM and 5±46ms for IRM pairs. Figure 1 shows cross-correlograms and CUSUMs of one IAM and one IRM MU pair.

Quantification of cross-correlogram peaks showed on average a higher relative amplitude (k_{intra} = 2.27±0.56 vs. k_{inter} = 2.08±0.36) and a higher peak mean (k'_{intra} = 1.69±0.28 vs. k'_{inter} = 1.55±0.22) for IAM MU pairs compared to IRM ones. The normalised peak areas of IAM pairs were larger (b_{intra} = 0.026±0.009 vs. b_{inter} = 0.019±0.007; CIS_{intra} = 0.45±0.20 vs. CIS_{inter} = 0.32±0.14). These differences were all significant ($P < 0.05$, one sided U-test). Peak widths remained in the same range (8±5 ms).

A higher occurrence of synchronisation was found as force increased (58% IAM, 37% IRM). Some pairs, showing a drop in peak amplitude, were also detected, even though the firing rate of both MUs increased with force.

The comparison between single- and multi-unit findings disclosed coincidences of MU synchronisation with global EMG coupling. They occurred in the same proportion with synergies in time and amplitude domain (26/114), and were more likely at higher forces.

DISCUSSION

The main finding of this investigation is the presence of short-term synchronisation between activated MUs of the same and different muscles during the generation of finely graded force. The occurrence rate of significant cross-correlogram peaks was more than

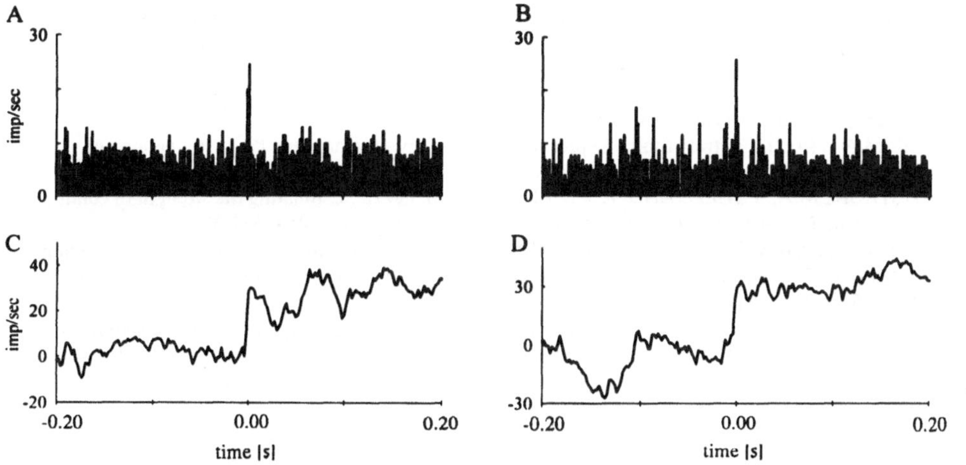

Figure 1. Cross-correlogram and CUSUM of IAM and IRM MU pairs. A, C: **1PI MU 2 * 1PI MU 3**; 2 N force level, 648 trigger spikes, peak at 0 ms, peak width 4 ms; synchronisation indices: k = 3.12, k' = 2.27, b = 0.036, CIS = 0.65. B, D: **EPL MU 1 * EPB MU2**; 3 N force level, 503 trigger spikes, peak at -2 ms, peak width 12 ms; synchronisation indices: k = 3.66, k' = 1.69, b = 0.034, CIS = 0.55. Top: Cross-correlograms, bin width: 2 ms; bin counts converted to impulses per sec. = (counts/bin) / (bin width x no. of triggers). Bottom: CUSUM, bin width 2 ms; control period starting at 0.2 s prior to trigger spike and lasting till 1 bin before peak onset. Abbreviations: 1PI, first palmar interosseus muscle; EPL, extensor pollicis longus muscle; EPB, extensor pollicis brevis muscle.

50%, but less than 88% (cf. Bremner et al., 1991). Generally, the strength of the synchronisation was relatively weak, but stronger for IAM MU pairs compared to IRM pairs. However, this difference was not as large as that reported by Bremner et al. (1991). We attribute the overall weakness to the shortness of the recording periods resulting in small spike counts and the unstable firing of the MUs due to the lack of MU activity feedback (Schmied et al., 1993). In addition, it is conceivable that the common drive to the MUs of the thumb and index finger muscles is reduced in favour of a more selective innervation necessary for fractionated muscle activity. Bremner et al. (1991) have also observed less MU synchronisation in these muscles than in those of the middle and ring finger. Two factors may explain the weak overlap of MU synchronisation and global EMG synergies: limitations in EMG decomposition and the sub-threshold synchronisation of MU pairs which would become significant at the global level only when all coupled units of the two muscles are pooled. It still remains to determine the extent to which the emergence of muscular co-activation in the amplitude domain can be traced back to MU synchronisation.

ACKNOWLEDGEMENTS

Supported by the Swiss National Science Foundation and the Slack-Gyr Foundation.

REFERENCES

BREMNER, F.D., BAKER, J.R. & STEPHENS, J.A. (1991). Variation in the degree of synchronization exhibited by motor units lying in different finger muscles in man. *J. Physiol.* **432**, 381-399.

DAVEY, N.J., ELLAWAY, P.H. & STEIN, R.B. (1986). Statistical limits for detecting change in the cumulative sum derivative of the peristimulus time histogram. *J. Neurosci. Meth.* **17**, 153-166.

HAAS, W. S. & MEYER, M. (1989). An automatic decomposition system for routine clinical examinations and clinical research: ARTMUP - automatic recognition and tracking of motor unit potentials. In *Clin. Neurophysiol. Updates, Vol. 2, Computer-aided Electromyography and Expert Systems.* ed. DESMEDT, J.E., pp. 67-81. Elsevier, Amsterdam.

MAIER, M.A., HEPP-REYMOND, M.-C. & MEYER, M. (1991). Force control in precision grip: existence of synergies? *Soc. Neurosci. Abstr.* **17**, 1110.

MAIER, M.A., (1992). Control of force in precision grip: an investigation into the participation and coordination of fifteen muscles. Ph.D. Thesis, University of Zürich.

NORDSTROM, M.A., FUGLEVAND, A.J. & ENOKA, R.M. (1992). Estimating the strength of common input to human motoneurons from the cross-correlogram. *J. Physiol.* **453**, 547-574.

SCHMIED, A., IVARSSON, C. & FETZ, E.E. (1993). Short-term synchronization of motor units in human extensor digitorum communis muscle: relation to contractile properties and voluntary control. *Exp. Brain Res.* **97**, 159-172.

SEARS, T.A. & STAGG, D. (1976). Short-term synchronization of intercostal motoneurone activity. *J. Physiol.* **263**, 357-381.

EFFECTS OF MOTOR ACTIVATION STRATEGIES ON THE RELATIONSHIP BETWEEN EMG AND FORCE

S.J. Day, M. Hulliger, W.M. Morrow,
B. Kacmar and M. Edström

Department of Clinical Neurosciences
University of Calgary, Alberta, Canada T2N 4N1

INTRODUCTION

It should be possible to predict muscle force from the electromyogram (EMG), but a general EMG-force relationship has not been identified (reviewed by Basmajian & De Luca, 1985). Apart from technical differences, it is conceivable that different motor activation strategies may result in genuinely different EMG-force characteristics (Lawrence & De Luca, 1983; Woods & Bigland-Ritchie, 1983). Motor unit (MU) action potential size may not be linearly related to twitch amplitude (see Milner-Brown & Stein, 1975). Therefore EMG amplitude will depend on whether a given level of force is generated by uniform activation of all MUs (predominant rate modulation) or by more selective activation of smaller MUs (predominant recruitment). To re-assess the role of different MU activation strategies, a preparation has now been used in which activation of MUs could be controlled precisely and in which finer details of MU activation patterns could also be manipulated.

METHODS

In anaesthetised cats, ventral roots supplying soleus were divided and regrouped into 10 filaments, containing either functionally single or groups of α motor axons, stimulation of which generated 3-20% of maximum tetanic tension (approximately 5-30 MUs). Filaments were stimulated independently in parallel with 12s pulse trains generated by a real-time micro-processor. Individual patterns were defined as pre-edited lists of intervals with Gaussian noise (coefficient of variation 12.5%) superimposed on a background component (Figure 1.1). Interval lists were edited to minimise synchronisation between channels. Background signals were either rectangular, to simulate recruitment alone (Figure 1A), or ramp-shaped, to simulate rate modulation, either on its own (Figure 1C) or in combination with recruitment (Figure 1B). Single- and multi-MU filaments were ordered in ascending tetanic force size to imitate MU recruitment. Force was monitored at the muscle tendon (at a length corresponding to an ankle joint angle of 90°). EMG was recorded with bipolar surface and indwelling wire electrodes (separation of 3-5 mm). Signals were sampled at 250/s (force) and 2000/s (EMG). Window averaging was used to calculate the averaged rectified EMG (AEMG).

RESULTS

Three strategies of activation have been investigated under isometric conditions: pure recruitment (Figure 1A), pure rate modulation (Figure 1C), and various combinations of recruitment and rate modulation (Figure 1B). When individual filaments were recruited at low rate (7/s) and motor drive was subsequently increased by rate modulation, the resulting force profiles were smooth (Figure 1.3, B & C). In contrast, when motor drive was increased by recruitment alone, both AEMG and force records revealed distinct steps upon recruitment of additional filaments (Figure 1A.3), which led to steps in the AEMG-force relationship. This was attributed to the technical limitation of the present method. To obtain smoother estimates of the AEMG-force characteristics, pure recruitment designs were expanded to include 4 shifted sets of 10-channel stimulation configurations to permit averaging of the records using a moving window average algorithm.

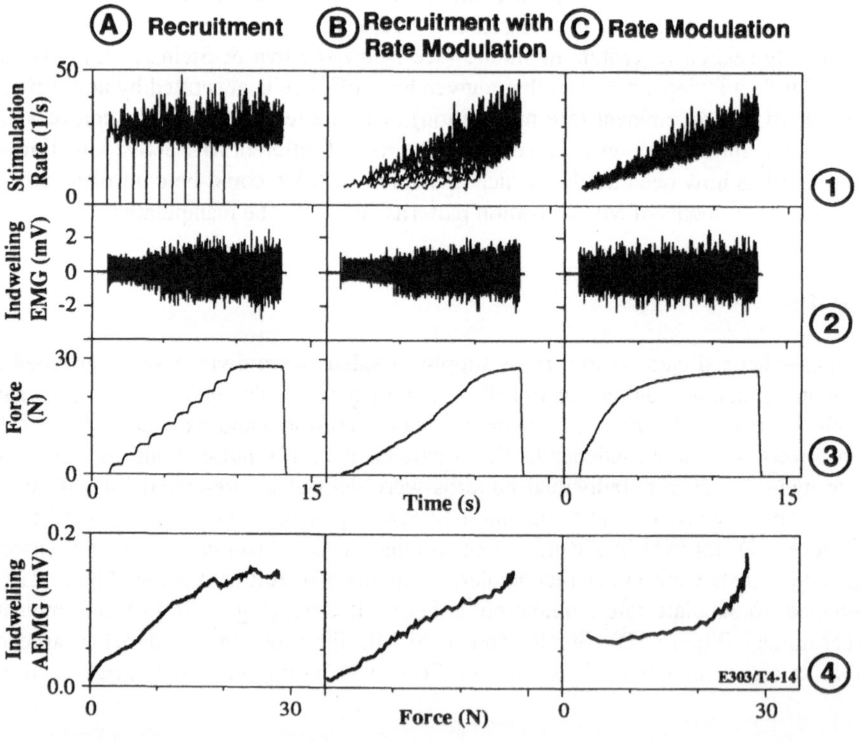

Figure 1. Raw EMG signals and forces generated by experimental simulations of 3 different activation strategies, spanning the range between the extremes of pure recruitment (A) and pure rate modulation (C). Data from a single experiment with the same indwelling EMG electrodes and 10-channel multi-MU filament constellation for all three strategies. Row 1, maximally desynchronised activation patterns with Gaussian noise. Row 4, AEMG-force relations computed from raw EMG (row 2) and force (row 3). See text.

AEMG-force characteristics varied considerably between experiments, mainly due to variations in the size of the EMG signals (attributable to variations in tissue attenuation). However, the general trends, which distinguished different activation strategies, were clearly recognised in individual experiments, and the variability was readily reduced by averaging AEMG and force data within corresponding time windows for each strategy.

AEMG-force relations revealed clear differences between the pure recruitment and pure rate modulation, while combined strategies had intermediate characteristics. In all cases the AEMG-force relations were monotonic, although some strategy-specific non-linearity occurred. Notably, wide-range recruitment combined with rate modulation (Figure 1B.4) most consistently revealed near linear input-output characteristics. Pure recruitment was characterised by gain compression (Figure 1A.4), pure rate modulation by gain expansion (Figure 1C.4) of the AEMG-force relation.

Were force to be controlled by recruitment alone, variations could only be achieved by varying recruitment rates. Effects were explored over a wide range (7-28/s) of rates. While maximum forces varied tenfold or more, the AEMG-force relations were nearly invariant.

Experiments were also conducted with multi-channel (10) activation of single MUs, using the same stimulation strategies. The preliminary data indicate that the same general features are encountered with multi-channel stimulation of single MUs as were found with activation of multi-MU filaments.

CONCLUSIONS

The present observations confirm that isometric AEMG-force relations are monotonic for different central MU pool activation strategies. However, significant differences were found between the extremes of pure recruitment and rate modulation, and are in remarkable agreement with recent computer simulations of EMG-force relations (Fuglevand, Winter & Patla, 1993). Only minor differences were observed among various mixed strategies, which combined rate modulation of MU firing with various recruitment strategies, and which were more plausible imitations of physiological activation patterns. This suggests that EMG still may be useful to estimate muscle force, although more work is needed to delineate the effects of fatigue and MU synchronisation on AEMG-force relations.

ACKNOWLEDGEMENTS

Supported by an AHFMR studentship award (SJD) and a Canadian MRC grant (MH).

REFERENCES

BASMAJIAN, J.V. & DE LUCA, C.J. (1985) *Muscles Alive* (5th edition). pp. 187-200. Williams & Wilkins, Baltimore.

FUGLEVAND, A.J., WINTER, D.J. & PATLA, A.E. (1993) Models of recruitment and rate coding organization in motor-unit pools. *J. Neurophysiol.* **70**, 2470-2488.

LAWRENCE, J.H. & DE LUCA, C.J. (1983) Myoelectric signal versus force relationship in different human muscles. *J. App. Physiol.* **54**, 1653-1659.

MILNER-BROWN, H.S. & STEIN, R.B. (1975) The relation between the surface electromyogram and muscular force. *J. Physiol.* **246**, 549-569.

WOODS, J.J. & BIGLAND-RITCHIE, B. (1983) Linear and non-linear surface EMG/force relationships in human muscles. *Am. J. Phys. Med.* **62**, 287-299.

RECRUITMENT PATTERNS OF INTRAFUSAL AND EXTRAFUSAL FIBRES IN MICE DURING PROLONGED SWIMMING

N. Fujitsuka, A. Yoshimura, Y. Shimomura,
T. Murakami, K. Kawakami, Y. Khobu
and C. Fujitsuka

Laboratory of Bioscience
Nagoya Institute of Technology
Gokiso-cho, Showa-ku,
Nagoya 466, Japan

Though recruitment patterns of fast and slow extrafusal fibres in exercise are well established (Armstrong, Saubert, Sembrovich, Shepherd & Gollnick, 1974), there have been no reports to date regarding the recruitment of different intrafusal fibres of mammalian muscle spindles during voluntary movement. We have used glycogen depletion during prolonged swimming to investigate recruitment patterns of the three intrafusal fibre types in mice. The degree of glycogen depletion was estimated by two methods, visual inspection (VI) and our newly developed optical scanning (OS) of periodic acid Schiff (PAS)-stained sections. Fibre typing was done by combining morphological, histochemical, and immunohistochemical characteristics, and we compared the results for intrafusal and extrafusal fibres.

We used 48 adult male mice in 8 groups of 6. Fibre typing of intrafusal and extrafusal fibres in soleus (SOL) and extensor digitorum longus (EDL) muscles was done in one group. Oxygen uptake during exercise (indicating exercise intensity) was measured in another, which then served to demonstrate a strong correlation (r = 0.96) between the glycogen concentration (mmol glucosyl units/kg, wet weight) on PAS-stained sections and their optical intensity (Vøllestad, Vaage & Hermansen, 1984). The other 6 groups were divided into a resting control group, and groups that swam for 0.5, 1, 2, 4, and 8 hours. Serial sections (12 μm) of muscles were cut with a cryo-microtome along the longitudinal axis of the muscle. Fibres were classified by means of hematoxylin staining, myosin ATPase staining, and an immunohistochemical method using chicken slow and fast myosin antibodies (Yoshimura, Fujitsuka, Kawakami, Ozawa, Ojala & Fujitsuka, 1992). By visual inspection under the microscope, fibres in PAS-stained sections were rated dark, light or negative. In the OS method, the images of PAS-stained sections were recorded on video tape with a CCD camera (Ikegami, Model ICD42DC), which made it possible to differentiate between 250 grades of optical intensity. The intensity was converted into the absolute glycogen concentration (mmol glucosyl units/kg, wet weight) according to the correlation we had established (Y = 98.6 - 0.67X, Y = mmol glucosyl units/kg, wet weight, X = intensity of PAS-stained section). The glycogen concentration in muscle sections was also measured biochemically, and the Student's t-test was employed to determine any significant differences between the two methods.

Intrafusal muscle fibres in serial section reconstructions of 135 SOL and 188 EDL spindles were morphologically classified as either longer nuclear bag (L-b), shorter nuclear bag (S-b), longer nuclear chain (L-c) or shorter nuclear chain (S-c) fibres. More than 80% of all examined SOL and EDL spindles consisted of one L-b, S-b, L-c, and S-c fibre each. Histochemical and immunohistochemical examinations identified the S-b fibre as a slow tonic fibre, the L-b fibre as an intermediate fibre, and the L-c and S-c fibres as fast twitch fibres. The S-b fibre has been shown to correspond to the nuclear bag_1 fibre, the L-b fibre to the nuclear bag_2, and the L-c and S-c fibres to the nuclear chain fibres in rat muscle spindles (Yoshimura et al., 1992). However, L-c fibres are not like the long chain fibres in cat spindles.

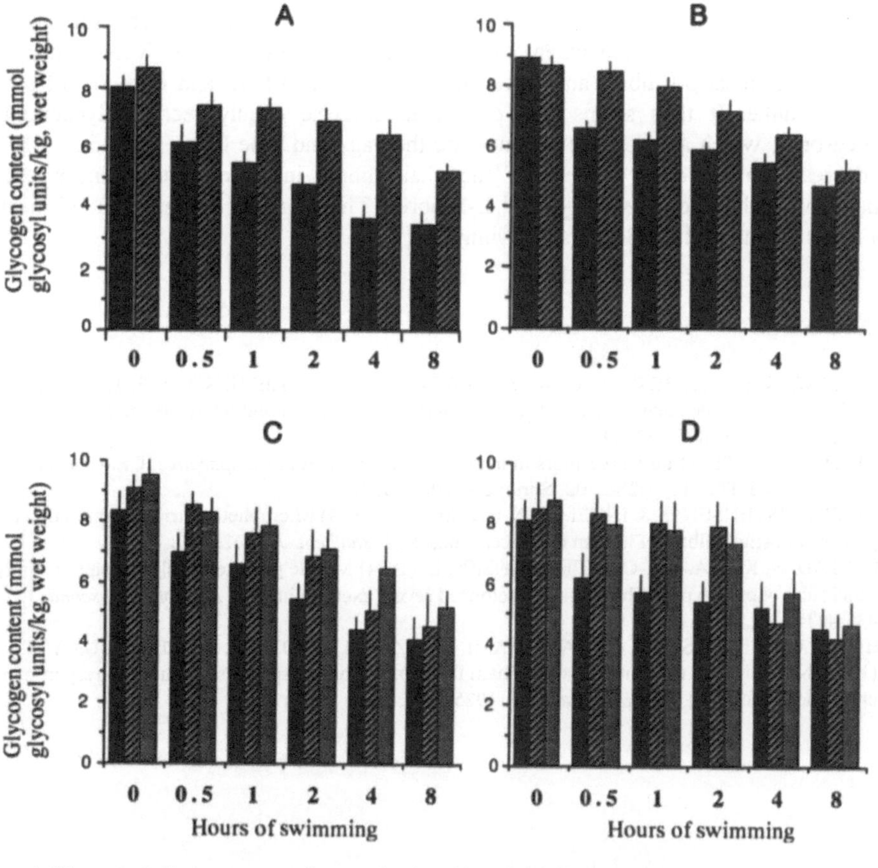

Figure 1. Glycogen depletion patterns for extrafusal and intrafusal fibre types. Above - extrafusal fibres (SOL - A, EDL - B); below - intrafusal fibres (SOL - C, EDL - D). In A and B black bars are for type I and hatched bars are for type II fibres. In C and D black bars are for S-b (bag_1), hatched bars are for L-b (bag_2) and stippled bars are for L-c and S-c fibres. Data are means ± SD.

Exercise intensity during swimming was kept in the range of 60-65% of mean VO_2max 136.6±11.8 ml/kg/min). The linearity of the correlation between glycogen content and optical intensity grading extended over a range of 1.4-11.8 mmol, but this was much smaller than the ranges of 60-120 mmol (Halkjær-Kristensen & Ingemann-Hansen, 1978) and 1-252 mmol (Vøllestad et al., 1984) previously reported for human muscles. The discrepancy seems to be a species-related, as the mean glycogen concentration in mice reported here is similar to that previously reported by others.

Figure 1 shows the glycogen depletion pattern for extrafusal and intrafusal fibres as estimated by the OS method. For extrafusal fibres, the VI and the OS methods clearly showed that mainly type I fibres were recruited at the beginning of the exercise in both SOL and EDL muscles. Type II fibres took over when type I fibre became fatigued. For intrafusal fibres, glycogen content decreased with increasing exercise duration in all three types, S-b (bag_1), L-b (bag_2), and L-c and S-c (chains). After 8 hours of swimming, the glycogen depletion in SOL muscle was 4.3 mmol/kg (i.e. a drop of 51% from the resting controls) in S-b, 4.4 mmol/kg (49%) in L-b, and 4.3 mmol/kg (45%) in L-c and S-c fibres. Almost the same amount of glycogen was utilized in the three fibre types. After 8 hours of swimming, the amount of glycogen depletion in EDL muscles was somewhat lower in S-b (3.5 mmol/kg; 44%) than in L-b (4.2 mmol/kg; 50%) and L-c and S-c (4.1 mmol/kg; 47%) fibres. A different glycogen depletion pattern was found in the three fibre types. From the start to 2 hours of swimming, S-b fibres were more depleted than the other two fibre types. From 2 to 4 hours of swimming, L-b fibres were most strongly depleted. This suggests that L-b fibres are chiefly recruited when S-b fibres become fatigued. The glycogen depletion patterns of intrafusal S-b fibres and extrafusal type I fibres in SOL and EDL spindles were very much alike. It thus seems that prolonged exercise mainly recruits dynamic beta motoneurones, which are thought to innervate the bag_1 and type I fibres (Gladden, 1992). The decrease in glycogen content of the chain fibres increased rather linearly when compared with the decrease in S-b and L-b fibres. This suggests a steady recruitment of chain fibres throughout the 8 hours of swimming.

REFERENCES

ARMSTRONG, R. B., SAUBERT IV, C. W., SEMBROWICH, W. L., SHEPHERD, R. E. & GOLLNICK, P. D. (1974) Glycogen depletion in rat skeletal muscle fibers at different intensities and durations of exercise. *Pflügers Arch* **352**, 243-256.

GLADDEN, M. H. (1992) Muscle receptors in mammals. In *Advances in Comparative & Environmental Physiology*, ed, ITO, F., pp281-302, Springer-Verlag, Berlin.

HALKJÆR-KRISTENSEN, J. & INGEMANN-HANSEN, T. (1979) Microphotometric determination of glycogen in single fibres of human quadriceps muscle. *Histochem. J.* **1**, 629-638.

VØLLESTAD, N. K., VAAGE, O. & HERMANSEN, L. (1984) Muscle glycogen depletion patterns in type I and subgroups of type II fibres during prolonged severe exercise in man. *Acta physiol. scand.* **122**, 433-441.

YOSHIMURA, A., FUJITSUKA, C., KAWAKAMI, K., OZAWA, N., OJALA, H. & FUJITSUKA, N. (1992) Novel myosin isoform in nuclear chain fibers of rat muscle spindles produced in response to endurance swimming. *J. App. Physiol.* **73**, 1925-1931.

HYSTERESIS OF MUSCLE CONTRACTION AND EFFECTS OF UNCERTAINTY IN PROPRIOCEPTIVE ACTIVITY AND MOTOR PERFORMANCE

A. I. Kostyukov, V. L. Cherkassky,
and A. N. Tal'nov

A. A. Bogomoletz Institute of Physiology
National Academy of Sciences of Ukraine
Bogomoletz str. 4, 252601 Kiev 24, Ukraine

Accuracy in estimation of mechanical properties of skeletal muscles is of primary importance for the adequate analysis of various problems of motor control. Based on a well-known family of length-tension curves obtained on a nerve-muscle preparation in isometric conditions, the spring model of muscle and the equilibrium point hypothesis were suggested for treatment of single-joint movements (Feldman, 1986). However, it is shown that efficiency of muscle contraction is strongly movement-dependent: a muscle generates stronger forces at lengthening vs shortening and such a hysteresis leaves long-term after-effects (Kostyukov, 1987). Some nonlinear effects in muscle dynamics have been evaluated in our study on the ankle extensors of anaesthesised cats (Figure 1). The muscles were activated by a distributed stimulation of the motor fibres supplying them; a servo-controlled linear motor was used as a muscle stretcher. Length-tension hysteresis was well expressed even at very slow movements and resulted in strong after-effects. In spite of the identity of stimulation parameters and applied loads, the steady lengths differed substantially depending on the direction of the previous movement (Figure 1A). At the same time pronounced uncertainty could be observed after the movements in the same direction as well (Figure 1B). In this case, powerful hysteresis after-effects connected with oppositely directed conditioning movements were not fully eliminated during the subsequent identical test load steps. The increase of these steps brought a divergence between the steady values of length down to its full disappearance.

We have shown that there are strong movement-dependent after-effects in the activity of primary and secondary endings of the muscle spindles (Kostyukov & Cherkassky, 1992). The after-effects were studied on the spindles of de-efferented muscles under load and length servo-control conditions, similar experiments were also made in the presence of stimulation of the static γ-efferents (Figure 2). With large-amplitude movements (more than 2-4 mm), the steady firing rates were always higher after preceding lengthening (loading) and lower after shortening (unloading) (Figure 2A, bottom traces; 2B). The difference in rate of the spindle ending steady firing at the same level of controlled parameter reached 15-20/s. The after-effects were usually less pronounced for small-amplitude movements (Figure 2A, top traces). With γ-stimulation, there were similar after-effects for large-amplitude test movements (Figure 2C). Hysteresis in the activity of muscle spindles amplifies the muscle hysteresis proper. This effect was demonstrated in our study of the stretch and unloading reflexes in decerebrate cats (Kostyukov, 1989). The divergence between the lengthening and shortening branches of the length-tension loops with externally induced movement of a muscle was always higher when it was subjected to the spinal reflex control as compared to the case of constant stimulation of the muscle. The similar powerful hysteresis seems to be present in awake animals as well. This effect was demonstrated in experiments on unanaesthesised cats, the simple stereotyped movements of which were evoked by a long-lasting microstimulation of the motor cortex (Kostyukov & Tal'nov, 1991). Joint stiffness in the presence of torque disturbances was shown to be dependent on the immediate past-movement history; it increased with changes in direction of the movement.

In conclusion, it should be pointed out that the muscle hysteresis itself seems to be increased significantly in spinal and supraspinal reflex circuits mainly due to the powerful hysteresis in activity of afferent terminals of muscle spindles and following central

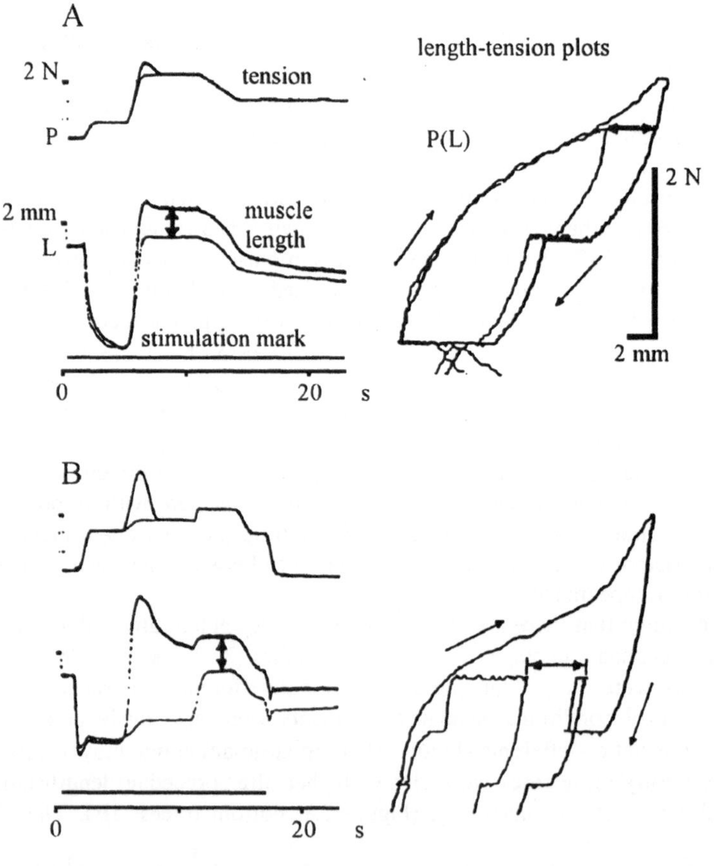

Figure 1. Effects of the movement-dependent uncertainty in equilibrium length of active muscle. The m. soleus of anaesthesised cat was activated by distributed stimulation of five distal filaments of the dissected L_7 - S_1 ventral roots, rate of a single channel stimulation was 6/s. The tension servo-control mode was used; the muscle movement trajectories for two different command signals and length-tension plots are superimposed in A and B fragments (thin and thick curves on experimental records).

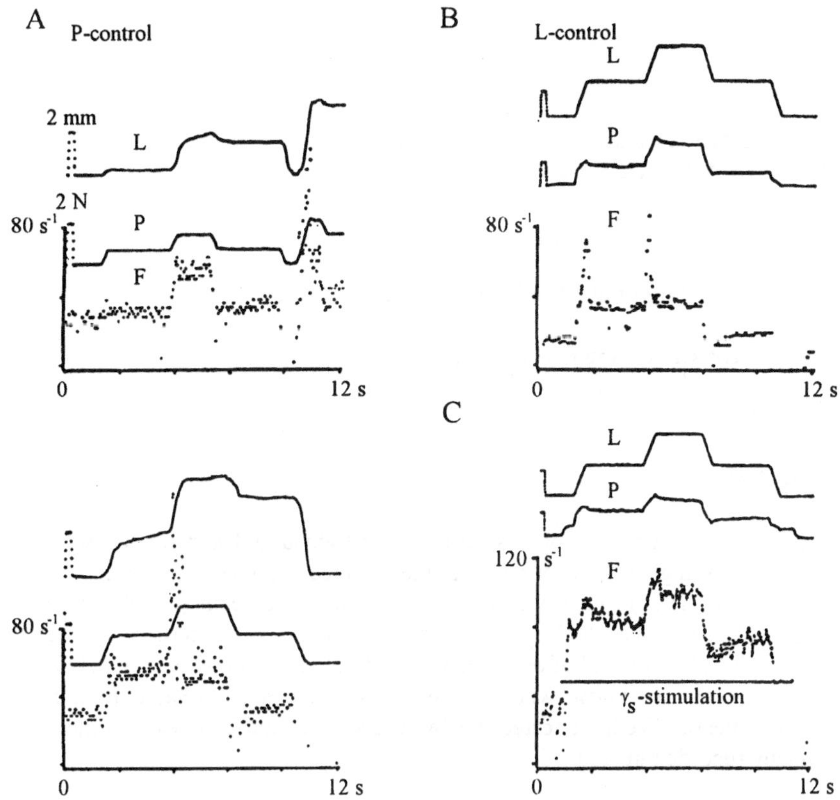

Figure 2. Response of the spindle primary ending from the m. gastrocnemius to servo-controlled changes in tension (A) or in muscle length (B & C). Instantaneous (A) and averaged (B & C) spindle discharge rates are given; stimulation of static γ-efferent with 50/s rate was applied in C.

transformation of the proprioceptive signals. This results in pronounced uncertainty in overall motor performance which is to be overcome in many real goal-directed movements. It can be supposed that elimination of the undesirable hysteresis-like uncertainties in motor behaviour might be possible only at a volitional level of motor control.

REFERENCES

FELDMAN, A.G. (1986) Once more on equilibrium-point hypothesis (λ–model) for motor control. *J. Mot. Behav.* **18**, 17-54.

KOSTYUKOV, A.I. (1987) Muscle dynamics: dependence of muscle length on changes in external load. *Biol. Cybern.* **56**, 357-387.

KOSTYUKOV, A.I. (1989) Dynamic properties of the stretch reflex. *Neirofisiologiya.* **21**, 589-597.

KOSTYUKOV, A.I. & CHERKASSKY, V.L. (1992) Movement-dependent after-effects in the firing of the spindle endings from the de-efferented muscles of the cat hindlimb. *Neuroscience.* **46**, 989-999.

KOSTYUKOV, A.I. & TAL'NOV, A.N. (1991) Effects of torque disturbances on elbow joint movements evoked in unanesthetized cats by microstimulation of the motor cortex. *Exp. Brain. Res.* **84**, 374-382.

THE EFFECTS OF HYPERACTIVITY ON H AND T RESPONSES IN AWAKE RATS

Chantal Pérot and Maria Izabel Almeida-Silveira

URA CNRS 858 Compiègne, France

INTRODUCTION

In man the soleus H reflex appears different for trained and non trained subjects. The H_{max}/M_{max} ratio is higher in endurance trained subjects (Ginet, Guiheneuc, Prevot & Vecchierini-Blineau, 1975) and lower in strength trained subjects (Casobona, Polizzi & Perciavalle, 1990). Moreover the H_{max}/M_{max} and T_{max}/M_{max} ratios increase when measured on the same subjects before and after an endurance training period (Pérot, Goubel & Mora, 1991). It is important to know whether the reflex changes are related to a fibre type transition phenomenon. We have tested this by recording H and T reflexes in the rat just as they are usually recorded in man.

METHOD

In rats, H and T reflexes of the ankle plantar flexors were recorded by means of surface electrodes, 2 mm in diameter, stuck on the depilated skin at the level of the gastrocnemius lateralis. A needle in the tail served as a reference electrode. The rat lay on the platform of an adapted ergometer and was lightly held in position. Stimulating electrodes were held in position over the sciatic nerve. Stimuli (0.2 ms in duration) were delivered when the rat was quiet, as shown by a very low level of the electromyogram (EMG) continuously observed on a control oscilloscope. The stimulus intensity varied between 3 and 20 mA. The EMGs were stored on a microcomputer (sampling frequency of 50 KHz; duration of acquisition 30ms). Taps on the Achilles tendon were applied by means of a small electromagnetic hammer, 1 mm distant from the skin. When switched on, the hammer struck the tendon for about 20 ms. The experiment was conducted on both legs, the right or the left being tested first at random. Some hours or days separated the two tests.

The reflexes were analysed in different groups of rats. The first group of 8 were trained to run inside a motorised wheel. The training period lasted 20 weeks. The training intensity increased progressively during the first 8 weeks and was kept as follows for the last 12 weeks : 5 sessions per week, 40 min per session at a velocity of 25m/min. Thus in the last 12 weeks the rats covered about 10 km per week. For a second group of 14 rats stretch-shortening cycles were repeatly imposed to the hindlimb muscles with the aid of an apparatus which hydraulically raised and lowered the cage. The stretch-shortening cycles were as follows : a 50 cm fall of the cage at a velocity of about 5 m/sec was suddenly arrested. This stretched the ankle plantar flexors, and was followed by a quick, short rise of the cage which caused the rats to jump producing a shortening of the pre-stretched muscles. Three times a week the rats went through a session of about a 100 of these cycles, a 3 minute rest being given after each 10 cycles. In each type of training the same number of age-matched rats served as control. The rats tolerated the procedures well, but any significant sign of stress led to the interruption of the test.

RESULTS

Examples of the recorded EMGs are given in Figure 1. The EMG detected at high intensity of stimulation (≅ 15 mA) had a mean latency of 0.94±0.09 ms and was identified as a M response. At lower intensities of stimulation it was possible to detect a response longer in latency (mean latency 7.14±0.49 ms), corresponding to an H reflex. The T responses obtained by the tendon taps had a mean latency of 9.53±1.09 ms. These EMGs may represent the reflex activation of not only the gastrocnemius lateralis, where the electrodes were placed, but also (and perhaps mainly) of the more reflexly active soleus. The reflex contribution of the soleus (deep muscle in the rat) was probably detected by cross-talk at the site of the electrodes. Such cross-talk has been observed between big human muscles (Türker & Miles, 1990), and cannot be excluded when testing EMGs in a small animal.

The mean peak to peak amplitudes of the M, H or T responses were calculated for each leg from the 10 largest stored responses, and H_{max}/M_{max} and T_{max}/M_{max} ratios calculated. More than half of the animals presented H reflexes significantly different on the right and the left sides. Thus, as previously proposed (Pérot & Almeida-Silveira, 1994), the legs were divided into 3 classes called "dominant, non-dominant and ambilateral". The comparisons between the reflexes obtained from trained and non-trained rats were made on the basis of

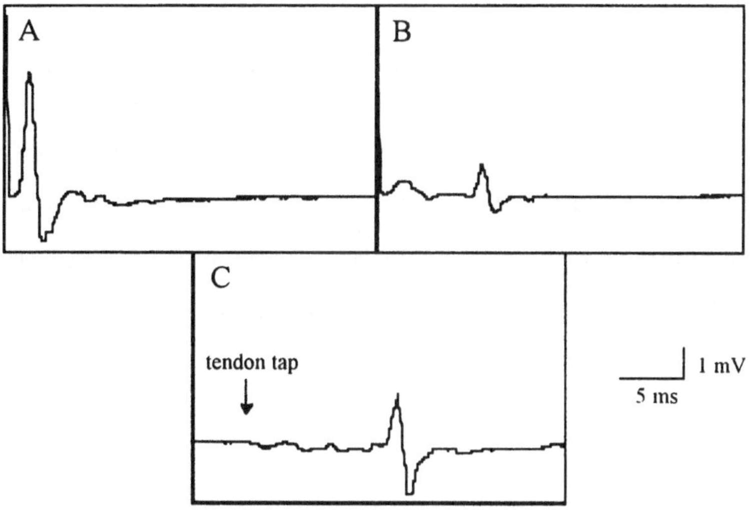

Figure 1. Examples of electromyographical activities detected in the ankle plantar flexors of awake rats during the 30 ms following electrical or mechanical stimuli. After a rather high sciatic nerve stimulation a M response was observed (A) ; a lower electrical stimulus allowed the recording of the H response (B) ; a tap on the Achilles tendon was followed by a T response (C). In A, B, and C the tracings lasted 30 ms.

this classsification. The H reflex was increased for the endurance trained rats but the difference was significant only for the ambilateral class. The T reflex appeared to be higher for the endurance trained rats than for the control animals, but the difference was not significant.

The soleus fibre type composition was determined by classical histochemical methods. The trained soleus presented a slight increase in type I fibres and a slight decrease in type II fibres. The results obtained for the muscles trained to stretch-shortening cycles were more convincing since all measured parameters were notably affected by training. The H reflex was significantly lower, the T reflex was slightly decreased and the number of type II fibres increased significantly at the expense of the type I fibres, the total number of fibres being unmodified. Thus the decrease in reflex excitability of the ankle plantar flexors may be due to the fact that repeated stretch-shortening cycles enhance the number of fast motor units, having a higher excitability threshold. In both types of training the T reflex changes were less marked than the H reflex changes. This result is explainable considering the fact that the T response would be affected not only by changes in the motoneuronal excitability but also by changes in the way the spindles are activated. The stiffness of musculo-tendinous elements in series with the spindle, and the sensitivity of the spindle itself would contribute to the reflex amplitude.

CONCLUSION

In awake rats, both H and T reflexes are easily obtained. These reflexes were modified in amplitude after training periods and their changes could be related to a fibre type transition phenomenon. Thus H and T reflexes give a neurophysiological index of the neuromuscular plasticity induced by hyperactivity. With the aid of H and T responses recorded in awake rats, neuronal excitability could be tested under physiological conditions regarding central synaptic transmission and descending controls.

REFERENCES

CASABONA, A., POLIZZI, M.C. & PERCIAVALLE, V. (1990) Difference in H-reflex between athletes trained for explosive contractions and non trained subjects. *Eur. J. App. Physiol.* **61,** 26-32.

GINET, J., GUIHENEUC, P., PREVOT, M. & VECCHIERINI-BLINEAU, F. (1975) Etude comparative du recrutement de la réponse réflexe monosynaptique de soléaire chez des sujets non entraînés et chez des sportifs. *Méd. Sport* **49,** 55-64.

PEROT, C. GOUBEL, F. & MORA, I. (1991) Quantification of T and H responses before and after a period of endurance training. *Eur. J. App. Physiol.* **63,** 368-375.

PEROT, C. & ALMEIDA-SILVEIRA, M.I. (1994) The human H and T reflex methodologies applied to the rat. *J. Neurosci. Meth.* **51,** 71-76.

TÜRKER, K.S. & MILES, T.S. (1990) Cross-talk from other muscles can contaminate EMG signals in reflex studies of human leg. *Neurosci. Lett.* **111,** 164-169.

MODIFICATION IN THE LOCOMOTOR ACTIVITY OF THE RAT SOLEUS MUSCLE IN RESPONSE TO HINDLIMB UNLOADING: PRELIMINARY RESULTS

M. H. Canu, D. Leterme, S. Fodili
and M. Falempin

Laboratoire de Physiologie
des Structures Contractiles,
Université des Sciences
et Technologies de Lille,
59655 - Villeneuve d'Ascq Cedex,
France

INTRODUCTION

In rats, previous studies have shown that hindlimb unloading (HU) modifies muscle characteristics, especially in antigravitational muscles such as the soleus (SOL). However, little is known about the neural effects of unloading. Some authors have found that the EMG activity in the rat SOL during HU is significantly reduced, and that the patterns of activity are changed (Blewett & Elder, 1993). Locomotion has not yet been studied after HU, but some clinical studies are of particular interest. It has been demonstrated that i) many locomotor disorders are due to a muscular deficit; ii) an adaptation of the locomotor pattern can occur in order to maintain the muscular system performance, in spite of muscle changes. This data supports the idea that the locomotor pattern may be modified, in response to HU and to its associated muscular transformation.

METHODS

Male Wistar rats (250 - 350g) were trained to walk on a treadmill at a low speed (0.20 to 0.30 m/s) before surgery. Pairs of Teflon coated multistranded stainless steel wire were implanted in the SOL and TA muscles under sodium pentobarbitone anaesthesia. On the day following the surgery, the rats were suspended for 14 days as previously described by Morey (1979). A sample of EMG data was recorded, during a 5 min locomotion session, i) before suspension (HU_0), ii) on the 7th and 14th day of suspension (HU_7 and HU_{14}). Raw EMGs were amplified (gain x2,000, band pass 10-3,000 Hz) and recorded on tape for off-line analysis with a microcomputer. The analysis program measured the burst duration, step cycle duration, burst surface (V.ms), electrical activity (burst surface/burst duration, V.ms/s) and mean peak amplitude. Data are presented as means ± SE. Results were analysed using the Student's t test.

RESULTS

Figure 1 shows an EMG recording at HU_0 and HU_7. In control rats (Figure 1A), a regular alternating pattern of bursts of activity and silence in SOL was observed. Bursts in TA appeared during the SOL silent periods. After suspension, this alternating pattern was not seen either in SOL or in TA. The rat had some difficulties in moving, and this was evident by frequent stops during which a tonic EMG activity was recorded in the SOL (Figure 1B). Some abnormal extension movements of the hindpaw were observed with a longer SOL EMG burst. Extrabursts were also recorded in TA (Figure 1C).

Figure 2A shows the ratio of burst duration to step cycle duration. It can be observed that at HU_7, the ratio was significantly weaker as compared to HU_0 (0.7680±0.0334 vs 0.8107±0.0237). Normal values were restored at HU_{14} (0.8090±0.0167). Burst duration was plotted against cycle duration and the slope of the regression line and the correlation coefficient were calculated ($r_0 = 0.94$; $r_7 = 0.83$; $r_{14} = 0.97$; $a_0 = 0.82$; $a_7 = 0.74$; $a_{14} = 0.85$). Note that the smallest values were obtained at HU_7.

Figure 2B shows burst electrical activity, i.e. burst surface for 1 s. A very significant decrease was observed at HU_7 (0.099±0.004 for HU_0; 0.070±0.007 for HU_7). Values at HU_{14} were slightly different from HU_0 (0.090±0.005). The regression line "burst surface against burst duration" was calculated. The slope was 71×10^{-6} at HU_0, 59×10^{-6} at HU_7 and 99×10^{-6} at HU_{14}. The correlation coefficient was smaller at HU_7 ($r_7 = 0.73$) than at HU_0 or HU_{14} ($r_0 = 0.91$; $r_{14} = 0.83$ respectively).

Mean peak amplitude (mV) was also calculated. Results are shown in Figure 2C. As before, amplitude decreased at HU_7 (-17%), and returned toward the normal value at HU_{14}.

DISCUSSION

Our preliminary results indicate that a disruption in the locomotor pattern of the rat

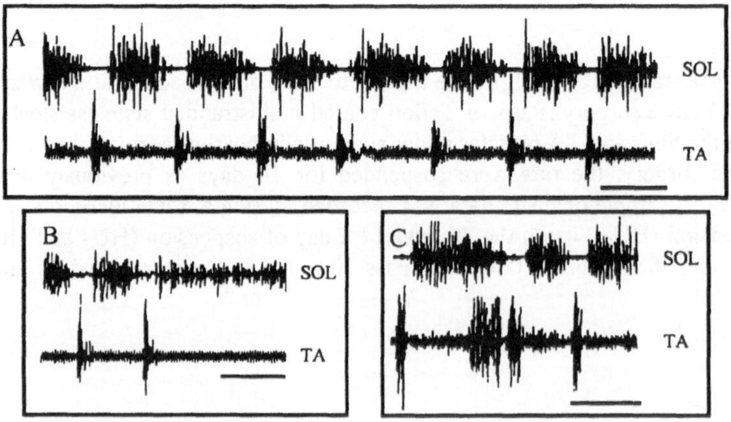

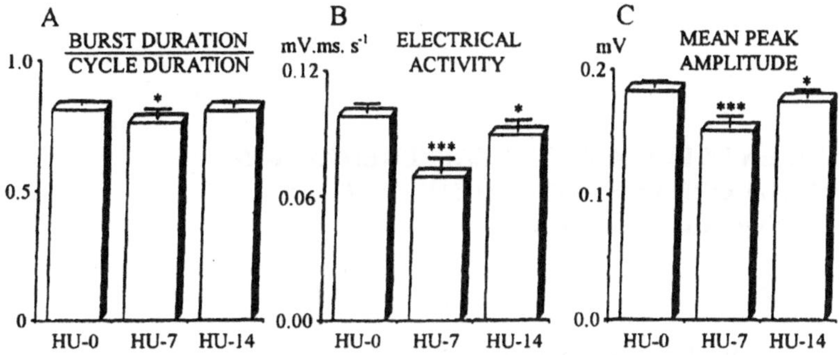

Figure 2. Burst duration/cycle duration, electrical activity and mean peak amplitude. Note the significant decrease at HU_7 in all cases, and restoration at HU_{14} (* $P < 0.05$; ** $P < 0.01$; *** $P < 0.001$).

soleus muscle during a walking test on a treadmill occurred after a 7 day period of HU. A burst EMG activity was always recorded during the stance phase, but the alternating pattern between the stance and swing phase during locomotion was modified. A decrease in the ratio of burst duration/step cycle duration was observed. This meant that there was a reduction of the stance phase. We also observed a decrease in the burst surface and in the mean peak amplitude of the EMG activity. These decreases probably reflect a decreased activation of the SOL, i.e. a decrease in the number of motor units recruited and/or a decrease in the firing rate of units already recruited.

These results indicate that a transitory disruption of the SOL motor pattern occurs within the first few days of HU, followed by a restoration of an almost normal pattern. Nevertheless, in spite of almost normal burst characteristics, the observation of the rat indicates that the locomotion is still disrupted. Since locomotion requires rhythmic alternate movements and coordination of limbs, and adequate coordination of muscles within one limb, we can hypothesize that one or several of these factors remain disrupted (work in progress).

ACKNOWLEDGEMENTS

This work was supported by grants from the Centre National d'Etudes Spatiales (CNES).

REFERENCES

BLEWETT, C. & ELDER, G.C.B. (1993) Quantative EMG analysis in soleus and plantaris during hindlimb suspension and recovery. *J. App. Physiol.* **74**, 2057-2066.

FALEMPIN, M., LECLERCQ, T., LETERME, D. & MOUNIER, Y. (1989) Time-course of soleus muscle change in and recovery from disuse atrophy. *The Physiologist* **33**, 588-589.

MOREY, E.R. (1979) Spaceflight and bone turnover : correlation with a new rat model of weightlessness. *Bioscience* **29**, 168-172.

STEVENS, L., MOUNIER, Y., HOLY, X. & FALEMPIN, M. (1990) Contractile properties of rat soleus muscle after 15 days of hindlimb suspension. *J. App. Physiol.* **68**, 334-340.

EARLY AND LATE EFFECTS OF DEAFFERENTATION UPON SPINAL CORD FUNCTION IN THE RAT - AN ELECTROMYOGRAPHIC STUDY

P. Hník[1], R. Vejsada[1] and Ruth Payne[2]

[1]Institute of Physiology
Czech Academy of Sciences
Prague, Czech Republic
[2]Department of Biomedical Sciences
University of Aberdeen
Aberdeen, UK

Deafferentation by dorsal root section has been extensively studied for assessing the significance of peripheral feedback information for the performance of movement (see e.g. Mott & Sherrington, 1895). Most authors have concluded that after an early depression of reflex and locomotor activity, the responsiveness of the deafferented spinal cord segments is enhanced (Drake & Stavraky, 1948). In this report we have attempted to obtain some more information about the changes in EMG activity occurring in extensor and flexor muscles of the rat shortly after dorsal root section and several weeks to months after the operation.

The experiments were performed on adult male rats. An EMG electrode array, consisting of a percutaneous connector, coiled steel wire leads (0.1mm in diameter) and Silastic (Dow-Corning) plate probes, was implanted under Nembutal anaesthesia (60 mg/kg i.p.). The plate electrodes were placed for bilateral recording from the soleus (SOL) and tibialis anterior (TA) muscles. For further details see Hník, Vejsada & Kasicki, (1981). After several days of EMG recordings, the animals were anaesthetised and the dorsal roots L1-S1 were sectioned proximal to the spinal ganglia. Before dorsal root section, both SOL muscles exhibit continuous EMG activity when the rat is sitting in the cage. This activity is of reflex myotatic origin and disappears when the rat is lifted up in the air, since the stretch of the SOL muscles is released. During the first 1-2 days after deafferentation, reflex myotatic activity in the SOL is absent. However, after this period "spontaneous" EMG activity appears in the deafferented SOL and is manifested by a tendency to extension of the deafferented limb during locomotion. This EMG activity can be turned off (or on) by various manoeuvres from the contralateral limb and by supraspinal influences (e.g. vestibulospinal). When "spontaneous" EMG activity had developed after deafferentation, it could be inhibited in the SOL (and also gastrocnemius muscle) by dorsiflexion of the ankle. This phenomenon was named "stretch-induced inhibition" (SII) of deafferented extensor muscles to stretch (Hník, Vejsada, Kasicki & Afelt, 1984). The EMG stereotype of walking or running appears to be unaffected by deafferentation, although the deafferented limb is ataxic with a tendency to extension. The EMG activity which appears in the SOL a few days after dorsal root section is apparently due to hyperexcitability of extensor motoneurones. It is tempting to speculate that extensor motoneurones receive a larger number of monosynaptic "boutons" on their surface than flexor neurones, so that section of dorsal roots then leads to degeneration of a greater number of synaptic endings on their surface causing their "partial denervation". This could make them hyperexcitable to neural inputs from various sources. No such symptoms of hyperexcitability were found in the deafferented TA, a flexor muscle. It is difficult to put forward even a tentative explanation of the SII phenomenon. Since dorsal roots L1-S1 were sectioned proximal to the ganglia, thus eliminating the sensory input from the whole limb, there remain two possible routes. Either ventral root afferents or side branches of sensory axons from spinal ganglia and their possible activation of Renshaw cells, which could thus inhibit the "spontaneous" EMG activity of deafferented extensor muscles.

REFERENCES

DRAKE, C.G. & STAVRAKY, G.W. (1948) An extension of the "Law of Denervation" to afferent neurones. *J. Neurophysiol.* **11**, 229-238.

HNÍK, P., VEJSADA, R. & KASICKI, S. (1981) Reflex and locomotor changes following unilateral deafferentation of rat hind limb assessed by chronic electromyography. *Neuroscience* **6**, 195-203.

HNÍK, P., VEJSADA, R., KASICKI, S. & AFELT, Z. (1984) Stretch-induced inhibition of spontaneous EMG activity in extensor muscles of the rat caused by chronic deafferentation. *Physiol. Bohemoslov.* **33**,139-145.

MOTT, F.W. & SHERRINGTON, C.S. (1895) Experiments upon the influence of sensory nerves upon movement and nutrition of the limbs. *Proc. R. Soc.* **57**, 481-488.

PART 3

GAMMA REFLEXES

REFLEX ACTIVATION OF γ_s- AND γ_d-MOTONEURONES OBSERVED IN ISOLATED MUSCLE SPINDLES OF CAT HINDLIMB MUSCLES

M.H. Gladden, M. Dickson and T. Lumsdon

Muscle Spindle Physiology Group
I.B.L.S., West Medical Building
University of Glasgow
Glasgow G12 8QQ, Scotland

INTRODUCTION

It is clear from even a cursory examination of the literature on muscle spindles that these are receptors of great sophistication and functional versatility. But a simple textbook 'role' for them in the serious business of motor control cannot be formulated realistically without taking into account that some information they signal is inconsequential to motor control, as for example when spindle afferents are influenced by the arterial pulse (Ellaway & Furness, 1977), and γ-motoneurones are influenced by laughing and mental arithmetic (Ribot, Roll & Vedel, 1986). Clearly their involvement is detachable, which presumes there are rules for engaging and disengaging them. Their participation is context specific; information from spindles must not only be interpreted in the context of any current motor act, but also be set against the gamut of information continually flowing into the spinal cord from other receptors - joint, cutaneous and all. The high stretch sensitivity of primary sensory endings sometimes must be essential to motor control, and at other times irrelevant and even dangerous, and will need to be excluded. This is not new (e.g. Graham Brown, 1911; Forssberg, Grillner & Rossignol, 1975; Taylor & Gottlieb, 1985).

The CNS may chose to ignore what is coming from the spindles by suitably setting the interneurones, but it has a very versatile set of 'interneurones' to control the spindles themselves, the γ-motoneurones, which, as Johansson (1981) has argued, are themselves a gathering ground for afferent information. Yet in wresting definitive and reproducible experimental results on γ-motoneurone activity the context is easily forgotten. For the methods almost obligatorily involve not only widespread denervation, but also extensive surgery which must activate pain afferents that in life would inhibit normal movement, including bone pain afferents from fixation of the limbs and axial skeleton. For these reasons it seems worthwhile to persevere in developing a method to follow γ–motoneurone activity which can leave the rest of the nervous system intact, particularly as there is no easily applicable, non-invasive technique available, similar to EMG recording for α-motoneurones.

METHODS

In anaesthetised animals, without denervation and with minimal surgery γ_s- and γ_d-motoneurone activity can be detected by observing the contractions of the separate types of intrafusal fibre that these motoneurones control in spindles which have been partially dissected within an exteriorised muscle (Dickson & Gladden, 1992). The muscle is usually a small one, like tenuissimus or the abductor digiti quinti medius muscle of the hind foot, but favourably-situated spindles in large muscles, even soleus, can be used. Recently we have increased the information these preparations yield by placing two recording electrodes on the spindle nerve supply in-continuity, one next to the spindle (Figure 1, middle trace), and the other further centrally. This means that potentials entering or leaving the observed spindle can be recognised by their individual shapes, and established as afferent or efferent by spike-triggered averaging from the distal to the more proximal electrode (Figure 2A; Dickson, Gladden & Yoshimura, 1992). By selecting the terminal spindle supplied by the intramuscular nerve, in our case the most distal spindle in tenuissimus, the number of γ-axons in the nerve branch was confined to the spindle supply, usually three to six. Individual video frames from movement sequences of intrafusal fibres can also be correlated directly with the neural discharges (Lumsdon, Dickson & Gladden, 1994; 1995).

γ–MOTONEURONES EXCITED BY MUSCLE STRETCH

Muscle stretch can be very potent in exciting γ_s-motoneurones of tenuissimus spindles

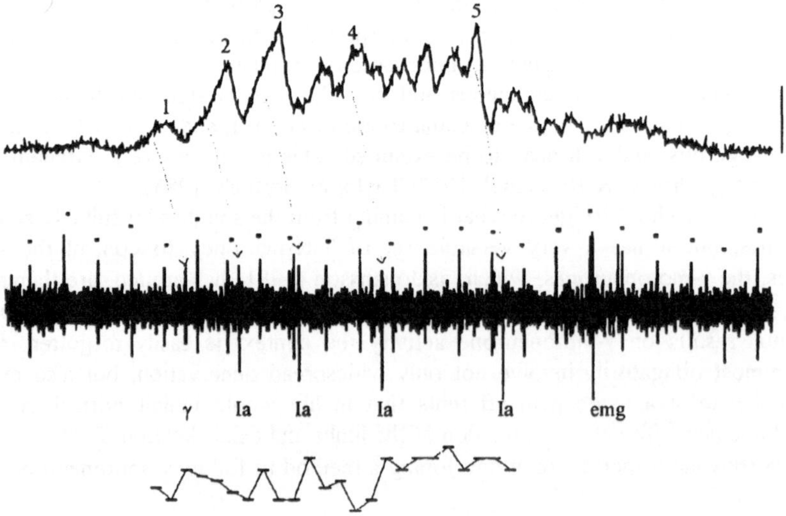

Figure 1. Record from nerve supply of most distal spindle in tenuissimus (middle trace) during repeated weak pulls on a thread attached to superficial fascia over the belly of the ipsilateral gastrocnemius muscle (upper trace). Ia spikes (Ia) follow pulls 2-5 as indicated by arrows and were preceded by γ-spikes - see Figure 2. A single γ-spike (γ) followed pull 1. (•): spikes from a secondary ending. emg: emg from extrafusal fibres, not necessarily associated with the reflex. Movement of chain fibres was scarcely detectable (lowest trace, chain fibre displacement in arbitrary units). Time scale 200ms; tension, 0.05N.

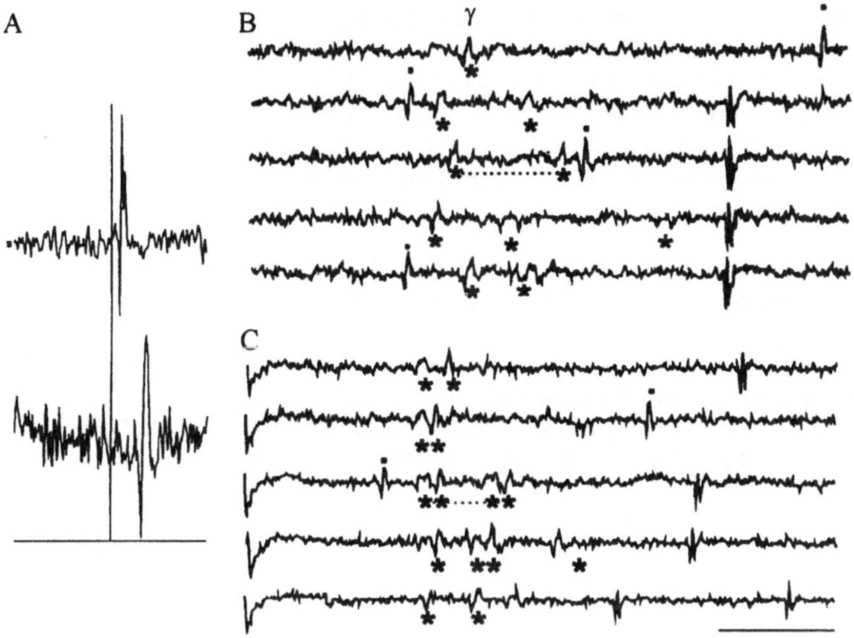

Figure 2. A. Averaged spikes from primary (upper) and secondary (lower) sensory endings of the most distal spindle in the tenuissimus muscle recorded from main tenuissimus nerve and triggered from spikes recognised by their shape in the spindle nerve, same experiment as Figure 1. B. Expanded records from Figure 1B triggered, except for trace 1, on the Ia spike to show the preceding γ-spikes (*). As in Figure 1 (•) indicates the secondary spikes. Stippled line joins recurring spike shapes i.e. couplets. C. Traces triggered on stimuli to biceps nerve branch at just α threshold, same experiment and same symbols as in Figures 1& 2A & B. Time scale: A, 20ms; B & C, 10ms.

under barbiturate anaesthesia. The effect was strongest early in experiments, and tended to wear off, and even disappear with time. Stretch of tenuissimus itself was effective, but so also was stretch of other muscles, biceps femoris, quadriceps, soleus and gastrocnemius, but the reflex was confined to the ipsilateral limb. Figure 1 shows an example of reflex contraction of intrafusal fibres in tenuissimus to weak pulls on a thread attached about half way along the ipsilateral gastrocnemius muscle surface. Stronger pulls caused obvious contractions in proximal and distal chain poles, and the bag_2 proximal pole, all contracting independently of eachother, so at this stage in the experiment it was possible to recruit three separate γ_s-motoneurones. With the weak pulls illustrated in Figure 1 (top trace) the reflex was at threshold, and the first γ-spike ('γ ' in Figures 1 & 2) failed to illicit a Ia spike (Figure 2B, first trace). Even with pulls 2-5 (Figure 1, top trace) only minute movements occurred in the spindle, and in different poles for each trial. This was because the γ-motoneurones were recruited in various combinations, as can be seen from the changing γ-spike shapes preceding each Ia spike in Figure 2B. For the chain fibre monitored in Figure 1 (bottom trace) definite movement was coincident with the fourth Ia spike only. The extrafusal EMG which followed the third Ia spike was not thought to result from the reflex activity, because similar EMG's occured irregularly in the absence of pulls at this stage in this experiment, and were not present during reflex activity which resulted from stronger pulls of the gastrocnemius muscle.

Cutaneous reflexes can be demonstrated easily in these preparations, and pulling on muscles can excite cutaneous receptors. However, reflex intrafusal contraction still occurred when care was taken to exclude cutaneous disturbances, and also when the skin was denervated locally by cutting cutaneous nerves. Spindle afferents must be involved in the reflex, because stretch of part of the tenuissimus containing only spindles was effective. Nevertheless, they may not be solely responsible in general. However, stimulation of muscle nerves at just α threshold (Figure 2C) recruited the same γ_s-motoneurones. Contraction of bag_1 fibres was seldom seen, which suggests that γ_d-motoneurones were only occasionally excited.

γ–ACTIVITY AND SECONDARY ENDINGS

Two observations were more conveniently investigated further in 'reduced' preparations. These were firstly that sometimes γ–spikes could occur in couplets (e.g in Figure 2; identical spike shapes follow each other in third traces of both B & C, couplets indicated by stippled lines), and secondly that secondary endings seemed virtually

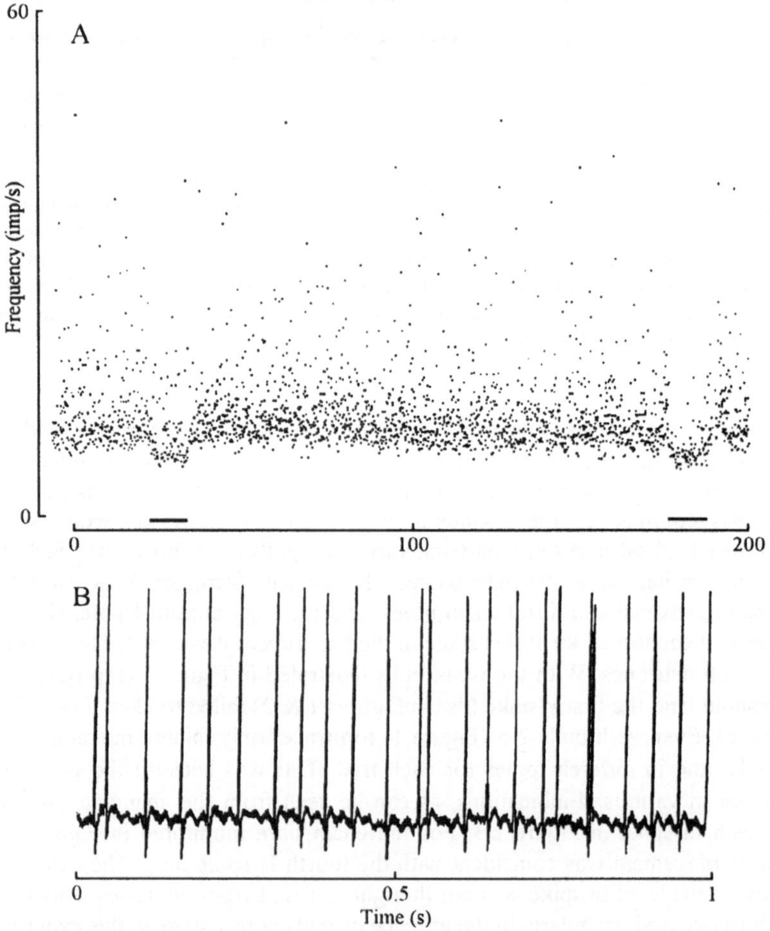

Figure 3. A. Spontaneous discharge of a γ_s-motoneurone recorded intracellularly from a long chain fibre of a tenuissimus muscle spindle in a precollicular decerebrate preparation. Bars indicate two periods of midbrain stimulation (coordinates A:1.5; L: 1.2; -2mm) of 0.3 and 0.35 mA at 500 pulses/s. B. Intracellular spikes from period between stimulations to illustrate couplet spikes, occuring here at about 0, 0.5 and 0.8s.

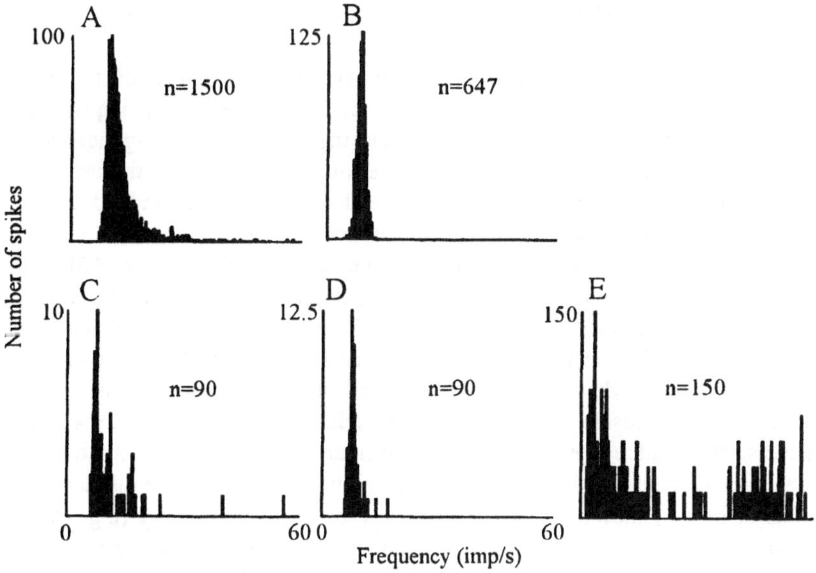

Figure 4. Distribution of frequencies of γ-discharge. A. Spontaneous activity of unit illustrated in Figure 3. B. γ-discharge induced by extracellular inotophoretic application of DL-homocysteic acid (Jankowska & Gladden, unpublished). C & D. Unit illustrated in Figure 3 and in A during the two periods of midbrain stimulation (C - first, D - second period). E. Unit recorded intracellularly from a typical chain fibre pole in a decerebrate preparation whose discharge included couplets of a limited frequency range.

unresponsive to γ-activity. Was the pattern of γ-activity preferentially exciting primary sensory endings, or was over-shortening of the spindles by dissection making the secondaries abnormally unresponsive? This was investigated by comparing the responses of primary and secondary endings of single tenuissimus spindles firstly to natural γ–activity and then to 'classical' testing applied to γ_s- and γ_d-axons which supplied the same spindle (Gladden & Lumsdon, 1995). The cats were decerebrate, and the spindles were undissected, with the normal relations of the muscle preserved. In these preparations γ-activity was difficult to evoke because the ipsilateral limb was totally denervated. However, it had a much stronger effect on the response of the primary than of the secondary ending. Nevertheless, when single γ–axons which supplied these spindles were stimulated at constant frequency in the usual ranges applied in classical tests (Emonet-Dénand, Laporte, Matthews & Petit, 1977) the secondaries could respond strongly. Comparison of afferent responses to natural γ-activity and to known constant frequencies of γ-stimulation suggested that the mean frequency of the natural γ-activity was low, but that this was punctuated by bursts at higher frequency.

This conclusion was supported by experiments in which the activity of individual γ-motoneurones was recorded in preparations with a similar central state to the experiments just described, that is, decerebrate, and with the ipsilateral hindlimb largely denervated. γ-activity was recorded intracellularly from individual intrafusal fibres in dissected spindles of the tenuissimus muscle. Recordings from chain fibres are useful because single chain fibre poles are rarely innervated by more than one γ_s-motor axon. An example is shown in Figure 3B. A fairly regular discharge appears to be interrupted by couplet spikes at higher frequency. Figure 4A shows that there was a continuum of frequencies in this case, but in another example the couplet frequency tended to be fixed at about 45 imp/s (Figure 4E).

γ–DISCHARGE PATTERNS

Functionally the bursting activity pattern is interesting because of the mechanical consequences in the spindle. Bursts at high frequency cause phasic twitching of chain fibres which preferentially excites primary endings, while the secondary ending signals are more related to the mean frequency, which can be quite low. The mean low frequency, together with the low frequency of the bursting, will tend to ensure that individual bursts are expressed in the primary discharge because this reduces the probability of interference from other on-going fusimotor spike trains which arrive at the same spindle. Although single stimuli in γ_S-axons supplying chain fibres can cause Ia spikes - the familiar phenomenon of driving - couplets or triplets may be better suited to the mechanical behaviour of chain fibres, which easily buckle on shortening, and so are less likely to be able to influence the primary sensory ending when the spindle is at all shortened during contraction of the surrounding extrafusal muscle. The couplets and triplets will tend to take up slack in them, and so increase the chance they can extend the primary terminals. The double pulse is also more likely to be successful in breaking through a high frequency resting discharge.

Notions of how γ_S-motoneurones would use the spindle machinery (e.g. Boyd, 1981) have been dominated by actions at high constant frequencies, and have not predicted this pattern of use. However, it is not known at this stage whether the bursting behaviour is common. So far it has been seen in a proportion of γ_S-motoneurones supplying the tenuissimus muscle. Differences in the activity patterns of single γ_S-motoneurones could explain the variations in reactivity of individual chain and bag_2 intrafusal fibres to ATP-ase staining, and to antibodies against myosin types, which occur not only between fibres, but also along the length of individual fibres (Yoshimura, Gladden & Dickson, 1992). These differences are not confined to intrafusal fibres of tenuissimus spindles, but are also found in spindles of other feline axial and hindlimb muscles (Yoshimura, personal communication). The phenomenon of couplet or triplet spikes would not tend to be noticed in the work of others on γ–motoneurones when γ–activity is averaged, for example in the method used by Bessou, Joffroy, Montoya & Pagès (1984).

THE ORIGIN OF COUPLETS AND THEIR SUPPRESSION

If a couplet occurs during a fairly regular γ–spike train the short interspike interval is frequently, though not invariably, followed by a longer interspike interval than the one immediately preceding the couplet (see Figure 3B). This suggests that the couplet has reset the rhythm, and that the burst is the consequence of some additional brief excitatory input to the γ–motoneurone. An observation which lends support to this idea is that no such bursts were found when γ-motoneurones were activated by iontophoretic application of L-glutamate or DL-homocysteic acid (Jankowska & Gladden, unpublished observations). The discharge was then irregular, but the frequencies were distributed approximately normally around a mean value set by the rate of release (Figure 4B). this suggests that the high frequency bursts reflect synaptic actions of excitatory interneurones on γ-motoneurones rather than characteristics of the responses of γ-motoneurones themselves. In these experiments, extracellular recordings were made from γ-motoneurones of peroneus tertius or brevis muscles in the motor nucleus of the L7 segment. They were differentiated from α-motoneurones by the 3.5 - 5.5 ms longer latency of their antidromic activation and a higher threshold (2.5 - 3.5 times threshold of the most sensitive fibres of the nerve). the range of conduction velocities was 29 - 40 m/s. It was not possible to identify them as static or dynamic. The recordings were made using glass micropipettes filled with a 0.5 M solution of glutamate or homocysteate in 0.3 M solution of KCl (pH 8.6). The amino acids were ejected by 5 -10 nA constant negative current passed through the recording electrode.

In the precollicular decerebrate preparation from which the example shown in Figure 3 was taken, midbrain stimulation could reduce the mean frequency of γ–discharge without affecting the bursting (Figure 4C). The γ-motoneurone must have been inhibited and was not the source of the couplets. Stronger stimulation still did not stop the bursts completely (Figure 4D).

The mechanical outcome of chain fibres responding to high frequency couplets with variable interspike intervals would be contractions with a pulsatile appearance, and this behaviour can be seen frequently in tenuissimus spindles in anaesthetised and decerebrate preparations. An example is shown in Dickson & Gladden (this volume). The frequency of these pulsing contractions, <10/s, corresponds to the γ–burst frequency illustrated in Figure 3B. Stimulation in some midbrain sites could not only stop these pulsing contractions in chains but simultaneously recruit other γ_s-motoneurones that in our experiments caused bag_2 fibres to contract (Dickson & Gladden, 1992 and this volume).

Recognition of the bursting pattern may help to reconcile apparent inconsistencies between recent work on the types of intrafusal fibres innervated by γ_s-axons. Boyd (1986) started a quest to explain why a single group of neurones should control two types of fibre together which individually have such strikingly different effects on spindle afferents (Boyd, 1981). He had attempted to resolve the problem by seeking evidence that these two fibre types had largely separate anatomical innervation from subgroups of γ_s-motoneurones (Boyd, 1986), but evidence accumulated to date on this point has not supported a simple dichotomy (Banks, 1991; Dickson, Emonet-Dénand, Gladden, Petit & Ward, 1993; Gladden, 1995). On the other hand, if some γ_s-motoneurones can be recruited and others simultaneously suppressed (Dickson & Gladden, 1992) the γ_s-motoneurone pool cannot be a homogeneous entity. Possibly the answer lies in functional γ_s-control patterns, not in the distribution of the communication lines, with the imprint of these patterns being left on the intrafusal fibres in the histological and immunohistological changes which Yoshimura describes. A more precise proposition is that γ_s-motoneurones might belong to different functional groups so far as CNS control is concerned, with activity patterns directed towards controlling, and in sympathy with the electrical properties and mechanical behaviour of either chain fibres or of bag_2 fibres (possibly related also to cell size, see Banks (1991) and Emonet-Dénand & Gladden, (1993)).

During the progress of natural movements afferent input to the cord from receptors without efferent control will be subject to gating and afferent gain control by interneurones, but the γ-system can adjust the spindle afferent contribution continuously at source. The contribution which work on isolated spindles might make in future endeavours to understand the control of movements is derived from the fact the technique reveals directly the interface where these efferents and afferents meet.

ACKNOWLEDGEMENT

This work was supported by the Wellcome Trust.

REFERENCES

BANKS, R.W. (1991) The distribution of static γ–axons in the tenuissimus muscle of the cat. *J. Physiol.* **442**, 489-512

BESSOU, P., JOFFROY, M., MONTOYA, R. & PAGES, B. (1984). Effects of triceps stretch by ankle flexion on intact afferents and efferents of gastrocnemius in the decerebrate cat. *J. Physiol.* **346**, 73-91.

BOYD, I.A. (1981). The action of the three types of intrafusal fibre in isolated cat muscle spindles on the dynamic and length sensitivities of primary and secondary sensory endings. In *Muscle Receptors and Movement.* eds. TAYLOR, A. & PROCHAZKA, A. pp. 17-32. Macmillan, London.

BOYD, I.A. (1986). Two types of static γ axon in cat muscle spindles. *Exp. Physiol.* **71**, 307-327.

DICKSON, M. & GLADDEN, M.H. (1992) Central and reflex recruitment of γ motoneurones of individual muscle spindles of the tenuissimus muscle in anaesthetised cats. In *Muscle Afferents and Spinal Control of Movement.* eds. JAMI, L., PIERROT-DESEILLIGNY,E. & ZYTNICKI, D. pp. 37-42. Pergamon, Oxford.

DICKSON, M., GLADDEN, M.H. & YOSHIMURA, A. (1992). γ–Reflex activity in cats anaesthetised with barbiturates. *J. Physiol.* **459**, 461P.

DICKSON, M., EMONET-DÉNAND, F., GLADDEN, M.H., PETIT, J. & WARD, J. (1993) Incidence of non-driving excitation of Ia afferents during ramp frequency stimulation of static γ-axons in cat hindlimbs. *J. Physiol.* **460**, 657-673.

ELLAWAY, P.H. & FURNESS, P. (1977). Increased probability of muscle spindle firing time-locked to the electrocardiogram in rabbits. *J. Physiol.* **273**, 92P.

EMONET-DÉNAND, F., LAPORTE, Y., MATTHEWS, P.B.C. & PETIT, J. (1977) On the subdivision of static and dynamic fusimotor actions on the primary ending of the cat muscle spindle. *J.Physiol.* **268**, 827-861.

EMONET-DÉNAND, F. & GLADDEN M.H. (1993) Type of fusimotor excitation induced in individual spindles by their fastest-conducting gamma axons. Proc. IUPS. XVIII (Congress XXXII, Glasgow), 322.6/P.

FORSSBERG, H., GRILLNER, S. & ROSSIGNOL, S. (1975) Phase-dependent reflex reversal during walking in chronic spinal cats. *Brain Res.* **85**, 103-107.

GLADDEN, M.H. (1995) Isolated muscle spindles, their motor innervation and central control. In *Neural Control of Movement.* eds. W. FERRELL & U.PROSKE. Plenum Press, New York (in press).

GLADDEN, M.H. & LUMSDON, T. (1995) γ–reflexes assessed from the activity of spindle afferents and intrafusal fibres in decerebrated cats. *J. Physiol.* (in press).

GRAHAM BROWN, T. (1911) Studies on the physiology of the nervous system. VIII: Neural balance and reflex reversal with a note on progression in the decerebrate guinea pig. *Q. J. Exp. Physiol.* **4**, 273-288.

JOHANSSON, H. (1981) Reflex control of γ–motoneurones. Umea University Medical Dissertations. New Series no. 70.

LUMSDON, T., DICKSON M., & GLADDEN, M.H. (1994) A frame-triggered counter to correlate image and data analysis. *J. Physiol.* **476**, 8P.

LUMSDON, T., DICKSON M., & GLADDEN, M.H. (1995) A system for aligning video-recorded visual data and high-quality multichannel recordings of simultaneously occuring signals. *J. Microscopy.* (in press).

RIBOT, E., ROLL, J.P. & VEDEL, J.P. (1986) The reflex activity of mammalian small nerve fibres. *J. Physiol.* **375**, 251-268.

TAYLOR, A. & GOTTTLIEB, S. (1985) Convergence of several sensory modalities in motor control. In *Feedback and Motor Control in Invertebrates and Vertebrates.* eds. BARNES, W.J.P. & GLADDEN, M.H. pp. 77-92. Croom Helm Ltd., London.

YOSHIMURA, A., DICKSON, M. & GLADDEN MH. (1992) Mechanical properties of chain fibres and regional variations in their histo- and immunohistochemical reactivity in tenuissimus muscles of anaesthetised cats. *J. Physiol.* **459**, 502P.

INFLUENCES ON THE γ-MUSCLE SPINDLE SYSTEM FROM JOINT MECHANORECEPTORS

P. Sjölander and H. Johansson

Division of Work Physiology
National Institute of Occupational Health
S-907 13 Umeå, Sweden

INTRODUCTION

The functional role of information mediated by sensory fibres from mechanoreceptors in joint tissue has been a controversial issue for several decades. Until recently it was generally believed that the principal function of joint mechanoreceptors were to initiate reflex activation of α-motoneurones as joint movements approach the limits of the joint's normal working range, and thereby to protect the joint against hyper-rotations. This view on the functional role of joint receptor afferents was to a large extent based on early findings showing that (1) during passively imposed limb movements, only a very small number of joint mechanoreceptors were active at intermediate joint angles, whereas the majority of joint afferents were most active close to the limits of the physiological working range of the joint, (2) α-motoneurones were influenced by activity in high threshold joint afferents, but not by activity in low threshold joint afferents, and (3) no clear-cut proprioceptive, kinaesthetic or movement deficits were induced after removal of joint receptor feedback. The latter findings together with the observations that muscle vibration could induce illusions of movement in stationary limbs and that muscle receptor afferents had cortical projections made it widely accepted that muscle spindles and Golgi tendon organs were of considerable importance for proprioception, kinaesthesia and reflex regulation of movements, leaving joint receptor afferents to provide nonspecific facilitation of spinal pathways and to initiate joint protective reflexes.

Over the last two decades, however, the traditional view of the functional role of joint mechanoreceptors has been re-examined. In a number of recent studies, experimental data have been presented which strongly suggest that joint afferents may participate in more sophisticated functions. For instance, the proportion of mid-range units (i.e. joint afferents active at intermediate joint positions) seems to be significantly larger during passive joint movements than originally believed (e.g. Ferrell, 1980; Baxendale & Ferrell, 1983) and, more importantly, the activity range of joint mechanoreceptors appears to be considerably larger during active as compared to passive joint movements (e.g. Marshall & Tatton,

1990). Moreover, by using novel techniques permitting measurement of the ability to detect small joint movements and changes in joint angles, convincing evidence has been presented for a significant contribution of joint receptors to movement and position sense (e.g. Barrett, Cobb & Bentley, 1991; Ferrell & Craske, 1992). In this context it should be recalled that low-threshold joint afferents have projections to different supraspinal structures, including the somatosensory cortex and the thalamus (for references, see Johansson & Sjölander, 1993). In man, microstimulation of single joint afferents produced sensations of joint displacement in some instances (Macefield, Gandevia & Burke, 1990), whereas microstimulation of spindle afferents evoked no specific sensations unless the stimulation caused muscle movement. Finally, recent investigations have shown that low threshold joint afferents can induce potent effects on α-motoneurones as well as on γ-motoneurones and different types of interneurones (for references, see Johansson & Sjölander, 1993), indicating that joint mechanoreceptors may take part in motor control via pathways other that those projecting directly to α-motoneurones.

Already in the 1960's there was speculation on the possibility that an important function of articular mechanoreceptors might be to provide the CNS with feedback, via reflex actions on γ-motoneurones, which is necessary for accurate co-ordination of muscle tone in posture and movements (Freeman & Wyke, 1967). Although this proposal was based on a line of indirect evidence, it has later on been experimentally showed that joint afferents may cause reflex effects on the γ-muscle spindle system which are both potent and frequent. The aim of this paper is to summarise briefly available data on effects from joint receptor afferents on the γ-muscle-spindle system and to discuss some possible functional interpretations of these effects.

ELECTRICAL STIMULATION OF JOINT AFFERENTS

Effects on γ-motoneurones

In the majority of studies aimed at investigating reflex effects from joint receptors on spinal neurones, graded electrical stimulation of the posterior articular nerve of the knee joint (PAN) in the anaesthetised or decerebrate cat has been used to activate joint afferents. With this method, Voorhoeve & van Kanten (1962) were the first to show that afferents emanating from joint tissue can induce reflex actions on γ-motoneurones. They reported excitatory effects on gastrocnemius γ-fibres at high intensity stimulation of the PAN. Later, Grillner, Hongo & Lund (1969) made micro-electrode recordings from γ-cells projecting to triceps surae (GS) and posterior biceps/semitendinosus (PBSt) muscles, and observed that both GS and PBSt γ-cells could be excited by electrical activation of high threshold PAN afferents (above twice the nerve threshold). In addition, some of the GS cells showed inhibitory effects during high intensity stimulation of the PAN. The first evidence of low threshold PAN effects (1.1-1.4 times the nerve threshold) on γ-motoneurones appeared in the mid 1980's (Johansson, Sjölander & Sojka, 1986). In this study micro-electrode recordings from γ-cells projecting to various extensor and flexor muscles in the cat hind limb were made. The cells were indirectly classified as dynamic or static based on their responses to electrical stimulation of the mesencephalic area for dynamic control. It was found that more than 90% of the static and the dynamic γ-motoneurones were responsive to stimulation of the PAN, some cells only from low- or high threshold PAN afferents and others from both low- and high threshold afferents. Excitatory and inhibitory effects were observed on both static and dynamic, and on flexor as well as on extensor, γ-motoneurones. Among flexor γ-cells excitatory effects predominated, while for extensor γ-cells excitation and inhibition occurred approximately equally frequently. The shortest latencies for excitatory effects were compatible with a trisynaptic spinal pathway, while inhibitory effects seemed to be mediated through longer pathways. The presence of both excitatory and inhibitory reflex actions on extensor γ-motoneurones at low stimulation intensities were recently confirmed in a study by Baxendale, Davey, Ellaway & Ferrell (1992), in which electrical stimulation of the PAN was applied during recordings from GS γ-efferents. Based on the latencies of the effects and the occurrence of complex response patterns on single γ-efferents (i.e., various combinations of late and early excitations and inhibitions), it was suggested that rapidly and slowly adapting group II PAN afferents may have different effects on individual γ-motoneurones.

Effects on muscle spindle afferents

The occurrence of reflex actions on γ-efferents and γ-cells upon electrical stimulation of the PAN do not necessarily imply that such effects would induce alterations in the sensitivity of muscle spindles, which of course is a prerequisite for considering these effects as functionally significant. In 1978, McIntyre, Proske & Tracey (1978a) reported that the discharge rate of primary spindle afferents from cat hind limb muscles was increased as a result of electrical stimulation of the PAN at stimulation intensities above twice the nerve threshold. Although it was suggested that the effects were mediated mainly via static fusimotor neurones, no safe conclusion could be made regarding the type of fusimotor neurone primarily involved since the experiments were made at constant length of the receptor bearing muscles. The observation that fusimotor reflexes evoked by electrical activation of knee joint afferents might be potent enough to change significantly the length sensitivity of muscle spindles was confirmed by Johansson, Sjölander & Sojka (1988). In this study, however, significant alterations in the spindle afferent sensitivity was documented at lower stimulation intensities (1.5 times the nerve threshold), and based on responses to sinusoidal stretching of the spindle containing muscle it was concluded that the effects were caused by reflex actions on predominantly dynamic fusimotor neurones.

In summary, by using electrical stimulation of the PAN in the anaesthetised or decerebrate cat, it has been shown that joint afferents can evoke both excitatory and inhibitory reflexes on static and dynamic γ-motoneurones to extensor as well as to flexor muscles. Moreover, it has been found that these reflex effects are sufficiently powerful to significantly alter the discharge rate of primary muscle spindle afferents. The observations that low intensity stimulation can evoke clear-cut reflexes may indicate that these effects are elicited from large diameter joint afferent fibres emanating from Golgi tendon organ-like endings, Pacin-form corpuscles and/or Ruffini endings. In this context it might be inferred that the PAN sometimes contains muscle spindle afferents and Golgi tendon organ afferents, originating from the popliteus muscle, which potentially might have caused the observed low threshold fusimotor effects. However, the proportion of muscle afferents in the PAN is most probably not as large as indicated based on succinylcholine induced activation of PAN afferents (Burgess & Clark, 1969), since it has been demonstrated that this substance also excites large diameter fibres emanating from articular tissue (Ferrell, 1980). It should also be noted that while the PAN appears to contain popliteus muscle afferents in about 50% of cats (McIntyre, Proske & Tracey, 1978b), Johansson et al. (1986) observed low threshold reflexes on γ-motoneurones in nearly all experiments and, furthermore, that Baxendale et al. (1992) found low threshold effects on γ-efferents in experiments where the popliteus muscle was destroyed.

PHYSIOLOGICAL STIMULATION OF JOINT RECEPTORS

Although graded electrical stimulation of peripheral nerves has been, and still is, a most valuable tool in attempts to reach a better understanding of the sensory-motor functions of different types of receptor afferents, the method has several distinct limitations. For instance, due to the considerable overlap in activation thresholds between afferents from different receptor types it is usually not possible to selectively activate afferents from specific receptor types. It may also be difficult, or impossible, to induce selective activation of receptor afferents within a certain tissue or structure. The PAN, for example, innervates the knee joint capsule, the intra-articular fat pads, the posterior cruciate ligament, the both collateral ligaments, the menisci, and sometimes the popliteus muscle. Another limitation is that electrical stimulation causes synchronous activation of a number of afferent fibres, often emanating from several types of endorgans, which is unlikely to occur during most natural limb movements. Physiological stimulation, on the other hand, permits natural activation of specific types of receptors as well as of endorgans within defined tissues or structures.

Effects on γ-efferents

In a study aimed at investigating effects on motoneurones during acute joint inflammation, He, Proske, Schaible & Schmidt (1988) recorded from PBSt γ- and α-axons before and after induction of acute knee joint inflammation in the anaesthetised cat. It was found that inflammation caused increased resting activity and responsiveness to passive knee joint movements on both γ- and α-efferents. Before induction of the inflammation, 41% of the γ-motoneurones and 14% of the α-motoneurones were responsive to the joint movements, indicating a higher reflex responsiveness of the γ-motoneurones as compared to α-motoneurones. However, since the hind limb was kept largely intact it was not possible to asses to what extent joint receptor afferents were accountable for the reflex effects. Yet, in a more recently study on the decerebrate cat evidence was found that selective physiological activation of joint receptors can evoke reflex effects on γ-efferents (Baxendale et al., 1992). In this study mechanosensitive receptors within the posterior aspect of knee joint capsule were excited by applying local capsular indentations by means of a small diameter probe driven by a vibrator coil. Peri-stimulus time histograms of the responses clearly showed that activation of small numbers of joint receptors might evoke both excitatory and inhibitory reflex actions on GS γ-efferents.

Effects on muscle spindle afferents

The first demonstration that physiological activation of mechanoreceptors in joint tissue can evoke fusimotor reflex effects which are powerful enough to significantly change the stretch sensitivity of muscle spindle afferents was made in our laboratory. In experiments on anaesthetised cats, muscle spindle afferent responses to sinusoidal stretching of the receptor bearing muscle (GS and/or PBSt), elicited during ongoing and in absence of physiological joint receptors stimulation, were recorded and compared in order to evaluate the size and the type (i.e., static, dynamic or mixed) of induced fusimotor reflex effects. By applying maintained pressure on different joint capsules in extensively denervated hind limbs it was shown that mechanosensitive joint receptors in the ipsilateral knee joint and in the contralateral knee and ankle joints potently influence the dynamic and the static sensitivity of muscle spindle afferents from the ipsilateral GS muscle (Appelberg, Hulliger, Johansson & Sojka, 1981; Johansson, Sjölander & Sojka, 1988). Subsequently, fusimotor reflex effects on GS and PBSt muscle spindle afferents evoked by selective stimulation of different ipsilateral knee joint ligaments were investigated (Johansson, Sjölander, Sojka & Wadell,

1989; Johansson, Lorentzon, Sjölander & Sojka, 1990; Sjölander, Djupsjöbacka, Johansson, Sojka & Lorentzon, 1994). Mechanosensitive ligament receptors were activated by stretching the intact ligaments (i.e., anterior and posterior cruciate ligaments, and lateral and medial collateral ligaments) by applying transverse traction forces, while simultaneous recordings were made from two to four single muscle spindle afferents. Figure 1 shows an example of fusimotor reflex effects induced on two simultaneously recorded spindle afferents (one GS primary and one PBSt secondary afferent) during loading of different knee joint ligaments. The reflexly evoked changes in mean rate of firing (fitted mean) and in depth of modulation of the sinusoidal responses of the spindle afferents indicate that the primary spindle afferents exhibited mixed dynamic and static fusimotor effects, with a predominance of the dynamic component, whereas the secondary spindle afferent showed fusimotor effects which most certainly were mediated via static fusimotor neurones (cf. Johansson et al., 1990; Sjölander et al., 1994). The most striking finding in this series of experiments was the remarkably high responsiveness to non-nociceptive stretch of individual knee joint ligaments (both cruciate and collateral ligaments). Moreover, although static, dynamic and mixtures of static and dynamic fusimotor effects were elicited by all four ligaments, a predominance of dynamic fusimotor reflex effects were encountered by stretch of the anterior cruciate ligament. On the other hand, mainly static reflex effects were evoked by stretch of the posterior cruciate ligament, and dynamic and mixed dynamic and static

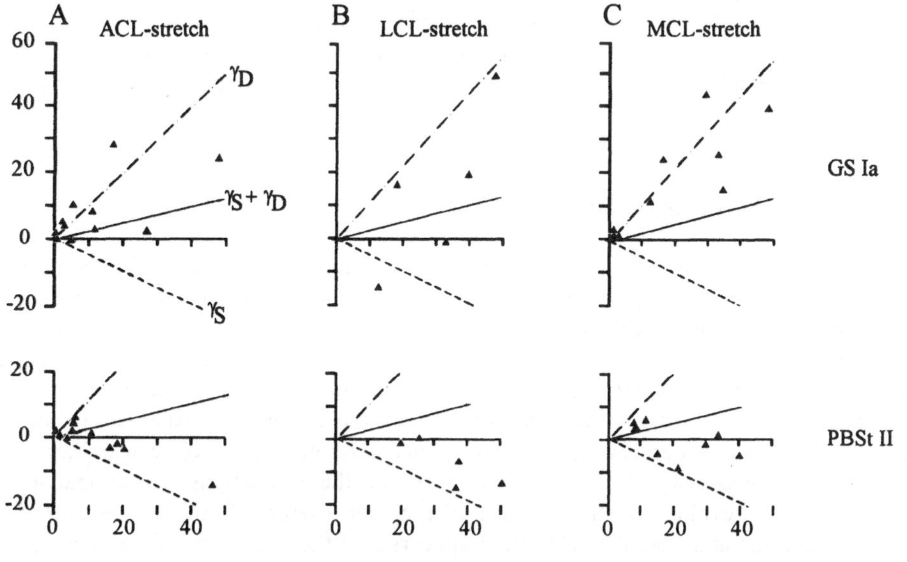

Figure 1. Fusimotor reflex effects simultaneously evoked on a GS primary and a PBSt secondary muscle spindle afferent by stretching the anterior cruciate ligament (A), the lateral collateral ligament (B) and the medial collateral ligament (C) of the knee joint in the anaesthetised cat. In each diagram, changes in sinusoidal responses for a number of control-test pairs (11 in A, 5 in B, 9 in C) are displayed. Each control and test response represented averaged values in mean rate of firing (fitted mean) and in depth of modulation calculated for 10 consecutive sinusoidal stretching cycles, at 1 Hz for the GS and at 0.9 Hz for the PBSt muscles (peak-to-peak amplitude = 2 mm; mean muscle length = 2 mm below maximum physiological length). The slopes of the reference lines (labelled γD, γS+γD and γS) displayed in A give the mean change in depth modulation/mean change in fitted mean of sinusoidal responses of soleus primary muscle spindle afferents elicited by selective and combined electrical activation of dynamic and static γ-efferents (Hulliger, Matthews & Noth, 1977). A and C are modified from illustrations in Johansson *et al.*, 1990 and Sojka, Sjölander, Johansson & Djupsjöbacka, (1991) respectively.

effects by stretch of the collateral ligaments. The high responsiveness and the potent fusimotor effects evoked by knee joint ligaments might be somewhat surprising since it would be expected that the ligaments contain a rather limited number of stretch sensitive mechanoreceptors. Nevertheless, these findings are in accordance with the demonstration of clear-cut reflex effects on γ-efferents of physiological activation of small numbers of joint receptors in the knee joint capsule (Baxendale et al., 1992), and seem to indicate that the synaptic coupling between joint mechanoreceptors and lumbar fusimotor neurones is impressively strong.

The observations briefly described above show that physiological activation of mechanoreceptors within different structures of the knee joint evoke static and dynamic fusimotor reflexes which result in clear-cut alterations in the sensitivity of primary and secondary muscle spindle afferents. The effects were elicited by applying light pressure on the joint capsule and by stretching the ligaments with small forces, suggesting that receptors with low threshold to mechanical stimuli were activated. Both the capsule and the ligaments in the cat knee joint are known to accommodate endorgans with low mechanical thresholds, i.e. Ruffini endings and Pacini-form corpuscles (for references, see Johansson & Sjölander, 1993). Ruffini endings are supposed to be largely slowly adapting whereas Pacini-form corpuscles are rapidly adapting. These receptor properties, together with the fact that the knee joint ligaments were stimulated by applying a constant load for 20-25 seconds, indicate that Ruffini endings most likely were accountable for the fusimotor effects obtained (Johansson et al., 1989; Johansson et al., 1990; Sjölander et al., 1994). In the investigation by Baxendale et al. (1992), repetitive indentations of the capsular surface were induced, which most certainly activated rapidly adapting Pacinian corpuscles and perhaps also Ruffini endings.

POSSIBLE FUNCTIONAL IMPLICATIONS

To what extent impairment in motor control, proprioception and kinaesthesia observed following disturbances in the feedback from joint mechanoreceptors is mediated via the γ-muscle spindle system or via other spinal neural systems and/or supraspinal structures remains conjectural. Nevertheless, if it is accepted that information mediated by muscle spindle afferents plays a significant role in co-ordination of normal movements and in the transmission of cues for movement and position sense it seems reasonable to assume that receptor afferents and descending pathways which through the γ-loop can influence the muscle spindle sensitivity also are of relevance for these functions. In comparison to fusimotor reflexes evoked by various types of receptor afferents from muscles, joints and skin in decerebrate and anaesthetised cats it appears as if the fusimotor effects elicited from joint mechanoreceptors are particularly potent (for review, see Johansson & Sjölander, 1993). Necessary requirements for considering these findings as functionally relevant are of course that the reflex pathways from joint afferents to the γ-muscle-spindle system are open under natural conditions and that the gain in these pathways is sufficiently high. In both animals and man it has been indicated that the gating of reflex pathways to γ-motoneurones from cutaneous afferents may vary according to the task that is performed (Murphy, Stein & Taylor, 1984; Burke & Gandevia, 1992). It has also been shown that the gain in sensory transmission is dependent on the novelty or difficulty of the task, and that the gain in one sensory pathway may change differently from that in another pathway when the task is altered (for review, see Prochazka, 1989). Although it has still to be experimentally demonstrated it seems likely that the spinal interneuronal networks can be controlled to predominantly transmit information from a certain type of receptor (e.g., joint, cutaneous or muscle receptors), or from particular combinations of various receptors, so as to establish sensory feedback optimal for muscle co-ordination, muscle stiffness, proprioception and kinaesthesia in different tasks and situations.

ACKNOWLEDGEMENTS

Supported by the Swedish Work Environment Fund and Centrum för Idrottsforskning.

REFERENCES

APPELBERG, B., HULLIGER, M., JOHANSSON, H. & SOJKA, P. (1981) Reflex activation of dynamic fusimotor neurones by natural stimulation of muscle and joint receptor afferent units. In *Muscle Receptors and Movements*, eds. TAYLOR, A. & PROCHAZKA, A., pp. 149-161. Macmillan, London.

BARRETT, D. S., COBB, A. G. & BENTLEY, G. (1991) Joint proprioception in normal, osteoarthritic and replaced knees. *J. Bone Joint Surg.* **73B,** 53-56.

BAXENDALE, R. H., DAVEY, N. J., ELLAWAY, P. H. & FERRELL, W. R. (1992) The interaction between joint and cutaneous afferent input in the regulation of fusimotor neurone discharge. In *Muscle Afferents and Spinal Control of Movement*, eds. JAMI, L., PIERROT-DESEILLIGNY, E. & ZYTNICKI, D., pp. 95-104. Pergamon Press, Oxford.

BAXENDALE, R. H. & FERRELL, W. R. (1983) Discharge characteristics of the elbow joint nerve of the cat. *Brain Res.* **161**, 195-302.

BURKE, D. & GANDEVIA, S. C. (1992) Selective activation of fusimotor neurones innervating human tibialis anterior. In *Muscle Afferents and Spinal Control of Movement*, eds. JAMI, L., PIERROT-DESEILLIGNY, E. & ZYTNICKI, D., pp. 151-156. Pergamon Press, Oxford.

CLARK, F. J. & BURGESS, P. R. (1975) Slowly adapting receptors in the cat knee joint: Can they signal joint angle? *J. Neurophysiol.* **38**, 1448-1463.

FERRELL, W. B. (1980) The adequacy of stretch receptors in the cat knee joint for signalling joint angle throughout a full range of movement. *J. Physiol.* **299**, 85-99.

FERRELL, W. B. & CRASKE, B. (1992) Contribution of joint and muscle afferents to position sense at the human proximal interphalangeal joint. *Exp. Physiol.* **77**, 331-341.

FREEMAN, M. A. R. & WYKE, B. (1967) Articular reflexes at the ankle joint: An electromyographic study of normal and abnormal influences of ankle joint mechanoreceptors upon reflex activity in the leg muscles. *Brit. J. Surg.* **54**, 990-1001.

GRILLNER, S., HONGO, T. & LUND, S. (1969) Descending monosynaptic and reflex control of γ-motoneurones. *Acta physiol. scand.* **75**, 592-613.

HE, X., PROSKE, U., SCHAIBLE, H.-G. & SCHMIDT, R. F. (1988) Acute inflammation of the knee joint in the cat after responses of flexor motoneurones to led movements. *J. Neurophysiol.* **59**, 326-340.

HULLIGER, M., MATTHEWS, P. B. C. & NOTH, J. (1977) Effects of combining static and dynamic fusimotor stimulation on the response of muscle spindle primary endings to sinusoidal stretching. *J. Physiol.* **267**, 839-856.

JOHANSSON, H., LORENTZON, R., SJÖLANDER, P. & SOJKA, P. (1990) The anterior cruciate ligament. A sensor acting at the γ-muscle-spindle systems of muscles around the knee joint. *Neuro-Orthopedics* **9**, 1-21.

JOHANSSON, H. & SJÖLANDER, P. (1993) Neurophysiology of joints. In *Mechanics of Human Joints. Physiology. Pathophysiology, and Treatment*, eds. WRIGHT, V. & RADIN, E. L., pp. 243-290. Marcel Dekker, New York.

JOHANSSON, H., SJÖLANDER, P. & SOJKA, P. (1986) Actions on γ-motoneurones elicited by electrical stimulation of joint afferent fibres in the hind limb of the cat. *J. Physiol.* **375**, 137-152.

JOHANSSON, H., SJÖLANDER, P. & SOJKA, P. (1988) Fusimotor reflexes in triceps surae muscle elicited by natural and electrical stimulation of joint afferents. *Neuro-Orthopedics* **6**, 67-80.

JOHANSSON, H., SJÖLANDER, P., SOJKA, P. & WADELL, I. (1989) Reflex actions on the γ-muscle-spindle systems of muscles acting at the knee joint elicited by stretch of the posterior cruciate ligament. *Neuro-Orthopedics* **8**, 9-21.

MACEFIELD, G., GANDEVIA, S.G. & BURKE, D. (1990) Perceptual responses to microstimulation of single afferents innervating joints, muscles and skin of the human hand. *J. Physiol.* **429**, 113-129.

MARSHALL, K. W. & TATTON, W. G. (1990) Joint receptors modulate short and long latency muscle responses in the awake cat. *Exp. Brain Res.* **83**, 137-150.

MCINTYRE, A. K., PROSKE, U. & TRACEY, D. J. (1978a) Fusimotor responses to volleys in joint and interosseous afferents in the cat's hind limb. *Neurosci. Lett.* **10**, 287-292.

MCINTYRE, A. K., PROSKE, U. & TRACEY, D. J. (1978b) Afferent fibres from muscle receptors in the posterior nerve of the cat's knee joint. *Exp. Brain Res.* **33**, 415-424.

MURPHY, P. R., STEIN, R. B. & TAYLOR, J. (1984) Phasic and tonic modulation of impulse rates in motoneurons during locomotion in premammillary cats. *J. Neurophysiol.* **52,** 228-243.

PROCHAZKA, A. (1989) Sensorymotor gain control: a basic strategy of motor control. *Prog. Neurobiol.* **33,** 281-307.

SJÖLANDER, P., DJUPSJÖBACKA, M., JOHANSSON, H., SOJKA, P. & LORENTZON, R. (1994) Can receptors in the collateral ligaments contribute to knee joint stability and proprioception via effects on the fusimotor-muscle-spindle system? An experimental study in the cat. *Neuro-Orthopedics* **15**, 65-80.

SOJKA, P., SJÖLANDER, P., JOHANSSON, H. & DJUPSJÖBACKA, M. (1991) Influences from stretch-sensitive receptors in the collateral ligaments of the knee joint on the γ-muscle-spindle systems of flexor and extensor muscles. *Neurosci. Res.* **11**, 55-62.

VOORHOEVE, P. E. & VAN KANTEN, R. W. (1962) Reflex behaviour of fusimotor neurones of the cat upon electrical stimulation of various afferent fibres. *Acta Physiol. Pharm. Neerland.* **10**, 391-407.

CUTANEOUS REFLEX CONTROL OF FUSIMOTOR NEURONES

P.R. Murphy and H.A. Martin

Division of Neurobiology, Medical School
University of Newcastle upon Tyne
Newcastle, UK

INTRODUCTION

The fusimotor (γ) system receives a wide variety of inputs from peripheral, spinal and supraspinal structures (for review, see Hulliger, 1984). Although many previous investigations, in animal preparations, have shown that cutaneous afferents produce potent reflex effects on hindlimb γ-motoneurones, no consistent pattern of response has emerged and a functional interpretation of such sensorimotor control has been difficult to achieve (Hunt, 1951; Eldred & Hagbarth, 1954; Hunt & Paintal, 1958; Voorhoeve & van Kanten, 1962; Grillner, Hongo & Lund, 1969; Catley & Pascoe, 1978; Bessou, Joffroy & Pagés, 1981; Johansson & Sojka, 1985; Davey & Ellaway, 1989; Murphy & Hammond, 1991, 1992). The reported responses have ranged from a "simple" flexor reflex pattern (i.e. flexor excitation, extensor inhibition; Hunt, 1951) to those that appeared individualised for a given γ-motoneurone (Johansson & Sojka, 1985). Various factors may have contributed to the diversity of effects in these studies including differences in preparation, afferent input and γ sampling. In addition, the presence of two functionally distinct types of fusimotor neurone, static and dynamic (Matthews, 1962), is likely to have contributed since direct studies of the reflex behaviour of the γ system have largely involved recordings from unclassified units. In the present article we review results from a decerebrate cat preparation in which cutaneous reflex responses to electrical stimulation have been investigated in the resting state and during locomotion.

METHODS

Full details of experimental procedures have been published recently (Murphy & Hammond, 1991; 1992) and will be summarised only briefly here. Adult cats were anaesthetised with halothane delivered in a mixture of 70% oxygen and 30% nitrous oxide. The nerve supply of the left hindlimb below the hip was sectioned, except for the common lateral gastrocnemius-soleus nerve. The animals were placed in a stereotaxic head holder over a treadmill with pins at the iliac crests and clamps on the left knee and ankle. Decerebration was performed by a section angled from just rostral to the superior colliculus to just in front of the mammillary bodies. Brain tissue above the section was removed and anaesthesia discontinued. The present preparation, unlike the classical intercollicular animal, is capable of generating periods of regular locomotor activity. Functionally single γ-efferents were recorded from the cut MG nerve on twin platinum wire electrodes. Units were identified as γ-motoneurones on the basis of their conduction velocities (< 40m/s) and discharge characteristics (Murphy, Stein & Taylor, 1984). EMG was recorded via a pair of silver wires inserted in the lateral gastrocnemius muscle.

In separate experiments, generally, the cut end of a cutaneous nerve was electrically stimulated either with single shocks or a brief train. In both cases stimuli (0.1ms width, up to 20T) were applied through bipolar platinum wire electrodes and the in-going volley was continuously monitored using a sciatic cuff electrode. Single shock stimulation was applied either to the sural or the medial plantar nerves in the resting state and a ditigal computer used to construct peri-stimulus time histograms (PSTHs) and their cumulative sums (CUSUMs; Ellaway, 1978). Stimulus trains were applied to the sural nerve at rest and during locomotion. For a given γ-efferent recording, stimulation (generally 5 stimuli, 50/s) was applied every 4s in both states at a range of stimulus intensities (1.5, 2, 3, 5, 10, 20T). Responses were assessed by calculating the change in mean γ rate (impulses/s) which occurred during the 100ms after stimulus onset compared either to the preceding 100ms, for responses at rest, or to the equivalent period of the preceding control step cycle, for responses during locomotion. During locomotion the onsets of the periods of measurement occurred at the same time (t) in the stimulated (S) and control (C) step cycles (Figure 1). γ-motoneurones were classified as static or dynamic indirectly, on the basis of their locomotor and/or resting discharge characteristics according to previously described criteria (Murphy et al., 1984; Murphy & Hammond, 1991, 1992).

RESULTS

Sural nerve

Single shock stimulation (up to 20T) of the sural nerve (which supplies the skin of the lateral posterior leg) in the resting state produced short latency excitation of MG static (n = 9) and dynamic (n = 7) γ-efferents. Although net excitation was the rule, with no sign of mixed or purely inhibitory effects, we cannot exclude the possibility of simultaneous inhibitory inputs which were outweighed by concomitant excitation. Effects appeared at low intensities (generally ≤ 2T) and increased with stimulus strength. The central delays of γ responses were estimated by subtracting afferent and efferent conduction times from the overall latency. The resultant values therefore represent intraspinal conduction time and synaptic delay. The central delays of the responses of static (3.0±1.1ms, mean ± SD) and dynamic (3.4±1.0ms) γ-motoneurones were not significantly different ($P > 0.1$, Students' t test) and are consistent with spinal oligosynaptic pathways, in agreement with previous results (Johansson & Sojka, 1985).

In the above experiments both types of fusimotor neurone showed net excitation to sural afferent stimulation in the resting state. To investigate the possibility that the reflex profile may alter during movement we examined the effect of sural stimulation on MG γ-efferents during locomotion. Brief trains of stimulation were applied during the step cycle at 1.5, 2, 3, 5 10 and 20T for each unit studied and responses during locomotion were compared to those at rest (see **METHODS**). As with the experiments involving single shocks we again found that sural stimulation produced graded, low threshold (≤ 2T) excitation of both types of γ-efferent in the resting state. During locomotion, however, the responses of static and dynamic γ-efferents differed. Dynamic neurones (n = 8) showed excitation (Figure 1) and there was no significant difference ($P > 0.1$) between mean responses at rest and those during locomotion (Figure 3). In contrast, the responses of static units were reduced, or even abolished, during locomotion compared to the resting state (Figures 2 & 3). An abolition of response during locomotion was observed for seven static γ -efferents. For the other eight units responses were reduced ($P \leq 0.05$) during locomotion, with stimuli ≤ 3T generally having no effect. Although the responses of static, but not dynamic, γ-efferents were task dependent, in neither case was their magnitude obviously related to step cycle phase, or γ rate, and responses occurring during or between homonymous EMG bursts were not significantly different ($P > 0.1$).

Medial plantar nerve

The previous results, concerning sural nerve, may be representative of the general pattern of cutaneous effects in the present preparation. On the other hand, it may transpire that the cutaneous reflex control of the fusimotor system is rather more specialised. We have begun to examine this problem by studying other cutaneous inputs. In recent experiments we have investigated the effects of stimulation (single shock, up to 20T) of the medial plantar nerve, which supplies the sole of the foot, on MG fusimotor neurones in the resting state. In marked contrast to our results with sural, we found complex patterns of response for both types of γ-efferent. Thus dynamic neurones (n = 8) showed powerful, short latency (15±1.2ms) spinal inhibition followed by weaker excitation. Static units showed two patterns of response. Some units (7) were purely excited at medium latency

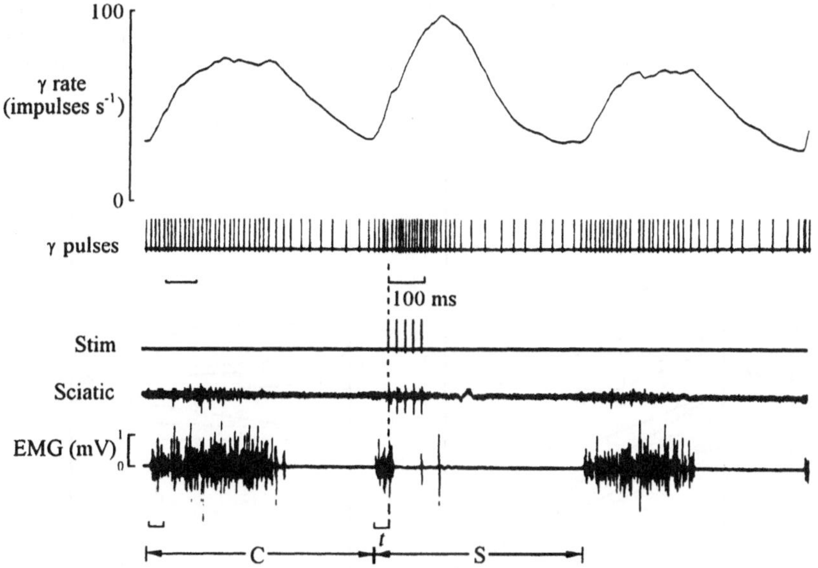

Figure 1. Response of a dynamic γ-efferent to stimulation (20T, 5 stimuli, 50/s) of the sural nerve during locomotion. Fusimotor excitation was accompanied by simultaneous inhibition of on-going EMG activity. The onsets of the measurement periods (see **METHODS**) occurred at the same time (*t*) in the stimulated (S) and control (C) step cycles. Fusimotor rate was produced by passing standard pulses to a leaky integrator (time constant = 100ms).

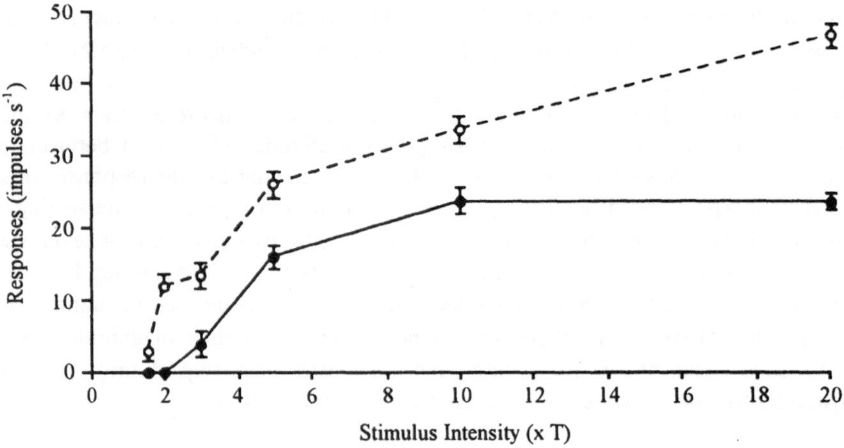

Figure 2. Example of a static γ-efferent in which mean responses to sural nerve stimulation (5 stimuli, 50/s) during locomotion (•) were reduced ($P < 0.05$ for 1.5T, otherwise $P < 0.01$) compared to the resting state (o) at each stimulus intensity. Vertical lines, S.E.M.

(39.9 ±12.2ms) while the remainder (8) showed mixed effects in which short latency (18± 3.6ms) spinal inhibition was followed by stronger excitation. Inhibitory and excitatory responses were generally present at 2T.

DISCUSSION

The functional implications of the present data will not be considered in detail here (see Murphy & Hammond, 1991, 1992; Murphy, Martin & Hammond, 1994); rather the discussion will focus on some general observations. Our results confirm earlier

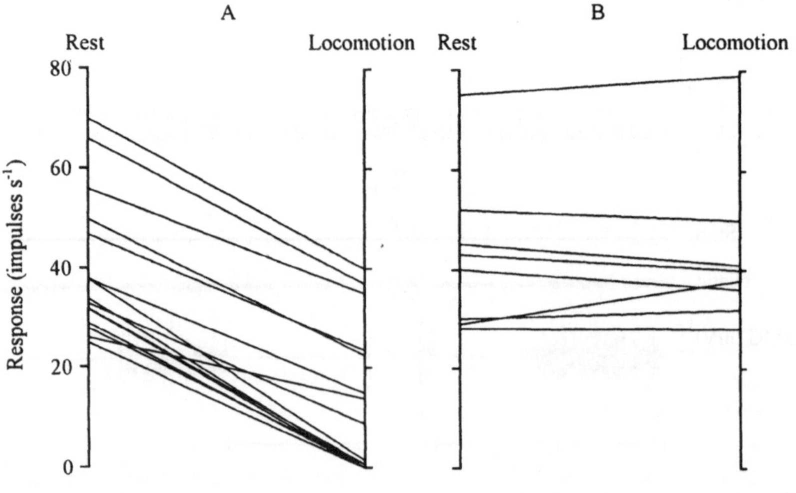

Figure 3. Mean responses of twenty-three individual static (A) and dynamic (B) γ-efferents at rest and during locomotion to sural nerve stimulation (20T, 5 stimuli, 50/s). Each line represents one unit. Responses of static, but not dynamic, fusimotor neurones were reduced, or even abolished, during locomotion compared to the resting state.

investigations (see **INTRODUCTION**), mainly of hindlimb muscles in non-behaving preparations, which indicate that low threshold (i.e. $\leq 2T$) cutaneous afferents are capable of exerting a significant level of control over the fusimotor system. However, previous studies have not revealed a consistent pattern of response. The present data, from the resting state, agree with this observation to the extent that ankle extensor static and dynamic γ-efferents did not show a uniform reflex profile to cutaneous inputs. Nevertheless, the variations in effect could be related to the source of the afferent input or the type of unit involved. Thus responses to sural nerve stimulation were relatively simple (excitation) while those to medial plantar were complex (excitation and mixed). Further, in the latter case, dynamic γ-efferents showed net inhibition while excitation dominated the responses of static neurones.

An additional factor in determining the cutaneous responses of fusimotor neurones appears to be behaviour. Thus, the reflex responses of static, but not dynamic, γ-efferents were task dependent, being reduced, or even abolished, during locomotion compared to the resting state. In this context it should be noted that, to date, there have been no direct recordings from identified and classified γ-efferents in freely behaving animals and we cannot be certain of the degree to which the pattern of cutaneous responses in reduced preparations is representative of the intact state. Most evidence relating to this problem comes from spindle afferent recordings from which fusimotor effects are inferred (for review, see Hulliger, 1984). Although most studies (e.g. Prochazka, 1983; Loeb, Hoffer & Marks, 1985) have stressed a lack of effect of innocuous stimulation this view may, in part, reflect the difficulties involved in deducing γ activity from spindle afferent discharge (e.g. Hoffer, Caputi, Pose & Griffiths, 1989). In addition, the present results indicate that the cutaneous control of fusimotor neurones may vary according to afferent source, γ type and behaviour. Thus evidence for such reflexes from spindle afferent recordings in freely moving animals may require specific testing procedures.

In conclusion, the present results suggest that the cutaneous reflex control of fusimotor neurones, though complex, is related to a number of functionally meaningful factors. Such diversity is consistent with a sophisticated control system and may have its basis in the "dual" capability of muscle spindle afferents (i.e. sensory/motor).

ACKNOWLEDGEMENTS

This work was supported by the MRC.

REFERENCES

BESSOU, P., JOFFROY, M. & PAGÉS, B. (1981) Efferents and afferents in an intact muscle nerve: background activity and effects of sural nerve stimulation in the cat. *J. Physiol.* **320**, 81-102.

CATLEY, D.M. & PASCOE, J.E. (1978) The reflex effects of sural nerve stimulation upon gastrocnemius fusimotor neurones of the rabbit. *J. Physiol.* **276**, 32P.

DAVEY, N.J. & ELLAWAY, P.H. (1989) Facilitation of individual γ-motoneurones by the discharge of single slowly adapting type 1 mechanoreceptors in cat. *J. Physiol.* **411**, 97-114.

ELDRED, E. & HAGBARTH, K.E. (1954) Facilitation and inhibition of gamma efferents by stimulation of certain skin areas. *J. Neurophysiol.* **17**, 59-65.

ELLAWAY, P.H. (1978) Cumulative sum technique and its application to the analysis of peri-stimulus time histograms. *Electroenceph. Clin. Neurophysiol.* **45**, 302-304.

GRILLNER, S., HOMGO, T. & LUND, S. (1969) Descending monosynaptic and reflex control of γ-motoneurones. *Acta physiol. pcand.* **75**, 592-613.

HOFFER, J.A., CAPUTI, A.A., POSE, I.E. & GRIFFITHS, R.I. (1989) Roles of muscle activity and load on the relationship between muscle spindle length and whole muscle length in the freely walking cat. *Prog. Brain Res.* **80**, 75-85.

HULLIGER, M. (1984) The mammalian muscle spindle and its central control. *Rev. Physiol. Biochem. & Pharm.* **101**, 1-110.

HUNT, C.C. (1951) The reflex activity of mammalian small nerve fibres. *J. Physiol.* **115**, 456-469.

HUNT, C.C. & PAINTAL, A.S. (1958) Spinal reflex regulation of fusimotor neurones. *J. Physiol.* **143**, 195-212.

JOHANSSON, H. & SOJKA, P. (1985) Actions on γ-motoneurones elicited by electrical stimulation of cutaneous afferent fibres in the hind limb of the cat. *J. Physiol.* **366**, 343-363.

LOEB, G.E., HOFFER, J.A. & MARKS, W.B. (1985) Activity of spindle afferents from cat anterior thigh muscles. III. Effects of external stimuli. *J. Neurophysiol.* **54**, 578-591.

MATTHEWS, P.B.C. (1962) The differentiation of two types of fusimotor fibre by their effects on the dynamic response of muscle spindle primary endings. *Q. J. Exp. Physiol.* **47**, 324-333.

MURPHY, P.R. & HAMMOND, G.R. (1991) The role of cutaneous afferents in the control of γ-motoneurones during locomotion in the decerebrate cat. *J. Physiol.* **434**, 529-547.

MURPHY, P.R. & HAMMOND, G.R. (1992) Short latency cutaneous reflex responses of γ-efferents in the decerebrate cat. *Exp. Brain Res.* **89**, 140-146.

MURPHY, P.R., MARTIN, H.A. & HAMMOND, G.R. (1994) The effect of medial plantar nerve stimulation on ankle extensor γ-efferents in the decerebrate cat. *J. Physiol.* **475**, 39P.

MURPHY, P.R., STEIN, R.B. & TAYLOR, J. (1984) Phasic and tonic modulation of impulse rates in γ-motoneurones during locomotion in premammillary cats. *J. Neurophysiol.* **52**, 228-243.

PROCHAZKA, A. (1983) Reflex regulation of fusimotor neurones: discussion. In *Reflex Organisation of the Spinal Cord and its Descending Control.* eds. PORTER, R. & REDMAN, S. pp25. Roche, Canberra City.

VOORHOEVE, P.E. & VAN KANTEN, R.W. (1962) Reflex behaviour of fusimotor neurones of the cat upon electrical stimulation of various afferent fibres. *Acta Physiol. Pharm. Neerland.* **10**, 391-407.

REGULATION OF THE DYNAMIC AND STATIC SENSITIVITY OF GASTROCNEMIUS-SOLEUS MUSCLE SPINDLE AFFERENTS BY JOINT AND CUTANEOUS AFFERENTS IN THE CAT

P.H. Ellaway[1], N.J. Davey[1], M.J. Ljubisavljevic[2] and N.P. Anissimova[3]

[1]Department of Physiology
Charing Cross & Westminster
Medical School, London W6 8RF, UK
[2]Institute for Medical Research
Dr Subotica 4, Beograd, Yugoslavia
[3]Pavlov Institute of Physiology
St. Petersburg, Russia

INTRODUCTION

The regulation of dynamic and static fusimotor drive to muscle spindles by cutaneous and articular somatic sensory receptors has been studied in the cat mainly by the use of electrical stimulation of peripheral nerves (see Hulliger, 1984). Since electrical stimulation excites axons of different sensory modalities the significance of the findings has been uncertain. Natural stimulation has provided some specific examples of regulation of the fusimotor system by skin and joint receptors. The cutaneous, slowly-adapting, type-1 mechanoreceptor is known to exert a powerful excitatory action on gamma motoneurones (Davey & Ellaway, 1989) but the dynamic or static nature of the effect is unknown. Stretch of the cruciate ligaments, which occurs naturally in movements of the knee joint, results in reflex regulation of both the dynamic and static sensitivity of spindle primary endings (Sojka, Johansson, Sjölander, Lorentzon & Djupsjöbacka, 1989; Johansson, Sjölander & Sojka, 1990). With reference to the gastrocnemius/soleus muscles, dynamic effects were more evident to stretch of the posterior cruciate ligament (Sojka et al., 1989) and static effects to stretch of the anterior ligament (Johansson et al., 1990). Stimulation of receptors within the knee joint, achieved by mechanical distortion of the posterior aspect of the joint capsule, was found to excite unclassified gamma motoneurones to the gastrocnemius/soleus muscles (Baxendale, Davey, Ellaway & Ferrell, 1990; 1992). In the same study, brushing the skin at the heel excited gamma motoneurones but exerted a profound inhibition of the excitatory action of electrical stimulation of the posterior articular nerve to the knee joint.

The diverse and interactive effects of peripheral afferent discharge on the fusimotor system make it impossible, as yet, to propose a clear model of these regulatory pathways. An additional complication is that the spinal reflex actions of specific afferent modalities appear to be influenced strongly by centres in the brain (see Hulliger, 1984; Baker, Catley, Davey & Ellaway, 1991). The present experiments were designed to improve our understanding of the influence of skin and joint afferents on dynamic and static gamma motoneurones and to examine the extent to which such regulation may be controlled by supraspinal activity. We chose to study the fusimotor supply of the gastrocnemius/soleus muscles and to stimulate afferents from skin at the insertion of those muscles (the sural nerve field at the heel) and joint afferents from the knee joint over which the gastrocnemius acts. These sensory and motor components will be integrated during normal use of the limb. A preliminary account of some of this work has been published (Ellaway, Davey Ljubisaljevic & Anissimova, 1994).

METHODS

Experiments were performed on cats either under α-chloralose anaesthesia (initial dose 80 mg/kg) or decerebrated at the mid-collicular level under halothane in oxygen anaesthesia. Full details of the preparation and afferent recording are given in Davey & Ellaway (1989). The left hind limb was denervated widely below the level of the hip. Nerves left intact were the sural nerve, the posterior articular nerve (PAN) supplying the knee joint and the nerve supply to either the gastrocnemius medialis, the gastrocnemius lateralis or the combined gastrocnemius lateralis and soleus muscle. The corresponding gastrocnemius/soleus (GS) muscle tendon was sectioned at its insertion on the calcaneum and tied to a servo-controlled electromagnetic puller.

The discharges of GS muscle spindle afferents were recorded from dorsal root ganglia using tungsten electrodes. This method of recording ensured that minimum disruption occurred to afferents from the skin of the heel (sural) or the knee joint (PAN). Muscle spindle afferents were identified as having high sensitivity to stretch of the parent muscle and evident spontaneous changes in discharge or reflex fusimotor action to pinna or skin stimulation. Primary spindle endings were distinguished from secondary endings by their faster axonal conduction velocity (> 70 m/s) and higher sensitivity to vibration.

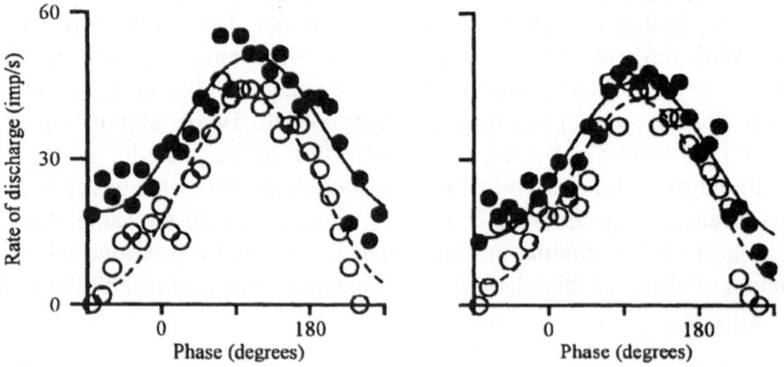

Figure 1. The effects of mechanical stimulation of the knee joint capsule, or of skin over the lateral aspect of the heel, on the averaged responses of a primary spindle afferent to sinusoidal stretch of the gastrocnemius lateralis muscle in a decerebrate cat with intact spinal cord. Open symbols: control. Closed symbols: stimulation. Note the increase in mean rate of discharge and reduction in peak-to-peak modulation caused by both stimuli.

The GS muscle was stretched sinusoidally by up to ± 1 mm at 0.98 Hz at a resting length of the muscle close to the maximum natural length. Electromyography ensured that the response of a muscle spindle was not influenced by the contraction of skeletal muscle motor units. Histograms of the response of spindle afferents to the sinusoidal stretch were constructed using 20 cycles. A least squares algorithm was used to fit a simple sine function to the experimental data in each histogram. The mean and half peak-to-peak modulation of discharge rate of the fitted curve were measured. Innocuous cutaneous stimulation within the left sural nerve field was achieved by light brushing of the skin. Mechanical stimulation of the knee joint was achieved by the use of a hand-held probe to the posterior capsule.

RESULTS

The responses of spindle afferents to sinusoidal stretch of the muscle were compared in the presence and absence of skin stimulation (sural field) or knee joint capsule stimulation. The patterns of response were examined both in cats with an intact spinal cord and in cats with the spinal cord severed at T9.

Decerebrate - spinal cord intact

Five of fifteen GS primary endings showed changes in their cycle histograms to stimulation of the skin or joint that were no greater than inter-control differences. The dominant effect of skin and joint stimulation in the other cases was an increase in the mean rate of discharge. In most cases there was also an obvious reduction in the peak-to-peak modulation of discharge in response to stretch. Figure 1 shows cycle histograms and best fit sine function curves for the discharges of a primary spindle afferent and the changes produced by stimulation of the knee joint capsule or the skin over the lateral aspect of the heel. Capsule stimulation caused an increase in mean rate of 12.3 impulses/s (imp/s) and reduction in modulation of 3.7 imp/s Skin stimulation increased the mean rate by 6.8 imp/s and reduced modulation by 2.6 imp/s There was little change in phase of the fitted curves as a result of either capsule or skin stimulation. Two secondary spindle afferents were studied and both showed increases in mean discharge rate and reductions in modulation to both capsule and skin stimulation.

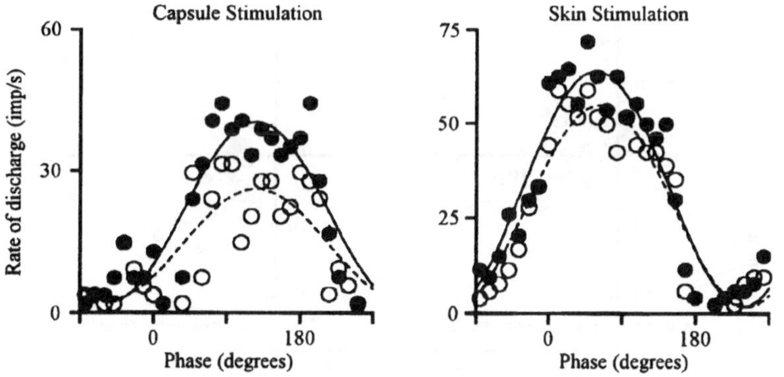

Figure 2. The effects of mechanical stimulation of the knee joint capsule, or of skin over the lateral aspect of the heel, on the averaged responses of two primary spindle afferents to sinusoidal stretch of the gastrocnemius lateralis muscle in a cat with a complete spinal cord section at T9. Open symbols: control. Closed symbols: stimulation. Note the increases in mean rate of discharge and peak-to-peak modulation caused by both stimuli.

Decerebrate - spinal cord severed

In cats with a complete spinal cord section at T9 over half (22 out of 40) of the spindle afferents studied had responses to stretch of the muscle that were not affected by capsule or skin stimulation. Of the remainder, skin stimulation tended to produce an increase in mean rate of discharge that was accompanied either by an increase or a decrease in modulation. In contrast, the dominant action of capsule stimulation was an increase in modulation accompanied by either an increase or decrease in mean rate of discharge. Figure 2 shows histograms and best fit curves for the discharges of two primary afferents in a spinal cat and illustrates the action of both capsule and skin stimulation. Capsule stimulation clearly increases both the modulation to stretch and the mean rate of discharge. Skin stimulation produced an insignificant reduction in mean rate and modulation in the same primary ending (not illustrated). However, in the same cat, skin stimulation produced a modest increase in both modulation and mean rate of another primary ending (Figure 2). The responses to stretch of two out of five secondary spindle afferents in this study were also influenced by joint or skin stimulation. In both cases, joint and skin stimulation caused an increase in mean rate but no change in modulation.

In neither of the preparations (cord intact, cord severed) was there any indication that spindle afferents from the different GS muscles reacted differently to peripheral stimulation. However, the relatively small numbers of afferents studied from the separate muscles preclude any further analysis related to the known functional differences between the soleus and gastrocnemius muscles (LaBella, Kehler & McCrea, 1989).

DISCUSSION

Figure 3 summarises the action of joint and skin stimulation for primary afferent endings in the two preparations. Our interpretation of the effects on the fusimotor system of stimulating joint and cutaneous afferents has been based on spindle afferent responses to electrical stimulation of individual dynamic and static gamma motoneurone axons (Hulliger, Matthews & Noth, 1977). To this end, the plots presented in Figure 3 are divided into sectors by dashed lines, computed from data published by Hulliger et al., (1977), that

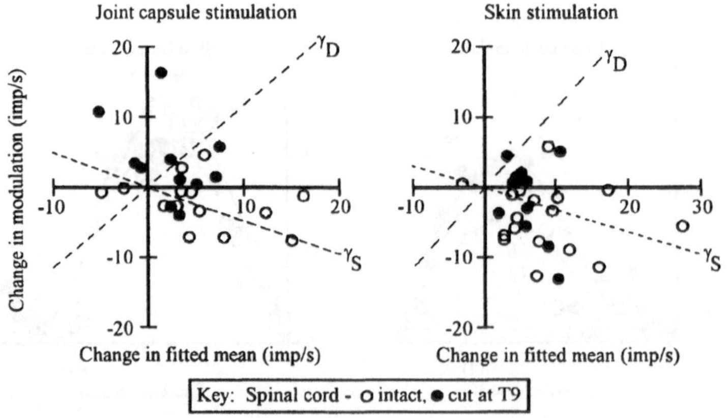

Figure 3. Change in peak-to-peak modulation plotted against the change in mean rate of discharge of GS primary spindle afferents evoked by mechanical stimulation of the knee joint capsule, or of the skin on the lateral aspect of the heel, in cats with intact and severed spinal cords. The dashed lines indicate the changes expected as a result of selective activation of static or dynamic gamma motoneurones (see text).

represent the expected effect of increasing or decreasing activity selectively in either dynamic or static neurones. In interpreting the actions of joint and skin stimulation it should be noted that the dashed lines in Figure 3 representing selective dynamic and static activity do not correspond to the vertical and horizontal axes and are not orthogonal. This may reflect the limited number of observations on which the data from Hulliger et al., (1977) is based and their conclusion that patterns of summation and occlusion of the actions of static and dynamic neurones can be asymmetrical. It should also be noted that some recent work has drawn attention to other problems in the interpretation responses to sinusoidal stretch (Rodgers, Durbaba, Taylor & Fowle, 1994).

We can conclude from our study that the effects of joint capsule and cutaneous stimulation were weak or absent for many spindles, especially in cats with spinal cord section. This confirms previous findings (Johansson, Sjölander, Sojka & Wadell, 1989; Johansson et al., 1990) and suggests that the strong effects that have been observed on gamma motoneurone discharge (Johansson & Sojka, 1985; Davey, Ellaway, Halliday & Rosenberg, 1990; Baxendale et al., 1992) may be directed at only a small proportion of the motoneurone pool. In addition, the data presented in Figure 3 suggest that stimulation of the joint capsule and skin at the heel rarely reduced the mean rate of discharge of spindle afferents. This result suggests either a lack of inhibitory input or that the background static fusimotor drive was low, which is certainly known to be the case for GS muscles in the spinal cat (Grillner, 1969). The exceptions were a few instances where capsule stimulation in the spinal cat caused a substantial increase in the modulation of discharge by stretch but little change in mean rate. However, the points lie clearly to the left of the expected selective dynamic line, implying an actual reduction in static gamma motoneurone activity. Reductions in modulation were more common than reductions in mean rate, especially in response to skin stimulation. The reductions in modulation may be accounted for either by a selective increase in static gamma motoneurone action (Crowe & Matthews, 1964) or inhibition of dynamic gamma motoneurone discharge, or both. The reduced modulation occurred in cats both with and without spinal cord section and may indicate that dynamic neurones have a background discharge in both preparations.

The dominant effects of both capsule and skin stimulation were undoubtedly an increase in the mean rate of spindle afferent ending reponses to stretch. However, as can be seen from Figure 3, stimulation could also result in an increase in modulation and this was seen more frequently with joint capsule stimulation. Thus, in terms of gamma motoneurone activation, joint stimulation evokes a stronger excitatory input to dynamic neurones and skin stimulation a stronger excitatory input to static gamma motoneurones.

Comparison of the effects of peripheral stimulation in cats with and without section of the spinal cord suggest that there was only a weak influence over the balance of reflex inputs from skin afferents between static and dynamic gamma motoneurones (see Figure 3). Skin stimulation had more pronounced effects on mean rate of discharge when the spinal cord was intact suggesting either that descending pathways may facilitate the excitatory skin input to static gamma motoneurones or that an additional long-loop excitatory pathway may operate. The distribution of data points in Figure 3 suggest a different organisation of the regulation of gamma motoneurones by joint afferents. Cutting the spinal cord appears to change the action of joint afferents from one which predominantly excites static neurones to one that excites dynamic gamma motoneurones.

This investigation has shown that joint and skin afferents from regions asscociated with the origin and insertion of the GS muscles exert an identifiable pattern of regulation over static and dynamic gamma motoneurones to those muscles, and that these reflex inputs can be controlled by supraspinal components of the central nervous system. The results are in broad agreement with previous investigations (Johansson et al., 1989; 1990) and stress that the influence of peripheral afferents on gamma motoneurones in this particular part of the limb are regulated by the brain and may differ according to the task to which the GS muscles contribute (Murphy & Hammond, 1991).

ACKNOWLEDGEMENTS

This work was supported by the MRC. MJL was supported by the European Science Foundation and NPA by the Royal Society and the Physiological Society. We gratefully acknowledge the expert technical assistance of Steven Rawlinson and Helen Thomas.

REFERENCES

BAKER, J.R., CATLEY, M.C., DAVEY, N.J. & ELLAWAY, P.H. (1991) Influence of the pontine and medullary reticular formation on the synchrony of gamma motoneurone discharge in the cat. *Exp. Brain Res.* **87**, 604-614.

BAXENDALE, R.H., DAVEY, N.J., ELLAWAY, P.H. & FERRELL, W.R. (1990) The influence of cutaneous receptors on the reflex pathways from knee joint afferents to gamma motoneurones in the decerebrated, low-spinal cat. *J. Physiol.* **429**, 34P.

BAXENDALE, R.H., DAVEY, N.J., ELLAWAY, P.H. & FERRELL, W.R. (1992) The interaction between joint and cutaneous afferent input in the regulation of fusimotor neurone discharge. In: *Muscle Afferents and Spinal Control of Movement*. ed. JAMI, L., PIERROT-DESEILLIGNY, E. & ZYTNICKI, D., pp. 95-104. Pergamon Press, Oxford.

CROWE, A. & MATTHEWS, P.B.C. (1964) Further studies of static and dynamic fusimotor fibres. *J. Physiol.* **175**, 132-151.

DAVEY, N.J. & ELLAWAY, P.H. (1989) Facilitation of individual gamma motoneurones by the discharge of single slowly-adapting type-1 mechanoreceptors in cats. *J. Physiol.* **411**, 97-114.

DAVEY, N.J., ELLAWAY, P.H., HALLIDAY, D.M. & ROSENBERG, J.R. (1990) Association between the discharges of gamma motoneurones during cutaneous stimulation revealed by coherence analysis in the cat. *J. Physiol.* **429**, 35P.

ELLAWAY, P.H., DAVEY, N.J., LJUBISAVLJEVIC, M. & ANISSIMOVA, N.P. (1994) Tonic supraspinal regulation of the action of joint and cutaneous afferents on the dynamic and static sensitivity of spindle afferents in the cat. *J. Physiol.* **479**, 30P.

GRILLNER, S. (1969) Supraspinal and segmental control of static and dynamic gamma motoneurones in the cat. *Acta physiol. scand.* **327**, 3-34.

HULLIGER, M. (1984) The mammalian muscle spindle and its central control. *Revs. Physiol. Biochem. & Pharm.* **101**, 1-110.

HULLIGER, M., MATTHEWS, P.B.C. & NOTH, N. (1977) Effects of combining static and dynamic fusimotor stimulation on the response of muscle spindle primary endings to sinusoidal stretching. *J. Physiol.* **267**, 839-856.

JOHANSSON, H., SJOLANDER, P., SOJKA, P. & WADELL, I. (1989) Effects of electrical and natural stimulation of skin afferents on the gamma-spindle system of the triceps surae muscle. *Neurosci. Res.* **6**, 537-555.

JOHANSSON, H., SJOLANDER, P. & SOJKA, P. (1990) Activity in receptor afferents from the anterior cruciate ligament evokes reflex effects on fusimotor neurones. *Neurosci. Res.* **8**, 54-59.

JOHANSSON, H. & SOJKA, P. (1985) Actions on gamma motoneurones elicited by electrical stimulation of cutaneous afferent fibres in the hind limb of the cat. *J. Physiol.* **366**, 343-363.

LA BELLA, L.A., KEHLER, J.P. & McCREA, D.A. (1989) A differential synaptic input to the motor nuclei of triceps surae from the caudal and lateral cutaneous sural nerves. *J. Neurophysiol.* **61**, 291-301.

MURPHY, P.R. & HAMMOND, G.R. (1991) The role of cutaneous afferents in the control of γ-motoneurones during locomotion in the decerebrate cat. *J. Physiol.* **434**, 529-547.

RODGERS, J.F., DURBABA, R., TAYLOR, A. & FOWLE, A.F. (1994) The use of sinusoidal and ramp stretch stimuli in characterizing fusimotor effects on cat muscles psindles. *Exp. Physiol.* **79**, 337-355.

SOJKA, P., JOHANSSON, H., SJOLANDER, P., LORENTZON, R. & DJUPSJOBACKA, M. (1989) Fusimotor neurones can be reflexly influenced by activity in receptor afferents from the posterior cruciate ligament. *Brain Res.* **483**, 177-183.

REFLEX CONTROL OF FUSIMOTOR NEURONES: WHERE DO WE STAND?

Manuel Hulliger

Department of Clinical Neurosciences
University of Calgary, Calgary, Alberta,
Canada T2N 4N1

Studies on the central control of γ motoneurones have revealed a bewildering variety of fusimotor reflexes whose functional interpretation remains elusive. The bewilderment partly arises from our expectation that there should be simple schemes permitting convenient incorporation of straightforward notions of fusimotor reflex control into more complex schemes of higher order motor control. The final verdict may well be that the true organisation is much more complex than is convenient for simplifying didactic notions, or that the underlying organisation does not even lend itself to schematic simplification at all.

A common rationalisation of the confusing diversity of fusimotor reflex actions is that it may be more apparent than real, with a great deal of it being attributable to methodological heterogeneity arising from the use of different preparations (spinal, decerebrate, anaesthetised, widely denervated, vs intact and freely moving animals), of electrical vs natural stimulation of sensory afferents, of direct vs indirect assessment of actions on γ motoneurones, and of invariably indirect methods of classification of fusimotor efferents as static or dynamic. Indeed, anaesthetised preparations suffer from often poor reflex responsiveness. The size of reflex actions therefore tends to be underestimated and certain reflex contributions may be overlooked altogether, if (polysynaptic) pathways are strongly depressed. The non-anaesthetised alternatives, like spinal or decerebrate preparations, especially when conditioned by convenient cocktails of drugs, are more rewarding for the study of handsome effects, yet reflex actions so unmasked may be exaggerated, reflecting disturbed balance rather than physiological control. Further, the desirability of rigorous control of reflex inputs would make electrical stimulation of sensory fibres the method of choice; yet it has long been recognised that this form of rigor may be more deceptive than illuminating, e.g. since electrical threshold categories poorly reflect functional modalities. Finally, the difficulty of direct recordings from adequately identified and classified γ-efferents has led to the use of indirect methods of γ motoneurone classification whose generality and accuracy is difficult to gauge. The conclusion from this assessment is that much more work based on carefully standardised techniques is required to identify the functional role of fusimotor reflexes. This is the sobering view.

Yet there may be an alternative interpretation of the apparent variability and inconsistency of fusimotor reflex effects. For the sake of the argument, a neural network model of the peripheral proprioceptive system of a limb could be set up, with interneurones and γ motoneurones as two separate layers of hidden units and, in somewhat unorthodox fashion, with joint angles and tactile stimuli serving input elements and with spindle afferent populations of different muscles as output units. To add some realism the model would have to build on limb geometry to transform joint angle into length variations. The spindle receptors would have to be equipped with a reasonable, if approximate, description of input-output behaviour and, in particular, responsiveness to both muscle length variations and alterations in fusimotor drive. Input functions would be limb movements, target behaviour would be spindle population responses coding length and velocity for various reasonable sensitivity settings dictated by a supervisor. Finally, the γ-spindle connectivity matrix would be reasonably constrained. Judging from what has emerged from network simulations of other neural systems (e.g. Robinson, 1994; Fetz, 1994), it would not be surprising to find (following network optimisation based on some meaningful learning rule) a rather significant degree of variability and inconsistency, including Robinson's (1994) "rogue" cells, even among γ-efferents supplying a single muscle. This might suggest that γ-reflex variability as such was nothing to despair about, as it might simply reflect highly distributed, parallelised and individualised γ motoneurone reflex organisation, with the important distinction, that - by design - this would nevertheless lead to well orchestrated and functionally meaningful sensory feedback. Yet the unacceptable would have to be accepted, namely that one would have to abandon claims of understanding γ reflex function at the level of single neurone recordings.

It is almost certainly premature to accept the above distributed network interpretation of γ reflex variability, until the phenomenon of single cell inconsistency is firmly established experimentally, i.e. until genuine variability can reliably be distinguished from variability due to methodological artefacts. Hence, for future experimental studies, the implication is nearly the same, namely that γ reflex variability has to be documented much more systematically. Again, further painstaking experimental work is unavoidable.

GLADDEN, DICKSON AND LUMSDON

The methodological developments outlined in the Gladden, Dickson and Lumsdon paper are most promising and, when fully developed and exploited, may lead to significant developments towards an understanding of the integrative interaction of peripheral and descending control of fusimotor neurones. While still applied under the umbrella of anaesthesia, the method minimises surgical trauma, can dispense with widespread denervation, and would appear to permit quasi-direct recordings from reliably identified static and dynamic fusimotor efferents (by recording potential trains from identified intrafusal fibres) during various forms of controlled natural stimulation.

The approach clearly is ambitious. It remains to be seen how widely it can be applied and whether it will permit comprehensive quantification of the magnitude of fusimotor effects on target spindle afferents. In particular, the reliability and wider applicability of the dual electrode recording technique for afferent spike train identification and quantification remains to be firmly established, the question being whether it will be possible to routinely record from small intramuscular branches (supplying multiple spindles) or whether it will remain restricted to single-spindle nerve ramifications. In the interest of standardisation of methodology in this field the technique should be portable to other laboratories, yet one wonders whether this will be practicable, as it seems to require so much skill and expertise.

The significance of bursting γ motoneurone firing patterns is more difficult to assess.

The short intervals (termed "couplets") might turn out to be useful markers of certain central pathways operating on γ motoneurones, but this requires that statistical identification procedures be established. Whether these short intervals have much impact in the periphery (in shaping spindle afferent firing patterns) may be doubted. They may be viewed as exaggerated extremes in what appear to be highly variable interval distributions (or γ firing patterns) anyway. Even with single γ axon activation most of this variability is likely to be strongly attenuated given the low-pass filter properties of the contractile fibres. How much of any burst-induced irregularity of encoder firing will persist if several γ efferents operate in parallel (and perhaps on the basis of competitive encoder interaction) can only, and indeed should, be explored in experiments with controlled electrical γ fibre stimulation, using pulse trains mimicking normal background variability of discharge, with and without short "couplet" intervals added. This would also provide the setting to test the notion of distinct functional groups of static γ motoneurones. On this view, static γ efferents would exert different actions depending on the characteristics of their discharge patterns. This proposal needs to be and can be tested in simple acute experiments. It implicitly rests on the assumption that frequency response characteristics of different intrafusal target fibres can differ significantly. The cautious view is that not too much should be expected, since even γ_d- and γ_s-operated intrafusal fibres appear to differ only marginally (Chen & Poppele, 1978; Hulliger, 1979).

SJÖLANDER AND JOHANSSON

By reviewing earlier work rather than presenting new material the authors largely avoid sticking out their necks for critique. Methodologically their work covers the ground from single γ cell electrophysiology to spindle Ia sensitivity testing during physiological activation of joint receptor afferents, which permits assessment of combined fusimotor action of several γ efferents converging onto a target afferent. Their published work demonstrates that even in widely denervated hindlimb preparations joint afferent stimulation is capable of eliciting appreciable fusimotor reflex excitation of Ia afferents. They suggest that, in addition to any direct input, a second component of joint receptor contribution to kinesthesia might be mediated through fusimotor reflex sensitisation and/or biassing of spindle afferents. However, the conclusive psychophysical experiments to distinguish between the two remain to be designed.

Beyond that the role of fusimotor reflex control of joint afferents remains fairly baffling. It seems difficult to ascribe a convincing role to the apparently preferential activation of dynamic fusimotor efferents from the anterior cruciate ligament and of static efferents from the posterior cruciate ligament. Also, although the pulling forces which elicited fusimotor reflexes were relatively modest, it is difficult (in the absence of chronic recordings of ligament forces) to assess whether ligament strain during natural movements would be sufficient to elicit such responses.

The challenges of future research might be, first, to attempt chronic recordings from hindlimb spindle afferents, using some realistic form of defined mechanical stimulation of joint capsules or ligaments and imposed perturbations for Ia sensitivity testing to assess the efficacy of fusimotor reflex pathways from joint afferents; second, mathematical models of the knee from the biomechanics literature might be adapted to explore likely physiological forces in cat ligaments; third, a more ambitious biomechanical model of the cat hindlimb might be adapted to incorporate proprioceptive feedback and fusimotor reflex pathways to explore the likely functional significance of what appear to be topographically specialised reflex actions on static and dynamic fusimotor efferents.

MURPHY AND MARTIN

The emphasis in the work summarised by Murphy and Martin is on organisational and functional aspects of fusimotor reflex modulation. They recorded from γ efferents, indirectly classified as static or dynamic, while cutaneous nerves were stimulated electrically during semi-fictive (isometric) locomotion. Of particular interest is their finding of state dependence of cutaneous reflex responses, in that activation of presumed static efferents was suppressed during locomotion, compared with the resting state, while dynamic fusimotor reflexes were unaffected. Broadly this ties in with a number of observations which have demonstrated cycle phase dependence of reflex activation of α motoneurones during stepping or cycling. In this context it would be interesting to learn whether the present example of reflex modulation was simply state dependent (locomotion vs rest) or whether in addition reflex gating also depended on the phase of the step cycle.

In evaluating this work, two concerns need to be considered. First, the authors somewhat lightly dismiss the observations that in chronically implanted and freely moving animals natural skin stimulation failed to elicit significant responses in spindle afferents. They blame the lack of specificity of the tests carried out in chronic animals. However, the authors should also accept the challenge of demonstrating that they themselves are not studying artefacts attributable to the use of electrical stimulation (see above), all the more as natural skin stimulation had previously been reported to be ineffective in activating γ efferents in the same preparation (Murphy, Stein & J. Taylor, 1984).

Second, the reliability and generality of the very indirect method of classification of γ efferents still remains to be demonstrated more convincingly. The main concern is that the distinguishing patterns (phasic vs tonic firing profiles) appear to be specific for certain muscles. In extensor muscles phasic γ discharge patterns are taken to identify dynamic efferents (Murphy et al., 1984), whereas in flexor muscles they are said to be typical of static efferents (Murphy & Hammond, 1993). The apparent lack of generality limits the method's applicability to a small number of selected muscles. Also, the original control experiments, where pattern-identified γ axons were stimulated in continuity, to identify the type of fusimotor effect they elicited, were based on a very small sample. Although the requisite experiments are extremely taxing, the data base of observations relating discharge pattern with type of action clearly has to be expanded, if this method of indirect classification is ever to gain wider acceptance.

ELLAWAY, DAVEY, LJUBISAVLJEVIC AND ANISSIMOVA

The authors make out a case for the importance of supraspinal control of fusimotor reflex circuits by demonstrating that in preparations with intact spinal cord, reflexes elicited by skin and joint stimulation were larger than after spinal section. Also, the organisation of supraspinal regulation of reflex pathways from cutaneous and joint afferents appeared to differ regarding the extent to which static or dynamic fusimotor efferents could be activated preferentially. Studies of this type will undoubtedly contribute to further detailed analysis of the spinal and supraspinal circuitry involved, and the present observations will help differentiate between genuine variability and artefactual fusimotor reflex effects (see above).

A note of caution not to expect too much in terms of strict hierarchical concepts of reflex control is nevertheless indicated. Neural network analysis and simulations have clearly demonstrated in principle that significant functional specificity can be achieved by remarkably distributed processing (Alexander, DeLong & Crutcher, 1994; Fetz, 1994). Further, it may be pointed out that the notion of supraspinal control tells us very little about what precisely supraspinal pathways control, and how they do it. Are they providing simple gain control, mediated by some tonic signal, or is phasic reflex strength modulated by concerted action of short- and long-loop pathways?

The method of classification of the present, rather modest, reflex effects also merits scrutiny. The use of sinusoidal stretching for sensitivity testing relies on reference data, which were originally obtained with high rate stimulation of single static or dynamic γ efferents eliciting mostly powerful fusimotor effects (Appelberg, Hulliger, Johansson & Sojka, 1981). From the outset this approach was limited by the implication that single fibre γ action on Ia sensitivity mimicked the more likely physiological mode of operation, where several efferents of the same type, all acting on the same spindle, may be activated simultaneously. The problem is that the effects of such multi-fibre activation have never been explored systematically. While for high rates of stimulation multi-fibre and single-fibre actions are likely to be similar, qualitatively, the effects of combined low-rate activation are much less predictable, as single fibre stimulation at low rates has been shown to elicit quite paradoxical effects on spindle sensitivity (Hulliger, Emonet-Dénand & Baumann, 1985; augmentation of sensitivity by low-rate γ_s activation). These concerns matter, since the majority of reflex actions on spindle sensitivity of Ellaway et al.'s and Sjölander & Johansson's papers clearly were rather weak, indicative of low-rate activation of fusimotor neurones. Therefore, this method of classification of fusimotor effects by sensitivity testing ought to be further validated in calibration studies documenting the actions of multiple low-rate activation of γ efferents of the same type on spindle sensitivity.

REFERENCES

ALEXANDER, G.E., DELONG, M.R. & CRUTCHER, M.D. (1994) Do cortical and basal ganglionic areas use 'motor programs" to control movement? In *Movement Control.* eds. CORDO, P. & HARNAD S. pp. 54-63. Cambridge University Press, Cambridge.

APPELBERG, B., HULLIGER, M., JOHANSSON, H. & SOJKA, P. (1981) Reflex activation of dynamic fusimotor neurons by natural stimulation of muscle and joint receptor afferent units. In *Muscle Receptors and Movement.* eds. TAYLOR, A. & PROCHAZKA, A. pp. 149-161. Macmillan, London.

CHEN, W.J. & POPPELE, R.E. (1978) Small-signal analysis of response of mammalian muscle spindles with fusimotor stimulation and a comparison with large-signal responses. *J. Neurophysiol.* **41**, 15-27.

FETZ, E.E. (1994) Are movement parameters recognizably coded in the activity of single neurons? In *Movement Control.* eds. CORDO, P. & HARNAD S. pp. 77-88. Cambridge University Press, Cambridge.

HULLIGER, M. (1979) The responses of primary spindle afferents to fusimotor stimulation at constant and abruptly changing rates. *J. Physiol.* **294**, 461-482.

HULLIGER, M., EMONET-DÉNAND, F. & BAUMANN, T.K. (1985) Enhancement of stretch sensitivity of cat primary spindle afferents by low-rate static gamma-action. In *The Muscle Spindle.* eds. BOYD, I.A. & GLADDEN, M.H. pp. 189-193. Stockton Press, New York.

MURPHY, P.R. & HAMMOND, G.R. (1993) The locomotor discharge characteristics of ankle flexor γ-motoneurones in the decerebrate cat. *J. Physiol.* **462**, 59-70.

MURPHY, P.R., STEIN, R.B. & TAYLOR, J. (1984) Phasic and tonic modulation of impulse rates in γ-motoneurons during locomotion in premammillary cats. *J. Neurophysiol.* **52**, 228-243.

ROBINSON, D.A. (1994) Implications of neural networks for how we think about brain function. In *Movement Control.* eds. CORDO, P. & HARNAD S. pp. 42-53. Cambridge University Press, Cambridge.

EXCITATORY AND INHIBITORY EFFECTS OF MIDBRAIN STIMULATION ON γ_s- AND γ_d-MOTONEURONES OF A CAT HINDLIMB MUSCLE

M. Dickson and M.H. Gladden

Muscle Spindle Physiology Group
I.B.L.S., West Medical Building
University of Glasgow
Glasgow G12 8QQ, Scotland

INTRODUCTION

Areas in the midbrain where stimulation preferentially affects γ_s- and γ_d-motoneurones have been reported (see Appelberg, 1981; Fowle, Taylor, Rodgers & Durbaba, 1992; Taylor, Durbaba, Rodgers & Fowle, 1993). In the tenuissimus muscle of anaesthetised animals we also found these effects in experiments in which the responses of Ia afferents to standard ramp and hold stretches were monitored during midbrain stimulation (Dickson & Gladden, 1990). Static effects tended to predominate. This was also the case in separate experiments in which activity of γ-motoneurones was detected by observing the contractions of the types of intrafusal fibre they control in spindles exteriorised under anaesthesia. Chain and/or bag_2 contractions indicate active γ_s-motoneurones, whereas active γ_d-motoneurones cause bag_1 contractions. In confirmation of Fowle et al. (1992) chain fibres could regularly be recruited with bag_2 (pure γ_s- activation), and these fibres were also recruited together with bag_1 (mixed γ_s- and γ_d-activation) in 14 experiments, including several in which more than one spindle was isolated. It proved difficult to recruit bag_1 fibres alone (pure γ_d-activation). Spontaneous γ_s-activity could also be inhibited. Here we document an example in which some γ_s-motoneurones were activated while others were inhibited.

METHODS

Tenuissimus muscles were exteriorised into an organ bath containing Krebs' solution, keeping the afferent and efferent nerve supplies and blood supply intact. Spindles were dissected sufficiently to expose intrafusal fibres in the capsular region. A row of three stainless steel or tungsten electrodes, 1.5mm apart, were inserted stereotaxically into the mesencephalon. The Horsley-Clarke co-ordinates for the stimulation in the case illustrated in Figure 1 were 3.4 A, 2.0 L ,-2 D. Stimulation was with 5 μA, 200 Hz, 0.2ms wide pulses to each electrode relative to a remote return. Video sequences were analysed with a Kontron (IBAS) system. Points or edges of the contracting fibres were followed through successive frames (25/s). Active movement of individual fibres could be distinguished easily from passive.

RESULTS

Chain fibre movement is not easily monitored when they are taut, at long spindle length, because their sarcomeres are fainter than bag fibre sarcomeres, and mistakes can be made because the plane of focus changes from one fibre to another during movement. For this reason observations of chain fibre contractions were made with unloaded fibres, at short spindle length, so that contraction in chain fibres straightened them out. Chain fibres buckle most obviously in the fluid space, between the sensory endings and the sleeve region, where there are no sensory endings or capsule cells to bind them to the bag fibres (Gladden, 1976). For the sequence analysed in Figure 1 the straightening out of a buckled region in a chain fibre by contraction is shown in the lower trace. Pulsatile movements due to spontaneous activity in a γ_s-motoneurone were halted by midbrain stimulation. The drift downwards of the trace during these contractions is due to background movement which has not been subtracted. In this region of the spindle the bag_2 always moves towards the pole if it contracts, that is, it moves towards the motor endings in the sleeve region. This movement towards the pole is shown in the upper trace of Figure 1. It increased at the onset of midbrain stimulation. The mechanical arrangement of the intrafusal bundle was such that any passive movement of chain fibres by bag_2 contraction would have straightened out all the chain fibres, because the equatorial region of the bag_2 fibre was virtually stationary, and at

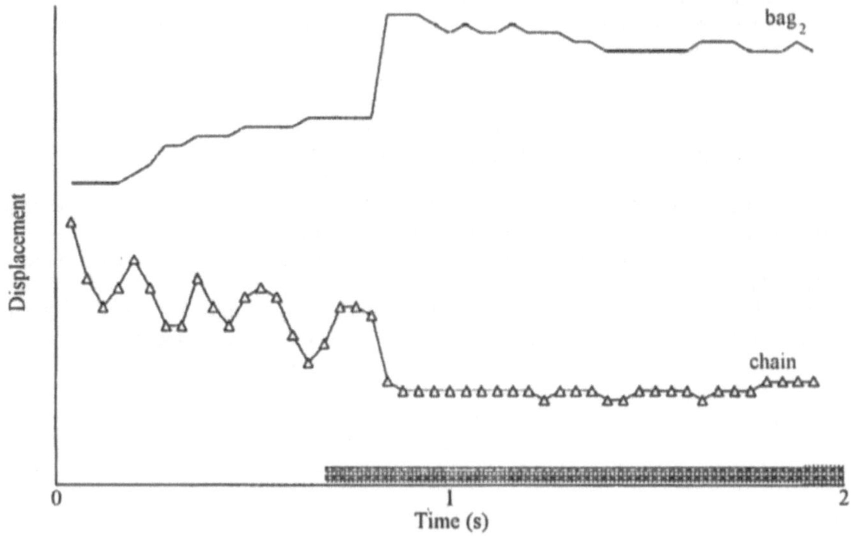

Figure 1. Contractions of a chain fibre (lower trace) due to spontaneous activity of a γ_s- motoneurone abolished by midbrain stimulation indicated by hatched box. Simultaneously another γ_s- motoneurone was activated, indicated by contraction of a bag_2 fibre (upper trace). Contraction is upwards for both. Displacement measured in arbitrary units (pixels) by video analysis.

the other side of the buckle in the chain fibres the bag_2 was moving to the pole. Therefore the relaxation of the chain fibre cannot have been a spurious effect due to passive unloading by bag_2 contraction.

DISCUSSSION

During these experiments we were surprised to observe that stimulation at specific sites could recruit some intrafusal fibres and simultaneously inhibit other fibres, but all these fibres were controlled by γ_s-motoneurones. This indicates that some γ_s-motoneurones were activated while others were inhibited. It is possible that the γ_s-motoneurones were selected unintentionally in some way by our experimental procedure. In order to demonstrate inhibition it was necessary to find spontaneously contracting intrafusal fibres, and there must have been other γ_s-motoneurones which were either silent, or discharging at a very low rate for which we could not establish any effect because contraction in bag_2 fibres is not obvious at frequencies >10-15/s. Nevertheless we conclude that the γ_s-motoneurone pool of the tenuissimus muscle does not respond as a homogeneous entity to midbrain stimulation.

ACKNOWLEDGEMENTS

This work is supported by the Wellcome Trust.

REFERENCES

APPELBERG, B. (1981) Selective central control of dynamic γ-motoneurones utilised for the functional classification of γ cells. In *Muscle Receptors and Movement.* eds. TAYLOR, A. & PROCHAZKA, A. pp 97-107. MacMillan, London.

DICKSON, M. & GLADDEN, M.H. (1990) Dynamic and static gamma effects in tenuissimus muscle spindles during stimulation of areas in the mes- and diencephalon in anaesthetised cats. *J. Physiol.* **423**, 73P.

DICKSON, M. & GLADDEN, M.H. (1992) Central and reflex recruitment of γ-motoneurones of individual muscle spindles of the tenuissimus muscle in anaesthetised cats. In *Muscle Afferents and Spinal Control of Movement.* eds. JAMI, L., PIERROT-DESEILLIGNY, E. & ZYTNICKI, D. pp. 37-42. Pergamon, Oxford.

FOWLE, A.J., TAYLOR, A., RODGERS, J.F. & DURBABA R. (1992). Mesencephalic and diencephalic areas for fusimotor control in the anaesthetised cat. *J. Physiol.* **446**, 230P.

GLADDEN, M.H. (1976) Structural features relative to the function of intrafusal muscle fibres in the cat. *Prog. Brain Res.* **44**, 51-59.

TAYLOR, A., DURBABA, R., RODGERS, J.F. & FOWLE, A.J. (1993) Reciprocal actions of midbrain stimulation on static and dynamic fusimotor neurones of the hindlimb in anaesthetized cats. *J. Physiol.* **473**, 15P.

FUSIMOTOR OUTFLOW TO MEDIAL GASTROCNEMIUS MUSCLE DURING ISOTONIC CONTRACTION OF ITS CLOSE SYNERGISTS IN DECEREBRATE CATS

M. Ljubisavljevic and R. Anastasijevic

Institute for Medical Research
Dr Subotica 4, Beograd, Yugoslavia

INTRODUCTION

It has been shown that fusimotor discharge increases during muscle fatigue due to isometric contractions (Ljubisavljevic, Jovanovic & Anastasijevic, 1992). The present report deals with experiments aimed at showing whether an increase in fusimotor discharge rate would occur also with isotonic contractions are they as common in everyday motor activity and seldom studied in experimental animals.

METHODS

Experiments were performed on decerebrate cats. Operative procedure, before decerebration, was carried out under halothane in oxygen anaesthesia. Fusimotor spikes were recorded from filaments of otherwise intact MG muscle nerve. The neurones were identified as fusimotor if their conduction velocity, determined by spike-triggered back-averaging of impulse traffic in the parent nerve, was in the range of 10 - 45 m/s. Muscle contractions were elicited by either continuous or interrupted repetitive electrical stimulation of LGS nerves. Pulses were delivered at 1.3x motor threshold, 0.2 ms duration applied at a rate of 40 Hz. The muscles were considered to be fatigued when, against a load (weight) equal to one third of the tension developed at the onset of an isometric contraction, they returned to the length at which the isometric contraction was elicited.

RESULTS

Changes in discharge rate of 26 fusimotor neurones to medial gastrocnemius muscle (MG) were studied during isotonic fatiguing contractions of lateral gastrocnemius and

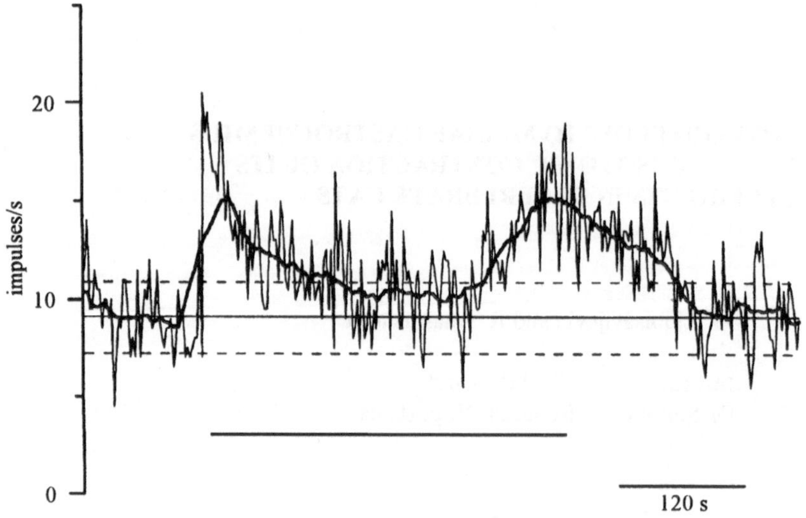

Figure 1. Changes in discharge rate (impulses/s, actual and smoothed by ten-points averaging) of a fusimotor neurone showing an early and a late long-lasting increase during isotonic muscle contraction. Horizontal continuous and broken lines - spontaneous firing rate, mean and SD. Lowermost continuous line indicates muscle contraction. The muscle loaded with 200 g, shortening from the onset to the end of the contraction by 6-2 mm.

soleus muscles (LGS) in decerebrate cats. At the onset of muscle contraction an increase in discharge rate, lasting for 5-220 s, occurred in all except one of the neurones. In 73% of the units a second increase developed in parallel with muscle fatigue outlasting the contraction for 5-180 s (Figure 1), while all but one of the remaining neurones exhibited a burst of spike discharges coincident with the end of contraction (Figure 2). Similar early and late increase in discharge rate developed during a train of short-lasting repetitive contractions.

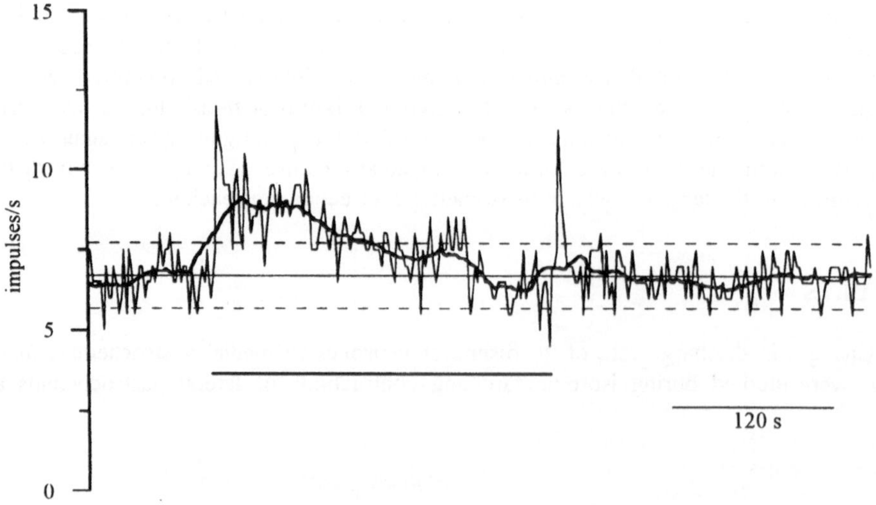

Figure 2. Changes in discharge rate of a fusimotor neurone showing an early long-lasting increase during isotonic muscle contraction and a sharp burst at its end. The muscle loaded with 200 g, shortening from the onset to the end of the contraction by 5-2 mm.

In comparison to changes in fusimotor discharge rate in an isometric regime (Ljubisavljevic et al., 1992) the early fusimotor responses were similar in amplitude, but tended to be longer lasting in isotonic contractions, while the opposite was true for the slowly developing late responses. When the contracting muscle was made ischaemic by clamping the femoral artery, the late increase lasted until the clamp was removed. It was supposed therefore to be a reflex response to afferent discharges from chemosensitive muscle afferents, activated by metabolic products liberated in muscle tissue. Short-lasting, brisk late responses, at the end of muscle contraction, which were not encountered previously in the isometric regime, are supposed rather to be due to discharges in mechanosensitive afferents, possibly sensitised by the metabolic products.

To test the effectiveness of the late increase in fusimotor discharge rate, spike discharges of muscle spindle primary afferents were recorded from dorsal root filaments during isotonic muscle contractions under similar experimental conditions. Post-contraction increase in discharge rate occurred in four out of eight units recorded. It was similar in both amplitude and duration to that established to be due to the late increase in fusimotor discharge rate in the isometric regime (Ljubisavljevic & Anastasijevic, 1994a,b).

DISCUSSION

From the functional point of view it is supposed that the early fusimotor responses might help to maintain muscle spindle responsiveness during isotonic, shortening contractions (e.g. Vallbo, 1973) which under these conditions might be more important than in an isometric regime. On the other hand the long-lasting late fusimotor responses might be less important if the blood flow is less impaired in an isotonic regime and the recovery from fatigue easier. The short-lasting late bursts, however, might provoke prolonged increase in spindle responsiveness (Brown, Goodwin & Matthews, 1969).

REFERENCES

BROWN, M. C., GOODWIN, G. M. & MATTHEWS, P. B. C. (1969). After-effects of fusimotor stimulation on the response of muscle spindle primary endings. *J. Physiol.* **205**, 677-694.

LJUBISAVLJEVIC, M. & ANASTASIJEVIC, R. (1994a) Fusimotor-induced post-contraction changes in muscle spindle afferent outflow and responsiveness in muscle fatigue in decerebrate cats. Effects of afferent inflow from outside the contracting muscle. *Acta Veterinaria* **43**, 343-352.

LJUBISAVLJEVIC, M. & ANASTASIJEVIC, R. (1994b) Fusimotor-induced changes in muscle spindle outflow and responsiveness in muscle fatigue in decerebrate cats. *Neuroscience.* **63**, 339-348.

LJUBISAVLJEVIC, M., JOVANOVIC, K. & ANASTASIJEVIC, R. (1992). Changes in discharge rate of fusimotor neurones provoked by fatiguing contraction of cat triceps surae muscles. *J. Physiol.* **445**, 499-513.

VALLBO, A. B. (1973). Muscle spindle afferent discharge from resting and contracting muscles in normal human subjects. In *New Developments in Electromyography and Clinical Neurophysiology*. ed. DESMEDT, J. E., pp. 251-262. Karger, Basel.

FUSIMOTOR OUTFLOW TO PRETIBIAL FLEXORS DURING ISOMETRIC AND ISOTONIC FATIGUING CONTRACTION OF TRICEPS IN DECEREBRATE CATS

R. Anastasijevic, M. Ljubisavljevic,
S. Radovanovic and I. Vukcevic

Institute for Medical Research
Dr Subotica 4, Beograd, Yugoslavia

INTRODUCTION

It has been shown that changes in discharge rate occur in fusimotor neurones to a muscle during fatiguing contractions of it or its close synergists during long-lasting isometric (Ljubisavljevic, Jovanovic & Anastasijevic, 1992a) and isotonic contractions (Ljubisavljevic & Anastasijevic, this volume). Effects are also seen in fusimotor neurones to a remote, partly synergistic non-contracting muscle group (Ljubisavljevic, Jovanovic & Anastasijevic, 1992b). These effects seem to be due to both mechanosensitive and chemosensitive small-diameter muscle afferents. These same muscle afferents can be expected to exert reflex effects also on fusimotor neurones to pretibial flexors (Appelberg, Hulliger, Johansson & Sojka, 1983a,b). Since the signals from muscle spindles in antagonist muscles may be as important for the control of movements as those from the agonist muscles, it seemed to us of interest to look for the changes in discharge rate of fusimotor neurones to pretibial flexors, acting primarily as antagonists to the triceps surae muscles, during long-lasting isometric and/or isotonic contractions of the latter.

METHODS

Experiments were performed on decerebrate cats. Operative procedure, before decerebration, was carried out under halothane in oxygen anaesthesia. Fusimotor spikes were recorded from nerve filaments dissected free from the peroneal nerve. The neurones were identified as fusimotor if their conduction velocity, determined by spike-triggered back-averaging of impulse traffic in the parent nerve, was in the range of 10 - 45 m/s. The contractions were elicited by electrical stimulation of the nerves to triceps applied, in isometric regime, until the muscle tension fell to approximately one third of its initial value. Pulses were delivered at 1.3x motor threshold, 0.2 ms duration applied at a rate of 40 Hz. In isotonic contractions the muscles were considered to be fatigued when, against a load (weight) equal to one third of the tension developed at the onset of the isometric contraction, returned to the length at which the isometric contraction was elicited.

RESULTS

Responses of the fusimotor neurones were diverse. In the isometric regime, 13 out of 38 units recorded exhibited a biphasic increase in discharge rate, the late one developing slowly and outlasting the contraction (Figure 1), i.e. similar to that found previously in fusimotor neurones to the contracting muscles (Ljubisavljevic et al., 1992a) and shown to be related to muscle fatigue. Another seven neurones, mainly silent at rest, exhibited burst-like discharges at the onset and the end of the contraction. In another 11 units the initial response was a decrease in discharge rate, lasting in six of them throughout the contraction (Figure 2A). Another six units exhibited a sustained plateau-like increase in discharge rate throughout the contraction, starting at its onset and maintained at a lower level, but still above the spontaneous firing rate, thereafter (Figure 2B). It should be mentioned also that many (about 20) silent neurones, mainly responding to manipulating the skin and the paw and/or stroking the fur, did not respond to triceps contractions.

Same patterns of changes in discharge rate, recorded in 31 of the 38 units, were encountered also during isotonic triceps contractions though the late responses tended to be less expressed.

DISCUSSION

It is supposed that the responses of the fusimotor neurones to pretibial flexors (antagonists) are evoked primarily by mechanosensitive muscle afferents from the

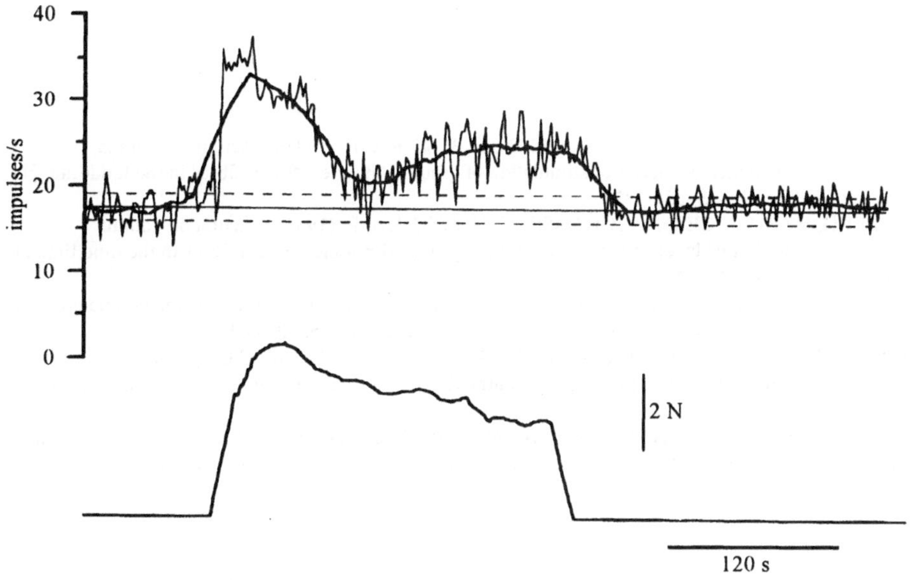

Figure 1. Changes in discharge rate (upper trace, impulses/s, actual and smoothed by ten-points averaging) of a fusimotor neurone showing an early and a late long-lasting increase during triceps muscle contraction (lower trace, N). Horizontal continuous and broken lines - spontaneous firing rate, mean and SD.

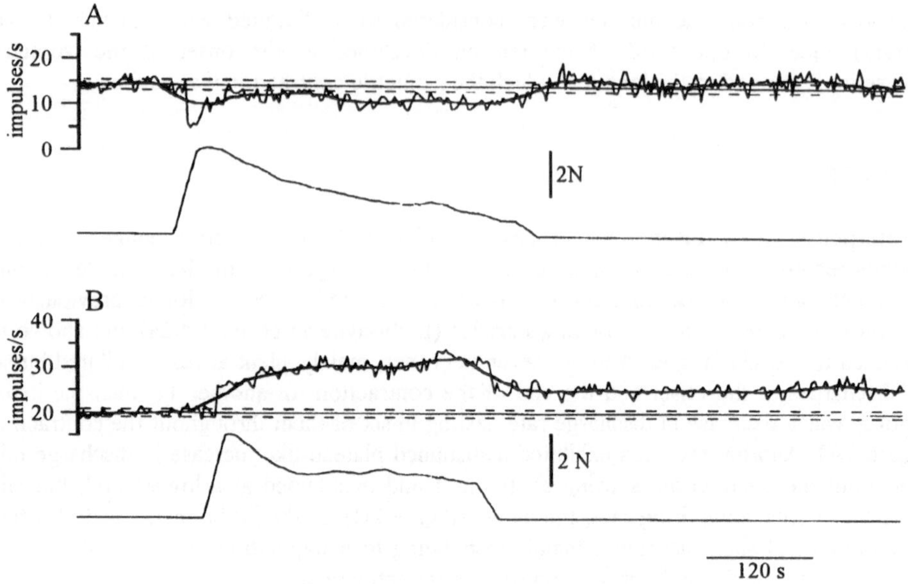

Figure 2. A - decrease in discharge rate of a fusimotor neurone during an isometric contraction. B - sustained increase in firing level of another fusimotor neurone.

contracting muscles (agonists) while the responses to chemosensitive muscle afferents or else to agonist muscle fatigue seem to be much weaker. Sustained shifts in firing level, encountered also in a few fusimotor neurones to hamstring muscles during fatiguing contractions of the triceps surae (Ljubisavljevic et al., 1992b), resembled the bistable behaviour which has been observed in skeletomotor neurones (Hounsgaard, Hultborn, Jespersen & Kien, 1984).

REFERENCES

APPELBERG, B., HULLIGER, M., JOHANSSON, H. & SOJKA, P. (1983a). Action on gamma motoneurones elicited by electrical stimulation of group II muscle afferent fibres in the hind limb of the cat. *J. Physiol.* **335**, 255-273.

APPELBERG, B., HULLIGER, M., JOHANSSON, H. & SOJKA, P. (1983b). Action on gamma motoneurones elicited by electrical stimulation of group III muscle afferent fibres in the hind limb of the cat. *J. Physiol.* **335**, 275-292.

HOUNSGAARD, J., HULTBORN, H., JESPERSEN, B. & KIEN, O. (1984). Intrinsic membrane properties causing a bistable behaviour of α-motoneurones. *Exp. Brain Res.* **55**, 391-394.

LJUBISAVLJEVIC, M., JOVANOVIC, K. & ANASTASIJEVIC, R. (1992a). Changes in discharge rate of fusimotor neurones provoked by fatiguing contraction of cat triceps surae muscles. *J. Physiol.* **445**, 499-513.

LJUBISAVLJEVIC, M., JOVANOVIC, K. & ANASTASIJEVIC, R. (1992b). Changes in discharge rate of cat hamstring fusimotor neurone during fatiguing contractions of triceps surae muscles. *Brain Res.* **579**, 246-252.

THE INTERACTION OF LOW AND HIGH THRESHOLD JOINT AFFERENTS IN RELATION TO REFLEX EFFECTS IN α- AND γ-MOTONEURONES

D.T. Scott, R.H. Baxendale and W.R. Ferrell

Division of Neuroscience and Biomedical Sciences,
The University, Glasgow G12 8QQ, UK.

INTRODUCTION

Group II (Aβ) joint afferents can exert powerful short latency excitatory reflex actions on alpha motoneurones (Baxendale, Ferrell & Wood, 1988; Ferrell, Rosenberg, Baxendale, Halliday & Wood, 1990) and these actions change when tested against a background of articular nociceptive activity (Baxendale, Ferrell & Wallace, 1989). Acute inflammation of the knee joint has a predominately excitatory action on flexor alpha motoneurones (He, Proske Schiable & Schmidt, 1988), increasing flexion reflex intensity and abolishing the modulation of the flexion reflex associated with joint angle (Baxendale & Ferrell, 1981; Ferrell, Wood & Baxendale, 1988).

Electrical stimulation of joint afferents has also been shown to elicit mixed excitatory and inhibitory actions in both flexor and extensor gamma motoneurones (Johansson, Sjölander & Sojka, 1986; Baxendale, Davey Ellaway & Ferrell, 1993). Acute knee joint inflammation enhances the spontaneous activity of flexor gamma motoneurones and their responsiveness to limb movement (He et al., 1988), while natural excitation of receptors by pressing on the capsule can excite gamma motoneurones (Appelberg, Hulliger, Johansson & Sojka, 1979, 1981; Ferrell et al., 1990; He et al., 1988). We will present data concerning the effect of high threshold (groups III & IV) joint afferent stimulation on extensor motoneurone excitation induced by of group II joint afferents. The nature of the interaction of the joint afferents is also examined.

GAMMA MOTONEURONES

Cats were decerebrated under gaseous anaesthesia, spinalised (T_{12}), paralysed and artificially ventilated. The posterior articular nerve (PAN) was dissected free and placed over two pairs of stimulating electrodes. The lateral gastrocnemius nerve was dissected free and recordings made from spontaneously discharging single units. Gamma motoneurones which had a period of excitation in response to stimulation of group II PAN afferents were used to examine the effect of a preceding maximal electrical stimulation of the PAN. Repetitive (0.5 to 2.0 Hz) electrical stimulation of the PAN, produced a short latency increase of gamma motoneurone firing with a latency ranging from 8 to 35ms. This facilitation was diminished or abolished if the first stimulus was preceded by a conditioning stimulus which excited the group IV afferents in the PAN. This depression of facilitation was dependent on the interval between the test and conditioning stimuli and the interval which produced the greatest depression was consistent with the effect being due to group IV afferents. Inhibition of the facilitation did not significantly alter the on-going discharge rate of the gamma motoneurones.

ALPHA MOTONEURONES

In similar preparations, PAN was dissected free and maintained in continuity with the capsule. Popliteus was tenotomised and recordings were made from soleus motor units. Repetitive mechanical indentation of the dorsal aspect of the cat knee joint capsule with a fine probe resulted in synchronisation of discharge of soleus motor units to the stimulus. This effect was mediated via articular afferents as it was reversibly abolished by the application of 2% lignocaine to PAN. This synchronisation was also reduced, and in some cases abolished, by application of 1% capsaicin suspension to PAN. This effect lasted a few minutes. As capsaicin is known to activate sensory C fibres, these findings indicate that excitation normally mediated by sensory receptors with low thresholds to mechanical stimuli can be abolished by activity in nociceptive afferents arising from the same joint. Conditioning electrical stimulation of PAN sufficient to activate group IV afferents prior to mechanical capsule indentation also produced inhibition, with a time course consistent with group III/IV effects.

INTERACTION OF JOINT MECHANORECEPTORS AND NOCICEPTORS

Decerebrated spinalised cats were prepared as before. PAN was dissected for stimulation and recording and the sural, gastrocnemius and contralateral PAN dissected for stimulation. A laminectomy was performed from the fifth lumbar to sacral segments, the dura opened and a tungsten microelectrode was inserted into the cord to establish the threshold for excitation of intraspinal terminals of joint afferents.

Changes in the excitability of the intraspinal terminals of group II afferent fibres from the PAN were used as a measure of polarisation induced by prior stimulation of group IV fibres (electrical and chemical) from the same nerve. Changes induced in the PAN by stimulation of group IV fibres in the sural, gastrocnemius and contralateral PAN were also examined.

The results show that both electrical and chemical stimulation of group IV PAN afferents result in hyperpolarisation of ipsilateral group II fibres. Stimulation of group IV sural and gastrocnemius fibres produces a weaker hyperpolarisation, while stimulation of the contralateral PAN had no effect.

ACKNOWLEDGEMENTS

This project was supported by the MacFeat Bequest.

REFERENCES

APPELBERG, B., HULLIGER, M., JOHANSSON, H. & SOJKA, P. (1979) Excitation of dynamic fusimotor neurons of the cat triceps surae by contralateral joint afferents. *Brain Res.* **160**, 529-532.

APPELBERG, B., HULLIGER, M., JOHANSSON, H. & SOJKA, P. (1981) Reflex activation of dynamic fusimotor neurons by natural stimulation of muscle and joint receptor afferent units. In *Muscle Receptors and Movement.* eds. TAYLOR, A. & PROCHAZKA, A. pp 149-161. MacMillan, London.

BAXENDALE, R.H., DAVEY, N.J., ELLAWAY, P.H. & FERRELL, W.R. (1993) The interaction between joint and cutaneous afferent input in the regulation of fusimotor neurone discharge. In *Muscle Afferents and Spinal Control of Movement.* eds. JAMI, L., PIERROT-DESEILLIGNY, E. & ZYTNICKI, D. pp 95-104. Pergamon, Oxford.

BAXENDALE, R.H. & FERRELL, W.R. (1981) The effect of knee joint afferent discharge on transmission in flexion reflex pathways in decerbrate cats. *J. Physiol.* **315**, 231-242.

BAXENDALE, R.H., FERRELL, W.R. & WOOD, L. (1988) Response of quadriceps motor units to mechanical stimulation of knee joint receptors in the decerebrate cat. *Brain Res.* **453**, 150-156.

BAXENDALE, R.H., FERRELL, W.R. & WALLACE, K. (1989) The effects of nociceptive input on the excitation of soleus motor units by knee joint mechanoreceptors in the decerebrate cat. *J. Physiol.* **420**, 48P.

FERRELL, W.R., ROSENBERG, J.R., BAXENDALE, R.H., HALLIDAY, D. & WOOD, L. (1990) Fourier analysis of the relation between the discharge of quadriceps motor units and periodic mechanical stimulation of cat knee joint receptors. *Exp. Physiol.* **75**, 739-750.

FERRELL, W.R., WOOD, L. & BAXENDALE, R.H. (1988) The effect of acute joint inflammation on the flexion reflex excitability in the decerebrate, low-spinal cat. *Q. J. Exp. Physiol.* **73**, 95-102.

HE, X., PROSKE, U., SCHIABLE, H-G. & SCHMIDT, R.F. (1988) Acute inflammation of the knee joint in the cat alters responses of flexor motoneurones to leg movements. *J. Neurophysiol.* **59**, 326-340.

JOHANSSON, H., SJOLANDER, P. & SOJKA, P. (1986) Actions on γ-motoneurones elicited by electrical stimulation of joint afferent fibres in the hind limb of the cat. *J. Physiol.* **375**, 137-152.

REGULATION OF THE γ-MUSCLE SPINDLE SYSTEM BY CHEMOSENSITIVE MUSCLE AFFERENTS

Mats Djupsjöbacka[1], Håkan Johansson[1], Mikael Bergenheim[1,2] and Per Sjölander[1]

[1]Division of Work Physiology
National Institute of Occupational Health
Box 7654, S-907 13 Umeå, Sweden
[2]Department of Physiology
University of Umeå, S-901 87 Umeå, Sweden

INTRODUCTION

Appelberg, Hulliger, Johansson & Sojka (1983a) has shown that electrical stimulation of group III muscle afferents evokes excitatory effects on both homo- and heteronymous γ-motoneurones. It is also well known that increased intramuscular concentrations of substances liberated during static muscle contractions and/or inflammation increase the activity of group III and IV muscle afferents. Furthermore, it has been shown that the activity in the primary muscle spindle afferents (MSAs) is most important for the reflex mediated muscle stiffness.

These observations have led Johansson & Sojka (1991) to suggest a mechanism which might be of importance for the genesis and spread of (occupational) muscle tension and pain. They hypothezised that increased activity in group III and IV muscle afferents might, via reflexes to the fusimotor-muscle spindle system, lead to increased muscle stiffness, which may lead to increased production of metabolites and further increased stiffness in primarily affected (homonymous) as well as heteronymous muscles. In this potential vicious circle the secondary MSAs were proposed to be of particular importance, since the secondary MSAs (unlike the primary) project back to the γ-muscle-spindle system, and may have potent effects on both dynamic and static γ-motoneurones (Appelberg, Hulliger, Johansson & Sojka, 1983b). Since the activity of the primary MSAs is influenced by fusimotor reflexes from secondary MSAs, this second positive feedback loop may have a key role in the regulation of the reflex mediated muscle stiffness.

Experimental support for this mechanism has been presented by Jovanovic, Anastasijevic & Vuco (1990) who have shown that injections of KCl, lactic acid, bradykinin or 5-HT into the arterial supply of the triceps surae muscle (GS) induce an increase in the activity of γ-efferents, probably via reflexes mediated by chemosensitive muscle afferents.

However, several questions of relevance for the mechanism hypothesised by Johansson & Sojka (1991) remain to be answered. For instance: (1) To what extent do the fusimotor reflexes, evoked by chemically stimulated muscle afferents, activate the MSAs? (2) What type of fusimotor neurons are activated by chemosensitive muscle afferents, and (3) are both primary and secondary MSAs influenced? (4) To what extent will chemical changes in a muscle influence the MSAs of surrounding muscles?

Thus, the aim of the present study was to investigate how increased concentrations of metabolites in one muscle influences the activity of primary and secondary MSAs from the chemically affected muscle as well as from surrounding muscles. A detailed account of the data presented in this paper has been given by Djupsjöbacka (1994).

METHODS

The experiments were performed in cats anaesthetised with α-chloralose (60 mg/kg IV). In both hind limbs the posterior biceps and semitendinosus (PBSt) and triceps surae (GS) muscles were prepared with intact nerve supplies. The rest of the limbs were denervated. The tendons of the ipsilateral PBSt and GS muscles were cut and subjected to sinusoidal stretches (1 mm, 1 Hz) at 2 mm below maximum muscle lengths. Injections of metabolites into the GS muscles were given via a cannula inserted in the sural artery. The injected volumes were 0.5-1.0 ml and the injection rate was 1 ml/min. The substances tested were l-lactic acid, KCl, arachidonic acid, bradykinin and 5-HT.

Simultaneous recordings of up to 12 GS and/or PBSt MSAs from separate dorsal root filaments were made with a multichannel hook electrode. The mean rate of firing and the modulation of the MSA responses to the sinusoidal stretching were used as quantitative measures of MSA sensitivity. Alterations in these parameters were also used for estimation of the type of fusimotor neurones involved in any reflex response.

RESULTS

Responsiveness

Injections of KCl (50-600 mM) or lactic acid (20-200 mM) into the ipsilateral GS muscle evoked responses on 88% of the MSAs (n = 32), while injections into the contralateral GS muscles evoked responses on 50% of the MSAs tested (n = 50). Injections of arachidonic acid (0.3 - 1.0 mg) into the ipsilateral GS muscle evoked responses on 86% of the MSAs (n = 36), while injections into the contralateral GS muscle evoked alterations on 45% of the MSAs (n = 22). The responsiveness of MSAs to ipsi- or contralateral GS-injections of bradykinin (9-100 (g) was 89% and 84% respectively (n = 37 and n = 19), while the responsiveness to injections into the ipsi- or contralateral GS muscles of 5-HT (25-150 μg) was 83% and 33% respectively (n = 23 and n = 15). There were no substantial difference in responsiveness between the primary and the secondary MSAs. Also, for injections into the ipsilateral GS muscles, the responsiveness of the GS (homonymous) MSAs was similar to the responsiveness of the PBSt (heteronymous) MSAs.

Type, size and duration of the effects

The alterations in MSA activity were in general, for both the GS and the PBSt MSAs, compatible with excitation of static or both static and dynamic fusimotor neurones, i.e. an increase in mean rate of firing accompanied by a decrease or no change in modulation.

Effects compatible with inhibition of fusimotor activity were rare. The size of the induced changes in mean rate of firing were in the range 10-140% of the mean firing rate prior to the injection. Usually, the duration of the alterations in MSA responses varied between 1-4 minutes. Yet, on some occasions the increase in MSA activity lasted for periods up to 60 minutes.

The reflex nature of the effects

It was concluded that the alterations in MSA activity, evoked by the injections of the test substances, were caused by reflexes to fusimotor neurones from chemically activated muscle afferents since: (1) Local anaesthesia of the nerve to the injected muscle always abolished or greatly diminished the effects on previously responsive MSAs. (2) Injections of the dissolving agents never induced any alterations in MSA activity. (3) Intravenous injections of the test substances only rarely gave rise to alterations in MSA responses.

DISCUSSION

The main finding of the present study is that increased intramuscular concentration of substances known to be produced during static muscle contractions and/or inflammation can induce sizeable increases in the static sensitivity of MSAs from the chemically affected muscle as well as from heteronymous muscles (including contralateral muscles), via fusimotor reflexes.

Since the present series of experiment show that activation of chemosensitive muscle afferents may induce activation of predominantly static fusimotor neurones, it seems conceivable that the short latency stretch reflex response of a muscle may be depressed in situations when the intra-muscular concentrations of metabolites are high. Also, the increased mean firing rate of the MSAs may lead to increased muscle stiffness during such conditions. Thus, the results lend further support to the mechanism behind genesis and spread of (occupational) muscle tension and pain proposed by Johansson & Sojka (1991).

REFERENCES

APPELBERG, B., HULLIGER, M., JOHANSSON, H. & SOJKA, P. (1983a) Actions on γ-motoneurones elicited by electrical stimulation of group III muscle afferent fibres in the hind limb of the cat. *J. Physiol.* **335**, 275-292.

APPELBERG, B., HULLIGER, M., JOHANSSON, H. & SOJKA, P. (1983b) Actions on γ-motoneurones elicited by electrical stimulation of group II muscle afferent fibres in the hind limb of the cat. *J. Physiol.* **335**, 255-273.

DJUPSJÖBACKA, M. (1994) Regulation of the γ-muscle-spindle system by chemosensitive muscle afferents and joint afferents. A conceivable mechanism behind onset and spread of increased muscle tension. Umeå University Medical Dissertations New series no. 394, Umeå, Sweden

JOHANSSON, H. & SOJKA, P. (1991) Pathophysiological mechanisms involved in genesis and spread of muscular tension in occupational muscle pain and in chronic musculoskeletal pain syndromes. A hypothesis. *Medical Hypotheses* **35**, 196-203.

JOVANOVIC, K., ANASTASIJEVIC, R. & VUCO, J. (1990) Reflex effects on γ-fusimotor neurones of chemically induced discharges in small-diameter muscle afferents in decerebrate cats. *Brain Res.* **521**, 89-94.

SYMPATHETICALLY-INDUCED CHANGES IN THE SPINDLE AFFERENT RESPONSE TO VIBRATORY STIMULI ELICITING THE TVR IN RABBIT JAW CLOSING MUSCLES

M. Passatore[1], F. Deriu[1], S. Roatta[1]
and C. Grassi[2]

[1]Department of Anatomy and Human Physiology
University of Turin, Turin, Italy
[2]Institute of Human Physiology
Catholic University, Rome, Italy

INTRODUCTION

Sympathetic nervous system activation can influence the stretch reflex in the jaw closing muscles of decerebrate rabbits through an action exerted at the peripheral level (Grassi, Deriu, Artusio & Passatore, 1993; Grassi, Deriu & Passatore, 1993). In particular, stimulation of the cervical sympathetic nerve (CSN), at frequencies within the physiological range, induces a marked depression, often preceded by a transient enhancement, of both the jaw jerk and the tonic vibration reflex (TVR) in the jaw closing muscles. This effect is mainly mediated by α_1-adrenergic receptors, though the noradrenaline co-transmitter NPY may make a small contribution. Indirect evidence supports the idea that the decrease in the stretch reflex induced by sympathetic stimulation is due to an action on muscle spindle afferents exerted by adrenergic mediators. We have tested this hypothesis by studying the sympathetically-induced changes in the spindle afferent response to vibratory stimuli eliciting the TVR.

METHODS

The experiments were performed on rabbits precollicularly decerebrated under urethane, ketamine-xylazine anaesthesia (400, 5 and 1.5 mg/kg respectively, IV). Some trials were carried out after having blocked neuromuscular junctions with pancuronium bromide or tubocurarine (0.3 and 0.25 mg/kg respectively, IV, repeated when necessary). Vibratory stimuli (5-50 μm at 170 Hz) lasting 5 s were applied to the mandibular symphysis at 15 s intervals through a length servo-controlled puller. Afferent discharge from jaw muscle spindles was recorded in the mesencephalic trigeminal nucleus by tungsten microelectrodes (5 MΩ impedance) and EMG activity of the masseter muscle was recorded by either belly-tendon copper leads or bipolar coaxial electrodes. Tension developed by jaw closing muscles was measured with a force transducer. The peripheral stump of the CSN was stimulated at 10/s (trains lasting 30 s - 2 min, 0.5 ms pulse duration, 4-8 V). Thyroid artery and femoral vein were cannulated for drug injection.

RESULTS

Stimulation of the CSN at 10/s consistently induced a marked depression of spindle afferent response to vibratory stimuli, lasting throughout the stimulation period. This was often preceded by a transient increase. In most of the tested units, parallel effects were observed both on EMG activity of the masseter muscle and on the reflexly developed tension (Figure 1). While the spindle afferent response to muscle vibration was still depressed by the stimulation, enhancing the vibration amplitude could restore spindle afferent activity as well as muscle activity recorded in the same units under control conditions.

The above described reduction in the spindle afferent response to 5 s of continuous vibration was superimposeded on sympathetically-induced changes of the background discharge seen during the intervals between successive periods of vibration.

Changes in background firing usually consisted of a 10-25 s latency increase in the discharge frequency, ranging from 20 to 400% of the control values. This was occasionally preceded by a period of decrease (see also Passatore, Grassi, & Filippi, 1985). The reduction of the spindle response to vibration was also present in those units in which no significant change or a decrease in the basal discharge occurred. The above described effects on spindle afferent activity were also present in curarised animals.

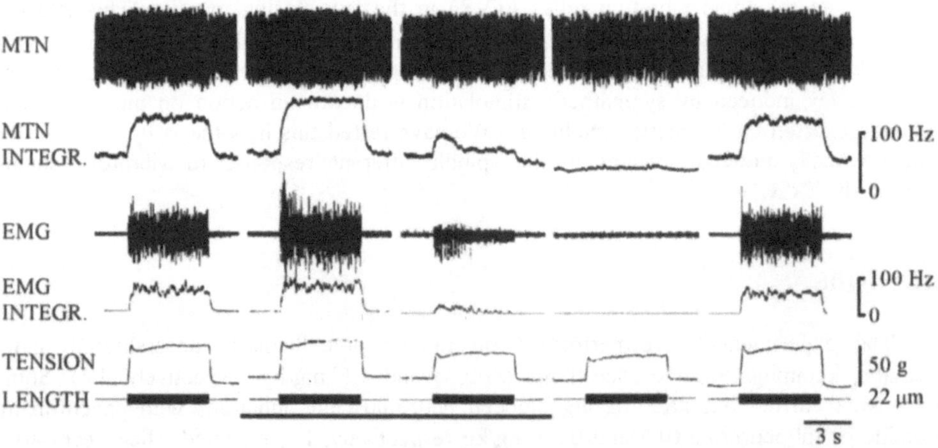

Figure 1. Effect of CSN stimulation on spindle afferent discharge (MTN), masseter EMG and developed tension during TVR in the jaw closing muscles. The sympathetically-induced decrease in both EMG activity and jaw muscle tension is associated with a parallel reduction of the unitary spindle afferent activity. This effect is often preceded by a transient increase of the reflex, during which the recorded parameters usually exhibit parallel changes. The TVRs are taken during control, 15 and 60 s after the beginning of CSN stimulation (signalled by bar), 10 s and 3 min after its interruption. CSN stimulated with a 10/s train lasting 90 s, 0.5 ms pulse duration, 8 V. Time constant of integrated signals, 200 and 500 ms for MTN and EMG, respectively.

The sympathetic effects on spindle afferent activity were mimicked by close arterial injection of the selective α_1-adrenoceptor agonist phenylephrine (3.5-7.0 μg/kg).

CONCLUSION

These data show that activation of the sympathetic can modulate both static and dynamic responses of muscle spindle afferents. These changes match the sympathetically-induced modifications of the TVR and are therefore probably responsible for the previously described changes in the stretch reflex in jaw closing muscles.

The gain of the feedback pathways controlling a particular movement exhibits large modulations, as shown in a variety of motor tasks (see Prochazka, 1989). It is suggested that the action exerted by the sympathetic supply on spindle afferent activity may be one of the mechanisms involved in adjusting a motor act to the context in which it is executed, such as states of physical and emotional stress.

ACKNOWLEDGEMENTS

Supported by CNR and MURST grants.

REFERENCES

GRASSI, C., DERIU, F., ARTUSIO, E. & PASSATORE, M. (1993) Modulation of the jaw jerk reflex by the sympathetic nervous system. *Arch. Ital. Biol.* **131**, 213-226.

GRASSI, C., DERIU, F. & PASSATORE, M. (1993) Effect of sympathetic nervous system activation on the tonic vibration reflex in rabbit jaw closing muscles. *J. Physiol.* **469**, 601-613.

PASSATORE, M., GRASSI, C. & FILIPPI, G. M. (1985) Sympathetically-induced development of tension in jaw muscles: the possible contraction of intrafusal muscle fibres. *Pflügers Archiv* **405**, 297-304.

PROCHAZKA, A. (1989) Sensorimotor gain control: a basic strategy of motor systems?. *Prog. Neurobiol.* **33**, 281-307.

PART 4

INNERVATION PATTERNS

EARLY DEVELOPMENT OF THE MUSCLE SPINDLE AND ITS DEPENDENCE ON NEUROTROPHIC FACTORS

Jan Kucera[1], Patrik Ernfors[2],
Jens Schwarze[1] and Rudolf Jaenisch[2]

[1]Department of Neurology
Boston University School of Medicine
80 East Concord Street
Boston MA 02118-2394, USA
[2]Whitehead Institute for Biomedical Research
and Department of Biology, MIT
Nine Cambridge Center
Cambridge MA 02142, USA

INTRODUCTION

Formation of muscle spindles in hindlimbs of rodents is dependent on contact between a sensory neurone and a developing muscle fibre (Zelená, 1957). This type of contact not only induces the formation of intrafusal fibres, but may also mediate the support of proprioceptive neurones in the dorsal root ganglia (DRGs) by muscle-derived trophic factors. Neurotrophins such as brain-derived neurotrophic factor (BDNF) or neurotrophin-3 (NT-3) support the survival of developing muscle sensory neurones *in-vitro* and can be retrogradely transported to DRGs from muscles (DiStephano, Friedman, Radziejewski, Alexander, Boland, Schick, Lindsay & Wiegand, 1992; Hory-Lee, Russell, Lindsay & Frank, 1993). Subpopulations of DRG neurones express the BDNF and NT-3 receptors, *trk*B and *trk*C (Mu, Silos-Santiago, Carroll & Snider, 1993).

In the present study, we examined the type and morphology of early nerve-muscle contacts as they relate to spindle precursors in rat hindlimbs. In addition, we have addressed the role of neurotrophins in the differentiation of proprioceptive neurones by examining spindle development in two strains of mutant mice carrying a deletion of either the BDNF or NT-3 gene.

METHODS

Fetal Sprague-Dawley rats were used to study the development of innervation in the soleus muscle at embryonic (E) days 15-21. The day of impregnation was considered as E0. BALB/c 126 outbred mice carrying a deletion of the BDNF or NT-3 gene were used to study the effect of neurotrophin deficiency on spindle development at postnatal (P) day 14. The mutant mice were generated by homologous recombination in embryonic stem cells (Ernfors, Lee, Kucera & Jaenisch, 1994a; Ernfors, Lee & Jaenisch, 1994b). Genotype was determined by Southern blot analysis of DNA. Hindlimbs and lumbar spinal cords were excised from homozygous, heterozygous recessive and wild type animals under sodium pentobarbital anesthesia (50mg/100g body weight). Tissues were fixed in 2.5% glutaraldehyde and 2% paraformaldehyde, embedded in eponate 12 and examined by light and electron microscopy. Transverse 0.5μm thick sections of L4 DRGs, L4 ventral roots and L4 dorsal roots were analyzed using the Zeiss Videoplan Image Analyzer to obtain frequency histograms of soma and nerve fibre sizes.

RESULTS

Early nerve-muscle contacts

Four categories of neuromuscular contacts were identified in the developing rat soleus muscle. Definitive motor nerve contacts were observed at E17-E18. They resembled motor endplates. They consisted of small to medium size axon terminals with synaptic vesicles. In addition, a distinct basal lamina was visible in a relatively wide synaptic cleft. Definitive sensory nerve-muscle contacts were identified at E18. They were present only on muscle fibres enclosed by the spindle capsule. Axon terminals were large and the synaptic

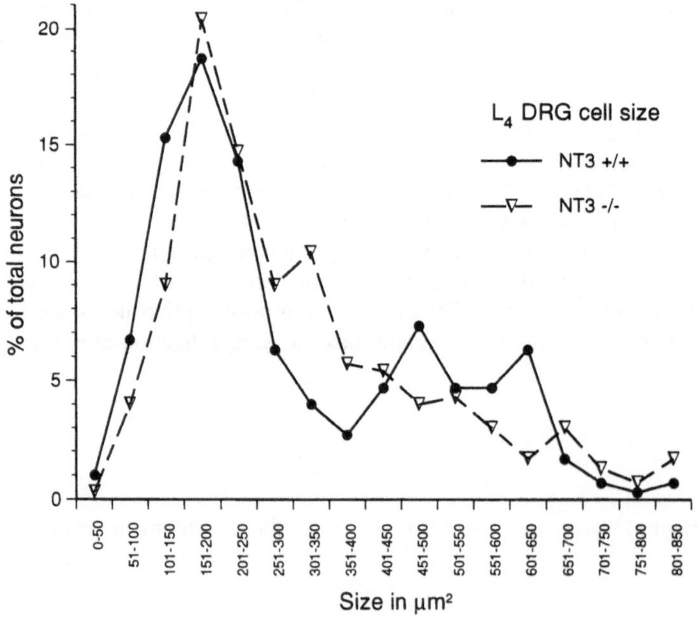

Figure 1. Sizes of 300 cells each in an L4 dorsal root ganglion from an NT-3 +/+ and an NT-3 -/- mouse at P14. Note the similarity in distribution of cell sizes, except for a slight relative deficiency of cells in the 400-650μm² range in the NT-3 mutant.

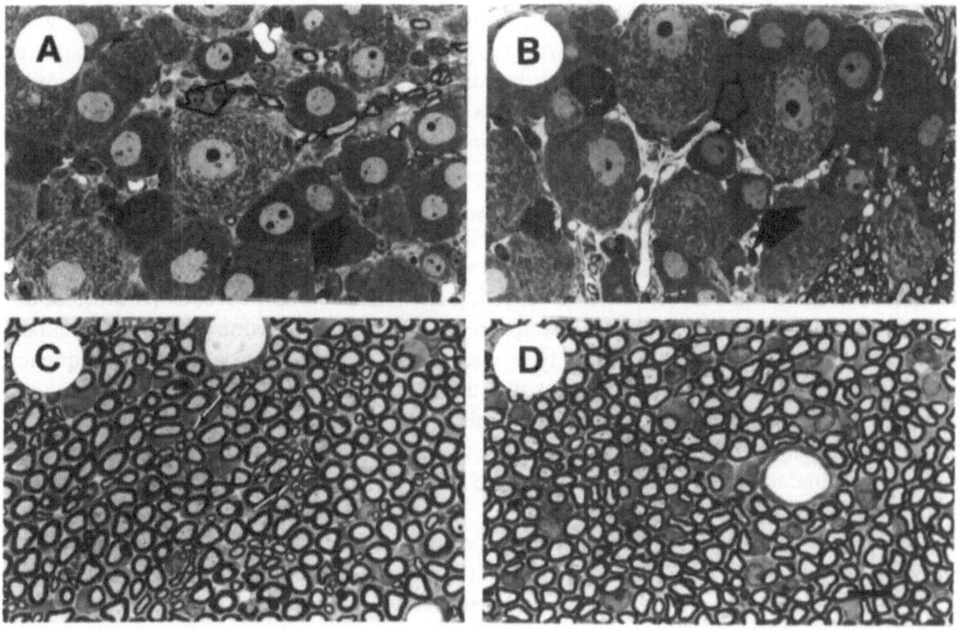

Figure 2. Sections of an L4 dorsal root ganglion (A & B) and an L4 ventral root (C & D) from an NT-3 -/- (A & C) and a wild type (+/+) mouse (B & D) at P14. Note that both large, light neurons (light arrows) and small, dark neurones (dark arrows) are present in both mutant and wild type. Also note that the small-caliber motor nerve fibres (arrows) are absent in the VR of the mutant mouse. Bar = 40μm (A & B) and 20μm (C & D).

cleft was relatively narrow; there was no interposed basal lamina. Probable sensory nerve-muscle contacts of E16-E17 muscles were ultrastructurally similar to the definitive sensory nerve-muscle junctions except for the absence of a spindle capsule. Undifferentiated nerve-muscle contacts were observed in E15-E16 muscles. They were derived from axons or bundles of axons whose features did not permit classification as either motor or sensory neurones. The cleft between the apposing sarcolemma and axolemma was narrow and devoid of basal lamina, although patches of basal lamina lined the adjacent regions of many of the myotubes. Axon profiles contained few or no vesicles. The contacts were frequently observed on the cells devoid of myofilaments (presumptive myoblasts), and on newly-formed primary myotubes.

NT-3 and BDNF mutant mice

Neonatal mice homozygous for a deleted NT-3 gene (-/-) were smaller than control littermates, and exhibited ataxic, pseudo-athetoid movements of the hindlimbs, with difficulty maintaining an upright posture (Ernfors et al., 1994a). Soleus (SOL) muscles of NT-3 -/- mice were entirely devoid of spindles, although SOL of heterozygotes for the NT-3 deletion (+/-) contained one-half the spindle complement (5.1±1.1, n = 10) of the SOL from wild type (+/+) mice (10.6±1.3, n = 11). An absence of spindles in NT-3 -/- mice was noted as early as at E17 and E18, thus the deficiency of spindles observed in heterozygous and homozygous mutants resulted from an inhibition of spindle formation rather than degeneration of newly-formed spindles. Probable sensory nerve-muscle contacts were observed in E15 NT-3 +/+ muscles, but not in E15 NT-3 -/- muscles.

Paucity of spindles in NT-3 -/- mutants was associated with a deficiency of nerve fibres in the spinal nerves at P14. In addition, both the sensory and motor nerve fibres were slightly smaller in mutant relative to wild type mice, presumably reflective of the smaller body size of NT-3 -/- animals. The distribution of cell sizes in the L4 DRGs was similar between the mutant and wild type mice, except for a slight relative deficiency of medium to large neurones in the mutants (Figure 1). However, no differences could be discerned in the types of neurones present in DRGs of NT-3 -/- and NT-3 +/+ mice upon examination by light microscopy (Figure 2a & b). Specifically, the light, type A as well as the dark, type B sensory neurones were present in both the wild type and mutant DRGs (Rambourg, Clermont & Beaudet, 1983). The L4 dorsal root (DR) contained half the normal complement of myelinated nerve fibres reflecting the loss of DRG neurones (Ernfors et al., 1994a). Fewer myelinated fibres were present in L4 ventral roots (VRs) of mutant (671$\pm$105, n = 3) compared to wild type mice (mean 1100$\pm$40, n = 2). The frequency histogram of the L4 VR nerve fibre sizes was unimodal in the mutants rather than bimodal, indicative of a gross deficiency of the small-caliber myelinated nerve fibres (Figure 3). The deficiency in this type of neurone was not due to a delay of myelination because only a few or no unmyelinated axons were observed in the VRs, thus the paucity was assumed to result from a deficiency in NT-3 production.

BDNF -/- mice had a normal complement of spindles in hindlimbs even though their L4 DRGs were showed a 30% loss of neurones (Ernfors et al., 1994b). Both the small- and large-caliber motor nerve fibres were present in the VRs of these mutants, similar to wild type mice.

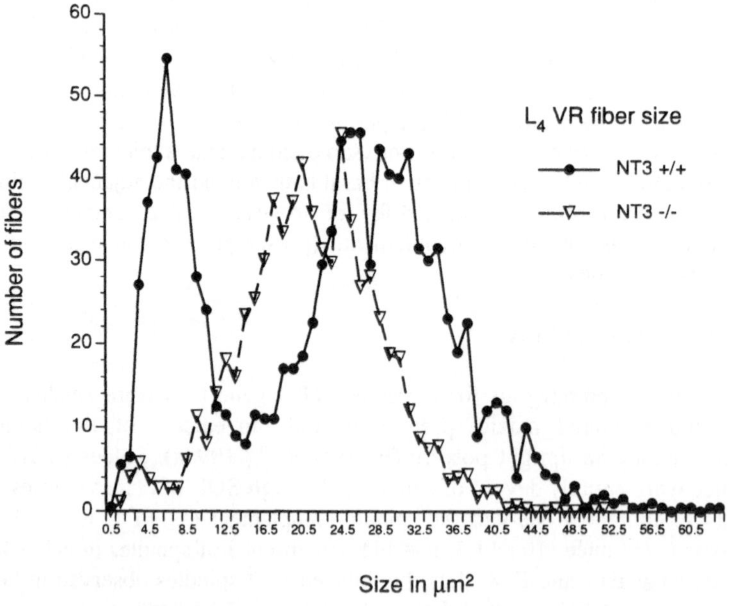

Figure 3. Histogram of mean nerve fibre sizes in L4 ventral roots of NT-3 -/- (n = 3) and wild type (+/+) mice (n = 2) at P14. All nerve fibres are slightly smaller in the mutant relative to the wild type and the nerve size distribution is unimodal in the mutant compared to bimodal in the wild type, reflective of the absence of a population of small-caliber nerve fibres.

DISCUSSION

The study shows that nerve-muscle contacts are present in rat hindlimbs several days prior to the formation of recognisable muscle spindles and motor endplates. The undifferentiated nerve-muscle contacts of E15-E16 muscles were devoid of basal lamina in the synaptic cleft, and therefore resembled the sensory contacts of established spindles. Whether all of these early contacts were sensory or whether the precursors of motor endplates can also be devoid of basal lamina in the junctional cleft was unclear. However, functional motor junctions must be present in early fetal muscles because nerve stimulation of E14 intercostal muscles triggers muscle contraction (Kelly & Zacks, 1969).

A striking feature of the undifferentiated nerve-muscle contacts was the involvement of muscle cells at the earliest stages of their development, including presumed myoblasts and newly-assembled myotubes. Thus, the contacts may be important for mutual recognition of specific neuronal and muscle cell populations, or exchange of neuroactive molecules between the apposing neurones and muscle cells.

The absence of muscle spindles in NT-3 -/- mice probably resulted from an absence of Ia afferent neurones, as suggested by the absence of projections of Ia axon collaterals to the spinal motor pools and a loss of neurones immunoreactive to parvalbumin in the DRGs (Ernfors et al., 1994a). Development of spindles requires sensory, but not motor, innervation (Kucera & Walro, 1992). The absence of Golgi tendon organs in NT-3 mutants may similarly reflect a paucity of Ib neurones (Ernfors et al., 1994a). Neither the Ia nor Ib neurones may correlate with a single morphological type of afferent neurone as the types of neurones represented in the mutant and wild type DRGs were similar.

NT-3 may be essential early in the development of spindles. The absence of spindles in NT-3 -/- embryos suggests that either the developing Ia sensory neurones never innervated the limb, or that they innervated the limb but did not establish permanent nerve-muscle junctions in the absence of NT-3.

Spindle number is highly predictable in rodent hindlimb muscles. Our study suggests that the availability of the appropriate types of myotubes is not the factor which limits spindle formation. Slow myotubes (recognised by the pattern of expression of myosin heavy chain isoforms) which could form intrafusal fibres were present in the mutant SOL and exceeded the number of spindles which form in wild type mice (Kucera, unpublished). Rather, numbers of Ia neurones most likely regulate spindle number. Their numbers may in turn be regulated by NT-3 in a dose-dependent manner. One-half the normal complement of spindles were present in NT-3 +/- mice, presumably because only half the normal amount of NT-3 was produced.

The small-caliber fibres in ventral roots are mostly fusimotor (γ) neurones, although a few preganglionic autonomic neurones are present. The paucity of small fibres in the VRs of NT-3 -/- mice suggests that most fusimotor (γ) neurones either did not develop or did not survive in the absence of spindles. Few γ axons survive nerve section in neonatal rats suggesting that developing γ neurones may depend on a muscle-derived factor for trophic support (Kucera, Walro & Gao, 1993). Whether this factor is NT-3 or some other neurotrophic agent associated with intrafusal fibres is unclear. However, mice lacking the *trk*C receptor that binds NT-3 also have fewer motor axons in the VRs, suggestive that some motor neurones are dependent on NT-3 for their survival (Klein, Silos-Santiago, Smeyne, Lira, Brambilla, Bryant, Zhang, Snider & Barbacid, 1994). The population of large-caliber VR fibres, reflective of skeletomotor (α) and skeletofusimotor (β) neurones was slightly diminished in the mutants, raising the possibility that β neurones are also depleted in the absence of NT-3.

Thus NT-3 may act in a coordinated fashion to support, either directly or indirectly, the development of each of the three classes of neurone -- Ia, Ib and γ -- that underlie the muscle proprioceptive system. The specificity of the NT-3 effect is underscored by the preservation of limb proprioceptive organs in BDNF-deficient mice.

ACKNOWLEDGEMENTS

The study was supported by PHS grant NS25796 to J.K., an NIH grant (5R35CA44339) ro R.J. and by the Amgen Corporation.

REFERENCES

DISTEPHANO, P.S., FRIEDMAN, B., RADZIEJEWSKI, C., ALEXANDER, C., BOLAND, P., SCHICK, C.M., LINDSAY, R. & WIEGAND, S.J. (1992) The neurotrophins BDNF, NT-3 and NGF display distinct patterns of retrograde axonal transport in peripheral and central neurons. *Neuron* 981-993.

ERNFORS, P., LEE, K.-F., KUCERA, J. & JAENISCH, R. (1994a) Lack of neurotrophin-3 leads to deficiencies in the peripheral nervous system and loss of limb proprioceptive afferents. *Cell* 77, 503-512.

ERNFORS, P., LEE, K.-F. & JAENISCH, R. (1994b) Mice lacking brain-derived neurotrophic factor develop with sensory deficits. *Nature* **368**, 147-150.

HORY-LEE, F., RUSSELL, M., LINDSAY, R.M. & FRANK, E. (1993) Neurotrophin-3 supports the survival of developing muscle sensory neurons in culture. *Proc. Natl. Acad. Sci. USA* **90**, 2613-2617.

KELLY, A.M. & ZACKS, S.I. (1969) The histogenesis of rat intercostal muscle. *J. Cell Biol.* **42**, 435-152.

KLEIN, R., SILOS-SANTIAGO, I., SMEYNE, R.J., LIRA, S.A., BRAMBILLA, R., BRYANT, S., ZHANG, L., SNIDER, W.D. & BARBACID, M. (1994) Disruption of the neurotrophin-3 receptor gene *trk*C eliminates Ia muscle afferents and results in abnormal movements. *Nature* **368**, 249-251.

KUCERA, J. & WALRO, J.M. (1992) Superfluousness of motor innervation for the formation of muscle spindles in neonatal rats. *Anat. Embryol.* **186**, 301-309.

KUCERA, J., WALRO, J.M., GAO, Y. (1993) Fusimotor-free spindles in reinnervated muscles of neonatal rats. *Neuroscience* **52**, 219-228.

MU, X., SILOS-SANTIAGO, I., CARROLL, S.L. & SNIDER, W.D. (1993) Neurotrophin receptor genes are expressed in distinct patterns in developing dorsal root ganglia. *J. Neurosci.* **13**, 4029-4041.

RAMBOURG, A.; CLERMONT, Y. & BEAUDET, A. (1983) Ultrastructural features of six types of neurons in rat dorsal root ganglia. *J. Neurocytol.* **12**, 47-66.

ZELENÁ, J. (1957) The morphogenetic influence of innervation on the ontogenic development of muscle-spindles. *J. Embryol. Exp. Morphol.* **5**, 283-292.

ORIGIN OF SPINDLES IN DEVELOPING RAT HINDLIMB MUSCLES

J.M. Walro and Jun Wang

Department of Anatomy,
Northeastern Ohio Universities
College of Medicine,
Rootstown OH 44272-0095, USA

INTRODUCTION

Whether intrafusal and extrafusal fibres arise from a common progenitor in developing rat skeletal muscle is unresolved. Pedrosa & Thornell (1990) proposed that the primary myotubes which differentiate into the bag_2 intrafusal fibre form from fusion of myogenic cells committed to differentiate into intrafusal fibres only. They further suggested that the precursors of bag_2 fibres are capable of attracting sensory innervation to the developing muscle spindles, and that all subsequent generation types of intrafusal fibre form by a similar process (Pedrosa & Thornell, 1990). In contrast, Kucera & Walro (1990) proposed that both intrafusal and extrafusal fibres originate from several pools of bipotential myotubes, and that the interaction between primary afferents and bipotential myotubes mediates transformation of these myotubes into the different types of intrafusal fibres.

The present study reports that an avian monoclonal antibody (mAb), S46, binds to a myosin heavy chain (MHC) isoform expressed by subpopulation of slow primary myotubes in rats. This population of myotubes, the oldest myotubes in hindlimb muscles, is bipotential in that it consists of precursors of both bag_2 intrafusal and type 1 extrafusal fibres.

MATERIALS AND METHODS

Animals

Fetal, neonatal and adult (60 day) offspring of ten timed-pregnant Sprague-Dawley rats were a source of muscles. The ages of rats were determined either from the day of impregnation (embryonic day 0 or EO) or the date of birth (postnatal day 0 or P0). Two rats each were examined at 24 h intervals from day E15 to E21, and at PO, P7 and P60.

Tissue processing

Rats were anaesthetised with sodium pentobarbital (50 mg/kg i.p.). Entire right lower legs (crura) of prenatal or early postnatal rats or individual leg muscles of adult rats were removed and frozen in isopentane. Midportions of the crural muscles were cut transversely into serial 8-μm sections in a cryostat. Sets of serial sections were reacted with a panel of four monoclonal antibodies: WBMHC-s, specific for slow-twitch MHC; MY32, specific for neonatal/adult MHC; ALD19, specific for slow-tonic MHC; and S46, specific for SM1 and SM2 (Page, Miller, DiMario, Hager, Moser & Stockdale, 1992). Binding of the primary antibodies was demonstrated using the avidin-biotin-complex (ABC) peroxidase method (Kucera, Walro & Gorza, 1992).

Identification of muscles and spindles

Although all crural muscles were examined, particular attention was given to the tibialis anterior (TA), a muscle predominantly composed of type 2 extrafusal fibres, and the soleus (SOL), a muscle predominantly composed of type 1 fibres. Myotubes were classified as primary or secondary based on size and chronology of development, and as slow, fast or mixed (slow/fast) based on their patterns of binding for WBMHC-s and MY32 (Kucera & Walro, 1990). Spindles were identified as encapsulations of small-diameter muscle fibres that contained one or two fibres reactive to the slow-tonic mAb, ALD 19 (Pedrosa & Thornell 1990). Intrafusal fibres were classified as nuclear bag_2, nuclear bag_1 and nuclear chain fibres on the basis of relative fibre size and patterns of MHC expression (Kucera & Walro, 1990). Regions of intrafusal fibres in P60 rats were classified as A (equatorial and juxta-equatorial), B (polar encapsulated) or C (extracapsular) based on established morphological criteria (Kucera, Dorovini-Zis & Engel, 1978).

RESULTS

Early in development at E14-16, the inner slow (red) and outer fast (white) regions characteristic of the mature TA muscle could not be distinguished by binding of WBMHC-s and MY32 because the newly-formed TA muscle contained a homogeneous population of primary myotubes which expressed both slow-twitch and neonatal/fast MHC isoforms. However, these two regions could be identified by differences in S46-binding as early as E16. Myotubes in the future slow region were S46-reactive, whereas myotubes in the future fast region did not bind S46. The two regions of the TA were first differentiable using WBMHC-s and MY32 on E17-18 when the primary myotubes in the outer region of the muscle began to lose their reactivity to the slow-twitch mAb, whereas the primary myotubes of the inner zone began to lose their reactivity to the neonatal/fast mAb. All primary myotubes bound S46 on E14 but by E18, the end of primary myogenesis, myotubes reactive to S46 were located only in the inner axial (future red) region of the TA muscle. Primary myotubes in the outer (future white) region of the TA, especially the lateral part of the TA that overlies the EDL muscle, did not bind S46. A band of myotubes moderately reactive to S46 separated the strongly reactive myotubes of the inner region from the unreactive myotubes of the outer region. No secondary myotubes bound S46 in the TA. Similar patterns of reactivity to mAb S46 were observed in other predominantly fast muscles.

Unlike the TA, the SOL muscle showed no distinct fibre regions at any age. All primary myotubes of the SOL bound the slow-twitch mAb WBMHC-s throughout fetal development. The predominantly slow SOL was the only muscle of the crus in which all primary myotubes, regardless of their location in the muscle, bound mAb S46. In addition,

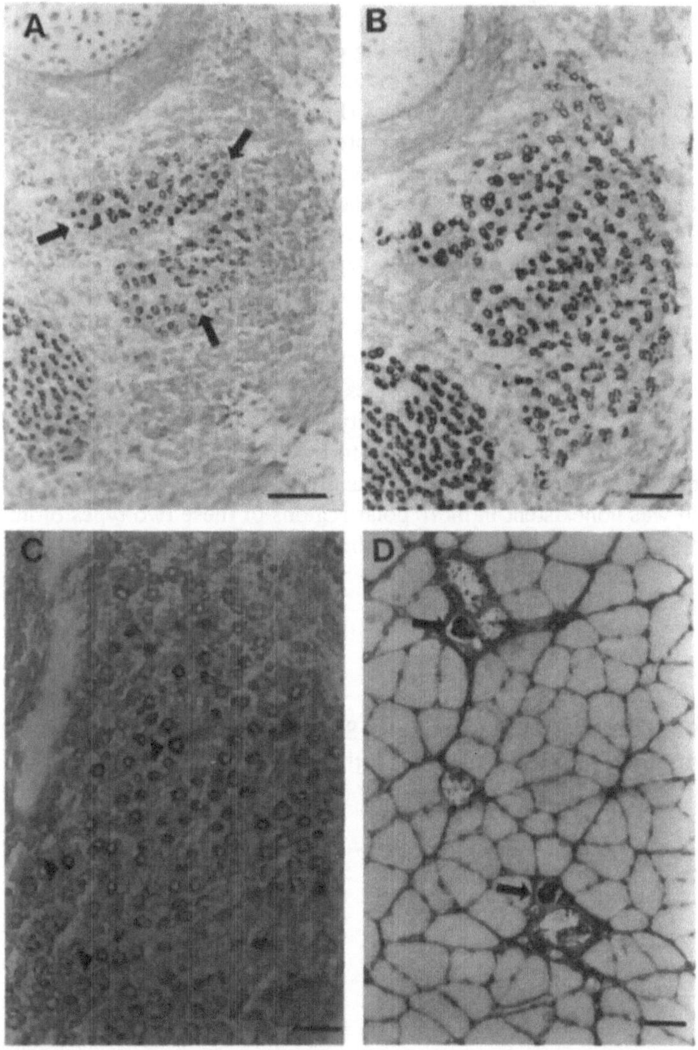

Figure 1. Tibialis anterior muscles from E17 (A & B), E19 (C), and P60 (D) rats. A, C, and D are reacted with the mAb S46 and B is reacted with mAb WBMHC-s. A & B: Myotubes which react with S46 (arrows) in A are a subset of the myotubes which react with the slow-twitch mAb in B. C: At E18, myotubes differentiating into bag_2 fibres strongly bind S46 (arrowheads), whereas S46-binding is dissipating in myotubes differentiating into type 1 extrafusal fibres. D: Nuclear bag intrafusal fibres in two adult spindles (arrows) bind S46 in the polar encapsulated region, analogous to that which would be predicted for mAb ALD19. Scale bars = 100 μm in A and B, and 50 μm in C and D.

the slow/fast secondary myotubes bound S46 in the SOL, but fast secondary myotubes did not.

Spindle formation begins on E17 when afferents contact some of the primary myotubes (Zelena 1957; Kucera, Walro & Reichler, 1989). Nascent spindles are then detectable in rat hindlimbs at E17.5-E18 as single myotubes binding ALD 19 (Pedrosa & Thornell, 1990). In the present study, the bag_2 fibre, the first intrafusal fibre to form, originated from the population of S46-reactive primary myotubes (Figure 1A), which was in turn a subset of slow primary myotubes (Figure 1B). Binding of S46, detectable in these myotubes as early as E14 in the TA and E15 in the SOL, intensified in some of the myotubes at E17.5-E18 located in the inner portions of both muscles (Figure 1C). These myotubes were assumed to be undergoing transformation into bag_2 intrafusal fibres consequent upon the afferent innervation (Kucera et al., 1989) because these myotubes developed equatorial aggregations of nuclei and became encapsulated 1-2 days later. Concurrently, S46 reactivity decreased in myotubes of the TA and SOL muscles that did not transform into intrafusal fibres (Figure 1C). As a result, strong binding of S46 to bag_2 intrafusal fibres contrasted with a residual weak binding of S46 by future type 1 extrafusal fibres in TA and SOL muscles at E21. Maturing type 1 extrafusal fibres in both TA and SOL muscles ceased to bind S46 by the end of the first postnatal week, thus only nuclear bag_2 and nuclear bag_1 intrafusal fibres of spindles reacted to S46 in adult (P60) muscles (Figure 1D).

Patterns of S46- and ALD19-reactivity were similar in the three types of adult intrafusal fibres. Both mAbs bound strongly to the A and inner B regions of the bag_2 fibre, and throughout the entire the A and B regions of the bag_1 fibre in adult rats. Binding rapidly dissipated and was nonexistent in more polar regions of these two types of intrafusal fibres. Both antibodies also bound to some chain fibres, but binding was limited to 1-2 sections at the equator of the fibre.

DISCUSSION

The isoform recognised by S46 is not SM-1, SM-2 or SM-3 subtypes of avian slow myosin. None of the antibodies specific for these isoforms such as NA1, NA2, NA3, NA7

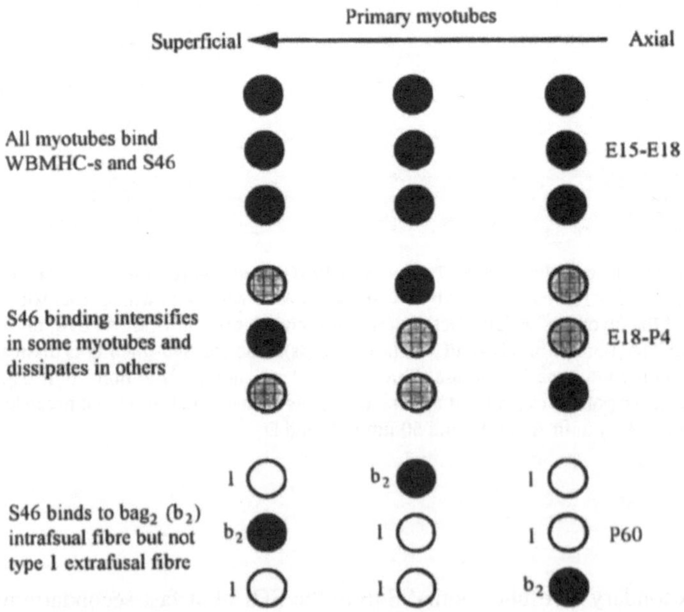

Figure 2. Development of the bag_2 fibre in a "slow" muscle such as the SOL. Age of myotubes/myofibres is shown at right. Intensity of shading denotes reactivity to the mAb S46. B_2 denotes the bag_2 fibre; and 1 denotes type 1 extrafusal fibres.

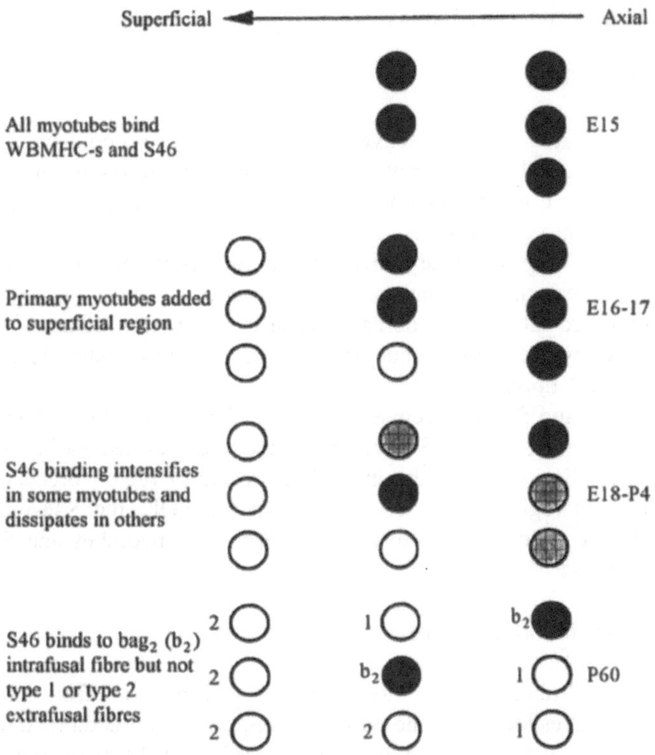

Figure 3. Development of the bag$_2$ intrafusal fibre in a "fast" muscle such as the TA. Age of myotubes/myofibres is shown at right. Intensity of shading denotes reactivity to the mAb S46. B_2 denotes the bag$_2$ fibre, and 1 and 2 denote type 1 and 2 extrafusal fibres.

and NA8 (E. Bandman, unpublished data), bound to any muscle fibres in the rat hindlimb. In addition, S46-reactive isoform is not one of the three developmental slow isoforms described recently in mammals (Hughes, Cho, Karsch-Mizrachi, Travis, Silberstein, Leinwand & Blau, 1993) because the duration of expression, and the types of fibres which express the isoform recognised by S46 differ from the three developmental slow isoforms. The isoform recognised by mAB S46 (henceforth referred to as MHC_{sd}) in rat hindlimb muscles is probably a developmental slow isoform, because S46 binds only myotubes which also bind slow-twitch mAbs and binds fetal myotube precursors of adult extrafusal fibres, but not to adult extrafusal fibres themselves.

Expression of MHC_{sd} is probably myogenic. Its expression is detectable as early as E14, concurrent with entry of nerves into the developing muscle masses, but prior to nerve-muscle cell contact. In addition, it is expressed in the absence of innervation in fetal rats treated with ß-bungarotoxin and in postnatal rats following sciatic nerve section (Kucera and Walro, unpublished data). However, the intensity of expression of MHC_{sd} by adult intrafusal fibres exceeded that of S46-reactive myotubes prior to E18. Thus afferents not only maintain, but also upregulate, expression of this isoform in nuclear bag intrafusal fibres.

The binding patterns of S46 and ALD19 were identical in adult spindles, thus both mAbs may recognise the same MHC isoform. However, S46 may be more sensitive than ALD19 because it binds to myotubes as early as E14, whereas ALD19-binding is first detectable at E17-18. However, the two antibodies differed in that S46 did not bind to type 2 extrafusal fibres, irrespective of dilution, whereas ALD19 cross-reacts with type 2 extrafusal fibres at high concentrations.

Development of extrafusal fibres

Two types of myotubes, S46-reactive and S46-unreactive, could be distinguished in fetal hindlimb muscles. All S46-reactive slow primary myotubes in the TA and SOL differentiated into type 1 myofibres only, whereas all fast secondary myotubes in both muscles differentiated into type 2 fibres. The fate of slow/fast myotubes differed in the two muscles. Mixed (slow/fast) secondary myotubes in the SOL differentiated into type 1 extrafusal fibres, whereas the slow/fast primary myotubes of the outer region of the TA must have differentiated into type 2 fibres because in this region of adult TA muscles is composed entirely of type 2 fibres. Thus MHC_{sd} and not slow-twitch MHC is a reliable marker for future type 1 fibres.

The S46-reactive and S46-unreactive myotubes differed not only in their ultimate fate, but also in the time of assembly. S46-reactive myotubes formed earlier than the S46-unreactive myotubes. Formation of the inner (S46-reactive) region preceded that of the outer (S46-unreactive) region in the TA muscle. In the SOL, the S46-reactive primary myotubes formed first, followed by S46-reactive secondary myotubes and S46-unreactive fast secondary myotubes.

Development of bag_2 intrafusal fibres

The first intrafusal fibre to form, the nuclear bag_2 fibre, forms from a subset of S46-reactive primary myotubes located in the deep axial regions of rat hindlimb muscles (Figures 2 & 3). All myotubes which developed into bag_2 fibres bound the mAb S46 and expressed slow-twitch MHC, analogous to those myotubes which gave rise to type 1 extrafusal fibres. Collectively this bipotential set of myotubes was the oldest set of myotubes in the muscle and they were located in the deep, axial regions of most muscles. Origin of both bag_2 and type 1 extrafusal fibres from this set of myotubes may be the mechanism responsible for the location of spindles in the deep (red) regions of hindlimb muscles.

Bag_2 intrafusal fibres arise from myotubes which initially express the same myosin heavy chain isoforms as myotubes which give rise to type 1 extrafusal fibres. Heterogeneity in MHC expression within this population of myotubes coincides with innervation of future bag_2 fibres by afferent neurones. If contacted by an afferent neurone, a S46-reactive myotube may differentiate into a bag_2 fibre; otherwise it differentiates into a type 1 intrafusal fibre. Whether this afferent-myotube contact is a random event is not known. Some primary myotubes which express MHC_{sd} may have receptors for afferent neurones whereas others may not. If so, the nature of such receptors remains to be elucidated.

ACKNOWLEDGEMENTS

ALD19 was a gift from Dr. J. Sawchak, S46 was a gift from Dr. F. Stockdale and NA1, NA2, NA3, NA7, and NA8 were gifts from Dr. E. Bandman.

REFERENCES

HUGHES, S.M., CHO, M., KARSCH-MIZRACHI, I., TRAVIS, M., SILBERSTEIN, L., LEINWAND, L.A. & BLAU, H.M. (1993) Three slow myosin heavy chains sequentially expressed in developing mammalian skeletal muscle. *Dev. Biol.* **158**, 183-199.

KUCERA, J., DOROVINI-ZIS, K. & ENGEL, W.K. (1978) Histochemistry of rat intrafusal fibers and their motor innervation. *J. Histochem. Cytochem.* **26**, 973-988.

KUCERA, J. & WALRO, J.M. (1990) Origin of intrafusal muscle fibers in the rat. *Histochemistry* **93**, 567-580.

KUCERA, J., WALRO, J.M., & GORZA, L. (1992) Expression of type-specific MHC isoforms in rat intrafusal muscle fibers. *J. Histochem. Cytochem.* **40**, 293-307.

KUCERA, J., WALRO, J.M. & REICHLER, J. (1989) The role of nerve and muscle factors in the development of muscle spindles. *Am. J. Anat.* **186**, 144-160.

PAGE, S., MILLER, J.B., DiMARIO, J.X., HAGER, E.J., MOSER, A., & STOCKDALE, F.E. (1992) Developmentally regulated expression of three slow isoforms of myosin heavy chain: diversity among the first fibers to form in avian muscles. *Dev. Biol.* **154**, 118-128.

PEDROSA, F. & THORNELL, L-E. (1990) Expression of myosin heavy chain isoforms in developing muscle spindles. *Histochemistry* **94**, 231-244.

ZELENA, J. (1957) The morphogenic influence of innervation on the ontogenetic development of muscle spindles. *J. Embryol. Exp. Morphol.* **5**, 283-292.

REINNERVATION OF CAT SKELETO-FUSIMOTOR UNITS AFTER NERVE SECTION

D. Barker[1], J.J.A. Scott[2] and M.J. Stacey[1]

[1]Department of Biological Sciences,
University of Durham, Durham DH1 3LE, UK.
[2]Department of Cell Physiology
and Pharmacology, University of Leicester,
Leicester LE1 9HN, UK.

Skeleto-fusimotor (β) axons form an integral part of the motor supply to mammalian skeletal muscles. In cat peroneus tertius (PT), for example, 31% of the motor axons innervating extrafusal muscle fibres also innervate intrafusal ones contributing to the motor innervation of 77% of the spindle population (Jami, Murthy & Petit, 1982; Scott, 1987). It has been established that, as in the case of γ axons, some β axons have a dynamic action on the primary-ending response (β_d axons) and others a static action (β_s axons). Also, the action is the same for all spindles innervated; there are usually more β_s axons in a muscle nerve than β_d; and whereas the slowest β axons are always dynamic and the fastest static, the conduction velocities (c.vs.) of the two kinds overlap (Bessou, Emonet-Dénand & Laporte, 1963; Harker, Jami, Laporte & Petit, 1977; Jami et al., 1979; Jami et al., 1982; see review by Banks, 1994). Of 81 β axons supplying nine PT muscles in three investigations, 49 were static and 32 dynamic (Jami et al., 1982; Scott, 1987; Emonet-Dénand, Petit & Laporte, 1992). Emonet Dénand et al. (1992) found that β_d axons supplying PT had c.vs. of 55-75 m/s as against 50-85 m/s for those supplying peroneus brevis (PB); these compared with c.vs. for β_s axons of 75-95 m/s (PT) and 75-90 m/s (PB).

The innervation of spindles by β axons is highly selective with regard to fibre type, β_d axons almost exclusively innervating bag_1 (b_1) fibres (Barker, Emonet-Dénand, Harker, Jami & Laporte, 1977), while β_s axons selectively innervate long chain (lc) fibres (Harker et al., 1977; Jami et al., 1979). The extrafusal component of β_d muscle units consists of slow-contracting (S) fibres (type I), or, rarely, fast-contracting fatigue resistant (FR) fibres (type IIA), whereas in β_s muscle units the extrafusal fibres are usually FR, occasionally fast fatiguable (type IIB), or, rarely, S (Jami et al., 1982). These identifications were made by determining the unit's fatigue index. In a serial-section glycogen-depletion study of four β muscle units we found the extrafusal composition to be a mixture of two fibre types, though predominantly type I in a β_d unit and type IIB in three fast fatiguable β_s units (Barker, Scott & Stacey, 1992).

There has been no detailed study of the motor reinnervation of spindles after nerve section, though it is known that functional β and γ axons are restored, and that the β innervation is the first to be re-established (Bessou, Laporte & Pagès, 1966; Brown & Butler, 1976; Scott, 1987).

METHODS

The experiments were performed on 7 adult cats. During the initial surgery to carry out nerve section the animals were anaesthetised with Alphaxalone supplemented with Halothane in oxygen. Under fully aseptic conditions the common peroneal nerve was exposed at the knee and, using an operating microscope, the fascicle that innervates PT and PB was freed from the main trunk, cut with fine scissors, and immediately repaired with 10/0 monofilament sutures. Because of the very limited nature of this lesion, there was no evidence of functional disability or discomfort following the operation. The animals were then maintained for six months to allow for reinnervation and subsequent stabilisation of the histochemical and physiological properties of the muscles.

At the end of this period the animals were prepared for acute recordings and glycogen depletion. Under sodium pentobarbitone anaesthesia the left hindlimb was denervated except for the nerve to PT, which was freed and hooked over a recording electrode. Care was taken to ensure that the vascular supply to the muscle remained intact. The distal tendon of PT was freed and attached to a feedback-controlled puller, the muscle length being set at 2 mm short of its physiological maximum. The S1 and L7 dorsal and ventral spinal roots were exposed and the dorsal roots subdivided to isolate filaments containing single Ia-like afferents from PT. The afferents were identified as being Ia-like on the basis of their responses to twitch contraction of the muscle and to ramp-and-hold stretch (Scott, 1990). The ventral roots were similarly subdivided to isolate single α axons. Confirmation of the isolation was obtained by recording an all-or-nothing, orthodromic action potential from the muscle nerve following stimulation of the ventral root filament (Figure 2A). Great care was taken to ensure that there were no γ axons in the filament by averaging the neurogram.

Recordings were made from the Ia-like afferents while the ventral root filaments were stimulated briefly with pulse trains of 50-100 Hz. A motor axon was initially identified as β if it caused acceleration in the discharge of one or more of the afferents. When this occurred, the axon was further tested to enable its positive identification as β. Tests included step changes in stimulation frequency and evaluation of the effect of stimulation on the afferent's firing rate during a ramp-and-hold stretch (Jami et al., 1982). The overall amount of stimulation was kept to a minimum throughout the identification procedures to prevent any inadvertent depletion of glycogen.

When a β axon was positively identified, its action on the Ia-like afferent(s) it activated was characterised by recording the responses to a range of stimulation frequencies applied during ramp-and-hold stretches of the muscle. The muscle unit was then depleted of glycogen by stimulating for 0.5s every second, the muscle being given a triangular stretch of 2 mm at 5 mm/s every 9s. The stimulation frequencies were varied between 10 and 50 Hz to ensure maintained activation of both the extrafusal and intrafusal components of the muscle unit. This regime was maintained for up to 3 hours until the extrafusal component was fully fatigued. During the last few minutes of the regime the stimulus frequencies were raised to between 100 and 150 Hz. At the end of the stimulation regime the muscle was rapidly excised, placed in a boat containing embedding compound (Cryo-M-Bed), and plunged into a bath of isopentane previously cooled to -160°C in liquid nitrogen.

Decorte, Emonet-Dénand, Harker, Jami & Laporte (1984) have shown that there is a risk of 'non-neural' depletion of the b_1 fibre during prolonged periods of stimulation that give rise to fatigue of the spindle response. It is unlikely that any such depletion occurred in our experiments since the periods of stimulation used were much shorter and the spindle response to the stimulus train was still marked, though reduced, even in the final period.

After fixation the muscle was serially cross-sectioned at 15μm in a cryostat at -20°C. The sections on every 8th and 9th slide were processed for fibre typing (mATPase profiles), the rest for observing the presence or absence of glycogen (PAS staining). Myofibrillar ATPase profiles were obtained after acid (pH 4.1 and 4.5) and alkaline (pH 10.1) pre-incubation following method A in Snow et al. (1982). All depletions, as well as the identity of the fibre type depleted, were independently assessed by D.B. and M.J.S. and a result recorded only if both depletion and fibre type were agreed. Depletions in IIB fibres < 130 μm long, the mean minimum length reported by Barker et al. (1992), were not recorded.

RESULTS

A single regenerated β axon was isolated in each experiment, and we shall refer to these axons as β1-7 They activated nine Ia-like afferents; no doubt they activated more, but it was not possible to isolate more than one or two in each experiment. Stimulation of the β axons generally gave rise to very low contractile tensions; only β1 and β2, with tetanic tensions of 2 grams force (gf) and 4 gf, respectively, produced tetanic tensions > 1 gf.

Table 1 summarises the actions of the β axons in terms of their effects on the resting discharge and dynamic index of the individual Ia-like afferents, and Figure 1 summarises the glycogen depletion produced in the activated spindles located in the PT muscles.

The Ia-like afferents activated by β1 and β3 all showed significant increases in the dynamic indices of their responses accompanied by only modest increases in the resting discharge (Figure 2C). These effects are presumed to indicate contraction of the b_1 fibre (Boyd, 1981), and this correlates well with the finding of depleted b_1 fibres in both experiments. It is noteworthy that the β axons had the relatively fast conduction velocities of 75 m/s (β1) and 60.5 m/s (β3).

Table 1. Actions of the regenerated β axons on the Ia-like afferents.c.v. = conduction velocity, d = dynamic, s = static. *biassing of the resting discharge was taken as positive when the afferent firing rate increased by >70 imp/s, without driving, in response to stimulation at 100 Hz.

β axon	c.v. m/s	Number Ia-like afferents activated	biassing*	1:1 driving	dynamic index	action using criteria for normal Ia afferents
1	75.0	1	-	-	↑	d
2	65.0	2 (i)	+	-	↑	d/s
		(ii)	+	-	↓	s
3	60.5	2 (i)	-	-	↑	d
		(ii)	-	-	↑	d
4	60.0	1	+	-	↑	d/s
5	55.0	1	+	-	↓	s
6	52.5	1	+	-	↓	s
7	46.0	1	-	+	↓	s

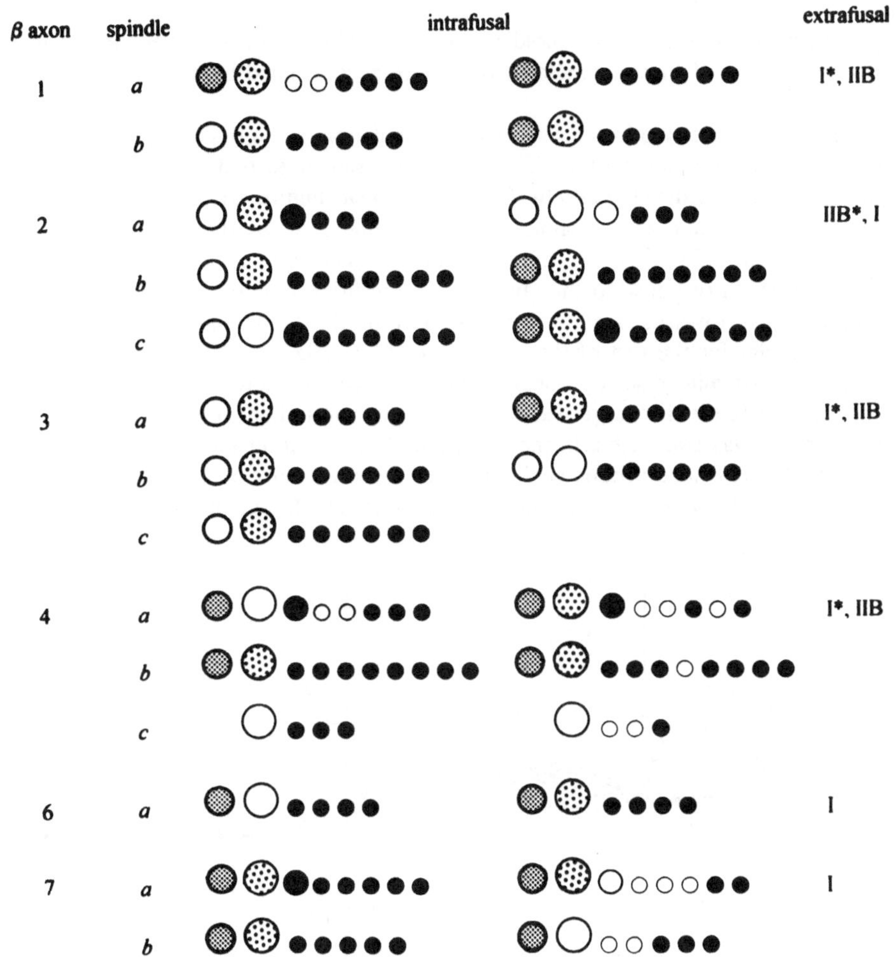

Figure 1. Summary of the glycogen depletion produced in activated spindles that were located in the reinnervated peroneus tertius muscles after stimulation of β axons 1-4, 6, 7 (no data available for β5). Medium-diameter circles with light shading represent non-depleted poles of b_1 fibres; large circles with coarse shading those of b_2 fibres; large filled circles those of long chain fibres, and small filled circles those of other chain fibres. Fibre types depleted in extrafusal component of muscle units also shown; asterisk indicates predominant fibre type.

In three instances (β2, afferent (ii); β5; and β6) stimulation of the β axon produced significant biassing of the afferent discharge, with no driving and a decrease in the dynamic index (Figure 2D). Such a pattern of response is comparable to that described by Boyd (1981) for contraction of the bag_2 (b_2) fibre and we may correlate this with the b_2 depletions produced by β2 in spindle *c* and β6 in spindle *a* (no data available for β5).

The Ia-like afferent activated by β7, which had the slowest conduction velocity (46 m/s), showed the only instance observed of 1:1 driving of the response at the resting length of the muscle (Figure 2B). The driving was secure at 50 and 70Hz stimulation, but was not maintained at 100Hz. Such patterns of driving indicate chain-fibre activation (see e.g. Boyd, 1981). However, during the ramp-and-hold stretch the response was not typical of that associated with chain-fibre contraction alone in that there was a marked response to the stretch with adaptation during the hold phase. Chain-fibre depletions were found in two spindles, and in one of these (spindle *b*) they were accompanied by a depletion of the b_2 fibre.

Finally, there were two instances in which stimulation of the β axon (β2 afferent (i); β 4) produced a mixed action on the afferent response similar to that observed by Emonet-Dénand, Laporte, Matthews & Petit (1977, see their Figure 6) during the combined stimulation of both static and dynamic γ axons. Figure 2 shows the action of β2 on the response of Ia-like afferent (i). At 50Hz stimulation there was a modest increase in the resting discharge. In response to the ramp stretch the firing frequency increased markedly, began to decrease towards a plateau at the start of the hold phase, and then became extremely irregular for the rest of the hold phase. At 100Hz stimulation the response was similar, but the irregular discharge began almost immediately after the end of the stretch itself, and at 150Hz the discharge was highly irregular throughout, though the response to the ramp stretch was still visible in the overall shift in the firing levels. The activation of spindle *a* might well have produced this type of response since it had both b_1 (dynamic action) and b_2 and long chain (static action) depletions. The Ia-like afferent activated by β4

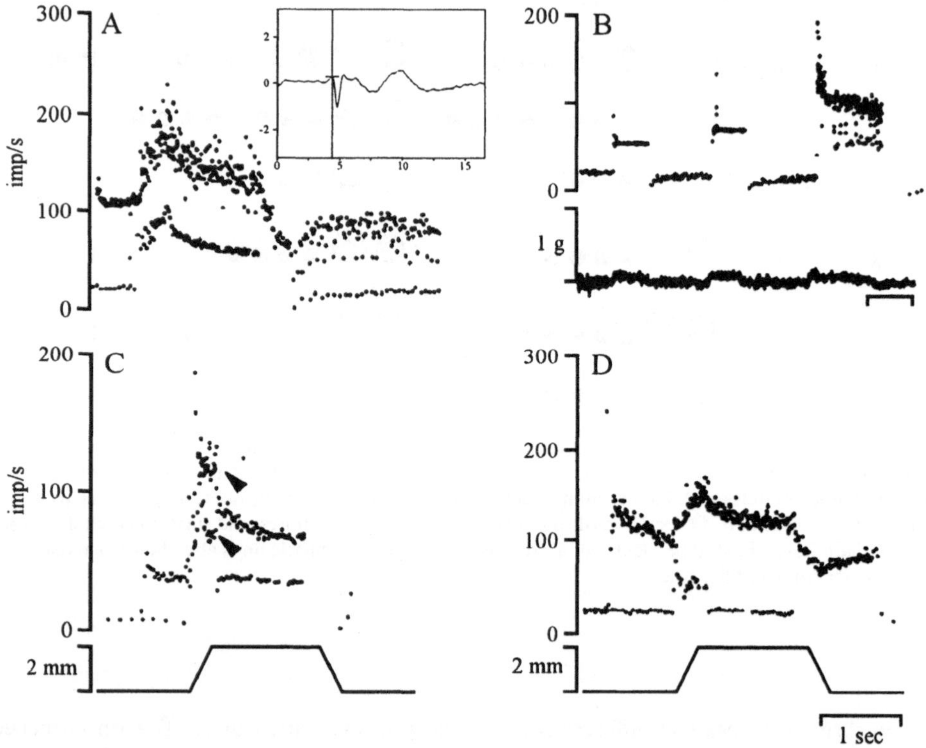

Figure 2. Example of responses of Ia-like afferents to stimulation of regenerated β axons. A. Instantaneous frequency records of a Ia-like afferent to ramp-and-hold stretch. Lower record shows the basic response; upper record the response to stimulation of β7 at 100Hz. Inset shows an averaged recording from the muscle nerve in response to supramaximal stimulation of the ventral root filament. There is a single neuronal action potential followed by a muscle action potential. B. The same afferent/β combination during stimulation at the resting length of the muscle at 50, 70 and 100Hz. Note the very small contractile tensions. Time bar represents 1s. C. Responses of a Ia-like afferent to ramp-and-hold stretch alone (lower record), and to stimulation of β3 at 150Hz (upper record). Arrowheads mark discharges at peak of the stretch in each case. D. Responses of a Ia-like afferent to stretch alone (lower record) and to stimulation of β2 at 100Hz (upper record).

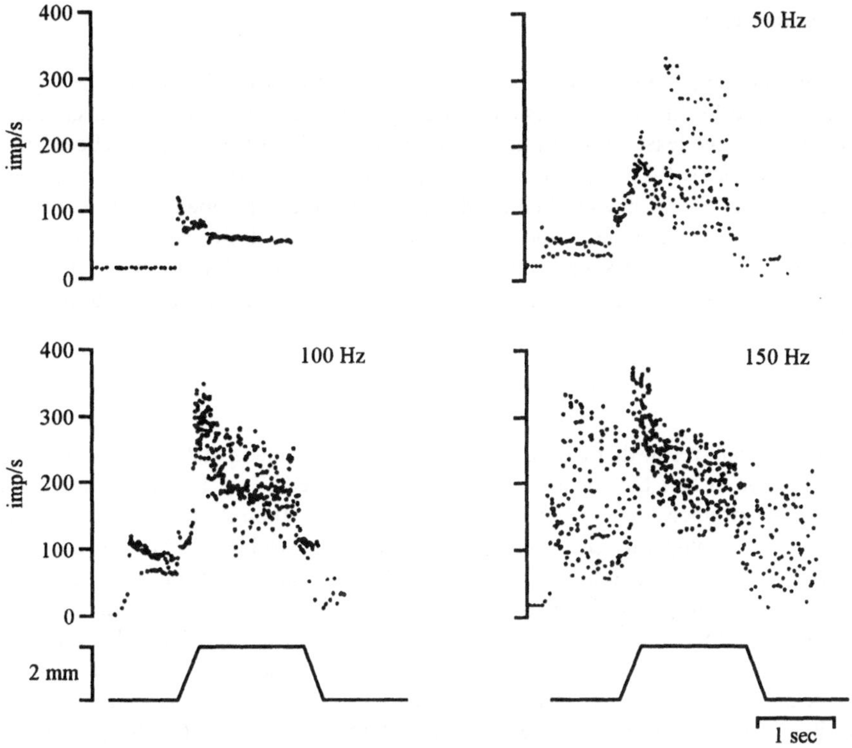

Figure 3. β2 appeared to have a mixed action in activating afferent (i) there being both dynamic and static elements in the response depending on frequency of stimulation and afferent firing rate.

Table 2. Summary of the results of correlating the actions of the β axons with the spindle depletion data and indicating which spindles were most likely to have been innervated by the activated Ia-like afferents.

β axon	Number of Ia-like afferents activated	Mode of action	Spindle(s) depleted	Fibre(s) depleted
1	1	d	b	b_1
2	2 (i)	d/s	a	b_1, b_2 & lc
	(ii)	s	c	b_1 & b_2
3	2 (i)	d	a, b & c	b_1
	(ii)	d	a, b & c	b_1
4	1	d/s	No correlation	
6	1	s	a	b_2
7	1	s	b	b_2 & 2 chains

also produced a response that appeared to be mixed in that there was significant biassing of the background discharge and increase in the dynamic index, but there was no marked irregularity in the discharge comparable to that produced by β2 in the response of afferent (i). Depleted b_2 fibres and chain fibres occurred among the depleted spindles but no b_1 fibres, so it is not clear which, if any, of the three located might have received the afferent that gave this response. These results are summarised in Table 2.

Extrafusal depletion

The muscle units depleted by β axons 1-4 were composed of a mixture of type I and IIB fibres (Figure 1). In all four units the mixture was that described by Barker et al. (1992) as 'moderate', i.e. the least common fibre type occurred with a frequency of about 20-30%. The unit innervated by β4 had a contraction tension of about 0.006 N and was small enough to enable all the depleted fibres to be traced and their positions mapped. It was composed of 27 extrafusal fibres, 22 type I and 5 type IIB. The type I fibres all occurred in the neighbourhood of the three spindles depleted except for four, which occurred in the same area as the IIB fibres. The muscle units depleted by β axons 6 and 7 were composed of type I fibres only.

COMMENT

The significant result that emerges from these experiments is that there were depletions of the bag_2 fibre in half the reinnervated spindles activated (7 of 14). Beta innervation of bag_2 fibres is extremely rare in normal adult cat spindles, though it does have a transient presence in developing rat spindles (Kucera, Walro & Reichler, 1988; Jones, Ridge & Rowlerson, 1987, see their Figure 1). It may be that this bag_2 rejection of β connexions is ensured by the Ia afferent, which plays such a key role in spindle development. If so, and if it exerts a similar control during spindle reinnervation, it may be significant that only half the primary endings supplied by annulospiral afferents to spindles reinnervated after nerve section belong to Ia afferents. Banks & Barker (1989) found that among 112 primary-ending sites reinnervated by such afferents, 52% were Ia, 28% Ib and 20% spindle II. They also showed that the non-Ia primaries responded to stretch like Ia primaries.

The c.vs. of the regenerated β axons showed a reversal of the normal slow dynamic/fast static sequence, but we doubt if this is significant in such a small sample. Low values of c.v. for β static axons might well result from the regenerating axons reaching the spindles by growing down Schwann tubes previously occupied by γ static axons. Similarly, it can be argued that the apparently selective reinnervation of bag_2 fibres by the β axons in our sample, especially β3, does not necessarily imply the existence of specific types of β axon, since these could have been α axons in the normal nerve that, after section, grew down tubes previously occupied by β_d axons.

Banks (1994) makes a good case for skeleto-fusimotor innervation being established by α axons that happen by chance to make intrafusal connexions, rather than by an intrinsically different type of motoneurone that has come to be known as β. Unfortunately our sample is too small to provide evidence either for or against this suggestion, but our interpretation of the results implies support.

ACKNOWLEDGEMENTS

We wish to thank the Medical Research Council (Grant G8814879N), and the Leverhulme Trust (award of an Emeritus Fellowship to D.B.) for financial support.

REFERENCES

BANKS, R.W. (1994) The motor innervation of mammalian muscle spindles. *Prog. Neurobiol.* **43**, 323-362.

BANKS, R.W. & BARKER, D. (1989) Specificities of afferents reinnervating cat muscle spindles after nerve section. *J. Physiol.* **408**, 345-372.

BARKER, D., EMONET-DÉNAND, F., HARKER, D.W., JAMI, L. & LAPORTE, Y. (1977) Types of intra- and extrafusal muscle fibre innervated by dynamic skeleto-fusimotor axons in peroneus brevis and tenuissimus muscles, as determined by the glycogen-depletion method. *J. Physiol.* **266**, 713-726.

BARKER, D., SCOTT, J.J.A. & STACEY, M.J. (1992) A study of glycogen depletion and the fibre-type composition of cat skeleto-fusimotor units. *J. Physiol.* **450**, 565-579.

BESSOU, P., EMONET-DÉNAND, F. & LAPORTE, Y. (1963) Occurrence of intrafusal muscle fibres innervated by branches of slow α motor fibres in the cat. *Nature* **198**, 594-595.

BESSOU, P., LAPORTE, Y. & PAGÈS, B. (1966) Observations sur la ré-innervation de fuseaux neuromusculaires de Chat. *Comp. Rend. Séances Soc. Biol.* **160**, 408-411.

BOYD, I.A. (1981) The action of the three types of intrafusal muscle fibre in isolated cat muscle spindles on the dynamic and length sensitivities of primary and secondary endings. In *Muscle Receptors and Movement.* eds. TAYLOR, A. & PROCHAZKA, A. pp. 17-32. Macmillan, London.

BROWN, M.C. & BUTLER, R.G. (1976) Regeneration of afferent and efferent fibres to muscle spindles after nerve injury in adult cats. *J. Physiol.* **260**, 253-266.

DECORTE, L., EMONET-DÉNAND, F., HARKER, D.W., JAMI, L. & LAPORTE, Y. (1984) Glycogen depletion elicited in tenuissimus intrafusal muscle fibres by stimulation of static γ-axons in the cat. *J. Physiol.* **346**, 341-352.

EMONET-DÉNAND, F., LAPORTE, Y., MATTHEWS, P.B.C. & PETIT, J. (1977) On the subdivision of static and dynamic fusimotor actions on the primary ending of the cat muscle spindle. *J. Physiol.* **268**, 827-861.

EMONET-DÉNAND, F., PETIT, J. & LAPORTE, Y. (1992) Comparison of skeleto-fusimotor innervation in cat peroneus brevis and peroneus tertius muscles. *J. Physiol.* **458**, 519-525.

HARKER, D.W., JAMI, L., LAPORTE, Y. & PETIT, J. (1977) Fast-conducting skeletofusimotor axons supplying intrafusal chain fibres in the cat peroneus tertius muscle. *J. Neurophysiol.* **40**, 791-799.

JAMI, L., LAN-COUTON, D., MALMGREN, K. & PETIT, J. (1979) Histophysiological observations on fast skeleto-fusimotor axons. *Brain Res.* **164**, 53-59.

JAMI, L., MURTHY, K.S.K. & PETIT, J. (1982) A quantitative study of skeletofusimotor innervation in the cat peroneus tertius muscle. *J. Physiol.* **325**, 125-144.

JONES, S.P., RIDGE, R.M.A.P. & ROWLERSON, A. (1987) The non-selective innervation of muscle fibres and mixed composition of motor units in a muscle of neonatal rat. *J. Physiol.* **386**, 377-394.

KUCERA, J., WALRO, J.M. & REICHLER, J. (1988) Innervation of developing intrafusal muscle fibers in the rat. *Am. J. Anat.* **183**, 344-358.

SCOTT, J.J.A. (1987) The reinnervation of cat muscle spindles by skeletofusimotor axons. *Brain Res.* **401**, 152-154.

SCOTT, J.J.A. (1990) Classification of muscle spindle afferents in the peroneus brevis muscle of the cat. *Brain Res.* **509**, 62-70.

SNOW, D., BILLITER, R., MASCARELLO, F., CARPENE, E., ROWLERSON, A. & JENNY, E. (1982) No classical typre IIB in dog skeletal muscle. *Histochem.* **75**, 53-65.

CRITIQUE OF PAPERS PRESENTED BY KUCERA ET AL., WALRO & WANG AND BARKER ET AL.

R.M.A.P. Ridge

Department of Physiology
School of Medical Sciences
University of Bristol
University Walk
Bristol BS8 1TD, UK

CRITIQUE OF KUCERA, ERNFORS, SCHWARZE AND JAENISCH

The work described here concerns the ontogenetic development of intrafusal muscle fibres and their innervation. The present view, derived from previous work from several laboratories, is that the bag_2 fibre of the muscle spindle develops from a primary myotube prenatally, its specialization from extrafusal precursors being associated with, and probably caused by, the termination on it of a primary afferent axon. Bag_1 and chain fibres appear later as secondary myotubes in close physical association with the bag_2 primary myotube. Temporally the arrival of the motor supply lags slightly behind the afferent supply and is not required for bag_2 fibre differentiation.

The first part of the present work is an attempt to examine again the pivotal role of the primary afferent in bag_2 differentiation, and the new observation is that undifferentiated nerve-muscle contacts appear at E15-El6, about a day earlier than previously reported in this muscle (rat soleus). As these contacts cannot be classified into motor or sensory on morphological grounds, there is the possibility that some are motor. Until very early selective motor denervation (for instance by β bungarotoxin *in utero*) has been used in a spindle development study, an early co-operative action between primary afferent and motor supply in bag_2 differentiation cannot be ruled out, though in the writer's opinion the present finding is not sufficient to cast serious doubt on the crucial importance of the primary afferent in this regard.

The main part of the work supports this view. Here the reverse experiment has been performed; that is, muscle development has been followed in the absence of primary afferents. The work concerns a fascinating new mutant mouse with a deletion of the neurotrophin 3 (NT-3) gene. (A second mutant in which the brain-derived neurotrophic factor gene was missing was found to have essentially normal spindle development). In the former mutant no spindles develop in soleus (they are absent from embryonic ages at which they would normally be appearing). Associated with this lack of spindles is a lack of primary afferents. In this, and in the abnormal movements made by the mice, the mutant is very similar to another recently described by Klein, Silos-Santiago, Smeyne, Lira, Brambilla, Bryant, Zhang, Snider, Barbacid (1994), that is defective for the NT-3 receptor (*trk*C receptor). It is therefore likely that the development of the primary afferent is NT-3 dependent, and that, as expected, the development of muscle spindles requires the primary afferent and does not take place in its absence. Recently the *trk*C mutant has also been shown to lack spindles in soleus and EDL muscles (Michael Chua, personal communication).

Kucera et al. compare normal and mutant dorsal root ganglion cell size and ventral root fibre size at L4. The most interesting aspect of these data as they stand at present is the gross deficit in small (γ) motor axons in the mutant. This could be a secondary loss following primary afferents, and consequent spindle, deficiency, but this should be confirmed by a study of muscle development with time in the mutant. On the sensory side, apparently about 50% of DRG cells are missing in the mutant, but this deficiency appears to be distributed across the entire spectrum of cell size in L4 DRG. The question of selectivity of afferent loss could be important, and needs to be examined in more detail in relation to cell size, and with specific cell markers (e.g. Lawson, Perry, Prabhakar & McCarthy, 1993).

The total absence of spindles in the muscles examined so far in homozygote NT-3 and *trk*C mutants is very interesting in relation to future developmental studies, and it is to be hoped that these mutants will become generally available. A previous mouse ("sprawling"; Duchen, 1975) and rat mutant ("mutilated foot"; Jacobs, Scaravilli, Duchen & Mertin, 1981) show large scale spindle deficits, but occasional spindles do occur, which makes them less clear-cut for developmental studies. Incidentally, the "sprawling" spindle deficit is much more marked in hindlimb muscles than in the forelimb. Perhaps it would be wise to examine forelimb muscles in the two new mutants. There is a large-scale deficit in γ motor axons in "mutilated foot".

A strange finding reported by Kucera et al is that the heterozygote mutant shows spindle counts in soleus reduced to half. If this is an NT-3 dose-dependency it raises basic questions about how NT-3 works in primary afferent development.

CRITIQUE OF WALRO AND WANG

Currently there is much discussion about the existence of different embryonic myoblast cell lines, and their possible significance in predetermining the fate of subclasses of the primary myotubes produced by them. Different cell lines are known to exist in avian muscle, but whether or not there is more than one cell line in rodents is not known (see Stockdale, 1992, for review). Separation of myoblast and primary myotube types is based so far on the types of myosin heavy chain present, mainly as detected using specific antibodies. In rats during development there are three slow myosin heavy chains (SMHC) known to make an appearance (Hughes, Cho, Karsch-Mizrachi, Travis, Silberstein, Leinwand, Blau, 1993), though only one is probably present in primary myotubes before secondary myotubes are generated.

In the present study a monoclonal antibody (S46) derived from avian muscle has been used in developing rat muscle. This antibody binds to some primary myotubes, but not others, in hind leg muscles, before secondary myotubes have appeared. Apparently the SMHC that binds the antibody is not one of the three types in the study by Hughes et al. since its time of appearance does not correspond with that of any of the three. However, this needs to be checked carefully, since in the Hughes study, as described, the development of primary myotubes was not concentrated upon in detail.

The main finding from this present work is that in leg muscles other than soleus there is a stage in early development when the S46 antibody binds to only a proportion of the primary myotubes (in soleus all are reactive). At this stage other antibodies reacting with SMHCs do not show this separation. The reactive subpopulation of primary myotubes then gives rise to intrafusal (b_2) and extrafusal (type I) muscle fibres. S46 binding then persists in the intrafusal fibres, and fades away during development of the extrafusal fibres.

This finding is interpreted as evidence in favour of primary myotubes being bipotential with regard to b_2 or extrafusal generation. It is indeed interesting that S46 divides the primary myotube population earlier in development than other antibodies, and that the division does not correspond to an intrafusal/extrafusal division. But this finding does not resolve the question of whether separate myoblast cell lines give rise to predetermined myotubes, the b_2 myotubes then attracting the primary afferent (and then depending on its successful contact for their survival), or whether truly bipotential myotubes are triggered to become intrafusal by the arrival of the primary afferent (either randomly or in a pattern generated elsewhere). If an attractant exists it is likely to be a surface molecule or diffusible substance which may or may not be associated with a specific SMHC intracellularly.

CRITIQUE OF BARKER, SCOTT AND STACEY

In this work connexions made by skeletofusimotor axons with intrafusal fibres were investigated in peroneus tertius muscle that had been denervated six months previously, and in which the nerve supply had regenerated. The β axons studied could originally have been β or α, and could have innervated peroneus tertius or peroneus brevis. On the basis of measured conduction velocities none were likely to have been γ fibres originally. The sensory axons from which recordings were made were selected on the basis of having Ia-like responses, but originally they could presumably have been any sort of afferent from either muscle (no data on conduction velocities or signal-to-noise ratios are given). It is known from previous work (Banks & Barker, 1989) that alien afferents will regenerate to spindles and respond to muscle stretch and fusimotor stimulation much like true spindle afferents.

The data were from physiological recording of spindle responses to motor stimulation and stretch, and from adjacent frozen sections processed for myosin ATPase and PAS (for fibre typing and glycogen depletion respectively). In the authors' interpretation the two sets of data are related to one another. The difficulty here is that for any one β fibre stimulated there are more spindles showing intrafusal glycogen depletion than yield physiological records, and the link can only be made on the basis of informed interpretation rather than proof. However, this does not matter here as the main conclusion can be based on glycogen depletion alone. In fact the interpretations based on the physiological records fit well with the glycogen depletion data.

The main finding is that in half of the spindles examined by glycogen depletion the bag_2 fibre has depleted in one or other (or both) poles. In normal adult spindles bag_2 fibres very rarely receive motor supply from β axons. Further, bag_2 depletion was, in three cases, associated with bag_1 depletion, thus contravening the usual discreetness of static (to b_2 and c) and dynamic (to b_1) innervation by γ fibres. Thus the usual rules of motor specificity with regard to intrafusal fibre type appear to have broken down. The authors hypothesise that such specificity (at least as far as the exclusion of β suppy from b_2 is concerned) is normally maintained in some way by the Ia afferent, and that in these aberrant cases an alien afferent had taken the place of the Ia afferent. However, an alternative explanation is that such specificity is imposed in some way during development only and is largely preserved in the normal adult because most connexions persist. In the experimental situation, of section and regeneration, all the connexions are changed and the factors imposing specificity of connexions are now weak or absent. What is needed to test the authors' interpretation is some marker that is specific for spindle Ia afferents. Unfortunately such a marker is not known at present.

The fact that the extrafusal component of four of these β units was of mixed fibre type (I, IIB) is interesting. Such mixed units have been described in normal (not regenerated) β units (Barker, Scott & Stacey, 1992), and other mixed extrafusal units (IIC/IIA) have been found in adult rat lumbrical, though it is not known if these are β units (Gates, Ridge & Rowlerson, 1991).

ACKNOWLEDGEMENTS

I thank Michael Chua (St Louis) and Anthea Rowlerson for useful discussion, and Bob Banks for giving me a preprint of his review (Banks, 1994).

REFERENCES

BANKS, R.W. (1994) The motor innervation of mammalian muscle spindles. *Prog. Neurobiol.* **43**, 323-362

BANKS, R.W. & BARKER, D. (1989) Specificities of afferents reinnervating cat muscle spindles after nerve section. *J. Physiol.* **408**, 345-372

BARKER, D., SCOTT, J.J.A. & STACEY, M.J. (1992) A study of glycogen depletion and the fibre-type composition of cat skeleto-fusimotor units. *J. Physiol.* **450**, 565-579.

DUCHEN, L.W. (1975). "Sprawling": a new mutant mouse with failure of myelination of sensory axons and a deficiency of muscle spindles. *Neuropath. App. Neurobiol.* **1**, 89-101.

GATES, H.-J., RIDGE, R.M.A.P. & ROWLERSON, A. (1991) Motor units of the fourth deep lumbrical muscle of the adult rat: isometric contractions and fibre type compositions. *J. Physiol.* **443**, 193-215.

HUGHES, S.M., CHO, M., KARSCH-MIZRACHI, I., TRAVIS, M., SILBERSTEIN, L., LEINWAND, L.A. & BLAU, H.M. (1993) Three slow myosin heavy chains sequentially expressed in developing mammalian skeletal muscle. *Dev. Biol.* **158**, 183-199.

JACOBS, J.M., SCARAVILLI, F., DUCHEN, L.W. & MERTIN, J. (1981). A new neurological rat mutant "mutilated foot". *J. Anat.* **132**, 525-543.

KLEIN, R., SILOS-SANTIAGO, I., SMEYNE, R. J., LIRA, S.A., BRAMBILLA, R., BRYANT, S., ZHANG, L., SNIDER, W.D. & BARBACID, M. (1994) Disruption of the neurotrophin-3 receptor gene trkC eliminates Ia muscle afferents and results in abnormal movements. *Nature.* **368**, 249-251.

LAWSON, S.N., PERRY, M.J., PRABHAKAR, E. & MCCARTHY, P.W. (1993). Primary sensory neurones: neurofilament, neuropeptides, and conduction velocity. *Brain Res. Bull.* **30**, 239-243.

STOCKDALE, F.E. (1992). Myogenic cell lineages. *Dev. Biol.* **154**, 284-298.

THE EXCEPTIONAL STRUCTURE OF MUSCLE SPINDLES IN SUPERFICIAL LUMBRICAL MUSCLES OF THE CAT HINDLIMB

R.W. Banks[1], L. Decorte[2], F. Emonet-Dénand[2], M.H. Gladden[3] & F. Sutherland[3]

[1]Department of Biological Sciences
University of Durham, Durham, UK
[2]Laboratoire de Neurophysiologie
Collège de France, Paris, France
[3]Muscle Spindle Physiology Group
I.B.L.S., West Medical Building
University of Glasgow
Glasgow G12 8QQ, Scotland, UK

INTRODUCTION

Histophysiological experiments using the superficial lumbrical muscle of the cat have provided the opportunity to analyse the structure of a large number of spindles (Decorte, Emonet-Dénand, Ginapé & Laporte, 1986). This muscle was initially chosen because of its small size, spindle content and known motor-unit properties (Emonet-Dénand, Hunt, Petit & Pollin, 1988).

Compared with the classical picture of the spindle, derived principally from tenuissimus, superficial lumbrical spindles show several important differences. In particular, a high proportion of lumbrical spindles possess long chain fibres (71% of poles, Decorte, Emonet-Dénand, Harker & Laporte, 1987) and greater individual variability in the number of bag fibres present (Decorte, Emonet-Dénand, Ginapé & Laporte, 1990).

Here we present further original features of superficial lumbrical (SL) spindles, emphasising the importance of the bag$_1$ (b$_1$) fibre, which is known to be innervated by both small (γ range) and large (α range) motor axons.

METHODS

The observations are based on serial cryostat sections, stained to demonstrate ATPase activity for fibre typing. Details of the method are given in Decorte et al. (1987).

RESULTS

The longest intrafusal fibre present was almost invariably a b_1 (26/30). The difference in length between b_1 (mean 3388μm) and bag_2 (b_2; mean 2320μm) fibres is especially notable. In only one case was a b_2 longer than a b_1, whilst in two cases the longest fibre was a (long) chain. Mean diameters of each fibre type were measured at regular intervals. The b_2 fibre was slightly smaller than b_1 fibre at the equator, and chain (c) fibres were significantly smaller than both. All types progressively decreased in diameter from region A through B, b_2 and most c fibres continued to decline through region C, whereas the mean diameter of long c fibres did not decrease further until beyond 2000μm from the equator. In contrast, the mean diameter of b_1 fibres increased markedly at the start of region C, and then again declined steadily. The b_1 fibres therefore were normally not only the longest, but considerably the largest, intrafusal fibres in the extracapsular poles.

In a few spindles there was only a single bag fibre, which was more often b_1 (2/4 complete spindles, or 7/11 poles) than is the case for similar spindles in tenuissimus. Perhaps related to this is the finding that in tandem-linked spindles continuous bag fibres were more often b_1 than b_2, which is again at variance to the familiar picture from tenuissimus. Some spindles appear to consist of two intrafusal bundles more or less conjoined so that the equator, and often one pole in addition are common to both.

CONCLUSION

Our observations add to the growing evidence that different muscles exhibit characteristic features of the structure and innervation of their proprioceptors. These are often of a statistical or quantitative nature, nevertheless they may be supposed to reflect adaptations to local control requirements.

ACKNOWLEDGEMENTS

We are very grateful to Dr. D.W. Harker, who participated in the initial histological analysis. RWB thanks the Wellcome Trust for Financial support.

REFERENCES

DECORTE, L., EMONET-DÉNAND, F., GINAPÉ, M. & LAPORTE, Y. (1986) Déplétion glycogénetique produite dans les fibres musculaires intrafusales à sac nucléaire par des étirements musculaires brefs de forte amplitude. *Comp. Rend. Séances l'Acad. Sci. Paris* **302,** 697-700.

DECORTE, L., EMONET-DÉNAND, F., HARKER, D.W. & LAPORTE, Y. (1987) High incidence of long-chain fibers in spindles of cat superficial lumbrical muscles. *J. Neurophysiol.* **57**, 1050-1059.

DECORTE, L., EMONET-DÉNAND, F., GINAPÉ, M. & LAPORTE, Y. (1990) Individual differences in multiple-bag spindles of cat superficial lumbrical muscles. *J. Anat.* **169**, 1-12.

EMONET-DÉNAND, F., HUNT, C.C., PETIT, J. & POLLIN, B. (1988) Proportion of fatigue-resistant motor units in hindlimb muscles of cat and their relation to axonal conduction velocity. *J. Physiol.* **400**, 135-158.

THE INNERVATION OF MUSCLE SPINDLES IN AN INTRINSIC MUSCLE OF THE HIND FOOT: THE SUPERFICIAL LUMBRICAL OF THE CAT

R.W. Banks[1], F. Emonet-Dénand[2],
M.J. Stacey[1] and D. Thiesson[3]

[1]Department of Biological Sciences
University of Durham, Durham, UK
[2]Laboratoire de Neurophysiologie
Collège de France, Paris, France
[3]URA CNRS 1448
Université de René Descartes
Paris, France

INTRODUCTION

An important, if largely unrecognised, feature of the design of the mammalian skeletomotor proprioceptive system is the characteristic variation from muscle to muscle in the provision of its components. This applies not only to the relative abundance of muscle spindles, but also to quantitative differences in their sensory ending complement (Banks & Stacey, 1988). Furthermore, Banks (1994) has shown that there is a linear correlation between the numbers of static γ and afferent axons supplied to each spindle, at least in the tenuissimus. Considerably more comparative information of this type is needed for a fuller understanding of the contribution that proprioception makes to particular motor tasks. Here we describe some preliminary observations on the innervation of the superficial lumbrical muscle of the cat's hind foot.

SENSORY INNERVATION

A sample of 53 teased, silver-impregnated spindles was analysed. The results from both 1st and 2nd superficial lumbrical muscles were combined, since there appeared to be no differences between the two. The spindles contained 55 primary (P) endings, two of which occurred singly in small, tandem-linked units that probably consisted only of bag$_1$ (b$_1$) and chain (c) fibres. All of the remaining units had the full complement of intrafusal fibre types, and almost all comprised single spindles, though a few were conjugated in parallel. In addition to a primary ending, each possessed from 0 to 4 secondary (S) endings with the following frequency distribution: P 23%; PS 55%; P2S 17%; P3S 4%; P4S 2%. Of the the 57 S endings 49 were in the S_1 position (next to P) and 8 in S_2 (next to S_1).

It is clear that a single Ia afferent is both necessary and sufficient to maintain (and probably to initiate) intrafusal fibre differentiation. Any additional afferent axons are usually of group II and supply secondary endings, as is the case in the superficial lumbrical described here. The mean number per spindle of these additional afferent axons in a particular muscle seems to be an important controlled variable of mammalian sensori-motor systems, since it has a characteristic value for that muscle while varying between different muscles. The value for the superficial lumbrical is lower (1.08) than that of any of a sample of 7 hind limb muscles of the cat (range 1.22, peroneus brevis, to 2.72, popliteus) studied by Banks & Stacey (1988). Precisely which spindles receive secondary endings appears to be a matter of chance, since the frequency of occurrence of spindles with different complements of afferent axons can be fitted by probabilistic statistics. The mean number of additional afferents per spindle is used to estimate the statistical parameters (Banks & Stacey, 1988). In most cases the two parameters (n, p) of the binomial distribution are necessary to describe adequately the observed distributions, but when n is large and p is small the distribution approximates to the Poisson. For the superficial lumbrical spindles that we have examined, binomial distributions with parameters $n = 2$, $p = 0.54$, or $n = 3$, $p = 0.36$ most closely approximate to the observed data. There are, however, insufficient degrees of freedom remaining after parameter estimation to test whether the theoretical distribution corresponding to either set of parameters differs significantly from the data.

MOTOR INNERVATION

The distributions of 131 intrafusal branches of motor axons within 29 poles of 16 teased spindles (5 P, 8 PS, 3 P2S) have been determined as follows: to bag_2 (b_2) or c or both, 56; to b_1, 75. Despite the presence of a high proportion of long-chain (lc) fibres in superficial lumbrical spindles (Decorte, Emonet-Dénand, Harker & Laporte 1987) only 2 axons to lc fibres showed characteristics of skeletofusimotor (β) innervation, a much lower incidence (0.13/spindle) than in tenuissimus (0.59/spindle; Banks, 1994). Conversely the incidence of b_1 innervation (4.7/spindle) that appeared to be dominated by p_1 plates and thus, potentially β innervation, was much greater than in tenuissimus (1.9/spindle). The β nature of some of this innervation was demonstrated by Barker, Emonet-Dénand, Laporte & Stacey (1980). We have studied this innervation using serial, cryostat sections of combined cholinesterase and silver-gold impregnation (Pestronk & Drachman, 1978), interspersed with alkaline- and acid-preincubated ATPase for fibre-typing. Our observations confirm that it almost invariably supplies the b_1 fibre.

The large majority (54/56) of the axons innervating b_2 and c fibres (separately or jointly) may be safely identified as purely fusimotor (γ). Mean values of the number of these axons per spindle, for each of the different sensory complements, were: P, 2.9; PS, 4.0; P2S, 4.4. Corresponding values for tenuissimus are 1.7, 3.0, and 4.4 respectively. A similar correlation between the numbers of afferent and γ axons may therefore exist in both muscles, though the constant of proportionality may differ.

CONCLUSION

We conclude that the pattern of innervation of superficial lumbrical spindles, although qualitatively similar to that of tenuissimus, differs quantitatively in several important respects. It seems possible, even likely, that such characteristics vary independently between different muscles.

ACKNOWLEDGEMENTS

RWB thanks the Wellcome Trust for financial support

REFERENCES

BANKS, R.W. (1994) Intrafusal motor innervation: a quantitative histological analysis of tenuissimus muscle spindles in the cat. *J. Anat.* **185**, 151-172.

BANKS, R.W. & STACEY, M.J. (1988) Quantitative studies on mammalian muscle spindles and their sensory innervation. In *Mechanoreceptors*, ed. HNÍK, P., SOUKUP, T., VEJSADA, R. & ZELENÁ, J. pp. 263-269. Plenum Press, New York.

BARKER, D., EMONET-DÉNAND, F., LAPORTE, Y. & STACEY, M.J. (1980) Identification of the intrafusal endings of skeletofusimotor axons in the cat. *Brain Res.* **185**, 227-237.

DECORTE, L., EMONET-DÉNAND, F., HARKER, D.W. & LAPORTE, Y. (1987) High incidence of long-chain fibers in spindles of cat superficial lumbrical muscles. *J. Neurophysiol.* **57**, 1050-1059.

PESTRONK, A. & DRACHMAN, D.B. (1978) A new stain for quantitative measurement of sprouting at neuromuscular junctions. *Muscle Nerve* **1**, 70-74.

SENSORY AND MOTOR INNERVATION OF THE ABDUCTOR DIGITI QUINTI MEDIUS MUSCLE OF THE CAT

R.W. Banks[1], M.H. Gladden[2],
F. Sutherland[2] and A. Yoshimura[3]

[1]Department of Biological Sciences
University of Durham, Durham, U.K.
[2]Muscle Spindle Physiology Group
I.B.L.S., West Medical Building
University of Glasgow
Glasgow G12 8QQ, Scotland, UK
[3]Department of Life Sciences
Institute of Technology, Nagoya, Japan

INTRODUCTION

The abductor digiti quinti medius (adqm) is an intrinsic muscle of the foot which abducts the fifth digit. Despite its small size (about 30mg) it contains all the elements of a complete skeletomotor and fusimotor system. It may therefore provide a practical model for studying integrated fusimotor and skeletomotor activity under a variety of conditions. Morphological investigation of the innervation of the adqm muscle of the cat hind limb was carried out as an adjunct to electrophysiological studies.

METHODS

Five adqm muscles from 3 cats were prepared for silver impregnation and teasing according to the method of Barker & Ip (1963).

One muscle was frozen in isopentane cooled with liquid-N_2, cut transversely at 10μm and reacted with antibodies raised against tonic and neonatal myosin (Rowlerson, Gorza & Schiaffino, 1985).

One muscle was fixed with glutaraldehyde, embedded in Araldite resin and serially sectioned transversely at 1 and 5μm. Sections were stained with toluidine blue for 30s on a hotplate.

Table 1. Numbers of spindles and tendon organs observed using the various techniques

Method	Number of spindles	Number of tendon organs
immunohistochemistry	4	7
toluidine blue	3	6
silver	17	26
total	23	37

All specimens were observed using light microscopy. Whenever possible, the numbers and identification of intrafusal fibres were noted, as were the number and position of sensory endings and the distribution of motor axons.

RESULTS

Bag fibres were distinguished from chain fibres by their greater diameters. The two types of bag fibre were distinguished histochemically by the marked presence of tonic myosin combined with a virtually total lack of reaction against neonatal myosin in the bag_1 fibre, as has been previously found in triceps (Rowlerson et al., 1985). In the resin sections, the bag_2 fibre had a more granular appearance than the bag_1 and had more elastic fibres surrounding its poles (Gladden, 1976). Also it often shared a capsular compartment with chain fibres. In the silver material the bag fibres were distinguished by the form of the sensory terminals and by the distribution of elastic fibres.

Each muscle contained 3 to 6 spindles and 3 to 7 tendon organs. Additionally, the silver material contained 4 paciniform corpuscles. Spindles contained up to 2 bag_1 fibres, 4 bag_2 fibres and 7 chain fibres, the highest numbers occurring in one spindle from the histochemical series that appeared to consist of two intrafusal bundles conjugated in parallel within the same capsule. This is a feature that has also been noted in the superficial lumbrical muscle (Banks, Emonet-Dénand, Stacey & Thiesson, this volume). Only one spindle appeared to be a b_2c unit. All 4 spindles studied by immunohistochemistry, 2 from the resin-embedded material, and at least 2 from the silver were linked in series with tendon organs. The overall sensory complements of the latter two associations were:

$$S_3S_2S_1PS_1S_2S_3\text{-TO and TO-}S_2S_1PS_1S_2\text{-TO}$$

In the overall sample the sensory complements of the b_1b_2c spindle units were:

P 2; PS 0; P2S 7; P3S 3; P4S 6; P5S 0; P6S 1

The mean number of secondary endings per spindle is 2.90. This is an unusually high average for a hind limb muscle, where mean values have previously been found to range from 1.08 in superficial lumbrical (Banks et al., this volume) to 2.72 in popliteus (Banks & Stacey, 1988).

In a preliminary physiological experiment we have separately recruited 4 Ia afferents and 2 paciniform corpuscles, recording spikes at two sites on the nerve supply to adqm so as to establish conduction velocities.

Motor endings as seen structurally in the resin-embedded material, and as indicated by acetylcholinesterase activity in the immunohistochemical material were located in reconstructions. The majority (about 94%) were intracapsular, although dynamic β innervation is known to occur regularly in this muscle (F. Emonet-Dénand, personal communication). Motor axons are predominantly unmyelinated at spindle entry.

CONCLUSION

Although this preliminary structural analysis demonstrates that the muscle contains all the sensory complement expected in a hind limb muscle, there are certainly some unusual features, notably the high incidence of secondary endings, and the in-series arrangement of many spindles and tendon organs.

ACKNOWLEDGEMENTS

Financial support from the Wellcome Trust is gratefully acknowledged.

REFERENCES

BANKS, R.W., EMONET-DÉNAND, F., STACEY, M.J. & THIESSON, D. The innervation of muscle spindles in an intrinsic muscle of the hind foot: the superficial lumbrical of the cat. (this volume).

BANKS, R.W. & STACEY, M.J. (1988) Quantitative studies on mammalian muscle spindles and their sensory innervation. In *Mechanoreceptors*, eds. HNÍK, P., SOUKUP, T., VEJSADA, R. & ZELENÁ, J. pp. 263-269. Plenum Press, New York.

BARKER, D. & IP, M.C. (1963) A silver method for demonstrating the innervation of mammalian muscle in teased preparations. *J. Physiol.* **169**, 73-74*P*.

GLADDEN, M.H. (1976) Structural features relative to the function of intrafusal muscle fibres in the cat. *Prog. Brain Res.* **44**, 51-59.

ROWLERSON, A., GORZA, L. & SCHIAFFINO, S. (1985) Immunohistochemical identification of spindle fibre types in mammalian muscle using type-specific antibodies to isoforms of myosin. In *The Muscle Spindle*, eds. BOYD, I.A. & GLADDEN, M.H. pp. 29-34. Macmillan, London.

EVIDENCE FOR γ_s INNERVATION OF LONG CHAIN FIBRES

R.W. Banks[1], F. Emonet-Dénand[2],
M.H. Gladden[3] and F.I. Sutherland[3]

[1]Department of Biological Sciences
University of Durham, Durham, UK
[2]Collège de France, Paris, France
[3]Muscle Spindle Physiology Group
I.B.L.S., West Medical Building
University of Glasgow
Glasgow G12 8QQ, Scotland, UK

INTRODUCTION

Nuclear chain fibres are classically regarded as being much shorter and thinner than bag fibres, either bag_1 or bag_2. However, Barker et al. (1976) noticed that some spindles possessed one or more chain fibres with unusually long poles, often of similar length to the bag fibres. As an estimate of the incidence of these 'long chain' fibres they assessed the proportion of spindles containing at least one chain fibre pole that extended for more than 1mm beyond the end of the capsule. Kucera (1980) formalised this practical criterion as a definition of a class of long chain fibres each of which possessed at least one long pole. Since that time, long chain fibres have often been treated as though they formed a distinct type of intrafusal fibre with a specific static beta (β_s) innervation (Kucera, 1982, 1984; Barker & Banks, 1986). Here we give evidence in detail of both static β and static γ innervation of long chain fibres.

METHODS

Two methods were used to examine the motor innervation of tenuissimus muscle spindles which were obtained from previous work (Banks, 1981; 1991; Arbuthnott, Ballard, Boyd, Gladden & Sutherland, 1982; Sutherland, Arbuthnott, Boyd & Gladden, 1985; Boyd, 1986). These were silver staining and teasing of muscle spindles and serial sectioning of resin-embedded muscles.

Table 1. Long chain fibre pole innervation. γs, βs were physiologically characterised static fusimotor axons.

No innervation	lch	lch + ch	lch + b_2	lch + b_1	Total
2	6 (1β_s) (2γ_s)	2	3 (1γ_s)	3	16

Tenuissimus muscles were silver stained and teased according to the method of Barker & Ip (1963) and the motor innervation traced. In addition, seven adult cat tenuissimus muscles were glutaraldehyde fixed, resin-embedded and serially sectioned in the transverse plane throughout at 1μm. Ultrathin sections were either taken every 10μm throughout the sleeve region (Arbuthnott et al., 1982) when photographs were taken with the Zeiss 109 electron microscope at x8,300 and x20,000 enlarged x2.5, or were taken when a batch of 1μm sections ended in a motor ending. Motor innervation of the spindles was in all cases reconstructed. In several instances of each method the function of the motor axons had been determined. Fusimotor axons were classified as static or dynamic from the response of the primary sensory afferent to muscle ramp and hold stretches and ramp stimulation between 0 and 150 Hz.

RESULTS

The intrafusal distributions of all motor axons that supplied at least one long pole of a chain fibre are given in Table 1, which also includes two instances of long chain poles that received no motor innervation. The most frequent distribution (6/16, 38%) was to long chain poles only, but long chain poles could also be supplied in common with bag_1, bag_2 or other chain fibres. The common innervation with bag_1 fibres is particularly noteworthy. Specific examples in which physiological properties of the motor innervation were known are considered in more detail below.

Long chain fibre pole innervated by known β static axon

In the spindle innervated by a known β axon, one chain fibre pole was long. The motor axon terminating on this long chain fibre pole had a conduction velocity of 85m/s and the primary sensory ending was driven, which is characteristic of static β axons (Jami, Petit & Scott, 1985). The two endings on this long chain fibre pole were p_1 plates 15μm apart and positioned inside the B region of the capsule.

Long chain fibre poles innervated by known γ static axons

In the specific example described here in which the spindle was reconstructed from serial sections there was a chain fibre, both poles of which were long. The motor innervation of these poles was totally contained within the spindle capsule sleeve. One static fusimotor axon (γ_{20} where 20 refers to the conduction velocity of the axon in m/s) terminated only on the proximal long chain fibre pole. On stimulation, this axon drove the primary sensory ending. A second static motor axon (γ_{18}) had an ending on the distal long chain fibre pole but also terminated on the bag_2 fibre in this pole. This axon showed bag_2 effects combined with some chain involvement (Boyd & Ward, 1982).

The form of the motor endings on each long chain pole was different. One was of the simple type and one of the complex type, as determined by Arbuthnott, Gladden & Sutherland (1992). The motor endings of the axon terminating on both bag_2 and long chain fibre poles were similar and of a simple type, i.e. axon terminals lying on a smooth post synaptic muscle membrane. The motor ending on the proximal long chain fibre pole was complex, i.e. the post synaptic muscle membrane was protruded into 'fingers', the axon terminals lying on or between these protrusions.

CONCLUSIONS

A long chain pole can be innervated by gamma or beta motor axons, and the incidence of long chain poles cannot be used to estimate the incidence of beta static innervation.

REFERENCES

ARBUTHNOTT E.R., BALLARD K., BOYD I.A., GLADDEN M.H. & SUTHERLAND F.I. (1982) The ultrastructure of cat fusimotor endings and their relationship to foci of sarcomere convergence in intrafusal fibres. *J. Physiol.* **331**, 285-309.

ARBUTHNOTT E.R., GLADDEN M.H. & SUTHERLAND F.I. (1992) Diversity and homogeneity within endplates associated with physiologically identified static γ-axons in cat tenuissimus muscle. *Exp. Physiol.* **77**, 443-453.

BANKS, R. W. (1981) A histological study of the motor innervation of the cat's muscle spindle. *J. Anat.* **133**, 571-591.

BANKS, R. W. (1991) The distribution of static γ axons in the tenuissimus muscle of the cat. *J. Physiol.* **442**, 489-512.

BARKER, D. & IP, M. C. (1963). A silver method for demonstrating the innervation of mammalian teased preparations. *J. Physiol..* **169**, 73P.

BARKER, D. & BANKS, R. W. (1986). The Muscle Spindle. In *Myology.* eds. ENGEL A.G. & BANKER, B.Q. pp 309-341. McGraw-Hill, New York.

BOYD I.A. (1986) Two types of static γ axons in cat muscle spindles. *Exp. Physiol.* **71**, 307-327.

BOYD, I.A. & WARD, J. (1982). The diagnosis of nuclear chain intrafusal fibre activity from the nature of the group Ia and groupII afferent discharge of isolated cat muscle spindles. *J. Physiol.* **329**, 17-18P.

JAMI, L., PETIT, J. & SCOTT, J.J.A. (1985) Driving of spindle primary endings by static β axons. In *The Muscle Spindle*. eds. BOYD, I.A. & GLADDEN, M.H. pp. 201-205. Macmillan, London.

KUCERA J. (1980) Histochemical study of long nuclear chain fibres in the cat muscle spindle. *Anat. Rec.* **198**, 567-580.

KUCERA, J. (1982). The topography of long nuclear chain intrafusal fibres in the cat muscle spindle. *Histochemistry.* **74**, 183-197.

KUCERA J. (1984) Histological identification of (static) skeletofusomotor innervation to a cat muscle spindle. *Brain Res.* **294**, 390-395.

SUTHERLAND F.I., ARBUTHNOTT E.R., BOYD I.A. & GLADDEN M.H. (1985) Two ultrastructural types of fusimotor ending on typical chain fibres in cat muscle spindles. In *The Muscle Spindle.* eds. BOYD, I.A. & GLADDEN, M.H. pp51-56. Macmillan, London.

RECURRENT INHIBITION AND SKELETOFUSIMOTOR INNERVATION OF PROXIMAL AND DISTAL FORELIMB MUSCLES OF THE CAT

M. Illert[1], H. Kümmel[1] and J.J.A. Scott[2]

[1]Physiologisches Institut
Universität Kiel, Kiel, Germany
[2]Department of Physiology
University of Leicester, Leicester, UK

Electrophysiological and morphological studies of the α-motoneurones of the long digit extensor muscles of the cat forelimb have revealed that they neither receive, nor give rise to recurrent inhibition (Hahne, Illert & Wietelmann 1988; Hörner, Illert & Kümmel, 1991). This contrasts with the α-motoneurones of the proximal muscles acting on the elbow (triceps medialis, TM; anconeus, An), which have been shown to have large recurrent collateral trees (Figure 1). The muscles of the wrist (extensor carpi ulnaris, ECU; extensor carpi radialis, ECR) are intermediate, with only a proportion of their motoneurones giving rise to collateral trees and these being relatively small.

The α-motoneurones to the long digit extensors are also characterised by a very specific pattern of monosynaptic Ia connections (Fritz, Illert, De La Motte, Reeh & Saggau, 1989), and retrograde labelling studies of the α-motoneurones to extensor digitorum communis (EDC) have shown that they have soma diameters in the range of small α-motoneurones (26-52μm). The small cell somata of the EDC motoneurones contrasts with the short after-hyperpolarisation (AHP, < 80 ms, $n = 19$) found in these neurones which corresponds to their contraction properties as fast units (Fritz & Schmidt, 1992).

The coincidence of these characteristics and the correlation in amphibians of the absence of recurrent inhibition (Székely, 1976) with the presence of skeletofusimotor or β-innervation (Katz, 1949) had led us to the hypothesis that the long digit extensor muscles of the cat forelimb might also have a high degree of β-innervation. This was tested by analysing the proportion of muscle spindle poles with p_1 endplates in silver impregnated EDC, TM, An, extensor digitorum longus (EDL) and triceps lateralis (TLa, medial and lateral parts). The results presented in Table 1 indicate that the EDC and EDL both have substantially larger numbers of p_1 plate endings than do the elbow muscles.

We suggest that the high degree of β-innervation in the digit muscles could support fast corrections of deviations from the intended muscle tension during manipulative movements.

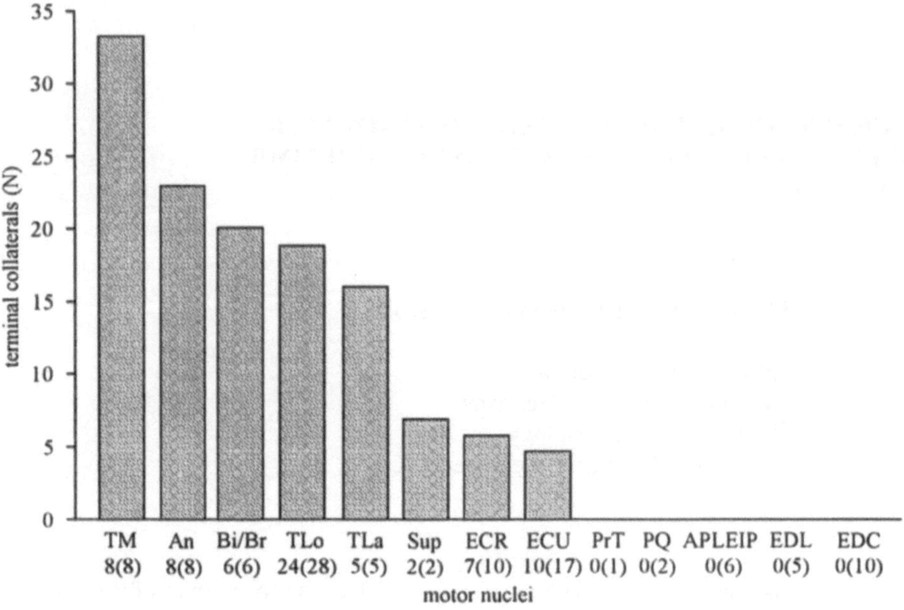

Figure 1. Average size of the collateral trees of motoneurones to different forelimb motor nuclei. The number of motoneurones with axon collaterals is given with respect to the number of neurones tested. TM: Triceps medialis, An: Anconeous, Bi: Biceps, Br: Brachialis, TLo: Triceps longus, TLa: Triceps lateralis, Sup: Supinator, ECR: Extensor carpi radialis, ECU: Extensor carpi ulnaris, PrT: Pronator teres, PQ: Pronator quadratus, APL: Abductor pollicis longus, EIP: Extensor indices proprius, EDL: Extensor digitorum lateralis, EDC: Extensor digitorum communis.

Contraction of the digit muscles against a resistance during a grip should activate the Ia afferents via the "efferent copy" to the intrafusal fibres and thus keep the involved motoneurones activated and maintain a stable muscular force. The parallel activation of motor and Ia axons found in humans during contraction of the EDC against a resistance (Hulliger et al., 1985) may be due to such a high degree of β-innervation. Recurrent inhibition might interfere with and disadvantage this function, by reducing the effect of the Ia EPSPs. The widespread Ia projections in the motor nuclei to the elbow muscles together with the distinct recurrent inhibition might allow the recruitment of motor units at low frequency (Kernell, 1976) and thus increase the muscle stiffness during their antigravity action (Grillner, 1973).

Table1. Proportion of spindle poles and estimated proportion of whole spindles with p_1 plate endings in long digit extensor and elbow muscles.

		p_1 - endplates (%)	
muscle	N	in spindle poles	whole spindles
EDC	3	61	71
EDL	5	67	72
TLa (med)	2	25	41
TLa (lat)	2	36	47
TM	2	40	45
An	3	35	41

REFERENCES

FRITZ, N., ILLERT, M., DE LA MOTTE, S., REEH, P. & SAGGAU, P. (1989) Pattern of monosynaptic Ia connections in the forelimb. *J. Physiol.* **419**, 321-351.

FRITZ, N. & SCHMIDT, C. (1992) Contractile properties of single motor units in two multi-tendoned muscles of the cat distal forelimb. *Exp. Brain Res.* **88**, 401-410.

GRILLNER, S. (1973) Muscle stiffness and motor control - forces in the ankle during locomotion and standing. In *Proceedings of the Second International Symposium on Motor Control*. ed. GYDIKOW, A.A., pp. 195- 215. Plenum Press, New York.

HÖRNER, M., ILLERT, M.& KÜMMEL, H. (1991). Absence of recurrent axon collaterals in motoneurones to the extrinsic digit extensor muscles of the cat forelimb. *Neurosci. Lett.* **122**, 183-186.

HAHNE, M., ILLERT, M. & WIETELMANN, D. (1988). Recurrent inhibition in the cat distal forelimb. *Brain Res.* **456**, 188-192.

KATZ, B. (1949) The efferent regulation of the muscle spindle in the frog. J. Exp. Biol. **26**, 201-217.

SZÉKELY, G. (1976) The morphology of motoneurons and dorsal root fibres in the frogs spinal cord. *Brain Res.* **103**, 275-290.

ULTRASTRUCTURE OF BAG AND CHAIN FIBRES AND 3-D RECONSTRUCTION OF THE SENSORY INNERVATION IN HUMAN LUMBRICAL SPINDLES

L.-E. Thornell[1], L. Carlsson[1], J. Fridén[2],
F. Pedrosa-Domellöf[1], U. Ranggård[1]
and T. Soukup[3]

[1]Departments of Anatomy and [2]Hand Surgery
Umeå University, S-901 87 Umeå, Sweden
[3]Institute of Physiology, Academy of Sciences
Prague, Czech Republic

INTRODUCTION

Morphologically, human muscle spindles conform, in principal, to those in other mammals. They contain bag_1, bag_2 and chain fibres and are innervated by sensory neurones and gamma motoneurones. However, the morphology of spindles seems to vary in different muscles in man more than in other mammals.

We have attempted to characterise spindles of the human first lumbrical muscle in detail, as this muscle is rather special. It originates from a flexor tendon and ends in the extensor aponeurosis of the first digit and is essential for the fine control of objects grasping between the first finger and thumb, as in writing. The diameters of the extrafusal fibres are small in comparison to those of limb extrafusal fibres and the muscle contains a large number of muscle spindles (Voss, 1937; Cooper & Daniel, 1963). We examined 21 lumbrical muscle spindles in detail. An extensive series of sections were stained for myofibrillar ATPase activity and antibodies against different myosin heavy chain (MHC) isoforms. These spindles contained on average 10 intrafusal fibres, which by their morphological, histo- and immunocytochemical characteristics corresponded to nuclear bag_1, nuclear bag_2 and nuclear chain fibres. The number of fibres of each type varied greatly: 1 - 5 bag_1, 1 - 4 bag_2 and 3 - 6 chain fibres per muscle spindle (Thornell, Pedrosa-Domellöf & Soukop, in preparation). Each intrafusal fibre type had a distinct MHC composition, but variations in the general pattern of MHC expression were common and showed a higher level of complexity than seen in limb muscle spindles (Lindmark, Pedrosa-Domellöf & Thornell, in preparation).

The present paper is a preliminary electron microscopic study of serially sectioned human lumbrical muscle spindles. We present the structural characteristics of bag_1, bag_2 and chain fibres and their afferent innervation. By computerised image analysis we have succeeded in getting a three dimensional view of the sensory innervation of the intrafusal fibres in one of the spindles.

MATERIAL AND METHODS

Lumbrical muscles were obtained during hand surgery from three patients aged 29, 50 and 67. The muscles were fixed at resting length in 2.5% glutaraldehyde in 0.1 M phosphate buffered saline (PBS). Three single spindles, about 5 - 7 mm long, were dissected out and were postfixed in 1% OsO_4 in PBS, dehydrated in an ascending series of acetone and then immersed in a mixture of acetone and Vestopal W. The spindles were finally embedded in glass capillary tubes using pure Vestopal W and then polymerised for 24 hours at 60°C.

Sections 70 nm thick were cut on a Reichert Ultracut and picked up on single-slot grids. They were stained with uranyl acetate and lead citrate with an LKB Ultrostainer and examined in a Jeol 1200 EX II electron microscope. Sections from the cross-sectioned blocks were photographed at every 1 - 8 µm interval both at low (x400) and high (x3000, x5000) magnifications, whereas sections from the longitudinal blocks were photographed at every second interval. Low magnification pictures were taken to get an outline of the spindles and served as a basis for the identification of the single fibres. Each fibre was subsequently analysed at high magnification to reveal sensory and motor terminals as well as

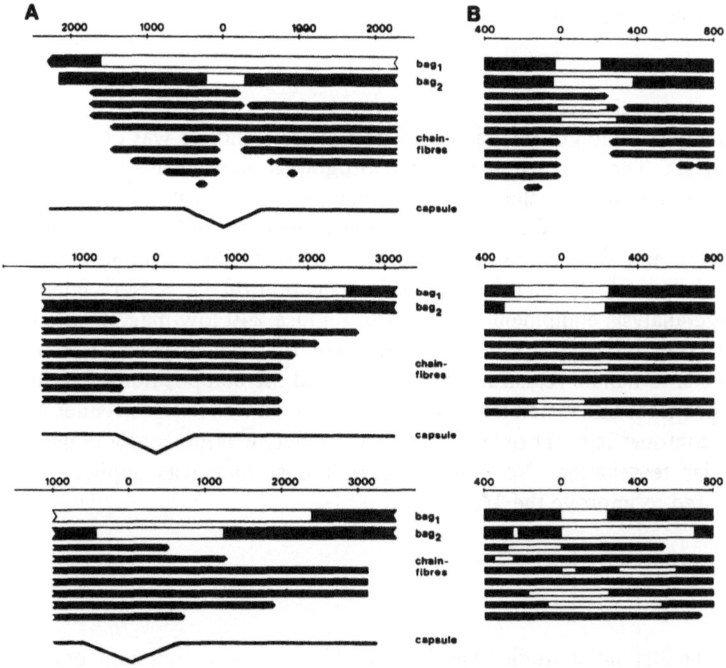

Figure 1. A. Schematic drawings depicting the organisation of three serially sectioned human lumbrical muscle spindles (1-3). Scale in micrometres, with centre of the equator at zero. Two nuclear bag fibres with larger diameter together with variable numbers of nuclear chain fibres having small diameters were present in each spindle. Notice also the large variability in number and length of the chain fibres. Solid areas mark the regions of fibres where myofibrils showed M bands. B. Schematic drawings of the sensory terminals in the equatorial region of the same spindles as in A. White bars indicate the sensory terminals. Note that some chain fibres lack sensory terminals.

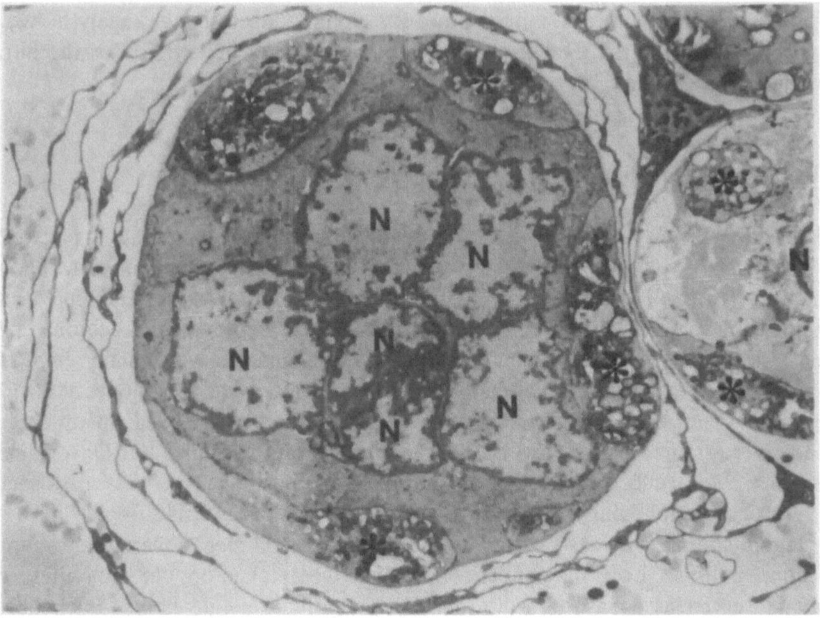

Figure 2. Electronmicrograph of a part of a muscle spindle cross-sectioned through its equatorial region. A bag_2 and a chain fibre are shown surrounded by inner capsular cells and bundles of elastic fibres. Nerve terminals (*) and nuclei (N). x 5000.

their internal structural organisation.

A 3-D reconstruction of the equatorial part of the spindle 1 was obtained with the help of image analysis. The low magnification photographs were analysed with respect to the capsule, intrafusal fibres, myelinated axons and vessels. The circumference of the outer capsule and of each intrafusal fibre were outlined manually and stored on an optical disk. Similarly, nerves and vessels were indicated and stored. High magnification photographs were used to help with the identification and indication of the sensory endings. An IBAS Kontron image analysis equipment was used to digitise 240 low magnification photographs out of the 15,000 slices. We developed software to trace manually the contour of the different objects, to handle registration of images and the composition of contours into 3-D objects. The program SunVision from Sun Microsystems was used to render the images. To compose the contours into 3D-objects, the surfaces between the contours were interpolated using triangular tessellation. No smoothing of the surface was applied. Finally, Phong shading was used to improve the 3D effect.

RESULTS

Each of the 3 spindles studied had two bag fibres, and varying numbers of chain fibres, identified on the basis of diameter and equatorial nucleation. A few fibres presented a small accumulation of nuclei but otherwise showed characteristics of chain fibres. The bag fibres differed in their ultrastructural appearance of the M-band in a characteristic way. In all 3 spindles one of the bag fibres lacked an M-band in the mid-equatorial region and in most of the encapsulated region. This fibre will be referred to as the bag_1 fibre. The other bag fibre in two spindles lacked an M-band in the equatorial region, for 0.5 and 1.5 mm respectively, whereas in one of the spindles an M-band was present throughout the whole length of the corresponding bag fibre. These fibres will be referred to as bag_2 fibres. Typical chain fibres and fibres which had a small accumulation of nuclei had a dense M-band throughout their whole length.

The number and the organisation of the intrafusal fibres and their relationship to the capsule, the equator and the periaxial space are shown schematically in Figure 1. In the equatorial region the diameters of bag_1 fibres were 17 - 20 μm, for bag_2 12 - 16 μm and for chain fibres 4 - 10 μm. In the non-equatorial encapsulated region, the bag_2 fibre often had the largest diameter, but the number and length of the chain fibres varied considerably. Some of the chain fibres ended in the equatorial region and, in one case, a chain fibre was interrupted by a short tendinous strand (Figure 1A). Although it is not apparent from the diagrams in Figure 1, the location of the intrafusal fibres in relation to each other varied also within the spindles.

Sensory nerve terminals

Abundant myelinated nerve profiles were present throughout the spindles. Myelinated axons that terminated on the densely nucleated equatorial region of the spindles were identified as branches of the primary afferent axon. The sensory terminals were characterised by the direct contact which they made with the intrafusal fibres. They were covered by the same basal lamina (Figure 2).

In spindle 1, sensory terminals were present on both bag fibres and on two chain fibres, as shown in Figure 1B. Four of the chain fibres totally lacked sensory innervation. The length of the primary endings on each fibre type varied. The bag_2 fibre had a much longer region of contact than the bag_1 and the two chain fibres. The number of axon terminal profiles per fibre in cross sections of spindle 1 was on average 3.9 (range = 1 to 8) for the bag fibres, and 1.4 and 1.5 for the two chain fibres (range = 1 to 4). As shown in Figure 3, the sensory terminals in 3-D appeared as longitudinal, somewhat curved strands with varicose swellings and irregular or curved branches. Although short stretches of terminals with an annulospiral configuration were seen on the bag fibres, this configuration was more commonly found on the chain fibres.

The density of terminals was highest close to the equator. In transverse sections the percentage of terminal area in relation to total fibre area was 12% for the bag_1, 8% for the bag_2 fibre and 18-20% for the chain fibres. Approximately 30% of the surface of each fibre was covered by sensory terminals except for the most equatorial region of the bag_2 fibre,

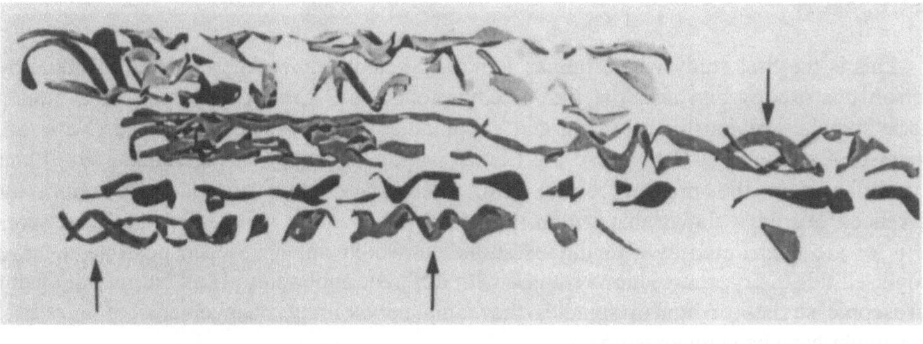

Figure 3. Computerised 3D image of the sensory terminals in the muscle spindle 1. The two bag fibres and two of the 6 chain fibres had afferent innervation. The sensory terminals appear mainly as longitudinal, somewhat curved strands with varicose swellings and irregular and curved branches. Short stretches of the terminals had annulospiral configuration (arrows).

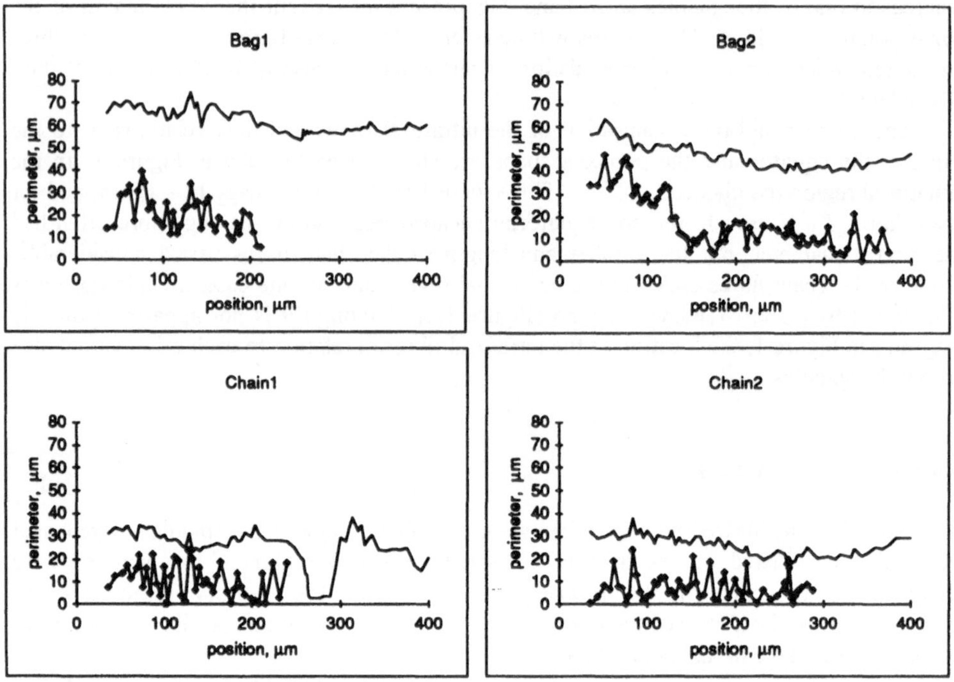

Figure 4. Graphs quantitatively illustrating the perimeter in the equator region of the four intrafusal fibres (upper line) and the sensory terminals (lower line) shown in Figure 2. One chain fibre (chain 1) was interspaced by a tendinous area 260-290 µm from the equator. Note the high proportion of terminals at the portion close to the equator of the bag_1 fibre.

where the density was approximately twice as high (Figure 4).

In spindle 2, the terminal region was similar in length in both bag fibres. Three out of 7 chain fibres had afferent innervation. In spindle 3, the length of the terminal region was much larger for the bag_2 fibre than for the bag_1 fibre. Furthermore, in the bag_2 fibre a small terminal area was seen separated from the main terminal region. Five out of 7 chain fibres had afferent terminals.

DISCUSSION

This is the first study where human lumbrical spindles have been studied in detail by electron microscopy and also the first attempt to get a 3-D reconstruction of a human muscle spindle. Previously, only frog and cat spindles have been reconstructed (Karlsson, Andersson-Cedergren & Ottoson, 1966; Banks, Barker & Stacey, 1982; Banks, 1986) and they differ from the present results. Our preliminary observations show important differences. We have shown that within the equatorial region a number of intrafusal fibres end in or are interrupted by a tendinous strand. It would not have been possible to have recognised this in cryostat sections stained with different antibodies. Thus, in previous light microscopic studies of human spindles the complexity of the arrangements of intrafusal fibres might have been underestimated.

As expected, three types of intrafusal fibres, bag_1, bag_2 and chain fibres, could be distinguished in human lumbrical spindles. It was not possible to distinguish with certainty between the two types of bag fibres on the basis of their diameters, in contrast to what has been suggested in other studies on human spindles. Kucera & Dorovini-Zis (1979) reported that bag_2 fibres generally were larger than bag_1 fibres. In our study, in one portion of the spindle, one of the bag fibres had a larger diameter than the other, but in another segment the situation was reversed. The differences in cross-sectional area between bag_1 and bag_2 fibres suggested by Gladden, Wallace & Craigen (1985) to be a helpful pointer in identifying bag_1 and bag_2 fibres were not apparent in our material. The M-band pattern and the accumulations of nuclei were used as the criteria for classifying the intrafusal fibres. Nuclear chain fibres and fibres with a small accumulation of nuclei always had dense M-bands along their entire length, whereas the nuclear bag fibres lacked an M-band in the equatorial region to a varying extent. The length of the fibre where an M-band was absent was used to distinguish between bag_2 and bag_1 fibres. One bag fibre in all 3 spindles had a substantially longer region, over 3 mm, which lacked an M-band pattern, whereas in the other bag fibre the occurrence of M-bands was more variable. An M-band pattern was absent over a distance of 0.5 mm in the first spindle. In the second spindle an M-band was seen along the whole bag fibre, whereas in the third spindle it was lacking over a distance of 1.5 mm.

From our previous correlative immunohistochemical and ultrastructural studies we know that lack of staining for MM-CK and M-protein, two proteins making up the M-band, correlates with the lack of M-band pattern in the myofibrils (Pedrosa, Butler-Browne, Dhoot, Fischman, & Thornell, 1989; Thornell, Carlsson, Kugelberg & Grove, 1987; Thornell, Carlsson & Pedrosa, 1990). As we have also shown that bag_1 fibres in human spindles lack staining for MM-CK and M-protein (Eriksson, Butler-Browne & Thornell, 1994; Thornell et al. in preparation), we conclude that the fibres mainly lacking an M-band in their intracapsular course are the bag_1 fibres. This difference in M-band appearance in different bag fibres and our way of typing the bag fibres are in accordance with the observations of Kucera & Dorovini-Zis (1979) on bag fibres in human intercostal muscles. Those authors observed that although both types of bag fibres had an identical M-band pattern (i.e. lack of a dense M-band), in the equatorial and juxta-equatorial regions a dense M-band was present in bag_1 fibres solely in the extracapsular region. They also found that the length of the fibre segment lacking an M-band in bag_2 fibres varied considerably. In one spindle, the nuclear bag_1 and bag_2 were identical up to 450 μm from the spindle equator, whereas in another spindle the myofibrils were observed to lack M-bands as far as 1830 μm from the spindle equator. Furthermore, there seems to be a general tendency in mammalian spindles for the bag_1 fibres to be those which lack an M-band pattern throughout most of their course (Banks, Harker & Stacey, 1977).

The functional significance of the differences in the composition of the myofibrillar M-band in the different types of intrafusal fibres can at present only be a matter of speculation. It is known that the MM-CK in the M-band is of importance for the reactivation of ATP, which is the prime fuel for the actin-myosin contraction (Wallimann, Wyss, Brdiczka, Nicolay & Eppenberger, 1992). It is also known that in general the bag_1 fibre differs from bag_2 and chain fibres in its MHC composition, e.g. in rat (Pedrosa et al., 1990), cat (Kucera & Walro, 1989) and human (Eriksson et al., 1994). The bag_1 fibre is probably the fibre which has the highest proportion of slow tonic myosin in its myofibrils as it lacks slow twitch myosin in its central region. In contrast, the bag_2 fibre contains slow twitch myosin in the whole capsular portion, in addition to slow tonic myosin (Thornell et al., in preparation). These morphological observations are in accordance with the different contraction features reported for bag_1 fibres. These fibres have the slowest rate of contraction and might therefore not need rapid reactivation of ATPase. This would be equivalent to slow tonic fibres in frogs and chicken which contain neither a dense M-band nor MM-CK at the centre of the myofibrillar A-band (Thornell et al., 1990).

The innervation of human spindles has been studied in both teased and sectioned material impregnated by silver nitrate and gold chloride (Kennedy, 1970; Swash & Fox,

1972; Swash, 1972) as well as in serial plastic 1 μm thick sections (Gladden, 1975; Kucera, 1986; Gladden et al., 1985) and by electron microscopy (Kennedy, Poppele & Staley, 1974; Gladden et al. 1985; Katto, Okamura & Yanagihara, 1987). Having established the identity of the intrafusal fibres we can conclude that human intrafusal fibres differ from those of other mammalian species in their afferent innervation. A main finding in the present study is that the sensory terminals were not distributed in the classical annulospiral pattern for most of their length, but formed longitudinal strands with some varicose swellings and irregular curved branches. A similar appearance was noted by Kennedy (1970) and Katto et al. (1987), whereas Swash (1982) described the primary endings as complex, loosely wrapped single or double spirals of unmyelinated axon terminals, which slightly indent the surface of the fibre from which flat, plate-like expansions arise.

Our observations that some chain fibres in human spindles lack sensory terminals are in accordance with those of Kucera (1986). The functional significance of this is unclear. Morphometric analysis of different objects such as volume of nuclei, myofibrils, mitochondria, leptomeres, cytoskeletal structures, axon terminals and area of contact between axon terminals and the intrafusal fibres may provide useful quantitative data. Hopefully, these will give new insights into the relationship between muscle spindle morphology and function.

ACKNOWLEDGMENTS

Supported by the Swedish Medical Research Council 12x-3934 and the Medical Faculty, Umeå University.

REFERENCES

BANKS, R.W. (1986) Observation on the primary sensory ending of tenuissimus muscle spindles in the cat. *Cell Tissue Res.* **246**, 309-319.

BANKS, R.W., HARKER, D.W. & STACEY, M.J. (1977) A study of mammalian intrafusal muscle fibres using a combined histochemical and ultrastructural technique. *J. Anat.* **123**, 783-796.

BANKS, R.W., BARKER, D. & STACEY, M.J. (1982) Form and distribution of sensory terminals in cat hindlimb muscle spindles. *Phil. Trans. R. Soc. Lond. B.* **299,** 329-364.

COOPER, S. & DANIEL, P.M. (1963) Muscle spindles in man; their morphology in the lumbricals and the deep muscles of the neck. *Brain* **86**, 563-586.

ERIKSSON, P-O., BUTLER-BROWNE, G.S. & THORNELL, L-E. (1994) Immunohistochemical characterization of human masseter muscle spindles. *Muscle Nerve* **17**, 31-41.

GLADDEN, M.H. (1975) Elastic fibres in human muscle spindles. *J. Anat.* **119**, 187-188.

GLADDEN, M.H., WALLACE, W. & CRAIGEN, M.L. (1985) Movement convergence and motor end plate location in a human muscle spindle. In *The Muscle Spindle*. eds. BOYD, I.A. & GLADDEN, M.H., pp. 115-120. Macmillan, London.

KARLSSON, U., ANDERSSON-CEDERGREN,. E. & OTTOSON, D. (1966) Cellular organization of the frog muscle spindle as revealed by serial sections for electron microscopy. *J. Ultrastruct. Res.* **14**, 1-35.

KATTO, Y., OKAMURA, H. & YANAGIHARA, N. (1987) Electron microscopic study of muscle spindle in human interarytenoid muscle. *Acta Otolaryngol.* **104**, 561-567.

KENNEDY, W.R. (1970) Innervation of normal human muscle spindles. *Neurology* **20**, 463-475.

KENNEDY, W.R., POPPELE, R.E. & STALEY, N.A. (1974) Isolation of viable human muscle spindles for electron microscopic and physiologic study. *Anat. Rec.* 179, 453-462.

KUCERA, J. (1986) Reconstruction of the nerve supply to a human muscle spindle. *Neurosci. Lett.* **63**, 180-184.

KUCERA, J. & DOROVINI-ZIS, K. (1979) Types of human intrafusal muscle fibers. *Muscle Nerve* **2**, 437-451.

KUCERA, J.& WALRO, J.M. (1989) Nonuniform expression of myosin heavy chain isoforms along the length of cat intrafusal fibers. *Histochemistry* **92**, 291-299.

PEDROSA, F., BUTLER-BROWNE, G.S., DHOOT, G.K., FISCHMAN, D.A. & THORNELL, L-E. (1989) Diversity in expression of myosin heavy chain isoforms and M band proteins in rat muscle spindles. *Histochemistry* **92**,185-94.

SWASH, M. (1972) Pathology of muscle spindle. In *Skeletal Muscle Pathology*. eds. MASTAGLIA, F.L. & WALTON, J., pp. 508-536. Churchill Livingstone, London.

SWASH, M. & FOX, K.P. (1972) Muscle spindle innervation in man. *J. Anat.* **112**, 61-80.

THORNELL, L-E., CARLSSON, E., KUGELBERG, E. & GROVE, B.K. (1987) Myofibrillar M-band structure and composition of physiologically defined rat motor units. *Am. J. Physiol.* **253**, 456-68.

THORNELL, L-E., CARLSSON, E. & PEDROSA, F. (1990) M-band structure and composition in relation to fiber types. In *The Dynamic State of Muscle Fibers*. ed. PETTE, D., pp. 369-383, Walter de Gruyter, Berlin and New York.

VOSS, H. (1937) Untersuchungen über Zahl, Anordnung und Länge der Muskelspindeln in den Lumbricalmuskeln des Menschen und einiger Tiere. *Zeitschrift für mikroskopisch-anatomische Forschung* **42**, 509-524.

WALLIMANN, T., WYSS, M., BRDICZKA, D., NICOLAY, K. & EPPENBERGER, H.M. (1992) Intracellular compartmentation, structure and function of creatine kinase isoenzymes in tissues with high and fluctuating energy demands: the phosphocreatine circuit for cellular energy homeostasis. *Biochem. J.* **281**, 21-40.

DISTRIBUTION OF SUBPOPULATIONS OF EFFERENT NEURONES INNERVATING COMPARTMENTS OF THE MASSETER MUSCLE IN THE RABBIT

M. Saad[1], R. Dubuc[1], C.G. Widmer[2],
K.G. Westberg[1] and J.P. Lund[1]

[1]Centre de recherche en sciences neurologiques
Université de Montréal, Montréal, Canada H3C 3J7
Département de Kinanthropologie
Université du Québec à Montréal
Montréal, Canada H3C 3P8
[2]Department of Oral Surgery
University of Florida, Gainesville,
FL 32610 USA

INTRODUCTION

It has long been known that there is a topographical organisation of trigeminal motoneurones innervating the muscles of mastication (e.g. Landgren & Olsson, 1976; Matsuda, Uemura, Kume, Matsushima & Mizuno, 1978; Mizuno, Matsuda, Sato, Konishi, Uemura & Matsushima, 1980). In all species studied so far, motoneurones of the jaw-closing muscles occupy the rostral and middle thirds of the trigeminal motor nucleus (motV). Within these parts, the motoneurone pool innervating the temporal muscle is represented dorsally, the medial pterygoid pool is ventral and the masseteric pool is found in a lateral position, although there is some overlap. The jaw-opening motoneurones are found ventromedially in the middle third and form the caudal third of the motV. In the rabbit, these motoneurones form a distinct cluster which is clearly segregated from the closer motoneurone pool in Nissl stained sections. The masseter and the digastric muscles of rabbits are also innervated by a group of small neurones with cell bodies located ventrolateral to motV (Donga, Dubuc, Kolta & Lund, 1992). This group of cells were initially described and named "cell group k" by Meessen & Olszewski (1949).

In the rabbit, the masseter muscle complex has at least 11 partitions that are innervated by separate nerve branches (Widmer, Lund & English, 1993). These partitions are arranged in superficial, intermediate and deep layers; three partitions in the superficial layer, two in each of the intermediate and deep layers. In addition, there are four partitions in the zygomaticomandibularis, which some classify as the deepest part of the muscle and others as a separate structure. The masseter nerve, which arises from the mandibular division of the trigeminal nerve, has 11-14 branches. The most proximal supply the deepest layer of the muscle, while the most distal branches go to the superficial layer (Widmer et al., 1993).

On the basis of the orientation of the fibres in each compartment, it has been suggested that the superficial oblique layer, which runs downwards and backwards to the jaw angle, is mainly responsible for the protraction of the mandible and contralateral movements, while closing force is generated by the vertical components of deeper layers (Schumacher, 1980). The variation in the expression of myosin isoforms throughout the muscle also suggest that there are different regional demands during function. The rabbit masseter muscle has a higher percentage of type I fibres in the rostral region, while type IIA and IM predominate in the caudal region. Variations are also observed as one moves from superficial to deep parts of the muscle (Bredman, Weijs, Moorman & Brugman, 1990). The arrangement of muscle fibres making up motor units into narrow columns within anatomical partitions is also consistent with the ability to control small regions of the muscle (Weijs, Jüch, Kwa & Korfage, 1993). In support of these anatomical findings, there is evidence that these compartments function somewhat independently during mastication (Herring, Wineski & Anapol, 1989; Weijs & Dantuma, 1981). On the working side (the side on which the food is crushed) EMG activity in the masseter muscle of the rabbit starts in the deeper parts, with a gradual transition towards the superficial ones. The compartments are recruited in reverse order on the other side (balancing side). Differences in recruitment order of the anterior and posterior deep masseter have also been observed (Weijs & Dantuma, 1981). The observed differences in onset time and activity levels between muscle compartments have profound effects on the resultant force vectors, which are functionally important in adapting the chewing pattern to different types of food (Weijs & Dantuma, 1981).

As a first step to uncover the rules governing the selection of trigeminal motoneurone pools, we have examinated the distribution of motoneurones supplying different compartments of the masseter muscle in the rabbit to see if these are also compartmentalised within motV and cell group k.

METHODS

The experiments were carried out in New Zealand rabbits (body weight 2-3kg). Animals were preanaesthetised with xylazine (5 mg, i.m.), followed one hour later with acepromazine maleate (20 mg, i.m.). Anaesthesia was induced with sodium pentobarbital (30 mg/kg, i.v.). Small doses of ketamine hydrochloride was used to maintain general anaesthesia during surgery. The masseter nerve was exposed after the removal of the zygomatic arch and two main branches (or groups of fine branches) were separated from the main nerve trunk. The branches were classified as superficial, intermediate, deep or zygomaticomandibular supplying the anterior, middle and posterior parts of the muscle. In most animals, two branches per side were cut. In some other animals, the whole trunk was isolated on one side, two branches on the other. The cut nerves were placed in small reservoirs of parafilm to prevent leakage of tracers. Crystals of Fluorogold (FG) were applied to the central end of one branch, and Fastblue (FB) crystals to the other. After 45 minutes, the branches were washed with saline and the wounds were closed with sutures.

After the animals had survived for 7 to 8 days, they were deeply anaesthetised with sodium pentobarbital (90 mg/kg, i.v.) and perfused transcardially with isotonic saline, followed by 4% paraformaldehyde in acetate buffer (0.1M, pH 6.5) then by 4% paraformaldehyde in borate buffer [0.1M, pH 9.5 (Gordon & Richmond, 1990)]. All Solutions were cooled to 4°C, and the head of the animal was surrounded with ice. After perfusion, the brain stem was removed and transferred through a graded series of sucrose solutions (10,20,30%) at 4°C, during 48 hours. Serial coronal sections of the frozen brain stem were cut on a cryostat at 30μm. The next day, the sections were dehydrated and coverslipped with Entellan. Slides were stored at -20°C to preserve their fluorescence.

Analysis

All sections covering the motV and cell group k were viewed with fluorescence microscopy under ultraviolet illumination, using a specific filter system to visualise FG and FB (UV 2A Nikon). The maximum and minimum diameters of each retrogradely labelled neurone was measured. Photomicrographs of some neurones were taken with Ektachrome film 400 ASA. The outline of each section was drawn from a projected image, then the positions of labelled neurones were transferred to the drawn sections by an x-y plotter

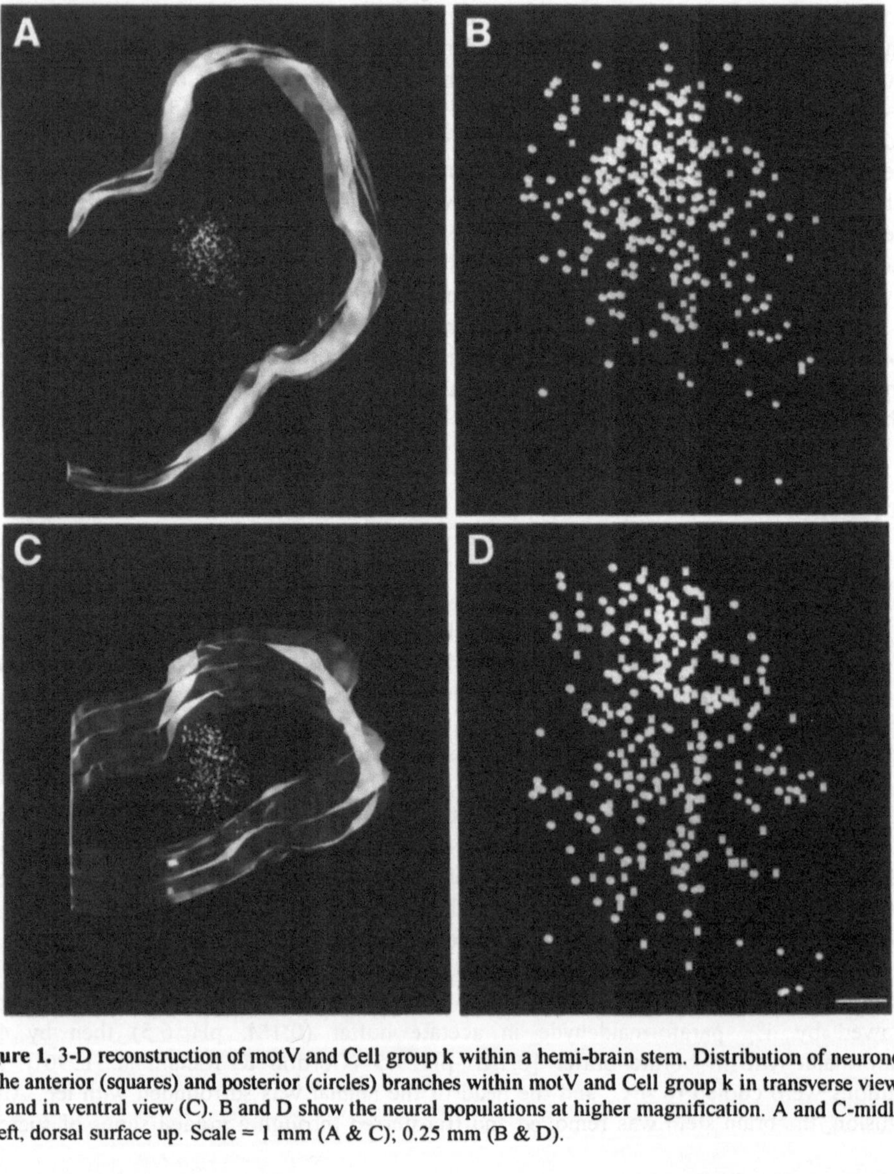

Figure 1. 3-D reconstruction of motV and Cell group k within a hemi-brain stem. Distribution of neurones of the anterior (squares) and posterior (circles) branches within motV and Cell group k in transverse view (A) and in ventral view (C). B and D show the neural populations at higher magnification. A and C-midline at left, dorsal surface up. Scale = 1 mm (A & C); 0.25 mm (B & D).

linked to linear potentiometers attached to the microscope stage. The three-dimensional positions of the neurones within the brain stem were plotted with a computer-based image analysis system (Autocad-Autoshade).

RESULTS

Motoneurones labelled with FG contain granules that appear golden-yellow under the microscope. Neurones labelled with FB had usually a uniform blue colour, although small silvery granules were seen in some cases. Both tracers filled the proximal dendrites.

Neurones sending axons to all the nerve branches studied were represented in both the motV and cell group k. Motoneurones supplying the superficial, intermediate and deep layers of the masseter muscle were found within the dorsal division of motV. Most of the cell group k neurones supplying the different muscle compartments were also found in the most dorsal part of this nucleus. Figure 1 shows the distribution of neurones in motV and cell group k innervating two superficial compartments. As can be seen, there is a complete overlap of the two populations in motV. This proved to be the rule even when the pools supplying deepest branches and superficial branches were compared in the same animals: there was no spatial representation of masseter compartments in motV.

Although it appears that one branch was represented more strongly in cell group k than the other (Figure 1B), this was not a general rule. All branches were supplied by cell group k and again, there was no separation of the various populations.

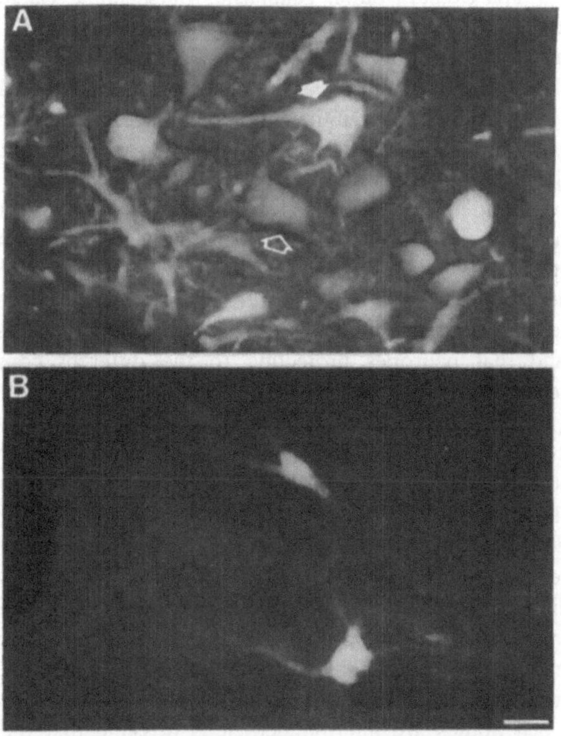

Figure 2. Photomicrographs showing the intermingling of motoneurones in the trigeminal motor nucleus (A) of the middle branch (FG) and anterior superficial branch (FB). The motoneurones labelled witn FG contain granules that appear golden-yellow (filled arrow) under the microscope. Neurones labelled with FB (empty arrow) were usually a uniform blue colour. B shows two neurones of cell group k that were labelled by FG applied to the posterior branch. Scale=50 μm.

There were significant differences in the shape and mean diameter of neurones in the two nuclei. The cell bodies of neurones in motV were usually round or ovoid (Figure 2A) with a mean diameter of 44 μm, while those of cell group k were fusiform (Figure 2B) and smaller in size (mean diameter = 31 μm).

DISCUSSION

Results from this study suggest that there is no topographical segregation in motV of motoneurones supplying different compartments of the masseter/zygomaticomandibularis complex. Motoneurones innervating the four layers and the different compartments within a layer intermingled extensively. The evidence suggests that there is a similar lack of topography in many of the motoneurone pools supplying complex muscles innervated by the spinal cord, although there does appear to be some separation of neurones supplying the gastrocnemius (Weeks & English, 1985) and trapezius muscle (Vanner & Rose, 1984). Gordon & Richmond (1990; 1991) and Gordon, Loeb & Richmond (1991) conducted a series of experiments to study the distribution of motoneurones of other compartmentalised muscles (diaphragm, sartorius and tensor fasciae latea and those of the neck). They showed that motoneurone subpopulations innervating different compartments of these muscles are extensively intermingled within the ventral horn, despite the fact that some muscles are supplied by more than one ventral root.

Our observations on the motoneurone distributions are not consistent with those of Matsuda et al., (1978) and Uemura-Sumi, Takahashi, Matsushima, Takata, Yasui & Mizuno (1982), who reported that the deep layer of the rabbit masseter muscle is represented by a separate pool of motoneurones from those of the superficial layer of the muscle. This discrepancy is probably due to differences in the experimental techniques used. In the studies of Matsuda et al. (1978) and Uemura-Sumi et al. (1982), the injections of the tracers were made directly into the muscle. The problem with this technique is that the exact location of the injections and their spread in the tissue is difficult to assess. In our experiments, we have avoided such problems by applying tracers to the ends of cut nerves.

Just as the masseter motoneurones in motV are found dorsal to those innervating the digastric, Donga et al. (1992) showed that the efferent neurones of cell group k which project to the masseter were located dorsal to those of the digastric muscle. In the present study we extend these observations by showing that cell group k innervates all parts of the masseter muscle and that neurones projecting to the different compartments overlap extensively with one another.

Donga et al. (1992) suggested that each masticatory muscle was innervated by two populations of motoneurones; one within motV and the other within the cell group k. In support of this hypothesis, we have preliminary results that these neurones are cholinergic. Cell group k neurones as well as classical motoneurones react strongly to antibodies raised against choline acetyltransferase (Saad, Widmer, Lepage, Umbriaco, Westberg, Dubuc & Lund, 1994). However, cell group k neurones are fusiform and smaller in size than motV cells, which could indicate that they are gamma motoneurones. This interpretation is not in line with the observation that the digastric muscle, which contains no muscle spindles in the rabbit, is also innervated by cell group k. The physiological properties of these neurones have not yet been studied and their function remains unknown.

Even though cell bodies are intermingled, it is possible that the dendritic trees of discrete motoneurone pools are oriented differently in order to attract particular inputs (Keirstead & Rose, 1983). Jaw closer motoneurones in the cat and the rat do have complex dendritic trees, but there is little evidence that these occupy separate spaces in the species studied (cat, Yoshida, Tsuru, Mitsuhiro, Otani & Shigenaga, 1987; rat, Lingenhöhl &

Friauf, 1991; Moore & Appenteng, 1991). Neurones with cell bodies located centrally in the motoneurone pool have spherical dendritic arbors that usually span most of motV, while motoneurones located at the periphery of the nucleus often have dendrites that project into the core of the nucleus (Lingenhöhl & Friauf, 1991). Distal dendrites of the motoneurones are also frequently found beyond the borders of the trigeminal motor nucleus. However, the dendrites of the closer motoneurones do not extend into the jaw-opening motoneurone subnucleus (Yoshida et al., 1987). The distribution of cell bodies and dendrites in the jaw-closing motoneurone pools suggests that topography is not an important determinant of synaptic connectivity. One alternative possibility is that the appropriate connections are specified during development by selective neurotrophic signals, that favour synapse formations between axon terminals and their partner motoneurones (Lichtman, Jhaveri & Frank, 1984).

CONCLUSION

In conclusion, our data and that of other groups working in the spinal cord clearly show that the subpopulations of motoneurones supplying most complex muscles do not have specific anatomical addresses because they have no spatial organisation that corresponds to the peripheral compartments. This seems to indicate that position of cell bodies is not an important factor in the selection of appropriate motoneuronal subpopulations during specific motor tasks.

ACKNOWLEDGEMENTS

We thank Mme. S. Lepage and M. Serge Dupuis for their valuable assistance. Supported by the Canadian MRC, the Swedish MRC and USPHS Grants DE10130 and DE00333.

REFERENCES

BREDMAN, J.J., WEIJS,W.A., MOORMAN, A.F.M. & BRUGMAN, P. (1990) Histochemical and functional fibre typing of the rabbit masseter muscle. *J. Anat.* **168**, 31-47.

DONGA, R., DUBUC, R., KOLTA, A. & LUND, J.P. (1992) Evidence that the masticatory muscles recieve a direct innervation from cell group k in the rabbit. *Neurosci.* **49**, 951-961.

GORDON, D.C,, LOEB, G.E. & RICHMOND F.J.R. (1991) Distribution of motoneurons supplying cat sartorius and tensor fasciae latae, demonstrated by retrograde multiple-labelling methods. *J. Comp. Neurol.* **304,** 357-372.

GORDON, D.C. & RICHMOND F.J.R. (1990) Topography in the phrenic motoneuron nucleus demonstrated by retrograde multiple-labelling techniques. *J. Comp. Neurol*. **292**, 424-434.

GORDON, D.C. & RICHMOND F.J.R. (1991) Distribution of motoneurons supplying dorsal suboccipital and intervertebral muscles in the cat neck. *J. Comp. Neurol.* **304**, 343-356.

HERRING, S.W., WINESKI, L.E. & ANAPOL, F.C. (1989) Neural organization of the masseter muscle in the pig. *J. Comp. Neurol.* **280**, 563-576.

KEIRSTEAD, S.A. & ROSE, P.K., (1983) Dendritic distribution of splenius motoneurons in the cat: comparison of mKotoneurons innervating different regions of the muscle. *J. Comp. Neurol.* **219,** 273-284.

LANDGREN, S. & OLSSON, K. Å. (1976) Localization of evoked potentials in the digastric, masseteric, supra- and intertrigeminal subnuclei of the cat. *Exp. Brain Res.* **26**, 299-318.

LICHTMAN, J.W., JHAVERI, S. & FRANK E. (1984) Anatomical bases of specific connections between sensory axons and motor neurons in the brachial spinal cord of the bullfrog. *J. Neurosci.* **4,** 1754-1763.

LINGENHÖHL, K. & FRIAUF, E. (1991) Sensory neurons and motoneurons of the jaw-closing reflex pathway in rats: a combined morphological and physiological study using the intracellular horseradish peroxidase technique. *Exp. Brain Res.* **83,** 385-396.

MOORE, J.A. & APPENTENG, K. (1991) The morphology and electrical geometry of rat jaw-elevator motoneurons. *J. Physiol.* **440**, 325-343.

MATSUDA, K., UEMURA, M., KUME, M., MATSUSHIMA, R. & MIZUNO, N. (1978) Topographical representation of masticatory muscles in the motor trigeminal nucleus in the rabbit: A HRP study. *Neurosci. Lett.* **8, 1-4**.

MEESSEN, H. & OLSZEWSKI, J. (1949) Cytoarchitektonicsher atlas des Rautenhirms des Kaninchens. Karger, Basel.

MIZUNO, N., MATSUDA, K., SATO, M., KONISHI, A., UEMURA, M. & MATSUSHIMA, R. (1980) Myotopical arrangement of masticatory motoneurons. In *Jaw Position and Jaw Movement.* eds. KUBOTA, K., NAKAMURA, Y. & SCHUMACHER, G.-H. pp. 198-206. Berlin. Veb Verlag Volk und Gesundheit.

SAAD, M., WIDMER, C.G., LEPAGE, S., UMBRIACO, D., WESTBERG, K.G., DUBUC, R. & LUND, J.P. (1994) Anatomical and immunocytochemical characterization of masseteric efferent neurons of Cell group k in the rabbit. Soc. Neurosci. Abs. **20**, 1585.

SCHUMACHER, G.-H. (1980) Comparative functional anatomy of jaw muscles. In *Jaw Position and Jaw Movement.* eds. KUBOTA, K., NAKAMURA, Y. & SCHUMACHER, G.-H. pp. 76-93. Veb Verlag Volk und Gesundheit, Berlin.

UEMURA-SUMI, M., TAKAHASHI, O., MATSUSHIMA, R., TAKATA, M., YASUI, Y. & MIZUNO, N. (1982) Localization of masticatory motoneurons in the trigeminal motor nucleus of the guinea pig. *Neurosci. Lett.* **29**, 219-224.

VANNER,S.J. & ROSE,P.K. (1984) Dendritic distribution of motoneurons innervating the tree heads of the trapezius muscle in the cat. *J. Comp. Neurol.* **226**, 96-110.

WEEKS,O.I.. & ENGLISH,A.W. (1985) Compartmentalization of the cat lateral gastrocnmius motor nucleus. *J. Comp. Neurol.* **235**, 255-267.

WEIJS, W.A. & DANTUMA, R. (1981) Functional anatomy of the masticatory apparatus in the rabbit (Oryctolagus Cuniculus L.). *Netherl. J. Zool.* **31(1)**, 99-147.

WEIJS, W.A., JÜCH, P.J.W., KWA, S.H.S. & KORFAGE, J.A.M. (1993) Motor unit territories and fiber types in rabbit masseter muscle. *J. Dent. Res.* **72**, 1491-1498.

WIDMER, C.G., LUND, J.P. & ENGLISH, A.W. (1993) Anatomical partitions and nerve branching patterns of the rabbit masseter. *Soc. Neurosci. Abs.* **19,** 153.

YOSHIDA, A., TSURU, K., MITSUHIRO, Y., OTANI, K. & SHIGENAGA, Y. (1987) Morphology of masticatory motoneurons stained intracellularly with horseradish peroxidase. *Brain Res.* **416**, 393-401.

CRITIQUE OF PAPERS BY THORNELL ET AL. AND SAAD ET AL.

Margaret H. Gladden

Muscle Spindle Physiology Group
I.B.L.S., West Medical Building,
University of Glasgow
Glasgow G12, 8QQ,
Scotland

The muscles used in these two studies, the masseter of the rabbit (Saed et al.) and the first lumbrical of the human hand (Thornell et al.) would seem to have very little in common. The first lumbrical has been intimately involved in some of the highest cultural achievements of human beings, since it is the muscle needed to apply pressure by the tip of the index finger against the thumb when holding a pen or paintbrush. However, one cannot be homocentrically dismissive about a muscle merely dedicated to lagomorph nutrition after reading the paper by Saed et al. The control of both these muscles must be closely co-ordinated not only with the control of several other muscles, but also with sensory information from the skin and joints of the hand in the lumbrical case, and from the buccal cavity and mandibular joints in the masseter case. But the rabbit masseter has an impressive versatility in addition, due to its internal organisation. As Saed et al explain it is divided by partitions into multiple compartments, from superficial to deep, capable of being activated successively, and possibly differentially even within each compartment. Surprisingly, however, this peripheral anatomical order was not reflected in a topographical organisation of the motoneurone cell bodies, which occur in two groups, in the trigeminal motor nucleus, and in cell group k, a separate collection of small fusiform neurones.

Saed et al. argue that the dendrites of these neurones are not organised into any special orientation either which might mirror the internal organisation of the muscle. Yet the control strategy must be complex in order to determine how small fractions of this muscle work independently for specialised functions, or co-ordinated together, while in addition ensuring that the forces generated are related to different food textures. Contraction of a portion of a muscle tends to stretch non-contracting areas. If muscle spindles were evenly distributed across all compartments their ensemble feedback might be difficult to interpret. Perhaps it is in order to keep the sensory feedback from the muscle simple that spindles in rabbit masster are confined to deep layers, and to the zygomaticomandibularis (Rowlerson et al., 1988). The control might rely more on the spacial distribution of sensory information from the buccal cavity than on information from within the muscle. However, spindles are similarly restricted to deep layers in mammals who use their mandible as a more straightforward hinge, without pronounced lateral movements (Lund et al., 1978; Rowlerson et al., 1988).

Thornell et al. found that muscle spindles of the first lumbrical muscle had primary sensory endings lacking an annulospiral arrangement, and chain fibres which are sometimes in discontinuous segments and sometimes without sensory contacts. One wonders how Rembrandt and Michaelangelo got by with such untidy spindles! These apparent structural abnormalities, however, need not have serious functional implications. Even in the cat the annulospiral form of the primary sensory ending is not completely standard, although this has become firmly entrenched in textbook descriptions. Ruffini himself wrote (1898) 'It offers a singular appearance not seen in any other nerve-ending. In some spindle its elegance and regularity are simply surprising.'.....but later....'Not every *primary ending* is so elegantly regular.' According to Banks et al., (1982) the primary terminals on both bag fibres are irregular at each end, and on bag_1 fibres only about 40% of the terminals on average are regular. The primary endings of b_2c units are mostly irregular.

In spindle 1 of Thornell et al. seven of the nine chain fibres terminate, or have breaks in continuity in the equatorial region (Figure 1A). The most likely explanation why so many terminate in the same position is that these chain fibres have ruptured. This could have occurred during some traumatic episode - we do not know why the patient came for hand surgery. If this is indeed the case, the fact that the bag fibres of this spindle remained intact suggests either that human chain fibres are more vulnerable to stretching, or that it was an abnormally violent contraction of the chain fibres themselves which caused them to rupture. In cat spindles it is the bag fibres which are more at risk from dissection damage than the chains. This is a consequence of their mechanical properties. Chain fibres are more perfectly elastic, while the viscoelastic properties of the bag fibres set up local internal stresses. There are also species differences in the susceptibility to damage from stretch. Rabbit spindles, for example, are more friable than cat spindles.

Thornell et al. found several chain fibres without sensory endings, two in spindle 1 (not counting those likely ruptured), three in spindle 2, and two in spindle 3 (Figure 1B). They could not distinguish between primary and secondary sensory endings, so this implies that these chains did not receive innervation from secondary endings if any were present, as well as primary sensory endings. Kucera (1986) also found one such fibre. This could be a consequence of development. Chain fibres are the last of the intrafusal types to develop (Milburn, 1973), and these non-innervated fibres may have developed too late, after the critical period for sensory innervation.

Chain fibres without sensory innervation are not necessarily a functionally redundant feature of these spindles, so long as they do have a motor innervation. Fusimotor activity without a direct connection to an afferent may, nevertheless, be capable of influencing the afferent. Sometimes spindle primary sensory endings respond to intrafusal contraction in close, neighbouring spindles even though these are not structurally integrated with them, except, presumably, by connective tissue linkages. Such effects are weak, but if the source were in the same spindle - i.e. a chain fibre without direct connection - the effect could be expected to be stronger. Presumably during development several chain fibres need to be produced in order to offer a sufficient number of potential sites for motor innervation, as human spindles can be innervated by considerable numbers of fusimotor axons. In an adult biceps brachii spindle Kucera (1986) traced 5 motor axons in one pole, and 8 in the other. In that spindle all 4 chain fibres were innervated at both poles. Lack of motor innervation in a chain fibre pole does seem to be a redundant feature, and this can occur in human spindles. In a rectus abdominis spindle from a child one of the 5 chain fibres of a pole was not innervated despite a large number of motor axons (7) supplying the pole (Gladden et al., 1985).

REFERENCES

BANKS, R.W., BARKER, D. & STACEY M.J. (1982) Form and distribution of sensory terminals in cat hindlimb muscle spindles. *Phil. Trans. R. Soc. B.* **299**, 329-364.

GLADDEN, M.H., WALLACE, W. & CRAIGEN M.L. (1985) Movement convergence and motor end plate location in a human muscle spindle. In *The Muscle Spindle.* eds. BOYD, I.A & GLADDEN, M.H., pp.115-120, Macmillan, New York.

MILBURN, A. (1973) The early development of muscle spindles in the rat. *J. Cell Sci.* **12**, 175-195.

KUCERA, J. (1986) Reconstruction of the nerve supply to a human muscle spindle. *Neurosci. Letts.* **63**, 180-184.

LUND, J.P., RICHMOND, F.J.R., TOULOUMIS, C., PATRY, Y. & LAMARRE, Y. (1978) The distribution of Golgi tendon organs and muscle spindles in masseter and temporalis muscles of the cat. *Neurosci.* **3**, 259-270.

ROWLERSON, A., MASCARELLO, F., BARKER, D. & SAED, H. (1988) Muscle-spindle distribution in relation to the fibre-type composition of masseter in mammals. *J. Anat.* **161**, 37-60.

RUFFINI, A. (1898) On the minute anatomy of the neuromuscular spindles of the cat, and on their physiological significance. *J. Physiol.* **23**, 1190-208.

MUSCLE SPINDLES IN THE MASTICATORY MUSCLES OF THE CAT AND OPOSSUM

M. Bosley[1], R. Burhanudin[2],
A. Rowlerson[1] and J.J. Sciote[3]

[1]Department of Physiology, UMDS
St. Thomas' Hospital, London SE1 7EH, UK
[2]Department of Orthodontics, UMDS
Guy's Hospital, London SE1 9RT, UK
[3]Basic Science Division
NYU College of Dentistry
345 East 24th St., New York, USA

INTRODUCTION

The morphology, location and intrafusal fibre type composition of spindles in the jaw-closer muscles of the cat and South American opossum (*Monodelphys domestica*) were identified histologically and by their myosin isoform composition.

METHODS

Frozen sections

Selected areas of the jaw-closer muscles were snap-frozen in composite blocks with samples of limb muscles for comparison. Serial 10 μm sections from these blocks were then stained for myosin ATPase activity, and with antibodies specific for a variety of myosin isoforms. The profiles of mATPase and antibody staining which characterise the extra- and intrafusal fibre types found in these muscles are summarised in Table 1.

Whole jaw-closer muscle blocks with skeletal attachments intact (cats) or whole heads (opossums)

These were fixed in modified Carnoy solution, decalcified in EDTA and embedded in paraffin wax for complete serial sectioning at 8 μm (opossum) or 10 μm (cat) in various planes in order to identify the exact position of spindles (method as described in Bosley, Rowlerson & Sciote, 1992). At intervals of not more than 400 μm throughout each series, a mini-series of sections was taken for Weigert-van Gieson staining (for general morphology) and for immuno-staining using the antibodies described in Table 1. Drawings were made from each 'level' (= mini-series), and the position of any spindles marked. Spindle counts were also made on the sections of cat jaw-closer muscles (Burhanudin, McDonald & Rowlerson, 1994). This was done only in muscle portions cut in transverse section, where each spindle was allotted a unique number and followed through the series.

RESULTS

Spindle morphology and intrafusal fibre type composition

Although the jaw-closer muscles of both species contain a predominant extrafusal fast fibre type (IIM) which is different from that in limb muscles (Table 1), intrafusal fibre types resembled those in limb muscle spindles. Most spindles were single spindles containing at least one bag_1, one bag_2 and two or more chain fibres. In the cat, paired spindles fused in parallel were quite common, and a small number of complex spindles (4 or more spindles fused together) was found in zygomaticomandibularis and masseter. In the opossum a few spindles were paired, but no complex units were found.

In six adult cats, the distribution of intrafusal fibres within spindles was determined for masseter and temporalis muscles. Mean numbers of intrafusal fibres in 60 individual spindles were: 1.1 bag_1, 1.0 bag_2 and 3.4 chains. The range of bag fibres was always between 1 and 2, but the chains were more variable, ranging from 0 to 8 (Sciote, 1993). There were no significant differences in intrafusal fibre distribution between simple and complex spindles.

Spindle location within the jaw-closer muscles

The co-localisation of slow (type I) fibres and muscle spindles was particularly striking in the jaw-closer muscles of the opossum; a similar but weaker association was observed in the cat. No spindles were observed in either digastric or lateral pteryogoid.

Medial pterygoid contained a few spindles in the cat, but not more than one in the opossum (none in one count, one in another). There was one spindle-rich area in masseter of both species, located moderately deep and anterior to the temporomandibular joint. In the cat this masseter area had about 50 spindles, and there was also another spindle-rich area with a similar number in zygomaticomandibularis.

Temporalis had only one spindle-rich region in the cat (more than 120 spindles in the deep anterior portion, between the coronoid process and the cranium). In the opossum, temporalis contained three spindle-rich regions. One was located in the most anterior part, and the other two extended posteriorly from the coronoid process, one medially towards the cranium and the other laterally along the temporalis tendon to a position superior to the zygomatic process of the temporal bone.

CONCLUSIONS

In jaw-closer muscles of the opossum, spindle morphology and intrafusal fibre type composition are very similar to that seen in the cat, except for the absence of complex spindles. The opossum, which is born in a very immature state, may therefore be useful for studying spindle development in these muscles.

Table 1. Myosin ATPase and immuno-staining profiles of extrafusal and intrafusal fibres in masticatory muscles of the cat and opossum. I. = slow-twitch; II = fast-twitch types IIA, IIB etc as found in limb muscles and digastric; IIM = fast-twitch type found in jaw-closer muscles of some species; * = Sigma clone MY32; # = specific for neonatal (& embryonic) myosins found in extrafusal fibres early in development.

	FIBRE TYPES					
	extrafusal			intrafusal		
	I	II	IIM	bag_1	bag_2	chain
mATPase:						
alkali	-	++	++	-	+	++
mild acid	++	+/-	++	+	++	-
strong acid	+	-	++	+	++	-
antibody:						
anti-I	++	-	-	+	++	-
anti-II*	-	-	-	-	+	++
anti-IIM	-	-	++	-	-	-
anti-TONIC	-	-	-	++	+	-
anti-NE#	-	-	-	-	+	++

The number of spindles found in the jaw-closer muscles of the cat was much higher than that found in a previous study by Lund, Richmond, Touloumis, Patry & Lamarre (1978), probably because the method we used allowed examination of adult material while avoiding any damage to the muscles.

The dentition and mandibular kinetics of the temporomandibular joint of the opossum are similar to those of the cat, suggesting that opossum mastication is also principally vertical, with very restricted lateral movement. However, in the opossum the anterior-posterior axis of the cranio-facial skeleton, and consequently the temporalis muscle, is relatively much longer than in the cat. This probably accounts for the additional spindle-rich areas in opossum temporalis, needed to ensure adequate proprioceptive feedback.

REFERENCES

BOSLEY, M., ROWLERSON, A. & SCIOTE, J.J (1992) Spindles in jaw-closer muscles of the cat. *J. Physiol.* **446**, 604P.

BURHANUDIN, R., MCDONALD, F. & ROWLERSON, A. (1994) Numbers and location of simple and complex spindles in the jaw-closer muscles of the cat. *J. Physiol.* **479**, 31P.

LUND, J.P., RICHMOND, F.J.R., TOULOUMIS, C., PATRY, Y. & LAMARRE, Y. (1978) The distribution of Golgi tendon organs and muscle spindles in masseter and temporalis muscles of the cat. *Neuroscience* **3**, 259-270

SCIOTE, J.J. (1993) Fibre type distribution in the muscle spindles of cat jaw-elevator muscles. *Archs. Oral Biol.* **38**, 685-688.

DIVERSE EXPRESSION OF EXTRACELLULAR MATRIX PROTEINS IN HUMAN MUSCLE SPINDLES

Fatima Pedrosa-Domellöf[1], Ismo Virtanen[2]
and Lars-Eric Thornell[1]

[1]Department of Anatomy
University of Umeå, Sweden
[2]Department of Anatomy
University of Helsinki, Finland

INTRODUCTION

Sensory transduction in muscle spindle involves distortion of the intrafusal fibres and of their sensory endings. In the equatorial region, where the sensory endings are found, the muscle spindle capsule is dilated and filled with fluid. It is generally accepted that the spindle capsule acts as a diffusion barrier and that the fluid-filled periaxial space provides both mechanical protection and a special internal millieu. Given these data, one would expect the extracellular matrix (ECM) to play an important role in muscle spindle function. In recent years, the number of ECM proteins and their proposed functions have expanded tremendously but data on the ECM of mammalian intrafusal fibres are scarse. We have therefore studied adult human muscle spindles with antibodies against the following ECM proteins: laminin, fibronectin and tenascin.

MATERIAL AND METHODS

The first lumbrical muscle was collected from the right hand of two human subjects at autopsy. The muscles were divided into smaller pieces and rapidly frozen in propane chilled in liquid nitrogen. Serial cross sections, 7 µm thick, were cut in a cryostat at -19°C. The sections were processed for immunocytochemistry with the following antibodies (Ab): Ab against laminin (E-Y Labs, USA), Ab 52BF12 against plasma fibronectin, Ab 143 DB7 (Locus-genex, Finland) which specifically binds the Mr 250,000 and 180,000 tenascin polypeptides. We used Ab SY38 against synaptophysin (Progen, Germany) and staining for acetylcholinesterase activity to label nerve endings. Antibody binding was visualised with the peroxidase anti-peroxidase method. A total of 36 muscle spindles were studied along series of over 200 cross-sections.

RESULTS

Fibronectin (Figure 1A) was present in all layers of the outer capsule and within the axial space. The level of staining by the Ab against fibronectin was highest in the outer layers of the capsule. The external lamina of the intrafusal fibres was weakly to moderately stained by the Ab against fibronectin along the whole length of the fibres.

Laminin (Figure 1B) was present in all layers of the outer and inner spindle capsule. The external lamina of the intrafusal fibres was strongly stained by the Ab against laminin along the whole length of the fibres.

Tenascin (Figure 1C) was present in the outer layers of the capsule but not in the inner capsule or in the axial space. In the equatorial region, the external lamina of the intrafusal fibres was strongly stained by the Ab against tenascin for a distance of approximately 50-500 μm which coincided closely with the area of distribution of the sensory endings (Figure 1D). Weak reactivity to the Ab against tenascin was very sporadically detected in intrafusal fibres along the B and C regions.

DISCUSSION

Our three main findings are: i) there is clear specialisation of the ECM within the capsule layers; ii) fibronectin is a component of the axial space and iii) the ECM of the intrafusal fibres is specialised at the site of sensory innervation and it is therefore likely to play an important role in muscle spindle function.

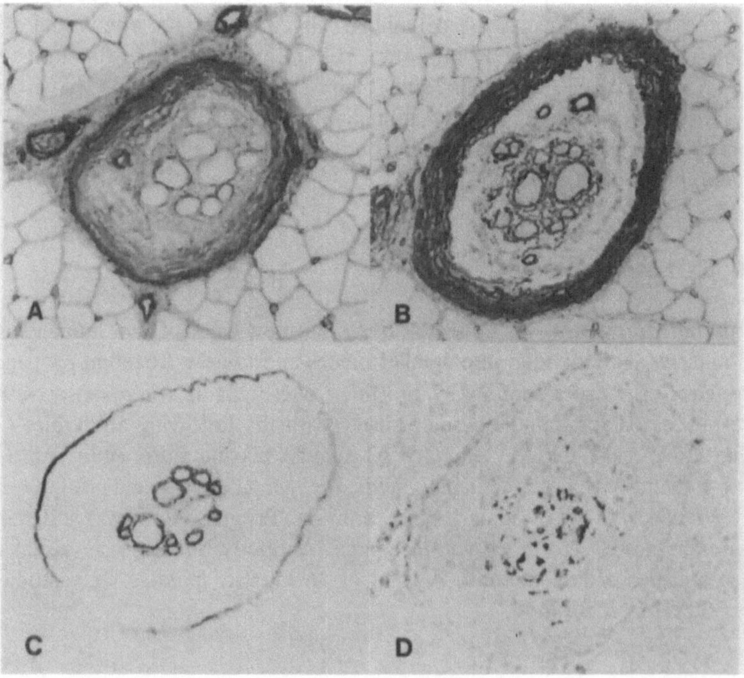

Figure 1. Muscle spindle stained with antibodies against fibronectin (A), laminin (B), tenascin (C) and synaptophysin (D). Notice the gradient in staining intensity with the antibody against fibronectin across the layers of the outer capsule (A). Tenascin (C) was present in the outer layers of the capsule and in a segment of the external lamina of the intrafusal fibres which coincided closely with the location of the sensory endings (D).

The muscle spindle capsule consists of an outer and an inner capsule (reviewed by Barker & Banks, 1986). The outer capsule is continuous with the perineurium of the spindle nerve and consists of multiple layers of flat cells arranged in a concentric tubular fashion, surrounded by a basal lamina. The inner capsule consists of a complex array of cells forming a meshwork around the intrafusal fibres and their sensory terminals (Ovalle & Dow, 1983). The presence of laminin in all layers of the outer capsule and in the inner capsule was expected because this multifunctional glycoprotein is present in all basement membranes. Laminin plays a role in adherence, growth, morphology, migration and differentiation of cells (reviewed by Engvall, 1993). In the present study, the staining obtained with the Ab against laminin was particularly useful to detect all layers of the outer spindle capsule. Fibronectin was also present in all layers, but a clear gradient was observed, with the highest staining intensity localised to the outermost layers. Taken together with the restricted location of tenascin to the outermost layers, these data indicate that the composition and therefore the function of the ECM vary within the layers of the spindle outer capsule. Whether this specialisation is mostly related to structure and support or to permeability can only be guessed. Fibronectins are glycoproteins found in most mesenchymal extracellular matrices and in blood plasma. They are also multifunctional proteins, promoting cell adhesion and affecting cell morphology, migration, differentiation and cytoskeletal organisation (reviewed by Hynes, 1993). Tenascin is a large glycoprotein found in the ECM (reviewed by Chiquet-Ehrismann, 1990). It is transiently expressed in many developing organs and, in the adult, it has been found in the perichondrium and periostium, ligaments, tendons and myotendinous junctions, as well as in smooth muscle.

The periaxial space is full of a highly viscous fluid, suggested to be hyaluronic acid (Brzezinski, 1961). The present results show that fibronectin is also present.

The localised expression of tenascin in the external lamina covering the intrafusal fibres and their sensory endings is most remarkable. This is, to our knowledge, the first study to show such an intimate spatial relation between tenascin and sensory nerve endings in human skeletal muscle. Although the exact functions of tenascin are still unclear, this protein probably plays a structural role within the ECM network. One can speculate that the ECM containing tenascin acts as a stabilising cuff around the intrafusal fibres and their sensory endings. Other roles of tenascin, related to regulation of cell function cannot be excluded.

REFERENCES

BARKER, D. & BANKS, R.W. (1986) The muscle spindle. In. *Myology*. eds. ENGEL, A.G. & BANKER, B.Q. pp 309-341, McGraw-Hill, New York.

BRZEZINSKI, D.K. (1961) Untersuchungen zur histochemie der muskelspindeln. *Acta histochem.* **12**, 75-79.

CHIQUET-EHRISMANN, R. (1990) What distinguishes tenascin from fibronectin? *FASEB J.* **4**, 2598-2604.

ENGVALL, E. (1993) Laminin. In. *Guidebook to the extracellular matrix and adhesion proteins* eds. KREIS, T. & VALE, R. pp 66-68, Oxford University Press, Oxford.

HYNES, R. (1993) Fibronectins. In. *Guidebook to the extracellular matrix and adhesion proteins* eds. KREIS, T. & VALE, R. pp 56-58, Oxford University Press, Oxford.

OVALLE, W.K. & DOW, P.R. (1983) Comparative ultrastructure of the inner capsule of the muscle spindle and the tendon. *Am. J. Anat* **166**, 343-357.

REGION DEPENDENT DISTRIBUTION OF MUSCLE FIBRE TYPES AND MUSCLE SPINDLES IN THE HUMAN MASSETER MUSCLE STUDIED IN CROSS-SECTIONS OF WHOLE MUSCLE

P.-O. Eriksson[1], A. Eriksson[2], I. Johansson[1], M. Monemi[1], and L.-E. Thornell[3]

Departments of [1]Clinical Oral Physiology, [2]Forensic Medicine and [3]Anatomy, Umeå University, S-901 87 Umeå, Sweden

INTRODUCTION

Human masseter is divided anatomical into superficial and deep parts, but regional specialisation in the distribution of fibre types and muscle spindles is organised differently. Thus, the anterior part is predominantly composed of type I fibres whereas type II fibres are concentrated in the posterior part. Spindles are most numerous and complex in the deep portion (Eriksson & Thornell, 1983; 1987). Previous studies concentrated on antero-posterior differences and were based on samples from restricted muscle regions. Since they may have missed medio-lateral variations the present study has been carried out on whole muscle transverse sections using enzyme- and immunohistochemical methods.

MATERIALS AND METHODS

The left masseter of a young adult male was dissected into its superficial and deep portions. Each was cut perpendicularly to the muscle fibres into slices which were placed flat on glass slides, covered with O.C.T compound (Tissue-Tek®, Miles Inc., USA.) and rapidly frozen in liquid propane chilled in liquid nitrogen. The superficial portion was sectioned at five different superior-inferior levels, the deep portion at three. Transverse sections were cut at 10µm across the whole muscle in a cryostat. They were processed for myofibrillar ATPase after alkaline and acid preincubations, mitochondrial enzyme, and routine histological staining. Adjacent sections were treated with antibodies (Abs) selective for slow, fast, embryonic, fetal, and slow tonic myosin heavy chain (MHC) isoforms, together with peroxidase antiperoxidase (PAP) staining.

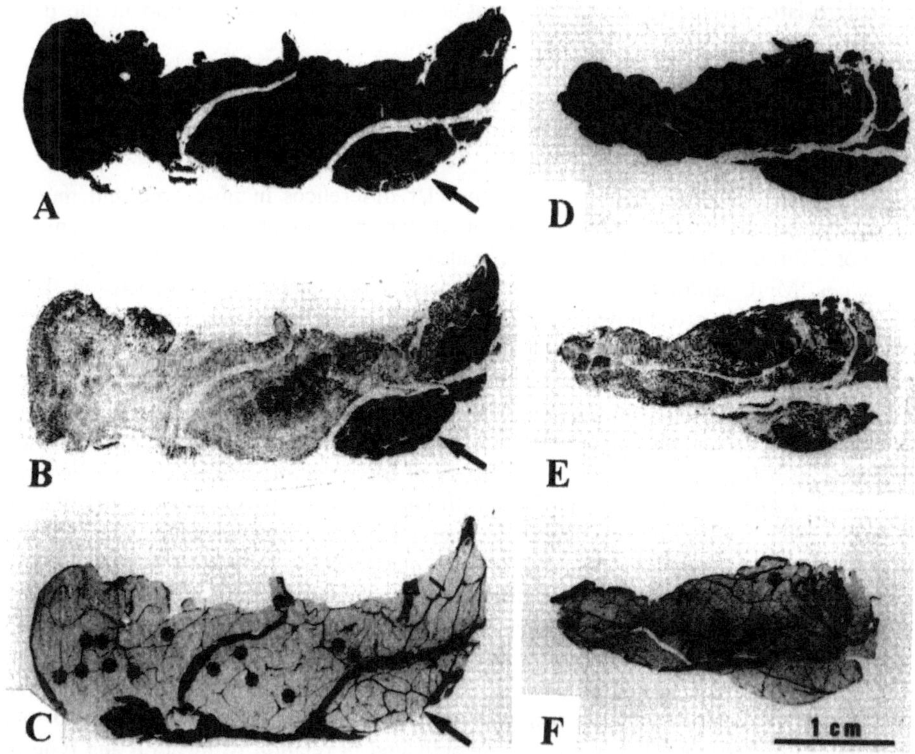

Figure 1. Serial cross-sections from the superior half of the superficial portion (A, B & C) and the superior third of the deep portion (D, E & F) of left masseter muscle. Left in the figure is anterior in the muscle, up is medial (near the bone). Sections are stained for myofibrillar ATPase at pH 4.3 (A & D), with an anti-fast MHC Ab and PAP staining (B & E) and Verhoeff-van Gieson (C & F). At pH 4.3 (A & D), dark staining represents type I muscle fibres and light staining (arrow) type II fibres. In the superficial portion, staining with the anti-fast MHC antibody is preferentially located in the postero-lateral part of the section (B, arrow). In the deep portion, anti-fast Ab staining is more variable though the posterior region shows the most intense staining. Single muscle spindle location within sections is marked (*; C & F). Note lack of muscle spindles in the postero-lateral part of the superficial portion.

RESULTS

The muscle fibres could be classified into type I (light staining at pH 10.3 and dark staining at pH 4.3), ATPase-IM (moderate at pH 10.3, dark at pH 4.3), IIC (dark at pH 10.3, moderate at pH 4.3) and IIB (dark at pH 10.3, unstained at pH 4.3). In the superficial portion, the antero-medial part contained predominantly type I fibres, whereas type IIB fibres were concentrated in the postero-lateral part of the muscle (Figure 1A). Type ATPase-IM and IIC fibres were distributed throughout the muscle cross-section. A similar intramuscular distribution of fibre types was seen in the deep portion (Figure 1D). However, in this portion the intramuscular variability in fibre type composition was considerable. All Abs stained the cross-sections in a heterogeneous non-random pattern (Figures 1 B & E). Type I fibres reacted with the anti-slow Ab, type ATPase-IM and IIC fibres with the anti-slow, anti-fast and anti-fetal Abs, and type IIB fibres with the anti-fast and anti-fetal Abs. Intrafusal fibres showed immunoreactivity with all Abs. The anti-embryonic and anti-slow tonic Abs exclusively reacted with intrafusal fibres. In the superficial portion, muscle spindles were frequently detected in the antero-medial region whereas none were observed in the postero-lateral region. The highest muscle spindle density was found in the deep portion (Figures 1 C & F).

DISCUSSION

The present findings of significant intramusclar differences in fibre type and muscle spindle compositions in the antero-posterior dimension corroborate previous studies (Eriksson & Thornell, 1983; 1987). This study also demonstrates intramuscular segregation of fibre types, MHC isoforms and muscle spindles in the medio-lateral dimension. Low threshold motor units are known to be composed of type I fibres. Therefore, the preponderance of type I fibres medially in the muscle is in line with the previous observation in the masseter that motor units with the lowest recruitment thresholds are located preferentially in the medial part of the muscle (Stålberg & Eriksson, 1987).

As a result of its complex multipennate architecture, the masseter muscle has a wide range of fibre orientations, and therefore a variety of force vectors and mechanical actions. The special fibre anatomy, in conjunction with the present and previous findings (Eriksson & Thornell, 1983; 1987), reflect functional segregation between as well as within the superficial and deep portions. These morphological characteristics suggest that each portion can be activated for a broad range of directions and levels of force, to fulfill separate functional demands. Further, high spindle density in the deep portion and lack of spindles in part of the superficial portion reflect how subregions rely on different proprioceptive support in control of particular motor tasks. Functional specialisation of subregions of the human masseter is supported by electromyographic studies which have demonstrated that motor units of different muscle regions can be activated in relation to location and type of activation, showing a "task-related" behaviour (Eriksson, Stålberg & Antoni, 1984; McMillan & Hannam, 1992).

ACKNOWLEDGEMENTS

Supported by the Swedish Medical Research Council (3934 and 6874), the Swedish Dental Society and the Faculty of Odontology, Umeå University, Sweden.

REFERENCES

ERIKSSON, P.-O. & THORNELL, L.-E. (1983) Histochemical and morphological muscle-fibre characteristics of the human masseter, the medial pterygoid and the temporal muscles. *Arch. Oral Biol.* **28**, 781- 795.

ERIKSSON, P.-O. & THORNELL, L.-E. (1987) Relation to extrafusal fibre-type composition in muscle-spindle structure and location in the human masseter muscle. *Arch. Oral Biol.* **32**, 483-491.

ERIKSSON, P.-O., STÅLBERG, E. & ANTONI, L. (1984) Flexibility in motor-unit firing pattern in human temporal and masseter muscles related to type of activation and location. *Arch. Oral Biol.* **29**, 707-712.

McMILLAN, A. S. & HANNAM, A. G. (1992) Task-related behaviour of motor units in different regions of the human masseter muscle. *Arch. Oral Biol.* **37**, 849-857.

STÅLBERG, E. & ERIKSSON, P.-O. (1987) A scanning electromyographic study of the topography of human masseter single motor units. *Arch. Oral Biol.* **32**, 793-797.

PART 5

SENSORY RECEPTOR PROPERTIES

SENSORY ENDINGS OF LIVING ISOLATED MAMMALIAN MUSCLE SPINDLES

Michael Chua and Carlton C. Hunt

Department of Cell Biology and Physiology
Washington University School of Medicine
St. Louis, MO, USA

We report here studies which are still in progress on the structure of living muscle spindle sensory endings, as visualised by differential interference contrast (Nomarski) microscopy and by fluorescence of vital dyes that are taken up by the sensory endings. Confocal microscopy has been used to create a 3-dimensional image of the sensory endings. Banks (1986) has previously used serial electronmicroscopic sections to reconstruct a 3-dimensional view of the sensory region of a cat tenuissimus muscle spindle.

Partially isolated muscle spindles in tenuissimus of cat were used by Bessou & Pagès (1975) and by Boyd, Gladden and their collaborators (Boyd, Gladden, McWilliam & Ward, 1977) to visualise intrafusal fibres and sensory endings during stimulation of single fusimotor axons in ventral root filaments in vivo, In contrast, our preparations have been totally isolated and studied in vitro. Spindles obtained from tail muscles of cat were used most often. They can be seen under the dissecting microscope and freed from extrafusal muscle fibres. Spindles have also been isolated from tenuissimus muscles. The capsule of the isolated spindles was often removed from the equatorial region by microdissection, in order to improve visualisation and to gain easier access to the sensory region. The isolated spindle was mounted in a chamber and viewed with Nomarski optics or epifluorescence. Confocal microscopy has been carried out using an Odyssey or Zeiss laser confocal system.

A Nomarski image of a primary ending on a bag fibre is shown in Figure 1. Such Nomarski images have a thin focal depth and high resolution. Bag and chain fibres can be readily identified in the equatorial region by their size and nucleation. By focusing at different levels the 3-dimensional extent of the annulospiral terminals of the primary endings can frequently be followed. The unmyelinated terminals often have a granular appearance, probably due to the numerous mitochondria they contain. Visualisation is often very clear but some areas remain poorly resolved, usually due to highly refractive elements, such as myelinated nerve branches above or below the plane of focus. Because of this, Nomarski images are not ideal for 3-dimensional reconstruction of the sensory ending. But images produced by fluorescent dyes, which appear to stain the totality of the sensory ending, were found to be more satisfactory.

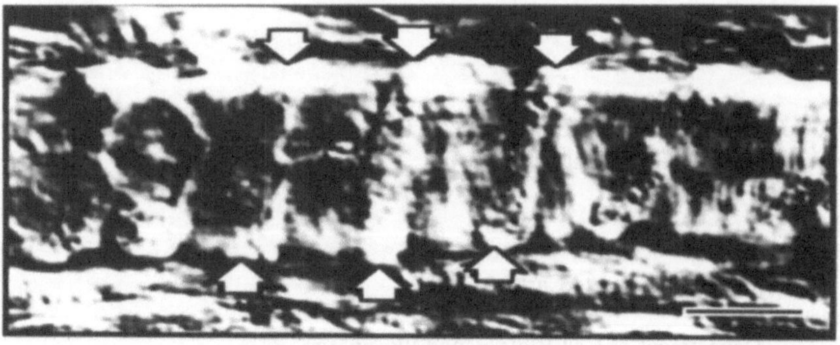

Figure 1. Nomarski image of a nuclear bag intrafusal muscle fibre in the equatorial region with overlying primary sensory terminals (arrows). Calibration bar is 10 μm.

In our initial studies of visualisation of sensory endings by fluorescent dyes, a small drop of solution containing lucifer yellow was applied to the axon of the sensory ending

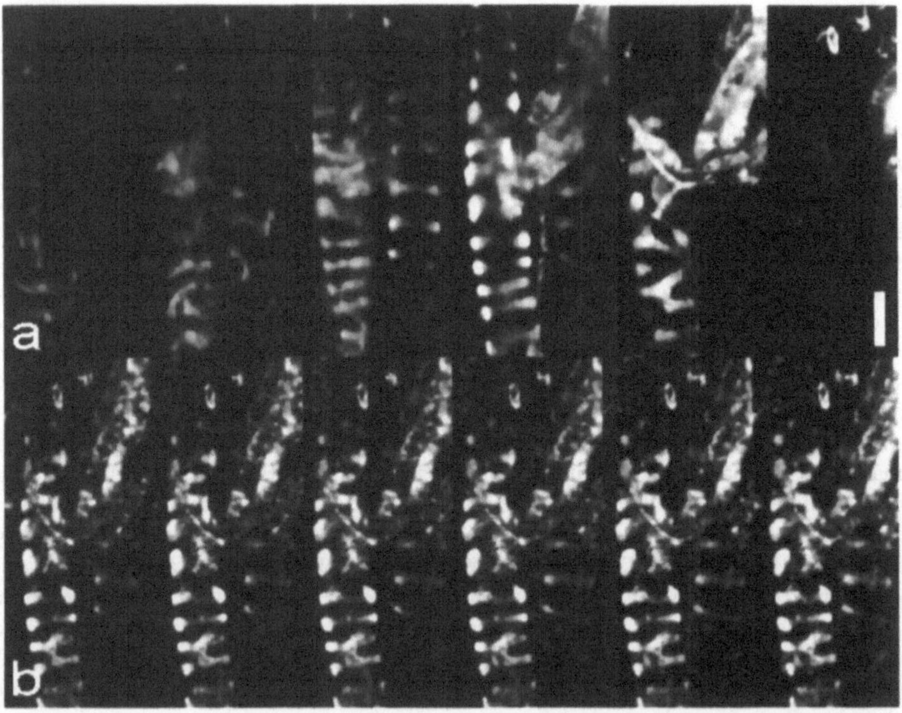

Figure 2. A. Original confocal optical sections from the primary sensory region of a cat muscle spindle stained with FM 1-43. Each frame is a sample at 8 μm steps from a stack of 53 sections spaced at 1 μm through the primary sensory region. Parts of the myelinated axon, preterminal unmyelinated region and sensory endings of the bag fibre are clearly visible in each optical section. B. Rendered views calculated from 1 μm confocal sections (sampled in A) and panned at 10 degree steps in the horizontal plane showing the full depth of the nerve terminals. The frame may be viewed like stereoscopic pairs (Focus with panels close to the face while relaxing ocular convergence. Then slowly move the panels to a normal view distance without readjusting convergence). Calibration bar is 20 μm.

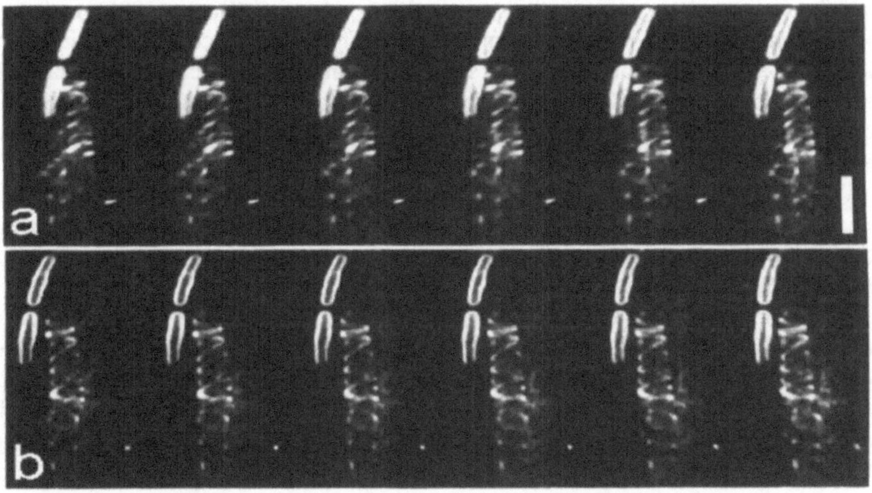

Figure 3. A. A spindle clearly showing the primary sensory ending on a bag fibre and several myelinated nerve segments. B. The spindle of A rendered with the spindle rotated showing the sensory endings upon several chain fibres, which in A are obscured by the bag fibre. Adjacent images are rendered with a 5 degree roll between images. A and B are rendered 40 degrees apart. Calibration bar is 40 μm.

raised in oil and current passed to move the dye into the terminals. However, styryl dyes, which are taken up by the endings spontaneously are more convenient. One of the most useful is FM 1-43. As Betz, Mao & Bewick (1992) have shown, this dye enters the motor terminals in an activity-dependent fashion; the uptake and release is thought to be associated with synaptic vesicle recycling. It is also avidly incorporated into the sensory endings of the spindle, both in snake and mammals. In contrast to the motor endings, activity, in the form of stretch or electrical nerve stimulation, does not appear to influence the uptake or release of these dyes in sensory endings.

The large 3-dimensional extent of the sensory endings of cat muscle spindles, particularly the primary endings, produces much out-of-focus information, which substantially degrades the image with conventional epifluorescence. Therefore we used confocal microscopy, in which the exciting light and fluorescent emission are focused through a pinhole in a secondary focal plane which excludes much of the out-of-focus information. Images were collected as averages at every 1 or 0.5 microns in the z-axis and stored digitally. A typical primary ending yields a stack of 50 or more optical sections (see Figure 2A) and a number of adjoining stacks are required to make a montage of the entire longitudinal extent. Images were collected as 512 by 480 pixel frames, each pixel being 8 bits of intensity. Figure 2B shows views calculated from the complete stack rendered at 10 degree intervals using the brightest point method. The myelinated axons approaching the spindle and the annulospiral endings are clearly evident on the bag fibres in this example.

Unmyelinated branches arise from each myelinated axon branch leading to an extensive array of unmyelinated terminals. Although individual spirals can be seen, many of the terminals are not complete spirals but annuli linked by short connectives, as noted by Banks (1986). There are additional interconnecting links between many of the unmyelinated subdivisions, some slender and others larger in diameter. The transducing elements, are therefore not only connected with the myelinated axon serially, but there are also parallel connectives. When modeling the electrotonic spread of potential in the endings, such shunts will have to be taken into account.

While primary sensory terminations are not all annulospiral (Banks, 1986), regions of a particular terminal may be helical for some length. From a stack of confocal sections, cross-sections of the intrafusal bundle with its sensory terminals may be computed. These show that there are many regions where the endings are spiral. The direction of the pitch in such a region can sometimes be seen to reverse. Rotation of the spindle is useful for showing the 3-dimensional structure of the primary ending. Examples of rotation about the longitudinal axis of the spindle are show in Figures 3A and 3B. The myelinated axon branches are clearly visible and can be traced from their termination close to the intrafusal muscle fibres, on which they have endings, back to their origin from the parent sensory axon. Nodes can be seen and internodal distances measured.

The advantage of 3-dimensional reconstruction from confocal fluorescent images over conventional microscopy are several. Collecting data is relatively fast and a number of spindles can be examined. A single spindle can be examined at several different lengths. Artifacts of fixation such as shrinkage are avoided. While serial electron micrographs have greater resolution, the detail provided by the confocal fluorescent images easily extends over the microscopically large expanse of a structure as large as the muscle spindle.

ACKNOWLEDGMENTS

Our thanks are due to Dr. Jeff Lichtman and Dr. Thomas Woolsey for assistance in the confocal microscopy. This research was supported by the National Institute of Neurological and Communicative Disorders and Stroke Grant NS07907.

REFERENCES

BANKS, R.W. (1986) Observations on the primary sensory ending of tenuissimus muscle spindles in the cat. *Cell Tiss. Res.* **246**, 306-319.

BESSOU, P. & PAGÈS, B. (1975) Cinematographic analysis of contractile events produced in intrafusal muscle fibres by stimulation of static and dynamic fusimotor axons. *J. Physiol.* **252**, 397-427.

BETZ, W. J., MAO, F. & BEWICK, G.S. (1992). Activity-dependent fluorescent staining and destaining of living vertebrate motor nerve terminals. *J. Neurosci.* **12**, 363-375.

BOYD, I.A., GLADDEN, M. H., McWILLIAM, P.N. & WARD, J. (1977). Control of dynamic and static nuclear bag fibres and nuclear chain fibres by gamma and beta axons in isolated cat muscle spindles. *J. Physiol.* **265**, 133-162.

PACEMAKER COMPETITION AND THE ROLE OF PRETERMINAL-BRANCH TREE ARCHITECTURE: A COMBINED MORPHOLOGICAL, PHYSIOLOGICAL AND MODELLING STUDY

R.W. Banks[1], M. Hulliger[2],
K.A. Scheepstra[2,3] and E. Otten[3]

[1]Department of Biological Sciences
University of Durham, Durham, UK
[2]Department of Clinical Neurosciences
University of Calgary, Calgary, Canada
[3]Department of Medical Physiology
University of Groningen, Groningen
The Netherlands

INTRODUCTION

Peripheral sensory endings often consist of groups of unmyelinated nerve terminals, specialised for transduction, that ultimately converge on the single afferent axon through a system of myelinated preterminal branches. At least in the mammalian muscle spindle the sensory terminals are in continuity only through the preterminal branches (Banks, 1986), so each may be supposed to have an associated spike-initiation (encoding) site, potentially able to act as a separate pacemaker. Interaction in such a system is expected to be highly competitive (Eagles & Purple, 1974) so that the final output of the afferent is in general a non-linear function of the activities of the separate encoders.

The primary sensory ending of the mammalian muscle spindle is a favourable example in which to study pacemaker interactions because it is frequently derived from two first-order branches that separately supply terminals on different types of intrafusal muscle fibre: bag_1 (b_1) fibre and bag_2 and chain (b_2 and c) fibres together (Banks, 1986). The greatly expanded axonal terminals, where mechanosensory transduction may be presumed to occur, are derived from short, unmyelinated preterminal branches. These in turn arise almost exclusively from the heminodes of the ultimate myelinated preterminal branches which may be of first to fourth order (Banks, 1986). The geometry of this arrangement indicates that the heminodes act as sites of spike initiation, a hypothesis that is supported by histochemical evidence from the cat showing that primary-ending heminodes share staining properties with known spike-initiation sites, such as motor axonal initial segments (Quick, Kennedy &

Poppele, 1980). The high degree of segregation between the dynamic and static fusimotor systems (reviewed by Banks, 1994) enables selective activation of the b_1 and b_2-c terminals respectively, by eliciting contractile activity in the appropriate muscle fibres.

Physiological studies have repeatedly shown that at least two pacemakers exist in primary endings of the cat and that they may be separately excited by dynamic and static fusimotor stimulation. Although competitive interactions predominate, a variable amount of summation may also occur (Hulliger & Noth, 1979; and other references therein). The possibility that the topology of the preterminal-branch tree architecture might influence, in part, the amount and pattern of summation has been supported by an unpublished modelling study (Otten, Hulliger & Schaafsma, in press). Here we extend those observations by determining, in the same spindle, both the nature of pacemaker interactions and the tree architecture of individual primary endings, and comparing the results with model simulations.

METHODS

The preparation was the tenuissimus of the anaesthetised cat. This facilitated precise location of the primary endings, an essential prerequisite for accurate histophysiological correlation. Details of the preparation, and of the data acquisition and control of experimental parameters, are given in Banks (1991) and Baumann & Hulliger (1991) respectively. Briefly, the left tenuissimus muscle was exposed by removal of biceps femoris, after extensive denervation of the sciatic distribution to the limb, sparing only the tenuissimus nerve. The muscle was freed from surrounding connective tissue for the greater part of its length, so as to allow effective transmission of stretch applied by an electromagnetic puller at its distal end. An efficient blood supply for the distal part of the muscle was maintained by creating a pedicle at the level of the popliteal fossa. Control signals for muscle stretch and fusimotor stimulation were taken from a hybrid signal generator that permitted reproducible synchronisation of electrical and mechanical stimulation patterns. Spikes obtained from single afferents were converted to TTL pulses, and stored on-line using a LS11/73 computer. Routinely, fusimotor actions were tested using both constant and triangularly modulated stimulation rates. Constant stimulation rates ranged from 25 to 150 Hz. During trapezoidal stretches, both static and dynamic axons were typically stimulated at 100 Hz. Triangularly modulated stimulation was applied with the muscles held at constant length. Maximal slope of the modulations was 50 Hz/s, with a peak stimulation rate normally not exceeding 150 Hz. This particularly facilitated diagnosis of the static effectors, using criteria established in a previous histophysiological study (Banks, 1991).

The proximo-distal locations of individual spindles were marked by epimysial stitches. At the end of the physiological part of each experiment the muscle was removed and fixed under light tension for about 7 hours in a Karnowsky fixative. In most cases, located spindles and their nerve supplies were teased out before or during secondary fixation in OsO_4, enabling the equatorial regions containing the primary endings to be identified and separately embedded in epoxy resin. Serial, 1μm sections were prepared, usually in the longitudinal plane, stained with toluidine blue, and used to reconstruct the afferent tree architecture.

Simulations were carried out using a model based on likely intrafusal fibre mechanical properties to produce a receptor potential associated with each type of fibre (Schaafsma, Otten & van Willigen, 1991). An ionic model of the action potential, based on modified Frankenhaeuser-Huxley kinetics (Otten, Hulliger & Scheepstra, in press), generated spikes at the heminodes of individual trees and propagated the spikes through the tree. The firing rate of each heminodal pacemaker might be influenced by the receptor potential of its own terminal, electrotonic spread from other terminals via the preterminal branches, and antidromic invasion by upstream nodal spiking.

A coefficient of interaction (C_i, see Figure 1) was calculated both for the real and simulated data, to provide a quantitative estimate of the amount of competition or summation that occurred between the dynamic and static pacemakers, and to allow correlations with parameters of tree structure to be estimated. The index was defined as the difference between the response to combined dynamic and static fusimotor stimulation (R_C) and the higher of the responses to separate stimulation (R_S or R_D), normalised with respect to the lower of the separate responses; formally

$$C_i = \begin{cases} (R_C - R_S)/R_D & R_S \geq R_D \\ (R_C - R_D)/R_S & R_D > R_S \end{cases}$$

C_i might be expected to vary between 0 (high competition, or complete occlusion of the momentarily slower pacemaker by the faster one) and 1 (complete summation, in which the overall output is the algebraic sum of the separate pacemakers).

If all heminodes are potential pacemakers, there is usually in a single tree more than one possible pathway linking dynamic and static pacemakers that could be active. We have selected the minimum path length in each case, estimating the path length quantitatively by the number of complete nodes (nMPL) contained in it.

RESULTS

Responses to fusimotor stimulation were obtained both during trapezoidal stretch and with the muscle held at a constant length. Here we give a preliminary account of the results

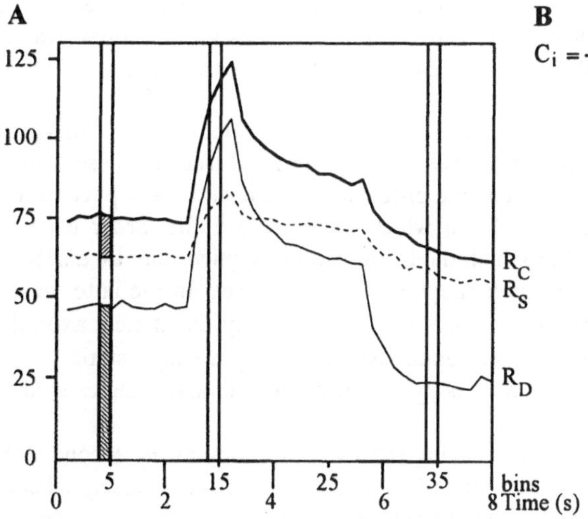

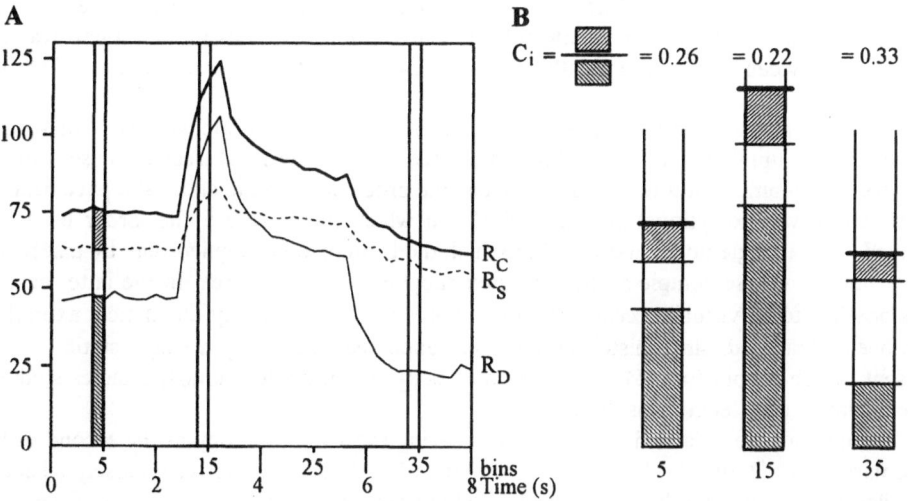

Figure 1. A. Population density plot showing all the responses to trapezoidal stretch, averaged in 200ms bins, during static (R_S), dynamic (R_D), and combined static and dynamic (R_C) fusimotor stimulation. Average stimulation rates were approximately 100 Hz in each case. B. Schematic representation of the calculation of the interaction index (C_i), as formally defined in the text, for the three representative bins (5, 15 and 35) shown in A.

Table 1. Mean coefficient of interaction (C_i) and the number of nodes in the minimum path lengths (nMPL) for dynamic and static pacemaker interactions in cat tenuissimus primary endings.

experiment/spindle	C_i	nMPL	experiment/spindle	C_i	nMPL
266/2	0.66	1	258/3	0.17	5
262/4	0.33	5	269/9	0.17	8
270/11	0.31	3	269/4	0.17	9
258/2	0.30	3	261/1	0.16	6
270/4	0.24	5	258/12	0.15	4
264/9	0.22	6	264/3	0.14	9
258/11	0.20	2	269/11	0.11	9
262/10	0.19	6	266/5	0.11	8
269/11	0.17	9	270/1	0.07	6

obtained using trapezoidal stretch, from eighteen different combinations of pairs of dynamic and static fusimotor action on 17 primary endings of 9 adult cats. C_i was calculated using data collected in 200ms bins. It usually varied according to the phase of the stretch, for example it often fell during the phase of muscle lengthening when the dynamic response greatly exceeded the static. A falling value of C_i indicates increasing competition. The effect proved to be a function of the difference in firing rate between the separate pacemakers, as shown by plotting the C_i against the difference between the separate dynamic and static responses for individual bins. The resulting scatter plot showed a clear tendency for C_i to reach maximum values (greatest summation) when the difference between the separate dynamic and static firing rates was 0.

Mean C_i, averaged from 40 bins, normally ranged from 0.07 indicating very strong competition, to 0.33 in which summation was evident but the response was still dominated by competition. Exceptionally, one primary showed very high summation (C_i = 0.66). The results are here tabulated in full (Table 1) according to decreasing order of mean C_i, together with the number of nodes in the minimum path length between the inferred dynamic (b_1) and static (b_2 or c) effectors. However, in most cases (13/17) there was no difference between the minimum path lengths for b_2 or c, and in no case was the difference greater than 1.

In most primary endings the Ia afferent branched dichotomously into first-order branches that supplied b_1 and b_2-c fibres separately. The preterminal trees of these endings possessed a minimum of 3 to 9 nodes between the probable dynamic and static pacemakers. There were two exceptional endings: 258/11 in which there were 5 first-order branches derived from a single node; and 266/2 in which the b_1 fibre was supplied only by a second-order branch, whose complementary division supplied several c fibres. In the latter case it was possible to be virtually certain that the c fibres were activated by the static γ axon that had been stimulated, since histologically there appeared to be only a single static γ axon present, which supplied all (6) c fibres, as well as b_2, in both poles in most spindles studied. The minimum path length was therefore 1.

Simulations were carried out using trees having the same topology as the second-order b_1-c part of 266/2 and the first-order b_1-c part of 266/5, since these were closely adjacent spindles in the same muscle and they exhibited virtually the most extreme values of mean C_i and minimum path length: 266/2, C_i = 0.66, nMPL = 1; 266/5, C_i = 0.11, nMPL = 8. Both showed considerably greater competition than the real endings, but whereas it was nearly complete in the 266/5 simulation, there was a much greater degree of summation in the 266/2 simulation.

DISCUSSION

Our results confirm previous reports (see Hulliger & Noth, 1979) in which it was concluded that mammalian primary endings contain multiple pacemakers, and that these pacemakers interact in a predominantly competitive manner. Detailed temporal analysis of the simulations showed that competition arises as predicted on theoretical grounds (Eagles & Purple, 1974) because of antidromic invasion of the momentarily less active pacemaker by spikes generated by the more active one.

A variable amount of summation can, however, be shown even by individual endings. There are several conceivable reasons for this variabilty, but we have now been able to demonstrate that in primary endings under comparable conditions of stretch and motor stimulation, a component is attributable to preterminal-branch structure. So far we have only considered topological features of the tree architecture and not the absolute dimensions of the trees. Those that we considered as potentially influential included the number of terminals, the proportion of branching nodes, and the number of nodes in the maximum path length. However, a multivariate analysis revealed that the only significant structural component was related to the number of nodes in the minimum path length. Moreover, the influence of this factor seems to decline very rapidly as minimum path length increases, as would be expected if the effect is mediated by the electrotonic spread of, say, receptor potentials. We may note that Hunt, Wilkinson & Fukami (1978) were able to record a compound receptor potential in the parent Ia axon after blocking Na^+ spikes with TTX.

The multivariate analysis also revealed that two other factors contributed to the variable amount of summation shown by individual endings: the overall firing rate, and the difference in rates when the dynamic and static pacemakers were separately activated. Summation increases with overall firing rate, a surprising observation and one not predicted by the simulations. However, it serves to remind us that competition normally predominates even, indeed especially, at low rates. This is perhaps explicable by supposing that the firing rate of an individual pacemaker is determined by its relative refractoriness, and that this can be counteracted by increasing the receptor potential. The increased summation seen at higher rates is more difficult to explain, but may be due to a shift in the effective pacemaker site to the next most proximal node if its threshold is exceeded by the electrotonically spread receptor potential. This would, of course, have the effect of reducing the minimum path length, and would therefore increase summation due to the structural factor. The histochemical study that showed heminodes to be potential pacemaker sites (Quick et al., 1980) also showed that at least some of the preterminal nodes shared the same staining properties, indicating a similar excitability.

The general similarity between the real and simulated data is encouraging, particularly in regard to the preponderance of competition and the effect of tree topology on summation. Nevertheless, it must be noted that overall the simulated results show very much greater competition than is exhibited by the real primary endings. The reasons for the discrepancy remain to be investigated, but presumably involve oversimplifications in the assumptions that form the basis of the model.

Finally we reflect on that feature of the primary ending with which we began: the dichotomous first-order division into b_1 and b_2-c parts. Despite, or perhaps because of, its regular occurrence in cat spindles, it seems to have been taken for granted when perhaps we should have been asking whether it must be so. Certainly it is one of the few quite constant features of the preterminal and terminal branch system of the primary, and the first-order branches often lie alongside each other for what seem to be extravagantly long distances (Banks, 1986). Could it be that the division is an adaptation to ensure that the interaction between separate dynamic and static pacemakers is highly competitve? We need to know more about the use made by the CNS of such a multiplexed signal in motor control, and whether summation would compromise that control. If so, an apparently unrelated fact of spindle development, which again has rather been taken for granted, becomes explicable: that the oldest of the secondary myotubes differentiates as the b_1 fibre (Milburn, 1984). The overall pattern of myotubal development is common to both intrafusal and extrafusal fibres (Kozeka & Ontell, 1981). Thus secondary myotubes separate from the growing cluster, centred on the primary myotube, in the order of their formation. This sequence will enhance the likelihood of a preterminal division of the Ia containing segregated dynamic and static pacemakers.

ACKNOWLEDEMENTS

We thank the wellcome Trust (RWB), the MRC of Canada (MH) and the Netherlands Science Foundation (KAS) for financial support.

REFERENCES

BANKS, R.W. (1986) Observations on the primary sensory ending of tenuissimus muscle spindles in the cat. *Cell Tiss. Res.* **246**, 309-319.

BANKS, R.W. (1991) The distribution of static γ-axons in the tenuissimus muscle of the cat. *J. Physiol.* **442**, 489-512.

BANKS, R.W. (1994) The motor innervation of mammalian muscle spindles. *Prog. Neurobiol.* **43**, 323-362.

BAUMANN, T.K. & HULLIGER, M. (1991) The dependence of the response of cat spindle Ia afferents to sinusoidal stretch on the velocity of concomitant movement. *J. Physiol.* **439**, 325-350.

EAGLES, J.P. & PURPLE, R.L. (1974) Afferent fibers with multiple encoding sites. *Brain Res.* **77**, 187-193.

HULLIGER, M. & NOTH, J. (1979) Static and dynamic fusimotor interaction and the possibility of multiple pace-makers operating in the cat muscle spindle. *Brain Res.* **173**, 21-28.

HUNT, C.C., WILKINSON, R.S. & FUKAMI, Y. (1978) Ionic basis of the receptor potential in primary endings of mammalian muscle spindles. *J. Gen. Physiol.* **71**, 683-698.

KOZEKA, K. & ONTELL, M. (1981) The three-dimensional cytoarchitecture of developing murine muscle spindles. *Develop. Biol.* **87**, 133-147.

MILBURN, A. (1984) Stages in the development of cat muscle spindles. *J. Embryol. Exp. Morphol.* **82**, 177-216.

OTTEN, E., HULLIGER, M. & SCHEEPSTRA, K.A. A model study on the influence of a slowly acivating potassium conductance on repetitive firing patterns of muscle spindle primary endings. *J. Theor. Biol.* (in press).

QUICK, D.C., KENNEDY, W.R. & POPPELE, R.E. (1980) Anatomical evidence for multiple sources of action potentials in the afferent fibers of muscle spindles. *Neuroscience* **5**, 109-115.

SCHAAFSMA, A., OTTEN, E. & VAN WILLIGEN, J.D. (1991) A muscle spindle model for primary afferent firing based on a simulation of intrafusal mechanical events. *J. Neurophysiol.* **65**, 1297-1312.

THE SENSITIVITY OF MAMMALIAN MUSCLE SPINDLES TO NEUROMUSCULAR BLOCKING AGENTS

U. Proske and R.W. Carr

Department of Physiology
Monash University,
Victoria 3168, Australia

INTRODUCTION

Ever since the first recordings were made from muscle spindles, many different drugs have been used to study their response properties. Two drugs in particular, both of which act at the neuromuscular junction, have been frequently employed, acetylcholine (ACh) and curare. In whole animal preparations succinylcholine (SCh) is commonly used instead of ACh since it has fewer systemic side effects and its action persists for longer. Similarly, the synthetic analog gallamine triethiodide is preferred rather than curare. Both drugs have helped significantly to advance our knowledge of the internal workings of the spindle. Here we briefly review their actions and include some recent observations that may help to resolve continuing controversies.

SUCCINYLCHOLINE

Succinylcholine or suxamethonium is a depolarising blocker of neuromuscular transmission. It opens the post synaptic receptor channels, producing a transient excitation of the muscle. SCh is broken down rather slowly and therefore its action persists until the channels progressively inactivate to block transmission. The excitatory action of SCh on muscle spindles was first described by Granit, Skoglund & Thesleff (1953). They proposed that SCh produced its effect through a combination of an intrafusal contraction and a direct depolarising action on the sensory terminal membrane. Even today there is still some uncertainty over the exact mechanism of action of SCh on muscle spindles.

The most easily interpreted observations of the actions of SCh are those of Gladden (1976) and Boyd (1985) using isolated muscle spindles. When ACh or SCh were added to the perfusate the two nuclear bag fibres contracted. Contraction threshold of the bag_1 fibre to ACh was lower than that of the bag_2 fibre (Gladden, 1976). The maximum contraction and consequent degree of opening of the spirals of the primary ending in the presence of

SCh was greater with the bag_2 fibre (34%) compared with the bag_1 fibre (8%). Nuclear chain fibres did not contract in the presence of ACh or SCh. However intrafusal neuromuscular transmission in chain fibres was blocked by both drugs. For bag fibres neuromuscular transmission was not blocked and the contraction evoked by fusimotor stimulation could be interpreted as being able to sum with that produced by the drug (Boyd, 1985).

The contraction produced in nuclear bag fibres by SCh is better described as a contracture. This property of intrafusal fibres to go into contracture is shared only by vertebrate tonic muscle, muscle which typically does not exhibit propagated action potentials and which has a distributed motor innervation. Twitch muscle will contract once and then inactivate in response to a depolarising current pulse. Like tonic muscle, intrafusal fibres do not show contractile inactivation and will contract for the duration of the pulse. The large increase in response of the spindle primary ending during stretch in the presence of SCh (Rack & Westbury, 1966) is due to the bag_1 contracture. The bag_2 contracture manifests itself as an increase in the level of "biassing" giving an overall increase in response (Boyd, 1985; Taylor, Rodgers, Fowle & Durbaba, 1992). In addition, an early increase in resting activity of primary endings of spindles in the presence of SCh has been attributed to a depolarising action of the drug on the sensory terminal membrane, either directly or as a result of the rise in extracellular levels of potassium ions released by the depolarised extrafusal junctions (Dutia, 1980).

The action of SCh on secondary endings of spindles remains the subject of controversy. It has been proposed that the entire response of 'true' secondary endings (axonal conduction velocity less than 60m/s) may be the result of a depolarising action, either direct or through rising potassium levels and that intrafusal contractions are not involved. Only 'intermediate' secondary endings (conduction velocity 60-75m/s), which are known to have significant sensory terminals on nuclear bag fibres include a component in their SCh responses from an intrafusal contraction (Dutia, 1980).

Here we would like to present two observations which contribute to the debate about the action of SCh on secondary endings. We have recorded the responses of secondary endings of the soleus muscle of the anaesthetised cat to ramp-and-hold stretches, before and after giving SCh intravenously (Figure 1, upper panel). We have quantified the response of each ending by counting the number of impulses generated by the stretch before and after SCh. As well as recording from secondary endings (38 afferents) we have included observations on a number of primary endings (30) and tendon organs (21) (Figure 1, lower panel). The mean increase in impulse count for secondary endings produced by SCh was 128 (± 9 s.e.m.) impulses compared with 455 (± 31 s.e.m.) impulses for primary endings and 2 (± 0.09 s.e.m.) impulses for tendon organs. Clearly, the responses of secondary endings, while less than those of primary endings, are significantly greater than for tendon organs. If the responses of secondary endings were the result of a direct depolarising action of the drug on the sensory terminal membrane, or a depolarisation produced by rising extracellular potassium levels, it might have been expected that tendon organs would show similar responses. The fact that they did not must therefore mean that either the sensory terminals of tendon organs differ in some fundamental way from the terminals of secondary endings or that the responses of secondary endings were not generated by a direct depolarising action but by an intrafusal contraction. A number of other observations made in our laboratory are consistent with the latter explanation (Taylor, Morgan, Gregory & Proske, 1994).

A second observation concerns the effects of fusimotor stimulation in the presence of SCh. We have recorded the responses of primary and secondary endings during stimulation of single identified fusimotor fibres, before and during intravenous infusion of SCh (200 μg/kg). For both endings the spindle responses to fusimotor stimulation were not blocked in the presence of SCh. Since secondary endings predominantly supply nuclear chain intrafusal fibres, if neuromuscular transmission was blocked in these fibres by SCh, (Boyd, 1985) it might have been expected that the responses to fusimotor stimulation would show a significant fall. In fact responses were potentiated. For both ending types the conclusion was that there was no evidence that SCh blocks intrafusal neuromuscular transmission in any of the intrafusal fibres. This finding agrees with some preliminary observations made by Rack & Westbury (1966). Boyd's observations may be explained by the fact that they were made on isolated spindles with 10 μg/l SCh added to the perfusate. We don't know what the actual concentration of SCh is at the level of the spindle in a whole animal preparation when SCh is given intravenously. It is possible that the local concentration is very much lower than 10 μg/l and insufficient to block intrafusal neuromuscular transmission.

One other point from these experiments was the apparent summation of responses to SCh and to fusimotor stimulation. However, if a sufficiently high stimulus rate was used the effects of fusimotor stimulation no longer summated with SCh responses. This result suggested that the size of the afferent response depended solely on the size of the intrafusal contraction, whether this was produced by SCh, fusimotor stimulation or both. A similar conclusion was arrived at by Boyd (1985) for the interacting effects of SCh and fusimotor

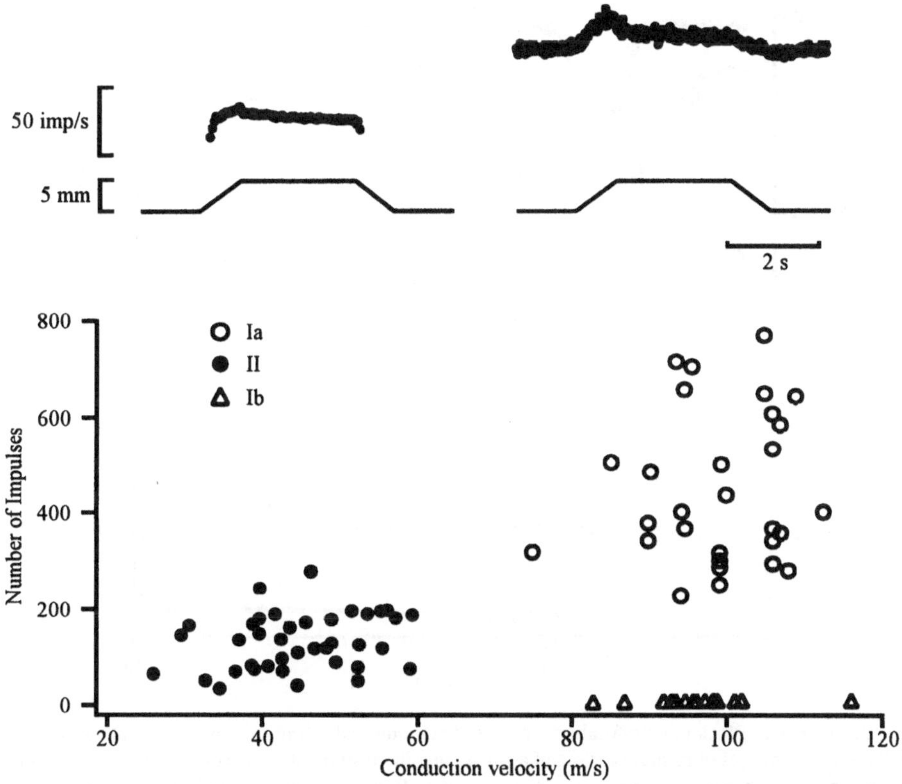

Figure 1. Responses of muscle receptors of the soleus muscle of the anaesthetised cat to i.v. SCh. Top panel, left, instantaneous frequency display of control response of a secondary ending to a ramp-and-hold stretch (5 mm at 5 mm/s). The response on the right was obtained 30 seconds after an intravenous dose of SCh (200 μg/kg). Lower panel graph shows the responses to SCh of the three kinds of muscle receptors. Open circles, primary endings of spindles, filled circles secondary endings, triangles tendon organs. Responses were quantified by counting the number of impulses produced by a ramp stretch in the presence of SCh minus the number from a control stretch. (Figure redrawn, in part, from Taylor et al, 1994)

stimulation on the bag_2 fibre of the isolated muscle spindle. For secondary endings if SCh had been exciting the endings by a direct depolarising action on the sensory terminal membrane some summation of SCh responses and fusimotor responses would have been expected under all conditions. This was not the case. To summarise, we propose that the majority of secondary endings including 'true' secondary endings respond to SCh as a result of an intrafusal contraction just as primary endings do and there is no direct action on the sensory ending, at least under the conditions of our experiment. Our preliminary evidence suggests that at the concentrations we have used, SCh does not block neuromuscular transmission in chain fibres. We are currently carrying out further experiments to test this point.

GALLAMINE

Like SCh, curariform drugs have been used on muscle spindles since the first recordings were made of their afferent responses. Curare is a competitive blocker of post synaptic receptor sites at neuromuscular junctions. It was reported by Katz (1949) that the

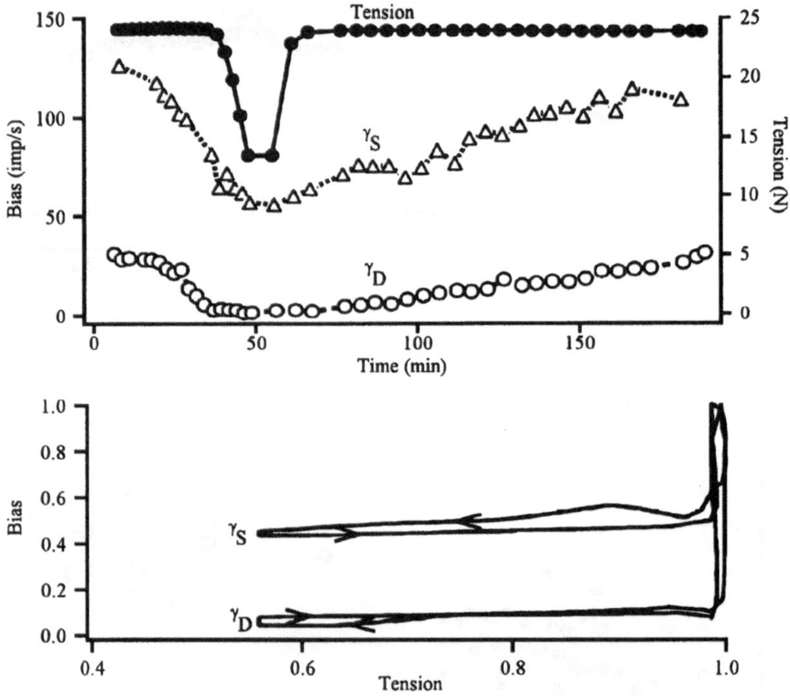

Figure 2. Top panel. Block of extrafusal and intrafusal neuromuscular transmission in the cat soleus muscle by gallamine. Responses of primary ending of a soleus spindle to stimulation at 100 pulses per second of a dynamic fusimotor fibre (γ_d open circles), a static fusimotor fibre (γ_s, open triangles) and whole muscle tension (filled circles) during a slow infusion of gallamine (0.15 mg/min) for 29 minutes, 10 minutes after recordings were begun. The solid bar below the time scale indicates the duration of the infusion. After infusion was stopped tension recovered rapidly while fusimotor responses recovered much more slowly. Lower panel, in a separate experiment spindle responses to dynamic (γ_d) and static (γ_s) fusimotor stimulation, each normalised with respect to their maximum values, plotted against normalised tension. The arrows in the two figures show the temporal sequence of the changes in biassing and tension during onset (arrow pointing to the left) and recovery (arrow to the right) from gallamine block. (Figures redrawn, in part, from Yamamoto et al, 1994).

intrafusal motor terminals of frog muscle spindles were more resistant to curare than the terminals on the adjacent extrafusal fibres. Since that time curare has been used as a tool to study the responses of spindles to motor nerve stimulation without the confounding mechanical effects of the accompanying extrafusal contraction. However some uncertainty remains about the exact difference in blocking threshold of intrafusal and extrafusal junctions in mammalian muscle. Hunt (1952) reported that in the cat, endplates were blocked by d-tubocurarine with doses which also blocked transmission to ordinary muscle fibres. Granit, Skoglund & Thesleff (1953), on the other hand, reported that it was possible to block most, if not all, extrafusal terminals, leaving intrafusal terminals substantially functional, although they admitted that a complete extrafusal block left action on spindles significantly reduced (Granit, Pompeiano & Waltman, 1959). Carli, Diete-Spiff & Pompeiano (1967) first used intravenous gallamine and found that some intrafusal junctions could be blocked as readily as extrafusal junctions while others were more resistant to the drug. We have recently re-examined the action of gallamine on intrafusal neuromuscular transmission (Yamamoto, Morgan, Gregory & Proske, 1994) and found, as had Carli *et al*, that transmission at some intrafusal junctions was blocked as readily as at extrafusal terminals while for others the threshold for a block was much higher.

The experiments were carried out on the soleus muscle of the anaesthetised cat. Gallamine was given with a constant infusion pump and the infusion rate was adjusted to be very slow, 0.15 mg/min. This was done to minimise any possible effects of diffusion barriers. The responses of primary endings of spindles were recorded during one second periods of stimulation of single, identified dynamic and static fusimotor axons. Whole muscle tension was also monitored by means of ventral root stimulation. Observations from one experiment are shown in Figure 2 (top). Here gallamine was infused 10 minutes after recordings had begun and the infusion was stopped 29 minutes later. The important observation was made that fusimotor responses showed evidence of a block, well before there was any sign of a drop in extrafusal tension. By the time extrafusal tension began to fall, the dynamic fusimotor response was fully blocked while the response to static stimulation had dropped from 130 to 60 impulses/s. After gallamine infusion was stopped tension recovered quite rapidly, within about 15 minutes. However, recovery of both fusimotor responses was much slower and they had still not returned to their control values 150 minutes later.

Our conclusion from this experiment is that some intrafusal junctions are more sensitive to gallamine than extrafusal junctions while others are less sensitive. Another way of representing the data of Figure 2 is to express tension and biassing as a fraction of their control values and plot one against the other (Figure 2, lower panel). For each of the two responses the figures drawn by this plot begin in the upper right-hand corner, move down and to the left, in the direction of the arrow, during onset of the block and back to the starting point during recovery as shown by the arrow in the reverse direction. If there were a diffusion barrier between the intrafusal fibres and the rest of the muscle by, for example, the spindle capsule, tension would drop before responses to stimulation and then recover earlier. This would generate a more open loop figure than the examples shown in Figure 2. On one or two occasions such open loops were observed but the majority of responses were like those illustrated here. We concluded that while diffusion barriers at times influenced responses to gallamine, with the low infusion rates used, such examples were in the minority. Furthermore all dynamic responses blocked early while for static effects there appeared to be a gallamine sensitive and a gallamine resistant component. Because each experiment lasted several hours we did not fully explore the gallamine resistant component. It would, of course, eventually block if gallamine infusion had been continued for long enough and we predict that it would then also recover more rapidly afterwards.

CONCLUSION

To conclude, we propose that some fusimotor endings are more sensitive to gallamine than extrafusal endings, while others are more resistant. The sensitive group includes dynamic and some static endings. Recovery from block at intrafusal junctions is also much slower than at extrafusal junctions. This second point is of some interest clinically. Muscle relaxants are used routinely in some surgical procedures. Our results suggest that when a patient recovers from anaesthesia, if a muscle relaxant has also been used, it is likely that muscle proprioceptive reflexes remain impaired for much longer than muscle force. Concerning the fusimotor effects which comprised gallamine sensitive and gallamine resistant components our current working hypothesis is that the sensitive terminals are those on nuclear bag fibres while the resistant terminals are on nuclear chain fibres. Experiments are under way to obtain further supporting evidence for this proposal.

REFERENCES

BOYD, I.A. (1985) Intrafusal muscle fibres in the cat and their motor control. In *Feedback and Motor Control in Invertebrates and Vertebrates*, eds. BARNES, W.J.P. & GLADDEN, M.H., pp. 123-144. Croom Helm, London.

CARLI, G., DIETE-SPIFF, K. & POMPEIANO, O. (1967) Mechanisms of muscle spindle excitation. *Arch. Ital. Biol.* **105**, 273-289.

DUTIA, M.B. (1980) Activation of cat muscle spindle primary, secondary and intermediate sensory endings by suxamethonium. *J. Physiol.* **304**, 315-330.

GLADDEN, M.H. (1976) Structural features relative to the function of intrafusal muscle fibres in the cat. *Prog. Brain Res.* **44**, 51-59.

GRANIT, R., POMPEIANO, O. & WALTMAN, B. (1959) The early discharge of mammalian muscle spindles at onset of contraction. *J. Physiol.* **147**, 399-418.

GRANIT, R., SKOGLUND, S. & THESLEFF, S. (1953). Activation of msucle spindles by succinylcholine and decamethonium. The effects of curare. *Acta physiol. scand.* **28**, 134-151.

HUNT, C.C. (1952) Drug effects on mammalian muscle spindles. *Fed. Proc.* **11**, 75.

KATZ, B. (1949) The efferent regulation of the muscle spindle in the frog. *J. Exp. Biol.* **26**, 201-217.

RACK, P.M.H. & WESTBURY, D.R. (1966) The effects of suxamethonium and acetylcholine on the behaviour of cat muscle spindles during dynamic stretching and during fusimotor stimulation. *J. Physiol.* **186**, 698-713.

TAYLOR, A., MORGAN, D.L., GREGORY, J.E. & PROSKE, U. (1994) The responses of secondary endings of cat soleus muscle spindles to succinyl choline. *Exp. Brain Res.* **100**, 58-66.

TAYLOR, A., RODGERS, J.F., FOWLE, A.J. & DURBABA, R. (1992) The effect of succinylcholine on cat gastrocnemius muscle spindle afferents of different types. *J. Physiol.* **456**, 629-644.

YAMAMOTO, T., MORGAN, D.L., GREGORY, J.E. & PROSKE, U. (1994) Blockade of intrafusal neuromuscular junctions of cat muscle spindles with gallamine. *Exp. Physiol.* **79**, 365-376.

CRITIQUE OF PAPERS BY CHUA AND HUNT, BANKS, HULLIGER, SCHEEPSTRA AND OTTEN AND PROSKE AND CARR

Anthony Taylor

Sherrington School of Physiology, UMDS,
St Thomas' Hospital, London SE1 7EH, UK

These three papers all deal with intricacies of structure and function of muscle spindles, but with such different aspects that they must be considered separately.

SENSORY ENDINGS OF LIVING SPINDLES

In the first paper Chua and Hunt describe some elegant and promising ways of visualising and reconstructing spindle afferent terminals. Examination of isolated spindles by Nomarski optics gives high resolution but very limited depth of field and myelinated axons, which are highly refractile, may obscure details. The alternative method used here was to apply styryl dyes which are taken up by the sensory endings independently of activity. Fluorescent images can then be captured by confocal microscopy with plane separations of 0.5 to 1.0 μm and computer processed into 3D form. These images were spectacularly demonstrated with video sequences which showed the power of the method for conveying spatial detail. For example, the viewpoint could be rotated so as to look down a primary ending and to see that the sense of the pitch sometimes reverses.

This imaging method is not only very much quicker than reconstruction from serial sections, but has the exceptional advantage of being applicable to living spindles. This opens up all kinds of new possibilities for correlating sensory function with detailed morphology. It should be possible to study the terminals of any axon, motor or sensory, with this method provided it can be selectively labelled with a fluorescent dye. It would, for example, be worth trying to penetrate axons close to a spindle so as to record or to stimulate then to inject an appropriate dye.

Another possibility, which links the interests of this paper with the following one, is that the images might be analysed to provide quantitative detail for sophisticated models. Once the images have been captured as digital files, as in this case, then with sufficient investment of effort it should be possible to process them to provide a description of the ending morphology in terms appropriate for biophysical analysis of impulse initiation.

PACEMAKER COMPETITION

This study by Banks et al. models the interactions between the pacemakers believed to be sited at the preterminal heminodes on spindle primary sensory endings. Other modelling studies reported previously have been set up to predict the consequences of competition between several impulse initiation sites, but this present work is exceptionally interesting because of the accompanying detailed histological reconstructions. These were performed by means of serial semi-thin sections and must have been very time-consuming. One wonders whether the method of Chua and Hunt could perhaps be used in any future extension of this work. The authors sought to relate predictions based on the observed morphology of the terminals of each primary afferent with recordings made from it. The predictions were based on a model with just two intrafusal muscle fibre types rather than three, since it was thought reasonable to replace bag_2 and chain fibres by a single fast fibre type. This is acceptable in that a single primary afferent axon branch usually supplies these two intrafusal fibre types together, but it ignores the fact that they have very different properties and hence differing actions. The rapid contraction of chain fibres can cause driving of afferent discharge while the slower contracture of bag_2 fibres causes simple biassing.

The firing rate of each hemi-node was made dependent upon the receptor potential of its own terminal together with electrotonic spread from others via the preterminal branches, and antidromic invasion from other active nodes. A coefficient of interaction (C_i) was defined which would tend to zero for maximal occlusion of one site by the firing of the other and which would tend to 1 for complete summation without competition. It would have been helpful to have had some computations done to show how C_i would be expected to vary with different specified proportions of competition and summation. Without this, the particular values obtained are a little difficult to assess. The interaction observed in the model on this basis was thought to be predominantly competitive and was more so than that observed in the natural case. The degree of summation observed was inversely related to the number of nodes in the minimum path length between preterminal hemi-nodes, which is appropriate for electrotonic spread of depolarisation.

In a critique such as this it is reasonable to raise the question as to the ways in which the subject is advanced by this type of study. The modelling described here has been done carefully and with a great deal of insight and is based on real data. Even so, the model can be seen to involve a number of approximations, notably grouping of bag_2 and chain fibres and assumptions regarding minimum path length. One has to decide how important they are. A longer paper would have allowed the authors to have explored in detail the effects of the various approximations, which is something they will no doubt do in the future. Ultimately, modelling must always involve simplifications otherwise the model is just as difficult to understand as the original system and does not lead to new insights. It is the failures of a model to reproduce natural behaviour which are its most interesting products, because it is these which lead to progressive refinement of knowledge of the system under study.

THE ACTION OF NEUROMUSCULAR BLOCKING AGENTS ON MUSCLE SPINDLES

Several rather contentious issues arise in this paper by Proske and Carr and it may be useful to start by restating some of the background from a slightly different point of view. There are two types of neuromuscular blocking drugs which are of interest for their action on spindles, acetylcholine (ACh) and its agonists, such as succinylcholine and competitive antagonists, such as curare and gallamine. Succinylcholine (SCh) has often been referred to as a depolarising blocker, but this is a little unfortunate because it has encouraged the idea that it might have a non-specific depolarising action on nerve endings. In fact, there is no evidence that, at the doses generally employed (200 μg/kg *in-vivo*), it has any significant action other than via ACh (nicotinic) receptors at neuromuscular junctions. As far as spindles are concerned the action of SCh is expressed solely by contracture of the two bag intrafusal fibre types, which leads to an increase in resting discharge (principally a bag_2 effect) and an increase in stretch sensitivity (principally a bag_1 effect). The first study in which such effects could be related to the two bag fibre types was by Gladden (1976) and in this case ACh was used *in-vitro.* Bag_2 fibres appeared to be less sensitive than bag_1 fibres and neuromuscular conduction to chain fibres was thought to be blocked. These observations were confirmed and extended by Boyd (1985) using SCh. Dutia (1980) provided an analysis of SCh effects *in-vivo* interpreted in terms of separate actions on the two bag fibre types, but at that time secondary endings were thought to be restricted to chain fibres. Therefore, when excitatory effects were observed on spindle secondaries they were interpreted as due to 'depolarisation' of the afferent terminals, either by a direct action of SCh or mediated by the liberation of potassium ions. This conclusion did not take into account the earlier observations (Ottoson, 1961) that ACh and various agonists had no excitatory effect on frog muscle spindle endings after intrafusal muscle fibres had been damaged by crushing. In intact spindles in cats an excitatory action is observed in a large proportion, but not all, secondary afferents (Scott, 1991; Taylor, Durbaba & Rodgers, 1992). The proportion showing this effect corresponded well with the proportion known from histology to have terminals upon bag_2 fibres (Banks, Barker & Stacey, 1982). Those showing no excitatory action can be taken to be those with endings restricted to chain fibres and in support of this they have the lowest conduction velocities. The smallest diameter axons terminate furthest from the equator largely on chains (Banks et al., 1982). Further evidence in favour of this interpretation has been set out in several recent papers (Taylor, Morgan, Gregory & Proske, 1994; Taylor, Durbaba & Rodgers, 1994; 1995) and the idea of there being a significant action of SCh directly on the sensory terminals or via potassium ions cannot now be supported. It is surprising therefore that Proske and Carr still maintain that the action of SCh on secondary endings remains the subject of controversy.

Where these authors do make a very interesting contribution is in their observation that a dose of 200 μg/kg SCh, which produces close to maximal excitatory effects on spindle afferents, does not block the response to fusimotor stimulation. This is the case with secondaries as well as with primaries, so it is thought most likely to be due to persisting transmission to chain fibres. An explanation of the conflict with Boyd's finding of paralysis of transmission to chain fibres (Boyd, 1985) is sought in terms of the different concentrations of the drug used in the different experiments, with the implication that the 10 μg/l in the *in-vitro* work was higher than that in the *in-vivo* experiments. However, an IV dose of 200 μg/kg would be expected to give something like 1000 μg/l in blood if it were evenly distributed. This suggests that there may be some barrier to the diffusion of the drug from the circulation into the interior of the spindles not seen *in-vitro*.

Another aspect of differential sensitivity of intrafusal fibres was seen in the effects of gallamine. When this was given as a slow IV infusion, blocking of dynamic and static fusimotor effects appeared before extrafusal transmission block, but while dynamic effects were completely blocked, static effects were reduced by only about 50%. This suggested that transmission to bag_1 and bag_2 fibres was more easily blocked than that to chain fibres. In this case, no evidence could be found for an important part being played by a diffusion barrier and it may therefore be that neuromuscular junctions on chain fibres are less sensitive than those on bag fibres to competitive antagonists as they are also to 'depolarising' blockers.

There is a long history of attempts to use graded doses of curare or its analogues to effect differential blocking of extrafusal and intrafusal contraction and the authors refer to this, but their work raises the other interesting possibility of differentially blocking transmission to bag fibres while leaving chain fibres still working. This could be very helpful in separating the chain and bag_2 components of the static fusimotor system. For example, when some procedure such as brainstem stimulation gives rise to static effects, a graded dose of gallamine should block the bag_2 component leaving unaffected that due to chain contraction.

REFERENCES

BANKS, R.W., BARKER, D., & STACEY, M.J. (1982) Form and distribution of sensory terminals in cat hindlimb muscle spindles. *Phil. Trans. R. Soc. B.* **99**, 629-644.

BOYD, I.A. (1985) Review: The internal working of muscle spindles. In *The Muscle Spindle.* eds BOYD, I.A. & GLADDEN, M.H. pp. 129-150. Macmillan, London.

DUTIA, M.B. (1980) Activation of cat muscle spindle primary, secondary and intermediate sensory endings by suxamethonium. *J. Physiol.* **304**, 315-330.

GLADDEN, M.H. (1976) Structural features relative to the function of intrafusal muscle fibres in the cat. *Prog. Brain Res.* **44**, 51-59.

OTTOSON, D. (1961) The effect of acetylcholine and related substances on the isolated muscle spindle. *Acta physiol. scand.* *53,* 276-287.

SCOTT, J.J.A. (1991) Responses of Ia afferent axons from muscle spindles lacking a bag_1 intrafusal muscle fibre. *Brain Res.* **543,** 97-101.

TAYLOR, A., DURBABA, R. & RODGERS. J.F. (1994) The site of action of succinylcholine on muscle spindle afferents in the anaesthetised cat. *J. Physiol..* **476**, 26P.

TAYLOR, A., DURBABA, R. & RODGERS, J.F. (1995) Quantitative aspects of the use of succinylcholine in the classification of muscle spindle afferents. In *Neural Control of Movement.* eds. FERRELL, W.R. & PROSKE, U. Plenum Press, New York. (in press).

TAYLOR, A., MORGAN, D.L., GREGORY, J.E. & PROSKE, U. (1994) The responses of secondary endings of cat soleus muscle spindles to succinyl choline. *Exp. Brain Res.* **100**, 58-66.

TAYLOR, A. RODGERS, J.F., FOWLE, A.J. & DURBABA, R. (1992) The effect of succinylcholine on cat gastrocnemius muscle spindle afferents of different type. *J. Physiol.* **456**, 629-644.

OSCILLATIONS IN THE DISCHARGE FREQUENCY OF PRIMARY CAT MUSCLE SPINDLE AFFERENTS DURING THE DYNAMIC PHASE OF A RAMP-AND-HOLD STRETCH

S.S. Schäfer

Department of Neurophysiology, OE 4230
Medical School of Hannover
30625 Hannover, Germany

INTRODUCTION

At the beginning of a ramp stretch, the muscle spindle Ia afferents show an initial frequency peak which is followed during the ongoing dynamic phase of stretching by second, third or even fourth peaks (Edin, 1991; Scott, 1992). The aim of this investigation is to compare the behaviour of the initial peak with the behaviour of the subsequent peaks.

METHODS

In 5 cats (anaesthetised with sodium pentobarbitone) 14 Ia afferents arising from the tibial anterior muscle, were isolated from the dorsal root L_7 or L_6. The host muscle was stretched with ramps from different prestretch levels (0 to 12 mm) and at different stretch rates (1 to 100 mm/s). Afferent spikes were displayed as instantaneous frequency plots. Each recording consisted at least four consecutive responses obtained under identical experimental conditions superimposed to demonstrate the discharge pattern. The magnitude and the latency of a peak were read from such records.

RESULTS

Oscillations were readily seen in the discharge frequency during a ramp stretch provided (i) a number of traces were superimposed, (ii) a medium prestretch of the muscle was used, (iii) the rate of ramp stretch was 10 mm/s or greater, and (iv) the variability of the interspike intervals was small. The behaviour of the initial peak was compared with the behaviour of the subsequent peaks under four experimental conditions. The first condition was an increase of the prestretch of the muscle. Under this condition, the initial peak and the subsequent peaks behaved in qualitatively the same manner. All peaks became clearer and occurred with shorter latency. Secondly, an increase of the stretch rate also made the peaks larger and shortened their latency. However, given the dependency of the magnitude of the a peak on its latency, for stretches at different rates, the dependency of the initial peak turned out to be a power function, whereas the dependency of the subsequent peaks was an exponential function. The difference in this relationship resulted from a higher sensitivity to the increasing stretch rate observed for the subsequent peaks than for the initial peak. The third experimental variable was a change in the time elapsing between successive ramp stretches. It was observed that when the waiting time was long (3 minutes), the initial peak was high. When the waiting time was short (7 seconds), the initial peak was small (cf. Morgan, Prochazka & Proske, 1984). However, the magnitude and the latency of the subsequent peaks was found to be independent of the waiting time. We varied the conditioning stretch preceeding the test stretch according to the procedure described by Gregory, Morgan & Proske (1987) (Figure 1A). The discharge pattern of Figure 1B was obtained under the experimental conditions 'keep long". The discharge pattern of Figure 1C was recorded under the condition 'keep short". The initial peak was strongly influenced by the conditioning, whereas the subsequent peaks were not. Figure 1D shows the mean magnitude (n = 7) of the initial peak and of the peak 2 and peak 3 under the conditions 'keep short" (open columns) and 'keep long" (hatched columns). The initial peak is significantly (p = 0.006) higher under the condition 'keep short" compared to the condition 'keep long". However, the peak 2 and peak 3 have the same magnitude (p > 0.2) under both experimental conditions.

DISCUSSION

The initial peak is explained by a largely synchronous opening of cross-bridges of the polar parts of the intrafusal muscle fibres (Morgan et al., 1984). Our results regarding the initial peak are in good agreement with this interpretation. On the otherhand, the subsequent peaks show a qualitatively different behaviour from the initial peak under three of the four

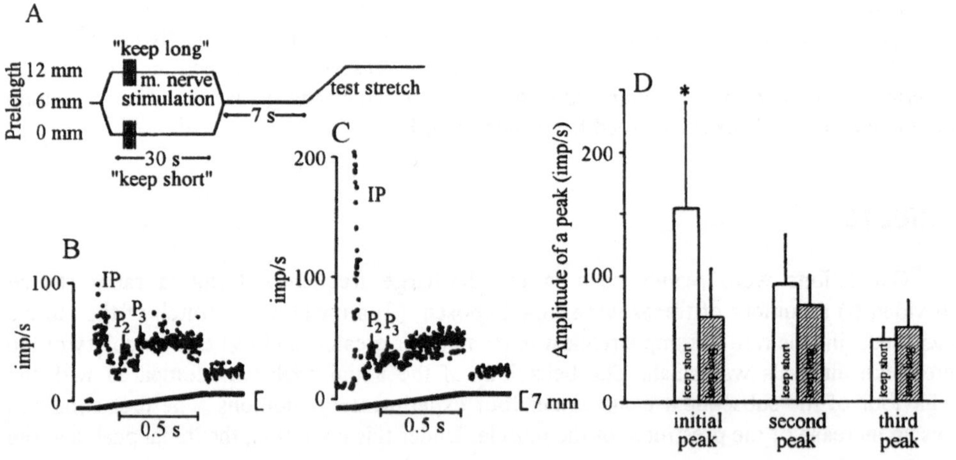

Figure 1. A. Experimental procedure used for obtaining the discharge patterns shown in B and C. Nerve stimulation over 2s with 30 stimuli/s. B. Discharge pattern of a Ia afferent obtained under the experimental condition "keep long". C. Dischrage pattern of the same Ia afferent under the experimental condition "keep short". D. The mean levels of the initial peak, peak 2 and peak 3 with SD under the experimental conditions "keep short" (open columns) and "keep long" (hatched columns). The initial peak decreases significantly (*: p<0.006) from the former to the latter condition, the subsequent peaks do not (p>0.2).

experimental conditions.

It appears that the subsequent peaks depend on factors different from those determining the intial peak. We assume that oscillatory changes in the receptor potential are the reason for the subsequent peaks. Oscillations in the receptor potential of mammalian receptors have been described for the hair cells of the cochlea and are have also been postulated for thermoreceptors. The oscillations in the membrane potential of these neurones are believed to depend on the alternation of an inward Ca^{++} current and a subsequent outward K^+ current, which is Ca^{++}-activated. Kruse & Poppele (1991) described for the isolated cat muscle spindle a Ca^{++} conductance that causes a Ca^{++}-activated K^+ conductance. Moreover, it is known, that the isolated muscle spindle is very sensitive to the intracapsular Ca^{++} concentration. In conclusion, we cannot exclude the possibility that the receptor potential of the muscle spindle tends to show intrinsic oscillations, which could be the cause of the subsequent peaks in the discharge frequency during a ramp stretch.

REFERENCES

EDIN, B.B. (1991) The 'initial burst' of human primary muscle spindle afferents has at least two components. *Acta physiol. scand.* **143**, 169-175.

GREGORY, J.E., MORGAN, D.L. & PROSKE, U. (1987) Changes in size of the stretch reflex of cat and man attributed to aftereffects in muscle spindles. *J. Neurophysiol.* **58**, 628-640.

KRUSE, M.N. & POPPELE, R.E. (1991) Components of the dynamic response of mammalian muscle spindles that originate in the sensory terminals. *Exp. Brain Res.* **86**, 59-366.

MORGAN, D.L., PROCHAZKA, A. & PROSKE, U. (1984) The after-effects of stretch and fusimotor stimulation on the responses of primary endings of cat muscle spindles. *J. Physiol.* **356**, 465 477.

SCOTT, J.J.A. (1992) The 'initial burst' of muscle spindle afferents with or without terminals on the bag_1 intrafusal muscle fibres. *Brain Res.* **585**, 327 329.

EFFECTS OF BRIEF PERIODS OF ACTIVATION OF BAG$_1$, BAG$_2$ AND CHAIN FIBRES ON THE RESUMPTION OF PRIMARY ENDING RESTING DISCHARGE

F. Emonet-Dénand, Y. Laporte and J. Petit

Collège de France, 75231 Paris Cedex 05, France

INTRODUCTION

The time at which primary endings resume discharging after quick muscle release is generally shortened by brief periods of repetitive stimulation of single γ axons immediately after the release. Significant differences in resumption times were seen by Proske, Gregory & Morgan (1991) when stimulating different γ axons supplying the same spindle, who suggested they were related to the type of intrafusal muscle fibre activated by each axon.

The present study aims to relate resumption times with the type of intrafusal muscle fibre activated. Identification of bag$_1$ fibre activation is easy since dynamic γ axons almost exclusively supply bag$_1$ fibres. However, static γ axons supply either chain or bag$_2$ alone or bag$_2$ and chain fibres together. Recently, a physiological method has been developed (Celichowski, Emonet-Dénand, Laporte & Petit, 1994) based on the responses of primary afferents during 30/s and 100/s stimulation of single static γ axons. Three types of activation by static γ axons could be distinguished: Fast, Slow and Mixed. Fast activation, ascribed to chains only, is characterised during stimulation at 30/s either by 1/1 driving or by an irregular increase in firing arising from a level close to that of the stimulation frequency and by the presence of significant peaks in correlograms of stimuli with afferent firing. Slow activation, ascribed to bag$_2$ fibres only, shows at 30/s a sustained and regular increase in firing with no peaks in the correlograms. Mixed activation, ascribed to chain and bag$_2$, shows at 30/s an irregular increase in firing which arises from a level distinctly higher than that of the frequency of stimulation and significant peaks in correlograms. The other purpose of this study was to determine whether differences in resumption times were influenced by muscle length.

METHODS

Experiments were carried out on peroneus tertius spindles of adult cats anaesthetised with pentobarbital sodium IP: 35 mg/kg, IV supplemented as required. In each of 5 experiments, the discharges of several primary endings were recorded from Ia fibres in dorsal root filaments at 5 muscle lengths (the minimal physiological length, L0, and 4 longer lengths, increased by steps of 0.5 mm). At each length, a 1 mm ramp stretch (rising and falling phases at 10 mm/s, duration of the hold phase: 2.3 s) was applied. The discharge of the ending stopped during muscle release and resumed a certain time after the end of the release. The time of resumption and the frequency of the resting discharge, 16 s after the end of the release, were systematically measured in passive conditions and after repetitive stimulation of single γ axons, at 20-100 Hz during 250 ms, 120 ms after the end of the muscle release. The effects of 36 γ axons (32 static and 4 dynamic), identified by the alterations of the responses to ramp-and-hold stretches, were studied on a total of 24 primary endings (2 γ on 10 spindles, 3 γ on 4 spindles, 4 γ on 6 spindles, 5 γ on 2 spindles and 6 γ on 2 spindles).

RESULTS

Resumption times, in active and passive spindles, varied from one spindle to the other and were longer for low frequencies. The resumption times as a function of resting discharge frequencies observed in a spindle in which the effects of 5 γ axons could be compared are shown in Figure 1. Of the 5 axons, 1 was dynamic (filled circles) and 4 static. One of these elicited a Mixed activation (filled squares) and 2 a Fast activation (open circles and triangles); the activation of the last one was identified as Fast (open squares) because during its stimulation at 30/s, driving was observed during the second half of the stimulation period; however, in this case, the classification was uncertain because during the first half of the stimulation the response presented a minimal value well above 30/s.

The resumption times observed after stimulation of the 2 static axons eliciting a Fast activation were very close to those of passive spindles whose values were distributed within the stippled area. For high resting discharges (approximately above 20 imp/s), no systematic difference in resumption times could be related to types of activation. For lower resting discharges, the shortest resumption time was observed after stimulating the static axon eliciting a Mixed activation. The stimulation of the static axon whose classification was uncertain also significantly reduced the resumption time.

The action of static axons eliciting a Slow activation (b_2) was studied in 5 instances (not illustrated). The resumption times observed after their stimulation were comparable to those observed after stimulating static axons eliciting a Mixed activation but much shorter than those measured after stimulating static axons eliciting a Fast activation (ch). The resumption times after stimulation of dynamic axons were generally similar to those observed after stimulation of static axons activating b_2 fibres, either alone or in association with chain fibres.

It was consistently found that the differences in resumption times ascribed to different intrafusal muscle fibres were observed only at shortest muscle lengths. At the longest, the times of resumption were very short and there was no systematic difference related to the type of activated fibres.

The frequency of primary ending discharges, measured 16 s after the end of muscle release, was unequally increased by the stimulation of different γ axons. At the shortest muscle lengths and low resting discharges, the increase observed after stimulation of static axons eliciting a Slow activation (bag_2) was greater than that elicited by axons with Fast activation (ch) and similar to that elicited by stimulation of static axons with a Mixed activation (b_2 + ch). Stimulation of dynamic axons had no effect. At the longest muscle lengths, the increases in frequency were very small and no differences between the actions of intrafusal muscle fibres could be observed.

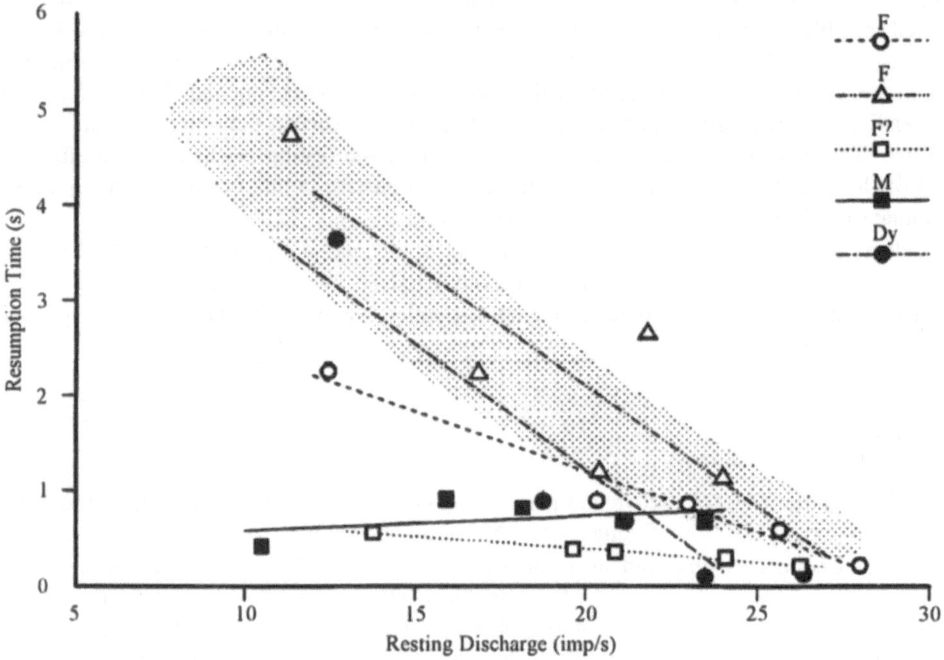

Figure 1. Resumption times of a primary ending discharge after stimulation of single γ axons as a function of resting discharge frequency. Dy: dynamic axon, M: static axon with a "mixed" activation of the ending, F: static axon with a "fast" activation, F ? : static axon with uncertain identification. Resumption times of passive spindles were distributed within the stippled area.

CONCLUSION

In summary, at shortest muscle lengths, bag_2 activation (either alone or in association with chain fibres) shortened the resumption time of primary ending discharges more than chain fibres alone. Bag_1 activation had a shortening effect comparable to that of bag_2. At the longest muscle lengths, resumption times were very short and no systematic difference could be ascribed to a particular type of intrafusal muscle fibres. The frequency of resting discharge measured 16 s after muscle release was increased, at the shortest muscle lengths, by bag_2 activation more than by bag_1 or chain activation.

These observations support the view that the state of bag_2 fibres may be the predominant factor which determines the frequency of primary ending resting discharge, as suggested by Proske et al. (1991), but at short muscle lengths only.

REFERENCES

CELICHOWSKI, J., EMONET-DÉNAND F., LAPORTE, Y. & PETIT, J. (1994) Distribution of static γ axons in cat peroneus tertius spindles determined by exclusively physiological criteria. *J. Neurophysiol.* **71**, 722-732.

PROSKE, U., GREGORY, J. E. & MORGAN, D. L. (1991) Where in the muscle spindle is the resting discharge generated? *Exp. Physiol.* **76**, 777-785.

TIME-DEPENDENT RESPONSES OF MUSCLE SPINDLE Ia-AFFERENTS ON A LONG LASTING SUCCINYLCHOLINE INFUSION

M.-A. Schoppmeyer and S.S. Schäfer

Department of Neurophysiology, OE 4230
Medical School of Hannover
30625 Hannover, Germany

INTRODUCTION

Succinylcholine (SCh) causes the bag_1 and bag_2 intrafusal fibres of cat muscle spindles to contract, while the chain fibres are paralysed (Gladden, 1976). In previous experiments Dutia (1980) delineated the time-dependent response of muscle spindle afferents during the first few minutes of SCh infusion. Taylor, Rodgers, Fowle & Durbaba (1992) described the behaviour of muscle spindle afferents following a single SCh injection.

METHODS

We studied the time-dependent behaviour of 49 Ia-afferents during IV SCh infusion for 15 minutes under four different concentrations (10-120 µg/kg/min) in order to show the different admixtures of static and dynamic fusimotor effects induced by SCh and their time-dependent changes. During the SCh infusions, repetitive ramp-and-hold stretches were applied to the tibialis anterior muscle of the cat. To characterise the time-dependence of the SCh induced effects we referred to representative discharge patterns obtained at the beginning and at the end of an infusion. We classified them into six categories ranging progressively from an apparently pure dynamic action (category I) to an apparently pure static action (category VI) based on Emonet-Dénand, Laporte, Matthews & Petit (1977).

RESULTS

Each of the 49 Ia-afferents changed their action continuously from a predominantly static effect at the beginning of the infusion to a more dynamic effect at the end of the infusion. The degree of change was specific for each Ia-afferent and Figure 1 illustrates this.

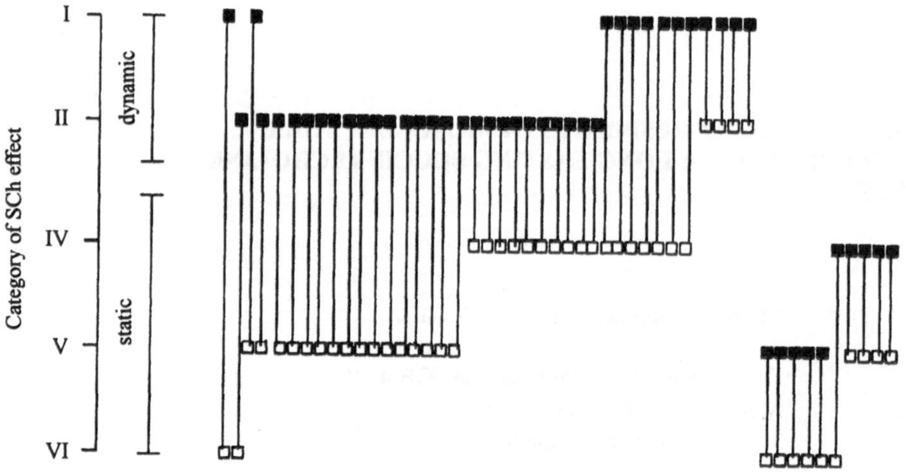

Figure 1. The degree of change of 35 Ia afferents (vertical lines) from a predominatly static effect at the beginning of the SCh infusion (open squares) to a more dynamic effect at the end of the infusion (closed squares). The y-axis gives the categories from I to VI.

The ordinates give the categories I to VI. The open squares at the bottom of each vertical line represent the category of a discharge pattern of an individual Ia-afferent observed at the beginning of the infusion, the closed squares given at the head of each vertical line represent the category of the same Ia-afferent at the end of infusion. 35 Ia-afferents changed their action from a predominantly static effect at the beginning of infusion (categories IV, V, VI) to a more prominent dynamic effect at the end of infusion (categories II, I). 4 Ia-afferents changed their action from a dynamic effect with suspected static modification (category II) to an apparently pure dynamic effect (category I) and 10 Ia-afferents changed their action from an apparently pure static effect (category VI) to a static effect with suspected or conceivable dynamic modification (categories IV, V).

DISCUSSION

We explain the observed SCh effects as a consequence of the simultaneous contraction of the bag_1 and bag_2 intrafusal fibres. Moreover, the receptor currents of the bag_1 and bag_2 terminals sum up in a non-linear fashion at the encoder site (Awiszus, 1994). Taking these two arguments into consideration, predominant static effects are observed at the beginning of infusion, because the bag_2 fibre contracts more strongly than the bag_1 fibre so that the receptor current of the bag_2 terminals has a stronger influence at the encoder site. At the end of an infusion, dynamic effects dominate because the contraction of the bag_2 fibre weakens so that the influence of the receptor current of the bag_1 fibre terminals dominate at the encoder site. This view explains the behaviour of the Ia-afferents and gives an understanding of the spindle specific degree of change we observe under a SCh infusion.

Taylor et al. (1992) classified 269 muscle spindle afferents into four different groups depending on the strength of influence of the bag_1 and the bag_2 fibres. Their results are consistent with our results at the beginning of SCh infusion. Nevertheless, we cannot classify our muscle spindles into four different groups because they are changing their behaviour at the end of infusion.

REFERENCES

AWISZUS, F. (1994) Repetitive activity of a branched Hodgkin-Huxley axon with multiple encoding sites. *Biol. Cybern.* **70**, 579-583.

DUTIA, M.B. (1980) Activation of cat muscle spindle primary, secondary and inter-mediate sensory endings by suxamethonium. *J. Physiol.* **304**, 315-330.

EMONET-DÉNAND, F., LAPORTE, Y., MATTHEWS, P.B.C. & PETIT, J. (1977) On the subdivision of static and dynamic fusimotor actions on the primary ending of the cat muscle spindle. *J. Physiol* **268**, 827-861.

GLADDEN, M.H. (1976) Structural features relative to the function of intrafusal muscle fibers in the cat. *Progr. Brain Res.* **44**, 51-59.

TAYLOR, A., RODGERS, J.F., FOWLE, A.J. & DURBABA, R. (1992) The effect of succinylcholine on cat gastrocnemius muscle spindle afferents of different types. *J. Physiol.* **456**, 629-644.

CORRELATION METHODS IN IDENTIFYING INTRAFUSAL MUSCLE FIBRE ACTIVITY

A. Taylor, R. Durbaba and J.F. Rodgers

Sherrington School of Physiology, UMDS
St. Thomas' Hospital, London SE1 7EH, UK

INTRODUCTION

An important current problem in muscle spindle physiology is to understand the extent to which the chain and bag_2 intrafusal muscle fibres are separately innervated by distinct classes of static γ-motoneurones. The ramp stimulation frequency test was introduced (Boyd & Ward, 1982) to distinguish the effects of chain and bag_2 fibre types on primary afferents. In this test a fusimotor axon is stimulated with a frequency rising smoothly from 0 to 150 Hz in 2.5 s and 1:1 driving or sub-harmonic afferent firing is taken to indicate a chain fibre effect. Lack of driving was taken to indicate that a given static fusimotor axon was restricted to bag_2 fibres (Boyd, 1986). It now seems that the ramp frquency test may over-estimate the incidence of bag_2 effects (Dickson, Emonet-Dénand, Gladden, Petit & Ward, 1993), because bag_2 activation suppresses the driving action of chain fibres. In that study and in a more recent one by Celichowski, Emonet-Dénand, Laporte & Petit (1994) correlation of afferent responses with stimulus trains was found to be a useful way of revealing chain innervation in the presence of bag_2. However, the use of regular or smoothly rising frequency stimulus trains gave rise to certain features which were difficult to interpret. We have therefore explored the use of random stimulus trains. An abstract of this work has been published elsewhere (Durbaba, Taylor, Rodgers & Fowle, 1993).

METHODS

Observations were made on spindles in the medial gastrocnemius muscle of cats deeply anaesthetised with pentobarbitone. Groups of 6 single spindle afferents were recorded simultaneously and the effects observed of stimulating single γ-axons in cut ventral root filaments. The afferents were characterised by conduction velocity and by the effect of succinylcholine on their responses to ramp stretches (Taylor, Rodgers, Fowle & Durbaba, 1992). Each γ-axon was tested at 50 - 100 Hz for its effect on the ramp stretch responses of primary afferents to determine its static or dynamic nature. Static units were then tested with stimulus ramps from 10 to 125 Hz in 2.5 s to look for driving. Finally, they were stimulated with Poisson distributed pulse trains at a mean frequency of 50 Hz (derived from a Geiger counter) and stimulus and afferent recordings made for 5 min periods, i.e. for 15,000 to 20,000 stimuli. Though β effects attenuated rapidly under these conditions, γ-fusimotor axon effects showed no sign of fatigue. Data were digitally recorded with the "Spike 2" package (Cambridge Electronic Design Ltd.) and cross-correlograms constructed with 1 ms bin widths and curves fitted with "Kaleidagraph" software (Abelbeck Inc.).

RESULTS

Driving γ-static units all yielded large, fast and simple correlogram peaks as shown in Figure 1A, with latency 11 to 13 ms and rise times of 2 to 2.5 ms. Attempts were made to fit the correlograms with various curves and by far the best fits were obtained with log-normal functions, as shown superimposed in Figure 1E. Many of the non-driving static γ-axons also produced distinct correlogram peaks, but they were smaller and slower than for driving units (Figure 1B). It was also evident that a single log-normal curve did not fit well, because of an additional slow component in the falling phase of the curve. This could be fitted with a second log-normal curve. The two components are shown together in Figure 1F. The simplest interpretation is that the fast component is due to chain fibre activation as in Figure 1A, and the slow component is due to bag_2 activation. This view is supported by the finding that occasionally non-driving static γ-axons generated only small, slow correlograms matching the above slow component (Figure 1C). Dynamic γ-axon stimulation gave correlograms similar to the above slow component, but of larger amplitude (Figure 1D).

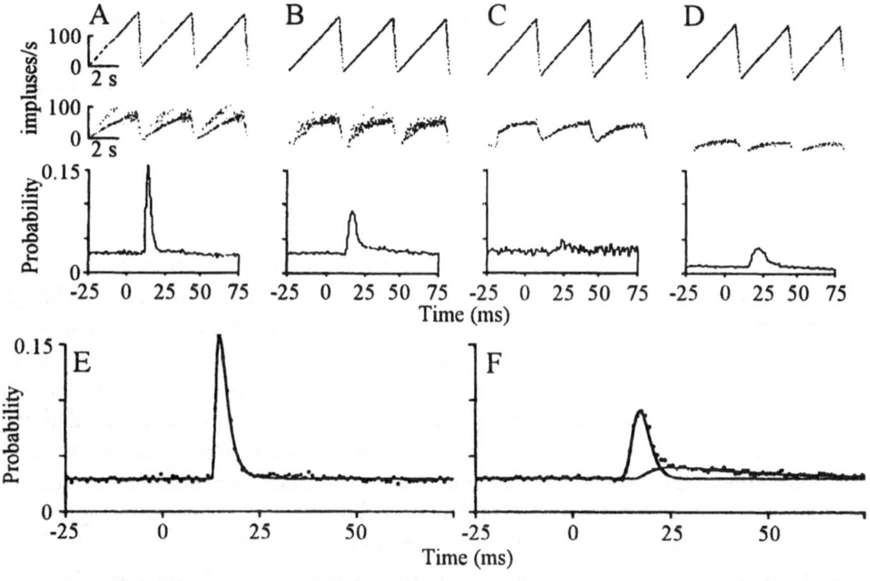

Figure 1. Responses of Ia afferents to ramp frequency and to random stimulus trains (Poisson, 50Hz mean) delivered to four different gamma efferent fibres. A - D. Top trace: stimulus ramp; middle trace: afferent response; bottom trace: cross-correlogram of the random stimuli with the afferent response based on 15 - 20 x 10^3 stimuli. The fusimotor fibres were A: driving static; B & C: non-driving static; D: dynamic. E is the fit of the correlogram in A with a single log-normal curve. F is the correlogram of B fitted with two log-normal curves.

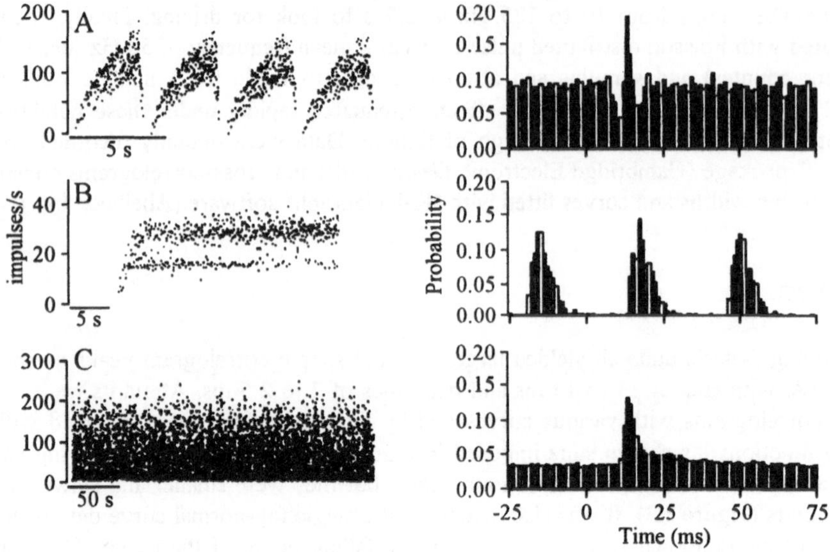

Figure 2. Response of a single Ia afferent to ramp (A), regular (B, 30Hz) and random (C, Poisson 50Hz mean) delivered to a non-driving static gamma efferent. All produce clear correlograms peaks, but only in the case of C could a second component be observed.

The random stimulus results contrast with correlograms derived from regular stimulation, as seen in Figure 2 in which ramp (A), regular (B) and random (C) stimulus patterns are compared for a non-driving static γ-axon. All yielded clear correlogram peaks, but only in the case of random stimulation were there two components, which could reasonably be separated by curve fitting. In ramp stimulation there is a marked trough on either side of the peak, while with regular stimulation the peak is repeated at the stimulus frequency.

DISCUSSION

In a linear system the result of cross-correlating a wide-band random input signal with the system output is the impulse response, which fully charaterises the system. Though the signal path from fusimotor input to afferent output of the spindle is no doubt far from linear, nevertheless some of the merits of this method of testing seem to be available. Thus, all frequencies are tested simultaneously and the reliability of the result depends only on the period for which data can be accumulated. Here we were able to continue stimulation for 5 mins. without any changes in response becoming apparent and the resulting curves were very smooth. Since the times of occurrence of the stimuli are completely statistically independent, there are no entrainment effects and the baseline of the correlogram is flat, which again facilitates measurement and the separation of components due to the chain and bag_2 fibres. It also seems likely that the natural resting firing of the afferent does not interfere with the test in any way. The usefulness of cross-correlation in studying fusimotor effects has been brought out in the recent papers referred to above (Dickson et al., 1993; Celichowski et al., 1994) and some of the additional potential advantages of using random stimuli have been appreciated before (Homma & Nakajima, 1985). The novelty of the present work is the construction of correlograms from very large numbers of impulses in response to *Poisson* distributed stimuli and the fitting of log-normal curves. The use of this method together with the ramp frequency test seems to make it possible to clearly distinguish cases in which a static fusimotor axon innervates chain fibres alone, bag_2 fibres alone or both together. This will be valuable in helping to decide on the degree to which these intrafusal fibres are separately accessible from the central nervous system.

REFERENCES

BOYD, I.A. & WARD, J. (1982) The diagnosis of nuclear chain intrafusal fibre activity from the nature of the group Ia and group II afferent discharge of isolated cat muscle spindles. *J. Physiol.* **329**, 17-18P.

CELICHOWSKI, J., EMONET-DÉNAND, F., LAPORTE, Y. & PETIT, J. (1994) Distribution or static γ axons in cat peroneous tertius spindles determined by exclusively physiological criteria. *J. Neurophysiol.* **71**, 722-732.

DICKSON, M.J., EMONET-DÉNAND, F., GLADDEN, M.H., PETIT, J. & WARD, J. (1993) Incidence of γ non-driving excitation of primary sensory endings using the ramp frequency stimulation test in the cat hindlimb. *J. Physiol.* **460**, 657-673.

DURBABA, R., TAYLOR, A., RODGERS, J.F. & FOWLE, A.J. (1993) Subclasses of fusimotor action on muscle spindles of the anaesthetized cat revealed by cross-correlation of firing with random stimulation. *J. Physiol.* **473**, 206P.

HOMMA, S . & NAKAJIMA, Y. (1985) Estimation of generator potential waveform elicited in muscle spindles by various inputs. In *The Muscle Spindle*. eds. BOYD, I.A. & GLADDEN, M.H. pp. 377-383. Macmillan, London.

TAYLOR, A., RODGERS, J.F., FOWLE, A.J. & DURBABA, R. (1992) The classification of afferents from muscle spindles of the jaw-closing muscles of the cat. *J. Physiol.* **456**, 609-628.

PART 6

ANALYSIS AND MODELLING

ANALYSIS OF ENCODING OF STIMULUS SEPARATION IN ENSEMBLES OF MUSCLE AFFERENTS

Håkan Johansson[1], Mikael Bergenheim[1,2],
Mats Djupsjöbacka[1] and Per Sjölander[1]

[1]Division of Work Physiology
National Institute of Occupational Health
Box 7654,S-907 13 Umeå, Sweden
[2]Department of Physiology, University of Umeå
S-90187 Umeå, Sweden

INTRODUCTION

In sensory physiology there have been two main theories relating to the processing of afferent information. These are the "labelled line" theory and the "ensemble coding" theory (cf. Ray & Doetsch, 1990a). The labelled line theory concentrates on the response properties of single receptors, while the ensemble coding theory places the emphasis on encoding in a neuronal population rather than by individual receptors or afferents (Erickson, 1968; Ray & Doetsch, 1990b).

Mechanical events in muscles seem most likely to be the subject of ensemble coding, since there are several types of mechano-receptors, with varying degrees of overlapping sensitivity. Also within a particular group, e.g. primary muscle spindle afferents (primaries), some degree of population coding might be expected, since they show very individualised or varied responses to different stimuli. Most current knowledge about the responses of muscle receptors is based on single sequential recordings and the disputable assumption is made that ensemble behaviour may be deduced from them (cf. Ray & Doetsch, 1990a). The present study involves simultaneous recording from a number of muscle spindle afferents, and a new method, based on Principal Component Analysis (PCA) has been developed for exploration of the capacity of ensembles for separating distinct amplitudes of sinusoidal muscle stretching.

METHODS

The experiments were conducted on 7 cats which were anaesthetised with α-chloralose (70 mg/kg). The left hindlimb was denervated save for the triceps surae. The animal was mounted in a metal frame with the pelvis and the left hindlimb rigidly immobilised. The triceps tendon was cut and tied to a computer-controlled electromagnetic puller.

In each experiment the muscles were stretched sinusoidally at 1 Hz at four or five different amplitudes (3 to 5mm peak-to-peak, in increments of 0.2 or 0.5mm). The stretches lasting 2s were superimposed on the plateau of a ramp-and-hold stretch (from -15 to -5mm, relative to maximal physiological length). This stretch sequence was repeated 5 times with 30s pauses, during which the muscles were fully relaxed.

Functionally single triceps primaries, with intact fusimotor supply, were recorded 4 to 9 at a time via a multichannel-hook electrode (Djupsjöbacka, Johansson, Bergenheim & Sandström, 1994) from dorsal root filaments (L7 or S1). A total of 43 primaries were studied. Their identification was based on maximal twitch tests (Hunt & Kuffler, 1951; Matthews, 1933) and conduction velocities using a dividing line at 72m/s (Boyd & Davey, 1968). The instantaneous firing frequency of the afferents was sampled on a PC-486 computer at a frequency of 6 kHz.

Analysis

Each set of primaries was tested with a number of sinusoidal stretches. For each test, a sequence of 1 second (i.e., one sine period, top to top) was selected, and the instantaneous firing frequency during every ms of the sequence was calculated and stored as variables in a data table.

Each row of such a table contains all the data from one test stimulation. Each afferent is represented by 1000 columns in the table, and the variables representing the different afferents follow each other successively in every row. Thus, the instantaneous frequency figures for the first afferent are organised as columns 1 - 1000 in the table, and the figures for the second afferent are represented by columns 1001 - 2000, and so on.

The variables of the table can be represented as coordinates in a p-dimensional system of coordinates (p = number of variables or columns). In such a system of coordinates each stimulus (row in the table) is defined by p coordinates. This multidimensional system of coordinates can be reduced to a three dimensional system of coordinates. The average distance between swarms representing different stimuli is taken as a quantitative measure of separation.

Mathematically, the calculation of the principal components can be described as follows (cf. Mardia, Kent & Bibby, 1979; Wold & Sjöström, 1977; Wold, 1982; Johansson, 1988; Johansson, Sjölander & Sojka, 1991): (1) the n objects (i.e. in our case the stimuli), which in the table correspond to n rows, is represented as a point swarm in a p-dimensional space. This is obtained by letting each variable define one orthogonal coordinate axis. Then (2), the swarm of points is represented by its mid-point. The coordinates of this point are the averages of the variables (x_k) which give the row vector x. The averages are (3) subtracted from the data. This gives the residuals e_{ik}, which become the elements in a new matrix E. Thus, the coordinate system is centred in the point x. The residual matrix is then denoted X, and then (4) the n points swarm in p-space is least-squares modelled by a line through the average x. The direction coefficients of the line are called "the variable loadings" (denoted p_{1k}, one for each variable) and form the row vector p_1, which defines the equation for the first principal component (PC). Thereafter, (5) each point is projected down on this line, and thereby the scores t_{i1} are obtained (i.e. the coordinate for the point i on the axis p_1). Then (6), t_{ip} is substracted from x_{ik}, which gives the residuals e_{ik} in the new residual matrix E. This (7) residual matrix (E) is then denoted X, and this new X is then used to fit a second straight line x to the data. This line will be orthogonal to the first line, and again the residuals are minimised with the least-squares method. In our analysis this was repeated three times, i.e. for the calculation of three PCs.

In our experiments three PCs always described more than 90% of the variance of the data set. The object score plots (see Figure 1A) may be complemented with a corresponding plot of the variable loadings (see Figure 1B), which shows the relations (distance in p-space) between the variables and in which the directions correspond to the directions of the object score plot. The significance of the computed PCs was checked with cross-validation techniques (Wold, 1978; Johansson, 1988; Johansson et al., 1991). Our analysis was performed with a program package run on a PC-486 computer.

Figure 1 illustrates the procedure of analysis for one experiment. First (1), PCA was made on all stimuli, and for all simultaneously recorded afferents (i.e. in Figure 1: 4 primaries, 5 Golgi-tendon-organ (GTO) afferents and 1 secondary). Figure 1A shows a graph of the object scores for the first two principal components, representing the Euclidean distance between the stimuli (i.e. sinusoidal stretches at 1Hz with a peak-to-peak amplitude of 0.3, 0.4, 0.5, 2.0, 2.5, and 3.0 mm). Each of the 6 stimuli was delivered 5 times, and thus the responses of the total population of afferents to the stimuli are represented in the object score plot of Figure 1A as 6 separate groups of 5 objects (dots). The next step (2) in the analysis was a comparison of the average variable loading for each afferent, in order to reveal how much each afferent contributed to the separation in the object score plot (Wold & Sjöström, 1977; Wold, 1982). This was achieved by averaging the 1000 consecutive variable loadings (1ms bins of instantaneous frequency figures) for each afferent (Figure 1B). Accordingly, in Figure 1B, the distance from origo for each afferent corresponds to the degree to which the afferent in question contributes to the resolution in the object score plot. As a third (3) step in the multivariate analysis, PCAs were calculated for each afferent and for each combination of afferents (i.e. every combination of 2 afferents, 3 afferents etc.) separately. For every afferent, and for every combination of afferents, (4) the degree of stimulus separation was then calculated.

Quantification of stimulus separation

An algorithm was designed to estimate the average separation of the stimuli (objects) in p-space. Since each type of stimulus is represented by a group of objects in the hyperplane, i.e. the 3-dimensional space spanned by the first 3 principal components, the algorithm should compute the distances between all the different object groups in the hyperplane. The algorithm would also have to take into account the spreading of the objects within the object groups. Thus an algorithm was used, where maximal separation would be obtained for maximal distances between object groups combined with minimal spreading within the object groups. The separation of 2 object groups was calculated according to the following equation (see Figure 1C):

$$\mathrm{Sep}(2-1) = \frac{|\mathrm{Mean2} - \mathrm{Mean1}|}{1.65 \times (\mathrm{S.D.1} + \mathrm{S.D.2})}$$

First, the separation for every pair of object groups was computed for each PC. Then the mean separation for all these pairs was calculated for each PC (below called: meanSepPCn). Finally, the separation of all object groups in the space of the three first PCs was calculated according to the following equation:

$$Separation = \sqrt{meanSepPC1^2 + meanSepPC2^2 + meanSepPC3^2}$$

Thus a relative measure of the degree of separation for a certain population of afferent could be obtained. It should be noted that comparison of such relative measures of separation can be made only for different sets of variables (i.e., afferent responses) determining the identical set of objects.

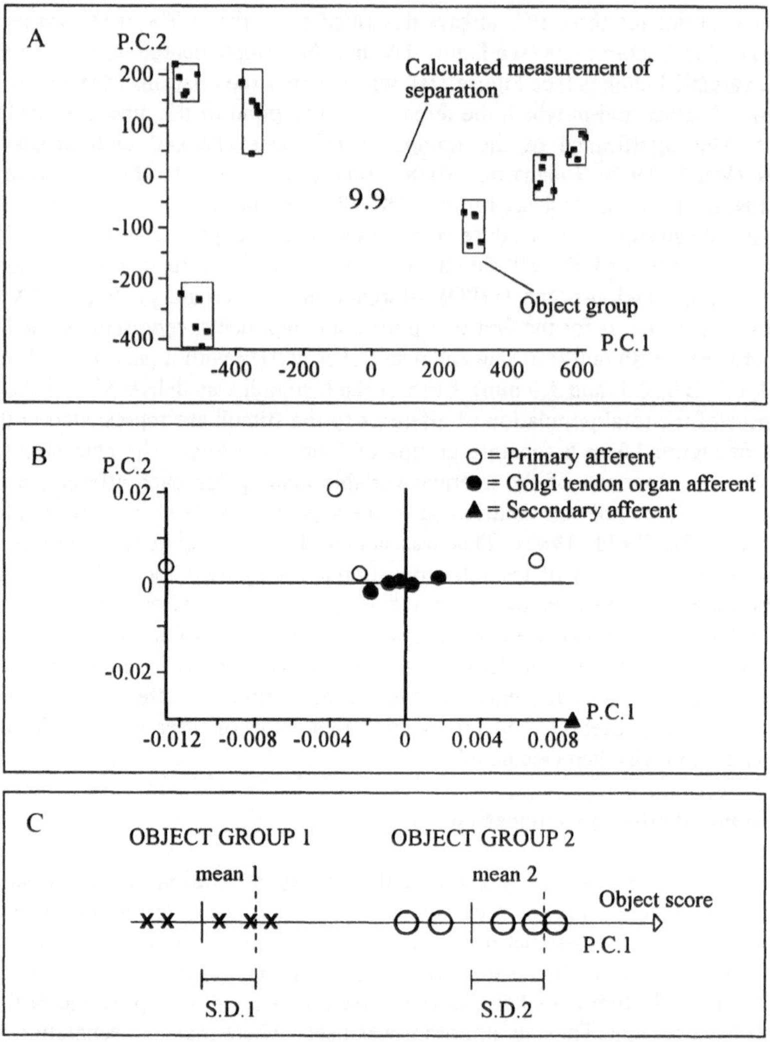

Figure 1. Procedure of analysis. *A*. Object score plot for the first two principal components of PCA of the response of 10 muscle afferents (i.e., 4 primaries, 1 secondary and 5 golgi tendon organ afferents) to 6 different stimuli (i.e., sinusoidal stretches at 1Hz with a peak-to-peak amplitude of 0.3, 0.4, 0.5, 2.0, 2.5, and 3.0 mm). Each type of stimulus was repeated 5 times. The 5 objects resulting from these identical stimuli is called an object group. The fitted square around each object group marks a 95% confidential interval of the spreading within the object group, and the figure in the middle of the graph is the calculated measurement of separation (c.f., Method) *B* Average variable loadings for each afferent for the first two principal components. Here, the distance from origo for each afferent reflects to what extent the afferent contributed to the resolution in the object score plot (1A). *C* Factors used for quantification of stimulus separation. Mean1 represents the mean object score for the 5 objects within object group 1, for the first principal component. S.D.1 is the standard deviation for these scores. Mean2 and S.D. 2 are the same values derived from object group 2.

RESULTS

In all experiments, the degree of stimulus separation increased with the size of the population of primaries. This is illustrated in Figure 2, which shows the average calculated stimulus separation for all combinations of 1, 2, 3, 4 etc. afferents (shaded columns; left hand scale), in an experiment with 8 simultaneously recorded primaries responding to 5 stimuli consisting of sinusoidal stretches with different amplitudes (3.0, 3.5, 4.0, 4.5 and 5.0 mm; 1Hz; superimposed on the plateau of a ramp stretch to 5mm below the maximal physiological length of the muscle). In addition, Figure 2 shows the figures of object group distances (open columns; right hand scale). Notably, the relative increase in object group distances with the number of primaries were greater than the corresponding increase in separation.

In all experiments there was a considerable variation in separation between the individual primaries. This is exemplified by the standard deviation (S.D.) given at the top of the first shaded column in Figure 2. Even though individual primaries sometimes could give a better separation than some populations of afferents, the best separation in all experiments was always obtained for a combination of several primaries. Interestingly, the S.D. for the separation decreased with increasing number of primaries in the populations (cf. Figure 2).

Furthermore, the diagram in Figure 2 seems to indicate that the curve of separation in relation to the number of primaries starts levelling at the right end of the diagram. This is further indicated by Figure 3, in which the results from 7 experiments (43 primaries) are pooled. The columns of Figure 3 show the average differences in separation between populations with successively increased number of primaries (i.e., from 1 to 3, 3 to 5, 5 to 7, and 7 to 9 primaries) expressed as percentage of increased separation while moving from a smaller to a larger population. Note the decrease in stimulus separation when moving from populations of 7 to 9 primaries, which however might be explained by the fact that the sample sizes decreased with the number of afferents.

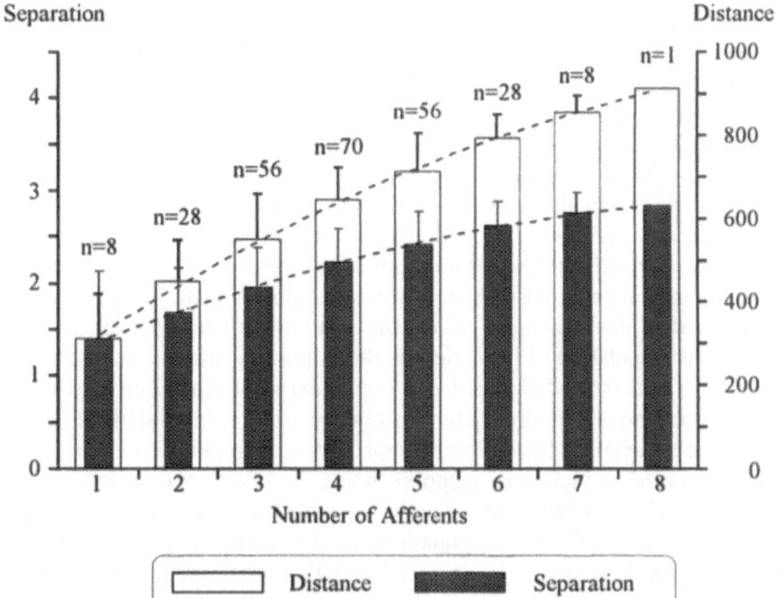

Figure 2. The average calculated stimulus separation (shaded columns; left hand scale), and the average distance (numerator in the equation in the methods section; open columns; right hand scale) between 5 different stimuli (i.e., 1 Hz sinusoidal stretches; peak-to-peak amplitude; 3.0, 3.5, 4.0, 4.5 and 5.0 mm; superimposed on the plateau of a ramp stretch to 5mm below the maximal physiological length of the muscle), for all populations of 1, 2, 3, 4 etc. afferents in an experiment with 8 simultaneously recorded primaries. *n* represents the number of permutations for each population size. The SD is given at the top of each column.

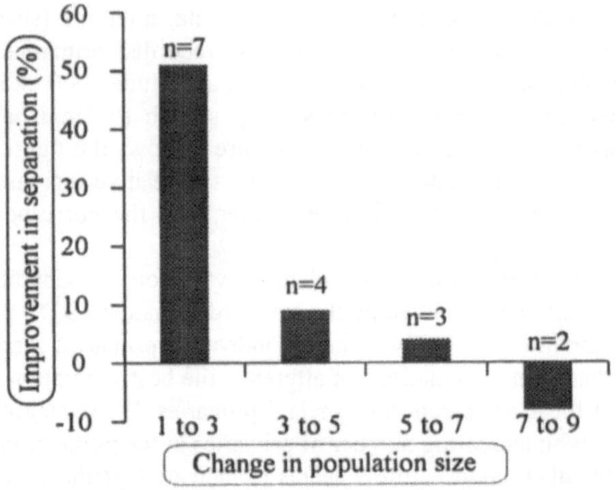

Figure 3. Differences in separation between populations with successively increased number of primaries (i.e., from 1 to 3, 3 to 5, 5 to 7, and 7 to 9 primaries). The diagram show the average improvement in percent in 7 experiments with different numbers of simultaneously recorded primaries (in total: 43 primary MSAs). *n* represents the number of experiments contributing to each column.

DISCUSSION

In this study we have used simultaneous recording of primaries to study ensemble coding. The analysis of an ensemble firing pattern should ideally be made with a method which assumes as little as possible about the unkown decoding (by the central nervous system) of the firing pattern, i.e. with a method which does not look for similarities between the representation of stimulus and the stimulus itself. Therefore, we developed a method based on PCA (Mardia et al., 1979; Wold & Sjöström, 1977; Wold, 1982; Johansson, 1988; Johansson et al., 1991). This method seems to have at least two advantages. First, it accesses stimulus discrimination and compares the relative discriminative ability of different populations of receptor afferents. Secondly, since the emphasis is put on the discrimination of stimuli, it involves relatively few assumptions about the decoding of the information.

Our results indicate that ensembles of primaries discriminate better between different muscle stretches than do individuals. Since the relation between the degree of separation and the size of the population of primaries reaches a plateau, the results also suggest that a limited number of units is needed in order to obtain maximal separation of stimuli. Whether this varies with the type of stimuli or afferents remains to be investigated. In this context the question might arise whether the increased separation of stimuli with increased population size simply reflects the reduction of response noise, i.e. a decrease in the variation of the responses of a certain population of primaries to a certain stimulus or a reduction of the denominator in the first of the two equations in the methods section. However, this is disproved by Figure 2, where it is demonstrated that the distances between the object groups (equation numerator) increase more than the separation (taking the denominator into account). We have also observed that the increase in separation with ensemble size is reduced or disappears completely after de-efferentation of the muscle spindles (Bergenheim, Johansson, Pedersen & Öberg, in preparation). This may indicate that the γ-system increases the ability of ensembles of spindle afferents to discriminate between different stimuli, perhaps by decorrelation of spindle afferent activity (cf. Inbar, Madrid & Rudomín, 1979).

Some of the teleological arguments in the discussion of ensemble vs. labelled line coding also gain support from this study. First, the so-called economy argument (Ray & Doetsch, 1990b) is bolstered by the fact that a higher degree of separation of stimuli implies a ability to encode a larger variety of stimuli. Secondly, the safety or reliability argument (Ray & Doetsch, 1990b) is supported by our observation that the loss of a single or of a few afferents probably means a relatively small loss in encoding (separation) capacity. Thirdly, the decrease in the S.D. of the separation with increasing ensemble size (Figure 2, shaded columns) confirms the theoretical argument (i.e. the stability argument, Ray & Doetsch, 1990b) that there is less variation in the discriminative ability of larger populations of afferents than in that of smaller populations, implying that with larger populations the choice of population for decoding is of less importance.

ACKNOWLEDGEMENTS

Supported by grants from The Swedish Work Environment Fund and The Swedish Sports Research Council (Centrum för Idrottsforskning).

REFERENCES

BOYD, I. A. & DAVEY, M. R. (1968) *Composition of Peripheral Nerve*, E & S Livingstone Ltd., Edinburgh and London,

DJUPSJÖBACKA, M., JOHANSSON, H., BERGENHEIM, M. & SANDSTRÖM. U. (1994) A multichannel hook electrode for simultaneous recording of up to 12 nerve filaments. *J. Neurosci. Meths.* **52**, 69-72.

ERICKSON, R.P. (1968) Stimulus coding in topographic and nontopographic afferent modalities: on the significance of the activity of individual sensory neurons. *Psychol. Rev.* **75**, 447-465.

HUNT, C.C. & KUFFLER, S.W. (1951) Stretch receptor discharges during muscle contraction. *J. Physiol.* **113**, 298-315.

INBAR, G., MADRID, J. & RUDOMIN, P. (1979) The influence of the gamma system on cross-correlated activity of Ia muscle spindles and its relation to information transmission. *Neurosci. Letts.* **13**, 73-78.

JOHANSSON, H. (1988) Rubrospinal and rubrobulbospinal influences on dynamic and static γ-motoneurones. *Behav. Brain Res.* **28**, 97-107.

JOHANSSON, H., SJÖLANDER, P. & SOJKA, P. (1991) Fusimotor reflex profiles of individual triceps surae primary muscle spindle afferents assessed with multi-afferent recording technique. *J. Physiol. (Paris)* **85**, 6-19.

MARDIA, K.V., KENT, J.T. & BIBBY, J.M. (1979) *Multivariate Analysis*, Academic, New York,

MATTHEWS. B. H. C. (1933) Nerve endings in mammalian muscle. *J. Physiol.* **78**, 1-53.

RAY, R.H. & DOETSCH, G.S. (1990a) Coding of stimulus location and intensity in populations of mechanosensitive nerve fibers of the raccoon. I. single fiber response properties. *Brain Res. Bull.* **25**, 517-532.

RAY, R.H. & DOETSCH, G.S. (1990b) Coding of stimulus location and intensity in populations of mechanosensitive nerve fibers of the raccoon. II. Across-fiber response patterns. *Brain Res. Bull.* **25**, 533-550.

WOLD, H. (1982) Systems under indirect observation. In *Systems under Indirect Observation.* eds. JORESKOG, K.G. & WOLD, H., pp. 1-54. North-Holland, Amsterdam.

WOLD. S. (1978) Cross validatory estimation of the number of components in factor and principal components analysis. *Technometrics* **20**, 397-406.

WOLD, S. & SJÖSTRÖM, M. (1977) SIMCA, a method for analyzing chemical data in terms of similarity and analogy. In *Chemometrics, Theory and Application*. ed. KOWALSKI, B., pp. 243-282. ACS Symposium Series, No. 52.

AN INTEGRATED MODEL OF THE MAMMALIAN MUSCLE SPINDLE

E. Otten[1], K.A. Scheepstra[1] and M. Hulliger[2]

[1]Department of Medical Physiology
University of Groningen
NL-9712 KZ Groningen
The Netherlands
[2]Department of Clinical Neurosciences
University of Calgary, Calgary, Alberta
Canada T2N 4N1

INTRODUCTION

A large number of experiments on mammalian muscle spindles, typically based on some combination of stimulation of dynamic and/or static gamma efferents with imposed length variations, have illustrated a range of functional properties and characterised muscle spindles as specialised mechanoreceptors (reviewed by Matthews, 1972; Hunt, 1990). A number of theories have been formulated to explain various aspects of experimental observations in terms of likely receptor mechanisms, which span a range from mechanical to ionic processes. Similar concepts have been applied in the analysis of other mechanoreceptors (Teorell, 1971). In most cases some combination of mechanical and ionic processes appears to give the most satisfactory general description of receptor behaviour. It can therefore be expected, that the same should apply for muscle spindles.

In addition to the extensive information available on global input-output properties, much is also known regarding the micro-anatomy and micro-physiology of muscle spindles, relating to possible receptor mechanisms. It is not too difficult to let these details be elements in some qualitative concept explaining the observed behaviour. It is, however, difficult to identify the relative contribution of individual receptor mechanisms to integral receptor behaviour, since there is much mutual dependence.

In this context, the design of an integrated mathematical model can be helpful, since in formulating a model it is necessary to make decisions on the relative weight of candidate mechanisms. Further, once individual processes are described quantitatively, experiments can often be designed to test predictions of the model. On the one hand, a mathematical model is no more than a quantitative theory, and as such needs to be corroborated or rejected. On the other hand, there are two advantages of models over ordinary verbal or diagrammatic theories. First, testing is straightforward, because the model is quantitative in nature and its predictions are in the form of numbers, which can be compared to measurements. Tests can be more critical than for verbal theories, which sometimes can be kept alive by mere shifts of emphasis. Second, internal coherence in formal models reduces the number of possible quantitative models compared with the number of reasonable but qualitative theories.

In this paper we emphasize general principles of operation of the muscle spindle. To be formulated mathematically and included in an integrated model, any concept had to meet three requirements: first, foundation in known micro-physiological processes and micro-anatomical data; second, ability to explain spindle properties; third, capability of predicting the outcome of new experiments, whose design was stimulated by observations on model behaviour. Four such principles are considered in the sections below. Each of these was first formulated quantitatively as a sub-model. These sub-models were then combined into a higher order model, which we refer to as the integrated model of the mammalian muscle spindle.

DYNAMICS FROM NON-LINEAR FORCE-VELOCITY PROPERTIES

It has repeatedly been postulated (Rudjord, 1970, Matthews, 1981) that important features of dynamic behaviour of mammalian muscle spindles originated in mechanical properties of its components. However, these properties are not well known: the only measurements available are those of fibre bundle (rather than individual fibre) tension (Hunt and Wilkinson, 1980) and of opening of spiral endings (Boyd, 1985)). It was therefore difficult to specify the parameters of a model simulating spindle mechanics with appropriate (i.e. spindle-derived) numerical values. Schaafsma, Otten & Van Willigen (1991) chose to rely on properties of extrafusal muscle fibres. But, interestingly, they had to modify them substantially to equip bag_1 fibres with dynamic properties that were sizable enough to account for dynamic responses of Ia afferents (by incorporating steep force increments during lengthening contractions). Figure 1 shows the difference in response to a trapezoidal

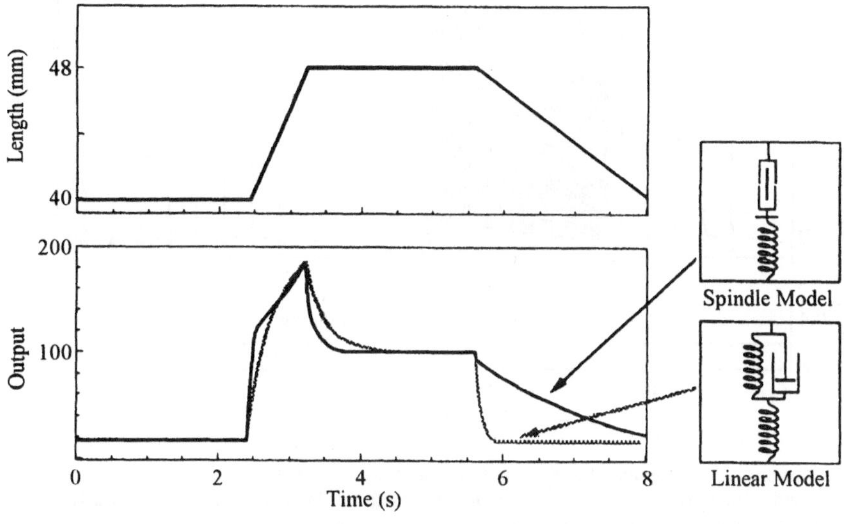

Figure 1. Comparison of mechanical properties of a linear visco-elastic Voigt/Maxwell model and of a muscle fibre model (spindle model of Schaafsma et al., 1991). Note the pronounced asymmetry in the spindle response (solid line) to stretch (lengthening contraction) and to release (shortening contraction).

length variation of a linear visco-elastic system and a system with muscle fibre properties. The much steeper increase in tension of the muscle system during stretch is evident.

Poppele and Quick (1981) reported evidence of stretch activation in the intrafusal fibre of a mammalian muscle spindle. Their observations showed a shortening of the juxta-equatorial part during a ramp stretch, while the length of the primary ending in series increased monotonically. Unless this shortening results from rearrangement of sarcomere lengths, stretch activation could be expected to help increase the sensitivity of the spindle. The data, however, showed that the shortening did not start until 1200 msec after the onset of the ramp, implying that it cannot contribute to the increase in dynamic sensitivity at the start of the ramp. It therefore seems most likely that non-linear force-velocity properties of the intrafusal muscle fibres are responsible for this phenomenon. It bears emphasis that the mechanical model of Schaafsma et al. (1991), which incorporated this particular non-linearity, was capable of simulating the pronounced dynamic sensitisation and the slow decay of Ia response during γ_d activation very adequately.

DYNAMICS FROM A SLOW K CONDUCTANCE IN THE SENSORY ENDING

One property which could not be simulated with muscle fibre mechanics only is post-release silence of spindle Ia responses. A model well tuned to simulate the dynamic peak during the ramp and the slow decay during the hold phase, still shows significant sensory elongation during the release (Schaafsma et al., 1991). If, in a mechanical model, sensory elongation is directly encoded as firing frequency, post-release silence cannot occur.

To overcome this, we simulated the transducer and encoder process using modified Frankenhaueser-Huxley (F-H) equations. The non-specific leak current and the contribution of the fast repolarising potassium system were reduced. In addition a slow potassium conductance was added (based on observations by Westbury, 1985). Figure 2 shows the

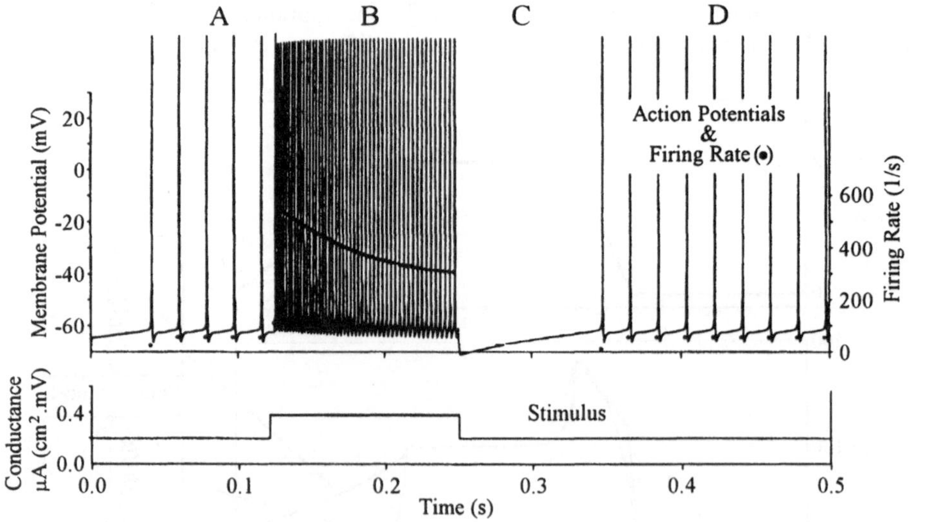

Figure 2. Response to rectangular stimulation of the sub-model of a spike generating site with slow potassium conductance. A. Steady state firing at low rate with minimal activation of the slow K conductance. B. High rate discharge during hold phase with progressive, but slow, activation of the (voltage-dependent) slow K conductance. C. Post-release silence due to the hyperpolarising action of the slow K conductance, whose activation diminishes only slowly. D. Return to the steady state firing at the same rate as in A, following return to a low level of activation of the slow K conductance.

behaviour of this model in response to a rectangular stimulus. During the hold phase the response revealed slow decay, which was due to the slowly activating and repolarising potassium conductance. The reverse occurred at the end of the rectangular stimulus (mimicking abrupt release), since the slow potassium conductance was still activated: following a period of post-release silence, firing slowly returned towards the resting level. Another consequence of adding a slow potassium conductance is that the Bode plots representing gain and phase of the receptor potential, as measured by Hunt and Wilkinson (1980) could be reproduced successfully and much more satisfactorily than with a mechanical model (Otten, Hulliger & Scheepstra, 1995).

The slow decay, which was introduced by the slow conductance, is very similar to the slow decay transients seen in several purely mechanical models, including that of Schaafsma et al. (1991; see above). This then leaves open the question, whether the slow decay in Ia firing following dynamic stretch is mediated by ionic or by mechanical processes, or by a combination of the two. Thus, without discriminative experiments, the relative contributions of mechanical and ionic receptor mechanisms cannot be identified. At present, experimental studies addressing this issue are under way (see Discussion).

COMPETITION AMONG TERMINALS BY SPIKE-GENERATOR RESETTING

Spike generator resetting and pacemaker switching are not new concepts in spindle or general mechanoreceptor research (Crowe & Matthews, 1964; Eagles & Purple, 1974; Hulliger & Noth, 1979). To illustrate their operation in a mathematical simulation, sub-models of a number of spiking locations were joined by myelinated axon segments. These segments only allowed axial currents, but not receptor potentials, to spread from one spiking location to another. Figure 3A shows the behaviour of a simple axon tree with two sensory endings, each stimulated by a trapezoidal input, when spike-generator resetting by

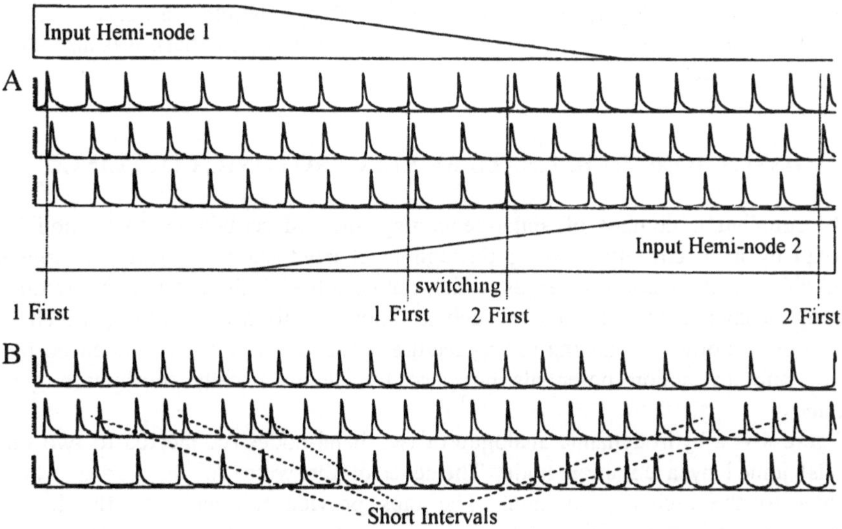

Figure 3. Sub-model of spike propagation through a simple terminal branch tree consisting of two hemi-nodes and a common node. Responses of each node are shown as trains of simulated action potentials during separate trapezoidal stimulation of each hemi-node. A. Spike-generator resetting by antidromic invasion enabled. B. Resetting disabled. In each case top trace: output hemi-node 1; middle trace: output common node; bottom trace: output hemi-node 2. Note that without resetting of the peripheral nodes unreasonably short inter-spike intervals are generated in the common node. See also text.

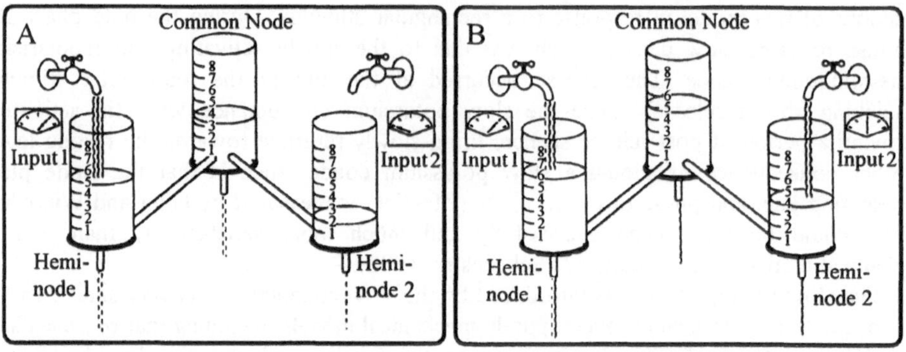

Figure 4. Electrotonic coupling between spike generating hemi-nodes, illustrated by an analogue model of water flow through pipes (axon segments) and cylinders (hemi-nodes and nodes). For details see text.

antidromically invading action potentials was enabled. When the two inputs are of equal size, pacemaker switching occurs. Otherwise, the ending with the smaller input lags behind, while firing at the same rate, as it is consistently invaded antidromically. If spike-generator resetting did not occur (Figure 3B), short inter-spike intervals would occur in the parent axon, whose minimum duration would be determined by the maximum firing rate of a node of Ranvier or by the peripheral conduction time (from the spiking site to the first common node). Since such short outlier intervals are not observed in Ia response patterns, it is safe to assume that resetting occurs.

Resetting results in total occlusion of the ending with the lower instantaneous firing rate, since it will simply follow the rate of the faster ending. Thus, when two or more endings project onto a common node, the discharge rate of the common node will be that of the ending with the highest instantaneous firing rate. However, total occlusion is rarely encountered (Banks, Hulliger, Scheepstra & Otten, 1995). Therefore a non-occluding form of interaction, i.e. coupling by electrotonic spread of receptor potentials, was also studied in a separate sub-model.

SYNERGISM AMONG TERMINALS BY ELECTROTONIC COUPLING

By coupling a number of spike-generating sites (described by their modified FH equations) by axial currents flowing through myelinated axon segments, a large set of coupled differential equations emerges, which can be solved with numerical approximations. The firing behaviour of the sites of such a simulation follow a pattern, which is best explained by analogy. The distinguishing feature is that a given site is influenced by other sites not only by action potentials but also by electrotonically transmitted generator potentials.

Figure 4 is a hydrodynamic analogue of two hemi-nodes (connected to two different terminals) joined by a common node. The level of the water in the graded cylinders is equivalent to the receptor potential. The taps provide the input to the hemi-nodes (equivalent to the depolarising current from mechano-sensitive channels). The pipes between the cylinders represent myelinated branches of an axon tree. The flow through the pipes is proportional to the difference in water level (equivalent to the potential difference between the two hemi-nodes). For steady-state flows the level at the common node is the average of the levels at the hemi-nodes. If hemi-node 1 has a large, and hemi-node 2 no input (Figure 4A), the resulting steady-state water levels are determined by input 1 and the leaks (proportional to the water levels in each cylinder). If, for the same input to hemi-node 1, hemi-node 2 also receives an input (albeit smaller; Figure 4B), the level at the common node increases. This results in a decrease of flow from node 1 to the common node, so that the level in node 1 will be higher. In summary, by turning on the input to node 2, the level at node 1 increases, even when its direct input (1) is larger than the parallel input (2).

Thus, electrotonic coupling is a synergistic mechanism, which results in interactions between endings somewhere between pure summation and pure occlusion. In computer simulations it was shown, that the amount of summation depends on the size of axon terminal branch trees, and that it decreases as the number of nodes between the hemi-nodes increases. This was a testable prediction of the model, which is dealt with elsewhere (Banks et al., 1995). Combining the sub-models of spike-generator resetting and electrotonic coupling permitted comprehensive simulations of interactions between endings in afferent trees of varying size.

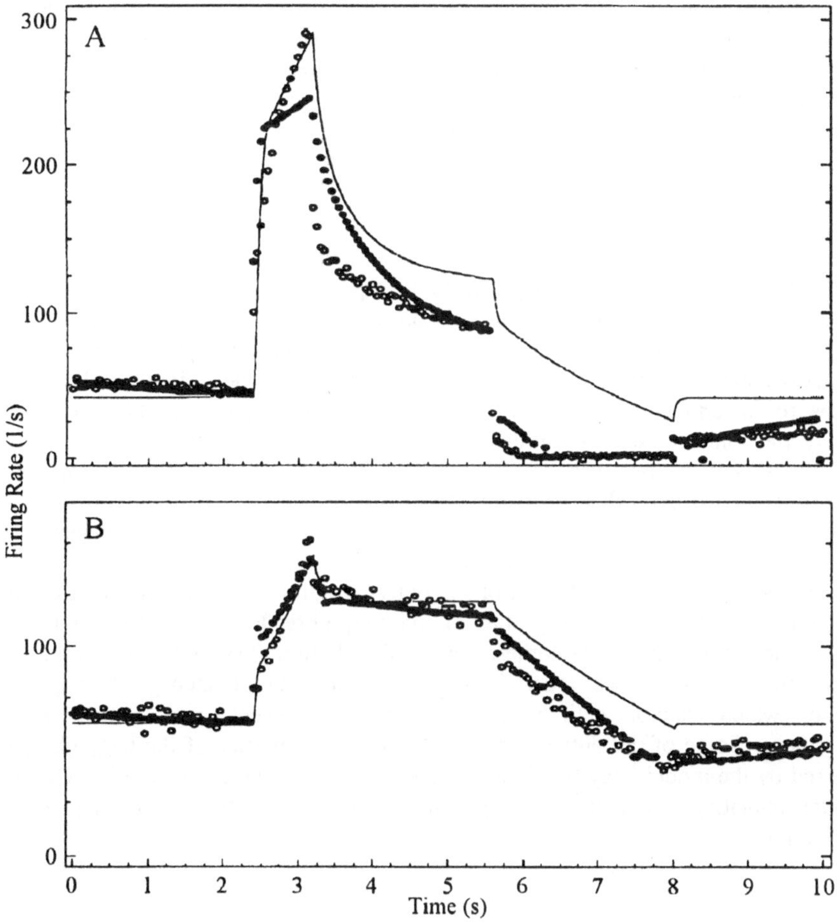

Figure 5. Behaviour of the integrated model of the mammalian muscle spindle compared with Ia population responses to trapezoidal stretch. A, γ_d stimulation at 100/s; B, γ_s stimulation at 100/s. Filled circles, response of integrated model; open circles, population responses of Ia afferents (Hulliger, Otten & Wang, unpublished data); thin solid line, length variations of the sensory ending as simulated by the model. Note in A (γ_d action) the mismatch of model and Ia response during the dynamic transients and the slow decay. Ongoing experimental and modelling studies are exploring the possibility of identifying additional ionic mechanisms to account for this discrepancy (see Discussion).

THE INTEGRATED MODEL

The integrated model was obtained by combining the four sub-models above. Figure 5 shows a comparison of integrated model responses with pooled experimental data. The elongation of the sensory endings (thin lines) of bag_1 (A) and bag_2 or chain (B) fibres during trapezoidal stretch permits identification of the contributions of mechanics and ionics to the integral response. For instance, the bag_1 fibre contribution (dynamic gamma action) owes its drop in firing rate during release almost exclusively to the slow potassium conductance, while the slow decay during the hold phase originates in mechanical creep, which is enhanced by the ionic system.

DISCUSSION

The formulation of above principles of operation in a mathematical model has permitted quantitative prediction of the outcome of experiments. For example, in the model slow decay of firing at the end of a ramp stretch originates in mechanical creep due to the force-velocity properties of the bag_1 fibre. It was predicted that abrupt release or termination of γ_d activation at the end of the ramp should abolish the slow decay. This was subsequently corroborated experimentally (Hulliger, Otten, Wang & Tabet, 1992, and in preparation). Further, the model predicts that spindles with large afferent trees will show stronger mutual occlusion of the responses of sensory endings than those with small trees. This was also confirmed experimentally (Banks et al., 1995).

The strategy of modelling used was to incorporate a candidate mechanism only if robust observations could otherwise not be reproduced. However, if several principles can generate comparable behaviour, the only way to select one over the other is to design discriminative experiments. An example is the capacity of both ionic and mechanical processes to introduce dynamics. One way of approaching this problem is to adapt selectively the ionic system by antidromic spike invasion (without engaging mechanical compartments) and to measure conditioning effects and recovery time constants. Any changes in dynamics must then be due to the conductances in the spike generating sites. Preliminary data from experiments along these lines indicate that ionic mechanisms indeed contribute to the dynamics of spindle Ia responses.

For several reasons redistribution of sarcomere length may be an important process to account for the distinguishing properties of the bag_1 fibre. First, their focal activation is bound to elicit non-uniform sarcomere length distribution. Secondly, the absence of a constant cross-sectional area must further compound this tendency. Thirdly, stretch activation, which cannot entirely be excluded at present, may cause additional inhomogeneity. Some of the outstanding force-velocity properties of the bag_1 fibre, which are inferred by the model, may be hidden in one or more of these possibilities. Currently a sub-model simulating sarcomere inhomogeneity is being studied to put these concepts on a firmer footing.

ACKNOWLEDGEMENTS

Supported by the Netherlands Science Foundation (KAS) and a Canadian MRC grant (MH).

REFERENCES

BANKS, R.W., HULLIGER, M., SCHEEPSTRA, K.A. & OTTEN, E. (1995) Pacemaker competition and the role of preterminal-branch tree architecture: a combined morphological, physiological and modelling study. (this volume).

BOYD, I.A. (1985) Muscle spindles and stretch reflexes. In *Scientific Basis of Clinical Neurology.* eds. SWASH, M. & KENNARD, C., pp. 74-97. Churchill Livingstone, London.

CROWE, A. & MATTHEWS, P.B.C. (1964) Further studies of static and dynamic fusimotor fibres. *J. Physiol.* **174**, 109-131.

EAGLES, J.P. & PURPLE, R.L. (1974) Afferent fibers with multiple encoding sites. *Brain Res.* 77, 187-193

HASAN, Z. (1983) A model of spindle afferent response to muscle stretch. *J. Neurophysiol.* **49**, 989-1006.

HULLIGER, M. & NOTH, J. (1979) Static and dynamic fusimotor interaction and the possibility of multiple pace-makers operating in the cat muscle spindle. *Brain Res.* **173**, 21-28.

HULLIGER, M., OTTEN, E., WANG, B. & TABET, M.S. (1992) On the nature of the γ_d-induced slow decay of cat spindle Ia firing following stretch. In *Muscle Afferents and Spinal Control of Movement.* eds. JAMI, L., PIERROT-DESEILLIGNY, E. & ZYTNICKI, D., pp. 63-69. Pergamon, Oxford.

HUNT, C.C. (1990) Mammalian muscle spindle: peripheral mechanisms. *Physiol. Rev.* **70**, 643-663.

HUNT, C.C. & WILKINSON, R.S. (1980) An analysis of receptor potential and tension of isolated cat muscle spindles in response to sinusoidal stretch. *J. Physiol.* **302**, 241-262.

MATTHEWS, P.B.C. (1972) *Mammalian Muscle Receptors and their Central Actions.* Arnold, London.

MATTHEWS, P.B.C. (1981) Evolving views on the internal operation and functional role of the muscle spindle. *J. Physiol.* **320**, 1-30.

OTTEN, E., HULLIGER, M. & SCHEEPSTRA, K.A. (1995). A model study on the influence of a slowly activating potassium conductance on repetitive firing patterns of muscle spindle primary endings. *J. Theoretical Biol.* (in press).

POPPELE, R.E. & QUICK, D.C. (1981) Stretch-induced contraction of intrafusal muscle in cat muscle spindle. *J. Neurosci.* **1**, 1069-1074.

RUDJORD, T. (1970) A second order mechanical model of muscle spindle primary endings. *Kybernetik* **6**, 205-213.

SCHAAFSMA, A., OTTEN, E. & VAN WILLIGEN, J.D. (1991) A muscle spindle model for primary afferent firing based on a simulation of intrafusal mechanical events. *J. Neurophysiol.* **65**, 1297-1312.

TEORELL, T. (1971) A biophysical analysis of mechano-electrical transduction. In *Handbook of Sensory Physiology*, Volume 1, ed. LOEWENSTEIN, W.R., pp. 291-339. Springer, Berlin.

WESTBURY, D.R. (1985) Evidence for the importance of calcium activated potassium conductance in frog muscle spindle sensory endings. In *The Muscle Spindle.* eds. BOYD, I.A. & GLADDEN, M.H., pp 359-363. Macmillan, London.

FORCE CODING BY POPULATIONS OF CAT GOLGI TENDON ORGAN AFFERENTS: THE ROLE OF MUSCLE LENGTH AND MOTOR UNIT POOL ACTIVATION STRATEGIES

M. Hulliger[1], P. Sjölander[1,2], U.R. Windhorst[1] and E. Otten[3]

[1]Department of Clinical Neurosciences
University of Calgary, Calgary, Canada T2N 4N1
[2]Division of Work Physiology, NIOH
S-913 07 Umeå, Sweden
[3]Department of Medical Physiology
University of Groningen, NL-9712 KZ Groningen
The Netherlands

INTRODUCTION

Golgi tendon organ (GTO) afferents respond very sensitively to forces generated by motor units (MUs) which are directly coupled with their receptor endings, but it has long been recognised that GTOs are poor indicators of whole muscle force. This can be attributed to their being operated by small subsets of MUs. For different strategies of MU pool activation these need not always be activated to the same degree at a given level of whole muscle force or motor drive. Individual afferents are therefore unlikely to be relied on exclusively for force feedback evaluation by the CNS (see review by Jami, 1992).

Alternatively, for a given muscle the ensemble activity of a population or subpopulation of GTOs, each contributing a "local" force estimate, might be a more faithful indicator of whole muscle force. This seemed borne out by statistical estimates of the firing rate-force relationship of a GTO ensemble, derived from individual GTO firing-force characteristics (Crago, Houk & Rymer, 1982), although these were based on measurements of reflex force which were not reproducible between trials. Yet the notion of GTO populations monitoring whole muscle force was challenged by Horcholle-Bossavit, Jami, Petit, Vejsada & Zytnicki (1990), who estimated population discharge from GTO responses to identical force transients. In their study large samples of GTOs from one and the same muscle were obtained, but only few MUs were activated and limited forces studied. The issue was re-investigated by Hulliger, Weytjens & Windhorst (1992): using multi-channel stimulation of small groups of MUs and exploring nearly the full physiological range of muscle force, they confirmed that GTO ensembles monitored muscle force remarkably faithfully. This study was limited in that only a single physiological MU activation strategy (an approximation of recruitment combined with rate modulation) was investigated, and that the question, whether population discharge-force characteristics depended on muscle length, was not explored.

In a subsequent study (Hulliger et al., in preparation) the role of different strategies of MU pool activation was investigated. Using similar multi-channel stimulation, strategies spanning the range between pure MU recruitment and pure rate modulation were simulated experimentally. The GTO ensemble firing-force characteristics of different activation strategies revealed small, but consistent differences. For large forces nearly identical sensitivity figures (around 4/s/N; mean population discharge/average force) were obtained. However, owing to different degrees of gain compression (response saturation with increasing force) over the entire range of force, small-force sensitivity varied over a nearly fivefold range, from 4/s/N for pure rate modulation to 18/s/N for pure recruitment, with a combined recruitment and rate modulation strategy revealing intermediate properties with a small-force sensitivity of 11/s/N.

The present study was undertaken to investigate whether population characteristics also depended on muscle length, given the well known length dependence of active force.

METHODS

In 9 pentobarbitone-anaesthetised cats, functionally single muscle afferents were recorded from fine dorsal root filaments. GTO afferents were identified on the basis of conduction velocity (60-110 m/s) and the characteristic response to supramaximal whole muscle twitch contraction. Responses of 41 GTOs were recorded during systematic exploration of various MU activation strategies, as described elsewhere (Hulliger et al., in preparation). For the present study a subset of 12 GTO afferents was investigated in more detail for a single activation strategy (recruitment combined with rate modulation; below), to investigate possible effects of muscle length.

The L7 and S1 ventral roots were extensively divided and then regrouped into eight filaments of widely ranging size (mean tetanic forces: 0.7-7 N), which together on average generated 90% of the maximum tetanic force of the soleus muscle. Individual filaments were estimated to control subsets of about 3-30 MUs. These 8 filaments were then ranked according to tetanic force and activated in parallel and independently, using electrical stimulation and computer-generated pulse trains which were reproducible between trials and experiments (inter-stimulus interval resolution 0.5 ms). The 8 stimulation patterns imitated an activation strategy based on recruitment combined with rate modulation (staggered recruitment at low rates, followed by ramp-shaped modulation of stimulation rate; Figure 1.3). The temporal order of filament activation was determined by their size (smallest first, largest last).

Whole muscle force was recorded conventionally at the tendon. Muscle length was calibrated during surgery, before detachment of the tendon from the heel, to permit recordings at defined and reproducible muscle lengths. Using a reference marker anchored in the tibia and a length marker on the tendon the relation between muscle length and ankle angle was determined. The length corresponding to an angle of 90° was chosen as zero reference.

Measurements were taken at fixed length, which was either intermediate (3 mm above the zero reference, corresponding to an ankle angle of 80°) or short (7 mm below the zero reference; 115° angle). Single unit responses are displayed as average frequency firing rate profiles (Figure 1.1). Population responses were computed as probability density histograms

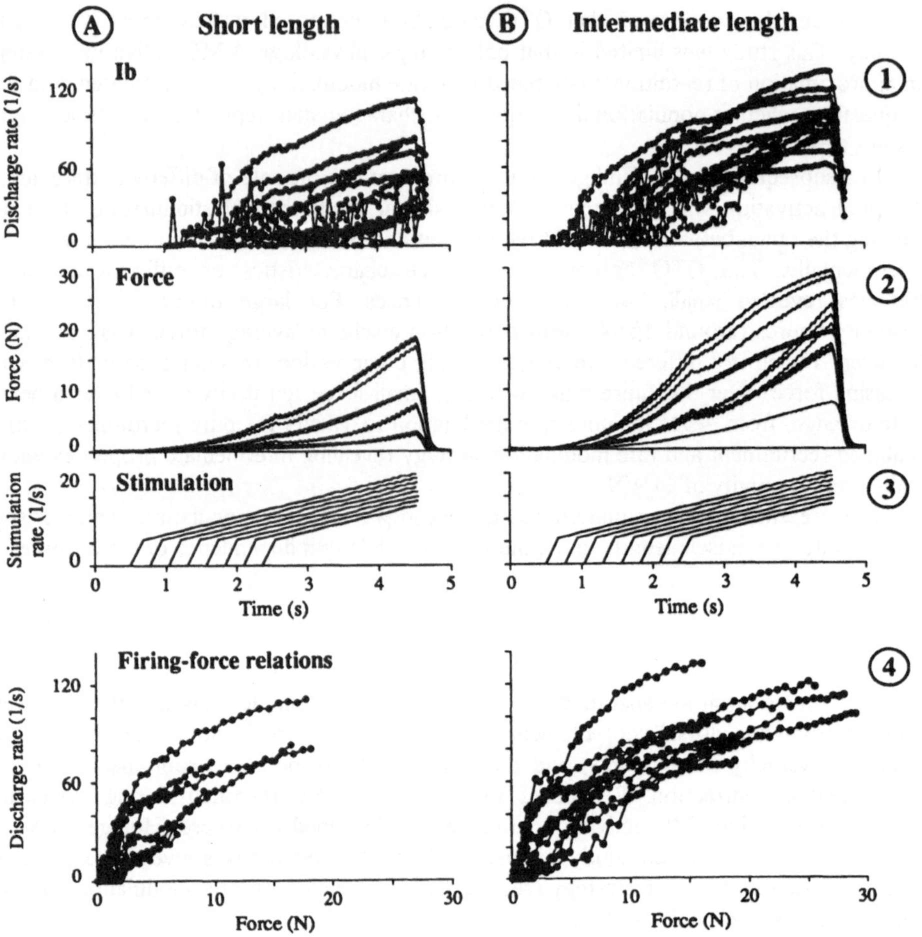

Figure 1. Range of individual GTO responses to force transients elicited by combined recruitment and rate modulation of 8 multi-MU filaments at short and intermediate muscle length. A, short length (-7 mm); B, intermediate length (+3 mm). Top, time course of average frequency firing rate of individual afferents, each recorded at short (A.1) and long (B.1) length; Second row, associated force transients; third row, 8-channel parallel stimulation patterns simulating a recruitment and rate modulation strategy. Bottom row, probability density estimates of firing rate vs force relationships, showing - for each afferent - the data of the 1st plotted against corresponding data of the 2nd row. For further detail see text.

(bin width 50 ms; Figure 2A). The firing rate-force relationships (Figures 1.4 & 2C) were calculated, for successive non-overlapping windows of the same duration, as probability density vs average force estimates.

RESULTS

Ventral root filaments encompassing MU groups of widely different size (3-30 MUs) exerted remarkably invariant biassing effects (average increase in GTO firing rate of 50/s), illustrating again the inadequacy of force feedback signals from individual GTO afferents. This was further emphasized by the wide range of individual response patterns illustrated in Figure 1 for 12 GTO afferents, which were recorded during activation based on recruitment

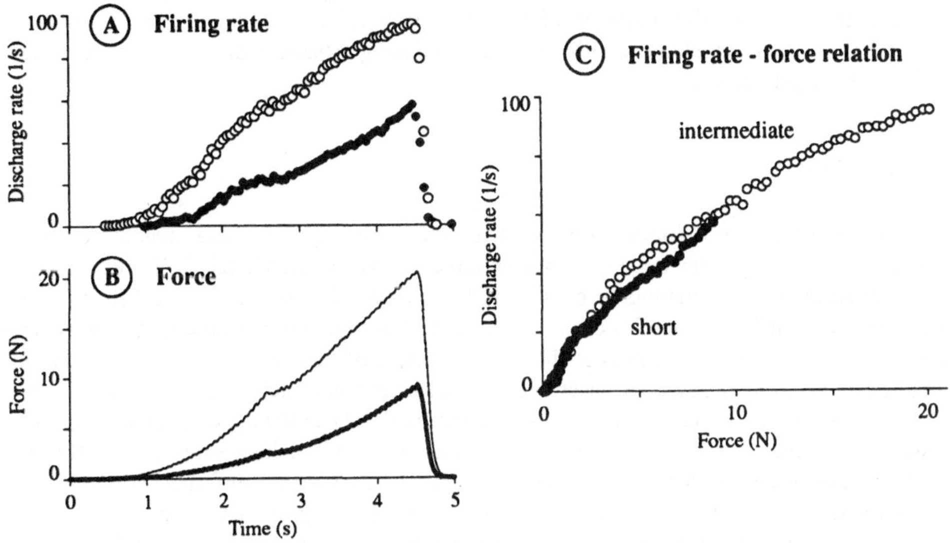

Figure 2. GTO ensemble responses at short and intermediate length. Average estimates of GTO subpopulation response and associated muscle force transient, based on the individual measurements of Figure 1. A, probability density estimates of ensemble response at short (filled circles) and intermediate (open circles) length. B, averaged force profiles at short length (thick line) and intermediate length (thin line). C, ensemble firing rate-force relations for the sub-population of GTO afferents studied at both lengths.

and rate modulation. Yet since force transients also varied appreciably (mainly between experiments; see Figure 1, A.2, B.2), the observation of GTO response variability did not *a priori* rule out consistency of firing rate *versus* force relations for individual afferents at different lengths.

However, firing rate-force relationships computed for individual GTO afferents still revealed pronounced variability. This is illustrated in Figure 1.4, which emphasizes the range of individual firing-force characteristics and the apparently only minor effects of muscle length regarding the general shape of the relationship. The main effects of increasing muscle length were, unsurprisingly, that muscle force increased appreciably (for invariant activation patterns) and that individual GTO afferents tended to respond at moderately higher firing rates (cf. Figure 1B with 1A). At both lengths gain compression was manifest (Figures 1.4 & 2C).

Upon averaging of individual measurements, ensemble responses revealed distinct trends: the smoothed GTO population responses (probability density averages) broadly followed the averaged force profiles (Figure 2A & B). The firing rate-force relations were monotonic, while still revealing gain compression, as was the case for a larger sample of afferents studied with different activation strategies (see Introduction). Figure 2C emphasizes that in spite of the pronounced effect of length on force (more than 50% reduction at short length) the GTO ensemble's input-output characteristics were only minimally affected.

The population response estimates of Figure 2 were based on a relatively small sample of 12 afferents studied at both lengths. A substantially larger number of GTO afferents was studied at the longer length, permitting comparison of ensemble response estimates based on samples of widely ranging size (3 to 41). It emerged, first, that sample size had remarkably little influence on the shape of the firing-force characteristics of a given activation strategy; secondly, that samples based on as few as 6 afferents consistently yielded ensemble characteristics very similar to those of larger samples and thirdly, that in particular the present sub-sample of 12 afferents provided an estimate of the firing rate-force relation at intermediate length, that was indistinguishable from that obtained with significantly larger samples.

DISCUSSION

The present observations confirm that the difficulty of force feedback signal interpretation arising from the non-representative nature of individual GTO responses is readily overcome by estimating force signals from ensemble responses of GTO afferents. Even relatively small sub-samples of GTOs yielded smooth and monotonic firing rate-whole muscle force relationships. However the characteristics of this relation are not entirely invariant. In a separate study three strategies of motor unit pool activation were explored in experimental simulations (Hulliger et al., in preparation). These strategies spanned the range between exclusive recruitment (at fixed activation rates) and exclusive rate modulation (of simultaneously recruited MUs). Over this wide range, appreciable differences in ensemble sensitivity were found for small (up to 10% of the physiological maximum), but not for large forces (above 50%). With exclusive recruitment small-force sensitivity was highest, for pure rate modulation it was lowest, and for the combined strategy it was intermediate. Yet activation of the motor pool by pure recruitment or pure rate modulation probably is the exception rather than the rule. For the more common combinations of recruitment and rate modulation the effects of activation strategy variations will probably be smaller.

If the extremes of MU pool activation strategies were to be employed by the CNS, the above differences may matter, in that the comparison of the force sensitivity of GTO ensembles for different activation strategies revealed a tradeoff between small-force sensitivity and linearity, which may be significant functionally: wide-range nearly linear monitoring of force appears best achieved with a pure rate modulation strategy. For high-sensitivity monitoring of small forces, a pure recruitment strategy seems best suited. For optimal tradeoff, the compromise of recruitment combined with rate modulation appears to be the strategy of choice.

Comparing the data for recruitment and rate modulation (obtained for filaments recruited according to size; Hulliger et al., in preparation), with the earlier data (Hulliger et al., 1992) on the same paradigm (but with non-size ordered filaments of uniform size), it emerged that the size of the filaments, or MU groups, was of only marginal importance, since the firing rate-force relations were nearly identical. Likewise, in spite of significant reduction of active muscle force at very short length, the firing-force relation was practically unaltered, further emphasizing the relatively general validity of this smooth, albeit non-linear relationship (Figure 2C).

The dependence of static force sensitivity on muscle length has previously been investigated for individual rather than GTO ensemble responses. The present data are in line with the observations of Crago et al. (1982), who for reflex-induced whole muscle activation (also soleus, at similar lengths) reported an average sensitivity of around 7/s/N. This is comparable with the small-force sensitivity of around 11/s/N and the large-force sensitivity of around 4.5/s/N of the ensemble relations of Figure 2C. Crago et al. also reported that static force sensitivity decreased marginally with increasing length. This was not seen in the present material (Figures 1.4 and 2C) and might be attributable to variations with length of reflex-induced MU pool activation strategies, although other factors may contribute. Yet this could not account for the observation of Stephens, Reinking & Stuart (1975) of a similar (30%) length-dependent reduction of static force sensitivity of gastrocnemius GTO afferents during activation of single MUs at a fixed rate. Although their observations on a mixed muscle studied at much longer length are not strictly comparable, it merits mention that one-point estimates of sensitivity for a given activation rate automatically reveal apparent sensitivity reduction, if input-output characteristics show saturation. On the evidence of Figure 2C, for the same maximum ensemble stimulation rates, the large-force sensitivity at short length was 6/s/N, while at long length it was reduced (by 25%) to 4.5/s/N. In summary, the present conclusion that the ensemble firing-force relation is not significantly influenced by muscle length does not entirely agree with previous interpretations of earlier single-unit data, although the individual properties of the present sample of GTOs appear identical with those described before.

The discrepancy between the present observations and those of Horcholle-Bossavit et al. (1990) is more intriguing and may require further experimental work to identify the relative importance of several possible factors. Probably most important is that in their study only a few selected MUs were activated at relatively high rates. In contrast, in the present investigation the full physiological range of force was explored and, in simulating MU pool activation strategies, physiological activation rates were studied, including recruitment of MUs at low rates (5/s). Further, the present observations only apply to a uniformly slow muscle (soleus), leaving it open whether similar simulations in a mixed muscle (as in Horcholle-Bossavit et al., 1990) would yield comparable results. Moreover, the principle estimates of ensemble characteristics of the present study were based on averages of data from different experiments, while Horcholle-Bossavit et al. (1990) relied on more homogenous samples of GTOs from individual muscles. However, this is of minor importance, since in the soleus material estimates from individual muscles revealed the same general features as the larger ensemble estimates across experiments.

A limitation of the present methodology is that for practical reasons independent activation of MUs was restricted to 8 channels and that, in order to explore the full physiological range of force, multi-MU filaments instead of individual MUs were stimulated. On limited evidence it seems unlikely that substantially different results would be obtained with multi-channel single-MU activation: the main concern is that some multi-MU filaments might have contained two or more MUs acting on a given GTO, and that these MUs were activated in non-physiological synchrony. Since soleus GTOs are operated by about 10 MUs (Houk & Henneman, 1967; Jami, 1992) this was statistically quite unlikely for the smaller, but a distinct possibility for the larger multi-MU filaments. However, GTO responses to activation of additional MUs typically reveal pronounced saturation (see Jami, 1992; review). Therefore it hardly matters whether activation of additional MUs is synchronised or not, since the incremental effects would be small in any case. This was supported by the observation that biasing effects were uniform over a tenfold range of tetanic forces of multi-MU filaments and of the same size (50/s) as reported for single MU activation (Gregory & Proske, 1979; Binder & Osborn, 1985).

The present estimates of ensemble discharge were based on simple calculation of probability density of discharge. While this seemed a reasonable estimate of likely postsynaptic events in interneurones receiving converging GTO input, there is at present no way of deciding whether an equivalent "algorithm" is used by the CNS circuits which decode force feedback signals. The signal processing in putative distributed neuronal networks, that may subserve this function, may or may not be determined by the event density of incoming sensory pulse trains. Only when theoretical work on signal processing in neuronal networks comes forward with more explicit concepts of decoding algorithms will it be possible to use more refined alternative measures to characterise ensemble discharge in populations of sensory afferents.

In summary, in contrast to individual afferents, GTO ensembles appear to be capable of monitoring whole muscle force faithfully over a wide range of force, regardless of whether force gradation arises from modulation of motor drive or from alteration in muscle length.

However, the firing rate-force relation, although generally monotonic, is not entirely invariant, being subject to subtle modification by different motor pool activation strategies, when these span a wide range of patterns. Force control by MU rate modulation extends the range of linearity of force monitoring at the expense of sensitivity, while recruitment augments small-force sensitivity at the expense of linearity.

ACKNOWLEDGEMENTS

Supported by an AHFMR fellowship award (PS) and a Canadian MRC grant (MH).

REFERENCES

BINDER, M.D. & OSBORN, C.E. (1985) Interactions between motor units and Golgi tendon organs in the tibialis posterior muscle of the cat. *J. Physiol.* **364**, 199-215.

CRAGO, P.E., HOUK, J.C. & RYMER, W.Z. (1982) Sampling of total muscle force by tendon organs. *J. Neurophysiol.* **47**, 1069-1083.

GREGORY, J.E. & PROSKE, U. (1979) The responses of Golgi tendon organs to stimulation of different combinations of motor units. *J. Physiol.* **295**, 251-262.

HORCHOLLE-BOSSAVIT, G., JAMI, L., PETIT, J., VEJSADA, R. & ZYTNICKI, D. (1990) Ensemble discharge from Golgi tendon organs of the cat peroneus tertius muscle. *J. Neurophysiol.* **64**, 813-821.

HOUK, J. & HENNEMAN E. (1967) Responses of Golgi tendon organs to active contractions of the soleus muscle of the cat. *J. Neurophysiol.* **30**, 466-481.

HULLIGER, M., WEYTJENS, J.L.F. & WINDHORST, U.R. (1992) Responses of cat soleus Golgi tendon organs to distributed ventral root stimulation at time-varying rates. *Soc. Neurosci. Abs.* **18**, 1407.

JAMI, L. (1992) Golgi tendon organs in mammalian skeletal muscle: functional properties and central actions. *Physiol. Rev.* **72**, 623-666.

STEPHENS, J.A., REINKING, R.M. & STUART, D.G. (1975) Tendon organs of cat medial gastrocnemius: responses to active and passive forces as function of muscle length. *J. Neurophysiol.* **38**, 1217-1231.

THE USE OF COHERENCE SPECTRA TO DETERMINE COMMON SYNAPTIC INPUTS TO MOTONEURONE POOLS OF THE CAT DURING FICTIVE LOCOMOTION

Thomas M. Hamm and Martha L. McCurdy

Division of Neurobiology
Barrow Neurological Institute
St. Joseph's Hospital and Medical Center
Phoenix, AZ, USA

INTRODUCTION

Our understanding of the organisation of the central commands directed to motoneurone pools during motor activity in mammals has been limited by the difficulty of determining the identity of segmental neurones that carry those commands to motoneurones. In rhythmic activities like locomotion, those commands are transmitted by interneuronal elements of spinal central pattern generators (CPG: Grillner, 1981; Gelfand, Orlovsky & Shik., 1988), whose identity and organisation remain to be determined. The present work was undertaken to investigate the organisation of the CPG for locomotion by determining which motoneurone pools receive a common synaptic input during locomotion, as indicated by patterns of synchronisation. Previous work suggests that synaptic inputs that are common to motoneurones are capable of producing a synchronised discharge of those motoneurones (Sears & Stagg, 1976; Kirkwood & Sears, 1978). The use of correlation methods in the time domain has demonstrated synchronisation between motoneurones in various forms of motor activity (Kirkwood, Sears, Tuck & Westgaard, 1982; Datta & Stephens, 1990).

Synchronisation between the discharge of neurones can be alternatively examined in the frequency domain by the use of coherence spectra (Rosenberg, Amjad, Breeze, Brillinger & Halliday, 1989). Coherence spectra provide a normalised measure of correlation that reflects synchronisation, the frequency composition of synchronisation, and easily computed confidence limits. Several studies have demonstrated its utility in detecting synchronisation (Bruce & Ackerson, 1986; Christakos, Cohen, Barnhardt & Shaw, 1991; Farmer, Bremner, Halliday, Rosenberg & Stephens, 1993). Comparison of correlation methods and coherence spectra show that the amplitudes of peaks in the coherence spectra are correlated with those in cross correlation functions (Farmer et al., 1993). In addition, comparison of power spectra and coherence spectra between medullary inspiratory neurones, phrenic motoneurones, and phrenic neurograms (Christakos, Cohen, See & Barnhardt, 1989; Christakos et al., 1991) demonstrates that high-frequency oscillations of medullary neurones produce correlated discharges in the phrenic motoneurone pool and neurogram, supporting the suggestion that significant coherence peaks in a particular frequency range indicates the frequency of discharge of the common presynaptic inputs to the synchronised motoneurones (Farmer et al., 1993).

METHODS

To determine which motoneurone pools received common synaptic inputs from central pattern generators, coherence spectra were computed for pairs of rectified neurograms collected during episodes of fictive locomotion. Fictive locomotion was produced in cats which had been decerebrated at the precollicular - postmammillary level while anesthetised with isoflurane anesthesia. While the cats were anesthetised, a laminectomy (L6-L7) and a hindlimb dissection was performed. The muscle nerves innervating the anterior and middle portions of biceps femoris (ABMB), posterior biceps femoris and semitendinosus (PBSt), medial gastrocnemius (MG), lateral gastrocnemius and soleus (LGS) and tibialis anterior (TA) were exposed and fitted with bipolar cuff electrodes for recording neurograms. The dura was opened and the L7 and S1 ventral roots were gently separated and placed on hook electrodes so that intramyelin recordings could be made from individual motor axons with glass microelectrodes (Hamm, Sasaki, Stuart, Windhorst & Yuan, 1987). Following the completion of surgery, anesthesia was removed, and the cat was immobilised by the administration of gallamine and was placed on a mechanical ventilator. A tungsten stimulating electrode was placed in the mesencephalic locomotion region (MLR) by using landmarks on the surface of the colliculi and then making several electrode tracks to find a site at which locomotion patterns could be produced by low-amplitude currents (< 140 μA, 20-25 Hz). Fictive locomotion was produced while recordings were made of the neurograms and the spike trains of one or two individual motor axons.

Data were collected on FM analog tape and analyzed off line. The neurogram signals were high-pass filtered (low frequency cutoff of 120 Hz), rectified, and passed through a sampled integrator, which was reset at the sampling rate for each neurogram channel (5kHz). Spike trains were passed through window discriminators and formed into trains of standard pulses having durations of one sampling period (0.2 msec) before being digitised with the neurograms. Bursts of activity in the rectified neurograms were selected using an interactive computer program. These selected bursts were passed through a modified cosine window, and the means were subtracted from each burst before power spectra and cross power spectra were computed by fast Fourier transform. Spectra for the standardised spike trains were computed using these same algorithms. Spectra from several episodes of locomotion were averaged before the coherence and partial coherence functions were computed, so that the coherence functions were based on numbers of data segments ranging from approximately 50 to 500.

A sample of fictive locomotion is illustrated in Figure 1A, which shows neurograms of ABMB, PBSt, MG and LGS, and recordings from two motor axons. In this presentation we will concentrate on the activity of these motor pools, all of which innervate extensor muscles, with the exception of PBSt, and which are coactive during locomotion (Engberg & Lundberg, 1969). In addition, the bifunctional muscles PB and St are usually classified as flexors, but are often coactive with extensors during fictive locomotion (Figure 1A; see Perret & Cabelguen, 1980), and display stance-phase activity at some speeds of locomotion (Engberg & Lundberg, 1969; English & Weeks, 1987). Consequently, the set of neurogram recordings that was analysed consisted of motor pools that are often coactive during fictive locomotion, and include motor pools acting at the same joint (MG and LGS), extensor motor pools acting at different joints (ABMB vs. MG and LGS), and a bifunctional motor pool (PBSt).

RESULTS

Characteristics of the coherence spectra that were often found in this study are illustrated by one example of coherence between ABMB and MG (Figure 1B). Significant coherence peaks were often found at low frequencies (5-10 Hz). These are below the rates of motoneurone discharge that we observed in these experiments, and most likely represent the low-frequency modulation of motoneurone discharge produced by the envelope of excitation during the burst. MLR stimulation usually produced a peak at 20-25 Hz and at harmonics of this frequency, as could be anticipated from the large EPSPs produced in motoneurones by this source (Shefchyk & Jordan, 1985). In some cases, this medium-frequency peak was broader than could be accounted for the frequency of MLR stimulation. In these cases, synchronisation of motoneurones at their frequency of discharge may contribute to the coherence peak. Inspection of the power spectra of individual motor axons showed that they often discharged in this mid-frequency range (15-40 Hz) although their rates of discharge could be somewhat higher than this range. In five of seven experiments, significant coherence was also seen at higher frequencies (45-170 Hz), as displayed prominently in Figure 1B. These coherence peaks occurred at frequencies that in most cases

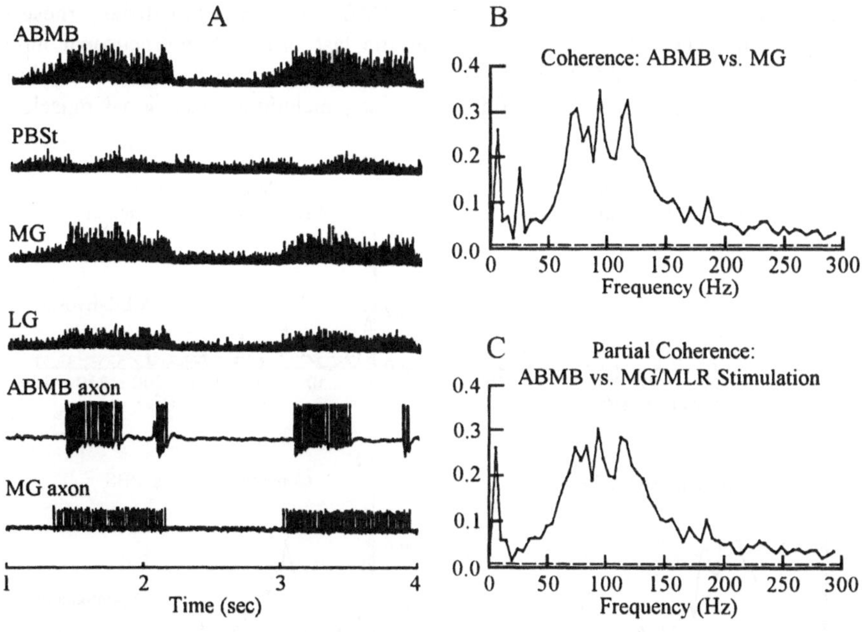

Figure 1. A. Rectified neurograms recorded from the muscle nerves to ABMB, PBSt, MG and LGS. Below these traces are two intramyelin recordings from ventral-root motor axons to ABMB and MG. Fictive locomotion was produced by stimulation of the MLR. B. Coherence spectrum calculated for the ABMB and MG rectified neurograms. A coherence value of one indicates that the neurograms are perfectly correlated at that frequency, while a value of 0 indicates that no correlation exits. The dashed line across the bottom of the figure indicates the value of the 99% confidence limits, above which there is only a 1% chance that the coherence value occurs by chance. C. Partial coherence function for the same neurogram pair, in which the contribution of the MLR stimulation near 25 Hz has been factored out.

were considerably higher than the rates of motoneurone discharge. Although coherence at these higher frequencies will be contributed by harmonics of the frequencies represented in the mid-frequency range, this expected contribution was insufficient to account for most of the high-frequency coherence. Based on these findings and the results and suggestions of the studies cited previously, we attribute these high-frequency components to the rates of discharge of common input to the motoneurone pools from interneurones associated with the spinal CPG.

Since the MLR stimulus is itself a synchronising influence, we wished to reduce the contribution of the MLR stimulus to the coherence functions as much as possible so that an estimate could be made of the contribution of common input from the CPG to the observed synchronisation. One advantage of the use of coherence functions is the ability to factor out the effect of a known synchronising influence that contributes to the coherence function between two signals. This can be done by the computation of a partial coherence function (Rosenberg et al., 1989), an example of which is shown in Figure 1C for the ABMB and MG neurograms. The fundamental MLR peak and its harmonics have been removed from the coherence function with little diminution in the magnitude of the high-frequency coherence peak. Except for one case of spontaneous locomotion (see below), all results described below will be based on partial coherence functions.

In the seven experiments we have analyzed, significant coherence peaks were observed between all four muscle-nerve neurograms illustrated in Figure 1A. In four of the five experiments with significant high-frequency coherence, the strongest coherence was found between ABMB and MG. Surprisingly, this was stronger than that found between the same-joint synergists MG and LG. In general, the weakest coherence in the high-frequency range was found between LG and PBSt and between ABMB and PBSt. Nonetheless, these were usually significant. These findings indicate that the last order of interneuronal input to motoneurones from the spinal CPG during fictive locomotion is distributed to several motor pools innervating muscles that act at different joints, including bifunctional muscles like

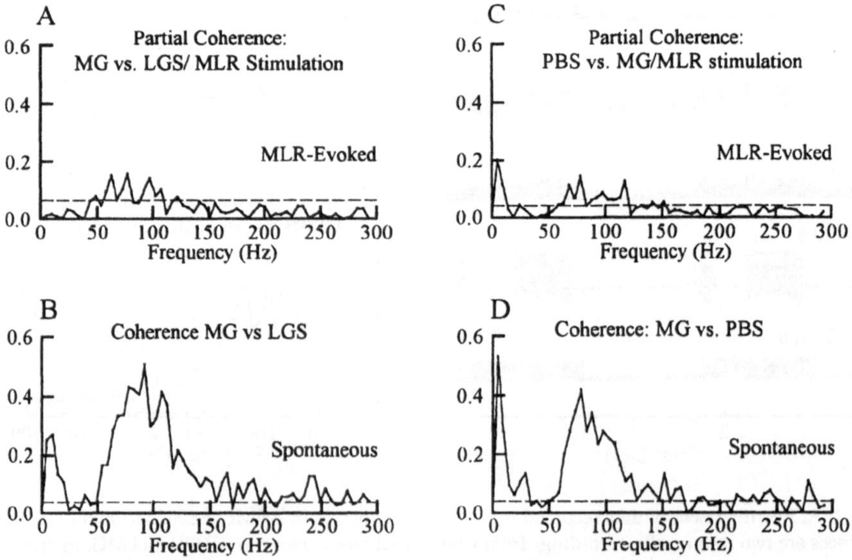

Figure 2. This figure shows partial coherence functions for two neurogram pairs recorded during MLR-evoked fictive locomotion (A & C). During episodes of spontaneous fictive locomotion that occurred in the same experiment, the coherence functions between these neurogram pairs increased dramatically over most of the frequency range (B & D).

PBSt. Furthermore, this finding implies that the spinal CPG is not organised for the control of motor nuclei that innervate muscles at individual joints, but rather is organised for the control of motor nuclei that innervate muscles at different joints and share periods of activity. Thus our findings are more compatible with CPG models with a half-centre organisation (Brown, 1911) than distributed models based on individual joints (Grillner, 1981).

Additional evidence that the high-frequency coherence reflects the contribution of interneuronal inputs to the motoneurone pools is revealed by an experiment in which significant coherence was observed during spontaneous fictive locomotion. Figures 2A & C show partial coherence functions computed for MG and LGS and for PBSt and MG from trials during MLR-evoked locomotion. These figures show coherence in the high-frequency range that is significant, but weak. Coherence functions computed for these neurograms during episodes of spontaneous locomotion in this experiment are shown in Figures 2B & D. Both records show striking increases in both low- and high-frequency coherence during spontaneous locomotion. Coherence between ABMB and MG also increased during the spontaneous locomotion. However, this increase was less than that observed between MG and LG. Moreover, the high-frequency peak for the MG-LG coherence function was broader than that of the ABMB-MG function during spontaneous fictive locomotion. By comparison to most instances of MLR-evoked locomotion, in which the ABMB-MG combination displayed the most prominent high-frequency coherence, this finding suggests that the organisation of the CPG may not be fixed, and may vary to some degree between episodes of locomotion initiated by different mechanisms.

In addition to coherence functions based on pairs of neurograms, we have also examined coherence functions derived from individual motor-axon spike trains paired with neurograms. An example is shown is Figure 3, which shows significant coherence between the activity of the MG neurogram and an ABMB motor axon in low-, mid- and high-frequency ranges. This coherence is weak in comparison to the neurogram-to-neurogram coherence observed in the same experiment (Figure 1C). In most instances, coherence functions between motor axons and neurograms were not significant, indicating that the neurogram-to-neurogram coherence function is a more sensitive measure of coherence in a population of active motor units than motor axon-to-neurogram function. Although coherence functions based on aggregate signals of neural activity have certain limitations in the interpretation of their amplitudes (Christakos, 1994), our results indicate that they can be a sensitive measure of synchronisation within a population of neurones.

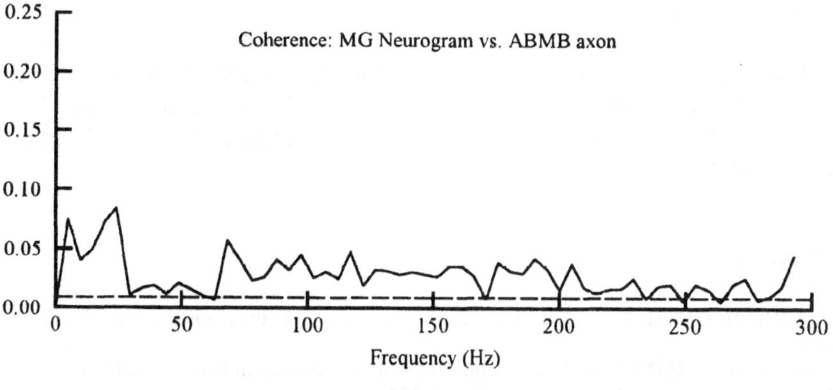

Figure 3. An example of a coherence spectrum calculated between the MG neurogram and an ABMB axon is shown in this figure. This coherence function exceeds the 99% confidence limits over much of the frequency range, although the coherence is weaker than for neurogram pairs (cf. Figure 1).

CONCLUSIONS

Several points should be stressed in summarising our results. We have found coherence functions computed for pairs of rectified neurograms to be sensitive measures of motoneurone synchronisation during fictive locomotion. Significant peaks in the coherence functions were found in low-, mid- and high-frequency ranges, of which we consider the last to be the most likely indicator of common synaptic input from the spinal CPG. Based upon the pairs of neurograms in which this high-frequency coherence is found, we surmise that motoneurones that innervate muscles at different joints, but are typically coactive during the stance phase of locomotion, receive common input from the CPG. This observation favors organisations for the CPG like those based on the half-centre hypothesis. In this context, our finding that PBSt exhibits activity that is coherent with that of hip and ankle extensors is consistent with the proposal of Perret & Cabelguen (1980) that St may receive commands from either an extensor or flexor module component of the CPG, dependent on segmental feedback. However, to distinguish between a half-centre model and some form of distributed model of the CPG, it will be necessary to determine the coherence functions between knee and hip or ankle extensors, other bifunctional muscles and flexors, and between muscles with idiosyncratic patterns of activity (e.g. O'Donovan, Pinter, Dum & Burke, 1982).

Even within the broad organisation indicated by the results of this study, the strength of common input to different motor pools may not be determined by rules of organisation that seem most intuitive. We found that the high-frequency coherence between ABMB and MG was usually stronger than that between the same-joint synergists MG and LG. Perhaps this is another indication that the organisation of segmental input to motoneurone pools is not necessarily organised according to mechanical action at a particular joint (cf. Labella, Kehler & McCrea, 1989). Finally, our anecdotal observation of changes in the coherence function during spontaneous fictive locomotion suggest that the patterns of input from the CPG to motoneurones may be flexible to some degree, depending upon the mechanism for initiating locomotion. Thus it may be instructive to determine patterns of coherence between motor pools under different conditions of locomotion. The techniques of analysis employed in this study can be applied to the investigation of EMG patterns obtained during treadmill or over-ground locomotion, so that the analysis of input to motor pools in a broader range of motor behaviors can readily be accomplished.

ACKNOWLEDGEMENTS

This work was supported by USPHS grants to T.M. Hamm (NS22454) and J.R. Bloedel (NS30013), a National Research Service Award to M.L. McCurdy (NS08773), and a training grant to the University of Arizona-Barrow Neurological Institute training program in motor control (NS07309).

REFERENCES

BROWN, T.G. (1911). The intrinsic factors in the act of progression in the mammal. *Proc. R. Soc. B.* **84**, 308-319.

BRUCE, E.N. & ACKERSON, L.M. (1986). High-frequency oscillations in human electromyograms during voluntary contractions. *J. Neurophysiol.* **56**, 542-553.

CHRISTAKOS, C.N. (1994). Analysis of synchrony (correlations) in neural populations by means of unit-to-aggregate coherence computations. *Neuroscience* **58**, 43-57.

CHRISTAKOS, C.N., COHEN, M.I., BARNHARDT, R. & SHAW, C.-F. (1991). Fast rhythms in phrenic motoneuron and nerve discharges. *J. Neurophysiol.* **66**, 674-687.

CHRISTAKOS, C.N., COHEN, M.I., SEE, W.R. & BARNHARDT, R. (1989). Changes in frequency content of inspiratory neuron and nerve activities in the course of inspiration. *Brain Res.* **482**, 376-380.

DATTA, A.K. & STEPHENS, J.A. (1990). Synchronization of motor unit activity during voluntary contraction in man. *J. Physiol.* **422**, 397-419.

ENGBERG, I. & LUNDBERG, A. (1969). An electromyographic analysis of muscular activity in the hindlimb of the cat during unrestrained locomotion. *Acta physiol. scand.* **75**, 614-630.

ENGLISH, A.W.M. & WEEKS, O.I. (1987). An anatomical and functional analysis of cat biceps femoris and semitendinosus muscles. *J. Morphol.* **191**, 161-175.

FARMER, S.F., BREMNER, F.D., HALLIDAY, D.M., ROSENBERG, J.R. & STEPHENS, J.A. (1993). The frequency content of common synaptic inputs to motoneurones studied during voluntary isometric contraction in man. *J. Physiol.* **470**, 127-155.

GELFAND, I.M., ORLOVSKY, G.N. & SHIK, M.L. (1988). Locomotion and scratching in tetrapods. In *Neural Control of Rhythmic Movements in Vertebrates*, eds. COHEN, A.H., ROSSIGNOL, S. & GRILLNER, S. pp. 167-199. John Wiley & Sons, New York.

GRILLNER, S. (1981). Control of locomotion in bipeds, tetrapods, and fish. In *Handbook of Physiology, The Nervous System: Motor Control, Sec.1, Vol.II, Pt.2,* ed. BROOKS, V.B. pp. 1179-1236. American Physiological Society, Bethesda.

HAMM, T.M., SASAKI, S.I., STUART, D.G., WINDHORST, U. & YUAN, C.S. (1987). The measurement of single motor-axon recurrent inhibitory post-synaptic potentials in the cat. *J. Physiol.* **388**, 631-652.

KIRKWOOD, P.A. & SEARS, T.A. (1978). The synaptic connexions to intercostal motoneurones as revealed by the average common excitation potential. *J. Physiol.* **275**, 103-134.

KIRKWOOD, P.A., SEARS, T.A., TUCK, D.L. & WESTGAARD, R.H. (1982). Variations in the time course of the synchronization of intercostal motoneurones in the cat. *J. Physiol.* **327**, 105-135.

LABELLA, L.A., KEHLER, J.P. & McCREA, D.A. (1989). A differential synaptic input to the motor nuclei of triceps surae from the caudal and lateral cutaneous sural nerves. *J. Neurophysiol.* **61**, 291-301.

O'DONOVAN, M.J., PINTER, M.J., DUM, R.P. & BURKE, R.E. (1982). Actions of FDL and FHL muscles in intact cats: functional dissociation between anatomical synergists. *J. Neurophysiol.* **47**, 1126-1143.

PERRET, C. & CABELGUEN, J.M. (1980). Main characteristics of the hindlimb locomotor cycle in the decorticate cat with special reference to bifunctional muscles. *Brain Res.* **187**, 333-352.

ROSENBERG, J.R., AMJAD, A.M., BREEZE, P., BRILLINGER, D.R. & HALLIDAY, D.M. (1989). The fourier approach to the identification of functional coupling between neuronal spike trains. *Prog. Biophys. & Molecul. Biol.* **53**, 1-31.

SEARS, T.A. & STAGG, D. (1976). Short-term synchronization of intercostal motoneurone activity. *J. Physiol.* **263**, 357-381.

SHEFCHYK, S.J. & JORDAN, L.M. (1985). Excitatory and inhibitory postsynaptic potentials in alpha-motoneurons produced during fictive locomotion by stimulation of the mesencephalic locomotor region. *J. Neurophysiol.* **53**, 1345-1355.

CRITIQUE ON THE PAPERS OF THE ANALYSIS AND MODELLING SECTION

Friedemann Awiszus

Klinik für Orthopädie
Otto-von-Guericke-Universität
Magdeburg
Germany

The four papers of this section deal with rather different subjects. Nonetheless, I will try to emphasize certain common points in the approaches.

The paper by Otten, Scheepstra & Hulliger deals with the internal mechanisms of a muscle spindle. The approach taken is that of "pure modelling", i.e. experimental work of others is used to construct a mathematical set of equations that behaves, in some way, similar to the real structure. The main task in constructing a muscle spindle model is to choose parameters for all the unknown mechanical and nerve membrane properties. Additionally, a choice for the relative contribution of each of the submodels has to be made.

Otten et al. made choices for all the parameters and gave reasonings for all of their choices which is an admirable effort. Both muscle mechanics and spike generation, represented by a Hodgkin-Huxley-type model, are taken into account along with some simple assumptions about pacemaker coupling along the afferent branching tree. One has to keep in mind, however, that direct experimental evidence of channel composition and kinetics at the pacemaker membrane is still lacking and that the firing characteristics of the submodel employed may differ considerably from that of the true encoder. Nevertheless, the model proposed may be viewed as a reasonable starting point for spindle modelling. Personally, I believe that although it is rather simplistic, this model has its greatest potential in allowing "inverse" modelling questions, i.e. to infer quantitatively spindle input from observed spindle output.

The other three papers of this section concentrate more or less on specific procedures for analysis of complex physiological data.

The papers of Hulliger, Sjölander, Windhorst & Otten and of Johansson, Bergenheim, Djupsjöbacka & Sjölander address the question as to what information can be obtained from an ensemble of afferents rather than a single afferent of a specific kind. These experimenters investigated an ensemble of peripheral receptors while controlling the input. In contrast to a direct modelling approach which attempts to simulate the receptor ensemble behaviour given the experimental input, these authors tried to obtain information about the input from the given ensemble output. Thus, this approach may in fact be regarded as "inverse modelling" of a receptor ensemble. An inverse model of a receptor ensemble represents a model for a decoder of afferent firing which in turn is an important part of the central nervous system. Little is known, however, about the afferent-decoding-neural network in the CNS and the approaches taken in the two papers may represent important first steps to obtain such information. Both papers find that the decoding process for a receptor ensemble is superior to that for a single afferent. Intuitively one would expect such a result and it is nice to see such an expectation borne out by experimental evidence.

The paper of Hulliger et al. deals with the ensemble of Golgi tendon organs in a particular muscle. The approach taken to decode the information in the afferent firng of such an ensemble is straightforward. First of all a feature extraction is performed on the input in so far as only the whole-muscle-force signal is investigated. Thereafter a feature extraction was performed on the afferent ouput (determination of firing density) and from this reduced output signal a reconstruction of the reduced input signal was attempted.

It was found that such a reconstruction was in fact possible. This does not mean, however, that the true "decoder" in the central nervous system behaves in a similar manner. Especially, there may be more information in the ensemble afferent firing of Golgi tendon organs than a simple coding of whole muscle force, and to what extent the real afferent decoder in the CNS is interested only in muscle force evaluation from the Golgi tendon organs has to remain an open question.

The approach taken by Johansson et al. is a more elegant one, as these authors do not restrict the interest of the proposed decoder to a single feature of the input given to the receptor ensemble. Principal component analysis of the ensemble afferent firing may be regarded as representing a decoder which is interested in identifying as many different stimuli given to the receptor ensemble as possible. Personally, I would expect that the real afferent decoder in the CNS attempts to differentiate between stimuli given to the receptor ensemble in a way similar to principal component analysis. The main disadvantage of this approach is that it is rather difficult to understand and thus to sell to the scientific comunity, as the principal components obtained represent features which have no straightforward physiological meaning on their own.

Both papers dealing with decoding of receptor ensemble firing restricted themselves to muscle receptors within the ensemble. This is not surprising as the experimental setup for such restricted ensembles is already extremly complicated. One should bear in mind, however, that the true afferent decoder within the CNS uses information obtained from non-homogenous receptor ensembles consisting not only of muscle spindle afferents and Golgi tendon organs but also joint afferents, skin afferents, and other muscle afferents. To what extent a full ensemble would be superior to the more homogenous ensembles studied in the two papers discussed cannot be answered with the present knowledge.

Finally, the paper by Hamm and McCurdy deals with an analysis task which is even more complicated. It is assumed that a central pattern generator provides input to several motoneurone pools involved in fictive locomotion. The authors measured the output of these motoneurone pools and tried to reconstruct the pool input (representing the unknown pattern generator output) from these data. Thus, the analysis presented may be also regarded as an inverse modelling of the motoneurone pool behaviour. The approach taken is to use the coherence function, which is the frequency-domain analogue of cross-correlation analysis in the time domain, which has been shown to represent a reliable approach to the inverse problem in motoneurone physiology. The results indicate that the motoneurone pools are activated in rather complex ensembles by the pattern generator which are not necessarily related to the function of the innervated muscles. Consequently, the organisation of the motor system even in such stereotyped movements as fictive locomotion is far more complex than appears necessary at first sight. One should bear in mind, however, that the motor system has to provide a huge variety of behaviours and thus this complicated innervation pattern might be necessary for other movement patterns.

In conclusion, the papers presented may be regarded as examples for pure and applied modelling which give some insight into the physiological basis of the processes studied. Especially, the "inverse modelling" approaches provide tools for the analysis of extremely complex physiological data which I believe will be important for future experiments in neurophysiology.

NON-LINEAR SUMMATION IN GTO RESPONSES: IMPLICATIONS FOR RECEPTOR MECHANISMS

P.I. Sjölander[1,2], M. Hulliger[1],
U.R. Windhorst[1] and E. Otten[3]

[1]Department of Clinical Neurosciences
University of Calgary, Canada T2N 4N1
[2]Division of Work Physiology, NIOH
Umeå, Sweden
[3]Department of Medical Physiology
University of Groningen, The Netherlands

INTRODUCTION

Golgi tendon organ (GTO) responses reveal pronounced non-linearities in that excitation by combined action of several motor units (MU) is less than the sum of the excitatory effects elicited by the individual MUs (see Jami, 1992; review). Such sub-linear summation is not caused by simple receptor saturation since the degree of non-linearity is independent of the level of firing rate (Gregory & Proske, 1979). It has been suggested that unloading effects elicited by MUs located in parallel to a GTO might account for the non-linear summation (Fukami, 1981) and, implicitly, for the gain compression in the firing rate-force relation. Another possibility is that the non-linearity results from interactions between multiple transducers and/or encoders within the receptor (Gregory, Morgan & Proske, 1985). The aim of the present study was to clarify the role of different receptor mechanisms in non-linear summation.

METHODS

Single soleus GTO afferents were isolated in pentobarbitone-anaesthetised cats. The L7 and S1 ventral roots were cut and divided into about 20 filaments. Those containing soleus α axons were regrouped into 8 multi-MU filaments (MMUFs; 3-30 MUs), which together generated up to 95% of maximum tetanic force (MTF). GTO responses were recorded during separate or combined stimulation of pairs or sub-sets of these MMUFs at constant muscle length (corresponding to 80° ankle flexion). Ramp (5 to 12 or 25/s) and rectangular (12, 17 or 30/s) stimulation patterns were generated by reading computer edited pulse trains. Whole muscle force was recorded isometrically at the tendon.

RESULTS

For each GTO afferent MMUFs were readily classified as excitatory (biasing action > 10/s) or non-excitatory (zero bias). Comparing GTO responses to ramp activation of all MMUFs with GTO responses to ramp activation of excitatory MMUFs alone, revealed no unloading effects from the non-excitatory MMUFs for forces from 0 to 95% of MTF. Instead, for all but 1 of the 9 GTOs studied, combined activation revealed loading effects. Some of these were large: in spite of generating potential unloading forces corresponding to 23-60% of MTF, non-excitatory MMUFs increased discharge rate by up to 20/s (mean = 8/s).

Comparison of discharge rate-force relations of ensemble GTO responses elicited by combined activation and by stimulation of excitatory MMUFs alone showed that activation of the non-excitatory MMUFs did not alter the overall monotonic relation, and reduced rather than augmented gain compression (minimum/maximum force sensitivity).

In order to evaluate possible encoder and transducer mechanisms, pairs of excitatory MMUFs were activated alone and together, using ramp stimulation for one MMUF and rectangular stimulation for the other. For each pair, the increment of firing upon combined activation and the difference between the discharge rates elicited by individual MMUFs were used to calculate a coefficient of interaction (C_i; see Banks, Hulliger, Scheepstra & Otten, this volume) and a normalised firing rate difference (ND; difference in firing rate/higher of the two individual firing rates). By definition C_i values range between 0.0 (complete occlusion) and 1.0 (linear summation).

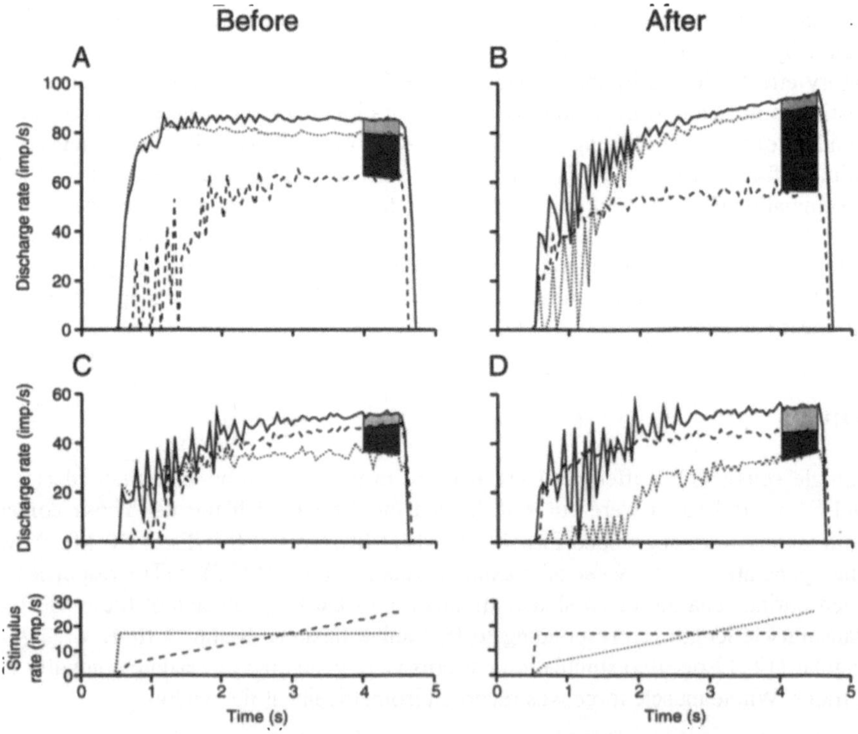

Figure 1. Effect on a GTO afferent of switching stimulation patterns between the individual multi-MU filaments in two different filament pairs (A-B and C-D). Dotted and dashed lines: stimulation of individual filaments; continuous lines: combined activation. All responses are displayed as average frequency estimates (bin width = 50 ms). Shaded areas indicate the increment of firing upon combined activation and the difference between the discharge rates elicited by stimulation of each filament individually.

Several MMUF pairs interacted as expected from predominant pacemaker competition, as significant occlusion was observed, which increased with increasing ND. However, some pairs showed interactions not compatible with pure pacemaker competition in simple terminal trees, when pronounced occlusion was found for very small values of ND.

Based on theoretical calculations (Otten, Scheepstra & Hulliger, this volume) pure pacemaker competition in a simple structure (one common node and two encoders) yields a broad bell-shaped and approximately inverse relation between C_i and ND, with a C_i maximum of 0.4 for ND = 0, and with C_i declining towards 0.1 for large values of ND. Thus, upon switching the stimulation patterns between the individual MMUFs C_i should remain unchanged if ND is unchanged, or decrease if ND increases (or vice versa). More than half of the MMUF pairs showed switching effects compatible with the above bell-shaped ND dependence of C_i. Figure 1A & B shows an example in which switching the stimulus pattern increased ND and decreased C_i. Yet for some MMUF pairs switching of activation patterns resulted in pronounced deviations from the inverse C_i-ND relation, either when considerable changes in C_i were observed without changes in ND, or when C_i and ND changed in parallel. The examples of Figure 1C & D illustrate a MMUF pair for which switching resulted in a significant increase in C_i, while ND was unaltered. Individual GTOs regularly revealed switching effects compatible with predominant pacemaker competition for some MMUF pairs, but not for others (Figure 1).

CONCLUSIONS

Mechanical unloading is unlikely to account for the sub-linear summation and gain compression in GTO responses. Some of the MMUF interactions of the present study are compatible with predominant pacemaker competition, whereas other types of interactions require additional mechanisms. These may include multiple transducer interactions (Gregory et al., 1985) that might arise from asymmetric compression of sensory terminals by contracting muscle fibres in series with the GTO receptor. To what extent combined operation of pacemaker and transducer mechanisms can fully account for the sub-linear summation of GTO responses is currently being investigated in model studies.

ACKNOWLEDGEMENTS

Supported by an AHFMR fellowship award (PS) and a Canadian MRC grant (MH).

REFERENCES

BANKS, R.W., HULLIGER, M., SCHEEPSTRA, K.A. & OTTEN, E. (1995) Pacemaker competition and the role of pre-terminal branch tree architecture: a combined morphological, physiological and modelling study. This volume.

FUKAMI, Y. (1981) Responses of isolated Golgi tendon organs of the cat to muscle contraction and electrical stimulation. *J. Physiol.* **318**, 429-443.

GREGORY, J.E. & PROSKE, U. (1979) The responses of Golgi tendon organs to stimulation of different combinations of motor units. *J. Physiol.* **295**, 251-262.

GREGORY, J.E., MORGAN, D.L. & PROSKE, U. (1985) Site of impulse initiation in tendon organs of cat soleus muscle. *J. Neurophysiol.* **54**, 1383-1395.

JAMI, L. (1992) Golgi tendon organs in mammalian skeletal muscle: Functional properties and central actions. *Physiol. Rev.* **72**, 623-666.

OTTEN, E., SCHEEPSTRA, K.A. & HULLIGER, M. (1995) An integrated model of the mammalian muscle spindle. This volume.

TOWARDS A MATHEMATICAL MODEL OF THE GOLGI TENDON ORGAN

E. Otten[1], M. Hulliger[2] and P. Sjölander[2,3]

[1]Department of Medical Physiology
University of Groningen, The Netherlands
[2]Department of Clinical Neurosciences
University of Calgary, Canada
[3]Division of Work Physiology
National Institute of Occupational Health
Umeå, Sweden

INTRODUCTION

The design of a comprehensive mathematical model of the Golgi tendon organ (GTO) is desirable, first, to provide a tool to explore the receptor's internal mechanisms of operation and, second, to provide a model module for incorporation into higher order models of motor systems circuitry. For the first application, models should include explicit formal descriptions of receptor processes. For the second, less analytical detail is required, but models must still be capable of simulating a wide range of receptor behaviour, including various non-linear properties. At present mathematical models meeting such requirements are not available. In the past, quantitative description was restricted to approximations by linear transfer functions (Houk & Simon, 1967; Anderson, 1974). But these offer little insight into receptor mechanisms and fail to account for the striking non-linearities of GTO responses.

Muscle spindles and GTOs have several features in common: like the Ia axon, the Ib axon also branches inside the receptor capsule to supply a number of sensory terminals. Further, both receptors have a muscular structure in series with, and providing the mechanical input to, the sensory terminals. However, for modelling purposes there are at least two distinct differences: first, a larger number of muscle fibres, independently controlled by separate motor units (MUs), act on a single GTO receptor. Second, Ib terminals are not necessarily exclusively excited by stretch; instead they may also be activated by lateral pressure and distributed distortion arising from activation of parallel intra-capsular muscle fibres (reviewed by Jami, 1992).

These differences are likely to result in different interactions between mechanical inputs in terms of firing behaviour. However, the complexity of the receptor precludes robust predictions regarding the type and size of interaction effects. Mathematical modelling seemed the method of choice to attack this problem. In the present study two types of non-linear behaviour were examined: gain compression (response saturation with increasing force) and sub-linear summation of the effects of separate excitatory MUs on an individual GTO (for references see Jami, 1992; Sjölander, Hulliger, Windhorst & Otten, 1995).

METHODS

Given the significant architectural similarities between spindles and GTOs, the design of a GTO model was based on the general structure of the muscle spindle model of Otten, Scheepstra & Hulliger (1995). Like the spindle model it features sub-models of muscle fibres, of the encoding process and of propagation of action potentials through the terminal branch tree of a sensory axon. Yet major structural adjustments reflecting the architectural differences also have to be implemented: first, parallel extrafusal muscle fibre inputs from limited numbers of motor units were incorporated; second, terminal deformation by lateral pressure was represented by equations describing volume-preserving distortions (Figure 1A). Simulated membrane strain was taken to determine terminal current, which in turn was fed into a modified Frankenhaeuser-Huxley model of an encoding site (see Otten et al., 1995).

RESULTS

Response saturation could be a manifestation of ionic properties of the terminal membrane alone, if its firing rate-current characteristics revealed significant gain

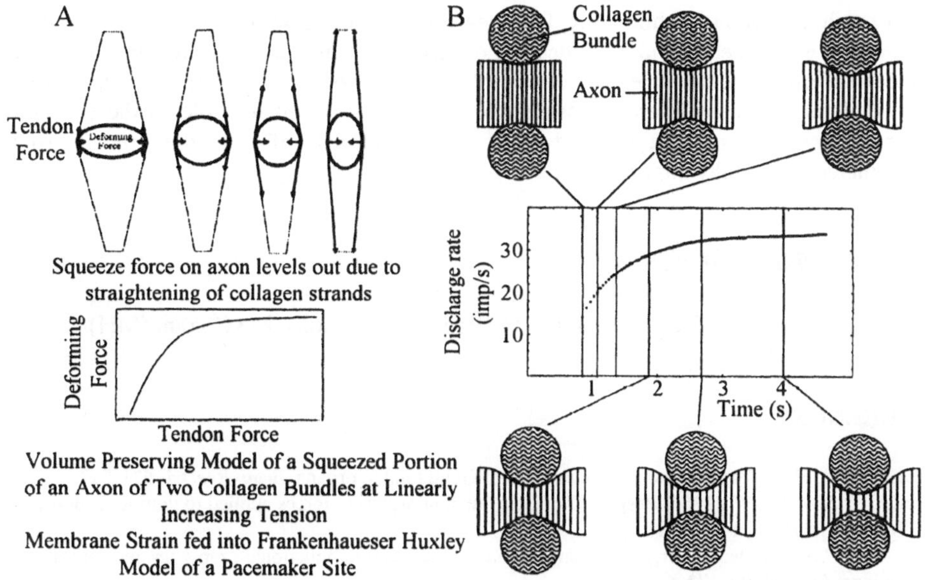

Figure 1. Saturation of GTO response to augmenting muscle fibre force, simulated in a sub-model of terminal deformation by lateral pressure. A, deforming force levelling off with increasing tendon fascicle force, as collagen strands straighten out. B, Magnitude of membrane deformation (flanking schematics), assumed to determine terminal current amplitude, was used as input to the encoder model. Centre panel, time course of simulated firing rate during progressive increase of muscle fibre force. See also text.

compression. However, it emerged that with realistic receptor membrane properties (see Otten et al., 1995) the magnitude of GTO response saturation could not be simulated satisfactorily. In contrast, when deformation of receptor terminals by lateral pressure was added the full extent of observed saturation was readily simulated (Figure 1B). In the present simulation saturation by terminal deformation arises from straightforward geometric principles. Figure 1A shows schematically how, as active muscle fibre force increases, initially slack tendon fascicles are pulled taut and straightened out, leading to deformation of the terminal. As force increases further, fascicles which are already taut cannot be straightened further. Hence deformation levels off, and the deforming force saturates.

Sub-linear summation of GTO response to activation of separate muscle fibres was in principle readily simulated when separate inputs activated different pace-makers in separate portions of the terminal branch tree. An occlusion component was attributable to simple pace-maker competition, while a residual (sub-linear) summation component was in principle accounted for by electrotonic coupling of receptor potentials (Otten et al., 1995). However, for GTO terminal branch trees of realistic size (see Jami, 1992) pace-maker competition combined with electrotonic coupling failed to account for experimental observations of near-total occlusion (Sjölander et al., 1995). It remains to be explored whether a combination of deformation-mediated saturation and pace-maker competition will permit more satisfactory simulations of such extremes.

DISCUSSION

The development of the present GTO model is still at an early stage. The receptor mechanisms simulated so far appear adequately founded in general morphological and physiological observations. Nevertheless, experimental confirmation of the size of their contribution to receptor behaviour is desirable. Some of the measurements required may be within reach of current methodology, like measurement of receptor membrane properties in patch clamp experiments, and identification of terminal branch tree architecture. Yet it will be more difficult to determine the site of action of interacting MUs, or the magnitude of terminal deformation by lateral pressure (to establish its postulated role as the saturation promoting mechanism. Last but not least, the sub-models of this study remain to be amalgamated into a fully integrated model.

ACKNOWLEDGEMENTS

Supported by an AHFMR fellowship award (PS) and a Canadian MRC grant (MH).

REFERENCES

ANDERSON, J.E. (1974) Dynamic characteristics of Golgi tendon organs. *Brain Res.* **67**, 531-537.

HOUK, J. & SIMON, W. (1967) Responses of Golgi tendon organs to forces applied to muscle tendon. *J. Neurophysiol.* **30**, 1466-1481.

JAMI, L. (1992) Golgi tendon organs in mammalian skeletal muscle: functional properties and central actions. *Physiol. Rev.* **72**, 623-666.

OTTEN, E., SCHEEPSTRA, K.A. & HULLIGER, M. (1995) An integrated model of the mammalian muscle spindle. This volume.

SJÖLANDER, P., HULLIGER, M., WINDHORST, U.R. & OTTEN, E. (1995) Non-linear summation in GTO responses: implications for receptor mechanisms. This volume.

MODELLING OF CHAOTIC AND REGULAR Ia AFFERENT DISCHARGE DURING FUSIMOTOR STIMULATION

K.A. Scheepstra[1], E. Otten[1],
M. Hulliger[2] and R.W. Banks[3]

[1]Department of Medical Physiology
University of Groningen
Groningen, The Netherlands
[2]Department of Clinical Neurosciences
University of Calgary, Calgary, Canada
[3]Department of Biological Sciences
University of Durham, Durham, UK

INTRODUCTION

Mammalian Ia afferents have a tendency to fire action potentials phase-locked to periodic stimuli, when certain static gamma axons are stimulated at constant frequency. This driving is thought to be mediated by chain fibres, which, in contrast to other intrafusal fibres, have fast contraction properties, comparable to type II extrafusal fibres. Intrafusal contractions are mechanically transmitted to the sensory ending, which transduces the movement into a receptor current. The ensuing oscillatory changes in receptor potential which accompany these rapid chain fibre contractions then cause the generation of one spike during each contraction cycle. If stimulation frequencies are high, or the bag_2 fibre is coactivated, the afferent may fire at subharmonic frequencies, giving 1:2 or 1:3 driving. Also, very irregular spike intervals can be observed, which gives the impression of an erratic firing response. Changes in the initial muscle length often cause transitions between subharmonic, 1:1 driving, and irregular firing (Boyd, Murphy & Mann, 1985; Banks, 1991).

The above explanation of the mechanisms underlying driving is plausible, but largely qualitative. Here we describe the simulation of driving and the transition to irregular firing with a detailed computer model. The model consists of two parts, the first describing intrafusal mechanical events (Schaafsma, Otten & van Willigen, 1991) and the second the ionic processes occuring in the sensory ending (Otten, Hulliger & Scheepstra, 1995). The ionic part is based on the Frankenhaeuser-Huxley equations, with an added slowly activating potassium conductance, which permits low frequency repetitive firing and also causes spike frequency adaptation.

METHODS

To simulate chain fibre contractions, the mechanical model was expanded to include a third fibre with intramuscular Ca^{++} dynamics similar to extrafusal type II fibre (Otten, 1987). All other parameters were taken from Schaafsma et al. (1991), except that passive damping was decreased 50 fold, to allow for rapid oscillations as observed in chain fibres.

The sensory elongation (SE), defined as the amount of stretch of the sensory ending, calculated by the mechanical part of the model was scaled to provide a depolarising input conductance (I) of the ionic part. We have assumed this process to be linear: $I = A \bullet SE + B$. Constants A and B calibrate both model parts with respect to each other, with A scaling the size of the contraction and B setting the initial stretch of the sensory ending. The sensory elongation was calculated every millisecond. The output of the ionic model consists of times of occurrence of action potentials.

RESULTS

Stimulation of the model chain fibre at 70/s resulted in an unfused tetanus, manifest as an oscillation with an amplitude of 5% of the total elongation produced by the fibre contraction. The ionic model generates action potentials repetitively during a maintained input conductance. We first determined the magnitude of constant input conductance needed to produce a steady state firing rate of 70 imp/s. The sensory elongation was then scaled (constant A) to give the same mean input conductance. Spikes generated by the model were now always time-locked to the stimulus pulses, i.e. the model showed driving. Even if the input was now increased or decreased by adding or subtracting a constant input conductance (constant B), which would normally change the firing rate by 10%, stable 1:1 driving persisted. When the input level was halved, a 1:2 driving pattern occurred. Intermediate patterns consisted of 1:2 driving with an occasional 1:1 spike or converse. Input conductances that were too high to maintain stable 1:1 driving caused irregular firing.

Frequencygrams (Bessou, Laporte & Pagès, 1968) were constructed to study irregular firing. During irregular firing the spikes accumulate into a pattern (Figure 1A, firing rates above the stimulation frequency), while random noise would show no pattern. Experimental data (Figure 1B) taken from an 8mm ramp and hold stretch during static fusimotor stimulation (53 Hz.) in cat soleus shows the same behaviour in real spindles.

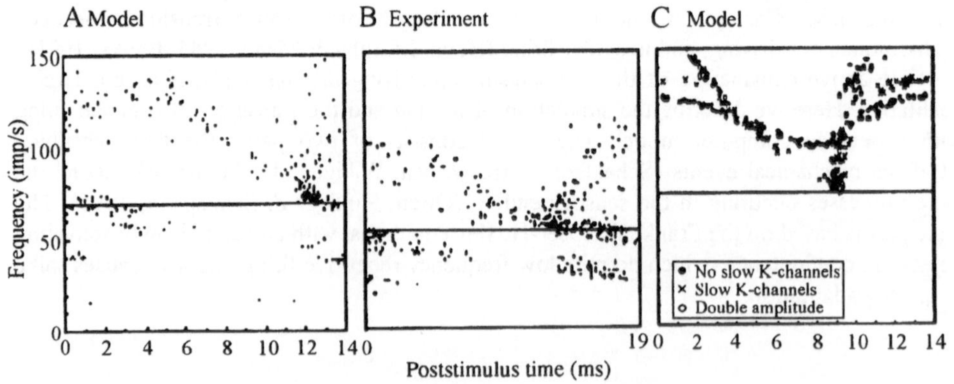

Figure 1. Frequencygrams constructed from model (A) and experimental (B) responses during static fusimotor stimulation, horizontal line indicates stimulation frequency. The pattern constructed from irregular firing can be altered by mechanical and ionic mechanisms (C).

This means that irregular firing results from a chaotic process. Increasing the amplitude of the oscillation clearly increased the tendency to fire irregularly (Figure 1C), but when the slowly activating potassium conductance was turned off, the tendency to fire irregularly was reduced. Thus irregular firing is not only caused by the intrafusal contraction, but also by the ionic processes in the sensory ending.

DISCUSSION

The present model study on driving by fusimotor stimulation demonstrates that the unfused tetanic contractions of chain fibres can cause driving, when sensory elongation varies by only 5%. Furthermore it was shown that chaotic firing patterns occur when the mean afferent firing rate exceeds the stimulation frequency.

However the model failed to reproduce the observation that changes in initial muscle length did not always alter the driving pattern (Boyd et al., 1985). This is because in the model the firing rate is primarily determined by the total amount of sensory elongation and not by the frequency of the oscillation.

With the present model the persistence of 1:1 driving at various muscle lengths can only be explained by assuming that the change in sensory elongation due to the length change was small (< 10%) compared to the total sensory elongation elicited by contraction, maintaining it inside the range of stable 1:1 driving. This implies that contraction levels of nuclear chain fibres are tuned to result in a sensory elongation that is needed to produce 1:1 driving at any stimulation frequency. Another explanation is that the encoder or tranducer is sensitive not only to the static level of sensory elongation, but also to the dynamic changes. This would allow afferent firing rates to be higher than predicted from the static level of sensory elongation alone and it would explain the generation of 1:1 driving over a wide range of sensory elongation levels.

Although the model has the property of eliciting stable driving it remains to be seen whether the phenomenon of driving can be fully accounted for.

REFERENCES

BANKS, R.W. (1991) The distribution of static γ-axons in the tenuissimus muscle of the cat. *J. Physiol.* **442**, 489-512.

BESSOU, P., LAPORTE, Y. & PAGÈS, B. (1968) A method of analysing the responses of spindle primary endings to fusimotor stimulation. *J. Physiol.* **196**, 37-45.

BOYD, I.A., MURPHY, P.R. & MANN, C. (1985) The effect of chain fibre 'driving' on the length sensitivity of primary sensory endings in the tenuissimus, peronius tertius and soleus muscles. In *The Muscle Spindle*. eds. BOYD, I.A. & GLADDEN, M.H., pp. 195-199. Macmillan, London.

OTTEN, E. (1987) A myocybernetic model of the jaw system of the rat. *J. Neurosci. Meth.* **21**, 287-302.

OTTEN, E., HULLIGER, M. & SCHEEPSTRA, K.A. (1995) A model study on the influence of a slowly activating potassium conductance on repetitive firing patterns of muscle spindle primary endings. *J. Theor. Biol.* (in press).

SCHAAFSMA, A., OTTEN, E. & VAN WILLIGEN, J.D. (1991) A muscle spindle model for primary afferent firing based on a simulation of intrafusal mechanical events. *J. Neurophysiol.* **65**, 1297-1312.

COURSEWARE FOR TEACHING MUSCLE SPINDLE PHYSIOLOGY

J.D. McGarrick and S. Durbaba

M & CAL Laboratory
Sherrington School of Physiology, UMDS
St. Thomas' Hospital, London SE1 7EH, UK

The basic principles of the operation of muscle spindles and tendon organs, and their contribution to stretch reflexes, form part of the teaching of preclinical medical and dental students. More advanced concepts may be introduced in BSc Neuroscience courses. Even at the most elementary level, it is difficult to instill a clear understanding of the close structure-function relationships of the spindle, and its relationship to movement, by traditional didactic methods. Some success has been reported in the use of physical models as visual aids to teaching (Brown, Harrison & Lab, 1979), but such models are not generally available. Mathematical models (eg. Schaafsma, Otten & van Willigen, 1991) may be very helpful to the specialist but are probably too sophisticated for average student teaching. Teaching films, later converted to video format, have been produced at the University of Glasgow (Boyd and Gladden & Ward) and are useful at both undergraduate and postgraduate level but suffer the disadvantage of limited accessibility. Clips from these films have been incorporated into the computer courseware described below.

Computer Aided Learning (CAL) lends itself particularly well to teaching dynamic concepts. The functional interrelationships of structures that move relative to one another are much easier to appreciate when explored by interactive animations. The operation of the muscle spindle is an excellent example of such a system.

We are developing a CAL tutorial which allows the student to explore both the structure and the function of the muscle spindle. For the student who is meeting the spindle for the first time, an introductory illustrated tutorial is offered, in which the explanatory text and diagram sequences are enriched with the appearance, in the appropriate contexts, of high quality images and video sequences (in Microsoft AVI format). Later in the tutorial, at a suitable entry point for the more advanced student, open exploration of the more detailed structure, with context-sensitive diagrams, animations and movie clips is provided (see illustration). By using a mouse to explore such 'active diagrams' that trigger text and graphical events, the structural and functional components become associated. The concepts of unloading and co-activation are illustrated by simple user-controlled animations of limb

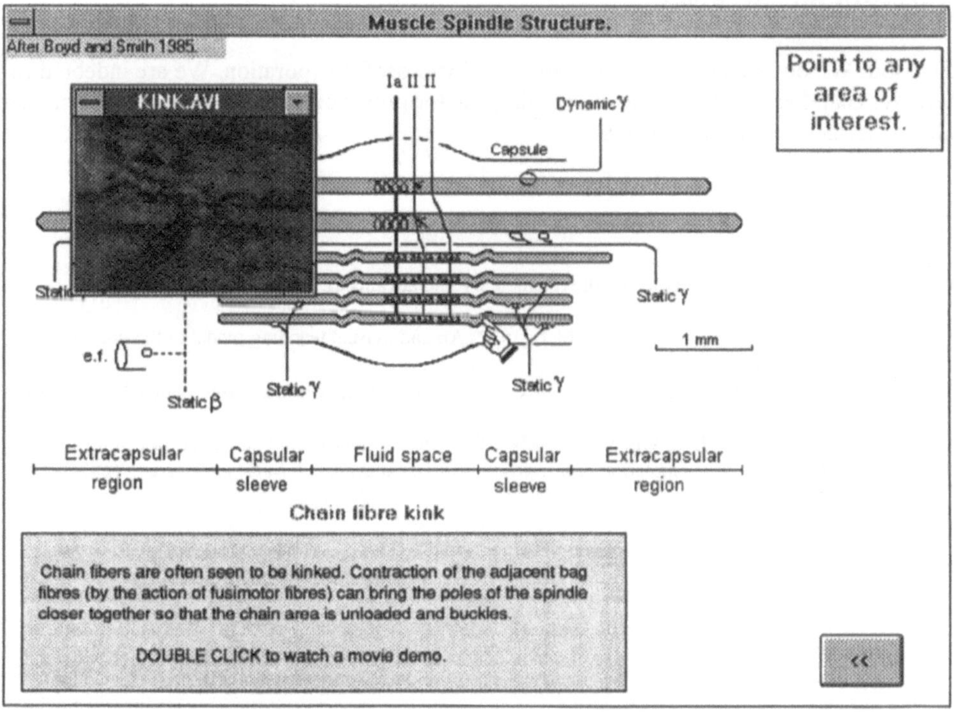

Figure 1. An example of a frame from the CAL tutorial. In the original the images are in colour and higher resolution.

movements with or without obstruction and with the option to eliminate fusimotor activity. The details of the involvement of static and dynamic fusimotor control of the intrafusal fibres are shown by animated overlays revealing how the primary and secondary regions respond to stretch and to fusimotor stimulation, with supportive video sequences.

A captured frame from the tutorial is illustrated above (Figure 1). In this example the user has double clicked on the chain fibre kinks in the diagram. This action has evoked an AVI sequence which shows how the chain fibres can buckle as they become unloaded by contraction of adjacent bag fibres. It should be noted that the quality of the moving image in the inset window is not well represented in this illustration.

This software is being developed in a PC Windows environment in Visual Basic 3.0 and will be transferred to the UMDS teaching network (currently running under Novell 4.02) for use and evaluation by students. In view of the file sizes associated with the sound and video sequences, final distribution is likely to be on CD ROM and will run on any Windows PC supporting 256 colour VGA with sound facilities. Other than this no special hardware is needed as Microsoft AVI format requires only the included software to display video sequences.

Part of this material is available as a World Wide Web document on the Internet. The URL (Universal Resource Locator) is:

http://www.umds.ac.uk/elsewhere/physiology/mcal/spinmain.html

ACKNOWLEDGEMENTS

Microsoft, Windows and AVI are trademarks of Microsoft Corporation. We are indebted to Margaret Gladden and the University of Glasgow for provision of and permission to use the video material referred to in the text.

REFERENCES

BOYD, I.A. The mammalian muscle spindle (elementary and advanced). Film / video. University of Glasgow, Glasgow.

BROWN, A.W., HARRISON, F.G. & LAB, M.J. (1979) An audiovisual teaching model of the muscle spindle. *Med. Education* **13**, 14-16.

GLADDEN, M.H. & WARD, J. Central and reflex activation of γ-motoneurones. Film / video. University of Glasgow, Glasgow.

SCHAAFSMA, A., OTTEN, E. & van WILLIGEN, J.D. (1991) A muscle spindle model for primary afferent firing based on a simulation of intrafusal mechanical events. *J. Neurophysiol.* **65**, 1297-1312.

THE MECHANICAL SPINDLE: A REPLICA OF THE MAMMALIAN MUSCLE SPINDLE

Pierre-Henry Marbot and Blake Hannaford

Biorobotics Laboratory
Department of Electrical Engineering FT-10
University of Washington, Seattle, USA

INTRODUCTION

The goal of the recently initiated Anthroform Arm Project is to understand the manipulation capabilities of the human arm through the development of a dynamically accurate replica arm. Previous attempts have focussed on manipulator kinematics, and have neglected dynamics and controller designs. Others have emphasized neural network control without attention to correct biomechanical modelling. Our manipulator and its controller are based on current knowledge of human biomechanics and neurophysiology respectively.

One key element is the mammalian muscle spindle which is responsible for position and velocity feedback. A spindle constitutes an active sensor, whose properties are altered by the central nervous system, according to its needs. For example, when a movement is being learned, gamma dynamic activity is high in the muscles used. This activity decreases when the trajectories have been learned. This paper describes features that we are attempting to copy and the mechanical and software aspects of our prototype. The model and prototype include active modulation of the spindle's non-linear response which models gamma effects.

THE MECHANICAL SPINDLE

The mechanical spindle is contained in an aluminium tube of 1 cm diameter and 11 cm in length. Within it is contained a non-linear force sensor in series with a spring, a leadscrew mechanism, a DC motor and a voltage-controlled oscillator (VCO) circuit which is used to convert the output of the strain gauges. Figure 1 shows schematic representation of the muscle spindle and its response to a quasi-static stretch. The motor is driven by Pulse Width Modulation (PWM), and its angular position feedback (q) is obtained from an optical encoder using quadrature signals. In this way, all the signals exchanged between the spindle and the processor are noise free because they are either frequency encoded or binary. A

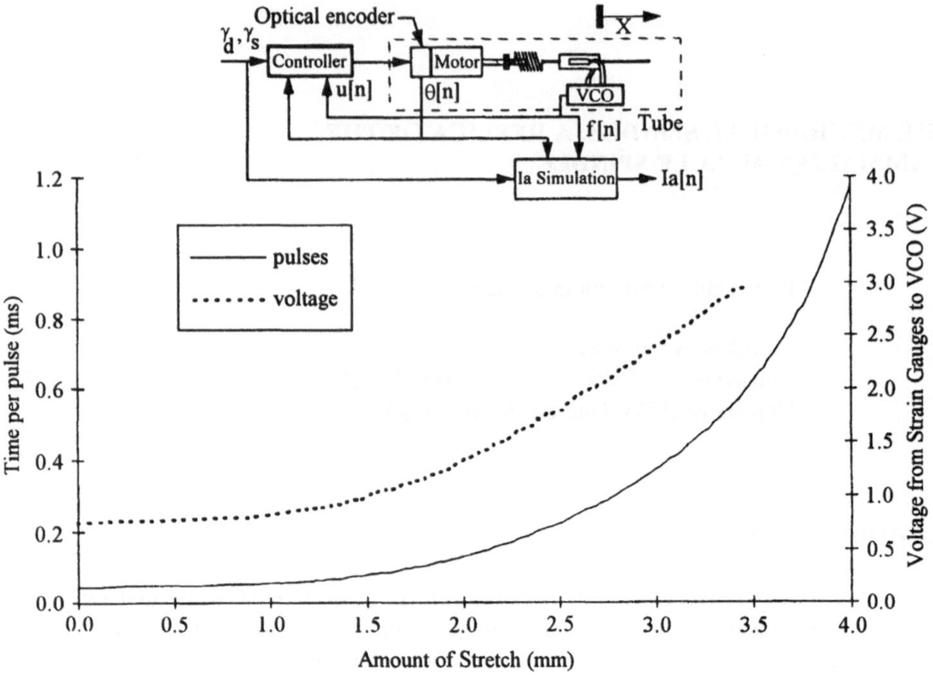

Figure 1. The mechanical spindle and its quasi-static response.

controller drives the motor such that it simulates the intrafusal muscle fibre. Finally, the Ia response software module matches the response of the real spindle.

CONTROLLING THE SPINDLE

A set of custom processor cards (Mac Duff & Veneima, 1992) based on the TMS320C30 Digital Signal Processor (DSP) and interconnected with a special bus is responsible for executing neural processing functions, handling sensor input and motor output, and communicating with a host. An I/O daughter card was developed to drive up to four mechanical spindles. Each "channel" is composed of a PWM output, an optical encoder input, and a frequency signal decoder. The latter (when associated with the VCO inside the spindle) constitutes a A/D channel with gain and offset adjustable by software, and a high immunity to noise.

The control law for the spindle motor uses an adaptive algorithm. The selected spindle motor is able to position the leadscrew along its travel (30 mm) in less than 0.5 seconds. The algorithm that was programmed to test the first spindle is the following (see Figure 1):

$$q_d[n] = q_d[n-1] + \begin{matrix} A & f[n] > f_0 + B \\ -A & f[n] < f_0 - B \end{matrix}$$

$$e[n] = q[n] - q_d[n]$$

$$u[n] = M * \mathrm{SIGN}(e[n] + a\,(e[n] - e[n-1]))$$

$$Ia[n] = g_s[n] * q[n] + g_d[n] * f[n]$$

where: Ia[n] are the Ia outputs, u[n] are torque commands to the motor; $g_s[n]$ and $g_d[n]$ are the static and dynamic gamma inputs; q[n] are the optical encoder positions, f[n] are strain gauge measurements; f_o, A and B are model parameters; M is maximum motor torque; a is automatically updated for maximum controller performance.

After considering models from Rudjord (1970), Poppele & Bowman (1970), Hasan (1983) and Ramos (1990), the best spindle model found to date was described in the paper from Schaafsma, Otten & van Willigen (1991). Its laws will be implemented as soon as the second version of the spindle has been fully characterised. The parameters will be adjusted to fit experimental data and outputs from one of the above model (when a relevant experiment is not available).

CONCLUSION AND FUTURE WORK

The mechanical muscle spindle introduces a new kind of sensor, original in two respects: it is active, and therefore capable of adjusting itself through a variable bias on the non-linear strain gauge; and it also uses two sensors in series, therefore giving it enhanced resolution. In the current design, the optical encoder gives a linear resolution of 16 microns, the strain gauges give 3 to 25 microns (their response is not linear). Within the range of the strain gauges (2.0 mm), the resolution of the spindle is 3 to 25 microns. The dual, independent sensor assembly permits velocity estimation. The motor/leadscrew can be used in conjunction with the non-linear sensing element to vary the natural frequency of the sensor for adjustable frequency selective response.

By trying to copy the human arm, we have generated a new technology for sensors. The second generation prototype (built to address reliability and fabrication issues) is complete, and is being tested.

ACKNOWLEDGEMENT

We gratefully acknowledge the efforts of Steve Venema and Melani Shoemaker, for their help on the processor hardware, and Professor Mark Binder for his expertise and guidance on muscle spindles. We thank the US National Science Foundation for support of this project through the Presidential Young Investigator Award to the second author.

REFERENCES

MACDUFF, I.G. & VENEIMA, S. (1992) The anthroform Neural Controller: an architecture for spinal circuit emulation. In *IEEE conference on Systems, Man and Cybernetics*, Volume 1. pp117-122. Chicago, USA.

POPPELE, R.E. & BOWMAN, R.J. (1970) Quantitative description of linear behavior of mammalian muscle spindles. *J. Neurophysiol.* **33**, 59-72.

RUDJORD, C.T. (1970) A second order mechanical model of muscle spindle primary afferent. *Kybern.* **6**, 205-213.

HASAN, Z. (1983) A model of spindle afferent response to muscle stretch. *J. Neurophysiol.* **49**, 989-1006.

RAMOS, C.F. (1990) Are detailed models of the muscle spindle appropriate for simulation studies of the stretch reflex? A general method for model comparisons. Ph.D. Dissertation. U.C. Berkeley.

SCHAAFSMA, A., OTTEN, E. & VAN WILLIGEN, J.D. (1991) A muscle spindle model for primary afferent firing based on a simulation of intrafusal mechanical events. *J. Neurophysiol.* **65**, 1297-1311.

THE ROLE OF PRIMARY AFFERENT DEPOLARISATION IN PRESYNAPTIC INHIBITION OF GROUP I FIBRES

B. Lamotte d'Incamps[1], M-L. Monnet[2],
C. Meunier[2] and D. Zytnicki[1]

[1]CNRS URA 1448, Université René Descartes
45 rue des Saints-Pères, 75270 Paris Cx 06, France
[2]CNRS UPR 014, Centre de Physique Théorique
Ecole Polytechnique, 91128 Palaiseau, France

In a previous study, declining inhibitory potentials, ascribed to the action of Ib afferents, were recorded in homonymous and synergic motoneurones during sustained subtotal isometric contractions of gastrocnemius medialis (GM) muscle (Zytnicki, Lafleur, Horcholle-Bossavit, Lamy & Jami, 1990). In a subsequent study, contraction-induced primary afferent depolarisations (PADs) were recorded intra-axonally from the intraspinal portion of homonymous Ib fibres during similar GM contractions (Lafleur, Zytnicki, Horcholle-Bossavit & Jami, 1992). Since PAD is known to be the electrophysiological correlate of presynaptic inhibition (Eccles, Schmidt & Willis, 1962), this observation supported the assumption that presynaptic inhibition of Ib fibres accounted for the decline of contraction-induced inhibition in motoneurones (see Zytnicki & L'Hôte, 1993). The aim of the present computer study was to evaluate the efficacy of the mechanism of presynaptic inhibition received by myelinated afferent fibres. Activation of an axo-axonic synapse may partly shunt the action potentials travelling along the fibre (Segev, 1990), but as recently suggested, the PAD might also contribute by itself to presynaptic inhibition by reducing the spike height (Graham & Redman, 1994). In the present paper, potentiation of the effects of an axo-axonic synapse by the PAD is demonstrated and explained on a simple model of the myelinated axon in terms of the dynamics of ionic currents.

A compartmental model of the terminal branch of a myelinated fibre was designed using a fully object-oriented simulator (Guillon, Meunier & Monnet, in preparation). The model incorporated a succession of 29 nodal zones and 30 internodal zones (see Figure 1A) whose dimensions were derived from the ultrastructural study of a Ia terminal arborisation in midlumbar spinal cord by Nicol & Walmsley (1991). Each nodal zone was modelled by a single compartment with Na^+ and K^+ membrane conductances exhibiting dynamics similar to the Hodgkin-Huxley model for the squid axon (Hodgkin & Huxley, 1952). Each internodal zone was subdivided into three passive compartments with high membrane resistivity and low capacitance in order to mimic the electrical properties of the myelin sheath. One active terminal zone carried the axo-axonic synapse (a passive terminal zone yielded qualitatively similar results) while a pulse of current was injected at the other terminal zone to generate an action potential. Activation of the axo-axonic synapse was modelled by a synaptic current $I_{syn}(t) = G_{max}.a(t).(V_{syn}-V)$ where G_{max} was the peak conductance, V_{syn} the reversal potential of the synapse, and the alpha function $a(t) = e.(t/\tau).exp(-t/\tau)$ described the time course of synaptic events (τ was the time constant). When the synaptic reversal potential was set at the resting membrane potential (typically -65 mV in agreement with experimental observations), a transient increase of the membrane conductance was the only effect of synaptic activation ('silent' synapse). When the synaptic reversal potential was set at a higher value than the resting membrane potential, the axo-axonic activation induced a PAD.

The effect of the activation of the axo-axonic synapse (peak conductance 50 nS, reversal potential -55 mV, i.e. 10 mV above rest) on an afferent action potential propagating down the axon is shown in Figure 1B. This action potential had an amplitude of 64 mV and was superimposed on a PAD of 7.5 mV amplitude. By comparison, a control action potential, elicited by a similar stimulation but in absence of synaptic activation, had an amplitude of 105 mV (thin line in Figure 1B). The synaptic activation reduced by 39% the amplitude of the action potential at the terminal. Most of the amplitude reduction was accounted for by a lower peak amplitude of the action potential during axo-axonic activation with respect to control conditions.

Then, the question arose whether the reduction in the peak amplitude of the action potential was only caused by the shunting effect of the synapse during activation or also by the PAD. Figure 1C shows the differences between action potential amplitudes during activation of a 'silent' synapse (open circles) and during activation of a PAD generating synapse (filled circles). For equivalent peak synaptic conductances, larger reductions of the spike height occurred in the presence than in the absence of PAD. The additional reduction elicited by the PAD was important for conductances under 100 nS. For instance, with a 60

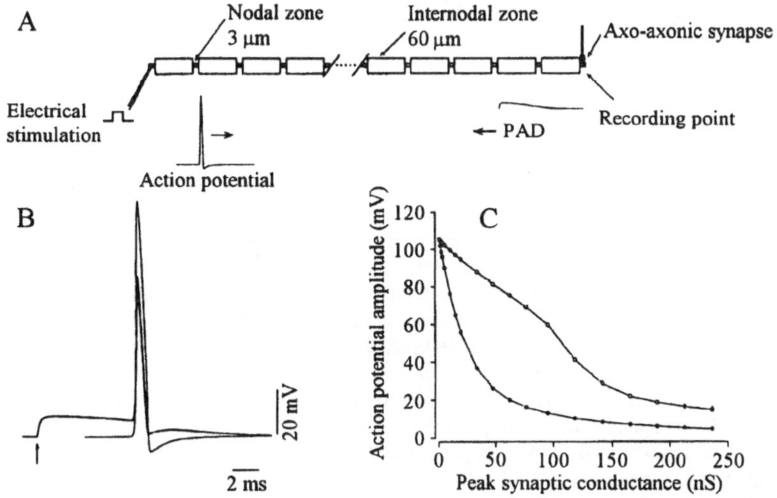

Figure 1. A. The Model. Action potential was elicited by electrical stimulation applied at the left end of the fibre and observed at the terminal zone on the right. PAD was induced by the activation of the axo-axonic synapse and diffused from right to left. B. Effects of the activation of the axo-axonic synapse on an afferent action potential. Thin line: control action potential; bold line: action potential superimposed on PAD. Arrow indicates the onset of synaptic activation: point indicates the stimulation. C. Reduction of action potential amplitude as a function of synaptic conductance. Open circles: silent synapse (V_{syn} = -65 mV); filled circles: V_{syn} = -35 mV.

nS peak conductance, the spike height fell from 75 mV to 20 mV in the presence of PAD. This effect of PAD tended to decrease for high conductances because of the shunt predominance. This computer study therefore strongly suggests that the PAD generated by an axo-axonic synapse located at the terminal of a myelinated fibre is much more than an electrophysiological correlate of presynaptic inhibition and may, by itself, contribute to the reduction of spike amplitude.

It is well known that the PAD increases the excitability of afferent terminal branches, as revealed by intraspinal microstimulation of the fibres (Eccles et al., 1962). It is therefore an apparent paradox that the PAD may at the same time reduce the amplitude of afferent action potentials. Analysis of ionic current perturbations at a distance from the axo-axonic synapse, where the spike height was reduced in the absence of shunting, provided an explanation for these paradoxical effects. PAD-induced perturbations in the inward sodium and outward potassium currents were examined by subtracting currents entering the compartment when the synapse induced a PAD from currents obtained during activation of a 'silent' synapse. At the onset of the spike, PAD induced a transient increase in the inward sodium current. This increase was responsible for a transient enhancement of the net current (balance between axial current entering the compartment and inward sodium, outward potassium, synaptic and membrane leak currents) and accounted for the transient hyperexcitability of the fibre induced by the PAD. Conversely, during the rising phase of the spike there was an increased inactivation of the inward sodium current and an increased activation of the outward potassium current. Together these two mechanisms were responsible for a large deficit in the net current flowing through the compartment and accounted for the reduction of the action potential amplitude.

In conclusion, our study suggests that the PAD may enhance the efficacy of axo-axonic synapses. Because of electrotonic diffusion, the PAD might also interact with the spike regeneration at some distance from the axo-axonic synapse. Such a remote effect of the PAD opens the possibility of interactions between axo-axonic synapses located on the same branch of the axon, or even at the end of different terminal branches. If this assumption proved correct the functional role of the PAD in presynaptic inhibition should be reexamined.

REFERENCES

ECCLES, J. C., SCHMIDT, R. F. & WILLIS, W. D. (1962) Depolarisation of central terminals of group I afferent fibres of muscle. *J. Physiol.* **160**, 62-93.

GRAHAM, B. & REDMAN, S. (1994) A simulation of action potentials in synaptic boutons during presynaptic inhibition. *J. Neurophysiol.* **71**, 538-549.

HODGKIN, A. L. & HUXLEY, A. F. (1952) A quantitative description of membrane current and its application to conduction and excitation in nerve. *J. Physiol.* **117**, 500-544.

LAFLEUR, J., ZYTNICKI, D., HORCHOLLE-BOSSAVIT, G. & JAMI, L. (1992) Depolarisation of Ib afferent axons in the spinal cord during homonymous muscle contraction. *J. Physiol.* **445**, 345-354.

NICOL, M. J. & WALMSLEY, B. (1991) A serial electron microscope study of an identified Ia afferent collateral in the cat spinal cord. *J. Comp. Neurol.* **314**, 257-277.

SEGEV, I. (1990) Computer study of presynaptic inhibition controlling the spread of action potentials into axonal terminals. *J. Neurophysiol.* **63**, 987-997.

ZYTNICKI, D., LAFLEUR, J., HORCHOLLE-BOSSAVIT, G., LAMY, F. & JAMI, L. (1990) Reduction of Ib autogenetic inhibition in motoneurons during contractions of an ankle extensor muscle in the cat. *J. Neurophysiol.* **64**, 1380-1389.

ZYTNICKI, D. & L'HÔTE, G. (1993) Neuromimetic model of a neuronal filter. *Biol. Cyber.* **70**, 115-121.

EFFECTS OF ELECTROTONIC SPREAD OF EPSPs ON SYNAPTIC TRANSMISSION IN MOTONEURONES: A SIMULATION STUDY

D.M. Halliday

Department of Physiology, Glasgow University
Glasgow G12 8QQ, UK

INTRODUCTION

Distal synaptic inputs to motoneurones produce Post Synaptic Potentials (PSP) at the soma with longer rise times and half widths than proximal inputs. This is due to electrotonic spread of the PSP. Previous studies have concentrated on the time course of a single compound EPSP at the soma for a variety of pre-synaptic input configurations (e.g. Segev, Fleshman & Burke, 1990). The EPSP represents only an intermediate variable in determining the output discharge of a cell. Simulation studies with compartmental models of motoneurones were used to investigate the functional consequences of synaptic location on synaptic transmission during repetitive firing.

METHODS

Two identical motoneurones were represented by compartmental models (Segev, Fleshman & Burke, 1989) assuming a soma diameter of 50 μm and one uniformly tapered dendrite with an initial diameter of 10 μm. Published values were used for passive membrane parameters and dendritic electrotonic length (Fleshman, Segev & Burke, 1988). These were membrane resistivity R_m = 11000 Ωcm^2, capacitance C_m = 1.0 $\mu F/cm^2$, cytoplasmic resistivity R_i = 70 Ωcm, and a total dendritic electrotonic length of λ_{tot} = 1.5. The dendrite had 15 compartments of equal electrotonic length (0.1) and assuming uniform taper of 0.5 μm per 100 μm length, λ_{tot} of 1.5 is equivalent to a physical length of 1.9 mm.

Afterhyperpolarisation conductances and threshold parameters were based on the model of Baldissera & Gustafsson (1974). EPSP conductances were modelled using an α function (Segev et al., 1990), with parameters set to give an EPSP of 100 μV, 0.35 ms rise time and 3.2 ms half width for a single pre-synaptic input activated from rest at the soma (Cope, Fetz & Matsumura, 1987). The same conductance activated at λ = 1.0 gave an EPSP of 38 μV, 2.0 ms rise time and 11.5 ms half width.

Simulations were performed with all the pre-synaptic inputs uniformly distributed firstly over the soma and first dendritic compartment and secondly about $\lambda = 1.1 \pm 0.1$. The mean discharge rate of each pre-synaptic input was set initially at 30/s, with a random (Poisson) inter spike interval distribution. The output firing rate was chosen to be about 13/s. To achieve this required 870 and 2160 active pre-synaptic inputs for each of the above simulations, giving an average of 26100 and 64800 EPSPs/s respectively. These numbers of pre-synaptic inputs were applied to both cells in the two simulations, using the same conductance as above, with 75% of the number of inputs inputs common to both cells.

Output spike times from each cell were recorded during 100 s runs. For analysis these were assumed to be realisations of stochastic point processes. The two cells showed a tendency for synchronous discharge, due entirely to the effects of the common inputs, thus pre- to post-synaptic transmission was characterised by time and frequency domain correlation estimates between the output discharges. In the time domain cumulant density functions, $q_{12}(u)$, were estimated and in the frequency domain coherence functions, $|R_{12}(f)|^2$, were estimated (Rosenberg, Amjad, Breeze, Brillinger & Halliday, 1989; Conway, Halliday & Rosenberg, 1993).

RESULTS

Figure 1A & B show cumulant estimates for the proximal and distal input cases respectively, with the corresponding coherence estimates in Figure 1C & D. Proximal inputs yielded a sharper central peak in the time domain than distal inputs. In the frequency domain

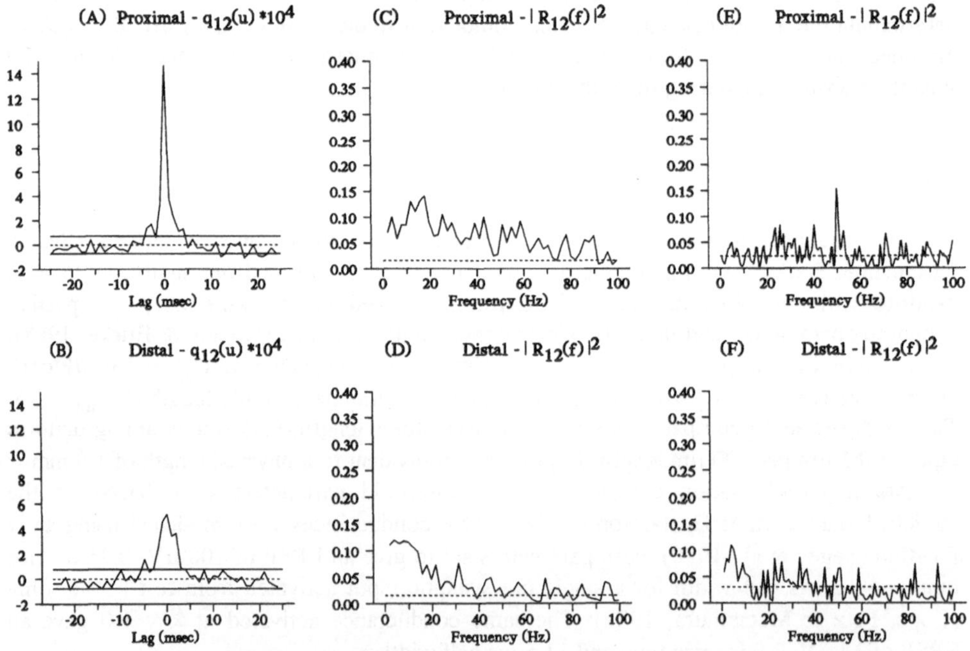

Figure 1. Cumulant estimates between output discharge of the two model cells with random stimuli for (A) proximal synaptic input location, and (B) distal synaptic input location. Horizontal lines represent the expected value (dashed), and the upper and lower 95% confidence limits (solid) for the case of independent processes. Corresponding coherence estimates are shown in (C) and (D) respectively. Coherence estimates between output of same model cells with 30/s periodic and 50/s regular stimuli for (E) proximal input location, and (F) distal input location. The horizontal dashed line in (C) to (F) represents the upper 95% confidence limit for the case of independent processes.

there was a corresponding broader range of significant coupling for proximal inputs than for distal inputs. These results suggest that one consequence of electrotonic lengthening of EPSPs on synaptic transmission is to reduce the bandwidth over which the post synaptic spike train is dependent on the frequency content of the pre-synaptic inputs.

This filtering effect was further investigated using a population model, based on the same configuration as above, but with 90% of the pre-synaptic inputs firing periodically at 30/s (normal inter spike interval distribution, mean = 33.3, S.D. 11.7 ms), and 10% firing regularly at 50/s (mean = 20.0, S.D. 0.8 ms). The inputs firing at 50/s were part of the 75% applied to both cells. Figure 1 E & F show the coherence estimates between the two outputs with the inputs for proximal and distal inputs respectively. Proximal inputs show regions of significant coherence around 30 Hz and a sharper peak at 50 Hz. The distal inputs gave a similar coherence around 30 Hz, but no peak at 50 Hz.

DISCUSSION

Simulation studies have been used to assess the functional significance of electrotonic spread. Results show that as the synaptic location becomes more distal there is a reduction in the bandwidth of transmission. This filtering does not affect the mean output firing rate which is determined by the general level of synaptic activity. The effect of electrotonic spread appears to be to smooth out the fluctuations in the membrane potential due to activation of EPSPs at shorter intervals, thus removing the higher frequency components these would transmit to the output discharge. This is particularly born out in the 50/s case. This suggests that one possible function for the dendritic tree, in terms of synaptic location, is to control the range of frequencies transmitted to the cell output discharge.

ACKNOWLEDGEMENTS

This research was undertaken initially at The Information Science Division, NTT Basic Research Labs, Musashino, Tokyo. Completion was supported in part by a Joint Research Council/HCI Cognitive Science Initiative grant. Presently supported by The Wellcome Trust.

REFERENCES

BALDISSERA, F. & GUSTAFSSON, B. (1974). Afterhyperpolarization conductance time course in lumbar motoneurones of the cat. *Acta physiol. scand.* **91**, 528-544.

CONWAY, B.A., HALLIDAY, D.M. & ROSENBERG, J.R. (1993). Detection of weak synaptic interactions between single Ia-afferents and motor-unit spike trains in the decerebrate cat. *J. Physiol.* **471**, 379-409.

COPE, T.C., FETZ, E.E. & MATSUMURA, M. (1987). Cross correlation assessment of synaptic strength of single Ia fibre connections with triceps surae motoneurones in cats. *J. Physiol.* **390**, 161-188.

FLESHMAN, J.R., SEGEV, I. & BURKE, R.E. (1988). Electrotonic architecture of type-identified α-motoneurones in the cat spinal cord. *J. Neurophysiol.* **60**, 60-85.

ROSENBERG, J.R., AMJAD, A.M., BREEZE, P., BRILLINGER, D.R. & HALLIDAY, D.M. (1989). The Fourier approach to the identification of synaptic coupling between neuronal spike trains. *Prog. Biophys. & Molecular Biol.* **53**, 1-31.

SEGEV, I., FLESHMAN, J.R. & BURKE, R.E. (1989). Compartmental models of complex neurons. In *Methods in Neuronal Modeling: From Synapses to Networks.* eds KOCH, C. & SEGEV, I. pp. 63-96. MIT Press, Cambridge and London.

SEGEV, I., FLESHMAN, J.R. & BURKE, R.E. (1990). Computer simulations of group Ia EPSP's using morphologically realistic models of cat α-motoneurones. *J. Neurophysiol.* **64**, 648-660.

PART 7

CENTRAL CONTROL

FLEXIBILITY IN THE RELATIONSHIP OF MONKEY CORTICO-MOTONEURONAL CELLS TO THEIR TARGET MUSCLES

R.N. Lemon and K.M.B. Bennett

Sobell Department of Neurophysiology
Institute of Neurology, London WC1N 3BG, UK

INTRODUCTION

For those interested in understanding how the central nervous system controls the excitatory drive to alpha motoneurones, the monkey cortico-motoneuronal (CM) system offers a unique advantage. CM neurones can be identified in a conscious working monkey, so that the use of reduced preparations, anaesthesia etc is avoided. Because the CM cell is displaced by only a single synapse from its target motoneurones, it is possible to gain a direct insight into how the discharge of a central neurone influences the pattern of muscle activity generated by these motoneurones. Although many different last-order neurones influence the alpha motoneurone, we know very little about the natural activity of most of them. The CM cell is an important exception. This article will discuss the apparent flexibility in the relationship between the CM cell and its target muscle.

CM cells are layer V pyramidal tract neurones located in the motor cortex and their axons form part of the corticospinal tract. The CM system is well developed in Old World monkeys (such as the macaque), apes and especially in man. In all these primates, it has a particularly strong influence upon motoneurones supplying hand and foot muscles (see Porter & Lemon, 1993, pp 186-193).

The spike-triggered averaging (STA) technique can be used to identify CM cells in the conscious monkey, trained to carry out a skilled, voluntary motor task (Fetz & Cheney, 1980), which in our studies has been the performance of a precision grip between the thumb and index finger (Lemon, Mantel & Muir, 1986). Rectified EMG recorded from selected hand and forelimb muscles is averaged with respect to the discharges of a spontaneously discharging cortical neurone. The identification of the CM cell relies upon the demonstration of a *post-spike facilitation (PSF)* of EMG activity in the STA that has an appropriate latency and duration.

CM cells active during the precision grip task produce PSF in a small group of hand and forearm muscles, the 'muscle field' of the CM cell (Fetz & Cheney, 1980). For many CM cells, the muscle field comprises muscles with synergistic actions (Buys, Lemon, Mantel & Muir, 1986; Lemon, Bennett & Werner, 1991).

CM CELL DISCHARGE CORRELATED WITH TARGET MUSCLE ACTIVITY

It might be expected that the direct nature of the CM connection would result in an inflexible relationship between a CM cell and its target motoneurone. However, early studies showed that this relationship was far from fixed. Fetz & Cheney (1980; 1987) and Muir & Lemon (1983) both reported a clear dissociation of CM cell discharge from target muscle activity according to the task performed. Our recent work suggests that there may be some flexibility in the relationship between a CM cell and its target muscle even within the different phases of the *same* task (Bennett, 1991; Bennett & Lemon, 1994).

We have analysed this relationship in two different ways. First, by looking at correlations between firing rate and level of EMG activity and secondly by measuring the amplitude of PSF under different conditions. Monkeys were trained to perform a precision grip task, in which they inserted the thumb and index finger through two small slots so as to gain access to two separate levers. Monkeys were rewarded for maintaining a target range of grip force (isometric condition) or lever displacement (auxotonic condition) for a period of around 1 s. We looked at two separate phases of this task: the initial *movement period*, in

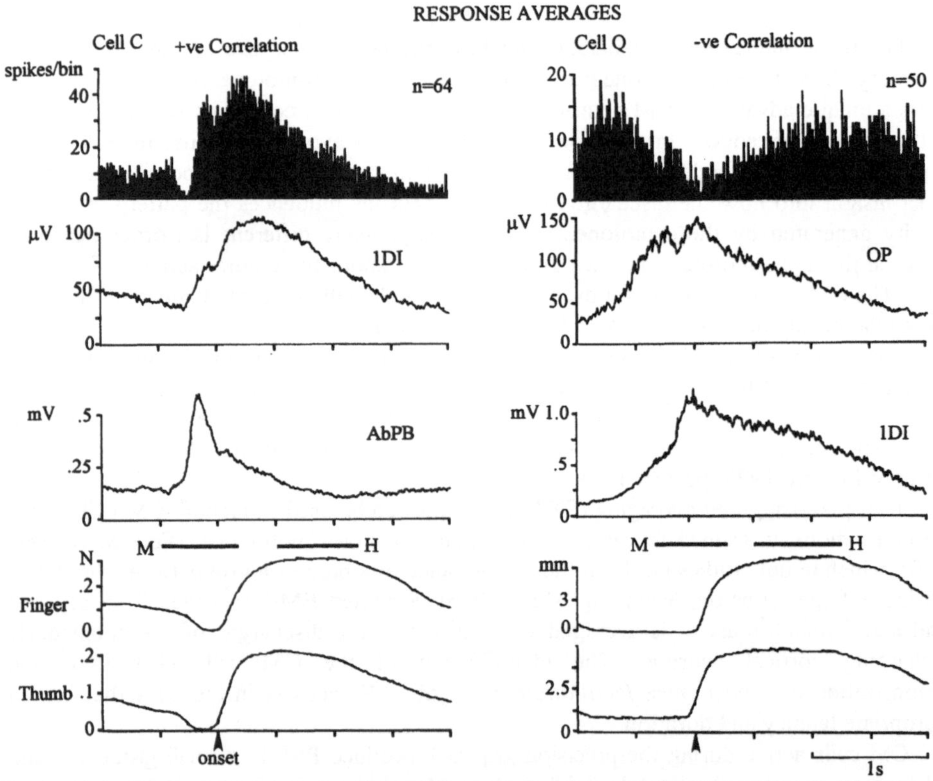

Figure 1. Response averages of 2 CM cells (C & Q) referenced to force (cell C) or movement (cell Q) onset (at arrow). n = number of averaged trials. Bars M and H indicate the periods referred to in the text as movement and hold, respectively. For each CM cell, the peri-event histogram of its discharge (bin width = 10 ms) is shown above averages of smoothed and rectified EMG activity of two target muscles, and the averaged force or movement traces of the finger and thumb. The discharge rate of cell C (left) was positively correlated to the activity of two target muscles, 1DI and AbPB. Firing in cell Q showed a negative correlation to the activity of two target muscles, OP (opponens pollicis) and 1DI. Reproduced with permission from Bennett & Lemon (1994).

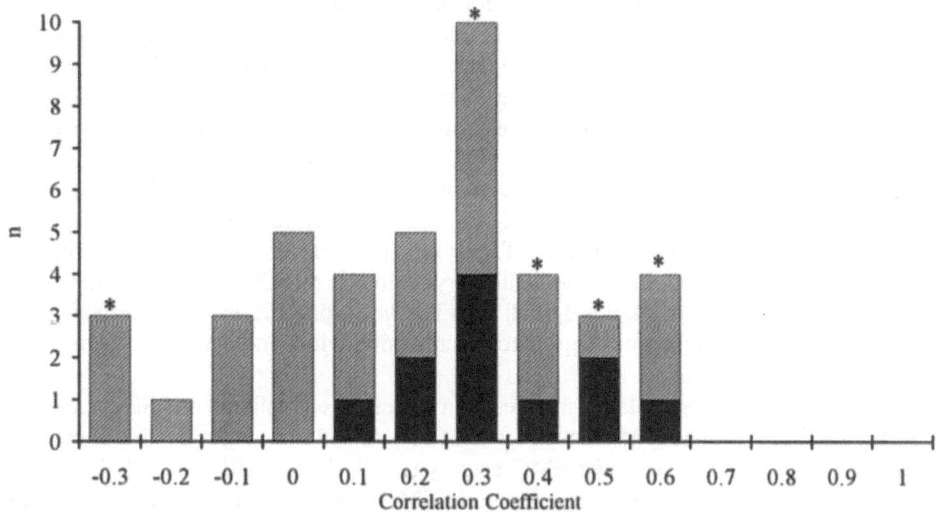

Figure 2. Distribution of correlation coefficients for a total of 42 CM cell-target muscle combinations. Cell activity (in spikes/s) and muscle activity (smoothed, rectified EMG) during the hold period of the precision grip task were subjected to linear correlation analysis. Filled and open columns indicate PSFs of >20%Mod and <20%Mod, respectively. * indicates a positive or negative correlation that was significant at the p <0.001 level (Student t test).

which the monkey positioned its digits on the two levers and began to exert force upon them, and the *hold period*, during which the monkey maintained the levers within the target force or displacement zone.

During the *movement period*, the hand muscles facilitated by a single CM cell often show a fractionated pattern of activity, with bursts of activity in one muscle out of phase with those in another. We found that 13/15 CM cells, each of which facilitated at least two hand muscles, fired at significantly ($p < 0.001$) higher rates when one of its target muscles was more active than the other. For example, cell C in Figure 1 showed a close relationship with the pattern of activity in first dorsal interosseus (1DI) but not with that in abductor pollicis brevis (AbPB), yet it produced PSF in both muscles. For 9/15 cells, the muscle with which the cell was more strongly coactivated was the one receiving the larger PSF. All cells fired at significantly ($p < 0.001$) higher rates during periods of fractionated activity than during the hold period.

During the *hold period*, EMG activity is characterised by a co-contracted pattern of activity in many different muscles (Smith, 1981), and it is assumed that it is the fine balance of ongoing activity within these muscles which is responsible for the level of force exerted by the monkey on the two levers (Maier, Bennett, Hepp-Reymond & Lemon, 1993). Bennett & Lemon (1994) carried out a detailed analysis of the relationship of CM cell discharge rate to the level of hold period EMG in target hand and forearm muscles. They analysed 23 CM cells (42 cell-muscle combinations) recorded from 4 monkeys. There were 41 PSFs and one case of post-spike suppression. In this study, the hold period EMG varied between 0.3% to 8.65% of the maximum activity recorded from the muscle during the task.

The mean CM cell discharge rates during the hold period ranged from 4 to 47 spikes/s (mean = 21.2 ± 11 spike/s). Figure 2 shows the distribution of correlation coefficients obtained; for 20/42 (48%) CM cell-muscle combinations, CM cell discharge rate increased with EMG activity ($r \geq 0.3$, $p < 0.001$). For five CM cells a correlation was observed with two target muscles. Data from one of these cells (cell C) is illustrated in Figure 1.

Figure 3 shows data from a positively correlated neurone (cell M), for which analysis was carried out at three different levels of EMG activity, high (H) medium (M) and low (L). Mean firing rate increased from 10 ± 0.5 SE spikes/s at low EMG levels to 22 ± 0.7 spikes/s at high EMG levels (see histograms on right of Figure 3). At the highest EMG level investigated, neurones with a positive correlation showed a mean increase of 253% above the rate at the lowest level.

But for 18/42 (43%) CM cell-muscle combinations there was no significant correlation of CM cell activity to its target muscle EMG. The discharge rate of three CM cells was negatively correlated to the activity level of their target muscles (see Figure 2), a result that was completely unexpected, given the facilitatory action of the CM synapse. Results from one such cell are shown in Figure 1 (cell Q). The large proportion of uncorrelated and negatively correlated cells did not appear to result from the low and narrow range of EMG occurring in the hold period.

There was a significant relationship between the relative amplitude of PSF (percentage modulation of background EMG level, or %MOD) and the strength of the correlation between CM cell discharge rate and EMG activity ($r = 0.41$, $p < 0.01$, dof 40). That is, an increase of CM cell discharge with target muscle EMG was more likely if the cell produced a large PSF in the muscle. Although correlated cell-muscle pairs ($r > 0.3$) had larger PSFs (mean %MOD 21.5 ± 13.9 SD, n = 20) than did uncorrelated ($r < 0.3$) pairs (mean %MOD

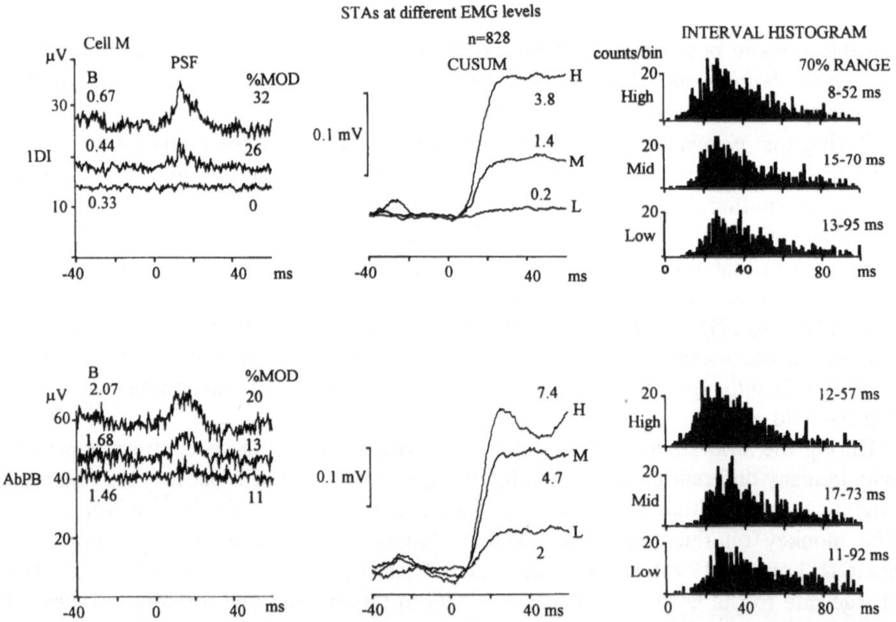

Figure 3. The activity of this CM cell was positively correlated with EMG from both its target muscles, 1DI and AbPb. Left: STAs of 1DI (above) and AbPB (below) at low (L), middle (M) and high (H) EMG levels of each muscle. 828 spikes were used for each average. The background EMG (B value) and relative amplitude of PSF (%MOD) both increased with EMG level. Centre: CUSUMs derived from the STA at each EMG level; CUSUM amplitude increased with EMG level. Right: interspike interval histograms of cell activity at the different levels; incidence of shorter intervals increased with EMG level. Values given in ms indicate range of intervals spanned by 70% of the spikes used to construct each histogram. Reproduced with permission from Bennett & Lemon (1994).

10.8 ± 7.3, n = 17), this difference was not significant. Large PSFs (>20 %Mod) are indicated by filled columns in Figure 2.

PSF AMPLITUDE CHANGES WITH LEVEL OF TARGET MUSCLE ACTIVITY

Movement period

There were pronounced differences in the strength of PSF exerted by the CM cells on their target muscles during the movement period. Our study addressed a group of 15 CM cells with at least two target muscles. For many of these cells, PSF of the muscle with which the CM cell was coactivated in the movement period was pronounced, while that of the other target muscle received little or no facilitation. Changes in the firing rate of a CM cell and in the degree of facilitation it exerted could combine to reinforce a fractionated pattern of activity in its target muscles (Bennett, 1991).

Hold period

Bennett & Lemon (1994) also investigated whether there were any changes in the strength of PSF at different levels of target muscle activity during the hold period. An increase in the *absolute* size of facilitation (measured from the CUSUM, see Figure 3) with an increase of EMG level was obtained for 93% (39/42) of CM cell-muscle combinations. From a low to a high level of EMG activity, the absolute amount of facilitation exerted by each CM cell increased by a factor of 1.2 to 32 (median value = 3.7). This increase in PSF occurred irrespective of the presence or absence of the correlation between CM cell discharge rate and target muscle activity referred to above. We also found that this increase was still present if spikes preceded by short intervals (<25 ms) were removed from the triggering spike train. Saturation effects were evident at the highest EMG level in some cases, and some cells did not produce detectable PSF at the lowest EMG levels.

Changes in the *relative* degree of facilitation (%Mod) were more variable. It remained constant in 9 combinations, increased in 13, and fell in the remaining 7. Interestingly, we found that the relative amplitude of PSF exerted by a *population* of CM cells, all of which facilitated the same hand muscle, remained constant over a range of EMG levels.

CONCLUSIONS

There is considerable flexibility in the relationship between the levels of activity in a CM cell and in its target muscle. A CM cell may exhibit a different relationship with a muscle during the movement and hold periods of the same task, and, during the hold period, different neurones may show a positive, negative or no correlation with EMG level.

This flexibility could arise for several different reasons. A weak correlation might suggest that the CM cell is providing only one of many excitatory inputs to the motoneurone pool. In this case, one could expect CM cells that produce large PSF to show a better correlation. We found a significant relationship between strength of PSF and the strength of correlation between cell and muscle activity. Most of the stronger CM effects (%Mod >20%) were found in cell-muscle combinations with significant positive correlations in their activity levels (see Figure 2). These large PSFs are underpinned by strong effects on individual motoneurones (Mantel & Lemon, 1987; see Porter & Lemon, 1993, pp. 177-186), and these CM inputs can provide a substantial proportion of the excitatory drive required to maintain steady discharge in these motoneurones. For these CM cells, the pattern of discharge is an important determinant of PSF amplitude (Lemon & Mantel, 1989).

However, there were some cases in which a CM cell did exert strong PSF and yet showed no significant correlation (Figure 2; see also Maier et al., 1993). This result still awaits explanation; perhaps it should prompt us to rethink the basic assumption that all central coding is related to changes in firing frequency (Lemon & Mantel, 1989). It must also be born in mind that both CM cells and target muscles show a high degree of trial-to-trial variability in their pattern of recruitment in the task. This probably reflects the many degrees of freedom involved in producing a given pattern of digit force or displacement. The absence of a correlation between a CM cell and its target muscle in averages compiled from many trials may conceal a very specific task-related engagement of the CM cell, but one which is not used (or,indeed, does not need to be used) in all trials (Maier et al., 1993).

Our analysis shows that the facilitation exerted by a CM cell generally increased with increasing EMG activity in the target muscle. In some cell-muscle combinations there was evidence of PSF increasing in proportion to the background EMG, a form of 'automatic gain compensation' (Gottlieb, Agarwal & Stark, 1970; Matthews, 1986). This may indicate that the CM cell, which is known to arborise widely within the ventral horn, may exert equally strong effects on the different motoneurones recruited for the precision grip task. For other combinations, the lack of PSF at low EMG levels, or the saturation of PSF at higher EMG levels tended to suggest that the cell exerted rather uneven effects on motoneurones of different sizes.

An important additional mechanism that may change the efficacy of the CM-EPSP is the 'setting' of the excitability of the motoneurone pool by other inputs (Crone, Hultborn, Kiehn, Mazieres & Wigström, 1988; Kernell & Hultborn, 1990). Inputs from other CM cells, and from other descending and segmental pathways could have contributed to changes at the motoneuronal level which caused PSF to increase with EMG activity for nearly all CM cells, irrespective of the correlation of the cell to the level of target muscle EMG activity.

We clearly need to know more about the operation of the CM synapse, and whether the CM system can, in fact, impose distinct patterns of EMG activity. Our results indicate that each CM cell-muscle combination operates in a particular fashion as the motoneurone pool becomes more active. Selective reinforcement of the excitatory drive at particular levels of muscle activity could be achieved by changing the firing rate within the population of CM cells facilitating the muscle, and this mechanism may be of strategic importance during fine control of precision finger movements.

ACKNOWLEDGMENTS

Rosalyn Cummings is thanked for her expert technical assistance. This work was supported by the MRC, Action Research and Felice Rosemary Lloyd Trust (Australia).

REFERENCES

BENNETT, K.M.B. (1991) Corticomotoneuronal control of muscle activity during the performance of precision grip tasks by the monkey. Ph.D. Thesis. Cambridge University.

BENNETT K.M.B. & LEMON R.N. (1994) The influence of single monkey cortico-motoneuronal cells at different levels of activity in target muscles. *J. Physiol.*, **477**, 291-307.

BUYS, E.J., LEMON, R.N., MANTEL, G.W.H. & MUIR, R.B. (1986) Selective facilitation of different hand muscles during voluntary movement in the conscious monkey. *J. Physiol.* **381**, 529-549

CRONE, C., HULTBORN, H., KIEHN, O., MAZIERES, L. & WIGSTRÖM, H. (1988) Maintained changes

in motoneuronal excitability by short-lasting synaptic inputs in the decerebrate cat. *J. Physiol.* **405**, 321-343.
FETZ, E.E. & CHENEY, P.D. (1980) Postspike facilitation of forelimb muscle activity by primate corticomotoneuronal cells. *J. Neurophysiol.* **44**, 751-772.
FETZ, E.E. & CHENEY, P.D. (1987) Functional relations between primate motor cortex cells and muscles: fixed and flexible. In *Motor Areas of the Cerebral Cortex.* eds. BECK, G., O'CONNOR, M. & MARSH, J. pp.98-117. Wiley, Chichester.
GOTTLIEB, G.L., AGARWAL, G.C. & STARK, L. (1970) Interactions between voluntary and postural mechanisms of the human motor system. *J. Neurophysiol.* **33**, 365-381.
KERNELL, D. & HULTBORN, H. (1990) Synaptic effects on recruitment gain: a mechanism of importance for the input-output relations of motoneurone pools? *Brain Res.* **507**, 176-179.
LEMON, R.N., BENNETT, K.M.B. & WERNER, W. (1991) The cortico-motor substrate for skilled movements of the primate hand. In *Tutorials on Motor Neuroscience.* eds STELMACH, G.E. & REQUIN, J. pp. 477-495. Kluwer Academic Publishers, Holland.
LEMON, R.N. & MANTEL, G.W.H. (1989) The influence of changes in discharge frequency of corticospinal neurones on hand muscles in the monkey. *J. Physiol.* **413**, 351-378.
LEMON, R.N., MANTEL, G.W.H. & MUIR, R.B. (1986) Corticospinal facilitation of hand muscles during voluntary movement in the conscious monkey. *J. Physiol.* **381**, 497-527.
MAIER, M., BENNETT, K.M., HEPP-REYMOND, M.-C. & LEMON, R.N. (1993). Contribution of the monkey corticomotoneuronal system to the control of force in precision grip. *J. Neurophysiol.* **69**, 772-785.
MANTEL, G.W.H. & LEMON, R.N. (1987) Cross-correlation reveals facilitation of single motor units in thenar muscles by single corticospinal neurones in the conscious monkey. *Neurosci. Lett.* **77**: 113-118
MATTHEWS, P.B.C. (1986) Observations on the automatic compensation of reflex gain on varying the pre-existing level of motor discharge in man. *J. Physiol.* **374**, 73-90.
MUIR, R.B. & LEMON, R.N. (1983). Corticospinal neurons with a special role in precision grip. *Brain Res.* **261**, 312-316.
PORTER, R. & LEMON, R.N. (1993) *Corticospinal Function and Voluntary Movement.* Oxford University Press. Oxford.
SMITH, A.M. (1981). The coactivation of antagonist muscles. *Can. J. Physiol. Pharm.* **59**, 733-747.

CEREBRAL CORTICAL CONTROL OF PRIMATE OROFACIAL MOVEMENTS: ROLE OF FACE MOTOR CORTEX IN TRAINED AND SEMI-AUTOMATIC MOTOR BEHAVIOURS

R.E. Martin[1], G.M. Murray[2] and B.J. Sessle[3]

[1]The Toronto Hospital, Toronto, Canada
[2]Faculty of Dentistry, University of Sydney Australia
[3]Faculty of Dentistry, University of Toronto Toronto, Canada

INTRODUCTION

Despite the extensive literature on the role of the sensorimotor cortex in limb motor control, only limited details were available until recently regarding the cortical mechanisms in orofacial motor behaviour in the primate. Over the past several years, we have used a combination of techniques to elucidate the organisation and neuronal properties of the primate lateral pericentral cortex in relation to orofacial sensorimotor function. In particular, we have shown that reversible cold block of the awake monkey's face motor cortex (MI), including tongue-MI, significantly reduces the successful performance of a trained tongue protrusion task, but causes little disruption of a biting task (Murray, Lin, Moustafa & Sessle, 1991). These findings are consistent with intracortical microstimulation (ICMS) data indicating a large tongue motor representation in MI but only a minor jaw-closing representation (see Huang, Sirisko, Hiraba, Murray & Sessle, 1988). They also agree with findings that single neurones located at ICMS-defined tongue-MI sites exhibit activity patterns selectively related to the tongue task, including directional specificity, whereas MI neurones with activity related to the biting task are comparatively scarce (Murray & Sessle, 1992 b,c). We have also shown that many tongue-MI neurones are characterised by a close spatial matching of somatosensory input and motor output (Huang, Hiraba & Sessle, 1989a; Murray & Sessle, 1992a). While these findings point to a role for many face-MI neurones in the generation and fine control of voluntary orofacial movements, our ICMS data also revealed that semi-automatic movements (such as those in mastication and swallowing) can be evoked from face MI (see Huang, Hiraba, Murray & Sessle, 1989b). Since this finding suggests that MI might participate in the initiation or control of semi-automatic as well as trained motor behaviours, further studies have been carried out to test this possibility.

METHODS

This study was carried out in 2 female monkeys (M. fascicularis) cared for in accordance with the Guiding Principles of the American Physiological Society and the guidelines of the Canadian Council for Animal Care. The methods have been extensively detailed elsewhere (Huang et al., 1988, 1989a,b; Murray et al., 1991; Murray & Sessle, 1992a,b; Moustafa, Lin, Murray & Sessle, 1994) and will therefore only be briefly outlined. Prior to implantation of a stainless-steel cylinder to allow transdural access of glass-coated tungsten microelectrodes to the sensorimotor cortex, each animal was trained to protrude its tongue to engage a small force transducer placed at a set distance from its mouth (see Murray et al., 1991; Murray & Sessle, 1992b). The transducer output moved a video screen cursor (in front of the animal), and the animal's task was to move the cursor into a target window that reflected a force level of 1.0N. A successful task trial was rewarded with 0.4ml of juice if the cursor remained within the baseline window during the pretrial period, exited the baseline window within 3s of target appearance, and remained within the target window for at least 0.5s. The electromyographic (EMG) activity of several orofacial muscles, monitored by chronically or acutely placed EMG electrodes, revealed that the genioglossus was the primary agonist for the tongue task (Murray et al., 1991; Moustafa et al., 1994). The EMG activity as well as neuronal activity (see below) were also recorded as the animal masticated and swallowed standard-sized pieces of apple or swallowed fruit juice. Orofacial movements were also recorded on video tape.

After a task success rate of 50-70% was achieved, a series of microelectrode penetrations were made in face MI, SI and more lateral cortical regions including the classical masticatory area (CMA) and swallow cortex. In each penetration, ICMS was systematically applied at neuronal recording sites to allow for a functional identification of the face MI, CMA and swallow cortex based on characteristic evoked movements and associated EMG patterns (e.g. Huang et al., 1989a,b; Murray & Sessle, 1992a). This report focuses on the properties of single neurones recorded at ICMS-defined output loci within face MI and their extracellularly recorded activity in relation to (a) the force-development and force-holding phases of the trained tongue task, (b) swallowing of the task juice reward, (c) mastication and swallowing of apple, and (d) sucking and swallowing of juice from a syringe. Superficial and deep sensory inputs to each neurone were also tested by the application of non-noxious mechanical stimuli to the orofacial tissues. A neurone was considered to have task, mastication, or swallow-related activity if its firing frequency during the relevant period was significantly different from that in the pretrial period. The significance of such a difference was determined by computer-assisted off-line analysis of usually 8 or more trials (repeated measures ANOVA and post-hoc Duncan comparisons, $p<0.05$). Recording and ICMS sites were histologically reconstructed and correlated with cytoarchitectonic areas of the cortex (e.g. Huang et al., 1989a; Murray & Sessle, 1992a).

RESULTS AND DISCUSSION

Our ICMS data confirmed earlier findings by ourselves and others of the multiple nested arrangement of the orofacial motor representation in face MI, and the extensive region of MI devoted to the tongue and facial muscles compared to the limited representation of the jaw musculature (see Huang et al., 1988; Murray & Sessle, 1992a). We also confirmed our previous findings (Huang et al., 1989b) that masticatory-like movements can be evoked by ICMS not only from CMA but also from sites within face MI (and SI), and additionally showed that ICMS at many intracortical loci within face MI and CMA could also or instead evoke swallowing. For example, of a total of 270 intracortical sites where swallowing was evoked by ICMS, orofacial muscle twitch was evoked at 50 (19%) of the sites; these 50 sites were located within the most lateral region of tongue-MI. Further, swallowing was evoked along with masticatory-like movements at 90% of the 270 sites, suggesting an extensive overlap of CMA and the swallowing cortex. Finally, increases in stimulus intensity at some intracortical sites shortened the latency of swallowing (from 3.5s to 0.5s) and increased the magnitude of swallow-related genioglossus EMG activity. This was consistent with earlier findings of the effects of changes stimulus parameters of ICMS on the properties of evoked masticatory-like movements (see Huang et al., 1989b).

The possibility that MI participates in the semi-automatic activities of swallowing and mastication, as well as in trained motor behaviours, was further borne out by our single neurone recordings in face MI. To date, a total of 72 tongue-MI neurones has been examined during both the tongue protrusion task and the swallowing of the juice reward. Of these, 46 (i.e. 64%) showed altered firing rates in relation to the tongue task and 12 (26%) of these task-related neurones exhibited modulation of firing in advance of the genioglossus activity associated with the task; these findings are consistent with our earlier data on the activity of tongue-MI neurones (Murray & Sessle, 1992b). In addition, of the 72 neurones tested, 51 (71%) showed altered firing in relation to swallowing of the juice reward. Moreover, of the 46 neurones that were significantly related to the tongue task, 34 (74%) also exhibited modulation of firing in relation to swallowing of the juice reward. The majority (approximately 80%) of these 34 neurones exhibited one of two firing patterns: 15 (44%) exhibited increased firing during both the tongue protrusion and juice reward swallow, while 12 (35%) showed increased firing during the tongue protrusion but decreased firing during the reward swallowing (e.g. Figure 1). Moreover, 71% of the 34 neurones had an intraoral mechanoreceptive field which, in the majority of cases (88%), was located on the superior tongue surface. In addition, at these 34 recording sites, ICMS evoked a variety of responses including lateral tongue movement (12/34), tongue protrusion (6/34), tongue retrusion (4/34), and more complex movements (12/34) such as depression of the tongue dorsum combined with elevation of the lateral tongue margins. These findings agree with our earlier MI data on spatial matching of input and output (Huang et al., 1989a; Murray & Sessle, 1992a) and reinforce our recent data (Murray et al., 1991; Murray & Sessle, 1992a,b) pointing to a major involvement of face MI in tongue motor control.

A total of 16 neurones that were examined during both the tongue task and juice reward swallowing were also studied in relation to chewing and swallowing of apple, and swallowing of juice from a syringe. Of these, 13 (81%) showed modulation of firing in relation to genioglossus EMG activity during mastication; in some cases, this activity was rhythmically related to the cyclic EMG activity associated with mastication while in others a more tonic pattern of activity was observed (e.g. Figure 1). In addition, 6 (46%) of these 13 also exhibited activity related to the tongue task, 11 (85%) showed altered activity during the task-related swallow, and 12 (92%) during apple or juice swallow. With regard to swallowing, 14 (88%) of the 16 neurones tested showed activity related to either swallowing of apple or swallowing of juice, and 11 to both types of swallowing. Of the 14 neurones that showed activity related to either swallowing of apple or juice, 7 (50%) also exhibited altered firing during the tongue protrusion task, while 13 (93%) of the 14 exhibited altered activity in relation to the reward swallow. Further, in 6 (43%) of the 14 neurones, the patterns of neuronal firing during the three types of swallowing were similar (i.e. increase in 3/6 and decrease in 3/6). Finally, 6 (38%) of the total of 16 neurones showed significant alterations of activity in relation to the tongue task, reward swallow, and the swallowing of apple or juice, and 5 (31%) showed activity in relation to mastication as well as these three motor behaviours (e.g. Figure 1).

Our finding that tongue-MI neurones may participate in semi-automatic orofacial movements as well as trained tongue motor tasks relates to the more general issue of

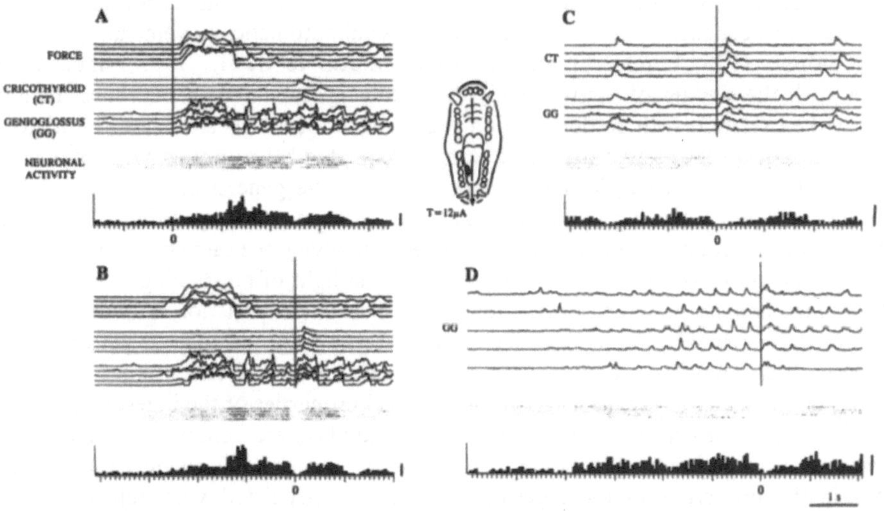

Figure 1. Activities of a single tongue-MI neurone during 5 successful trials of tongue protrusion task (A and B), swallowing of juice from a syringe (C), and chewing and swallowing of apple (D). Traces in A and B have been aligned to onset of significant increase in GG EMG activity associated with the tongue protrusion (A) and reward swallow (B); lower traces have been aligned to swallow onset (C) and swallow onset immediately following mastication (D). The neurone showed significant increase in activity during tongue protrusion (A), significant decrease in relation to reward swallow (B), juice swallow (C) and apple swallow (D), and significant increase during mastication (D). Vertical calibration bar for histogram corresponds to 10 impulses (50-ms bins). The lingual mechanoreceptive field of the neurone also is shown, as well the direction (arrow) of tongue movement produced by ICMS (at threshold intensity, T = 12 μA).

movement fractionation and the view that MI can cause muscle groups that would normally act in synergy to be independently activated (Evarts, 1986; Lemon, 1993). Certainly our face MI data (Murray & Sessle, 1992b) reveal a high selectivity of neurones for one trained motor behaviour (e.g. tongue protrusion task) over another (e.g. biting task). However, consistent with the many sites within MI from which swallowing and different types of chewing can be evoked with ICMS, our neurone recording data from face MI reveal many task-related neurones also showing activity during swallowing and/or rhythmic jaw movements carried out by the monkeys. The possibility that MI neurones are involved in both task and semi-automatic movements is of particular interest in relation to information on the role of limb MI in locomotion. For example, studies in awake cats have shown that some limb-MI neurones have activity only weakly correlated with self-paced (treadmill) locomotion but exhibit substantially greater modulations of firing when the animal adjusts its gait in order to negotiate obstacles during locomotion (e.g. Drew, 1988). Similarly, some limb-MI neurones that are activated in time with the step cycle show rapid, larger alterations of activity in response to an unexpected, externally imposed perturbation of the step cycle (Marple-Horvat, Amos, Armstrong & Craido, 1993). These findings have been taken to support the view that limb MI may modify the step cycle in light of incoming sensory inputs that reflect changing environmental conditions by superimposing a transcortical modulation of subcortical locomotor networks in order to achieve context-appropriate modifications in certain parameters of muscle activation patterns during the step cycle, without disrupting the overall locomotor rhythm. In a similar fashion, face MI has been thought to superimpose a repertoire of specific motor synergies principally on subcortical circuits responsible for the basic timing and sequencing of muscle activity of semi-automatic movements (Dubner, Sessle & Storey, 1978; Lund & Enomoto, 1988; Martin & Sessle, 1993). Indeed, in agreement with the finding of Drew (1988) that many pyramidal tract neurones do not exhibit a marked relation with cat locomotion except when obstacles are being negotiated, Hoffman & Luschei (1980) found that most biting task-related MI neurones may not exhibit a strong relation with chewing, leading Evarts (1986) to the general conclusion that MI plays a major role in controlling operantly conditioned movements (e.g. biting task) and a minor role in the control of semi-automatic movements involving the same muscles.

Such a view, however, may need to be reassessed in light of our finding of significant alterations in the activity patterns of many tongue-MI neurones in relation to both task and semi-automatic movements. During mastication and swallowing, unexpected perturbations and changing environmental conditions would appear to be the rule, rather than the exception. For example, during mastication, the physical properties of the ingested material, including its shape, consistency, temperature, and frangibility, are continually altered over successive chew cycles in preparation for swallowing, and there is evidence that such variations in the properties of the ingested material are associated with adjustments in certain aspects of the motor activities occurring in mastication and swallowing (Dubner et al., 1978; Hamlet, 1989; Lund & Enomoto, 1988; Martin & Sessle, 1993). This, combined with our previous and present findings of prominent somatosensory inputs to face MI, and the close spatial matching of sensory inputs and motor outputs (Huang et al., 1989a; Murray & Sessle, 1992a), suggests that MI may have a significant role in the control of adaptive modifications of the motor sequences of mastication and swallowing. It must also be kept in mind that a role also exists for the "classical" CMA and swallow cortex areas in the descending control of these semi-automatic behaviours (Dubner et al., 1978; Huang et al., 1989b; Lund & Enomoto, 1988; Martin & Sessle, 1993). Indeed, consistent with earlier findings (e.g. Lund & Lamarre, 1974), we have obtained preliminary data indicating that some neurones recorded in these areas at sites where ICMS evokes chewing-like movements and/or swallowing also exhibit alterations of activity in relation to these two semi-automatic behaviours and receive orofacial afferent inputs. Future studies will need to clarify the relative importance of MI versus these more lateral areas in the cortical control of mastication and swallowing.

It could be argued that the task-related and semi-automatic movement-related activity patterns of the tongue-MI neurones examined in the present study reflect reafferentation. The extent to which movement-related activity patterns of limb MI neurones reflect inputs from central sites or peripheral receptors (i.e. reafferentation) remains unclear, but we have found some neurones in which significant changes in firing preceded task-related EMG activity, suggesting the possibility that at least the task-related activity patterns of some tongue-MI neurones occur independently of peripheral inputs. The present findings also suggest a prominent representation of intraoral afferent inputs to tongue-MI neurones active during both trained and semi-automatic movements, consistent with the view that oral mechanosensory inputs may be important in the regulation of these orofacial movements. As noted above, the importance of sensory inputs in the control of mastication and swallowing is well established. Furthermore, it is possible that, as suggested for limb MI, afferent inputs to face MI are modulated during orofacial movements so that only selected inputs useful in guiding the movement or in adapting the movement to an altered orofacial environment gain access to MI; we have already demonstrated "gating" of sensory inputs to face SI neurones during the monkey's performance of tongue-protrusion or biting tasks (Lin & Sessle, 1994). The extent to which orofacial afferent inputs access face MI during movement and contribute to the activity patterns of MI neurones and descending control of orofacial movements will be the subject of our future studies.

ACKNOWLEDGEMENTS

This study was supported by Canadian M.R.C. grant MT-4918 to B.J.S.

REFERENCES

DREW, T. (1988) Motor cortical cell discharge during voluntary gait modification. *Brain Res.* **457**, 181-187.

DUBNER, R, SESSLE, B.J. & STOREY, A.T. (1978) *The Neural Basis of Oral and Facial Function.* Plenum Press, New York.

EVARTS, E.V. (1986) Motor cortex output in primates. In *Cerebral Cortex. Sensory-Motor Areas and Aspects of Cortical Connectivity.* eds. JONES, E.G. & PETERS, A. pp. 217-241. Plenum Press, New York.

HAMLET, S.L. (1989) Dynamic aspects of lingual propulsive activity in swallowing. *Dysphagia* **4**, 136-145.

HOFFMAN, D.S. & LUSCHEI, E.S. (1980) Responses of monkey precentral cortical cells during a controlled jaw bite task. *J. Neurophysiol.* **44**, 333-348.

HUANG, C.-S., SIRISKO, M.A., HIRABA, H., MURRAY, G.M. & SESSLE, B.J. (1988) Organization of the primary face motor cortex as revealed by intracortical microstimulation and electrophysiological identification of afferent inputs and corticobulbar projections. *J. Neurophysiol.* **59**, 796-818.

HUANG, C.-S., HIRABA, H. & SESSLE, B.J. (1989a) Input-output relationships of the primary face motor cortex in the monkey (Macaca fascicularis). *J. Neurophysiol.* **61**, 350-362.

HUANG, C.-S., HIRABA, H., MURRAY, G.M. & SESSLE, B.J. (1989b) Topographical distribution and functional properties of cortically induced rhythmical jaw movements in the monkey (Macaca fascicularis). *J. Neurophysiol.* **61**, 635-650.

LEMON, R.N. (1993) Cortical control of the primate hand. *Exp. Physiol.* **78**, 263-301.

LIN, L.-D. & SESSLE, B.J. (1994) Functional properties of single neurons in the primate face primary somatosensory cortex. III. Modulation of responses to peripheral stimuli during trained orofacial motor behavior. *J. Neurophysiol.* (in press).

LUND, J.P. & ENOMOTO, S. (1988) The generation of mastication by the mammalian central nervous system. In *Neural Control of Rhythmic Movements in Vertebrates.* eds. COHEN, A.H., ROSSIGNOL, S., & GRILLNER, S., pp. 41-72. Wiley, New York.

LUND, J.P. & LAMARRE, Y. (1974) Activity of neurones in the lower precentral cortex during voluntary and rhythmical jaw movements in the monkey. *Exp. Brain Res.* **19**, 282-299.

MARPLE-HORVAT, D.E., AMOS, A.J., ARMSTRONG, D.M. & CRAIDO, J. M. (1993) Changes in the discharge patterns of cat motor cortex neurones during unexpected perturbations of on-going locomotion. *J. Physiol.* **462**, 87-113.

MARTIN, R.E. & SESSLE, B.J. (1993) The role of the cerebral cortex in swallowing. *Dysphagia* **8**, 195-202.

MOUSTAFA, E.M., LIN, L.-D., MURRAY, G.M. & SESSLE, B.J. (1994) An electromyographicanalysis of orofacial motor activities during trained tongue-protrusion and biting tasks in monkeys. *Arch. Oral Biol.* (in press).

MURRAY, G.M., LIN, L.-D., MOUSTAFA, E. & SESSLE, B.J.(1991) Effects of reversible inactivation by cooling of the primate motor cortex on the performance of a trained tongue-protrusion task and a trained biting task. *J. Neurophysiol.* **65**, 511-530.

MURRAY, G.M. & SESSLE, B.J. (1992a) Functional properties of single neurons in the face primary motor cortex of the primate. I. Input and output features of tongue motor cortex. *J. Neurophysiol.* **67**, 747-758.

MURRAY, G.M. & SESSLE, B.J. (1992b) Functional properties of single neurons in the face primary motor cortex of the primate. II. Relations with trained orofacial motor behavior. *J. Neurophysiol.* **67**, 759-774.

MURRAY, G.M. & SESSLE, B.J. (1992c) Functional properties of single neurons in the face primary motor cortex of the primate. III. Relations with different directions of trained tongue protrusion. *J. Neurophysiol.* **67**, 775-785.

THE REPRESENTATION OF BREATHING MOVEMENTS IN THE CAT INFERIOR OLIVE

T.A. Sears[2], S. Magzoub[1,2] and C. Seers[1]

[1]Institute of Neurology, Queen Square
London WC1, UK
[2]Sherrington School of Physiology, UMDS
St. Thomas' Hospital, London SE1 7EH, UK

INTRODUCTION

The rib cage is a complex shell-like structure which undergoes rhythmical changes in configuration due to intercostal and related muscle activities. These automatic breathing movements, in concert with those of the diaphragm and abdominal wall displacements ventilate the lungs. As well as participating in other respiratory reflex activities, such as coughing, the rib cage provides axial stability and through its attachments to the upper limb girdle contributes to setting the postural stage for the execution of skilled fore-limb movements. It also powers phonation and by its fine regulation of sub-glottal pressure determines prosodic features of the human voice in speech and song. To mediate these diverse motor strategies the rib cage itself needs structural stability and to this end comprehensive information about the configuration and movements of the rib cage is signalled to the CNS by intercostal muscle spindles, tendon organs and costo-vertebral joint mechanoreceptors. At segmental level the spindle inputs subserve locally organised stretch reflexes which operate in cascade due to the multi-segmental organisation. Furthermore, their functional operation is closely linked to the spatially discrete regions of the rib cage activated by the current, central command for rib cage movements, either automatic, dictated by the prevailing chemical drive and behavioural state (awake, REM or non-REM sleep), or voluntary as in speech. These different skeletomotor activities range from the 'most automatic' to the 'most voluntary'. The former have the special attribute of occurring without volition in the sleeping and awake states but obtrude abruptly into consciousness in the event of an unexpected mechanical load, so that breathing movements provide a unique example of sensori-motor coordination (Sears, 1971). Indeed, Newsom Davis & Sears (1970) found it necessary to invoke a complex centrifugal control of afferent transmission from chest wall proprioceptors to account for both the sequence of inhibitory and excitatory reflex responses of voluntarily activated human intercostal muscles to unexpected mechanical loads and for the different way these reflexes were modified according to 'prior instruction', known otherwise as 'set'. Their explanation conjoined α-γ co-oactivation, error detection and servo assistance for the execution of learnt movements in relation to predictable loads (including learning as an aspect of development and the automation of habitual motor skills) with the occurrence of an unexpected mechanical load and proprioception resulting therefrom, known otherwise as 'event detection'. It might be expected, therefore, that proprioceptive information from the chest wall would also be directed to the cerebellum there to be involved in motor learning of the movement strategies underlying the different tasks (see Ito, 1984) including their automation. However, little relevant information is available about such mechanisms save for the reports of respiratory modulation of both Purkinje cell simple spike discharge and of intercostal-evoked climbing fibre responses in the thoracic b zone of the cerebellar vermis (Baker, Seers & Sears, 1993). In view of the suggested role of the inferior olive as a 'comparator' of intended and actual movement achieved (Oscarsson, 1969) and the plasticity conferred on mossy fibre transmission to Purkinje cells by climbing fibre inputs (see Ito, 1989), we have sought evidence for the representation of breathing movements in the inferior olive. The central respiratory rhythm generator shows a high level of autorhythmicity in the experimental preparation, but still subject to normal chemical and reflex controls. Study of such fictive movements under paralysis thus offered the prospect of gaining insights as to the nature of the central command for movement fed to the olive, as hypothesized by Oscarsson (1980) and also as inferred from the Purkinje cell complex spikes (Andersson & Armstrong, 1987) or olivary discharges (Gellman, Gibson & Houk, 1985) which occur in response to perturbed locomotion in awake animals.

METHODS

Experiments were carried out on cats anaesthetised with pentobarbitone, paralysed with gallamine triethiodide and artificially ventilated at a low stroke volume and high rate (40 - 60/min) to produce hypocapnic apnoea, typically at an end-tidal CO_2 fraction (ET_{CO2}) of 2.0 - 2.5%. Carbon dioxide could then be added to bring the ET_{CO2} to either the threshold for rhythm generation (ET_{CO2}trg), to the same level it had during spontaneous breathing, or to any other level. To determine the phasing and output of the central respiratory rhythm generator, α and γ motoneurone discharges were recorded from external (inspiratory) and internal (expiratory) intercostal nerve filaments (Sears, 1964a). Extracellular recordings were made from the inferior olive using glass-coated platinised tungsten electrodes (exposed tips of 5 or 25 μm) after removal of the posterior vermis to expose the obex. The recording sites were verified histologically in paraffin sections stained with Luxol Fast Blue and Cresyl Violet.

RESULTS

Segmental components of breathing

Before probing the olive the ET_{CO2}trg was determined. The state of hypocapnia is characterised by the tonic discharge of expiratory α and γ motoneurones (Sears, 1964a) and, as CO_2 is progressively incremented in steady state steps of 3min duration, rhythm generation can be detected by the onset of phasic inspiratory activity or, more sensitively, by the onset of phasic inhibition of tonic expiratory activity (Bainton, Kirkwood & Sears, 1978). Such tonic firing of the lowest threshold expiratory motoneurones is detectable in filaments innervating the most proximal sector of each intercostal space; and its topographic distribution across the rib cage corresponds to that of the low threshold, phasic expiratory motoneurone discharge during rhythm generation (see Sears, 1990). Provided that the segmental afferent innervation is substantially intact these expiratory motoneurones are reflexly activated during the inflation phase of the pump. This is due to stretch of internal intercostal muscle spindles and the monosynaptic excitation of homonymous and heteronymous internal intercostal motoneurones by afferents from primary and secondary (Kirkwood & Sears, 1974) endings supported probably by oligosynaptic pathways (Kirkwood & Sears, 1982). This unique manifestation of a sustained stretch reflex in the anaesthetised preparation owes its presence to a CO_2 dependent, brainstem-mediated drive conveyed by expiratory bulbospinal neurones (Bainton & Kirkwood, 1979) which monosynaptically excite the motoneurones (Kirkwood & Sears, 1974). With the onset of rhythm, as CO_2 is increased, this stretch reflex becomes increasingly confined to the expiratory phase, being gated out during inspiration by the deepening postsynaptic inhibition which is linked reciprocally to the intensity of the central inspiratory drive (Sears, 1964b). Similar considerations broadly apply to the chest wall deflation reflex. This reflex can be demonstrated during the inspiratory phase when rhythm generation is present and although it is often absent during hypocapnic apnoea (when the corresponding inspiratory bulbospinal neurones are mostly inactive; Cohen, 1964), can appear when the ET_{CO2} is held just below the threshold for rhythm generation. Similarly, the 'inspiratory shift' produced by a peripheral chemoreceptor drive during apnoea (Sears, Berger & Phillipson, 1982), can be used to bring the deflation reflex of the external intercostal motoneurones to threshold (Sears, unpublished observations). As shown below, the use of the fast rate of artificial ventilation was crucial in disclosing, and making sense of, the signals encountered in the olive as it allowed the central and proprioceptive components to be distinguished.

Recordings from the inferior olive

Systematic exploration of the inferior olive was carried out at an ET_{CO2} of usually

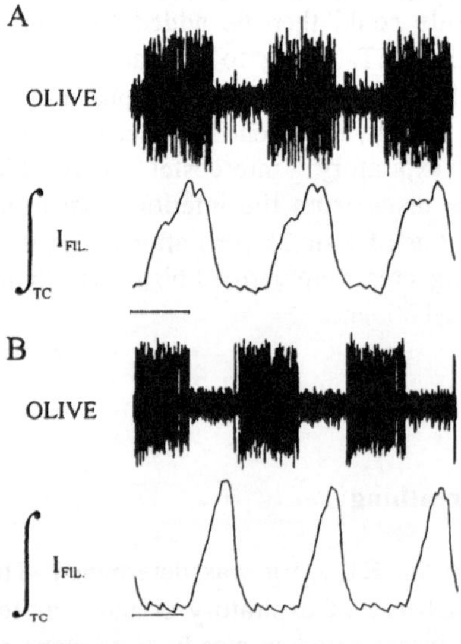

Figure 1. A and B, upper traces, extracellular recordings from cDAO; lower two traces, integrated inspiratory (IFil) motoneurone activities recorded from external intercostal nerve filaments. Note inspiratory (A) and expiratory (B) phasing of high frequency olivary discharge and also the presence of inspiratory phased low voltage mass activity in B.

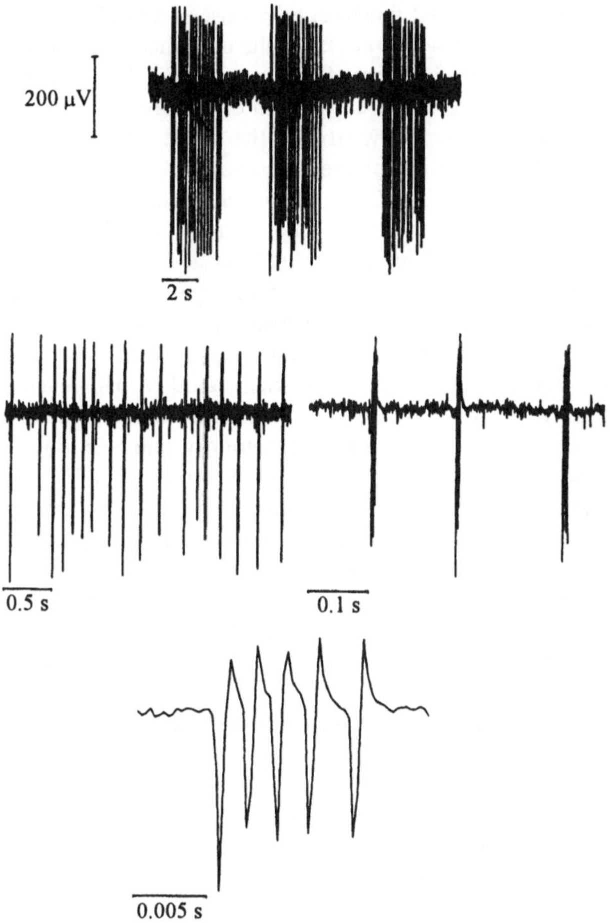

Figure 2. Expiratory phased olivary neurone discharge recorded at different sweep speeds to show respiratory, low-frequency (4 - 6 Hz) and burst (500 Hz) components of the single unit discharge.

0.5% above its eupnoeic level, such that phasic inspiratory and expiratory motoneurone activity was present. Under these experimental conditions two principal types of 'respiratory' phased activity were encountered,each comprising two sub groups, and all were located in the dorsal accessory olive (DAO). One type, as illustrated in Figure 1, consisted of low-amplitude, diphasic discharges of 1 - 2 ms duration forming mass activity in phase with either inspiration (A) or expiration (B). When resolved into single units the discharges showed strong respiratory modulation, the pattern resembling that of inspiratory or expiratory bulbospinal neurones of the ventral respiratory group (see Merrill, 1974) but discharging at lower frequencies, augmenting from 25 - 60 Hz, for example, within the half cycle. Induction of hypocapnic apnoea invariably led to the loss of the mass inspiratory phased activity, but there were some indications of a persisting, but poorly discriminated tonic discharge at the recording sites.

The other type of respiratory phased activity consisted of high amplitude, long duration (2 - 4 ms) spikes which could discharge either singly and sporadically, but with a preferred inspiratory or expiratory phasing; or repetitively at low frequencies (2 - 15 Hz) throughout a phase (Figure 2). Interestingly, as in Figure 2, a low frequency (4 - 5 Hz) discharge of a single unit would be found on a fast timebase to consist of a brief duration, high frequency (400 - 500 Hz) burst discharge(1 - 6 impulses) of the unit, this complex repeating at the low frequency of 4 - 5 Hz for the duration of the respiratory phase. Such burst discharges described here in relation to each impulse of a train of low frequency discharges of a single unit are identical to those which follow usually the single discharge of olivary neurones evoked by electrical stimulation of periperal nerves, as described by several authors (Armstrong & Harvey, 1966; Crill, 1970). However, its occurrence in association with sustained, repetitive firing of the single unit implies a sustained, probably strong synaptic drive although intrinsic electrical properties (Llinás & Yarom, 1989) might play an important role in maintaining olivary neurone rhythmicity under these in vivo conditions. In the example of Figure 2 the phasing was unambiguously expiratory, but more commonly, during rhythm generation, classification of the phase was represented by a greater probability of discharge in one phase rather than another. However, when CO_2 was withdrawn to create hypocapnic apnoea, such records invariably simplified to reveal low-frequency firing locked to the phasing of the respiratory pump, during either inflation or deflation. The recordings illustrated in Figure 3 were taken with the ET_{CO2} held close to the threshold for rhythm generation. Periods of central rhythm generation (left side, first two cycles) alternated with periods of apnoea (centre) during which the pump-locked, chest wall deflation and inflation reflexes occurred in external (middle trace) and internal (lower trace) intercostal motoneurones, respectively, to be replaced by central rhythm generation (right side, last two cycles). During central rhythm generation the olivary discharge was sporadic and of indeterminate phasing. During apnoea it became clearly locked to the deflation phase of the pump, at which time external intercostal muscle spindles become increasingly active (Kirkwood & Sears, 1982), but with the resumption of central rhythm generation it again fired sporadically.

Olivary location of repiratory phased activities

The left-hand panel in Figure 4 shows the spatial distribution of the different categories

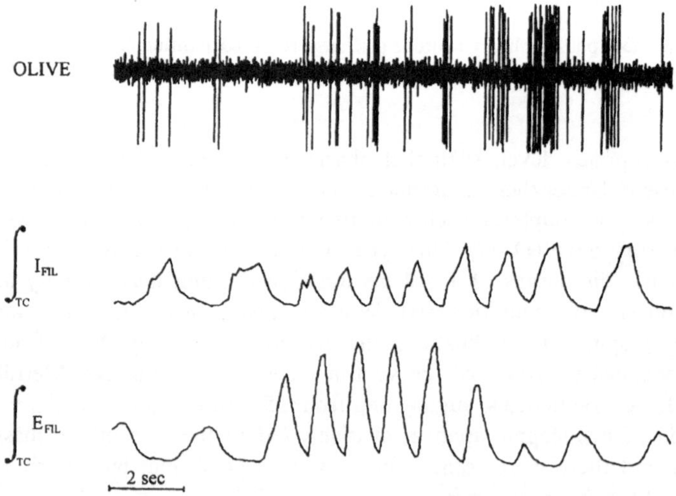

Figure 3. Upper trace, low frequency olivary neurone discharge showing transition from sporadic firing during respiratory rhythm generation to deflation phased, pump-locked activity during apnoea and converse with resumption of rhythm generation. Lower two traces, integrated neurograms of inspiratory (IFil) and expiratory (EFil) motoneurone activities recorded from external and internal intercostal nerve filaments, respectively.

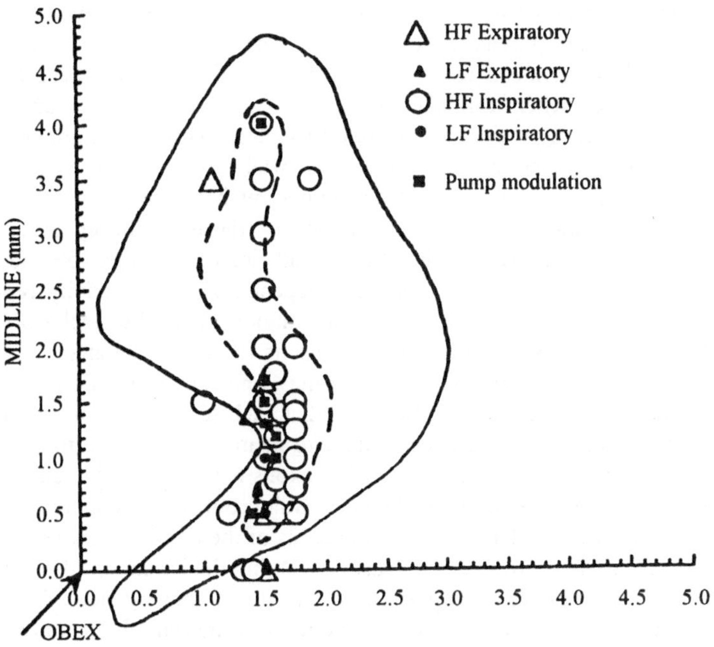

Figure 4. Distribution of all respiratory related activities in dorsal accessory olive. Dotted line designates termination sites of spino-olivary fibres from Matsushita et al. (1992).

of respiratory phased activities plotted in relation to the spatial outline of the dorsal accessory olive (DAO) as determined from our own histological material. The locations form a column extending throughout the DAO with a preponderance of points in the caudal half. Only at the end of this study did we become aware of the work of Matsushita, Yaginuma & Tanami (1992) who had determined the somatotopic termination of the spino-olivary fibres in the cat, segment by segment, using anterograde transport of wheat germ agglutinin-horseradish peroxidase. The dotted line in the right-hand panel in Figure 4 shows the location of the thoracic segments from that work, superimposed on our results to reveal a remarkable degree of correspondence. It should be noted that this map, based on the constellation of points representing the central respiratory drive and chest wall proprioceptor inputs, has a preponderantly caudal distribution in the DAO, compared to the more rostral placing of the trunk in the work of Gellman, Houk & Gibson (1983), based mainly on cutaneous inputs.

DISCUSSION

In a previous investigation it was established that both the intercostal-evoked climbing fibre responses and the Purkinje cell simple spike activity in the vermal, cerebellar thoracic b zone, are modulated during the fictive respiratory cycle (Baker et al., 1993). Those findings indicated that the central respiratory drive is directed towards the spino-olivary and the spino-cerebellar components of the cerebellar afferent systems. Concerning the latter system, this conclusion has now been directly confirmed by the demonstration in the paralysed cat of respiratory phased activity in some thoracic spinocerebellar tract (SCT) neurones with crossed ascending axons, i.e., from neurones corresponding to those of the ventral spinocerebellar tract subserving the hindlimbs (Tanaka & Hirai, 1994). Such activity was out of phase with the phrenic nerve discharge and this phasing determined whether chest expansion (due to artificial respiration) evoked a discharge of these crossed SCT neurones. Furthermore, when artificial respiration was stopped the units continued discharging, albeit less strongly, during the expiratory phase. However, in the non-paralysed preparation the inspiratory movement of the chest wall now occurred in phase with the central inspiratory discharge, the net outcome being a much reduced modulation of the unit activity during the respiratory cycle. The pattern of these responses of crossed SCT neurones closely parallel those of the chest wall inflation reflex of the expiratory motoneurones and the corresponding inflation - dependent olivary activity reported here, with regard to both the proprioceptive and central components of the respiratory drive. The simplest interpretation of these and our previous results is that the command for respiratory movement is fed to two sets of spino-olivary neurones, one conveying information about the inflation reflex, the other the deflation reflex, and that each set receives inhibition reciprocal to the antagonistic drive. The most likely candidate neurones would be those forming the ventral funiculus system whose crossed ascending axons have termination sites in the DAO corresponding to those reported here for the chest wall proprioceptive inputs. Alternatively, the input to the olive might be derived from collaterals of the crossed SCT system described above, which has the desired properties at least concerning the inflation reflex. In either case the relevant spinal neurones would appear to be the site of convergence of proprioceptive inputs and inspiratory or expiratory phased components of the central respiratory drive.

With regard to the respiratory-phased, high frequency discharges within the olive also described here, this finding would seem to require the presence of interneurones because the olivary neurones giving rise to the climbing fibres characteristically discharge at low frequencies with long duration spikes. To date, some GABAergic interneurones have been reported to occur within the olive along with the abundant GABAergic terminals found there (Nelson & Mugnaini, 1989) the former possibly mediating the inhibitory effects of rubrospinal inputs on olivary neurones (Weiss, Houk & Gibson, 1990). It must be recognised, however, that our experiments provide no direct evidence as to the immediate source of the inputs,spinal or supraspinal, although one from the latter is possible as the roof nuclei have now been shown to receive direct inputs from the rostral VRG, at least in the rat (Nunez-Abades, Pasaro & Sears; submitted for publication). Furthermore, recording in awake cats has revealed the presence of respiratory phased activity in the cerebellar roof nuclei (Gruart & Delgardo-Garcia, 1992), so that nuclear-olivary pathways are another possible source for the central respiratory drive fed to the inferior olive, as here newly described.

In relation to the general theme of the symposium, the lineage of the present work has its beginnings in the 60's and 70's. It stems from ideas relating to the length servo control of movements and the role of the muscle spindle, under gamma activation, as a comparator of the commanded and achieved change in muscle length. For the respiratory system, it became evident that the command for breathing movements, at least for the intercostal muscles, draws upon alpha and gamma coactivation as demonstrated by the direct recording of the efferent discharges in intercostal nerve filaments and the acceleration of spindle discharge during movement. However, this latter input did not constitute the entire command for movement, as would be required by a length follow-up servo mechanism driven solely through the fusimotor system. Instead, intracellular recording from intercostal motoneurones revealed 'central respiratory drive potentials' (CRDPs), which represent the central (brainstem mediated) command for respiratory movement being fed directly to the alpha motoneurones (Sears, 1964b). Its distribution clearly underlies the characteristic, topographic pattern of EMG activity in the rib cage (cf. Taylor, 1960) for the obtaining level of chemical drive, but also a sub-liminally excited fringe of motoneurones readily brought to threshold by an additional external mechanical load, including that imposed by phonation (Newsom Davis & Sears, 1970). Such a mechanism is of particular relevance for the rib cage where the in-series arrangement of the corresponding muscles in each segment, and the spindles therein, allow muscles in otherwise inactive segments (or sectors within them) to respond reflexly to changes in internal (pulmonary) or external mechanical loading. The intracellular studies also provided evidence for the reciprocal inhibition of the antagonistic (inspiratory or expiratory) alpha motoneurones as an important feature of the central command for respiratory movement The present results strongly suggest that these centrifugal controls also regulate afferent transmission of proprioceptive information in the spino-olivary-cerebellar pathways. Furthermore, the co-extensive distribution of central and proprioceptive signals indicates that the thoracic microzones of the dorsal accessory olive display a spatial mapping of the current distribution of the central command for breathing movements across the chest wall and thus the current status of interneuronal circuits locally subserving the central respiratory drive and the chest wall proprioceptive reflexes. Both inputs could derive from the spinal cord if the descending drives themselves activated the ascending pathways, as now established for thoracic neurones projecting to the cerebellum over the crossed spinocerebellar pathway. Another possibility is that the central respiratory drive could reach the olive directly from the ventral or dorsal respiratory groups although no such direct projection appears to have been described for the dorsal accessory olive. With regard to the functional significance of these results, one can only speculate. Through the projections of the DAO to the different parasagittal zones of the cerebellar cortex, namely rostral DAO to c1, c3 and d1, and hence to nucleus interpositus, and caudal DAO to the b zone and hence to the lateral vestibular nucleus (see Ito, 1984 for summary), information about the prevailing command for respiratory movement is made widely available for integration with limb movement and postural demands, respectively. Furthermore, these connections and the adaptive motor learning attributed to conjunctive activity of mossy and climbing fibre inputs to Purkinje neurones could contribute to the sensori-motor automation of breathing movements, whether subserving solely metabolic needs in sleep or wakefulness (eupnoea), as asssociated movements during changes in posture, or to power phonation in speech and song, the most voluntary of human motor activities requiring decades of motor learning.

ACKNOWLEDGEMENTS

The support of the Brain Research Trust is gratefully acknowledged.

REFERENCES

ANDERSSON, G. & ARMSTRONG, D.M. (1987) Complex spikes in Purkinje cells in the lateral vermis (b zone) of the cat cerebellum during locomotion. *J. Physiol.* **385**, 107-134.

ARMSTRONG, D.M. & HARVEY, R.J. (1966) Responses in the inferior olive to stimulation of the cerebellar and cerebral cortices in the cat. *J. Physiol.* **187**, 553-574.

BAINTON, C.R. & KIRKWOOD, P.A. (1979) The effects of carbon dioxide on the tonic and rhythmic discharges of expiratory bulbospinal neurones. *J. Physiol.* **296**, 219-314.

BAINTON, C.R., KIRKWOOD, P.A. & SEARS,T.A. (1978) On the transmission of the stimulating effects of carbon dioxide to the muscles of respiration. *J. Physiol.* **280**, 249-272.

BAKER, S., SEERS, C. & SEARS, T.A. (1993) Respiratory modulation of afferent transmission to the cerebellum. In *Respiratory Control. Central and Peripheral Mechanisms.* eds. SPECK, D.F., DEKIN, M.S., REVELETTE, W.R. & FRAZIER, D.T. pp 95-99. University Press of Kentucky, Kentucky, USA.

COHEN, M.I. (1964) Respiratory periodicity in the paralysed, vagotomised cat: hypocapnic polypnoea. *Amer. J. Physiol.* **206**, 847-864.

CRILL, W. (1970) Unitary multiple-spiked responses in cat inferior olivary nucleus. *J. Neurophysiol.* **33**, 199-209.

GELLMAN, R., GIBSON, A.R. & HOUK, J.C. (1985) Inferior olivary neurones in the awake cat: detection of contact and passive body displacement. *J. Neurophysiol.* **54**, 40-60.

GELLMAN, R.S., HOUK, J.C. & GIBSON, A.R. (1983) Somatosensory properties of the inferior olive in the cat. *J. Comp. Neurol.* **215**, 228-243.

GRUART, A. & DELGARDO-GARCIA, J.M. (1992) Respiration-related neurons recorded in the deep cerebellar nuclei of the alert cat. *Neuroreport.* **3**, 365-368.

ITO, M. (1984) *The Cerebellum and Neural Control.* Raven Press, New York.

ITO, M. (1989) Long-term depression. *Ann. Rev. Neurosci.* **12**, 85-102.

KIRKWOOD, P.A. & SEARS, T.A. (1974) Monosynaptic excitation of motoneurones from secondary endings of muscle spindles. *Nature.* **252**, 243-244.

KIRKWOOD, P.A. & SEARS, T.A. (1982) Excitatory post-synaptic potentials from single muscle spindle afferents in external intercostal motoneurones of the cat. *J. Physiol.* **322**, 287-314.

LLINÁS, R. & YARON, Y. (1986) Oscillatory properties of guinea pig inferior olivary neurones and their pharmacological modulation; An *in vitro* study. *J. Physiol.* **376**, 163-182.

MATSUSHITA, M., YAGINUMA, H. & TANAMI, T. (1992) Somatotopic termination of the spino-olivary fibres in the cat studied with the wheat germ agglutinin-horseradish peroxidase technique. *Exp. Brain. Res.* **89**, 397-407.

MERRILL, E.G. (1974) Finding a respiratory function for the medullary respiratory neurons. In *Essays on the Nervous System. A Festschrift for Professor J.Z. Young.* eds. BELLAIRS, R. & GRAY, E.G. pp. 451-486. Clarendon Press, Oxford.

NELSON, B.J. & MUGNAINI, E. (1989) Origins of GABAergic inputs to the inferior olive. In *Exp. Brain Res.* Series 17, 86-107.

NEWSOM DAVIS, J. & SEARS, T.A. (1970) The proprioceptive reflex control of the intercostal muscles during their voluntary activation. *J. Physiol.* **209**, 711-738.

OSCARSSON, O. (1969) The sagittal organisation of the cerebellar anterior lobe as revealed by the projection patterns of the climbing fibre system. In *Neurobiology of Cerebellar Evolution and Development.* ed. LLINÁS, R. pp. 525-532. Am. Med., Chicago.

OSCARSSON, O. (1980) Functional organisation of olivary projection to the cerebellar anterior lobe. In. *The Inferior Olivary Nucleus: Anatomy and Physiology.* eds. COURVILLE, J., DE MONTIGNY, C. & LAMARRE, Y. pp. 279-289. Raven Press, New York.

SEARS, T.A. (1964a) Efferent discharges in alpha and fusimotor fibres of intercostal nerves of the cat. *J. Physiol.* **174**, 295-315.

SEARS, T.A. (1964b) The slow potentials of thoracic respiratory motoneurones and their relation to breathing. *J. Physiol.* **175**, 404-424.

SEARS, T.A. (1971) Breathing: A sensori-motor act. In. *The Scientific Basis of Medicine Annual Reviews.* pp. 131-147. Athlone Press.

SEARS, T.A. (1990) Central rhythm generation and spinal integration. *Chest.* **97**, 45S-50S.

SEARS, T.A., BERGER, A.J. & PHILLIPSON, E.A. (1982) Reciprocal tonic activation of inspiratory and expiratory motoneurones by chemical drives. *Nature.* **299**, 728-730.

TANAKA,Y. & HIRAI, N. (1994) Physiological studies of thoracic spinocerebellar tract neurones in relation to respiratory movement. *Neurosci. Res.* **19**, 317-326.

TAYLOR, A. (1960) The contribution of the intercostal musclesto the effolrt of respiration in man. *J. Physiol.* **151**, 390-402.

WEISS, C., HOUK, J.C. & GIBSON, A.R. (1990) Inhibtion of sensory responses of cat inferior olive neurons produced by stimulation of red nucleus. *J. Neurophysiol.* **64**, 1170-1185.

DEVELOPMENT OF CORTICAL CONTROL IN BABIES AND CHILDREN

J.A. Eyre and S. Miller

Department of Child Health
University of Newcastle upon Tyne
Newcastle upon Tyne NE2 4HH, UK

INTRODUCTION

In the perinatal period rapid maturation of neurones in the central and peripheral nervous system takes place. Broadly, there is maturation of the cell body, with increasing size of the soma and extension of the dendritic tree. In parallel, the axon enlarges in length and diameter. However, the most striking changes, and perhaps the most significant for movement control, concern synaptic projections within the central nervous system. Axonal projections become more focused with withdrawal of collateral branches. At the same time a redistribution of synapses over the soma-dendritic surface occurs. The overall density of synaptic terminals initially rises and then falls, but the relative proportion of synapses on dendrites to soma increases. The morphological types of synaptic terminals also change with age, with an increasing density, particularly on the soma, of terminals ascribed to inhibitory function. These remarkable organisational changes (for references see Eyre & Miller, 1992) are reflected in developing motor control and in neurophysiological measurements, both in the corticospinal projection and in excitatory and inhibitory spinal reflexes (Koh & Eyre, 1988; Eyre, Miller & Ramesh, 1991; O'Sullivan, Eyre & Miller, 1991; McDonough, Eyre, Kelly, Metcalfe & Miller, 1993; Issler & Stephens, 1983). The process continues to adulthood, but is most rapid during the first five postnatal years. The present paper reports changes in central motor conduction delays, their relevance for the organisation of manual skills and the synaptic linkage of the corticospinal pathway during development.

EXPERIMENTAL APPROACH

Three studies on human subjects from birth to adulthood are reported and full details of subjects and methods are given in the papers referenced: measurements of motor evoked responses and somatosensory evoked potentials (Eyre et al., 1991); synaptic linkage of the corticospinal pathway in neonates and adults (Conway, Kelly, De Kroon, Miller & Eyre,

1992); development of the pincer grasp and its relationship to the development of adult corticospinal delays (Watts, Eyre, Kelly & Ramesh, 1992). In all studies ethical approval and written, informed consent from subjects or their parent(s) were obtained.

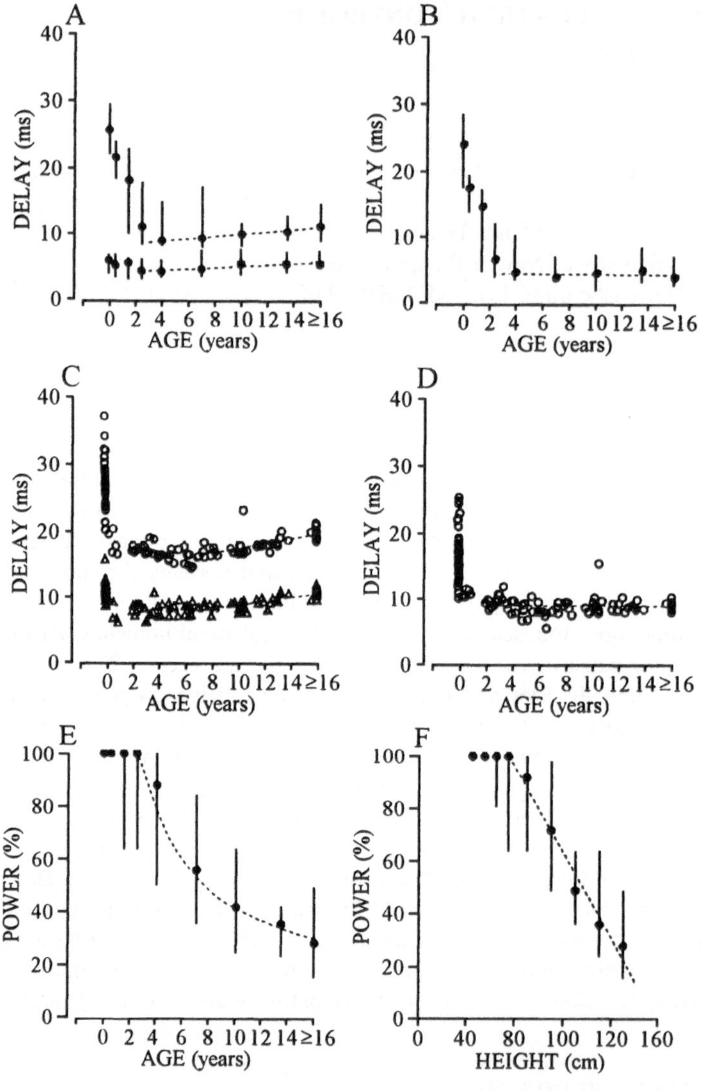

Figure 1. A & B. Motor evoked potentials in biceps brachii in relation to age in 308 subjects from 32 weeks gestation to 55 years. The upper curve of A shows the conduction delay following magnetic stimulation of the cortex, the lower curve the delays following magnetic stimulation of cervical spinal roots. The curve in B shows the central motor conduction delay, obtained by subtracting in A the lower from the upper curve. C & D. Somatosensory evoked potentials, evoked by electrical stimulation of the median nerve, in relation to age in 149 subjects aged from 34 weeks gestation to 52 years. The upper curve (open circles) in C shows the delay to the N_1 cortical response, the lower curve (filled triangles) the delay to Erb's point. The curve in D shows the central somatosensory delay obtained by subtracting in C the lower from the upper curve. E. Threshold power of magnetic stimulation for motor evoked responses in biceps brachii in relation to age. Power is expressed as the square of the voltage output of the stimulator, scaled to 100% as the maximum output. F. Threshold power (%) in relation to the height of the subject. In A, B, C & D the symbols provide the medians and the bars the 10th to 90th centile ranges. In A, B & F the regression lines were calculated on raw data for each subject over the age range 4-16 years. (Redrawn from Eyre et al., 1991).

ONTOGENY OF CENTRAL MOTOR AND SOMATOSENSORY TRANSMISSION

Sensorimotor integration plays an important role in the acquisition of motor skills in animals and in man (for references, see Eyre et al., 1991). Such integration involves not only peripheral reflex pathways, but also occurs at many levels within the central nervous system. In engineering systems, real-time controllers with feedback loops become unstable in the presence of significant transmission delays, particularly if they are variable or unpredictable. The question is therefore raised how stable, sensorimotor control is achieved during childhood, when stature increases by up to four times and the rate of growth varies over a 20-fold range (Tanner, Whitehouse & Takaishi, 1966). The nervous pathways underlying sensorimotor integration consequently undergo proportional increases in length and in their rate of change of length. In the maintenance of stability of control during childhood, the nervous system is likely to have either constant conduction delays in sensory and motor pathways or mechanisms for estimating and adjusting to changing conduction delays. In a cross-sectional study of 457 subjects, aged from newborn to adults and ranging in height from 40 - 180 cm, the conduction delays were measured in central and peripheral motor and somatosensory pathways to the upper limb (Eyre et al., 1991). Magnetic stimulation was used to determine central and peripheral conduction in motor pathways from the cortex to biceps brachii and hypothenar muscles, with responses recorded in the surface EMG. Somatosensory potentials were recorded at Erb's point and over the scalp in response to electrical stimulation of the median nerve at the wrist. The conduction delays in the central components of both motor and somatosensory pathways were found to decrease rapidly during the first two years after birth and thereafter to remain *constant* at adult values (Figure 1A-D). These observations apply only to the fastest conducting fibres. However, if this developmental phenomenon also holds for fibres of slower conduction velocities, it may provide stability of timing for such processes as internal feedback and efference copy, which have been proposed to play major roles in motor control and learning. In contrast, the conduction delays in the peripheral components of both motor and somatosensory pathways also decreased initially, but then from the age of five years progressively increased in proportion to arm length. It is surprising that the process of maturation is different in peripheral nerves, where early attainment of maximum fibre diameter (about 5 years, Gutrecht & Dyck, 1970) leads to progressive increase of conduction delay with subsequent growth.

The threshold stimulus intensity for evoking responses following magnetic stimulation of the cortex was initially high and fell progressively until the age of 16 years, showing a linear relationship with height for the range 70 - 180 cm (Figure 1E & F). Studies of corticospinal volleys and single corticospinal axons in the monkey have demonstrated that magnetic stimulation of the cortex excites corticospinal axons directly and that the threshold is linearly related to axonal diameter (Edgley, Eyre, Lemon & Miller, 1990, and in preparation). The linear relationship between threshold for a motor evoked response and height in the human studies provides further evidence that corticospinal axons continue to grow in diameter in proportion to increase in pathway length until adult height is achieved.

DEVELOPMENT OF THE PINCER GRASP AND ITS RELATIONSHIP TO THE DEVELOPMENT OF ADULT CORTICOSPINAL DELAYS

The human newborn infant has little voluntary control of arm and hand movement and develops dexterity in reaching and in performing relatively independent finger movements within the first postnatal year (Illingworth, 1983; Watts et al., 1992). Skilled movements of the hand and upper limb in primates are mediated by direct monosynaptic corticospinal projections to α-motoneurones (see Porter & Lemon, 1993). From studies in monkeys, Kuypers (1962) suggested that direct corticomotoneuronal projections are not present at birth and that the adult pattern of monosynaptic connections develops by the 6th to 8th postnatal month. In support of this conclusion, Lawrence & Hopkins (1976) found in monkeys that the adult level of skill in performing relatively independent finger movements also did not appear until 6 - 8 months and that sectioning the pyramidal tract within a few weeks of birth prevented the maturation of skilled hand and finger movements. A rapid reduction in central motor conduction delay occurs in infants between the ages of 12 - 18 months (see Figure 1B, and also Eyre et al., 1991). During this period children acquire a pincer grasp and relatively independent finger movements. The hypothesis was that the rapid reduction in central motor delay might reflect the establishment of monosynaptic corticomotoneuronal connections and would therefore precede the development of skilled hand use (Watts et al., 1992).

An 18 month longitudinal study was performed on 20 healthy children who were studied at 2 - 6 month intervals. Central motor conduction delay was measured using magnetic stimulation of the motor cortex and cervical spinal motor roots (Eyre et al., 1991). Responses were recorded in the surface EMG of biceps brachii and hypothenar muscles. The development of the pincer grasp and relatively independent finger movements was assessed using a modified Kluver board with different sized wells (Lawrence & Kuypers, 1968), from which the infants had to retrieve small, brightly coloured chocolate buttons.

The central motor conduction delay was prolonged in those aged <6 months and remained relatively constant (hypothenar muscle, median 17.3 ms, range 14.8 - 20.6 ms). The delay to both muscles shortened rapidly in individuals at varying ages between 6 - 15 months after birth, to reach adult values (hypothenar muscles, median 5ms, range 3.4 - 6.8

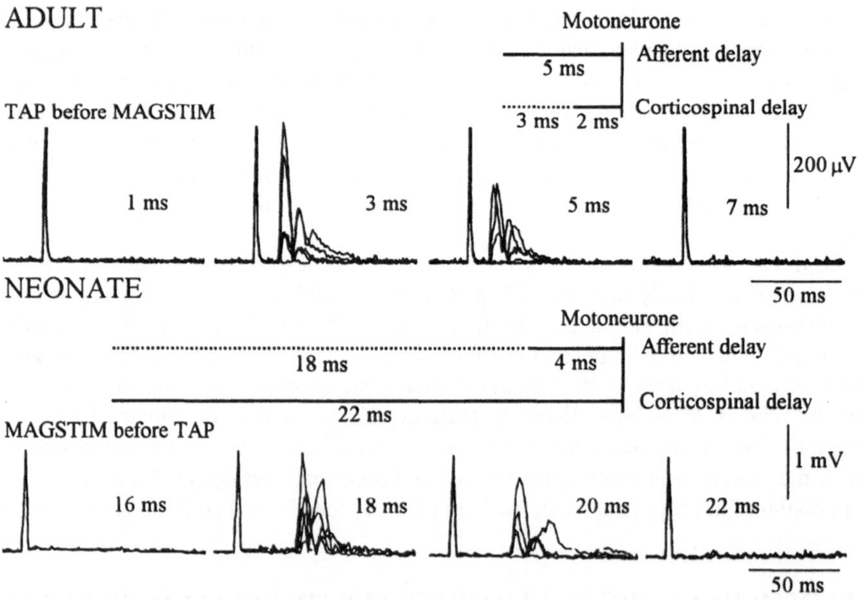

Figure 2. Trials of spatial facilitation recorded in the surface EMG of biceps brachii at different intervals between a subthreshold magnetic stimulus to the cortex (MAGSTIM) and a subthreshold tap to the tendon (TAP). The inset diagrams show the estimated time of convergence of afferent peripheral and corticospinal volleys at the motoneurones. The horizontal solid lines indicate the delay in each pathway, with each stimulus being delivered at the left hand edge of the line. The broken horizontal lines indicate the interstimulus delays. For the adult subject illustrated the TAP is given 3 ms before MAGSTIM and for the newborn subject MAGSTIM is delivered 18 ms before TAP.

ms). All but one subject developed a skilled pincer grasp and the majority developed relatively independent finger movements *before* the rapid reduction in central motor delay. If a direct projection is prerequisite for skilled hand use, as proposed by Lawrence & Hopkins (1976), the observation that skilled hand use developed without a significant change in central motor conduction delay suggests that direct monosynaptic corticomotoneuronal connections may be present in the human newborn.

EVIDENCE FOR A MONOSYNAPTIC CORTICOSPINAL PROJECTION TO MOTONEURONES OF BICEPS BRACHII IN THE HUMAN NEWBORN

The aim of this study was, therefore, to examine the synaptic linkage of the fastest conducting corticospinal axons with alpha motoneurones by estimating the rise time of cortically evoked motoneuronal EPSPs, using the technique of spatial summation (Uemura & Preston, 1965). Our hypothesis is that if the corticospinal projection is monosynaptic in the newborn then the rise time will be brief, of the order of 1 - 2 ms (monkey; Jankowska, Padel & Tanaka, 1975), and, since the projection is monosynaptic in the human adult (Boniface, Mills & Schubert, 1991), the rise time in the newborn will not be significantly longer than that observed in adult subjects. If the projection is polysynaptic then the rise time will be of the order of 5 - 10ms and will be significantly longer than that observed in adults (for references see Phillips & Porter, 1977).

The study was performed on 11 adult subjects and 11 newborn babies (Conway et al., 1992). A brief mechanical tap of the tendon of biceps brachii was used to bring the motoneurones close to threshold for a period, which was long in relation to the rise time of the EPSP evoked by subthreshold cortical stimulation (tendon tap; Burke, Gandevia & McKeon, 1983; primate monosynaptic corticomotoneuronal EPSP, Jankowska et al., 1975). Percutaneous magnetic stimulation of the cortex was used to excite a corticospinal volley subthreshold for evoking a motor response in the surface EMG of biceps brachii. The interstimulus intervals were varied to define the time course of spatial facilitation, which under these experimental conditions will define predominantly the time course of the corticomotneuronal EPSP (Conway, de Kroon, Miller & Eyre, under submission). In Figure 2 examples are shown of the responses in the EMG of biceps brachii following spatial facilitation of subthreshold stimuli. The time from onset to peak of the spatial summation was brief in both subjects groups, being 1 ms or less for 9/11 adults and 9/11 newborns, thus providing evidence to support monosynaptic corticomotoneuronal linkage in the newborn. This conclusion is supported by the observation in longitudinal studies that skilled manipulation, including relatively independent finger movements, develop at a time when central motor conduction delay is prolonged and not changing. The *subsequent* rapid fall in central motor conduction delay is therefore likely to reflect myelination of corticospinal axons, rather than a change from polysynaptic to monosynaptic corticospinal linkage.

ACKNOWLEDGEMENTS

Support from the Wellcome Trust, M.R.C., Spastics Society and Northern Region Health Authority is gratefully acknowledged.

REFERENCES

BONIFACE, S.J., MILLS, K.R. & SCHUBERT, M. (1991) Responses of single spinal motoneurones to magnetic brain stimulation in healthy subjects and patients with multiple sclerosis. *Brain* **114**, 643-662.

BURKE, D., GANDEVIA, S.C. & McKEON, B. (1983) The afferent volleys responsible for spinal proprioceptive reflexes in man. *J. Physiol.* **339**, 535-552.

CONWAY, E., KELLY, S., DE KROON, J., MILLER, S. & EYRE, J.A. (1992) *J. Physiol.* **452**, 274P.

EDGLEY, S.A., EYRE, J.A., LEMON, R.N. & MILLER, S. (1990) Excitation of the cortico-spinal tract by electromagnetic and electrical stimulation of the scalp in the Macaque monkey. *J. Physiol.* **425**, 301-320.

EYRE, J.A. & MILLER, S. (1992) The assessment of motor pathways. In *The Physiological Examination of the Newborn Infant.* ed. EYRE, J.A. pp124-154. MacKeith Press, London.

EYRE, J.A., MILLER, S. & RAMESH, V. (1991) Constancy of central conduction delays during development in man: investigation of motor and somatosensory pathways. *J. Physiol.* **434**, 441-452.

GUTRECHT, J.A. & DYCK, P.J. (1970) Quantitative teased fibre and histological studies of human sural nerve during postnatal development. *J. Comp. Neurol.* **138**, 117-130.

ILLINGWORTH, R.S. (1983) *The development of the infant and young child - normal and abnormal.* Churchill Livingstone, London.

ISSLER, H. & STEPHENS, J.A. (1983) The maturation of cutaneous reflexes studied in the upper limb in man. *J. Physiol.* **335**, 643-654.

JANKOWSKA, E., PADEL, Y. & TANAKA, R. (1975). Projections of pyramidal tract cells to alpha motoneurones innervating hind-limb muscles in the monkey. *J. Physiol.* **249**, 637-667.

KOH, T.H.H.G. & EYRE, J.A. (1988) Maturation of the corticospinal tract from birth to adulthood measured by electromagnetic stimulation of the motor cortex. *Archives of Disease in Childhood* **63**, 1347-1352.

KUYPERS, H.J.G.M. (1962) Corticospinal connections: postnatal development in the Rhesus monkey. *Science* **138**, 161-184.

LAWRENCE, D.G. & HOPKINS D.A. (1976) The development of motor control in the Rhesus monkey: evidence concerning the role of corticomotoneuronal connections. *Brain* **99**, 235-254.

LAWRENCE, D.G. & KUYPERS, H.J.G.M. (1968) The functional organization of the motor system in the monkey. Part 1 - The effects of bilateral pyramidal lesions. *Brain* **91**, 1-14.

McDONOUGH, S.M., EYRE, J.A., KELLY, S., METCALFE, A.V. & MILLER, S. (1993) Occurrence of short latency inhibition from triceps brachii to biceps brachii in human neonates. *J. Physiol.* **467**, 113P.

O'SULLIVAN, M.C., EYRE, J. A. & MILLER S. (1991) Radiation of phasic stretch reflex in biceps brachii to muscles of the arm in man and its restriction during development. *J. Physiol.* **439**, 529-543.

PHILLIPS, C.G. & PORTER, R. (1977) *Corticospinal neurones. Their role in movement.* Academic Press, London.

PORTER, R. & LEMON, R. (1993) *Corticospinal function and voluntary movement.* Clarendon Press, Oxford.

TANNER, J.M., WHITEHOUSE, R.H. & TAKAISHI, M. Standards from birth to maturity for height, weight, height velocity and weight velocity; British children. *Archives of Disease of Childhood* **41**, 454-613.

UEMURA, K. & PRESTON, J.B. (1965). Comparison of motor cortex influences upon various hind-limb motoneurons in pyramidal cats and primates. *J. Neurophysiol.* **28**, 398-412.

WATTS, C., EYRE, J.A., KELLY, S. & RAMESH, V. (1992) Development of the pincer grasp and its relationship to the development of adult corticospinal delays. *J. Physiol.* **452**, 34P.

THE EVOLVING GAMMA LOOP

A. Taylor, R. Durbaba and J.F. Rodgers

Sherrington School of Physiology, UMDS
St Thomas' Hospital, London SE1 7EH, UK

BACKGROUND

The term gamma loop was introduced by Granit to refer to the activation of alpha motoneurones indirectly through the effect of gamma efferent drive to the muscle spindles. Merton went on to popularise the notion that this type of action was essential to achieve negative feedback control of voluntary movement. However, subsequent development of detailed knowledge of the static and dynamic fusimotor systems and recordings from spindles during natural movements combined to show that the true significance of the gamma loop must be much more subtle. This theme was dealt with in penetrating fashion by Matthews in his 1981 review lecture. He pointed out that though the idea of servo control of movement through gamma action had provided a great stimulus to thought and to experiment, it was based on much too simple a knowledge of spindle structure and function. The decade from 1960 to 70 saw great advances with the recognition of the primary and secondary afferents and the division of fusimotor actions into static and dynamic. The elucidation of detail continued in the next decade with the description of the three types of intrafusal muscle fibres, but there was also increasing interest in the broader issues of spindle function in movement control. This was particularly dependent on the recording of spindle afferent activity during natural movements in man by microneurography (Vallbo & Hagbarth, 1968; Vallbo, 1973) and in animals by chronic electrode implantation (e.g. Taylor & Cody, 1974; Goodwin & Luschei, 1975; Prochazka, 1975; Prochazka, Westerman & Ziccone, 1976). In fact, in 1981 Matthews noted that this approach had "quite properly led to a shift of interest away from the fascinations of the muscle spindle itself towards the wider questions of its functional role in the body". It is therefore surprising to see the extent to which interest in the details of spindles has persisted since then and how few workers have involved themselves in the study of the 'wider questions'. A positive reason for this is that knowledge of some details has been seen to be incomplete and it is not very attractive or useful to construct theoretical models of function in these circumstances. The less worthy reasons are that chronic experiments on sophisticated animals require a lot of committment and resources. The present wave of intolerance of animal experiments, whipped up by extremists posing as liberal thinkers, is also impeding progress in this subject as in others.

THE CURRENT PROBLEM - ONE OR TWO STATIC FUSIMOTOR SYSTEMS?

The most important missing detail of spindle function referred to above is the question as to whether there is enough specific and separate innervation of bag_2 and chain intrafusal fibres to justify dividing the static fusimotor system into two subclasses. Boyd (1985) was very clearly of the opinion that essentially two separate groups of static fusimotor neurones controlled the bag_2 and the chain fibres. He based this conclusion on observing the effects of stimulating individual static fusimotor axons on the discharge of several spindles within tenuissimus. Chain action was diagnosed by the driving effect on primary afferents, while bag_2 action was inferred from a biassing action without apparent driving. Boyd did not maintain that there was a strict anatomical segregation, such that an axon producing driving effects (γ_{sc}) never contacted bag_2 fibres or that an axon producing non-driving static effects (γ_{sb}) never contacted chain fibres, but he did feel that the evidence pointed strongly to an effective functional distinction. Gladden and her colleagues have shown various situations in which chain and bag_2 fibres could be activated separately and sometimes reciprocally by brain stimulation (Gladden & McWilliam, 1977; Gladden, 1981; Dickson & Gladden 1990; Asgari-Khozankalaei & Gladden, 1990 a, b). It is true that the technique which they used of visualising the intrafusal fibres in a single spindle in exteriorised tenuissimus could not decide whether a static axon which had one type of termination in that spindle, terminated in the same way in any others which it supplied. However, the fact that the action on the two fibre types was sometimes reciprocal means that there must be two types of static fusimotor neurone with different central synaptic connections.

It would be remarkable if such specialisation were not backed up by a degree of systematic separate innervation of the chain and bag_2 fibres. Banks (1991) reviewed this subject carefully and, alone and with others, has carried out combined functional and morpholgical studies of the distribution of static fusimotor innervation. Clear evidence was produced in tenuissimus that some static axons were restricted to chain fibres in a number of spindles and they were generally the slowest conducting. The faster axons were more widely distributed and could innervate bag_2 fibres alone or could branch to chain fibres as well. Although the possibility of some degree of separate central control of bag_2 and chain fibres was not rejected, the separation of biassing, driving and indeterminate effects was so incomplete that differential activation under natural conditions was thought most likely to be restricted to what might be attributed to orderly recruitment according to the size principle.

Very recently the distribution of static gamma axons has been studied in peroneus tertius spindles using physiological criteria (Celichowski, Emonet-Dénand, Laporte & Petit, 1994). In this, as in another recent study (Dickson, Emonet-Dénand, Gladden, Petit & Ward, 1993), the ramp frequency test for chain effects was supplemented by cross correlating the afferent response spikes with the stimulus trains. This appears to be a sensitive way of detecting even minimal branching of gamma axons to supply chain fibres (see also Durbaba, Taylor, Rodgers & Fowle, 1993). The conclusion was that the distribution of individual static axons to bag_2, chain or both fibre types could be explained as an outcome of random chance. This might on the face of it induce us to lay the subject to rest except that it conflicts so much with the compelling observations made by Gladden's group referred to above, in which brain stimulation caused reciprocal effects on the two fibre types. It should also be remembered that Wand & Schwarz (1985) by microinjections of picrotoxin into the substantia nigra inhibited the static fusimotor bias of primary spindle afferents, but not of secondaries. The only obvious interpretation is that there must be separate activation of chain and bag_2 intrafusal fibres. With these findings in mind it is essential to look very closely at the evidence based on studying the distribution of static fusimotor axons with driving or biassing actions. It may be that in the experiments of Celichowski et al. (1994) the correlation test is so sensitive to chain effects that it indicates a mixed (non-specific) function for axons which would actually have predominant actions via bag_2 fibres under natural conditions. Plainly, this matter is not yet satisfactorily decided, but we would like to take the view for the moment that a substantial degree of independent control of the three intrafusal muscle fibre types is available to the CNS and to see whether we can fit this to the observations which have been made of spindle afferent activity patterns in man and chronic animals and of spindle and fusimotor discharge in reduced preparations.

FUSIMOTOR ACTIVITY PATTERNS - AN HYPOTHESIS

As spindle research has progressed there have been a succession of proposals for general schemes by which the potential independence of fusimotor action developed in mammals might best be exploited to give an advantage over the shared intrafusal and extrafusal innervation characteristic of lower vertebrates. The protagonists of each scheme tend to imply that a single unitary explanation might be found and that the latest idea therefore displaces previous ones. The truth is more likely to be that the separate fusimotor innervation gives just the sort of independence which would be needed to operate in different ways according to circumstances. A lengthy review would be needed to do justice to the various ideas that have been put forward and reference can be made to Matthews (1972, 1981), Prochazka (1980; 1981; 1986) and Prochazka & Hulliger (1983). The present purpose is best served by coming straight to an hypothesis which builds on the previous ones, but adds the possibility of separate control over bag_2 and chain fibres.

1. Mechanisms exist in the CNS which can separately engage gamma motoneurones supplying bag_1, bag_2 and chain fibres and alpha motoneurones.

2. For any given movement or postural task a certain level of tonic drive is directed to the bag_1 fibres to set an appropriate primary afferent sensitivty to displacement.

3. Similarly, a tonic discharge is directed to the bag_2 fibres to establish an appropriate bias of spindle discharge and hence of excitatory input to the alpha motoneurones.

4. During active shortening contractions there will be a fluctuating drive to chain fibres, the chief function of which is to maintain afferent firing.

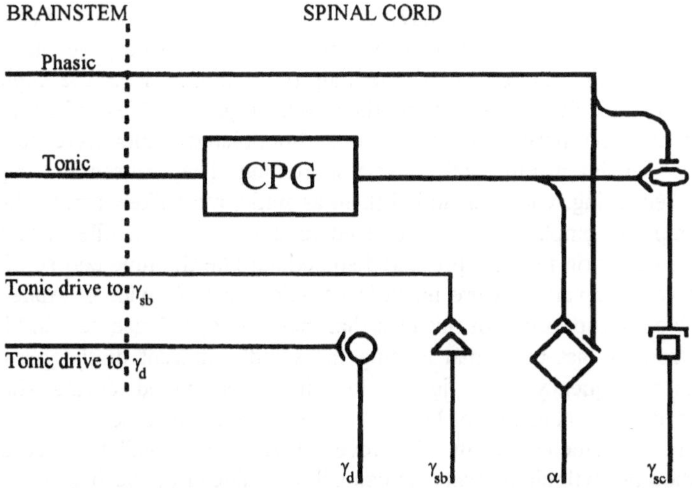

Figure 1. Diagram to represent a hypothesis for the way in which the three types of intrafusal fibre may be controlled. For simplicity, interneurones in the descending pathways, reflex connections, β-innervation and non-specific fusimotor innervation have been omitted. CPG represents the locomotor spinal pattern generator.

Figure 1 is a diagram representing these points. It is not proposed that all the features will be evident in all movements, rather it describes the resources which can be drawn upon in varying degree according to the situation. Now we will set out the supporting evidence (some admittedly circumstantial and even a little teleological) to see whether the hypothesis is viable.

For the first point there is ample evidence from reduced preparations that CNS stimulation can produce independent changes in alpha and in static and dynamic gamma motoneurones (see e.g. Matthews, 1972; Appelberg, 1981; Taylor & Donga, 1989) and so the necessary separate pathways exist. Chronic recordings in animals (reviewed by Prochazka, 1980; 1981; 1986) also attest to the existence of great flexibility in the relationship of alpha to static and dynamic gamma activity in natural movements. The question of independence of bag_2 and chain control has been dealt with above and we assume for the present that there is enough evidence in favour to accept it provisionally.

Tonic drive to bag_1 fibres at different levels according to circumstances is clearly implied by the recordings from jaw muscle spindles in naturally behaving cats (Taylor & Cody, 1974; Cody, Harrison & Taylor, 1975; Taylor & Appenteng, 1981). It is also backed up by recordings of fusimotor axons in masseter nerve in reduced preparations (Appenteng, Morimoto & Taylor, 1980; Donga & Taylor, 1993; Taylor, Appenteng & Morimoto, 1981). The concept of "fusimotor set", based on a large number of observation of cat hindlimb movements (Prochazka, Hulliger, Zangger & Appenteng, 1985), also entails a tonic activity in dynamic fusimotor neurones adjusted to the circumstances. That bag_1 activity should be tonic rather than modulated with movements also seems appropriate in view of the slow contractile properties of bag_1 fibres (*pace* Murphy, Stein & J. Taylor, 1984).

The idea that the bag_2 fibres should also be activated in a tonic fashion is once again in accord with the concept of fusimotor set as worked out in cat hindlimb studies. The "simulation" studies of Hulliger & Prochazka (1983) and Hulliger, Zannger, Prochazka & Appenteng (1985) have never shown any advantage of introducing alpha-linked modulation of static fusimotor firing in fitting natural data. It may be that the static fusimotor axons which they selected for stimulation were generally non-driving and so directed to bag_2 fibres, but this is not clear from the original accounts.

The fourth point above claims that when gamma firing is seen to be modulated in time with phasic movements it is in axons directed to chain fibres. The effect of this would be to tend to prevent unloading and silencing of the afferents. For this purpose chain fibres would be far more effective than either of the bag type fibres, since the rapid contractile characteristics of chain fibres would allow them to keep pace with extrafusal contraction. In recordings of masseter nerve fusimotor axons, "modulated" units were observed to fire bursts at up to 120 Hz in time with muscle shortening. They caused striking increases in secondary afferent firing, which identified them as static, most likely acting via chain fibres. The term "temporal template" was introduced to describe the profile of static fusimotor firing, on the assumption that it represented something like the time course of the intended movement. Two observations were made in conscious cats (Taylor & Appenteng, 1981), which are so clearly supportive of this idea that they are reproduced here in Figure 2. First, it is evident from A that in normal lapping the spindle afferent discharges at a low and almost constant frequency, with just a few impulses missed during shortening (the insensitivity to displacement cannot be due to the presence of tonic static fusimotor firing, because the mean frequency is low). However, in B after a small IV dose of barbiturate anaesthetic, though rhythmic movements can still be induced by placing fluid in the mouth, the pattern of spindle discharge is completely changed. The movements are slower and smaller than previously and there is a conspicuous burst of spindle firing now during the muscle shortening. This must be due to an underlying burst of static fusimotor firing, which was originally not quite enough to entirely prevent unloading, but now is far more than enough. Similar effects were also seen when normal chewing of soft food was impaired by unexpected tough pieces of food. The other striking observation in the alert cat (Figure 2 C & D) was that as the cat became satisfied with lapping milk, lapping and rhythmic spindle discharge stopped. However, after a short pause the animal took interest in the milk again and there ensued a period of rhythmically modulated spindle discharge without EMG or jaw movement before lapping recommenced and the previous pattern was restored. Evidently the central pattern generator started working and sending an output to the (static) fusimotor neurones before the excitability of the alpha motoneurones was sufficient to make an overt expression of the rhythm. It appears generally that alpha motoneurone activity is more readily suppressed than gamma by anaesthesia. It may be wondered why a separate group of gamma motoneurones should be needed for this task of delivering a train of motor impulses to the spindle, effectively in parallel with the alpha discharge, when this was already available in the linked system in amphibia. The answer presumably lies in the need to avoid spindle excitatory input on the fusimotor neurones, which would otherwise cause instability due to positive feedback.

EXPERIMENTAL TESTING

Some experimental evidence in favour of these ideas has been set out above. Much more evidence exists, some for and some against, but none conclusive. The way forward now must be to devise new experiments, because retrospective analysis of data is so vulnerable to the pitfalls of selectivity. Ideally we would record during natural movements

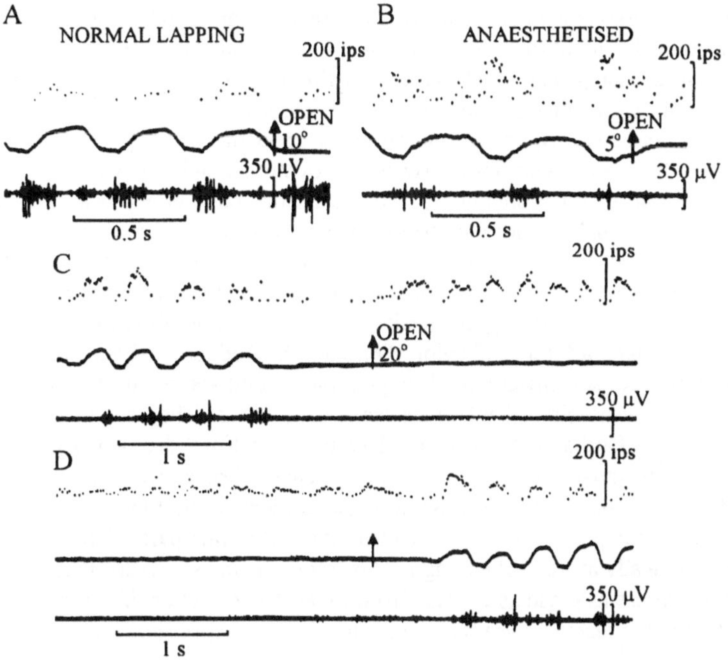

Figure 2. The firing of a primary spindle afferent from a jaw closing muscle in a chronically prepared cat (modified from Taylor & Appenteng, 1981), during lapping movements. In each record the upper trace is instantaneous frequency; middle, jaw movement and lower, masseter EMG. In A, C & D the animal was alert and lapping normally. In B a small dose of pentobarbitone was given IV and movements induced by placing milk in the mouth.

directly from fusimotor neurones and devise reliable means of diagnosing which intrafusal fibre type each controlled. This goal is probably a long way off, but fusimotor recording in a variety of movements in lightly anaesthetised or decerebrated animals is quite feasible.

Just two basic problems exist. First, it is necessary to know to what extent the peculiar conditions of the preparation degrade the normal motor control organisation. We have already seen how even very light anaesthesia can depress alpha activity so that a pattern of fusimotor discharge, which normally almost exactly cancelled mechanical unloading, became excessive to the point of making a large increase in afferent discharge during muscle shortening. It is also known that decerebration leads to a large tonic fusimotor background firing. Another consideration, the importance of which is only just emerging from the work of Gladden's group is the need to avoid noxious stimulation as far as possible and of preserving most of the normal afferent input. In the exposed *in situ* tenuissimus preparation, which allows these conditions, fusimotor reflex effects are obvious which previously have been unobtainable. Clearly, we shall need to devise new experimental approaches, moving away from the classical laminectomy with its massive acute trauma.

The other main problem is to identify fusimotor neurones as γ_d, γ_{sb} or γ_{sc}. Ideally one would record from their cell bodies or from their axons in continuity and characterise them by the effects on spindle discharge of stimulating them. An alternative, which has been used in jaw (Donga, Taylor & Jüch, 1993) and hindlimb (Rodgers, Durbaba, Fowle & Taylor, 1993; Taylor, Durbaba. Rodgers & Fowle, 1993) muscle studies is to find a region in the brainstem at which electrical stimulation causes a pure static or dynamic effect on spindles in the muscle in question. If, for example, a dynamic effect is produced, then any fusimotor neurones excited by this stimulus may be presumed to be dynamic. Another potentially useful method which we have been trying recently is to make use of cross-correlation of random stimuli to γ-axons with afferent firing (Durbaba et al., 1993). Four different types of correlogram could be identified. Static driving units, which it is agreed represent chain effects, gave large, fast peaks which were well fitted by a single log-normal curve. Non-driving static units gave smaller, more prolonged curves or a very small slow curve. The former type required two log normal curves for a good fit and were thought to represent mixed effects on bag_2 and chain fibres. Responses of the second type were fitted by a single slow log-normal curve and probably represented a pure bag_2 effect. An extension of this approach, which may make it possible to say when chain fibre activation is occurring is to look for cross-correlations between the firing of different spindle afferents from one muscle. We have been able to show that random stimulation of single static driving gamma axons can sometimes give clear correlogram peaks between primary afferents and even between primary and secondary afferents (Taylor, Durbaba & Rodgers, 1994a and this volume). Finally, by testing with succinylcholine it is possible to find some muscle spindle secondary afferents which show no sign of influence from bag_2 fibres. They are usually the slowest conducting (Taylor, Durbaba & Rodgers, 1994b). It follows that any sign of fusimotor activity in these units must indicate the presence of chain fibre contraction.

This list of methods which can now be used to identify the functional type of gamma motoneurones is incomplete, but is sufficient to show that this part of the problem may be soluble. The technically more challenging part is to devise ways of recording gamma activity in situations which are as close as possible to normal. Nevertheless, if sufficient interest can be generated, then the next few years may see the hypothesis presented here put to the test.

ACKNOWLEDGEMENTS

We are grateful for the support of Action Research and the Research Endowments of St. Thomas' Hospital.

REFERENCES

APPELBERG, B. (1981) Selective central control of dynamic gamma motoneurons utilised for the functional classification of gamma cells. In *Muscle Receptors and Movement.* eds. TAYLOR, A. & PROCHAZKA, A. pp. 97-107. MacMillan, London.

APPENTENG, K., MORIMOTO, T. & TAYLOR, A. (1980) Fusimotor activity in masseter nerve of the cat during reflex movements. *J. Physiol.* **305**, 415-431.

ASKGARI-KHOZANKALAEI, A. & GLADDEN, M.H. (1990a) Recruitment of intrafusal muscle fibres from the sensorimotor cortex in tenuissimus muscles of cats under barbiturate anaesthesia. *J. Physiol.* **420**, 26P.

ASGARI-KHOZANKALAEI, A. & GLADDEN, M.H. (1990b) Simultaneous excitation and inhibition of tenuissimus static gamma-motoneurones during cortical stimulation in the anaesthetized cat. *J. Physiol.* **429**, 9P.

BANKS, R.W. (1991) The distribution of static gamma-axons in the tenuissimus muscle of the cat. *J. Physiol.* **442**, 489-512.

BOYD, I.A. (1985) Intrafusal muscle fibres in the cat and their motor control. In *Feedback and Motor Control in Invertebrates and Vertebrates.* eds. BARNES, W.J.P. & GLADDEN, M.H. pp. 123-144. Croom Helm. London.

CELICHOWSKI, J., EMONET-DÉNAND, F., LAPORTE, Y. & PETIT, J. (1994) Distribution of static γ axons in cat peroneus tertius spindles determined by exclusively physiological criteria. *J. Neurophysiol.* **71**, 722-732.

CODY, F.W.J., HARRISON, L.M. & TAYLOR, A. (1975) Analysis of activity of muscle spindles of jaw-closing muscles during normal movements in the cat. *J. Physiol.* **253**, 565-582.

DICKSON, M., EMONET-DÉNAND, F., GLADDEN, M.H., PETIT, J. & WARD, J. (1993) Incidence of non-driving excitation of Ia afferents during ramp frequency stimulation of static γ-axons in cat hindlimbs. *J. Physiol.* **460**, 657-673.

DICKSON, M. & GLADDEN, M.H. (1990) Dynamic and static gamma effects in tenuissimus muscle spindles during stimulation of areas in the mes- and diencephalon in anaesthetized cats. *J. Physiol.* **423**, 73P.

DONGA, R., TAYLOR, A. & JÜCH, P.J.W. (1993) The use of midbrain stimulation to identify the discharges of static and dynamic fusimotor neurones during reflex jaw movements in the anaesthetised cat. *Exp. Physiol.* **78**, 15-23.

DURBABA, R., TAYLOR, A., RODGERS, J.F. & FOWLE, A.J. (1993) Subclasses of fusimotor action on cat muscle spindles revealed by cross-correlation of firing with random stimulation. *J. Physiol.* **473**, 206P.

GLADDEN, M.H. (1981) The activity of intrafusal muscle fibres during central stimulation in the cat. In *Muscle Receptors and Movement.* eds. TAYLOR, A. & PROCHAZKA, A. pp. 109-122. MacMillan, London.

GLADDEN, M.H. & McWILLIAM, P.W. (1977) The activity of intrafusal muscle fibres during cortical stimulation in the cat. *J. Physiol.* **273**, 28P-29P.

GOODWIN, G.M. & LUSCHEI, E.S. (1975) Discharge of spindle afferents from jaw-closing muscles during chewing in alert monkeys. *J. Neurophysiol.* **38**, 560-571.

HAGBARTH, K.E. & VALLBO, Å.B. (1968) Discharge characteristics of human muscle afferents during muscle stretch and contraction. *Exp. Neurol.* **22**, 674-694.

HULLIGER, M. & PROCHAZKA, A. (1983) A new simulation method to deduce fusimotor activity from afferent discharge recorded in freely moving cats. *J. Neurosci. Meth.* **8**, 197-204.

HULLIGER, M., ZANGGER, P., PROCHAZKA, A. & APPENTENG, K. (1985) Simulations reveal large variations in fusimotor action in normal cats: 'fusimotor set'. In *The Muscle Spindle.* eds. BOYD, I.A. & GLADDEN, M.H. pp. 311-315. MacMillan, London.

MATTHEWS, P.B.C. (1972) *Mammalian Muscle Receptors and their Central Actions.* London, Edward Arnold.

MATTHEWS, P.B.C. (1981) Evolving views on the internal operation and functional role of the muscle spindle. *J. Physiol.* **320**, 1-30.

MURPHY, P.R., STEIN, R.B. & TAYLOR, J. (1984) Phasic and tonic modulation of impulse rates in γ-motoneurones during locomotion in premammillary cats. *J. Neurophysiol.* **52**, 228-243.

PROCHAZKA, A. (1975) A pause in spindle afferent discharge during rapid active muscle shortening in the freely moving cat. *Pro. Australian Phys. Pharm. Soc.* **6**, 183-184.

PROCHAZKA, A. (1980) Muscle spindle activity during walking and during free fall. In *Spinal and Supraspinal Mechanisms of Voluntary Motor Control and Locomotion.* ed. DESMEDT, J.E. pp 282-293. Karger, Basel.

PROCHAZKA, A. (1981) Muscle spindle function during normal movement. *Int. Rev. Physiol.* **25**, 47-90.

PROCHAZKA, A. (1986) Proprioception during voluntary movement. *Can. J. Physiol. Pharmacol.* **64**, 499-504.

PROCHAZKA, A. & HULLIGER, M. (1983) Muscle afferent function and its significance for motor control mechanisms during voluntary movements in cat, monkey and Man. In *Motor Control Mechanisms in Health and Disease.* ed. DESMEDT, J.E.. pp. 93-132. Raven, New York.

PROCHAZKA, A., HULLIGER, M., ZANGGER, P. & APPENTENG, K. (1985) "Fusimotor set": new evidence for alpha-independent control of gamma motoneurones during movement in the awake cat. *Brain Res.* **339**, 136-140.

PROCHAZKA, A., WESTERMAN, R.A. & ZICCONE, S.P. (1976) Discharges of single hindlimb afferents in the freely moving cat. *J. Neurophysiol.* **39**, 1090-1104.

RODGERS, J.F., DURBABA, R., FOWLE, A.J. & TAYLOR, A. (1993) Flexor and extensor muscle fusimotor activation from midbrain stimulation in the anaesthetized cat. *J. Physiol.* **459**, 462P.

TAYLOR, A. & APPENTENG, K. (1981) Distinctive modes of static and dynamic fusimotor drive in jaw muscles. In *Muscle Receptors and Movement.* eds. TAYLOR, A. & PROCHAZKA, A. pp. 172-192. Macmillan, London.

TAYLOR, A., APPENTENG, K. & MORIMOTO, T. (1981) Proprioceptive input from the jaw muscles and its influence on lapping, chewing, and posture. *Cand. J. Physiol. Pharmacol.* **59**, 636-644.

TAYLOR, A. & CODY, F.W.J. (1974) Jaw muscle spindle activity in the cat during normal movements of eating and drinking. *Brain Res.* **71**, 521-530.

TAYLOR, A. & DONGA, R. (1989) Central mechanisms of selective fusimotor control. *Prog. Brain Res.* **80**, 27-35.

TAYLOR, A., DURBABA, R. & RODGERS, J.F. (1994a) Correlated discharges of muscle spindle afferents related to shared fusimotor input in the anaesthetised cat. *J. Physiol.* **475**, 37-38P.

TAYLOR, A., DURBABA, R. & RODGERS, J.F. (1994b) The site of action of succinylcholine on muscle spindle afferents in the anaesthetised cat. *J. Physiol.* **476**, 26P.

TAYLOR, A., DURBABA, R., RODGERS, J.F. & FOWLE, A.J. (1993) Reciprocal actions of midbrain stimulation on static and dynamic fusimotor neurones of the hindlimb in anaesthetized cats. *J. Physiol.* **473**, 15P.

VALLBO, Å. (1973) Muscle spindle afferent discharge from resting and contracting muscle in normal human subjects. In *New Developments in Electromyography and Clinical Neurophysiology.* ed. DESMEDT, J.E. pp. 251-262. Karger, Basel.

WAND, P. & SCHWARZ, M. (1985) Two types of cat static fusimotor neurones under separate central control? *Neurosci. Lett.* **58**, 145-149.

CRITIQUE OF PAPERS BY LEMON AND BENNETT, EYRE AND MILLER, SESSLE ET AL, TAYLOR ET AL. AND SEARS ET AL.

J.C. Rothwell

MRC Human Movement & Balance Unit
The Institute of Neurology, Queen Square
London WC1N 3BG, UK

Lawrence and Kuypers' (1968) observation that relatively independent movements of the digits disappear after section of the pyramidal tract is one of the most influential observations in motor physiology. When extended to man, the observation assumes an even greater, philosophical importance. The pyramidal tract is the evolutionary development which has given us a superior ability to manipulate objects and fashion tools from our surroundings.

Lemon's work is a description of a small proportion of the pyramidal tract system in monkey: the large diameter monosynaptic projection to motoneurones innervating muscles of the forearm and hand (the cortico-motoneuronal projection). Although this represents only a fraction of the total number of pyramidal fibres, calculations (Fetz, Cheney, Mewes & Palmer, 1989) show that in the hand and forearm, the cortico-motoneuronal projection is capable of supplying up to 40% of the synaptic input to discharge motoneurones at frequencies of 15 to 20Hz. At least for these muscles, the cortico-motoneuronal input should be reasonably representative for pyramidal tract control.

The first observation of this report confirms earlier work showing that cortico-motoneuronal cells are very active during independent movements of the digits whilst their activity is not so great in gross movements which involve forceful contraction of many muscles such as the power grip. My first point is that this simple, but remarkable observation is quite unexpected and deserves more than a passing mention. After all, why should the cortico-motoneuronal system, which can generate about 40% of the motoneuronal input actually be excluded from contributing fully to powerful contractions? The answer is not known, yet this is possibly one of the most intriguing clues as to the operation of the cortico-motoneuronal system that we have. This becomes clear if we speculate on the possible reasons for the result. The simplest approach is to say that it is *because* the cortico-motoneuronal system is designed to take part in relatively independent finger movements that it is less suited to perform powerful contractions. It may be that independent digit movement depends on a combination of excitatory and inhibitory inputs to the spinal cord so that some muscles are activated, whereas others are prevented from contracting. If the cortico-motoneuronal system could *only* provide this combination of positive and negative input then clearly the negative effect would interfere with production of high levels of contraction in many muscles and this could be the reason why the system is not very active during performance of the power grip. We come to the conclusion that rather than being an all powerful evolutionary novelty, there are limitations in the function of the pyramidal system. Whether these are caused by limitations in the inputs which drive it, or by limitations in the anatomy of the projections is not known.

The anatomy of the cortico-motoneuronal projection forms the basis of the second part of the results. Cortico-motoneuronal fibres to the hand have a very limited muscle field, and the discharge of any one neurone is greatest when the animal activates the muscle with which it forms the strongest anatomical projection. The hypothesis is that the ability to produce fractionated movements of the fingers is a direct consequence of the form of the electro-anatomical projection. However, the idea that independent finger movements *require* isolated projections to individual muscles is not self-evident at all. The types of finger movement which are associated with high levels of activity in the cortico-motoneuronal system are produced by the activity of many muscles acting in concert. Winkling out a raisin from a well is not the action of the first dorsal interosseous. It involves all the muscles which act on the index as well as those on the other fingers, which are held flexed away from the well. Relatively limited excitatory muscles fields for each cortico-motoneuronal axon are not an obvious advantage in producing the complex pattern of EMG activity needed to make independent movements of the digits. Indeed, if it were necessary to have small muscle fields to product fractionated movement, then, given the importance of forearm muscles in controlling the digits, we would expect the muscle fields of cortico-motoneuronal cells projecting to the wrist and forearm to be as small as those which project to the hand. In fact, the former are approximately twice as large as the latter. It is not even a prerogative of the cortico-motoneuronal system to produce fractionated movements. Walking and swallowing consist of a sequence of highly fractionated muscle activities none of which need cortico-motoneuronal input to occur. The conclusion is that independent finger movements do not require the small muscle fields so characteristic of the cortico-motoneuronal system. Examining the connectivity of the cortico-motoneuronal cells may not provide all the answers to the question of how the brain controls independent finger movement.

Why then are the projections of cortico-motoneuronal cells so limited? Perhaps we should turn to the other characteristic of movements produced by this system. That is their enormous repertoire and adaptability. The number of different combinations of finger movements that we can make is enormous, and more can probably be learnt relatively easily. Maybe small muscle fields are needed to allow flexibility into independent digit control.

The paper by Miller and Eyre examines the development of the cortico-motoneuronal system in infants. The results rely on the use of non-invasive transcranial magnetic stimulation to activate descending motor pathways in man with little or no discomfort. The authors show that it is possible to activate muscles from the brain using this technique in new-born babies, children and adults. There are two major results of this work. First, the latency of the EMG response is longest in new-born children and decreases to a minimum at the age of 4 to 5 years, and increases thereafter as the child grows. The second result is that the threshold for stimulation is highest in the new-born, is lower in children, and lowest in adults. The data are interpreted in terms of the development of the cortico-motoneuronal component of the pyramidal tract.

Unfortunately, this is a bold assumption which might fail to convince a sceptic. Given the difference in stimulation parameters needed in the new-born versus the adult, it would have been more parsimonious to suggest that two different pathways were being activated at different ages. In the new-born, a very high threshold, slow conducting pathway is recruited, whereas in children and adults, a lower threshold, faster conducting pathway is recruited. Indeed, given the very high intensities used to produce muscle responses in the new-born (up to the maximum output of a high power magnetic stimulator), the stimulus may spread deep below the cortex and perhaps activate structures in the brain stem. If this were the case, then the pattern of projections to different body muscles might differ from that expected from the cortico-motoneuronal system. However, the distribution of responses in different muscles at birth has not been investigated systematically.

The latter part of the paper argues that magnetic stimulation over the scalp produces monosynaptic facilitation of motoneurones in subjects of all ages. If correct, then this is a powerful argument that the same, probably cortico-motoneuronal pathways are studied from birth to adulthood. The results indicate that the biceps tendon jerk can be facilitated by a suitably timed magnetic stimulus, and that the time course of this effect is characterised by a short, and rapidly rising initial facilitation which then declines. Because the peak facilitation is reached so quickly, then the result is compatible with a rapidly rising EPSP at biceps motoneurones which might well be monosynaptic in origin. Unfortunately, although this may be a reasonable conclusion in adults, the results are quite unexpected in children. If magnetic stimulation does activate a cortico-spinal system with a very slow conduction velocity in the new-born, then we would expect there to be considerable temporal dispersion of excitation arriving at the spinal cord in the new-born compared with adults. Why then is there so little difference between the time course of facilitation of the tendon jerk in infants and adults? One possibility is that the time course consists of two phenomena; an initial excitation followed by a later inhibition. Thus, the time to peak facilitation of the tendon jerk represents a difference in timing between arrival of the excitatory and inhibitory inputs. If this interval is brief, the time to peak facilitation of the response will be brief, since it will be cut short by the later-arriving inhibition. The result is that the duration of the peak tells us nothing about the number of synapses in the excitatory pathway. The duration would relate only to the difference in the number of synapses (or relative conduction times) in the pathways to excitatory and inhibitory inputs. Monosynaptic excitation is not necessary.

The paper by Sessle and colleagues addresses the topical question of the role of primary motor cortex in control of automatic movements. Movements such as walking, swallowing, chewing etc. usually are assumed to be governed by activity in low level systems. Recently, it has become evident that besides initiating and terminating such rhythms, higher centres may continuously monitor the output and intervene when the task is unexpectedly changed. The situation is analogous to an air line pilot taking over from the more usual computer control. This has been described for walking movements in the cat where cortical activity has been shown to be very small during normal gait, but increases suddenly when unexpected obstacles are encountered (Marple-Horvat, Amos, Armstrong & Craido, 1993).

The present work is on chewing and swallowing, and postulates that the primary motor cortical areas which project to tongue and masticatory muscles, and which normally are active during the performance of fine motor tasks such as controlled biting or tongue protrusion, also are concerned with semi-automatic orofacial movements. Unfortunately, this is only a postulate; proof is lacking. The results demonstrate only that cells in these areas change their firing pattern when monkeys swallow or chew in a relatively uncontrolled manner. The question is whether the firing of these cells provides any useful input for the movement, or whether it simply reflects feedback in afferents from the moving part of the body. Indeed, it seems likely that the latter is the case since the authors have shown previously that cooling the motor cortex has little effect on automatic movements of the jaw.

The work is obviously preliminary. However, if the motor cortex is to be shown to have a useful contribution to semi-automatic movements, then it must be shown that disruption of motor cortical activity also disrupts the movement. Given the results in cat walking, perhaps the contribution would be most evident as a lack of adaptation to an unexpected disturbance of automatic rhythmical movement. At present, the evidence is only circumstantial.

The paper by Taylor and colleagues is a development of ideas of the use of muscle spindles. Their main point is that the three types of intrafusal fibre in the spindles may be under separate fusimotor control. This would produce a spindle in which we could control separately the sensitivity to stretch, the resting level of discharge in the afferents, and also allow the system to maintain afferent firing during muscle shortening. It is a most elegant model of the spindle and a very useful attempt to combine some of the more recent data on the fine structure of the spindle. It is unfortunate that the β-fusimotor innervation is not discussed, but whether this would make a great difference to the model is debatable.

Ever-increasing information about the fine structure of the muscle spindle may be intriguing in itself, but it is surely time to ask whether or not this information is of use to the final end user like myself who studies movement control in intact humans. The answer is probably no. As the authors have noted in their Introduction there are very few new ideas about the wider questions of the functional role of the muscle spindle in the body. Indeed, it is perfectly possible to construct attractive and thought-provoking theories on how we move with no more than minimal knowledge of the muscle spindle. Examples are Bizzi/Feldman hypothesis of trajectory control, or the work of Soechting and colleagues on the arm movement co-ordinate system used in pointing (Flanders & Soechting, 1990). The fact is that we have yet to discover the real role of muscle spindle in movement of any kind. Stretch reflexes exist but are hardly essential. They are large in some subjects small in others, and inherently variable from trial to trial (Marsden, Merton, Morton, Rothwell & Traub, 1981). Muscle spindle input may also contribute to a sense of joint position, but is only one of many inputs involved. In addition, there appears to be a virtually no pathophysiology of the muscle spindle which would help us to understand movement disorders. The few recordings that have been made from the human muscle spindle in neurological disease have shown that in all cases, the spindle behaves as in normal subjects (Burke, 1982). If the spindle never goes wrong, then is there any need to study it?

Clearly, these are extreme views. I have no doubt myself that the muscle spindle has a very important role in the control of movement, and that there are pathologies in which control of the spindle, or the structure of the spindle is abnormal. My plea is therefore, not for less work on the fine structure of the spindle, but for a good deal more on its function.

The final paper by Sears et al concerns the details of the central and peripheral inputs to the olive during breathing. This is somewhat outside my own area of expertise. However, the paper also has a more general message which is highly relevant to all of the preceding papers. That is the problem of how efferent motor commands might control the flow of afferent input to the central nervous system, and perhaps by other mechanisms, regulate the effectiveness of reflex pathways. The modulation, at different phases of the respiratory cycle, of reflexes in intercostal muscles evoked by sudden loading is an elegant example of the latter effect. But the crux of the matter is as always to prove that the "gating" of reflexes is a useful effect. Would respiration be greatly different if the reflexes were not closely controlled?

It is not clear whether this has yet been sorted out for the respiratory system. In the limbs, though, it is now a question of some importance. The reason is that one of the major effects of damage to central motor pathways, as seen, for example, in patients after stroke, is failure to regulate the effectiveness of reflex pathways during movement. This has been observed in almost every spinal pathway which has been studied (e.g. Ia reciprocal inhibition, Renshaw inhibition, presynaptic inhibition), so that it is almost possible to imagine that the reflexes we observe in patients are not abnormal in themselves, rather they are examples of normal reflexes "stuck" at a particular level of excitability. At some point during movement, similar reflex excitability would be encountered in normal subjects, but for them the effect would be transient, rather than permanent., If this reasoning is correct, then it provides us for once with an insight into the effectiveness of reflex control of movement, and may even begin to reveal the misty secrets of the muscle spindle.

REFERENCES

BURKE, D. (1982) A critical examination of the case for or against fusimotor involvement in disorders of muscle tone. In *Brain & Spinal Mechanisms of Movement Control in Man.* ed. DESMEDT, J.E. pp. 133-150. Raven Press, New York

FETZ, E.E., CHENEY, P.D., MEWES, K. & PALMER S. (1989) Motor cortex control of finely graded forces. *Prog. Brain Res.* **80**, 437-449.

FLANDERS, M. & SOECHTING, J.F. (1990) Parcellation of sensorimotor transformations for arm movements. *J. Neurosci.* **10**, 2420-2427.

LAWRENCE, D.G. & KUYPERS, H.G.J.M. (1968) The functional organisation of the motor system in the monkey. Parts I and II. *Brain.* **91**, 1-36.

MARPLE-HORVAT, D.E., AMOS, A.J., ARMSTRONG, D.M. & CRAIDO, J.M. (1993) Changes in the discharge patterns of cat motor cortex neurones during unexpected posturbation of on-going locomotion. *J. Physiol.* **462**, 87-113.

MARSDEN, C.D., MERTON, P.A., MORTON, H.B., ROTHWELL, J.C. & TRAUB, M.M. (1981) Reliability and efficacy of the long-latency stretch reflex in the human thumb. *J. Physiol.* **316**, 47-60.

TOPOGRAPHICAL ORGANIZATION OF NEURONES IN GROUP II REFLEX PATHWAYS OF THE RAT SPINAL CORD

J.S. Riddell and M.R. Hadian

Institute of Physiology, University of Glasgow
Glasgow G12 8QQ, UK

INTRODUCTION

In the cat, interneurones mediating the reflex actions of group II afferents of different hind-limb muscles are concentrated in different segments of the lumbosacral spinal cord. A particularly clear example of this is that group II afferents in the nerves to the quadriceps and pretibial flexor muscles produce their strongest actions on interneurones at the rostral end of the lumbosacral enlargement (in midlumbar segments, Edgley & Jankowska, 1987a,b) while group II afferents of gastrocnemius and posterior biceps-semitendinosus mainly contact interneurones at the caudal end of the lumbosacral enlargement (in sacral segments, Jankowska & Riddell, 1993, 1994). We have now investigated the topographical organisation of neurones in group II reflex pathways of the rat with the intention of providing some basic information necessary for studies in this species.

METHODS

Male Wistar rats were anaesthetised with sodium pentobarbitone (50 mg/kg i.p.) supplemented with 10 mg/kg i.v. as required. The segmental locations of group II-activated neurones were investigated by recording cord dorsum potentials evoked by electrical stimulation of muscle nerves while their distribution within the grey matter was defined by mapping the sites at which group II field potentials could be recorded with glass microelectrodes. At the end of experiments, marking electrodes were left at recording sites and the animal perfused intra-arterially with formol saline. The segmental location of recording sites was confirmed by dissection of spinal roots and blocks of spinal cord containing the electrodes cut into 50μm serial transverse sections and stained with Cresyl Violet. These were used to reconstruct the sites within the grey matter where group II field potentials were recorded and to relate the rostro-caudal position at which maximal cord dorsum potentials were recorded to the pudendal motor column or Clarke's column.

RESULTS

Stimulation of the quadriceps (Q), tibialis anterior-extensor digitorum longus (TA-EDL) and gastrocnemiussoleus (GS; or the lateral branch alone, LGS) nerves all produced cord dorsum potentials which grew substantially in amplitude as stimulus intensities were raised from 2 to 4-5 times threshold (T) for the most excitable afferents in the nerve; example records are shown in Figure 1A & B. For Q these potentials were largest over the L1 and L2 segments (see Figure 1C) and for TA-EDL they were distributed about one segment more caudally (i.e. L2 and L3). In contrast, the cord dorsum potentials evoked by stimulation of the GS nerve occurred at the opposite end of the lumbar enlargement being largest within the caudal half of the L5 segment (see Figure 1D). When histologically recognisable groups of neurones were used as landmarks, cord dorsum potentials evoked by group II afferents of Q were maximal 1-2 mm caudal of the most caudal cells of Clarke's column while cord dorsum potentials evoked by group II afferents of GS were maximal within 1 mm rostral of the pudendal motor column.

The distribution of field potentials evoked by group I and group II afferents of Q is shown in Figure 2B. Group II potentials were largest (>300μV) in the ventral half of the dorsal horn (see middle record of Figure 2A) but were also present in more dorsal parts of the ventral horn where they were sometimes preceded by shorter latency potentials produced by group I afferents (see lower records of Figure 2A). Field potentials evoked by group II afferents of GS were mainly encountered in the dorsal horn but smaller potentials evoked by group I and group II afferents were sometimes encountered more ventrally.

CONCLUSIONS

The results show that in the rat, as in the cat, interneurones in group II reflex pathways from different muscles are concentrated in particular spinal segments with those contacted

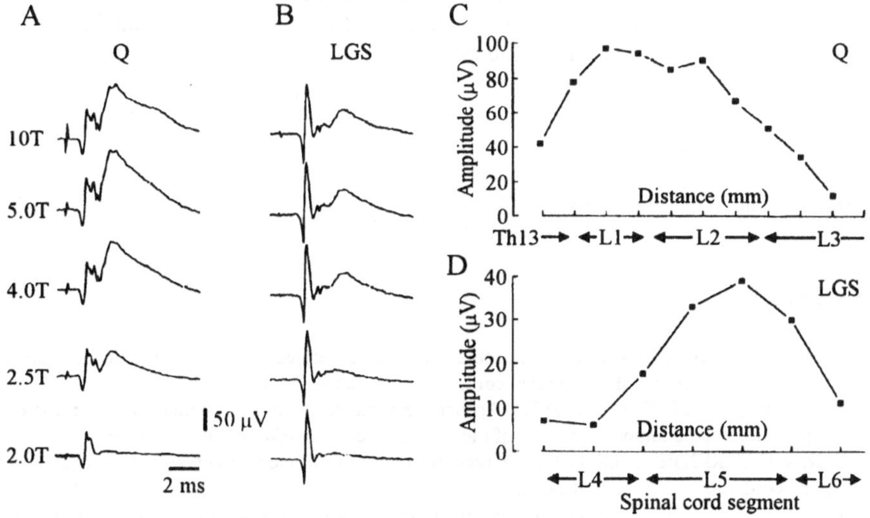

Figure 1. Cord dorsum potentials evoked by stimulation of group II muscle afferents of the quadriceps (A) and lateral gastrocnemius-soleus (B) nerves (averages of 64 sweeps) at different strengths. Note that stimuli of 2T were largely ineffective. C & D, plots showing the rostro-caudal distribution of cord dorsum potential amplitude (recorded at 1mm intervals) with respect to the spinal segments: C. Potentials produced by quadriceps (L1/L2) and D potentials produced by lateral gastrocnemius-soleus afferents (L5).

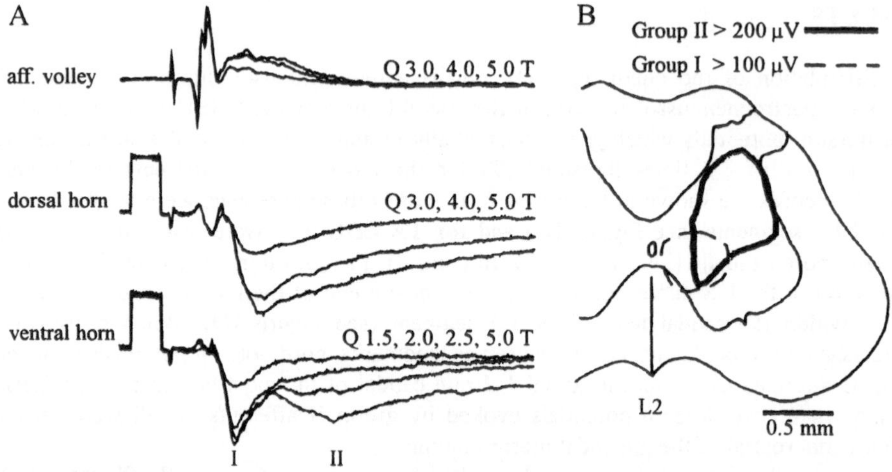

Figure 2. Field potentials evoked by group II muscle afferents. A, afferent volley (upper records) and field potentials recorded in the dorsal horn (middle records) and ventral horn (lower records) of the L2 segment. The records consist of superimposed potentials evoked by stimuli applied to the quadriceps nerve at strengths between 1.5 - 5T, as indicated. In the lower records potentials evoked by group I and group II afferents are denoted I and II respectively. Each record is an average of 64 sweeps. The calibration pulse in the middle and lower records is 200μV and 1ms. B, distribution within the grey matter of the L2 segment of field potentials evoked by group I and group II afferents of the quadriceps nerve.

by Q and TA-EDL afferents being located at the rostral end and those activated by GS group II afferents at the caudal end of the lumbar enlargement. Furthermore, as in the cat (Jankowska & Riddell, 1993), group II potentials produced by afferents of GS were found to be largest in close proximity to the rostral end of the pudendal motor column (McKenna & Nadelhaft, 1986), an internal landmark common to both species. The results suggest therefore that the topographical organisation of neuronal systems mediating the reflex actions of group II muscle afferents in the rat is similar in principle to that of the cat.

ACKNOWLEDGEMENTS

Supported by The Wellcome Trust.

REFERENCES

EDGLEY, S.A. & JANKOWSKA, E. (1987a). Field potentials generated by group II muscle afferents in the middle lumbar segments of the cat spinal cord. *J. Physiol.* **385**, 393-413.

EDGLEY, S.A. & JANKOWSKA, E. (1987b). An interneuronal relay for group I and group II muscle afferents in the middle lumbar segments of the cat spinal cord. *J. Physiol.* **389**, 675-690.

JANKOWSKA, E. & RIDDELL, J.S. (1993). A relay for Group II muscle afferents in sacral segments of the cat spinal cord. *J. Physiol.* 465, 561-578.

JANKOWSKA, E. & RIDDELL, J.S. (1994). Interneurones in pathways from group II muscle afferents in sacral segments of the feline spinal cord. *J. Physiol.* **475**, 455-468.

McKENNA, K.E. & NADELHAFT, I. (1986). The organization of the pudendal nerve in the male and female rat. *J. Comp. Neurol.* **248**, 532-549.

THE BASIS OF VARIABILITY IN MAGNITUDE OF THE RESPONSE OF MUSCLES TO TRANSCRANIAL MAGNETIC STIMULATION OF THE MOTOR CORTEX IN MAN

N.J. Davey[1], P.H. Ellaway[1], D.W. Maskill[1]
and N.P. Anissimova[2]

[1]Department of Physiology
Charing Cross and Westminster Medical School
Fulham Palace Road, London W6 8RF, UK
[2]Pavlov Institute of Physiology
St. Petersburg, Russia

INTRODUCTION

The magnitude of the compound motor evoked potential (cMEP) response of skeletal muscles to transcranial magnetic stimulation (TMS) in man may show a high degree of variability (Davey, Ellaway, Maskill, Anissimova, Rawlinson & Thomas, 1994). The variability occurs despite using a constant stimulating intensity and the efforts of the operator to maintain a constant stimulating coil position over the cranium. We have now investigated the basis of this variability.

METHODS

Four mormal male subjects (ages 34 -51 years) were investigated, with local ethical approval. Surface electromyographic (EMG) recordings were made from the thenar muscles of the hand and extensor digitorum communis (EDC) muscles of the forearm on both sides of the body. Subjects were seated comfortably and instructed to keep their muscles relaxed; they were provided with audio and visual feedback of EMG to help them achieve this state. TMS was delivered using two stimulators (Magstim 200) each connected to a 7 cm (mean diameter) figure-of-eight coil. The coils were hand-held by independent operators standing on either side of the subject. The coils were placed as close together as possible on either side of the cranium so that the centre of the cross-over region of each lay about 5 cm lateral to the vertex. The initial current induced in the brain by each of the coils flowed medially. The intensities of TMS were selected so as to evoke similar mean amplitudes of responses in homonymous muscles on the left and right sides. In general, such stimulation was approximately 5% of the maximum stimulator output above threshold. A stimulus was delivered through each coil every 5 seconds. One stimulator was discharged 1 ms after the other to avoid interaction between the two magnetic fields that evidently reduced the size of responses when the coils were discharged simultaneously. Each response was rectified and the mean, standard deviation and coefficient of variation of the amplitude of the 50 responses were calculated. When the data were grouped by muscle, the range in coefficient of variation for each muscle was: right thenar 0.25-0.87, right EDC 0.22-0.41, left thenar 0.3-1.32, left EDC 0.41-0.79. When the data were grouped by subject, the ranges of coefficient of variation were 0.25-0.41, 0.3-0.44, 0.41-1.32 and 0.22-1.12. It is clear from these ranges that no particular muscle or individual subject tended to exhibit consistantly more or less variability than the others.

RESULTS

In each subject, regression analysis was undertaken to look for any correlation between the amplitudes of the individual responses of the six possible pairs of muscles. In three of the four subjects, linear regression analyses showed that there was a significant positive correlation between responses in all instances. Figure 1 shows correlations between the mean cMEP amplitudes of right EDC and right thenar (A) and right and left EDC (B) for one of these three subjects. In a fourth subject, the responses of three muscles also showed positive linear correlations but a significant correlation between the responses of the right EDC and other muscles was absent ($P>0.05$). For the group of four subjects, the correlations between pairs of ipsilateral muscles were no stronger than between pairs of contralateral muscles, either homonymous or heteronymous.

CONCLUSION

The results indicate that fluctuations in amplitude of the response to TMS may be correlated between muscles on both sides of the body. Experiments are in hand to test whether the basis of the variability may be related to gross changes in synaptic excitability at the cortical or spinal cord level, or physical changes in brain position, for example as a result

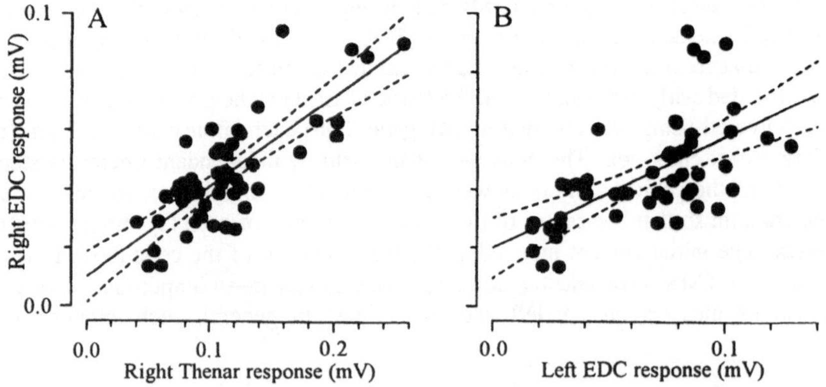

Figure 1. Linear regression analyses of mean cMEP amplitudes in response to 50 bilateral TMS delivered through each coil at 45% of maximum stimulator output. The dotted lines mark the 95% confidence limits for the regression. A. Right EDC and right thenar ($r=0.788$, $P<0.001$), B. Right and left EDC ($r=0.611$, $P<0.001$).

of the blood pressure pulse. However, our results show that variability cannot be due solely to variation in coil position relative to the cranium.

REFERENCE

DAVEY, N.J., ELLAWAY, P.H., MASKILL, D.W., ANISSIMOVA, N.P., RAWLINSON, S.R. & THOMAS, H.S. (1994) Variability in the amplitude of skeletal muscle responses to bilateral transcranial magnetic stimulation in man. *J. Physiol.* **476**, 33P.

THE SITE OF FACILITATION DURING TRANSCRANIAL DOUBLE PULSE MAGNETIC BRAIN STIMULATION IN MAN

D.A. Ingram[1] and D.N. Rushton[2]

[1]Department of Clinical Neurophysiology
The Royal London Hospital
[2]Department of Rehabilitation
London Hospital Medical College
Whitechapel, London, UK

INTRODUCTION

Magnetic transcranial brain stimulation excites the corticospinal tract by a mechanism which, at or near threshold, is believed to be mainly trans-synaptic. There is therefore scope for synaptic spatiotemporal summation at cortical level. This contrasts with the effect of threshold electrical transcranial stimulation, which seems to excite the initial segment of the pyramidal tract axon. However, in both cases there is also scope for spatiotemporal summation at spinal segmental (anterior horn cell) level.

Conventional single-pulse magnetic brain stimulation allows measurement of extrinsic reinforcing processes such as the Jendrassik manoeuvre, but double-pulse stimulation also allows study of the time-course and other properties of facilitation in the pyramidal tract. Near threshold, this may be seen as a phase of facilitation of the effect of the test stimulus by a preceding conditioning stimulus.

During double-pulse stimulation, the inter stimulus interval (ISI) and stimulus strength show a complex relationship between excitatory and inhibitory effects (Valls-Solé, Pascual-Leone, Wassermann & Hallett, 1992). In intrinsic hand muscles, the strongest facilitatory effect was seen just above activation threshold, with an ISI of 20 - 40 ms. By contrast, when the ISI is short (< 6 ms), and the conditioning pulse is just subthreshold, the net effect of the conditioning pulse is strongly inhibitory in the relaxed condition (Ridding et al., this meeting).

These considerations prompt the question: at what level is the temporal facilitation occurring? Is it cortical (which we might term F_c), spinal (F_s), or a combination of the two (F_{c+s})? We report here experiments which attempt to answer this question by assessing excitability changes of the target motoneurone pool in leg muscles.

The overall facilitation was measured as changes in the amplitude of responses to a cortical test stimulus, induced by a conditioning cortical magnetic pulse. The spinal element

(Fs) was measured as the change of H reflex amplitude, again following a threshold conditioning cortical stimulus. The purely cortical element (F_c) of facilitation could thus be inferred by subtracting F_S from F_{c+s}, or perhaps by dividing F_S into F_{c+s}.

METHODS

Muscle responses to midline magnetic brain stimulation and tibial nerve stimulation were recorded (with local Ethical Committee approval) from the right soleus in 5 informed normal males (aged 32 - 49 years), using surface electrodes and a Dantec Counterpoint EMG machine. Soleus was chosen for the reliability of H reflexes elicited in the relaxed state, and for the accessibility of this muscle to magnetic brain stimulation. In order to minimise extraneous sources of response variation, subjects were instructed to relax, and background EMG activity was continuously monitored.

Magnetic brain stimuli, adjusted to be just suprathreshold, were delivered via two Magstim 200 machines linked by a Bistim module to a 120 mm figure-8 right-angled coil. The coil was centred in the midline, just anterior to the vertex, at a site adjusted to optimise lower limb responses. The Bistim arrangement permitted the ISI to be adjusted between 1-999 ms. Subjects were studied relaxed and semi-recumbent. Test H-reflex muscle responses were adjusted to a stable control amplitude of between 0.5 and 1.0 mV.

Results in the test conditions were expressed as percent of control amplitudes. ISI's were expressed as the interval between two cortical magnetic pulses, or the interval between a cortical magnetic pulse and an H reflex stimulus. It should be noted that the conduction time from cortex to lumbar cord is approximately (but not necessarily exactly) equivalent to the conduction time from the popliteal fossa to the lumbar cord. Thus, ISI's given below in the RESULTS are approximately equivalent in the two conditions.

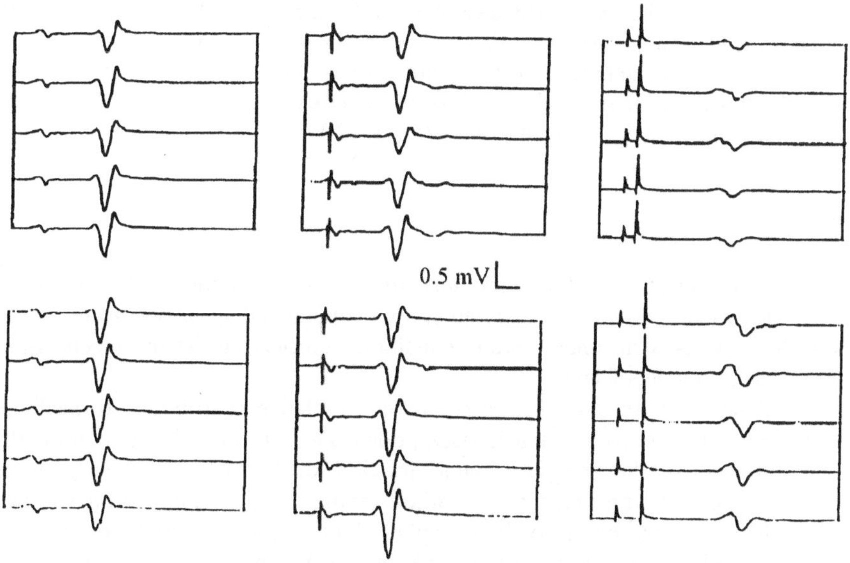

Figure 1. Representative series of test and conditioning stimuli in a single subject, compared with control H reflex responses, for each interstimulus interval (ISI) at 5 msec (upper traces) and 10 msec (lower traces) to show facilatation. Left column: Control H reflexes. Middle column: Conditioning magnetic brain stimulus followed by test H reflex stimulus. Right column: Conditioning magnetic brain stimulus followed by test magnetic brain stimulus. Horizontal calibration bar is 10 msec.

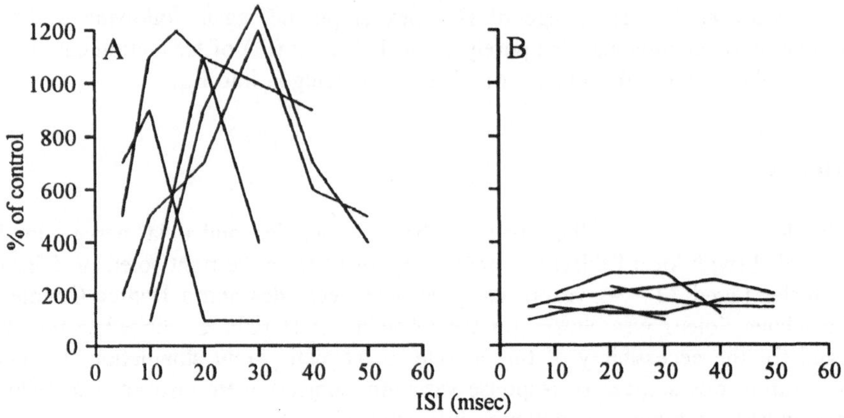

Figure 2. Time-course of facilitation of the response to the test cortical magnetic stimulus by the preceding (conditioning) stimulus. Interstimulus interval (ISI) varied between 5-50 msec. In the upper part (A) the test stimulus is applied at cortical level. In the lower part (B), the test stimulus is segmental (H reflex).

RESULTS

All subjects showed marked facilitation of the response to the second cortical stimulus (F_{c+s}). All subjects showed a much smaller facilitation of the H reflex response (F_s). The time-course of F_s was similar to that of F_{c+s}, with the peak of facilitation occurring with an ISI between 10 - 40 ms. Figure 1 shows a representative set of data from one subject.

The ISI giving the largest facilitation varied somewhat between subjects. In Figure 2 the time-course of facilitation is plotted for each of the 5 subjects. The range of ISI giving facilitation is similar to that previously described for hand muscles (Valls-Solé et al., 1992).

Percent facilitation for each experimental condition was obtained following calculation of the mean peak-peak amplitude of five consecutive muscle responses at each ISI (see Figure 1). As shown in Figure 2, the maximal F_s never exceeded 280% (range 120-280%). By contrast, the maximal F_{c+s} ranged from 900% to 1200%.

CONCLUSION

These results suggest that for paired, just suprathreshold stimuli, applied at the optimal ISI, facilitatory effects can be demonstrated at both cortical and spinal segmental levels. The results further suggest that, under certain conditions, cortical facilitatory mechanisms may outweigh those at segmental level.

However, attempts to quantify and compare the relative contribution to facilitation at each level are limited by certain implicit assumptions. These include the assumption that the H reflex stimulus is addressing the same fraction of the motoneurone pool as the corticospinal volley, and, further, that the earliest recruited alpha motoneurones in the pool respond in the same size order to these unphysiological stimuli. The responses at these relatively low levels of stimulation are also assumed to be approximately linear, and the H reflex response size is chosen to be in the most linear part of its range. The cortical stimuli were placed in the just-suprathreshold range to avoid some sources of non-linearity, such as response saturation. Facilitation at the two sites could be additive or multiplicative, or related by some other function; our data do not allow us to determine this.

REFERENCE

VALLS-SOLÉ, J., PASCUAL-LEONE, A., WASSERMANN, E.M. & HALLETT, M. (1992). Human motor evoked responses to paired transcranial magnetic stimuli. *EEG & Clin. Neurophysiol.* **85**:355-364.

MOVEMENT IDEATION AND ITS EFFECTS ON THE SHORT-LATENCY SOMATOSENSORY POTENTIALS EVOKED FROM THE MEDIAN NERVE

Malcolm Lidierth and
Vicente Gradillas

Sherrington School of Physiology
UMDS, St Thomas' Hospital
London SE1 7EH, UK

INTRODUCTION

The somatosensory evoked potentials (SEPs) produced by electrical stimulation of the median nerve are reduced during movement (Lee & White, 1974). This reduction arises partly because of interaction between the experimental stimulus and movement-related sensory feedback and partly because of central regulation of sensory excitability. We have attempted to isolate a component of the central regulation by examining the effects of mentally rehearsing a complex finger movement on the SEPs. The movement sequence was that described by Roland, Larsen, Lassen & Skinhoj (1980) who also showed that its mental rehearsal produced an increase in cerebral blood flow through the supplementary motor areas of cerebral cortex. These regions provide cortico-cortical projections to the peri-Rolandic cortex as well as contributing to the pyramidal tracts and may therefore regulate sensory excitability at both cortical and sub-cortical levels.

METHODS

Healthy male volunteers lay on a bench with their arms outstretched to the sides and supported. Electrical stimuli were delivered to the median nerve at the wrist through Ag/AgCl disc surface electrodes fixed in place with collodion. Similar electrodes were used to record the evoked potentials from the median nerve above the elbow using a pair of electrodes as well as from Erb's point and the scalp overlying the sensory cortex (c3') both referenced to the anterior frontal scalp (aFz). The recordings were amplified and filtered (3 Hz to 3kHz) and digitally averaged over 500 stimulus presentations. For each subject, five control and five rehearsal periods were recorded.

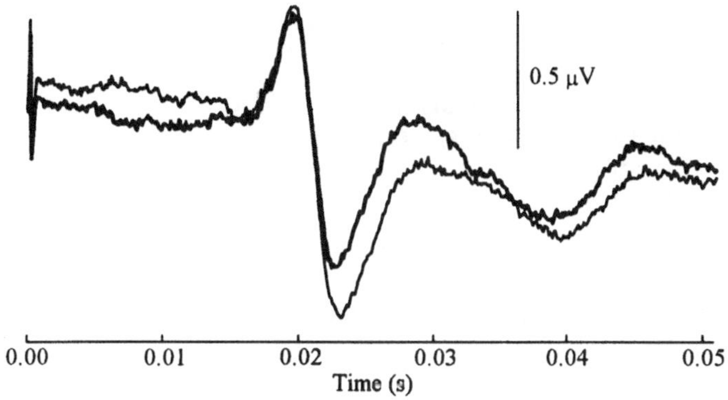

Figure 1. Averaged somatosensory evoked potentials recorded in a subject at rest (thin line) and during rehearsal of the movement (bold line). The potentials were recorded between c3' and aFz and averaged over 500 stimuli in each case.

RESULTS

Eleven subjects have been examined so far. Figure 1 shows the average scalp potential recorded in one subject at rest (thin line) and during rehearsal of the movement task (bold line). Note that the initial cortical response, N20, is unaffected by the rehearsal while the subsequent P22 is clearly reduced. N20 represents the sensory cortical response to the stimulus while P22 probably arises pre-centrally in the motor areas (see Jones, 1993 for discussion). The results illustrated in Figure 1 are typical; nine of ten subjects in whom the P22 was clear exhibited a reduction in its amplitude during mental rehearsal of the movement task.

DISCUSSION

The N20 was not consistently affected during mental rehearsal of the finger movement. Therefore, we have found no evidence of regulation of sensory excitability through descending control of the sensory pathways or of the postcentral sensory cortical areas. The reduction in ampltitude of P22 is therefore most probably attributable to corticocortical regulation of pre-central excitability to the peripheral nerve stimulus.

REFERENCES

JONES, S.J. (1993) Somatosensory evoked potentials I: Methodology, generators and special techniques. In *Evoked Potentials in Clinical Testing.* ed. HALLIDAY, A.M. pp 383-419. Churchhill-Livingstone, London.

LEE, R. G. & WHITE, D. G. (1974). Modification of the human somatosensory evoked response during voluntary movement. *EEG & Clin. Neurophysiol.* **36**, 53-62.

ROLAND, P. E., LARSEN, B., LASSEN, N. A. & SKINHOJ, E. (1980). Supplementary motor area and other cortical areas in organization of voluntary movements in man. *J. Neurophysiol.* **43**, 118-136.

BALANCE RECOVERY IN CEREBELLAR PATIENTS: EVIDENCE FOR A CONTRIBUTION OF THE FUSIMOTOR SYSTEM

M.C. Do[1], B. Bussel[2], Ph. Thoumie[1]
and O. Remy-Neris[2]

[1]Laboratoire de Physiologie du Mouvement
ERS CNRS 102, Université de Paris-Sud
Orsay, France
[2]Service de Rééducation Neurologique
Hôpital R. Poincaré, Garches, France

INTRODUCTION

It was suggested that the role of the cerebellum in locomotion includes a contribution to adjustments of the timing of locomotor movements of the limbs, an involvement in interlimb coordination and a modification of various reflex activities. In particular, it has been suggested that the cerebellum participates in the control of the sensitivity of muscle spindle stretch receptors through the fusimotor system (Gilman, 1969). However, Gorassini, Prochazka & Taylor (1993) have shown that the cerebellar nuclei (interpositus and dentate nuclei) are not primarily responsible for fusimotor control in cat locomotion.

In this study we have examined the motor behaviour of cerebellar patients in an experimental paradigm involving the recovery of balance which includes two motor programs, a compensatory reaction and a locomotor program. More precisely, to test the role of fusimotor system, we examined the soleus EMG activity of cerebellar patients and healthy subjects in which the soleus group I afferents were blocked.

METHODS

The compensatory reaction and the locomotor activity were induced by a forward fall obtained by a disequilibrium of an initial inclined posture. The subject, leaning forward (approximately 20°), was wrapped with an abdominal belt connected to an electro-mechanical device with a steel cable. The cable was released at a moment unknown to the subject, who was instructed to step to recover the balance. In the following description muscles of the leg which steps are indicated by the suffix m (mobile) and the others by s (stance).

12 healthy subjects and 12 cerebellar patients participated in the experiment. The static cerebellar syndrome was of vascular, tumoral or traumatic origin, without pyramidal syndrome or sensory disturbance. The cerebellar patients were selected on the basis of their ability to walk without help and to rise on their toe tips. In 4 healthy subjects, ischemia of the stance leg was induced (the pneumatic cuff was placed above the knee). The variables analysed were the EMG activity of right and left soleus (Sol) and tibialis anterior (TA), the kinetics of the centre of gravity and the ankle angular displacement.

RESULTS

It has been shown that the above perturbation of an initial inclined posture brings into play successively a reproducible compensatory reaction and stepping to recover balance (Do, Breneire & Brenguier, 1982; Do, Bouisset & Breniere, 1988). The compensatory reaction started with the activation (latency = 65 ms) of both soleus muscles (Figure 1, CTL) which braked the fall (an inversion of the vertical acceleration, G_Z). The end of Sol EMG activity (in the mobile foot, Sol_m) generally coincided with the positive peak of G_Z. The activation of TA_m followed, then the step occurred (TO). The first step was carried out with a downward acceleration and ended at HC (heel-contact) time. During the step

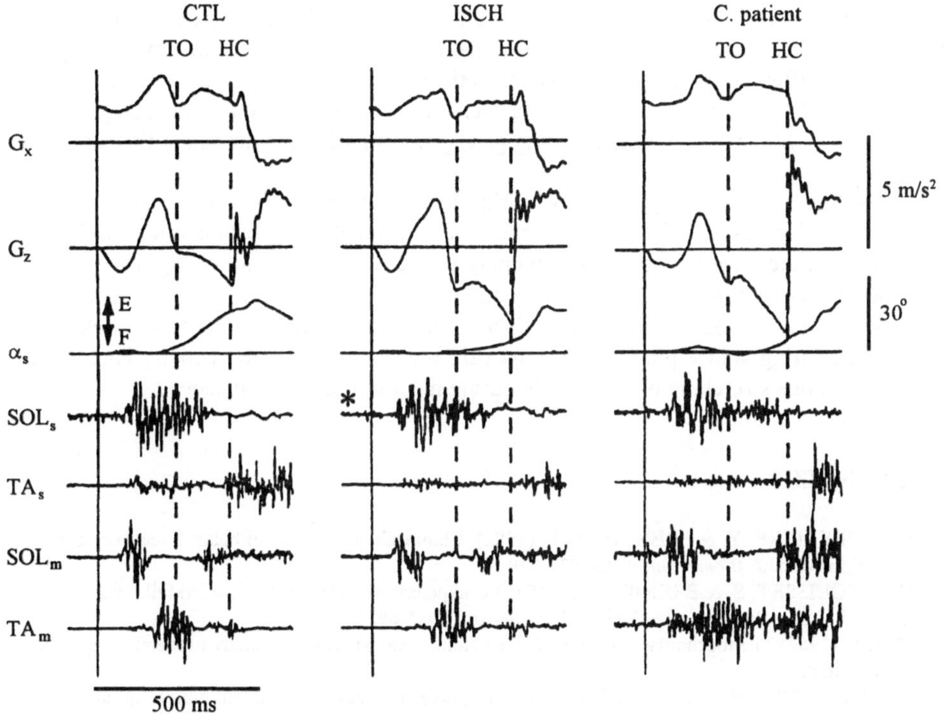

Figure 1. Single recordings of motor activity and biomechanics during forward fall in different situation. CTL, control healthy subject; ISCH, healthy subject with ischaemia of stance leg (*), same subject as in CTL; C. patient, cerebellar patient. G_x, G_z; acceleration of centre of gravity following the progression and vertical axes; as, ankle angular displacement (E, extension; F, flexion). Sol_s, Sol_m, TA_s, TA_m raw EMG activity of soleus and tibialis anterior in stance (s) and starting (m) leg. TO, HC, toe-off and heel-contact.

execution phase, the Sol_S EMG activity (stance foot) continued over the stance phase, and stopped just before heel-contact (HC); the ankle of the stance foot was continuously in extension.

Most of the cerebellar patients showed compensatory reactions appearing at a latency similar to that of healthy subjects. However, the Sol EMG pattern during the compensatory reaction was more variable and sometimes a co-contraction of TA and Sol was observed. During the unipedal stance phase (TO-HC), the Sol_S EMG activity (Figure 1, C. patient), the ankle extension and the vertical CG acceleration at HC were smaller in cerebellar than in healthy subjects. Furthemore, the step execution was impaired so that the cerebellar patients could not completely recover their balance.

In healthy subjects, ischemia of the stance leg (Figure 1, ISCH) did not induce any modification of Sol latency. The features of mechanical and EMG traces were similar to those observed in cerebellar patients. The comparison between the control and ischemia situations showed significant decreases of the positive peak of G_Z and EMG activity of Sol_S.

DISCUSSION

The absence of modification of Sol latency after ischemia suggests that the compensatory reaction to forward fall was not triggered by the Sol group I afferents. In contrast, the significant decrease of the Sol_S EMG activity after ischemia evidences the participation of Sol group I discharge during the unipedal stance phase of the balance recovery movement. It seems unlikely that the decrease in Sol_S EMG activity could be due to an effect of blockade of Ib afferents of the TA_S. Since the TA_S did not show phasic activity the Ib afferents would not have been activated. In other words, the decrease of Sol_S activity would be due to a decrease of Ia facilitation. This is in line with Yang, Stein & James (1991) who have shown that during gait, group Ia afferents modulate the reflex gain of Sol.

If the Sol_S EMG activity observed during the stance phase is Ia dependent, it seems likely that the decrease of EMG activity observed in cerebellar patients is due to a decrease of γ drive. In other words, during the stepping to recover balance following a forward fall the stance Sol activation is dependent on γ drive. The profile of ankle angular displacement could explain the purpose of the γ drive, since in the control situation the Sol muscle length is shortened during stance phase. In normal gait, the stance Sol EMG activity is reinforced by the continuous stretching of the muscle during most of the stance phase.

REFERENCES

DO, M.C., BRENIERE, Y. & BRENGUIER, P. (1982) A biomechanical study of balance recovery during the fall forward. *J. Biomechanics* **15**, 933-939.

DO, M.C., BOUISSET, S. & BRENIERE, Y. (1988) Compensatory reactions in forward fall : are they initiated by stretch receptors? *EEG & Clin. Neurophysiol.* **69**, 448-452.

GILMAN, S. (1969) The mechanism of cerebellar hypotonia. An experimental study in the monkey. *Brain* **92**, 621-638.

GORASSINI, A., PROCHAZKA, A. & TAYLOR, J.L. (1993) Cerebellar ataxia and muscle spindle sensitivity. *J. Neurophysiol.* **5**, 1853-1862.

YANG, J.F., STEIN, R.B. & JAMES, K.B. (1991) Contribution of peripheral afferents to the activation of the soleus muscle during walking in humans. *Exp. Brain Res.* **87**, 679-687.

FUNCTIONAL ORGANISATION OF THE INTERMEDIATE CEREBELLUM

M. Garwicz, C.-F. Ekerot, H. Jörntell
and J. Schouenborg

Department of Physiology and Biophysics
University of Lund
Sweden

INTRODUCTION

The uniform organisation of the neuronal circuitry throughout the cerebellar cortex suggests a uniform mode of operation and thus emphasises the importance of local afferent and efferent connections in determining the function of a particular part of the cortex. Based on the organisation of these connections the cerebellar cortex of the cat is divided into about ten sagittally oriented *zones* (see Ito, 1984 for references). A zone is anatomically defined by its projection to a restricted part of the intracerebellar or vestibular nuclei and its climbing fibre input from a circumscribed part of the inferior olive. Some of the zones are functionally coupled in that they receive branching collaterals from common olivary neurones and in turn project to the same subdivision of the intracerebellar nuclei. Since each part of the inferior olive receives input from a specific set of spino-olivary pathways, the zones can be electrophysiologically identified by the latencies and receptive fields of climbing fibre responses evoked on peripheral stimulation. The organisation of olivary afferent and nuclear efferent connections suggests that each zone, or in some cases an ensemble of zones, controls specific motor systems.

The intracerebellar nucleus interpositus anterior receives cortical input from the C1, C3 and Y zones in the overlying pars intermedia and in turn projects to the magnocellular red nucleus and, via thalamus, to the motor cortex, thus controlling the rubrospinal and corticospinal tracts. Studies of this cerebellar system in awake animals have indicated its importance for the execution of voluntary arm/forelimb movements requiring coordination of several limb segments, such as reaching for a target. Inactivation of nucleus interpositus has been shown to cause severe action tremor or dysmetria during reaching in monkeys and cats, respectively. Furthermore, single unit recordings in pars intermedia of the cerebellar cortex and in nucleus interpositus have revealed significant modulation of neuronal discharge patterns during different types of reaching movements.

SPATIAL ORGANISATION OF CLIMBING FIBRE RECEPTIVE FIELDS

Recent electrophysiological investigations in our laboratory, using mainly barbiturate anaesthetized cats, have been concerned with the detailed functional organisation of the olivo-cortico-nuclear connections in the forelimb part of this cerebellar control system. In an initial study, the climbing fibre input was characterized in terms of spatial organisation of peripheral receptive fields and patterns of termination in the cortex (Ekerot, Garwicz & Schouenborg, 1991). Responses in climbing fibres on nociceptive and tactile stimulation of the forelimb skin were recorded as complex spikes in single Purkinje cells in the C3 zone. The noxious pinch stimulus evoked unusually vigorous, sustained responses which greatly facilitated the mapping. In general, the proximal borders of the receptive fields were located close to joints and the area from which maximal responses were evoked was usually located eccentrically within the receptive field. Based on their spatial characteristics the receptive fields could be divided into eight classes, which in turn were tentatively divided into about thirty subclasses. The climbing fibre input to the C3 zone had a detailed topographical organisation, with different classes of receptive fields represented in different parts of the zone. Furthermore, climbing fibres belonging to the same subclass of receptive fields terminated within narrow, sagittal strips of cortex, thus forming *microzones* defined by their homogeneous climbing fibre input.

CORTICAL PROJECTION TO NUCLEUS INTERPOSITUS

In the second study, the topographical organisation of the cerebellar cortical projection to nucleus interpositus anterior was investigated using a new method (Garwicz & Ekerot, 1994). This was based on the observation that stimulation of the forelimb skin evoked positive extracellular field potentials in the nucleus. The positive potentials reversed polarity in the dorsal part of the nucleus, had response characteristics similar to cortical climbing fibre field potentials and were reversibly reduced in amplitude when a local anaesthetic was applied to the cerebellar surface. These properties showed that the positive potentials reflected local inhibitory synaptic currents generated by climbing fibre activated Purkinje cells. Thus, determining the area on the skin from which positive potentials could be evoked at a particular site in the nucleus was almost equivalent to mapping the climbing fibre receptive fields of Purkinje cells projecting to that nuclear site and this approach could be used to investigate systematically the organisation of the cortico-nuclear projection. The projection was found to be topographically organised and the receptive fields identified in nucleus interpositus had characteristics similar to those of single climbing fibres projecting to the C3 zone, thus suggesting that there was little convergence in the cortico-nuclear projection between microzones with different climbing fibre input. On the other hand, microzones with similar climbing fibre receptive fields in the C1, C3 and Y zones were proposed to converge on common groups of nuclear neurones. These olivo-cortico-nuclear *modules* were suggested to constitute the functional units of this part of the intermediate cerebellum (Garwicz, 1992).

RELATION BETWEEN CLIMBING FIBRE INPUT AND MOTOR OUTPUT

In a concluding study, the relation between climbing fibre input and motor output of the modules was investigated in decerebrated cats (Ekerot, Jörntell & Garwicz, 1995). As in the previous study, the climbing fibre receptive field was established by recording positive field potentials in the nucleus evoked on natural stimulation of the forelimb skin. The electrode was then switched to stimulating mode and the movement of the forelimb evoked on local electrical microstimulation in the nucleus was assessed. The movements usually involved more than one limb segment. Shoulder retraction and elbow flexion were frequently evoked, whereas elbow extension was rare and shoulder protraction never observed. Flexion and extension movements at the wrist were diverse, depending on activation of muscles with radial, central or ulnar insertions on the paw. Movements of the digits consisted mainly of dorsal flexion of central or ulnar digits. A comparison of climbing fibre receptive fields and associated movements indicated a general specificity of the input-output relationship of the modules in this cerebellar control system. Several findings suggested that, assuming the limb was in a free position, the movement evoked from a particular site would act to withdraw the area of the skin corresponding to the climbing fibre receptive field of the afferent microzones. For example, sites with receptive fields on the dorsum of the paw were

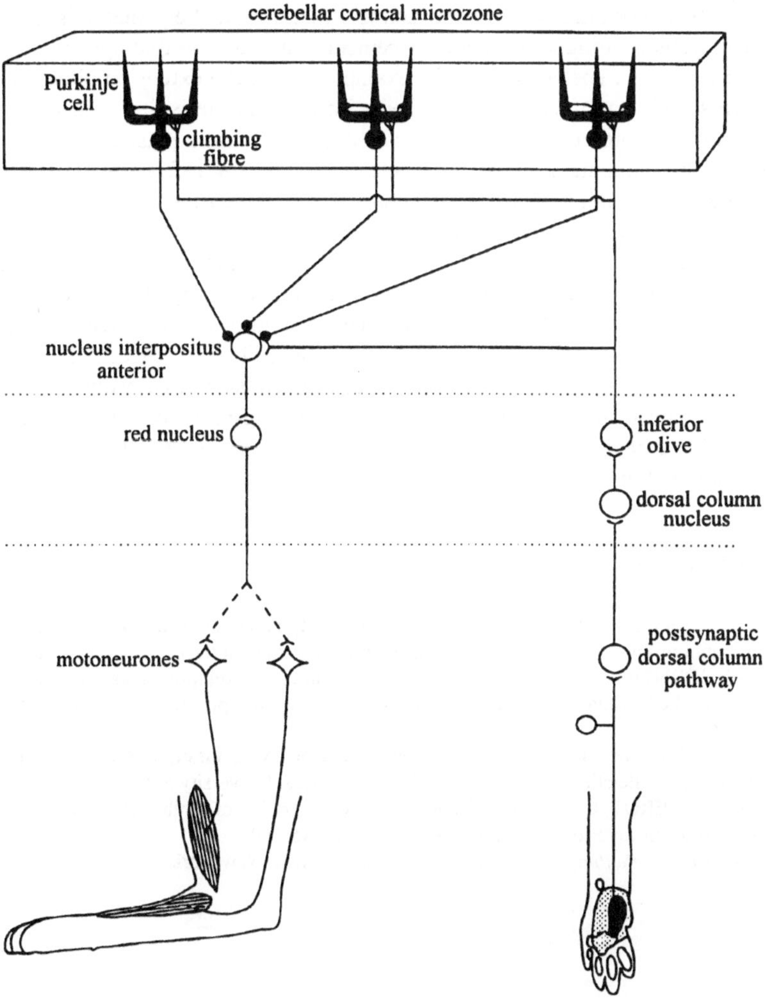

Figure 1. Diagram of the proposed modular organisation of cerebellar control of forelimb movements via the rubrospinal tract. Note that only one microzone is shown for simplicity. Filled cells are inhibitory. The cutaneous receptive field of the climbing fibre afferents to this particular module is depicted on the ventral side of the paw. Centre of receptive field shown in black. The associated movement is composed of a dorsiflexion of the paw and a flexion at the elbow. Dashed lines indicate indirect connections.

frequently associated with palmar flexion at the wrist, whereas sites with receptive fields on the ventral side of the paw and forearm were associated with dorsiflexion at the wrist (Figure 1). Correspondingly, receptive fields on the lateral side of the forearm and paw were often related to flexion at the elbow, whereas sites with receptive fields on the radial side of the forearm were associated with elbow extension.

CONCLUSIONS

Our results suggest that the part of the intermediate cerebellum which controls forelimb movements via the rubrospinal and corticospinal tracts has a modular organisation. Each module receives a homogenous climbing fibre input and appears to control a specific movement of the limb. The general bias for flexion movements in our material and some of the results from investigations in awake animals make it unlikely that this system is the driving motor system during explorative movements. Instead, the findings are suggestive of a "braking action" of cerebellar output, for example during the extension phase of reaching-like movements. This notion can be reconciled with the hypothesis that climbing fibres carry information about "motor errors" and the evidence for synaptic plasticity in the cerebellar cortex induced by climbing fibre activity (see Ito, 1984 for references). In a context of motor learning, unintended skin contact during the extension phase of reaching-like movements would give rise to an "error signal" in microzones with climbing fibre receptive fields overlapping the affected skin area. This would in turn induce long-lasting reduction of transmission between the parallel fibres which were active during the movement and their target Purkinje cells, eventually causing a reduced inhibition of the corresponding efferent neurones in nucleus interpositus anterior. Since climbing fibre receptive fields appear to be associated with output acting to withdraw the affected skin area from a stimulus, this disinhibition of nuclear neurones in the context of reaching-like movements would enhance "braking" of the extension movement of appropriate limb segments, thus reducing or eliminating the initial motor error.

REFERENCES

EKEROT, C.-F., GARWICZ, M. & SCHOUENBORG, J. (1991) Topography and nociceptive receptive fields of climbing fibres projecting to the cerebellar anterior lobe in the cat. *J. Physiol.* **441**, 257-274

EKEROT, C.-F., JÖRNTELL, H. & GARWICZ, M. (1995) Functional relation of corticonuclear input and movements evoked on microstimulation in cerebellar nucleus interpositus anterior in the cat. *Manuscript.*

GARWICZ, M. (1992) Cerebellar control of forelimb movements: Modular organisation revealed by nociceptive and tactile climbing fibre input. *Doctoral Thesis*, University of Lund.

GARWICZ, M. & EKEROT, C.-F. (1994) Topographical organization of cerebellar cortical projection to nucleus interpositus anterior in the cat. *J. Physiol.* **474**, 245-260

ITO, M. (1984). *The Cerebellum and Neural Control.* Raven Press. New York.

GATING OF SPINO-OLIVOCEREBELLAR PATHWAYS IN THE AWAKE CAT

Richard Apps and Nick Hartell

Department of Physiology, School of Medical Sciences
University of Bristol, Bristol, UK

INTRODUCTION

The spino-olivocerebellar pathways (SOCPs) are vital in motor control and in the cat, include paths which terminate as climbing fibres in the c_1, c_2 and c_3 zones in the paravermal cerebellar cortex. To date, little is known about any functional differences that may exist between zones, particularly in terms of their climbing fibre input and associated SOCPs. The possibility of operational differences between zones has therefore been investigated in the awake cat. In the present study the aim was to investigate those SOCPs that target the paravermal zones and determine if changes in excitability occur during rest and locomotion.

METHODS

At an initial aseptic operation under general anaesthesia an array of 25 microwires was chronically implanted into the electrophysiologically identified c_1, c_2 and c_3 zones within superficial folia of lobules V/VI of the cerebellar cortex in 3 cats. Additional leads were implanted to record EMG activity in a number of forelimb muscles (including triceps brachii) as well as stimulating and recording cuffs which were implanted around the superficial radial (SR) nerve in the ipsilateral forelimb. Following recovery from the operation, non-noxious electrical stimulation of the SR nerve was used to evoke, via the SOCPs, extracellular climbing fibre field potentials in the paravermal zones. Measurement of the size (mV.ms) of these responses was used to monitor excitability in the SOCPs at rest and during steady walking on an exercise belt.

RESULTS

Chronically implanted microwires yielded responses characteristic of the c_1, c_2 and c_3 zones at 15, 14 and 7 sites respectively. At each recording site the mean size of responses varied systematically during locomotion and such changes were not paralleled by changes in nerve volley size (not illustrated). At c_1 and c_3 sites, responses were consistently largest during late swing (Figure 1A) whereas least responses were more varied in timing but occurred most frequently during early stance (Figure 1B). In contrast, at most c_2 sites the largest responses during locomotion occurred during mid/late stance (Figure 1C) whereas least responses tended to occur during swing or early stance (Figure 1D).

For each recording site the mean size of responses evoked during rest was compared with the size of responses evoked by a similar stimulus during walking. At c_1 sites the largest responses during locomotion were often considerably larger (13/15 sites > 125% of rest; Figure 2A, stippled bars) while the smallest responses were usually smaller (14/15 sites < 75% of rest; Figure 2A, open bars). Similar findings were obtained for c_3 sites (largest responses: 5/7 sites > 125% of rest; smallest responses: 6/7 sites < 75% of rest; see Figure 2B, stippled and open bars respectively). By comparison, at c_2 sites the largest responses during locomotion were only rarely larger (5/14 sites > 125% of rest; Figure 2C, stippled bars) while the smallest responses were consistently smaller (14/14 sites < 75% of rest; Figure 2C, open bars).

CONCLUSIONS

The results confirm and extend the more limited findings of Apps, Lidierth & Armstrong (1990) and Lidierth & Apps (1990). During locomotion, excitability in SOCPs to the paravermal zones varies with the phase of the step cycle and can be either facilitated or depressed relative to rest. In contrast to the c_2 zone, the pattern of step-related

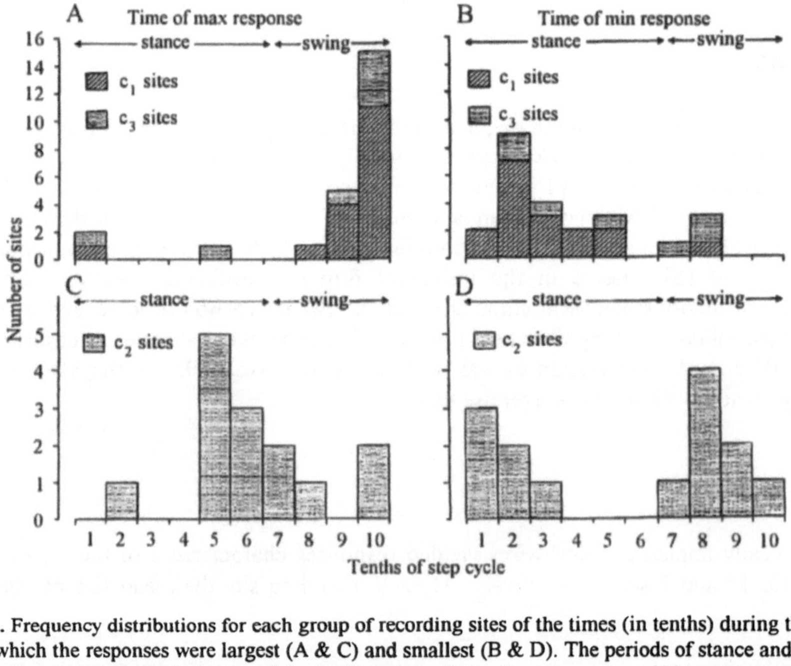

Figure 1. Frequency distributions for each group of recording sites of the times (in tenths) during the step cycle at which the responses were largest (A & C) and smallest (B & D). The periods of stance and swing are approximate timings for trajectory of the ipsilateral forelimb. Since a number of histograms were obtained for each site, the sample with the greatest difference between the largest and smallest response has been selected. As step-related fluctuations in response size at c_1 and c_3 sites were similar, the results are grouped together in A & B. Hatched bars c_1 sites (n = 17); stippled bars c_3 sites (n = 7). C & D c_2 sites (n = 14).

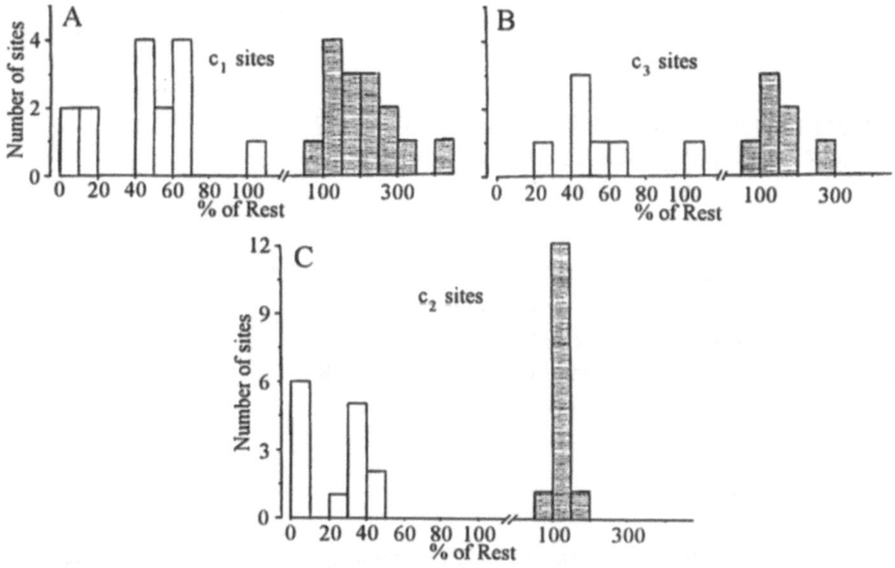

Figure 2. Frequency distributions for each recording site of response size during locomotion as compared to rest. For each site the largest and smallest mean response sizes ever encountered during locomotion are shown (stippled and open bars respectively) expressed as a percentage of mean response size at rest animal. A, c_1 sites (n = 15); B, c_3 sites (n = 7); C, c_2 sites (n = 14).

modulation is similar at c_1 and c_3 sites. The present data show that sensory traffic in the ipsilateral SR nerve is most likely to be relayed via the SOCPs to the c_1 and c_3 zones during the swing phase of the step cycle and least likely to be conveyed during the stance phase. By comparison, sensory traffic in the ipsilateral SR nerve is most likely to be relayed via the SOCPs to the c_2 zone during the stance phase of the step cycle and least likely during swing/early stance. Overall, the data add force to the suggestion that at least some zones of the cerebellum have different functional responsibilities in motor control.

ACKNOWLEDGEMENTS

Supported by the Wellcome Trust.

REFERENCES

APPS, R., LIDIERTH, M. & ARMSTRONG, D.M. (1990) Locomotion related variations in the excitability of spino-olivocerebellar paths to the cerebellar cortical c_2 zone in the cat. *J. Physiol.* **424**, 487-512.

LIDIERTH, M. & APPS, R. (1990) Gating in the spino-olivocerebellar pathways to the c_1 zone of the cerebellar cortex during locomotion in the cat. *J. Physiol.* **430**, 453-469.

THE SOMATOSENSORY ORGANISATION OF THE INFERIOR OLIVARY NUCLEUS OF THE RAT

B. Cappi

Sherrington School of Physiology, UMDS
St Thomas' Hospital, London SE1 7EH, UK

INTRODUCTION

Sensory input to the cerebellum arises from climbing fibres and mossy fibres. Mossy fibres supply the cerebellum from a wide range of brain structures, whereas the inferior olivary nucleus (ION) is the sole source of climbing fibres. As well as providing input to the cerebellar nuclei, climbing fibres exert a strong influence on the Purkinje cells in the cerebellar cortex in which they evoke complex spikes (Eccles, Llinás & Sasaki, 1966). Despite its apparent importance in sensorimotor control, the physiology of ION has not been studied in great detail.

Most studies of the olivocerebellar system have been concerned with recording complex spikes from Purkinje cells. However, this provides only indirect information about the organisation of sensory input to the inferior olivary nucleus. In this study the somatosensory receptive fields of single olivary cells have been examined using tactile stimulation and the field potentials evoked in the olive have been mapped following electrical stimulation of peripheral nerves.

METHODS

Wistar rats (300 - 350 g male) were anaesthetised with urethane (1.25 g/kg i.p.) and the cerebellum was exposed to allow a dorsal approach. For unitary studies, extracellular recordings were made in the ION using varnished stainless steel microelectrodes (1 - 2MΩ at 1 kHz) inserted at an angle of 22° to the vertical. Olivary cells were recognised by their slow spontaneous firing rate of approximately 1 Hz, and their characteristic "popping" sound on an audio monitor.

For evoked potential studies, the left sciatic nerve was dissected and mounted for stimulation and the cerebellum was exposed to gain access to the contralateral olivary nucleus. Evoked potential recordings were made using identical electrodes to those previously described. The sciatic nerve was stimulated at 1 Hz with a pulse width of 0.1 ms.

The evoked response in the olive was filtered (1 Hz - 1 kHz), and averaged over 32 stimulus presentations. The averaged data were further processed by taking an arbitrary point on each of the averaged traces (3 ms post stimulus) and used as a baseline to which all other readings were compared. Amplitude measurements were taken at 5, 6, 7, 8, and 9 ms post stimulus. Histological reconstruction of electrode tracks was aided by the placing iron marks (20 μA for 20s, tip positive).

RESULTS

In total, 32 olivary cells have been isolated and recorded. Sixteen showed no response to cutaneous stimulation. These neurones were more numerous in the medial parts of the inferior olivary nucleus. Sixteen cutaneous receptive fields associated with olivary cells were mapped. Those neurones located in the lateral parts of the inferior olive received input from the contralateral surface. Three of these cells were supplied from the forelimb, 5 from the hindlimb, 4 from the face, and 4 cells were supplied from both forelimb and hindlimb. Results are summarised in Figure 1. Four of these neurones responded only to light tapping (arrows) with no observable response to light brushing of the receptive field.

Field potentials were seen very clearly with latency between 5 and 8ms. Serial electrode penetrations were made to further localise these potentials. They were found to be located laterally in the dorsal accessory olive. Figure 2A shows the averaged data obtained for a single track. Note that some single action potentials may be superimposed onto the field potential. Figure 2B is a 3D surface graph representing this series of tracks.

DISCUSSION

It appears that there is a high level of somatotopic organisation within the ION of the

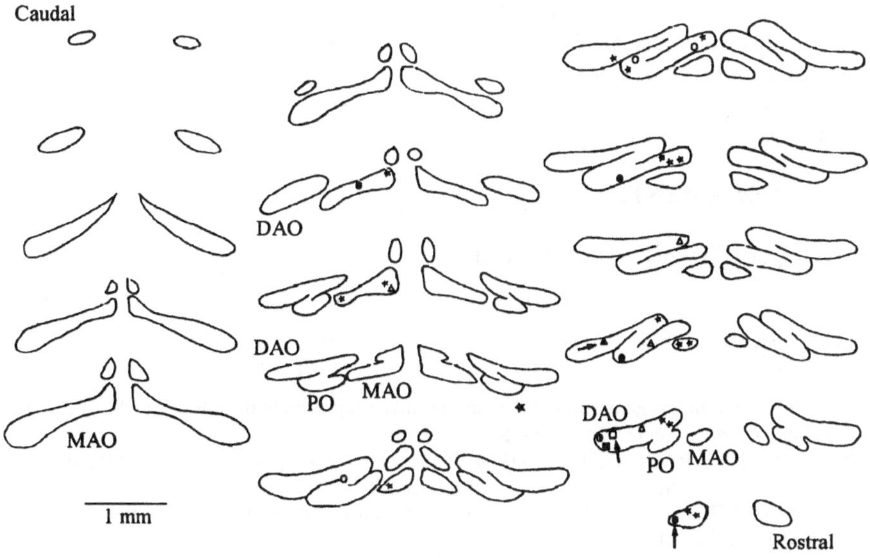

Figure 1. Standard outline drawings of transverse sections through the inferior olivary nucleus taken at 200 μm intervals in the rostro-caudal axis (taken from Apps, 1990). The symbols indicate the position of isolated cells and their receptive field location. (MAO: medial accessory olive; DAO: dorsal accessary olive; PO: principal olive).

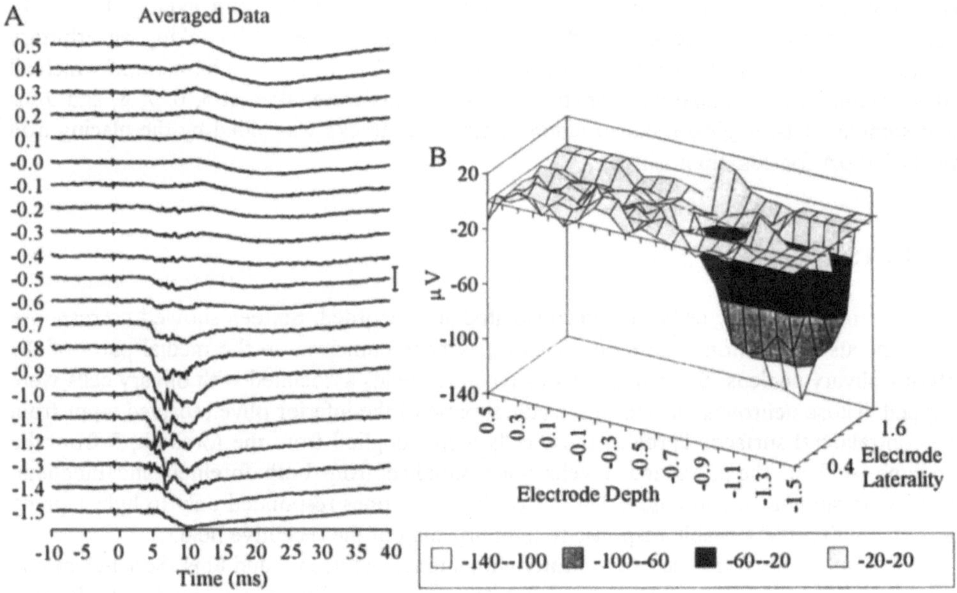

Figure 2. Field potentials recorded in the inferior olivary nucleus in response to electrical stimulation. A. data obtained at different depths from a single electrode penetration. Scale bar = 93.6 μV. B. Data at 9 ms latency, obtained from a series of tracks. Axes labelled "Electrode Depth" and "Electrode Laterality" indicate stereotaxic co-ordinates.

rat. The pattern emerging from the single unit study suggests that the lateral parts of the olivary complex receive greater input from limbs, while more medial parts have a greater input from the face. In very medial regions cells more often have no receptive fields. However more cells need to be isolated from wider areas of the olivary complex.

Thus far this study has failed to isolate cells that have a bilateral input or deep input as found by Gellman, Houk & Gibson (1983) in the cat. However those cells which exhibited a response to tapping only may have been responding to a small movement induced by the tapping process.

ACKNOWLEDGEMENTS

B.C. was supported by the Wellcome Trust.

REFERENCES

APPS, R. (1990). Columnar organisation of the inferior olive projection to the posterior lobe of the rat cerebellum. *J. Comp. Neurol.* **302**, 236-254.

ECCLES, J.C., LLINÁS, R. & SASAKI, K. (1966). The excitatory synaptic action of climbing fibres on the Purkinje cells of the cerebellum. *J. Physiol.* **182**, 268-296.

GELLMAN, R., HOUK, J.C. & GIBSON, A.R. (1983). Somatosensory properties of the inferior olive of the cat. *J. Comp. Neurol.* **215**, 228-243.

EFFECTS OF REDUCING THE TOOTH HEIGHT ON JAW-CLOSING MUSCLE ACTIVITY, BITE FORCE AND MUSCLE SPINDLE DISCHARGES DURING MASTICATION IN THE RABBIT

T. Morimoto[1], O. Nakamura[1], Y. Masuda[1],
K. Yoshikawa[2], T. Maruyama[2] and K. Takada[3]

Departments of Oral Physiology[1]
Prosthetic Dentistry[2] and Orthodontics[3]
Osaka University Faculty of Dentistry
1-8, Yamadaoka, Suita, Osaka, 565 Japan

INTRODUCTION

Electromyographic (EMG) activity of masticatory muscles and jaw movements are influenced by orofacial sensations arising during chewing foods of various properties (Thexton, Hiiemae & Crompton, 1980). Previous studies have suggested that both periodontal receptors and muscle spindles in the jaw closing muscles are primarily responsible for these influences (Lavigne, Kim, Valiquette & Lund, 1987; Morimoto, Inoue, Masuda & Nagashima, 1989). It was demonstrated that after lesioning of the mesencephalic trigeminal nucleus (MesV) in the anesthetised rabbit, the facilitative response of the masseter muscle to application of a test strip between the opposing teeth during mastication was greatly reduced. This result may be attributed to loss of muscle afferents from the jaw-closing muscles. The present study aims to obtain further evidence that the jaw-jerk reflex participates in the control of the biting force during mastication.

METHODS

Ten male rabbits (≈ 2.5 kg) were anaesthetised with IV α-choloralose (60 mg/kg) and urethane (500 mg/kg), supplemented with halothane in oxygen. The animal's head was fixed in a stereotaxic apparatus. Mastication was induced by intracortical microstimulation (0.2 ms, 30 Hz, 40 - 100 μA for 7 s) of the cerebral masticatory area (CMA). EMG activity was recorded, using pairs of 150 μm enamelled wires, from masseter and digastric muscles of both sides during mastication and compared before and after grinding the teeth. Jaw movements were recorded by an optoelectronic method. In a few animals, axially directed chewing force was recorded by a small force-displacement transducer fixed at the molar region.

RESULTS

Although various patterns of rhythmic jaw movements were induced by CMA stimulation, the crescent shaped movements with a distinct power phase were mainly employed here since they resemble jaw movements during natural chewing in awake rabbits. After grinding the teeth, the jaw could be raised by the amount of the reduced tooth height during the power phase. When the reduced height exceeded 0.7 mm, the masseter EMG activity during jaw closure was significantly decreased. The magnitude of EMG reduction appeared to be proportional to the amount of tooth grinding, and the EMG activity could be reduced to less than 20% of the value recorded before tooth grinding. The bite force was also reduced by tooth grinding. Similar effects were found in the animals with loss of oral sensations by trigeminal de-afferentation. These results indicate that both the masseter activity and the bite force induced during mastication were affected by the change in the intermaxillary distance. Since muscle spindles in the jaw-closing muscles are affected by the

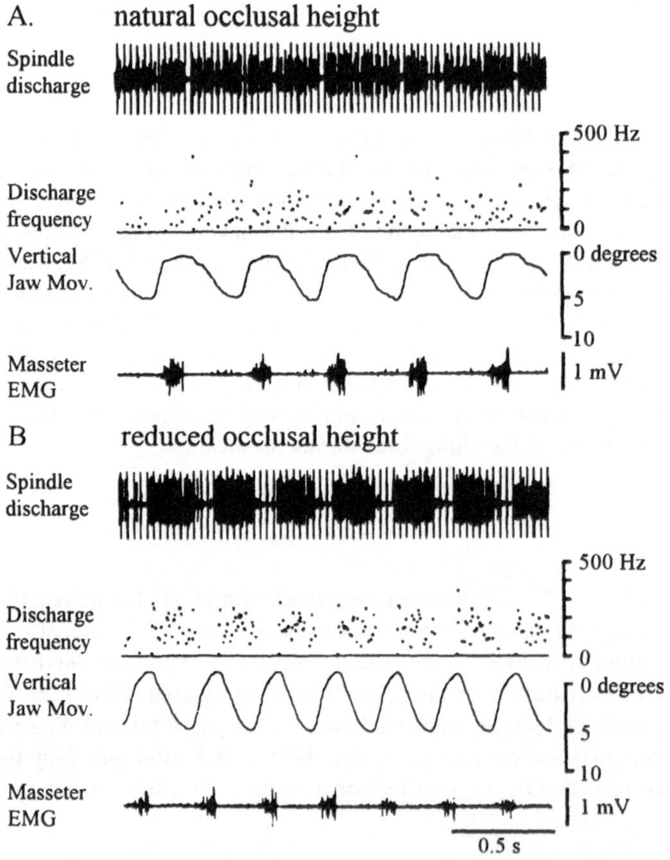

Figure 1. Muscle spindle response during mastication at two different occlusal heights. A. natural occlusal height. B. occlusal height reduced by 1.2 mm. Note that the discharges disappeared during jaw closing and power phases in B.

intermaxillary distance, we recorded muscle spindle discharges in the MesV during mastication after cutting the maxillary and inferior alveolar nerves. The spindle discharges were then compared at natural occlusal height and at the reduced occlusal height. An example of the MesV neurone activity is shown in Figure 1. This neurone was activated by passive jaw opening with ramp and hold stretches and ceased firing during jaw closure. Furthermore, the response increased after IV injection of succinylcholine, indicating that the neurone was likely to be a muscle spindle afferent. As shown in Figure 1A, this presumed spindle fired throughout the masticatory cycle when the teeth occluded at the natural tooth height, although the rate of firing was high during jaw-opening phase and low during jaw-closure. In contrast, the spindle discharges were greatly reduced or disappeared during the power phase after grinding the teeth (Figure 1B).

DISCUSSION

Several studies have been carried out on spindle discharges during mastication in awake animals or during peripherally induced rhythmic jaw movements in anesthetised animals (Taylor & Cody, 1970; Goodwin & Luschei, 1975). These studies demonstrated that when the teeth contact food and the movements of the mandible slow down, the discharges of primary spindle afferent increase with an increase in the EMG activities of jaw-closing muscles. These findings suggest that the increased spindle discharges during jaw-closure can enhance the jaw-closing muscle activities through the jaw-jerk reflex. Although food was not masticated in the present study, the muscle spindle was found to be activated during the power phase of the cortically induced mastication. When the tooth height was reduced by grinding, the spindle discharges were decreased or disappeared during the power phase. These findings may be accounted for by the same mechanism. The decrease of EMG activity of the jaw-closing muscles at the reduced tooth height may be due to the reduced spindle input to the jaw-closing motoneurones through the jaw jerk reflex arc. In the present study, the EMG activity of the jaw closing muscle recorded after tooth grinding could be reduced to less than 20% of the value before tooth grinding. This finding suggests that the EMG activity of the jaw closing muscles is produced mostly by the reflex mechanism and little by direct inputs from the central pattern generator.

ACKNOWLEDGMENT

Supported by the Grants from Japanese Ministry of Education (No. 05454500).

REFERENCES

GOODWIN, G.M. & LUSCHEI, E.S. (1975) Discharge of spindle afferents from jaw-closing muscles during chewing in alert monkeys. *J. Neurophysiol.* **38**, 560-571.

LAVIGNE, G., KIM, J.S., VALIQUETTE, C. & LUND, J.P. (1987) Evidence that periodontal presso-receptors provide positive feedback to jaw closing muscles during mastication. *J. Neurophysiol.* **58**, 342-358.

MORIMOTO, T., INOUE, T., MASUDA, Y. & NAGASHIMA, T. (1989) Sensory components facilitating jaw-closing muscle activities in the rabbit. *Exp. Brain Res.* **76**, 424-440.

TAYLOR, A. & CODY, F.W.J. (1970) Jaw muscle spindle activity in the cat during normal movements of eating and drinking. *Brain Res.* **71**, 523-530.

THEXTON, A.J., HIIEMAE, K.M. & CROMPTON, A.W. (1980) Food consistency and bite size as regulator of jaw movement during feeding in the cat. *J. Neurophysiol.* **44**, 456-474.

MODULATION OF THE JAW-OPENING REFLEX EVOKED BY TOOTH-PULP STIMULATION IN THE CAT

D. Banks

Sherrington School of Physiology, UMDS
St. Thomas's Hospital, London SE1 7EH, UK

INTRODUCTION

In lightly anaesthetised or conscious cats, tooth-pulp stimulation produces reflex EMG activation of the digastric muscle at a latency of 7-10 ms, resulting in a jaw-opening reflex (JOR). In many experimental animals this jaw opening is regarded as analogous to the limb flexion withdrawal reflex although the association between pain experience and the occurrence of a flexion withdrawal reflex may not be wholly appropriate for the jaw. The jaw-opening reflex can be elicited as a high threshold response to noxious stimulation of the teeth or as a low threshold response to non-noxious stimulation of the teeth and other intra-oral structures. The jaw-opening reflex is thought to involve interneurones in the rostral trigeminal spinal nucleus, in particular subnucleus oralis. In experiments on acute, anaesthetised preparations a variety of CNS structures have been shown to exert an influence on both the JOR and putative reflex interneurones including; somatosensory cortex, anterior pretectal nucleus, nucleus raphe magnus, periaqueductal grey and rostral ventromedial medulla. These central structures probably contribute to the behavioural depression of the JOR which accompanies a variety of activities.

Except for behavioural studies, most experimental animals are surgically traumatised during preparation for stereotaxic recording despite anaesthesia. In these preparations there is a long lasting depression in the excitability of neurones in the trigeminal sensory nuclei and an increase in the threshold of the JOR in response to tooth-pulp stimulation (Clarke & Matthews, 1990). In the present case we have avoided the effects of surgical trauma and investigated a variety of stimuli which modulate the JOR and putative reflex interneurones in subnucleus oralis in the chronically prepared, conscious or anaesthetised cat.

METHODS

Male cats were initially prepared under general anaesthesia (alphaxalone/alphadolone, 18 mg/kg I.M. and 5 - 8 mg/kg/h I.V.) for chronic recording as described previously

(Banks, Kuriakose & Matthews, 1993a; Boissonade, Banks & Matthews, 1991). An occipital craniotomy was made, sealed with elastomer and covered with a headpiece (Banks et al., 1993a). An anterior connector block was fixed to the skull over the frontal sinuses (Boissonade et al., 1991). An indwelling cannula was inserted into an external jugular vein. Six stainless steel wires were passed subcutaneously from the connector block to the right digastric muscle for recording EMG and Ag/AgCl fillings were placed in the upper and lower right canine teeth for electrical stimulation. Following recovery, observations were made when the cats were either re-anaesthetised or when conscious.

RESULTS

In previous studies using these techniques in deeply anaesthetised cats (Banks, Kuriakose & Matthews, 1992) we found that high intensity electrical conditioning stimuli applied to the ipsilateral forepaw at sufficient intensity to stimulate Aδ fibres (6mA/1ms/10Hz for 10 minutes) produced an increase in the thresholds of trigeminal brainstem neurones and of the jaw opening reflex to tooth-pulp stimulation, with the jaw opening reflex thresholds being elevated significantly more and for longer than those of the neurones. In experiments on acute preparations (Cadden, 1985), it was shown that a similar effect was also produced by stimulating Aδ fibres. In contrast, in chronically prepared, anaesthetised cats selective stimulation of C-fibres produced by topical applications of 5% mustard oil to the skin of the forearm decreases the thresholds of trigeminal neurones and the JOR to tooth-pulp stimulation (Banks, Kuriakose & Matthews, 1993b). We have established that in the chronically prepared, anaesthetised cat, the size of the JOR and the activity of neurones in the trigeminal spinal (pars oralis) nucleus following tooth-pulp stimulation can be modulated by a variety of sensory inputs outside the oral cavity.

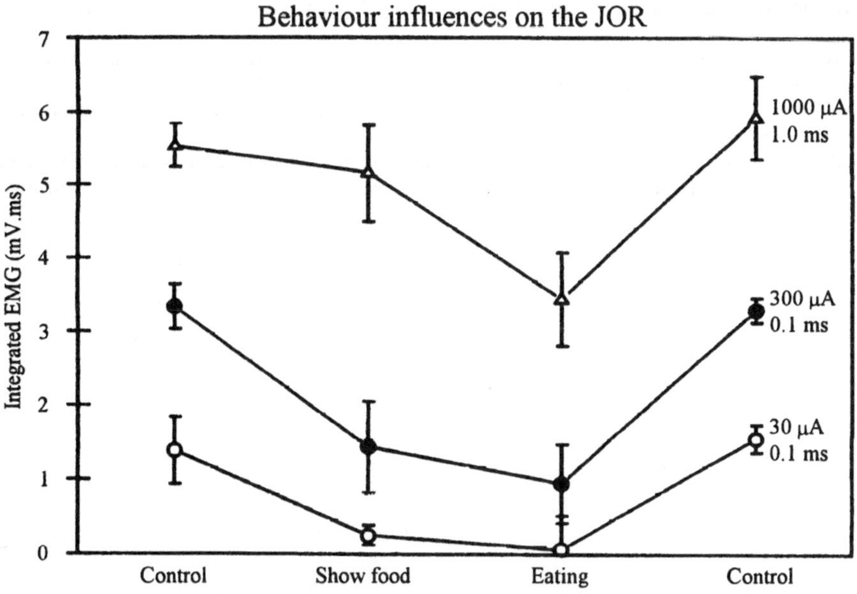

Figure 1. Variations in the size of the digastric EMG at three different intensities of electrical stimuli (open circles = 30 μA, 0.1 ms; filled circles = 300 μA, 0.1 ms; triangles = 1000 μA, 1.0 ms) applied at 0.1 Hz to the canine tooth-pulp of a conscious cat when sitting still (Control), distracted (Show Food) or when eating a sample of fish (Eating).

An example of the effect of behavioural modifications in the size of the JOR is shown in Figure 1. In this, the size of the digastric response (mean ± 2S.D., n = 10) to tooth-pulp stimulation applied at 0.1Hz and at three different intensities (shown on the right hand side as 1000μA, 1ms; 300μA, 0.1ms; 30μA, 0.1ms) is reduced from its resting value when an animal is shown food and during eating.

DISCUSSION

During eating the reduction in size of the JOR may be due partly to sensory inputs originating in and around the oral cavity. However this depression may originate from structures centrally since the site or smell of food is sufficient to inhibit the jaw-opening reflex. The size of this centrally originating depression is more pronounced when a single tooth-pulp stimulus is in the range 30 - 300 μA, 0.1 ms duration than at higher intensities. Thus sensory inputs from structures within or outside the oral cavity, in addition to the central nervous system are able to modulate the size of the jaw-opening reflex in response to tooth-pulp stimulation.

REFERENCES

BANKS, D., KURIAKOSE, M. & MATTHEWS, B. (1992) Modulation by peripheral conditioning stimuli of the responses of trigeminal brain-stem neurones and of the jaw opening reflex to tooth-pulp stimulation in chronically prepared, anaesthetized cats. *Exp. Physiol.* **77**, 343-349.

BANKS, D., KURIAKOSE, M. & MATTHEWS, B. (1993a) A technique for recording the activity of brain-stem neurones in awake, unrestrained cats using microwires and an implantable micromanipulator. *J. Neurosci. Meth.* **46**, 83-88.

BANKS, D., KURIAKOSE, M. & MATTHEWS, B. (1993b) Effect of cutaneous application of mustard oil on responses of trigeminal brainstem neurones to tooth pulp stimulation in cats. *Soc. Neurosci. Abstr.* **19**, 111.

BOISSONADE, F.M., BANKS, D. & MATTHEWS, B. (1991) Methods for recording the jaw-opening reflex to tooth-pulp stimulation in awake cats. *J. Neurosci. Meth.* **38**, 35-40.

CADDEN, S.W. (1985) The digastric reflex evoked by tooth-pulp stimulation in the cat and its modulation by stimuli applied to the limbs. *Brain Res.* **336**, 33-43.

CLARKE, R.W. & MATTHEWS, B. (1990) The thresholds of the jaw-opening reflex and trigeminal brain-stem neurones to tooth-pulp stimulation of acutely and chronically prepared cats. *Neurosci.* **36**, 105-114.

IS THE SUPRATRIGEMINAL AREA INVOLVED IN JAW-OPENING REFLEX ACTIVITY?

P.J.W. Jüch, R.F. Minkels,
F. Klok and J. IJkema-Paassen

Medical Physiology, University of Groningen
Bloemsingel 10, Groningen, The Netherlands

INTRODUCTION

Stimulation of the inferior alveolar nerve (IAN) elicits the so-called jaw-opening reflex (JOR) which is associated with excitation of the jaw-opening muscles and inhibition of the jaw-elevator muscles. High threshold afferents with their cell bodies located in the trigeminal ganglion (tg) are responsible for the activation of the jaw-opening muscles, whereas low threshold afferents with their cell bodies in the mesencephalic trigeminal nucleus (Me5) are involved in the inhibition of the jaw-elevator muscles (Dessem, Iyadurai & Taylor, 1988). The central neuronal routing of the JOR is not fully understood yet. High threshold short latency JOR activity is mediated through a disynaptic pathway involving interneurones in the trigeminal brainstem nuclear complex (TBNC). However, the location of interneurones involved in high-threshold long latency JOR activity is still disputed. This study indicates that the supratrigeminal area (Su5), adjacent to the Me5, contains elements of a neuronal pathway involved in reflex activation of the jaw-opening muscles.

METHODS

Rats were anaesthetised with halothane in oxygen. The IAN was stimulated with wire electrodes inserted into the mandibular canal. Bipolar wires were used for EMG recordings of the ipsilateral anterior digastric muscle. Intracellular recordings were made with glass microelectrodes, extracellular recordings with epoxy coated stainless steel electrodes. The Su5 was stimulated with bipolar stainless steel electrodes.

Neuroanatomical experiments comprised iontophoretic injections (day 0) of the anterograde tracer Phaseolus vulgaris leucoagglutinin in the Su5 and pressure injections (day 7) of the retrograde tracer Cholera Toxin subunit B in the left and right anterior and posterior digastric muscles. The animals were perfused through the heart (day 10) and cryosections were processed according to (Bruce & Grofova, 1992).

RESULTS

Stimulation of the Su5 with a short train (2 - 15 pulses, 50, 70 and 90 Hz) of low intensity stimulus pulses (150 μA, 0.2 ms duration) resulted in anterior digastric EMG responses following each stimulus pulse. Stimulus frequencies below 50 Hz did not produce clear EMG signals. The latency of the first response (at 50 Hz) varied between 2.0 and 2.8 ms; the second response followed between 8.9 and 12.2 ms. At higher stimulus frequencies the latency of the second response decreased to about 8.0 ms (at 90 Hz) whereas its amplitude doubled; the amplitude and latency of the first response remained approximately unaltered. Occasionally it was observed that the responses could not follow all stimulus pulses of one stimulus train.

Extracellular responses of high threshold units were recorded in the Su5 following single pulse stimulation of the IAN. The latencies ranged between 1.0 and 1.2 ms with threshold values equal to that of the simultaneously recorded jaw-opening EMG (150-200 μA). These high threshold units followed stimulus frequencies up to 600 Hz.

Intracellular recordings in the Su5 revealed cells (n = 2) that fired a single action potential (latency about 6.5 ms) superimposed on an EPSP following single shock stimulation of the IAN (see Figure 1). The threshold for firing an action potential was the same as that found for the simultaneously recorded digastric reflex activity (latency about 8.0 ms). At firing threshold, the EPSPs (latency 1.4 ms) had a rise time (onset to peak) of 5.7 ms whereas the time constant (peak-to-half decay) of the falling phase was 3 ms.

The neuroanatomical experiments revealed a few ipsi- and contralateral Su5 projections to labelled anterior and posterior digastric motoneurones. Labelled boutons were restricted to the dendrites of these motoneurones.

DISCUSSION

It was demonstrated that stimulation of the Su5 with a stimulus burst induces a short and a long latency digastric response. As single pulse stimulation did not evoke digastric activity we conclude that a certain amount of drive is needed to activate these Su5 elements. The finding that the short latency response showed no considerable variability in timing upon an increase in stimulus frequency indicates that this response was evoked from structures that synapse directly on the jaw-opening motoneurones. The second response, however, did show considerable variability in timing and amplitude indicating a synaptic

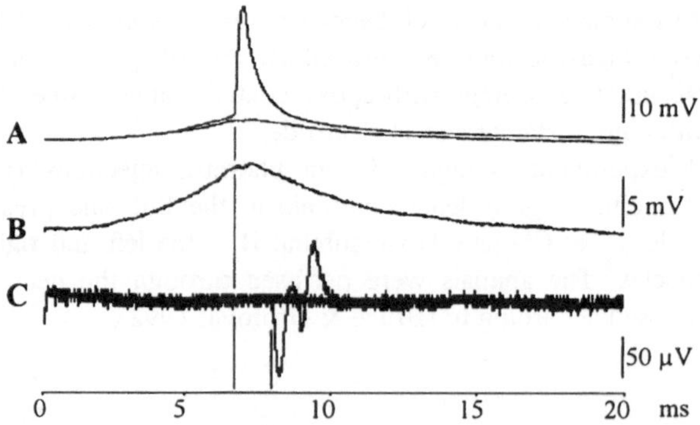

Figure 1. A. Intracellular recording of a unit (resting potential -65 mV) in the Su5 and simultaneously recorded anterior digastric EMG (C) in response to stimulation of the IAN at threshold (70 μA, 0.2 ms). B. Amplification of EPSP in A.

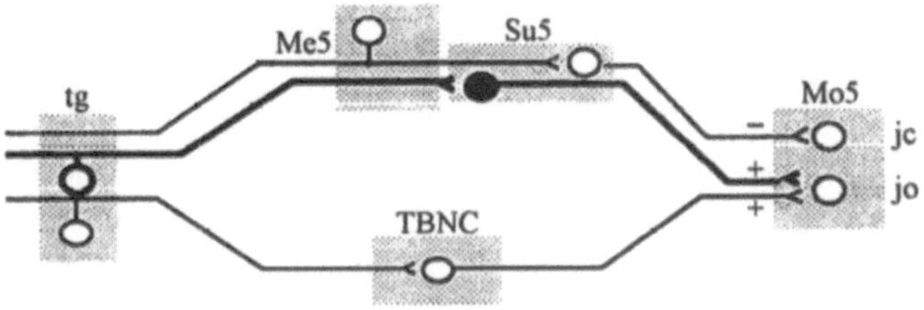

Figure 2. Diagram showing the central pathways for JOR activity. The structures indicated by the thick lines and the filled cell indicate the proposed pathway responsible for the long latency high threshold opening reflex activity as described in this study. (Abbreviations: jc, jaw-closing motoneurone; jo, jaw-opening motoneurone; Mo5, trigeminal motornucleus; further abbreviations, see text).

drive from IAN afferentation. If we assume that these pre- and postsynaptic events are linked in the same reflex arc, then excitatory interneurones receiving input from IAN afferents that have direct projections to the jaw-opening motoneurones can be found in the Su5. The intracellular recordings revealed units that are good candidates for these postulated interneurones for the following reasons: 1) the neurones responded with an EPSP or action potential to stimulation of the IAN, indicating a pre synaptic drive from IAN afferents, 2) the threshold for firing an action potential was equal or slightly higher than the threshold for the simultaneously recorded digastric reflex activity, indicating a link to the jaw-opening motoneurones, and 3) the difference in latency of the action potential and the jaw-opening EMG is sufficient for one synaptic delay which indicates that these interneurones project directly onto the jaw-opening motoneurones. As the extracellularly recorded Su5 units were able to follow fairly high stimulus frequencies they can be considered as responses recorded from axons of high threshold IAN afferents. These afferents then might serve as pre synaptic input for the Su5 interneurones because 1) the afferents had an equal threshold as found for the simultaneously recorded opening reflex activity, 2) a time difference of 4.5 ms between the latency of the units and the onset of digastric activity indicates that input from these afferents to the digastric motoneurone pool is relayed by at least one synaptic transmission, and 3) a time difference of 0.4 ms between the latency of these units and the latency of the recorded EPSP described above is adequate to allow one synaptic delay. The late positive deflection of the EPSPs recorded in the jaw-opening motoneurones, which is indicative for a temporal summation of afferent signals with longer latencies, might reflect the Su5 mediated high threshold IAN afferentation.

These results indicate that the Su5 contains neuronal elements that are involved in high threshold long latency JOR activity. As we were able to make only two recordings of Su5 interneurones in many electrode penetrations and the fact that a few ipsi- and contralateral Su5 projections to labelled anterior and posterior digastric motoneurones were found indicate that the Su5 contains only a limited number of interneurones involved in opening reflex activity. A central routing for the proposed pathway involving Su5 structures is depicted in Figure 2.

REFERENCES

BRUCE, K. & GROFOVA, I. (1992) Notes on a light and electron microscopic double-labelling method combining anterograde tracing with Phaseolus vulgaris leucoagglutinin and retrograde tracing with cholera toxin subunit B. *J. Neurosci. Meth.* **45**, 23-33.

DESSEM, D., IYADURAI, O.D. & TAYLOR, A. (1988) The role of periodontal receptors in the jaw-opening in the cat. *J. Physiol.* **406**, 315-330.

PART 8

PHARMACOLOGY OF CENTRAL CONTROL

EXPRESSION OF GLYCINE RECEPTORS BY IDENTIFIED ALPHA AND GAMMA MOTONEURONES

Robert E.W. Fyffe, Francisco J. Alvarez,
Deborah Harrington and Dianne E. Dewey

Department of Anatomy
Wright State University
Dayton, Ohio 45435, USA

INTRODUCTION

The neurotransmitter role of glycine is well established and has recently been reviewed by Aprison (1990). Glycine is a major inhibitory neurotransmitter in the mammalian spinal cord, and it exerts its postsynaptic effects via increases in the chloride permeability of the postsynaptic membrane. Glycine receptors, like $GABA_A$ receptors, are multimeric ligand-gated chloride channels whose structural and functional properties have been elucidated in great detail in recent years (Betz, 1991). Glycine receptor subunit mRNAs are expressed in widespread areas of the central nervous system (CNS; e.g. Malosio, Marquèze, Kuhse & Betz, 1991; Kirsch, Malosio, Wolters & Betz, 1993a). Advances in understanding of the molecular composition of glycine receptors have also led to the use of monoclonal antibodies against subunits of the glycine receptor complex (Pfeiffer, Simler, Grenningloh & Betz, 1984) to study the general localisation of glycine receptors in the CNS (e.g. Triller, Cluzead, Pfeiffer, Betz & Korn, 1985; Triller, Cluzead & Korn 1987; van den Pol & Gorcs, 1988). One of the antibodies used for such immuno-localisation studies is directed against gephyrin, a 93 kDa glycine receptor-associated, tubulin binding protein which is required for glycine receptor clustering and localisation at synaptic specialisations (Kirsch, Wolters, Triller & Betz, 1993b). In the mammalian ventral horn, gephyrin immunoreactivity has been shown to be localised to the sub-membrane cytoplasm, coincident with immunoreactivity against ligand binding subunits of the glycine receptor, and, importantly, is localised at sites corresponding to synaptic specialisations (see Triller et al., 1985 and unpublished observations from our laboratory). These properties make gephyrin a particularly useful marker for use in probing the organisation and distribution of glycine receptors in identified central neurones, as we will describe in this presentation.

Synaptic inputs to spinal motoneurones have been studied intensively for many years, with the most complete data currently available concerning Ia afferent input (e.g. Brown & Fyffe, 1981; Segev, Fleshman & Burke, 1990) and inhibitory synapses from Renshaw cells

(Fyffe, 1991a). Renshaw cells, which mediate recurrent inhibition of α-motoneurones (Renshaw, 1946; Eccles, Fatt & Koketsu, 1954) are probably glycinergic as their axon terminals appear to contain elevated levels of glycine (Fyffe, 1991b). However it should be noted that some Renshaw cells may use GABA as their neurotransmitter (Cullheim & Kellerth, 1981; Schneider & Fyffe, 1992) and there is mounting evidence for coexistence of glycine and GABA in spinal neurones (e.g. Todd & Sullivan, 1990). There are of course numerous other neural elements which contribute to the population of glycinergic neurones and axon terminals in the ventral horn, including the interneurones which mediate reciprocal inhibition from Ia afferents (see review by Jankowska, 1992).

In the present study we are attempting to gain insights into the overall topographical organisation of synapses on motoneurones because it is believed that the efficacy of synaptic transmission between neurones in the CNS depends on complex interactions between a variety of pre-and post-synaptic mechanisms (e.g. Burke, 1987) and therefore the manner in which different neurotransmitter receptors and synapses are distributed over the surface a neurone is likely to be of significance in determining the functional role of these synaptic inputs. We focus here on glycinergic synapses but, in general terms, knowledge of synaptic location, receptor density and active zone size for specified inputs systems should help us in understanding various aspects of synaptic mechanisms and plasticity (Redman, 1990; Bailey & Kandel, 1993).

METHODS

Experiments were performed on cats anesthetised with Nembutal (40mg/kg i.p., then 5-10 mg/kg i.v. as required for maintainance of deep anesthesia). Intracellular recordings were made in the lumbar spinal cord with micropipette electrodes filled with 8% Neurobiotin (Vector) in Tris buffer, bevelled to have a tip resistance of 15-25 MΩ. Various hindlimb muscle nerves, and dorsal roots L7 and S1 were dissected and mounted on bipolar silver/silver chloride electrodes for stimulation of motor axons or afferent fibres respectively. α- and γ-motoneurones were identified by antidromic conduction velocity and response to dorsal root stimulation (see e.g. Moschovakis, Burke & Fyffe, 1991). Neurobiotin was injected into the motoneurones by passing depolarising current pulses for 1 - 10 minutes, while constantly monitoring the condition of the cell. At the end of the experiment animals were perfused with 4% paraformaldehyde, the spinal cord was removed and processed histologically the next day. Neurobiotin-stained cells were visualised using 7-amino-4-methylcoumarin-3-acetic acid (AMCA)-labelled avidin (Vector). Gephyrin immunocytochemistry was performed using mouse mAb 7a (dilution 1:100; Boehringer Mannheim), and revealed for immunofluorescence using goat antimouse IgG coupled to fluorescein isothiocyanate (FITC; Jackson). After extensive fluorescence microscopy, the Neurobiotin-labelled cells were "solidified" for brightfield inspection by processing the sections with a Vector ABC kit, using diaminobenzidine (Sigma) as substrate for the peroxidase reaction; this procedure permitted accurate reconstruction of the labelled neurones and subsequent quantitative measurements of the three dimensional structure of the cell for correlation with the fluorescent micrographs (computer aided morphometric measurements were performed with a Eutectic Neuron Tracing System).

For electron microscopic immunocytochemistry, one cat was perfused with 4% paraformaldehyde and 0.4% glutaraldehyde in 0.1M phosphate buffer. Spinal cord blocks were postfixed in the same fixative overnight. 50 μm vibratome sections were obtained from spinal cord blocks and treated with 1% $NaBH_4$ (Sigma). Thereafter a standard "pre-embedding" procedure was performed using mAb 7a (dilution 1:100) followed by the ABC method to reveal the immunoreactive sites. The sections were then osmicated, dehydrated, and flat embedded in Epon/Araldite. Areas of interested were cut and mounted in EM capsules, then serially sectioned and observed in a Philips EM300 electron microscope at 60 or 80 KV. 3-D computer reconstructions from series of sections obtained through selected terminals apposed to glycine receptors were performed with a Eutectic Neuron Tracing system. These reconstructions demonstrated that gephyrin immunoreactivity corresponded with the morphology of the synaptic active zone.

RESULTS AND DISCUSSION

Fourteen α- and 6 γ-motoneurones were analysed for gephyrin immunoreactivity. Gephyrin immunoreactive patches (representing glycine receptor clusters) associated with the membrane of motoneurones varied considerably in size and shape, from small punctae less than 0.5 μm in diameter to large, more complicated arrangements usually 1-2 μm in

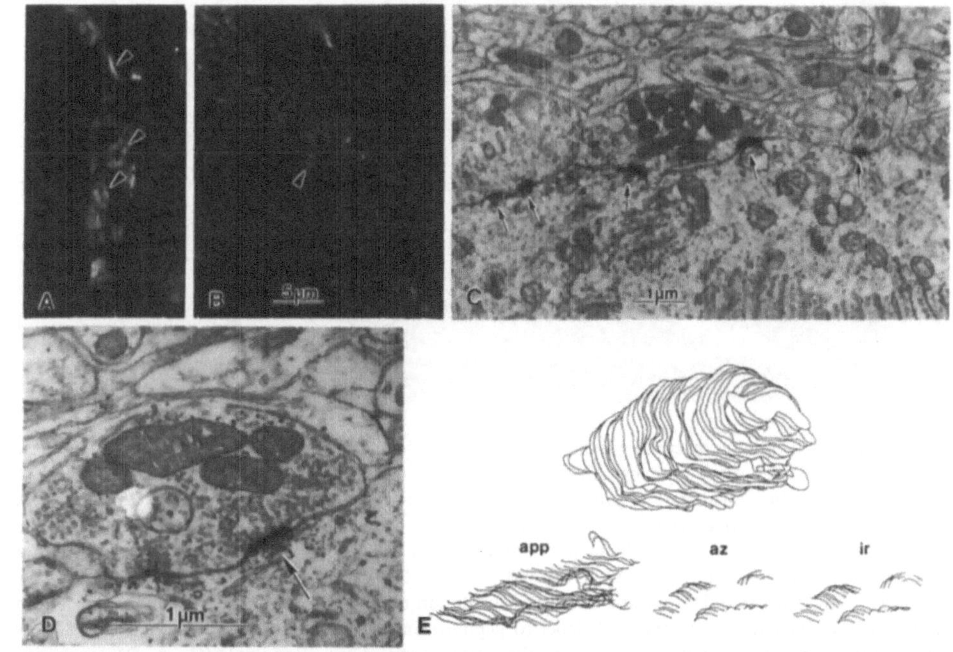

Figure 1. Subcellular organisation of gephyrin-immunoreactivity in the ventral horn of the cat spinal cord. A & B; two unidentified dendrites reveal different patterns of gephyrin cluster-immunofluorescence. A. Large gephyrin-immunofluorescence clusters form scalloped profiles, which when viewed "en face" (lower two arrowheads) appear to have central perforations (the upper arrowhead indicates a similar immunoreactive patch viewed from the side). B. Most of the immunofluorescence is revealed as consisting of small dots, with only occasional larger arrangements (e.g. arrowhead). C. Electronmicrograph of a portion of an α-motoneurone somatic membrane covered by various terminals establishing small (0.5 μm or less) gephyrin-immunoreactive synapses (arrows). D. Higher magnification of a terminal containing flattened synaptic vesicles and establishing a gephyrin-immunoreactive synaptic contact on an α-motoneurone cell soma. Gephyrin-immunoreactivity (arrow) is specifically located postsynaptic to the terminal active zone. The extent of the gephyrin-immunoreactivity accurately reflects the organisation of the synaptic active zone(s) and correlates well with the shapes and sizes of immunofluorescent structures described in A & B. E. Computer-aided serial section reconstruction of a gephyrin-immunoreactive synaptic contact on an α-motoneurone cell soma. In serial sections multiple active zones and their corresponding postsynaptic gephyrin-immunoreactivity can be easily identified. The bottom row shows the zone of apposition between the terminal and the cell soma (app), the extent of the active zones (az) and the postsynaptic immunoreactivity (ir), all in the same orientation as the reconstructed terminal (top row).

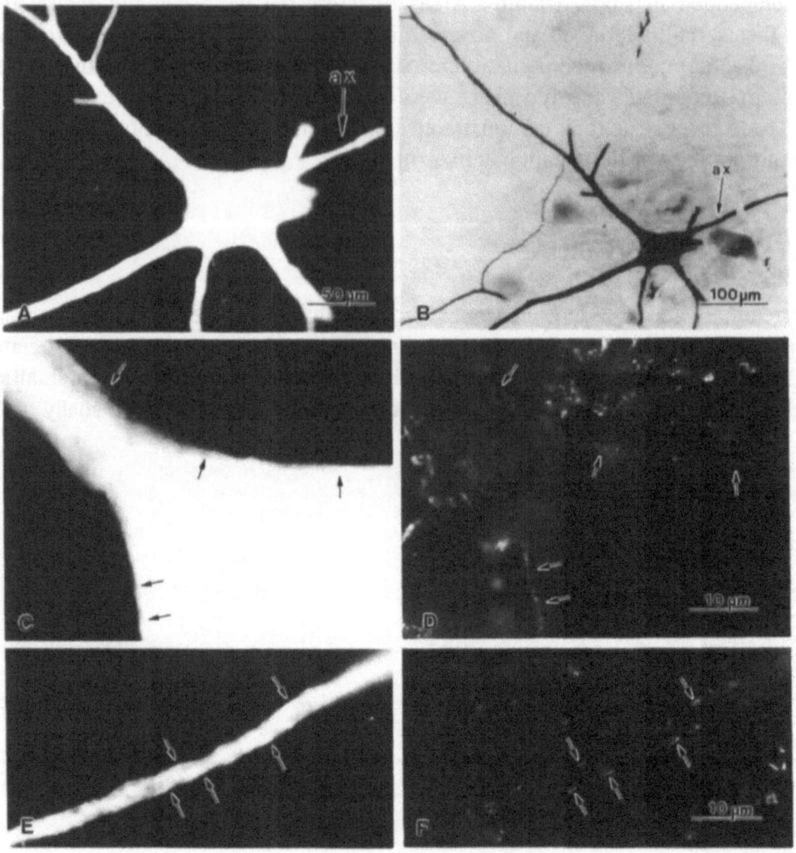

Figure 2. Gephyrin-immunoreactivity in an identified α-motoneurone. A. Medium magnification view (using fluorescence microscopy to reveal AMCA-avidin bound to the intracellularly injected neurobiotin) of a section containing the cell soma and proximal dendrites. The origin of the axon is indicated (ax) in A & B. B. Lower power brightfield view of the same motoneurone after "solidification" with ABC-peroxidase complex. C & D and E & F, are high magnification pairs of immunofluorescence micrographs showing portions of the α-motoneurone soma (C & D) and a segment of dendrite (E & F). The complementary micrographs show the same field viewed under UV excitation to visualise the AMCA-labelled neurobiotin (C & E), or blue excitation to reveal FITC labelled gephyrin-immunoreactivity (D & F). Matching arrows mark approximately the same locations on the micrograph pairs. Proximal (C & D) gephyrin-immunoreactive clusters are relatively densely packed, and of small size whereas distally (E & F), gephyrin-immunoreactive cluster density diminishes but the clusters increase in size and become more immunoreactive (brighter fluorescence). The dendritic segment shown in E was located at around 300 μm from the cell soma.

their longest axis (Figure 1A & B). Gephyrin immunoreactivity was located at the cytoplasmic side of the postsynaptic membrane and its extent closely matched the extent of the apposed presynaptic active zone (Figure 1C, D & E). The size and shape of gephyrin/glycine receptor aggregates measured in the electron microscope closely matched the configurations observed with immunofluorescence. Some gephyrin aggregates form small punctae; groups of 5 - 10 such punctae often outlined small ellipsoids on the neurone surface. This pattern corresponded ultrastructurally to the apposition of a single terminal establishing synaptic contact through multiple small, spot-like active zones. The larger and more complex conformations corresponded with similar active zone configurations at each of the synapses. Notably, perforated patches represented perforated synaptic active zones. For the purpose of describing the data in this study, gephyrin-immunoreactive patterns are simply divided into those that form simple, small punctae, and those which appear as larger complex structures (regardless of their actual configuration).

Gephyrin immunoreactivity was displayed throughout the dendritic trees of α-motoneurones, with the highest density of immunoreactivity appearing on the soma and proximal regions (up to a distance of 150 - 250 μm) of the dendrites. Indeed, almost half (49%) of the terminals in contact with α-motoneurone somas displayed postsynaptic gephyrin immunoreactivity. In contrast, 4 of the 6 γ-motoneurones had little or no gephyrin immunoreaactivity on their cell bodies.

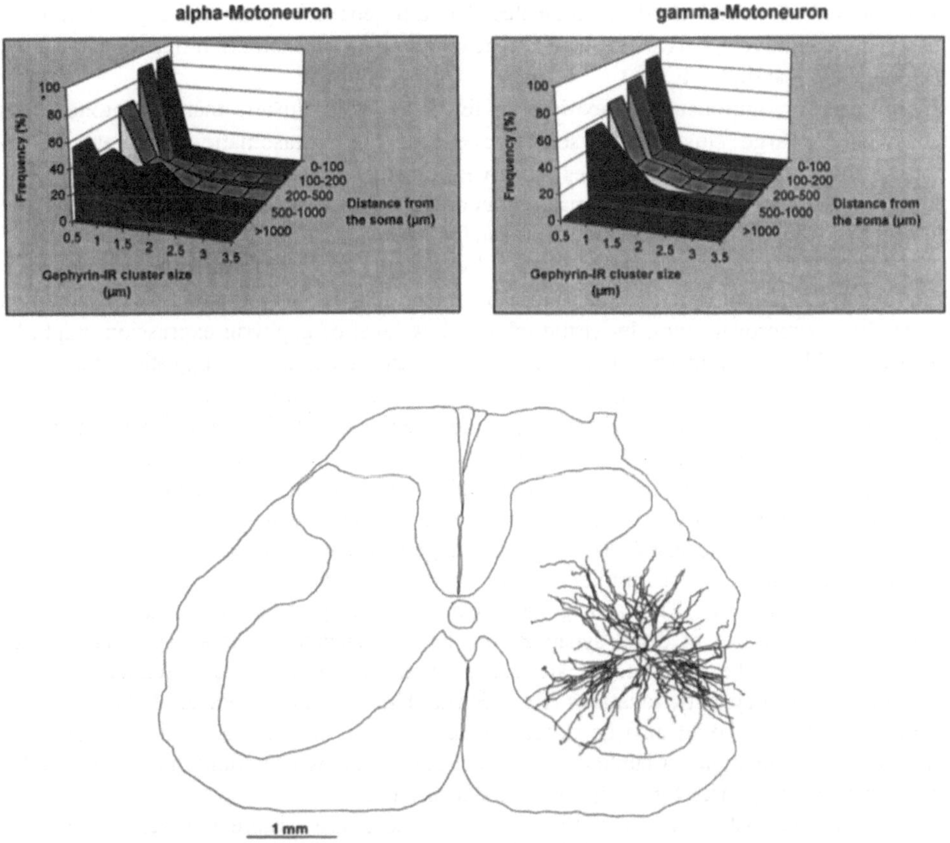

Figure 3. Quantitative analysis of gephyrin immunoreactive aggregates. The graphs depict the size distribution of immunoreactive clusters in the dendritic arbor of one α- and one γ-motoneurone at different distances from the cell soma. Cluster sizes were grouped in bins of 0.5 μm increments with the first bin (0.5) containing all immunoreactive spots measuring 0.5 μm or less. Bin grouping overcomes most of the error incurred when measuring small sizes from the fluorescence material. The graphs were generated from 2,654 and 947 measurements of independent clusters in the respective neurones. The graphs show that the size of immunoreactive gephyrin-immunoreactive clusters gradually increases distally from the cell soma. This is represented as a decrease in the frequency (Y-axis) of small clusters vs. larger clusters (size is represented in the X-axis) at increasing distances from the cell soma (Z-axis). Note that there is considerable variability in cluster size within distal dendritic compartments, while close to the cell soma most glycine receptor clusters measured around 0.5 μm. The lower illustration is a computer aided reconstruction of the full dendritic field of the α-motoneurone depicted in Figure 2 and in the graph.

Juxtasomatic glycine receptor clusters on α-motoneurones, including those on the axon hillock, were predominantly of the small punctate type (Figure 2C & D). At greater distances from the soma, the presence of gephyrin immunoreactivity progressively diminished, but the clusters that were present were larger and of more complicated morphology (Figure 2E & F). Proximally, very few complex aggregates were observed. At the most distal locations (>1,000 μm from the soma, and in the tapering segments of dendrites in the white matter), where cluster density was very low, the ratio of complex to simple clusters remained high, but on average the complex clusters decreased in size slightly (many were <1.0 μm in diameter at the most distal locations). Although dendritic spines are rare on spinal motoneurones, gephyrin immunoreactivity was observed on these structures as well as on the dendritic shafts (and also on varicose regions of some motoneurone dendrites). Despite the presence of a somato-dendritic gradient of cluster sizes, on average (see Figure 3), there was considerable variability in cluster morphology along dendritic shafts of mid- to distal dendritic branches. No differences in gephyrin expression were observed, qualitatively, between F- and S-type α-motoneurones although this aspect has not been pursued quantitatively.

In general, γ-motoneurones express far fewer gephyrin clusters than α-motoneurones and, whereas large clusters are seldom seen on the juxtasomatic membrane of α-motoneurones, they are sometimes observed on the most proximal dendrites of γ-motoneurones. Overall, however, small immunoreactive spots predominate proximally, with clusters becoming larger but less frequent distally.

Clearly, synaptic responses mediated by glycine receptors must be a significant component of the information that must be integrated by α-motoneurones, although it appears that γ-motoneurones, by virtue of their low level of gephyrin expression, might be less susceptible to glycinergic influences. A corollary of the latter suggestion may lie in observations that recurrent inhibition of γ-motoneurones is rather weak (Ellaway, 1971). However, because of the relatively small sample size this conclusion should be treated with caution and note should be made that a minority of the γ-motoneurones displayed gephyrin-immunoreactivity approaching (but never matching) that seen in α-motoneurones. For both types of motoneurone, however, the organization of glycinergic synaptic input is interesting because of the proximo-distal size gradient of gephyrin immunoreactive clusters, which represents, in effect, a size gradient of an important synaptic feature - the size of the presynaptic active zone. This arrangement would probably result in, on average, larger postsynaptic responses at distal synapses and help to enhance their synaptic efficacy by compensating for the electrotonic attenuation that affects synaptic potentials that are generated at distal synaptic locations (Rall, Burke, Holmes, Jack, Redman & Segev, 1992; Redman, 1990). It should also be noted that although a size gradient exists on average, there is also considerable variability in receptor cluster size, which could underlie variability of postsynaptic responses for synapses at similar distances.

It could be asked whether the fact that there are different types of synaptic arrangements on proximal vs distal dendrites of motoneurones simply reflects inputs from different types of glycinergic interneurones impinging preferentially on particular regions of the dendritic tree. However, although there is some degree of segregation of inhibitory inputs on motoneurones (Fyffe, 1991a) there is considerable overlap in synaptic distributions. Furthermore, the gephyrin immunoreactivity at synapses formed by axon terminals of individual interneurones is also extremely variable (not shown). This variability could underlie the observation that although there is a systematic trend over the neurone surface, there is also considerable morphological variability of gephyrin immunoreactive clusters on any region of the motoneurone dendrites. We have observed that gephyrin expression in other spinal neurones displays a similar systematic localisation pattern (see also Triller, Seitanidou, Franksson & Korn, 1990), suggesting that this type of size gradient may be an important factor in the regulation of receptor cluster distribution. Finally, the results suggest that although the number of glycine receptors present, and the complexity of the clusters they form, may be determined by both pre- and postsynaptic factors, receptor cluster features are more related to the position of the synaptic sites in the dendritic arbor than to the source of the presynaptic input.

ACKNOWLEDGEMENTS

This work was supported by NIH grant NS25547.

REFERENCES

APRISON, M.H. (1990) The discovery of the neurotransmitter role of glycine. In *Glycine Neurotransmission.* eds OTTERSEN, O.P & STORM-MATHISEN, J. pp1-23. Wiley, Chichester.

BAILEY, C.H. & KANDEL, E. (1993) Structural changes accompanying memory storage. *Annu. Rev. Physiol.* **55**, 397-426.

BETZ, H. (1991) Glycine receptors: heterogeneous and widespread in the mammalian brain. *TINS* **14**, 458-461.

BROWN, A.G. & FYFFE, R.E.W. (1981) Direct observations on the contacts made between Ia afferent fibres and alpha motoneurones in the cat's lumbosacral spinal cord. *J. Physiol.* **313**,121-140.

BURKE, R.E. (1987) Synaptic efficacy and the control of neuronal input-output relations. *TINS* **10**, 42-45.

CULLHEIM, S. & KELLERTH, J.-O. (1981) Two kinds of recurrent inhibition of cat spinal motoneurones as differentiated pharmacologically. *J. Physiol.* **312**, 209-224.

ECCLES, J.C., FATT, P. & KOKETSU, K. (1954) Cholinergic and inhibitory synapses in a pathway motor axon collaterals to motoneurones. *J. Physiol.* **126**, 524-562.

ELLAWAY, P.H. (1971) Recurrent inhibition of fusimotor neurones exhibiting background discharges in the decerebrate and spinal cat. *J. Physiol.* **216**, 419-439.

FYFFE, R.E.W. (1991a) Spatial distribution of recurrent inhibitory synapses on spinal motoneurons in the cat. *J. Neurophysiol.* **65**, 1134-1149.

FYFFE, R.E.W. (1991b) Glycine-like immunoreactivity in synaptic boutons of identified inhibitory interneurons in the mammalian spinal cord. *Brain Res.* **547**, 175-179.

JANKOWSKA, E. (1992) Interneuronal relay in spinal pathways from proprioceptors. *Prog. Neurobiol.* **38**, 335-378

KIRSCH, J., MALOSIO, M.-L., WOLTERS, I. & BETZ, H. (1993a) Distribution of gephyrin transcripts in the adult and developing rat brain. *Eur. J. Neurosci.* **5**, 1109-1117.

KIRSCH, J., WOLTERS, I., TRILLER, A. & BETZ, H. (1993b) Gephyrin antisense oligonucleotides prevent glycine receptor clustering in spinal neurons. *Nature* **366**, 745-748.

MALOSIO, M.-L., MARQUÈZE-POUEY, B., KUHSE, J. & BETZ, H. (1991b) Widespread expression of glycine receptor subunit mRNAs in the adult and developing rat brain. *EMBO J.* **10**, 2401-2409.

MOSCHOVAKIS, A.K., BURKE, R.E. & FYFFE, R.E.W. (1991) The size and dendritic structure of HRP-labeled gamma motoneurons in the cat spinal cord. *J. Comp. Neurol.* **311**, 531-545.

PFEIFFER, F., SIMLER, R., GRENNINGLOH, G. & BETZ, H. (1984) Monoclonal antibodies and peptide mapping reveal structural similarities between the subunits of the glycine receptor of rat spinal cord. *Proc. Natl. Acad. Sci. USA* **81**, 7224-7227.

POL VAN DEN, A.N. & GORCS, T. (1988) Glycine and glycine receptor immunoreactivity in brain and spinal cord. *J. Neurosci.* **8**, 472-492.

RALL, W., BURKE, R.E. HOLMES, W.R., JACK, J.J.B. , REDMAN, S.J. & SEGEV, I. (1992) Matching dendritic neuron models to experimental data. *Physiol. Rev.* **72**, S159-S185.

REDMAN, S.J. (1990) Quantal analysis of synaptic potentials in neurons of the central nervous system. *Physiol. Rev.* **70**, 165-198.

RENSHAW, B. (1946) Central effects of centripetal impulses in axons of spinal ventral roots. *J. Neurophysiol.* **9**, 191-204

SCHNEIDER, S.P. & FYFFE, R.E.W. (1992) Involvement of GABA and glycine in recurrent inhibition of spinal motoneurons. *J. Neurophysiol.* **68**, 397-406.

SEGEV I., FLESHMAN, J.W. & BURKE, R.E. (1990) Computer simulation of group Ia EPSPs using morphologically realistic models of cat α-motoneurons. *J. Neurophysiol.* **64**, 648-660.

TODD, A.J. & SULLIVAN, A.C. (1990) Light microscope study of the coexistence of GABA-like and glycine-like immunoreactivities in the spinal cord of the rat. *J. Comp. Neurol.* **296**, 496-505.

TRILLER, A., CLUZEAUD, F. & KORN, H. (1987) Gamma-aminobutyric acid-containing terminals can be apposed to glycine receptors at central synapses. *J. Cell Biol.* **104**, 947-956.

TRILLER, A., CLUZEAUD, F., PFEIFFER, F., BETZ, H. & KORN, H. (1985) Distribution of glycine receptors at central synapses: An immunoelectron microscopy study. *J. Cell Biol.* **101**, 683-688.

TRILLER, A., SEITANIDOU, T., FRANKSSON, O. & KORN, H. (1990) Size and shape of glycine receptor clusters in a central neuron exhibit a somato-dendritic gradient. *The New Biologist.* **2**, 637-641.

NORADRENERGIC LOCUS COERULEUS INFLUENCES ON POSTURE AND VESTIBULOSPINAL REFLEXES

O. Pompeiano

Dipartimento di Fisiologia e Biochimica
Via S. Zeno 31, I-56127 Pisa, Italy

INTRODUCTION

In addition to the lateral vestibular nucleus (LVN), which exerts a direct excitatory influence on tonic α- and static γ-motoneurones innervating the limb extensors, there are other structures, located in the dorsolateral pontine tegmentum, which exert a modulatory influence on posture (Pompeiano, Horn & d'Ascanio, 1991). These structures include a distinct cluster of neurones, located in the locus coeruleus (LC) and subcoeruleus (SC), which are not only noradrenergic (NA) but also norepinephrine (NE) sensitive, due to the existence of self-inhibitory synapses acting on α_2-adrenoceptors through mechanisms of recurrent and/or lateral inhibition (cf. Barnes & Pompeiano, 1991). This paper will review various experimental studies designed to analyse the mode of action of the LC and SC. It will be shown that the LC-complex controls posture by utilizing either a direct coeruleospinal (CS) projection or an indirect projection passing through the dorsal pontine reticular formation (pRF) and the related medullary inhibitory reticulospinal (mRS) system (Figure 1).

NORADRENERGIC LOCUS COERULEUS INFLUENCES ON THE SPINAL CORD

Direct projection

The CS projection forms a plexus of thin fibres terminating in part at least in the ventral horn, where they surround both large and small cells, presumably α- and γ-motoneurones, as well as interneurones (cf. Commissiong, 1991). Stimulation of the LC-complex in cats facilitated the monosynaptic reflexes in the ipsilateral hindlimb and also produced EPSPs in spinal motoneurones (Barnes, Fung & Pompeiano, 1989). Moreover, microiontophoretic application of NA increased extracellularly recorded firing of rat spinal motoneurones, and also produced a slowly developing small-amplitude depolarization in intracellularly recorded units from rats and cats. This neuromodulatory effect outlasted the ejection period and was associated with an enhancement of glutamate-evoked motoneuronal firing (cf. White, Fung & Barnes, 1991). Some of the effects described above were found to be mediated by α-(presumably α_1) adrenoceptors (Barnes & Pompeiano, 1991; White et al., 1991).

In addition to a direct facilitatory influence on spinal motoneurones, the CS system depressed the activity of inhibitory Renshaw (R)-cells, thus leading to disinhibition of α-motoneurones. In particular, we studied the effects of LC stimulation on recurrent inhibition in the lumbar spinal cord of decerebrate cats (Fung. Pompeiano & Barnes, 1988). Monosynaptic reflexes, recorded from a split bundle of the L_7 ventral root following stimulation of the gastrocnemius-soleus (GS) or common peroneal (CP) nerve, were conditioned with a single volley delivered to the remaining bundle of the L_7 ventral root at the appropriate interval (6 - 18 msec) to elicit the most prominent recurrent inhibitory effect. The ventral root-induced recurrent inhibition was regularly counteracted by LC preconditioning stimuli (3 - 4 cathodal pulses of 0.7 ms, at 770/sec). Such a decrease in recurrent inhibition, involving both extensor (GS) and flexor (CP) motor nuclei, depended on a NA mediated CS suppression of R-cell activity. The evidence for this is: 1) recurrent inhibition was reduced in some cases, even in the absence of any measurable facilitation of the monosynaptic reflex attributable to a direct depolarising influence of the CS projection on spinal motoneurones; 2) the magnitude of LC-induced disinhibition was significantly correlated with that of the ventral root-induced recurrent inhibition, i.e. the greater the recurrent inhibition, the larger the disinhibition; 3) the high frequency discharges of R-cells recorded intracellularly in response to ventral root volleys were inhibited by LC conditioning. It appeared also that the R-cell activity was inhibited by iontophoretically applied NA (cf. Fung et al., 1988). No attempt has been made so far to investigate whether this effect was mediated by β-adrenoceptors (see Figure 1).

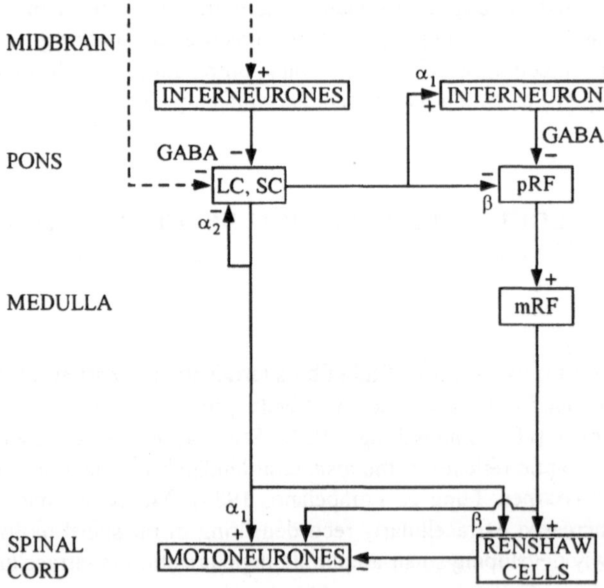

Figure 1. Descending projections from the NA LC-SC nuclei to the spinal cord. They include a direct coeruleospinal projection, as well as an indirect projection passing through the dorsal pontine reticular formation (pRF) and the related medullary reticular formation (mRF), from which inhibitory reticulospinal systems originate. α_1, α_2 and β refer to adrenoceptors; + and - indicate postsynaptic effects on various targets. Dashed lines refer to supramesencephalic descending systems which presumably inhibit the LC-SC nuclei either directly, or through interposed neurones.

In addition to R-cells, the NA system could also inhibit interneurones which convey information from flexor reflex afferents (FRA) to motoneurones. This was shown originally in spinal cats after systemic administration of L-DOPA, a precursor of NE (cf. Marshall, 1983). Also in anesthetized cats, properly timed stimuli of the LC complex, as well as locally applied NE, depressed the monosynaptic focal synaptic field potentials of group II origin, but not of group I origin, in the intermediate and ventral horn regions of midlumbar segments. These stimuli also selectively depressed the disynaptically evoked reflex actions of group II afferents in the hindlimb motoneurones (inhibition in extensor and excitation in flexor motoneurones; cf. Jankowska, Riddell, Skoog & Noga, 1993). Some of the effects described above were reduced but not suppressed by NA α-antagonists (cf. Jankowska et al., 1993). However, the possible contribution of β-adrenoceptors to these effects was not investigated.

Indirect projection

NA and NE-sensitive LC-complex neurones also send afferents to the dorsal pRF, on which they exert an inhibitory influence (cf. Barnes & Pompeiano, 1991). This influence, was mediated at least in part directly by β-adrenoceptors (cf. Pompeiano et al., 1991) and in part was attributable to activation through α_1-adrenoceptors of inhibitory GABAergic interneurones (Cirelli, d'Ascanio, Horn, Pompeiano & Stampacchia, 1993). On the other hand, dorsal pRF neurones, which are partly cholinergic and cholinosensitive, project to the inhibitory regions of the medial medulla, on which they exert an excitatory influence (cf. Barnes & Pompeiano, 1991). The corresponding mRS projections may actually inhibit spinal motoneurones by activating R-cells (cf. Pompeiano, 1984).

LOCUS COERULEUS INFLUENCES ON POSTURE

LC-complex neurones exert a prominent facilitatory influence on posture. In particular we have shown that in decerebrate cats electrolytic lesions of the LC complex, or microinjection into this structure of the α_2-adrenergic agonist clonidine, which inhibits the discharge of the NA LC neurones, decreased or suppressed postural activity particularly in the ipsilateral limbs (cf. Pompeiano et al., 1991). Similar results were also obtained after local injection into the LC-complex of $GABA_A$ or $GABA_B$ agonists, which inhibit LC neurones postsynatically (unpublished). These postural changes were attributed not only to inactivation of CS neurones, but also to disinhibition of the dorsal pRF neurones. In fact, a decrease or suppression of postural activity could be obtained after microinjection into the dorsal pRF of either the β-noradrenergic antagonist propranolol (cf. Pompeiano et al., 1991) or the α_1-adrenergic antagonist prazosin (Cirelli et al., 1993), which blocked the direct or indirect inhibitory influences exerted by the NA LC neurones on dorsal pontine reticular neurones. Activation of the pRF neurones by local injection of the cholinergic agonist carbachol or bethanechol (a muscarinic agent) also decreased or suppressed the postural activity in decerebrate cats, an effect which was attributed to postsynaptic inhibition of extensor motoneurones. On the other hand, lesions of these tegmental reticular structures suppressed episodes of postural atonia induced by systemic injection of an anticholinesterase (cf. Pompeiano et al., 1991).

The hypothesis that LC neurones and the related pRF neurones are critically involved in the maintenance of the decerebrate rigidity is supported by experiments utilising unit recording, showing that in decerebrate cats postural activity was present as long as the NA LC-complex neurones fired regularly, thus keeping pRF neurones under inhibitory control. However, as soon as the LC neurones ceased firing, e.g. after systemic injection of an anticholinesterase, the discharge rate of the presumably cholinergic and cholinosensitive pRF neurones as well as of the inhibitory mRS neurones increased, thus suppressing posture (Pompeiano & Hoshino, 1976). Reciprocal changes in firing rate of the LC-complex neurones as well as of the pRF neurones, similar to those described above, occurred also in intact, unanesthetised cats. In fact, the spontaneous discharge of the LC neurones decreased or disappeared during the episodes of postural atonia typical of desynchronised sleep, while that of the pRF neurones and the related inhibitory mRS neurones increased (Hobson & Steriade, 1986).

It is of interest that the resting discharge of the LC neurones, which is very low (1-2 imp/s) in the intact animal during quiet waking (Hobson & Steriade, 1986), increased to about 10 imp/s after decerebration (Pompeiano & Hoshino, 1976). We proposed that this increased discharge contributes to γ-rigidity, since the LC neurones exert not only a direct facilitatory influence on α-extensor (and flexor) motoneurones (Barnes et al., 1989), but also an inhibitory influence on R-cells (Fung et al., 1988), thus leading to disinhibition of tonic α- as well as static γ-motoneurones innervating extensor muscles (cf. Pompeiano, 1984). Indeed there is evidence that administration of L-DOPA may activate these two populations of motoneurones (cf. Commissiong, 1981). The increased discharge of the LC neurones could also contribute to decerebrate rigidity by suppressing the activity of dorsal pRF neurones and related inhibitory mRS neurones.

The increased discharge of LC neurones after decerebration can be attributed to interruption of a supramesencephalic pathway, exerting a tonic inhibitory influence on the NA LC neurones. This effect could be mediated either by an indirect projection acting through GABAergic inhibitory neurones, some of which are located close to or within the LC complex (cf. Barnes & Pompeiano, 1991), or by a direct GABAergic inhibitory projection from the substantia nigra (pars reticulata) to the pontine tegmentum (cf. Schwartz, Santag & Wand, 1984). The cascade of inhibitory systems involved in the LC control of postural activity is schematically illustrated in Figure 1.

LOCUS COERULEUS INFLUENCES ON VESTIBULOSPINAL REFLEXES

Observations made in decerebrate cats have shown that rotation about the longitudinal axis of the animal, leading to sinusoidal stimulation of labyrinth receptors, produces a contraction of limb extensors during side-down tilt and relaxation during side-up tilt. This reflex depends on the activity of vestibulospinal (VS) neurones, since most of them, particularly projecting to the lumbosacral cord, were excited during side-down and depressed during side-up tilt (Pompeiano, 1992). We discovered that most of the presumably inhibitory mRS neurones as well as of the LC-complex neurones including the CS neurones projecting to the lower segments of the spinal cord, responded to animal tilt with a predominant response pattern which was opposite in sign to that of VS neurones (Pompeiano, 1992). However, due to reciprocal interaction between LC-complex neurones and pRF units, activity of the excitatory CS neurones could from time to time predominate over that of the inhibitory mRS neurones or *vice versa*.

In precollicular decerebrate cats, in which a high resting discharge of LC neurones tonically inhibited that of dorsal pRF neurones and related inhibitory mRS neurones (Pompeiano & Hoshino, 1976), the amplitude of the EMG responses of limb extensors to labyrinth stimulation was quite small (Pompeiano et al., 1991). This finding was attributed to the fact that the R-cells linked with limb extensor motoneurones fired in phase with the corresponding motoneurones during side-down animal tilt, due to a reduced discharge of the CS neurones, leading to disinhibition of the related R-cells (Pompeiano, wand &

Srivastava, 1985). This would increase the functional coupling of these inhibitory interneurones with their own extensor motoneurones, thus reducing the gain of the VSR.

Different results, however, were obtained when in the same preparations activity of the LC neurones was partially or completely impaired, while that of pRF neurones and related inhibitory mRS neurones increased, due to some of the neurochemical procedures described in the previous section. In all these experiments, the increased discharge of the pontine and the medullary reticular neurones could be either moderate so as to decrease postural activity (*primary state*) or so prominent as to suppress posture (*secondary state*). During the *primary state* the gain of the VSR increased (cf. Pompeiano, 1992). This finding was attributed to the fact that the R-cells linked with limb extensors motoneurones now fired out of phase with respect to the corresponding motoneurones during side-down tilt, due to a reduced discharge of the RS neurones leading to disfacilitation of the related R-cells (Pompeiano et al., 1985). This would decouple these inhibitory interneurones from their own extensor motoneurones, thus increasing the response gain of limb extensors to labyrinth stimulation. During the *secondary state*, however, the increased discharge of pRF neurones and related inhibitory mRS neurones was so prominent as to suppress not only posture but also the VSR, due to intense postsynaptic inhibition of the extensor α-motoneurones (cf. Pompeiano, 1992).

In conclusion, it appeared that the increased discharge of the NA LC neurones which occurs after decerebration, enhanced postural activity but actually reduced the gain of the VSR. In contrast, a progressive decrease in discharge of the same neurones leading either to a decrease or to suppression of posture increased or decreased, respectively, the gain of the VSR. The neuronal circuit made by NA LC neurones and the related cholinergic and/or cholinoceptive pRF neurones may thus act as a variable regulator of the VSR gain (cf. Pompeiano, 1984). Since the resting discharge of NA LC neurones undergoes spontaneous fluctuations during the sleep-waking cycle (Hobson & Steriade, 1986), we postulate that this system exerts a prominent role in adapting the gain of the VSR to the state of arousal of the animal.

REFERENCES

BARNES, C.D., FUNG, S.J. & POMPEIANO, O.(1989). Descending catecholaminergic modulation of spinal cord reflexes in cat and rat. *Annals New York Acad. Sci.* **563**, 45-58.

BARNES, C.D. & POMPEIANO, O. (1991). *Neurobiology of the Locus Coeruleus*. Elsevier, Amsterdam.

CIRELLI, C., D'ASCANIO, P., HORN, E., POMPEIANO, O. & STAMPACCHIA, G. (1993). Modulation of vestibulospinal reflexes through microinjection of an α_1-adrenergic antagonist in the dorsal pontine tegmentum of decerebrate cats. *Arch. Ital. Biol.* **131**, 275-302.

COMMISSIONG, J.W. (1981). Spinal monoaminergic systems: an aspect of somatic motor function. *Fed. Procs.* **40**, 2771-2777.

FUNG. S.J., POMPEIANO, O. & BARNES, C.D. (1988). Coeruleospinal influence on recurrent inhibition of spinal motonuclei innervating antagonistic hindleg muscles of the cat. *Pflügers Archiv.* **412**, 346-353.

HOBSON, J.A. & STERIADE, M. (1986). Neuronal basis of behavioural state control. In *Intrinsic Regulatory Systems of the Brain. Handbook of Physiology. Section I. The Nervous System. Vol.IV.* ed BLOOM, F.E., pp 701-823, American Physiological Society, Bethesda.

JANKOWSKA, E., RIDDELL, J.S., SKOOG, B. & NOGA, B.R. (1993). Gating of transmisssion by stimuli applied in the locus coeruleus and raphe nuclei of the cat. *J. Physiol.* **461**, 705-772.

MARSHALL, K.C. (1983). Catecholamines and their actions in the spinal cord. In *Handbook of the Spinal Cord. Vol. 1. Pharmacology*, ed. DAVIDOFF, R.A. pp 275-328. Marcel Dekker Inc., New York.

POMPEIANO, O. (1984). Recurrent inhibition. In *Handbook of the Spinal Cord. Vol. 2 & 3. Anatomy and Physiology*, ed. DAVIDOFF, R.A. pp 461-557. Marcel Dekker Inc., New York.

POMPEIANO, O. (1992). The role of the noradrenergic locus coeruleus neurons in the dynamic control of posture during the vestibulospinal reflexes. In *Vestibular and Brain Stem Control of Eye, Head and Body Movements.* eds. SHIMAZU, M. & SHINODA, Y. pp. 91-110. Karger, Basel.

POMPEIANO, O., HORN, E. & D'ASCANIO, P. (1991). Locus coeruleus and dorsal pontine reticular influences on the gain of the vestibulospinal reflexes. *Prog. Brain Res.* **88**, 435-461.

POMPEIANO, O. & HOSHINO, K. (1976). Central control of posture: Reciprocal discharge by two pontine neuronal groups leading to suppression of decerebrate rigidity. *Brain Res.* **116**, 131-138.

POMPEIANO, O., WAND, P. & SRIVASTAVA, U.C. (1985). Influences of Renshaw cells on the gain of hindlimb extensor muscles to sinusoidal labyrinth stimulation. *Pflügers Archiv.* **404**, 107-118.

SCHWARZ,M., SANTAG, K.-M. & WAND, P. (1984). Non-dopaminergic neurons of the reticular part of the substantia nigra can gate static fusimotor action on flexors in cat. *J. Physiol.* **354**, 333-344.

WHITE, S.R., FUNG, S.J. & BARNES, C.D. (1991). Norepinephrine effects on spinal motoneurons. *Prog. Brain Res.* **88**, 343-350.

ENKEPHALINERGIC CONTROL OF SEGMENTAL REFLEX PATHWAYS AND DESCENDING PATHWAYS IN THE CAT

H. Steffens and U. Fronhöfer

Institute of Physiology, University of Göttingen
D-37073 Göttingen, Germany

INTRODUCTION

Flexor reflex afferents (FRA; Eccles & Lundberg, 1959) belong to one of the systems involved in segmental motor control (Lundberg, 1979). There is a wide spread convergence of nociceptive cutaneous and muscle afferents on interneurones of the segmental FRA pathways (Steffens & Schomburg, 1993) and therefore these afferents also participate in segmental motor control. Consequently, peptides like enkephalins which are released after nociceptive input under the control of higher centres not only suppress the ascending nociceptive pathways, but also have a strong influence on spinal motor control (Schmidt, Schomburg & Steffens, 1991). However, the site of enkephalinergic action is still the subject of speculation. The higher concentration of μ-receptors in the dorsal horn than in the ventral horn of the spinal cord (Yaksh & Noueihed, 1985) gives a clue as to where the action may take place. The aim of this study was therefore to investigate further the role of segmental and descending pathways concerning enkephalinergic motor control.

METHODS

The experiments were carried out on adult cats weighing 2.7-4.5 kg. Under general anaesthesia the cats were anaemically decapitated (cf. Kniffki, Schomburg & Steffens, 1981). Subsequently, the cat was artificially ventilated, spinalised at C1 and paralysed (Pancuronium 'Organon', about 0.15 mg/kg every hour i.v.). End-expiratory CO_2 concentration was regulated at 4%. The mean arterial blood pressure was maintained above 80 mmHg, if necessary by infusion of a dextran solution. Rectal temperature was kept close to 38°C.

A laminectomy was performed and the ventral roots L5-S1 on the left side and the nerves of the left hind limb were sectioned and mounted for electrical stimulation: quadriceps (Quad), posterior biceps and semitendinosus (PBSt), anterior biceps and semimembranosus (ABSm), flexor digitorum and hallucis longus (FDL), plantaris (Pl), peroneus longus, brevis and tertius (SPM), tibialis anterior and extensor digitorum longus (DP), gastrocnemius and soleus (GS), suralis (Sur), saphenous (Saph), and the cutaneous part of the superficial peroneal nerve (SPC). The plantar division of the tibial nerve (Tib), which is a mixed nerve supply to intrinsic foot muscles and to the skin of the foot pads, was left intact for radiant heat stimulation of the skin (for details see Steffens & Schomburg, 1993). A second laminectomy was performed at Th10 for electrical stimulation of the dorsolateral funiculus with its descending tracts. To avoid antidromic activity in the dorsal columns running down to the lumbar level, the dorsal columns were sectioned caudally to the Th10 stimulation site.

Electrical stimulation was performed with single 0.1 ms square pulses. Stimulus strength is expressed in multiples of threshold (T). Monosynaptic reflexes were recorded from the ventral roots of L7 and S1. Intracellular recordings from motoneurones were made with 2M potassium citrate microelectrodes in the L7 segment. Motoneurones were identified by antidromic invasion following electrical stimulation of the ventral roots and the monosynaptic response to stimulation of group Ia afferents of a muscle nerve. The afferent volley to electrical nerve stimulation was recorded near the dorsal root entry zone. Single interneurones of the dorsal horn were recorded extracellularly if they could be identified tentatively as FRA interneurones, i.e. if they were excited by muscle afferents (flexors and extensors) with 2T or higher and by all low to high threshold cutaneous afferents.

Conditioning effects of descending pathways and of pathways from cutaneous afferents on monosynaptic reflexes of the flexor PBSt and of the extensor GS were tested. Conditioning effects of both descending pathways and segmental pathways from cutaneous afferents were tested together. Convergence of descending pathways and segmental pathways from cutaneous afferents were tested with intracellular recordings. All recordings were made before, during and after the application of the μ-agonist DAGO ([D-Ala^2,N-Me-Phe^4,Gly^5-ol]-enkephalin, 1.3-1.6 mg/kg i.v.).

RESULTS

Basic effects of DAGO on monosynaptic reflexes

As expected with the high spinal animal, the basic reflex effects evoked by cutaneous afferents were characterised by the flexor reflex pattern, i.e. the monosynaptic reflex of the flexor was facilitated and the monosynaptic reflex of the extensor was inhibited (see Eccles & Lundberg, 1959). As shown earlier (cf. Schmidt et al., 1991), the μ-agonist DAGO has a strong depressive effect on the monosynaptic reflex of the flexor and on facilitatory effects from cutaneous afferents. These effects were less pronounced on the monosynaptic effect of the extensor, but the suppression of the inhibitory effect from cutaneous afferents on the reflex of GS was also pronounced. Accordingly, naloxone had the same relative strength of antagonising effects. With conditioning by stimulation of descending pathways, the monosynaptic reflexes were influenced in the same direction as with cutaneous stimulation, but again the effect for GS was less pronounced than for PBSt. The conditioning effect could be reduced by i.v. injection of DAGO, which could be antagonised by naloxone.

Interaction of descending pathways and segmental pathways from cutaneous afferents to motoneurones

As is already known, descending pathways converge on spinal interneurones within segmental pathways from FRA to motoneurones (Hongo, Jankowska & Lundberg, 1969;

Baldissera, ten Bruggencate, Lundberg, 1971; Baldissera, Hultborn & Illert, 1981). This convergence should be evident in intracellular recordings from motoneurones when conditioning the response from one pathway with the other one, and it should also be visible when conditioning the monosynaptic reflex response with both pathways. However, timing between the two conditioning stimuli together with timing of the test stimulus is difficult. This problem was solved with the aid of computerised stimulation. The muscle nerve was stimulated with a fixed delay. The delays of conditioning stimulation of descending pathways and of the skin nerve were systematically shifted by 1 ms from 8 ms to 0 ms before the test stimulus, performed with combined conditioning and single conditioning; control responses to test stimuli were also taken. Thus, three dimensional profiles of monosynaptic reflex amplitudes conditioned with both descending and segmental cutaneous stimulation could be calculated.

Before the application of DAGO, the monosynaptic reflex of PBSt was slightly facilitated by descending stimulation. Facilitation appeared with the test stimulus delayed by 8 to 2 ms after the conditioning stimulus (Figure 1A, left side). Similarly the monosynaptic reflex of PBSt was facilitated by stimulation of the Sur, with a maximum facilitation at -6 ms delay (Figure 1A, left side). With combined conditioning, the monosynaptic reflex amplitude profile did not follow simply as expected from control conditionings in the x- and z-directions, but showed a dip located with the co-ordinates given by x (descending delay -5 to -1 ms) and z (Sur delay -4 ms), extending to z -3 ms at x -3 ms (Figure 1A, right side). After the application of DAGO the facilitation of the PBSt monosynaptic reflex disappeared almost completely. Only between -4 to -3 ms delay did the cutaneous nerve stimulation facilitate the monosynaptic reflex (Figure 1B, left side). The three dimensional profile of reflex amplitudes conditioned with both descending stimulation and cutaneous stimulation followed more or less perfectly the single conditioning curves in the x- and z-directions (Figure 1B, right side).

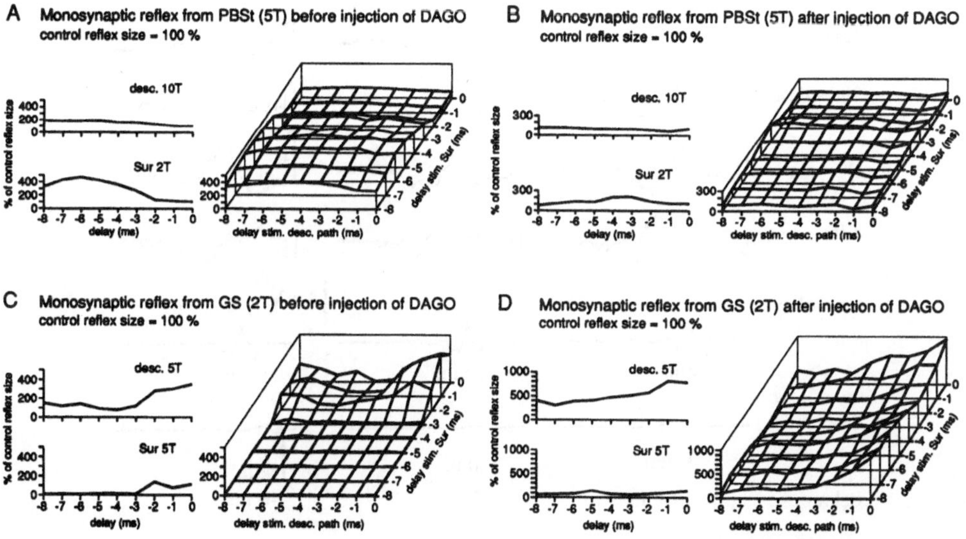

Figure 1. Three dimensional profiles of monosynaptic reflex amplitudes in response to combined conditioning with both stimulation of Sur and desc. with changing delays, related to the stimulation time of the homonymous nerve. Amplitudes of PBSt (A, B) and GS (C, D) monosynaptic reflexes before (A, C) and after (B, D) the application of 1.5 mg/kg DAGO. On the left side of each part control amplitude curves with only one conditioning stimulation. All amplitudes were measured from averaged responses (8 single responses each).

When the monosynaptic reflexes of the extensor (GS) were conditioned with both descending and cutaneous stimulation, the controls showed a weak facilitation of the reflex with a delay of -8 to -6 ms and a strong facilitation with a delay of -3 to 0 ms following conditioning stimulation of descending pathways (Figure 1C, left side). Conditioning of the monosynaptic reflex of GS with cutaneous stimulation (Figure 1C, Sur 5T) showed a complete inhibition for delays of -8 to -3 ms, and a weak facilitation at shorter delays. With combined conditioning the whole area of x (descending delay -8 to 0 ms) and z (Sur delay -8 to -3 ms) is flat because of the very strong inhibition by the cutaneous nerve. The rest of the profile follows both controls in x- and z-directions, except for the area at the 0-end for the delay of stimulation of Sur and -4 to -3 ms for the delay of descending stimulation (Figure 1C, right side). After the application of DAGO (1.5 mg/kg i.v.), a pronounced facilitation of the GS monosynaptic reflex was seen over the whole range of delays for descending conditioning, and the valley between -5 and -3 ms delay, which was there before application of DAGO, was missing (Figure 1D, left side). With single conditioning by stimulation of Sur the inhibition was more or less completely gone (Figure 1D, left side). As for the PBSt, the three dimensional profile of the double conditioned monosynaptic reflex of the GS followed both control curves in the x- and z-directions without any marked irregularities (Figure 1D, right side).

Intracellular recordings from motoneurones confirmed the existence of common interneurones in the pathways from segmental cutaneous afferents and descending pathways to motoneurones. An example is shown in Figure 2. As might be predicted from the reflex recordings with double conditioning, detection of this convergence was dependent on the correct timing of stimulation of the pathways. The PBSt motoneurone (Figure 2) showed

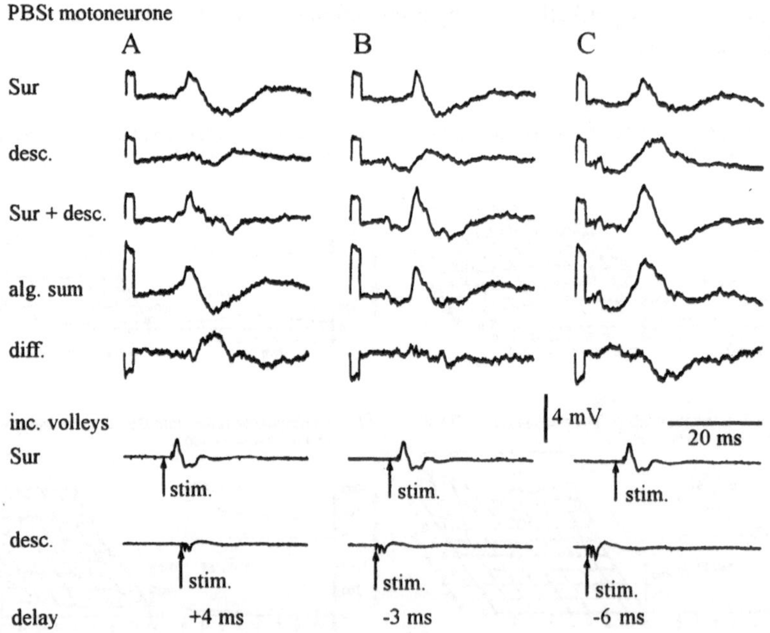

Figure 2. Spatial interaction between the segmental cutaneous pathway and the descending pathway. Intracellular recording of a PBSt motoneurone. Stimulation strength of Sur 5T (A, B, C) and of desc. pathways 5T (A, B) or 10T (C). 1st and 2nd line single stimulation of Sur or desc.; 3rd line combined stimulation of Sur and desc.; 4th line algebraic sum of row 1 and 2; 5th line difference (line 3 - line 4). The different delays between stimulation of Sur and desc. are indicated at the incoming volley recordings. All responses are averaged from 8 single responses each.

mixed effects after stimulation of Sur and of the descending pathways. When stimulating both the descending pathways and Sur together, the difference between the response after double stimulation and the algebraic sum of the control responses should indicate a convergence on common interneurones. With -3 ms delay between stimulation of the descending pathways and Sur stimulation (zero point is Sur stimulation), there was no difference between the algebraic sum and the double stimulation response (Figure 2B). With +4 ms delay there was an additional facilitation (Figure 2A, diff.) and with -6 ms delay an additional inhibition (Figure 2C, diff.) which indicated the convergence on common interneurones.

Motoneurone PSPs and the effect of DAGO and naloxone

Intracellular recordings of motoneurones revealed distinct effects after the i.v. application of DAGO and naloxone. Figure 3 shows an example for recordings of a PBSt motoneurone before, during and after the injection of naloxone (1 mg/kg i.v.) antagonising the effect of DAGO. The recordings began 45 min after the i.v. injection of 1.6 mg/kg DAGO, i.e. when the effect was maximal. Naloxone had no effect on the monosynaptic EPSP (Figure 3, PBSt 2T). PSPs from cutaneous afferents were mainly suppressed. However, small portions of the responses with short delay to cutaneous stimulation seemed to be stable. The PSPs from cutaneous afferents could be re-established by the injection of naloxone (Figure 3, Sur 2T). The PSPs evoked by stimulation of descending pathways seemed to undergo only a form of modulation by the drugs (Figure 3, desc. 10T). In this example an EPSP (the 2nd EPSP in response to desc. 10T) was replaced by inhibition after the application of naloxone. Additionally, convergence was tested before, during and after

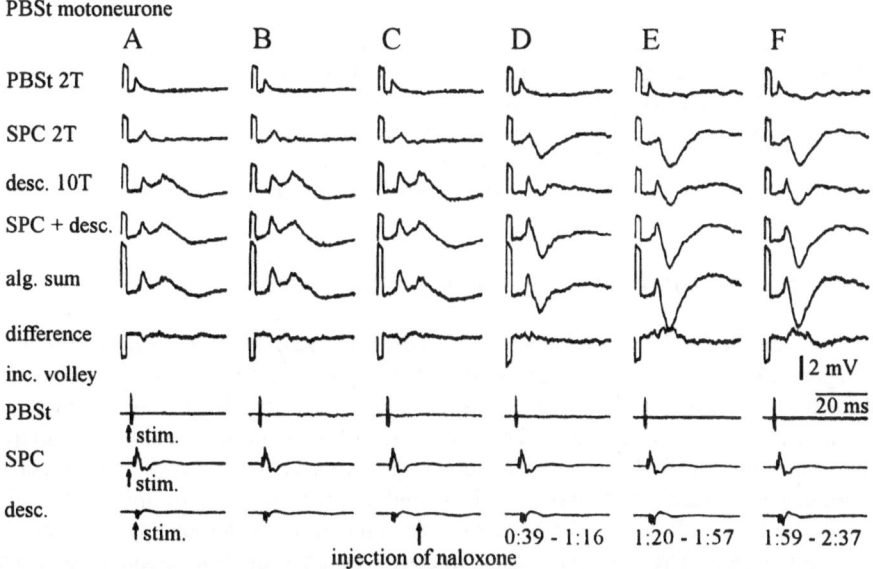

Figure 3. The effect of naloxone (1 mg/kg) on postsynaptic potentials of a PBSt motoneurone evoked by stimulation the homonymous muscle nerve (PBSt 2T) or the sural nerve (Sur 2T) or the descending pathways (desc. 10T). Spatial interaction between the segmental cutaneous pathway and the descending pathways are also shown (technique as in Fig. 2). Naloxone was given during the recording of column C. The recording times after the application of naloxone are indicated under the columns D, E and F. Stimulus times are indicated by arrows incoming volleys at column A. All responses were averaged from 8 single responses each. Recordings 45 min after i.v. injection of DAGO (1.5 mg/kg).

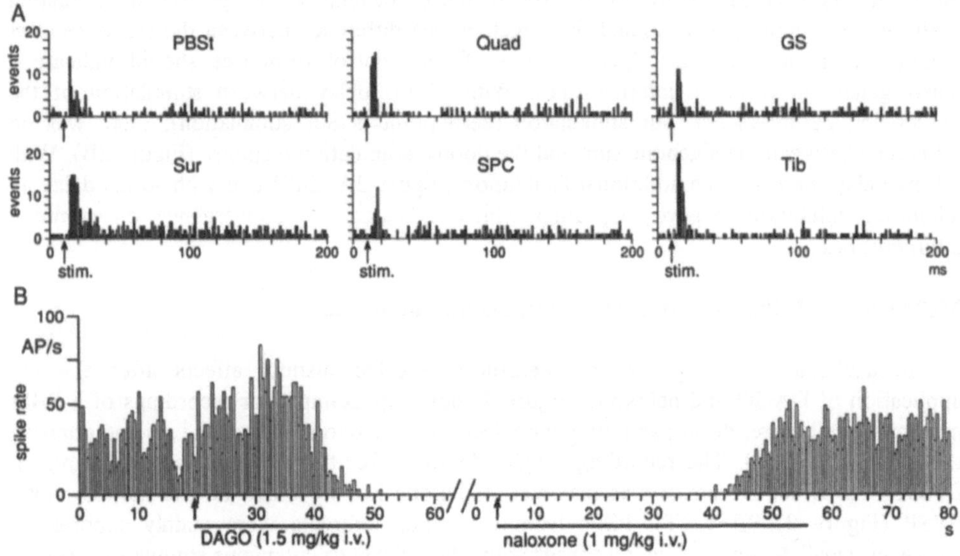

Figure 4. Response behaviour of an extracellularly recorded dorsal horn interneurone (1.2 mm depth of microelectrode tip) to stimulation of different nerves with 5T (A) before the application of DAGO, and change of spontaneous activity of the same interneurone (B) after the application of DAGO (1.5 mg/kg) and naloxone (1 mg/kg). Naloxone was applied 25 min after the application of DAGO. (A) peri stimulus time histograms with 20 responses summed up in each histogram, except for the PSTH from stimulation of Sur (16 responses); time scale 200 ms, 1 ms/bin. (B) frequency histogram, 25 bins/s, time scale in seconds; injection times are indicated at the bottom.

application of the drugs. Differences between the response to double stimulation and the algebraic sum of responses to single stimulation disappeared more or less completely with the application of DAGO, and reappeared with the injection of naloxone (cf. Figure 3, "difference"). Monosynaptic EPSPs in response to descending stimulation proved to be as resistant to DAGO and naloxone injection as those from the homonymous muscle nerve (cf. Figure 3, desc. 10T).

FRA interneurones and the effects of DAGO and naloxone

While microelectrodes were being advanced in the search for motoneurones, stable recordings were obtained from interneurones at a depth of 1 to 1.5 mm, and they were tested for their response to stimulation of all the prepared nerves. They were assumed to be FRA interneurones if they were excited by stimulation of medium threshold muscle afferents and low to medium threshold cutaneous afferents (Figure 4A). These types of interneurones were spontaneously active and were also facilitated by nociceptive input applied by radiant heat to the foot pad. However, their response to stimulation of descending pathways was weak or even absent. The maximum instantaneous frequency of their spontaneous activity was 700 Hz, but the calculated mean spike rate was about 25 to 75 Hz. This spontaneous activity stopped completely within one minute following the application of DAGO (Figure 4B), and this effect was antagonised by an injection of naloxone. After the application of DAGO the interneurone could no longer be excited by electrical stimulation of the FRA. Even stimulation strengths of 10T or nociceptive stimulation did not succeed in exciting these neurones. After application of naloxone the response pattern to electrical stimulation returned to that seen before application of DAGO.

DISCUSSION

Several main results from this investigation should be emphasised. By means of the method of double conditioning of monosynaptic reflexes which result in three dimensional profiles of the reflex amplitudes, it was shown that the descending pathway is less influenced by DAGO than is the segmental FRA pathway. The existence of convergence from descending and segmental FRA pathways on common interneurones within the pathways to motoneurones was confirmed (cf. also Hongo et al., 1969; Hongo, Jankowska & Lundberg, 1972; Baldissera et al., 1971; 1981). Positive deviations of the three dimensional profiles from the controls should indicate areas where the timing of stimuli was optimal for convergence. Signs of convergence disappeared after the application of DAGO. To show this with the method of conditioning an exact timing between the different stimuli was necessary. Intracellular recordings confirmed these patterns of convergence. In addition to the projections to interneurones, monosynaptic projections from descending pathways to motoneurones (Baldissera et al., 1981; Shapovalov, 1975) were also confirmed.

The application of DAGO and naloxone during intracellular recordings of motoneurones revealed a strong depression of the segmental FRA pathways but a modification of the descending pathway. The intracellular recordings confirmed that convergence of descending pathways and segmental FRA pathways on common interneurones vanishes with the application of DAGO and is re-established with the application of naloxone.

The results of this study may be considered in relation to previous work. Though the descending pathways which were stimulated contained small numbers of hypothalamospinal and pontospinal fibres and part of the descending tract from the nucleus raphe magnus (Leichnitz, Watkins, Griffin, Murfin & Mayer, 1978; Basbaum & Fields, 1979), the main pathway stimulated must have comprised rubrospinal fibres (Hongo et al., 1972; Kuypers & Huisman, 1982). Monosynaptic EPSPs from both the descending tract and from the homonymous muscle nerve were not affected by DAGO or naloxone. Therefore, it may be assumed that DAGO acts via the interneurones. Since the descending pathways were not suppressed in the same way as the segmental pathways from FRA, but were only modified, and also signs of convergence on common interneurones disappeared with the application of DAGO, we may further assume that the effect is only on segmental first-order FRA interneurones. The descending fibres may in part converge on the segmental pre-motor interneurones, including inhibitory interneurones. This would not however, explain the modifications of effects evoked by descending stimulation. The explanation for this may be that the spontaneous tonic drive from FRA interneurones is abolished by DAGO. These interneurones also fulfil the condition of not being excitable by descending stimulation.

From the point view of spinal and descending motor control, we must accept the existence of facilitation of segmental effects by descending pathways and the reverse effect from convergence on common interneurones (Lundberg, 1979; Baldissera et al., 1981; Schomburg, 1990). Though the release of enkephalins may interfer with certain features of segmental motor control, descending systems appear to be able to retain sufficient power to ensure efficient motor function.

ACKNOWLEDGEMENTS

Supported by the Deutsche Forschungsgemeinschaft (Scho 37-3/3).

REFERENCES

BALDISSERA, F., TEN BRUGGENCATE, G. & LUNDBERG, A. (1971). Rubrospinal monosynaptic connexion with last-order interneurones of polysynaptic reflex paths. *Brain Res.* **27**, 390-392.

BALDISSERA, F., HULTBORN, H. & ILLERT, M. (1981). Integration in spinal neuronal systems. In *Handbook of Physiology, Sect I, The Nervous System, Vol. II, Motor Control, Part 1.* ed. BROOKS, V.B. pp509-595. American Physiological Society, Bethesda.

BASBAUM, A.I. & FIELDS, H.L. (1979). The origin of descending pathways in the dorsolateral funiculus of the spinal cord of the cat and the rat: further studies on the anatomy of pain modulation. *J. Comp. Neurol.* **187**, 513-532.

ECCLES, R.M. & LUNDBERG, A. (1959). Synaptic actions in motoneurons by afferents which may evoke the flexion reflex. *Arch. Ital. Biol.* **97**, 199-221.

HONGO, T., JANKOWSKA, E. & LUNDBERG, A. (1969). The rubrospinal tract. II. Facilitation of interneuronal transmission in reflex paths to motoneurones. *Exp. Brain Res.* **7**, 365-391.

HONGO, T., JANKOWSKA, E. & LUNDBERG, A. (1972). The rubrospinal tract. IV. Effects on interneurones. *Exp. Brain Res.* **15**, 54-72.

KNIFFKI, K.-D., SCHOMBURG, E.D. & STEFFENS, H. (1981). Effects from fine muscle and cutaneous afferents on spinal locomotion in cats. *J. Physiol.* **319**, 543-554.

KUYPERS, H.G.J.M. & HUISMAN, A.M. (1982). The new anatomy of the descending brain pathways. In *Brain Stem Control of Spinal Mechanisms, Vol. 2.* eds. SJÖLUND, B. & BJÖRKLUND, A. pp29-54. Elsevier, Amsterdam.

LEICHNITZ, G.R., WATKINS, C., GRIFFIN, G., MURFIN, R. & MAYER, D.J. (1978). The projection from nucleus raphe magnus and other brainstem nuclei to the spinal cord in the rat: a study using the HRP blue reaction. *Neurosci. Lett.* **8**, 119-124.

LUNDBERG, A. (1979). Multisensory control of spinal reflex pathways. *Prog. Brain Res.* **50**, 11-28.

SCHMIDT, P.F., SCHOMBURG, E.D. & STEFFENS, H. (1991). Limitedly selective action of a d-agonistic leu-enkephalin on the transmission in spinal motor reflex pathways in cats. *J. Physiol.* **442**, 103-126.

SCHOMBURG, E.D. (1990). Spinal sensorimotor systems and their supraspinal control. *Neurosci. Res.* **7**, 265-340.

SHAPOVALOV, A.I. (1975). Neuronal organisation and synaptic mechanisms of supraspinal motor control in vertebrates. *Revs. Physiol. Biochem. Pharm.* **72**, 1-54.

STEFFENS, H. & SCHOMBURG, E.D. (1993). Convergence in segmental reflex pathways from nociceptive and non-nociceptive afferents to α-motoneurones in the cat. *J. Physiol.* **466**, 191-211.

YAKSH, T.L. & NOUEIHED, R. (1985). The physiology and pharmacology of spinal opiates. *Ann. Rev. Pharm.* **25**, 433-462.

DESCENDING CONTROL OF A CROSSED SPINAL REFLEX BY SEROTONIN

N.C. Aggelopoulos[1], R.W. Clarke[2]
and S.A. Edgley[1]

[1]Department of Anatomy, Downing Street
Cambridge CB2 3DY, UK
[2]Department of Physiology and
Environmental Science
University of Nottingham
Sutton Bonnington Campus
Loughborough LE12 5RD, UK

INTRODUCTION

Many different oligo- or polysynaptic spinal reflexes have been described. A general rule seems to be that multiple reflex pathways can exist, and that under different circumstances specific pathways can be selected (Lundberg, 1981). The mechanisms underlying reflex selection are not well understood, but descending pathways from the brainstem have been implicated (see Lundberg, 1981). Here we describe the control of specific patterns of crossed reflexes evoked from group II afferents which seems to be dependent on a descending serotonergic system acting via a specific receptor subtype.

The reflexes in question are evoked by stimulation of specific nerves at intensities which can activate group II afferents. Many neurones in the midlumbar segments of the cat spinal cord have very strong inputs from group II afferents (Edgley & Jankowska, 1987) and may mediate these reflexes. It has been demonstrated that secondary muscle spindle afferents contribute to these inputs at least for some of the cells. (Harrison, Jami & Jankowska, 1988). Latencies of the EPSPs evoked by group II afferents in midlumbar neurones indicate a minimal monosynaptic linkage. These neurones contribute to spinal reflex actions since many have projections to the lower lumbar motor nuclei and spike-triggered averaging has demonstrated monosynaptic excitation or inhibition evoked from them in lower limb motoneurones (Cavallari, Edgley & Jankowska, 1987).

In investigating these neurones we found an unusual pattern of crossed reflexes evoked by the same afferents which excite midlumbar neurones. Stimulation of group II afferents evokes excitation of ipsilateral flexors and inhibition of ipsilateral extensors (a flexion reflex). On the contralateral side such stimulation evokes not a crossed extension reflex (excitation of contralateral extensors), but inhibition of contralateral extensors (Arya, Bajwa & Edgley, 1991). However, this crossed inhibition is abolished and a crossed extension reflex can be evoked after spinal section, suggesting the existence of a tonic descending control of the reflex pattern. When the spinal cord is intact this pathway facilitates the crossed inhibition and suppresses the crossed excitation. This may have implications for interlimb coordination during movement or postural adjustments.

A further interesting property of the crossed IPSPs is their short latency: the minimal central latencies were a little over 3.0 ms from the arrival of group I volleys at the L6/L7 border. This is comparable to the latencies of EPSPs and IPSPs from the same afferents in ipsilateral motoneurones and raises the possibility of a simple disynaptic pathway involving a single commissural neurone (Arya et al., 1991). Such neurones have been shown to exist by transneuronal labelling from hindlimb motoneurones (Harrison, Jankowska & Zytnicki, 1986) and to have the appropriate inputs (Jankowska & Noga, 1990).

CENTRAL ANATOMY OF THE CROSSED INHIBITION FROM GROUP II AFFERENTS

The similarity of the nerves of origin of the crossed inhibition seen with the spinal cord intact (Arya et al, 1991) to the nerves causing excitation of midlumbar neurones (Edgley &

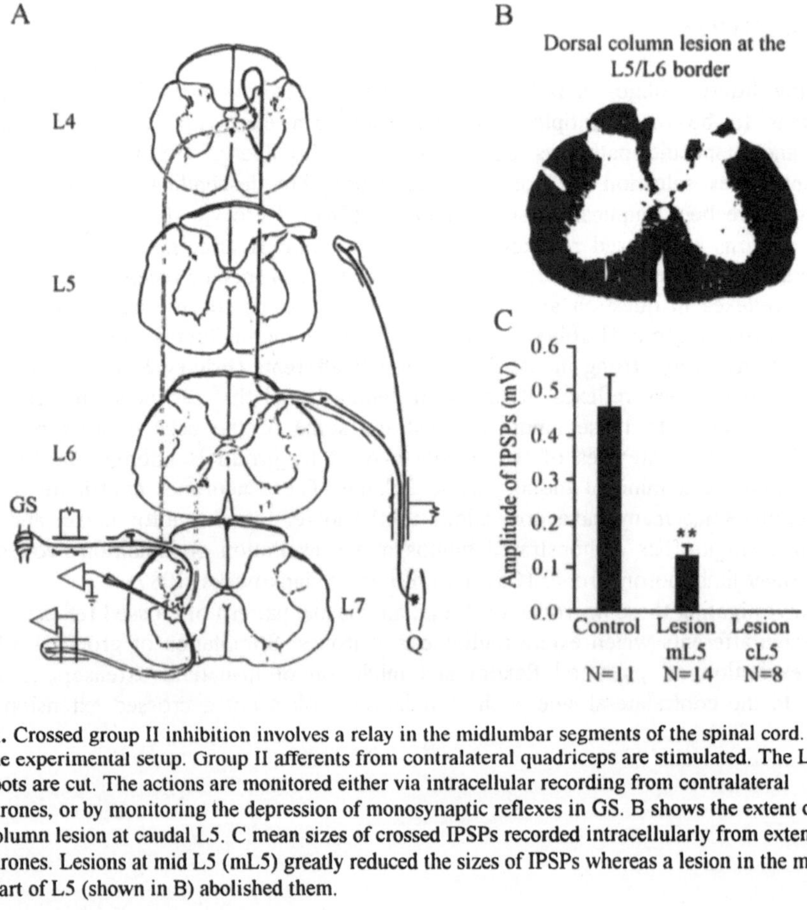

Figure 1. Crossed group II inhibition involves a relay in the midlumbar segments of the spinal cord. A shows the experimental setup. Group II afferents from contralateral quadriceps are stimulated. The L5 dorsal roots are cut. The actions are monitored either via intracellular recording from contralateral motoneurones, or by monitoring the depression of monosynaptic reflexes in GS. B shows the extent of the dorsal column lesion at caudal L5. C mean sizes of crossed IPSPs recorded intracellularly from extensor motoneurones. Lesions at mid L5 (mL5) greatly reduced the sizes of IPSPs whereas a lesion in the most caudal part of L5 (shown in B) abolished them.

Jankowska, 1987) would suggest that the crossed inhibition might be mediated via the latter. Recently however, evidence has been provided that neurones exist in the caudal lumbar segments with a convergence pattern appropriate for the mediation of these actions (Harrison & Riddell, 1989). We have therefore examined the central anatomy of the crossedinhibition with reversible (lidocaine) or irreversible physical lesions of the afferent fibres in the dorsal columns at different levels, monitoring the crossed inhibition by intracellular recording from motoneurones or by monosynaptic reflex testing. A consistent finding was that interruption of the dorsal columns in the caudal part of the L5 segment abolished the crossed inhibition, suggesting that the interneurones mediating the crossed inhibition are located rostral to this point (Figure 1). Interruption in mid- or rostral L5 reduced the inhibition, suggesting that the intervening neurones are distributed along the cord, the most caudal being in caudal L5, others being in rostral L5 or L4 (Aggelopoulos & Edgley, 1994).

This finding has implications for the central linkage of the crossed inhibition: as pointed out above, the short latency of the crossed inhibition suggests a short pathway. The observation that the pathway involves a relay in the midlumbar segments strengthens the possibility that the pathway is disynaptic (ie. a single intervening neurone).

LOCALISATION OF THE DESCENDING FIBRES CONTROLLING CROSSED INHIBITION

A consistent characteristic of the crossed inhibition was its absence after spinal section. As a first step to identifying the fibres concerned, we made limited funicular lesions of the spinal cord in the low thoracic region (Burton & Edgley, 1993). The crossed inhibition was abolished by lesions of the dorsal part of the lateral funiculus ipsilateral to the motoneurones (contralateral to the afferents), as illustrated in Figure 2. In subsequent experiments we verified that the crossed inhibition was still present if the entire dorsal funiculus, contralateral lateral funiculus and ventral funiculus were transected, but the ipsilateral lateral funiculus was left intact.

EVIDENCE FOR THE INVOLVEMENT OF 5-HT$_{1A\ OR\ 7}$ RECEPTORS

The possibility of involvement of 5-HT$_{1A\ or\ 7}$ in crossed group II reflexes was examined with intracellular recording from extensor motoneurones. Examples of recordings from GS motoneurones are illustrated in Figure 3. In these experiments we first verified that the crossed group II inhibition was present with the spinal cord intact, by examining a sample of extensor motoneurones (gastrocnemius-soleus, GS, or semimembranosus & anterior biceps, SMAB). In all experiments this was the case. Following this the spinal cord was sectioned in the low thoracic region. A second sample of extensor motoneurones was then tested to verify that the crossed inhibition had been abolished; in recordings begun ten minutes after spinal section, crossed group II IPSPs were only seen in (7%) of GS and SMAB motoneurones. As has been reported previously (Arya et al., 1991) this was accompanied by the appearance of later EPSPs (eg. a crossed extension reflex) in many motoneurones, but this usually required more than one stimulus.

To test the possible involvement of 5-HT$_{1A\ or\ 7}$ receptors, we subsequently administered a 5-HT$_{1A\ or\ 7}$ receptor agonist, 8-hydroxy 2-di-propylamino tetraline hydrobromide (8-OH-DPAT) intravenously at doses between 0.3 and 1.0 mg/kg. In all experiments 8-OH-DPAT at these doses restored the crossed inhibition in many motoneurones within 20 minutes (Figure 3). After 8-OH-DPAT, crossed group II IPSPs were seen in the large majority of extensor motoneurones. The mean size of the IPSPs after DPAT was not statistically significantly different from that before spinalisation (Students t-test). Furthermore, the reappearance of the inhibition was accompanied by a loss of the later EPSPs which had appeared after spinalisation. To some extent this may reflect shunting by the IPSPs, but this is unlikely to have been the whole story since the EPSP latencies were much greater than those of the IPSPs.

Finally, to verify further that the actions of 8-OH-DPAT were specific to 5-$HT_{1A\ or\ 7}$ receptors we administered the selective antagonist WAY-100135 (Wyeth Research UK). In all 4 experiments performed, WAY-100135 was able to antagonize the effects of 8-OH-DPAT by reducing the incidence of the crossed group II IPSPs. As was the case after spinal section, abolition of crossed IPSPs was accompanied by the appearance of later EPSPs, in line with the crossed extension reflex (Figure 3).

Monitoring the effects of these manipulations on the ipsilateral reflex actions of group II afferents provides a control of sorts, which allows us to consider that these effects are specific and to discount that they are the result of a global change in the excitability of the motoneurones, or of the spinal cord in general.

CONCLUSIONS

The results demonstrate a tonic descending facilitation of a crossed reflex originating from group II afferents. This is unusual since descending systems are usually seen as suppressing reflex actions.

It is well established that 5-HT has powerful effects on motor systems, in particular via its actions on motoneurones (see Hounsgaard, Hultborn, Jespersen & Kiehn, 1988). The site(s) of action of 5-HT cannot be determined from our experiments. However, since the pathway responsible for the crossed IPSPs is likely to be simple with a single commissural neurone, and since the pathway controlling the reflex descends in the dorsolateral funiculus ipsilateral to the motoneurones, the obvious sites are either at the motoneurones or on the interneurone terminals. An action on the afferent terminals is unlikely since these are contralateral to the motoneurones. An action on the motoneurones is unlikely since similar

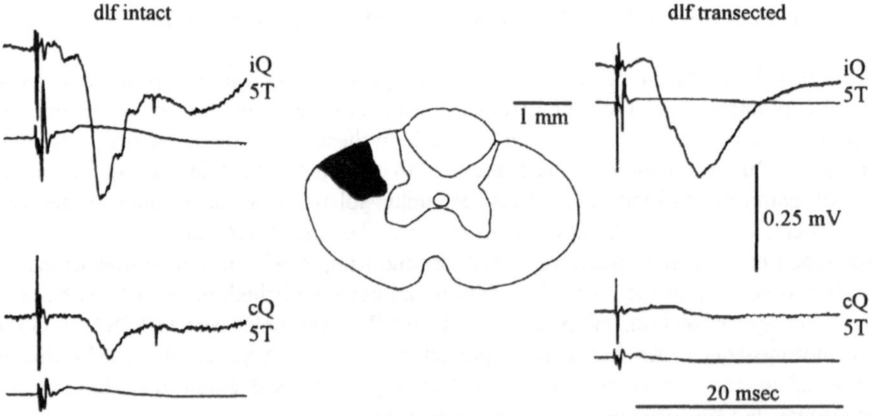

Figure 2. A pathway descending in the dorsolateral funiculus controls crossed group II inhibition. The figure shows IPSPs recorded intracellularly from GS motoneurones before and after the lesion of the dorsolateral funiculus (dlf) illustrated in the centre panel. With the cord intact (left), IPSPs are evoked by both ipsilateral and contralateral group II afferents. After the lesion, the ipsilateral IPSPs are still present, but the contralateral IPSPs are much reduced: larger lesions abolish the contralateral IPSPs.

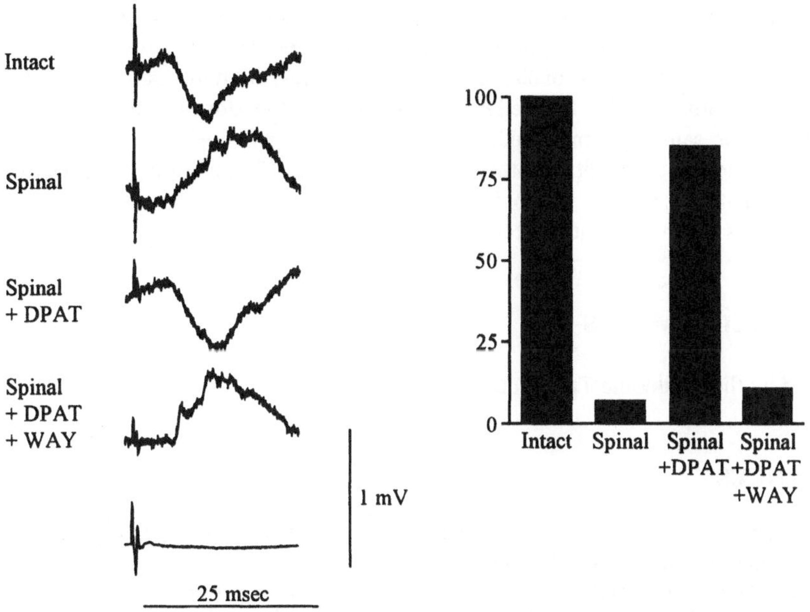

Figure 3. Changes in crossed group II reflex actions in GS motoneurones. The left hand panel shows intracellular recordings of the responses evoked by stimulation of the contralateral quadriceps nerve at 5T from 4 different GS motoneurones with an example of the simultaneously recorded cord dorsum potential below. From top to bottom: with the cord intact crossed IPSPs are evoked. After spinal transection the IPSPs are abolished and EPSPs appear. Administration of 8-hydroxy-DPAT restores the IPSPs, which are again abolished by WAY. The histogram on the right shows the frequency of crossed IPSPs in a population of GS motoneurones.

IPSPs evoked in the extensor motoneurones by ipsilateral group II afferents are not greatly affected by spinalisation or 8-OH-DPAT. Given that the pathway may be disynaptic and, if so, the interneurones should be on the same side as the afferents, this raises the possibility that the action of the serotonergic fibres is at the presynaptic terminals of the interneurones.

Obvious questions are what is the functional role of the crossed group II inhibition and what is the significance of this action of serotonin on the reflex. These experiments merely demonstrate a connectivity, rather than a function. Nevertheless we may speculate that the reflex pattern seen with the spinal cord intact is likely to promote bilaterally symmetrical actions in extensors, whereas the pattern of actions after the descending control is removed, is more appropriate to reciprocal actions on the two sides. Serotonin is often seen as having global actions in the spinal cord, but here its actions could be seen as selecting a specific reflex pattern. In parallel with the modifications of the crossed IPSPs in extensor motoneurones, the crossed EPSPs with a longer latency were seen to alter conversely. The latter observation is in keeping with the actions of a 'dorsal reticulospinal system' which mediates tonic suppression of spinal reflexes as described by Engberg et al. (1968). At the time of the initial observations, the dorsal reticulospinal system was thought not to be serotonergic on two counts. i) the descending fibres were rapidly conducting (up to 60 m/s), dorsally located and the cells of origin were outside the raphe, ii) the effects were not mimicked by administration of the serotonin precursor 5-hydroxytryptophan (Engberg et al. 1968; Lundberg, 1982). It has subsequently been shown first that serotonergic fibres originating from the reticular formation lateral to the raphe nuclei are rapidly conducting and descend in the dorsolateral funiculus, and secondly that a whole plethora of serotonin receptors exists (see Jacobs & Amzitia, 1992). Administration of a non-specific agonist may have unpredictable, sometimes opposing actions at different sites in a reflex pathway. Thus it is a possibility that a system of fast conducting serotonergic fibres acting via 5-$HT_{1A\ or\ 7}$ receptors is responsible for the reflex suppression by the dorsal reticulospinal system. Furthermore, the same system would, when active, not only suppress certain reflexes (eg. crossed extension) but facilitate others (crossed group II inhibition of extensors). The observation that this system is rapidly conducting (up to 60 m/s) is intriguing, since it suggests that its operation requires temporal specificity.

ACKNOWLEDGEMENTS

Supported by the Wellcome Trust and MRC.

REFERENCES

AGGELOPOULOS, N.C. & EDGLEY, S.A. (1994) Segmental localisation of the relays mediating crossed group II inhibition in the cat. *J. Physiol.* **475**, 40P.

ARYA, T., BAJWA, S. & EDGLEY, S.A. (1991) Crossed reflex actions from group II muscle afferents in the lumbar spinal cord of the anaesthetized cat. *J. Physiol.* **444**, 117-131.

BURTON, M.J. & EDGLEY, S.A. (1993) Fibres in the dorsolateral funiculus of the spinal cord control crossed reflexes from group II afferents in the anaesthetized cat. *J. Physiol.* **459**, 435P.

CAVALLIARI, P., EDGLEY, S.A. & JANKOWSKA, E. (1987). Postsynaptic actions of mid-lumbar interneurones of hindlimb muscles in the cat. *J. Physiol.* **389**, 675-689.

EDGLEY, S.A. & JANKOWSKA, E. (1987). An interneuronal relay for group I and II muscle afferents in the midlumbar segments of the cat spinal cord. *J. Physiol.* **389**, 647-674.

ENGBERG, I., LUNDBERG, A. & RYALL, R.W. (1968) Reticulospinal inhibition of transmission in reflex pathways. *J. Physiol.* **194**, 201-223.

HARRISON, P.J., JAMI, L. & JANKOWSKA, E. (1988) Further evidence for synaptic actions of muscle spindle secondaries in the middle lumbar segments of the cat spinal cord. *J. Physiol.* **402**, 671-686.

HARRISON, P.J., JANKOWSKA, E. & ZYTNICKI, D. (1986) Lamina VIII interneurones interposed in crossed reflex pathways in the cat. *J. Physiol.* **371**, 147-166.

HARRISON, P.J. & RIDDELL, J.S. (1989) Group II activated lumbosacral interneurones with an ascending projection to midlumbar segments of the cat spinal cord. *J. Physiol.* **408**, 561-570.

HOUNSGAARD, J., HULTBORN, H., JESPERSEN, B. & KIEHN, O. (1988) Bistability of α-motoneurones in the decerebrate cat and the acute spinal cat after intravenous 5-hydroxytryptophan. *J. Physiol.* **405**, 345-368.

JACOBS, B.L. & AMZITIA, E.C. (1992) Structure and Function of the Brain Serotonin System. *Physiol. Rev.* **72**, 165-229.

JANKOWSKA, E. & NOGA, B. (1990) Contralaterally projecting lamina VIII interneurones in the middle lumbar segments in the cat. *Brain Res.* **535**, 327-330.

LUNDBERG, A. (1981) Multisensory Control of Spinal Reflex Pathways. *Prog. Brain Res.* **50**, 11-28.

LUNDBERG, A. (1982) Inhibitory control from the brainstem of transmission from primary afferents to motoneurones, primary afferent terminals and ascending pathways. In *Brain Control of Spinal Mechanisms.* eds. SJÖLUND, B. & BJÖRKLUND, A. pp 179-224. Elsevier Biomedical Press. Amsterdam.

PHARMACOLOGY OF LOCOMOTION IN CHRONIC SPINAL CATS

S. Rossignol[1], C. Chau[1] and H. Barbeau[2]

[1]Department of Physiology, CNRS
Faculty of Medicine, Université de Montréal
[2]School of Occupational Therapy, McGill University
Montréal, Canada

This short review will summarize some of the work we and others have performed in the field of pharmacology of locomotion in cats and indicate the potential benefits of such an approach in clinical situations where we believe that a rational locomotor pharmacotherapy can be developed.

GENERATION AND MODULATION OF LOCOMOTION BY NORADRENERGIC DRUGS

In 1931, Hinsey, Ranson & Zeiss (1931) showed that the i.v. injection of ephedrine increased the ability of acute spinal cats (C1) to stand and even to display rhythmic movements of the forelimbs and hindlimbs. Although the possibility exists that the effects of ephedrine could be explained by an increase in blood pressure, the authors suggested an independent central action unrelated to cardiovascular effects. Comparing the behaviour of high spinal cats with and without ephedrine, they concluded that "the reflex pathways necessary for these responses are already present and ephedrine, whatever its action may be, is instrumental only in making possible their demonstration". This appears to be one of the earliest studies suggesting an important role of catecholamines in motor control and their potential role in facilitating spinal circuits important for stepping. In 1967, classical papers on the effects of l-DOPA on the spinal cord suggested that, through the release of noradrenaline, DOPA could activate spinal pathways capable of generating alternating bursts in flexor and extensor motor nerves as during locomotion (Jankowska, Jukes, Lund & Lundberg, 1967a,b). After potentiating DOPA with a monoamine oxidase inhibitor, nialamide, Grillner and Zangger (1975;1979) established that the spinal cord was capable of generating, in the absence of phasic afferent feedback, a detailed locomotor pattern with a characteristic ratio of flexor/extensor burst durations and a bilaterally organised alternating activity with muscles being activated at their respective characteristic times within the cycle.

This was confirmed in different preparations by several authors (Baev, 1977, Fleshman, Lev-Tov & Burke, 1984, Pearson & Rossignol, 1991).

The α_2 noradrenergic agonist clonidine was then used by Forssberg and Grillner (1973) to induce locomotion successfully in acute spinal cats. We have also used intraperitoneal clonidine to induce treadmill locomotion of the hindlimbs within the first week after spinalisation at the last thoracic segment (T13) of adult cats chronically implanted with EMG electrodes (Barbeau & Rossignol, 1991). We found that within that week the locomotor pattern evolved rather rapidly. When given during the first few days after spinalisation, the locomotor pattern is often not as well developed as later on. A stronger perineal stimulation is often needed and the rhythmic movements of the hindlimbs are often performed with the hips more extended than in the normal. This progresses rapidly so that after the first week, the locomotor movements have a greater amplitude and the hip excursion is more in the normal locomotor range. Similarly, the fictive locomotor pattern induced by DOPA or clonidine in paralysed cats (Pearson & Rossignol, 1991) changes with time after spinalisation. Indeed, in early-spinal cats, the rhythmic pattern (recorded in peripheral nerves with nerve cuffs) can be more rudimentary with flexor and extensor bursts often of similar duration. In late-spinal cats, the flexor bursts are typically much shorter than the extensor bursts, as in the normal walking cat.

The above results suggest that there is some degree of plasticity in the locomotor circuitry which evolves as a function of time after spinalisation. We have taken advantage of

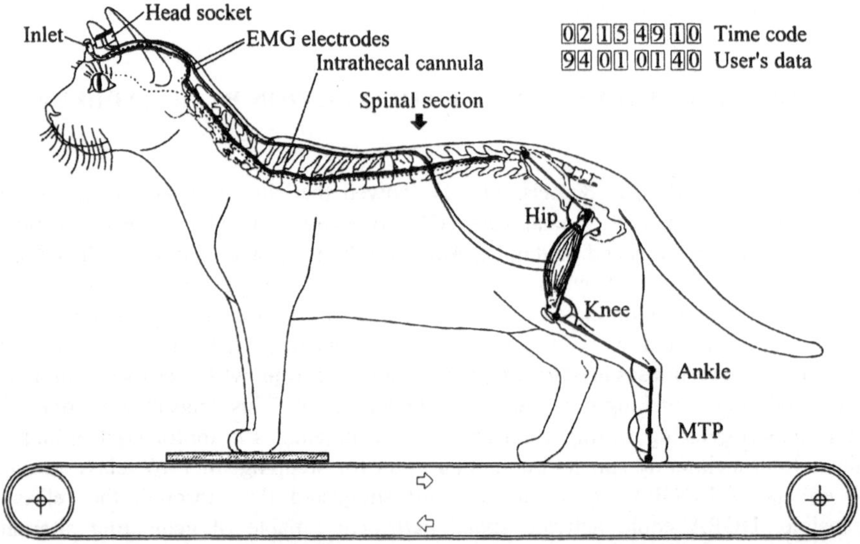

Figure 1. Methodology for EMG and video recordings in chronic spinal cats implanted with an intrathecal cannula. The cats are chronically implanted with bipolar EMG electrodes in selected hindlimb muscles and the leads are connected to a multipin head socket cemented on the skull. The intrathecal cannula consists of a Teflon tubing (24 LW) inserted through the atlanto-occipital ligament with its port of entry cemented on the head next to the EMG socket. The tip of the cannula is inserted in the intrathecal space down to approximately L4. The locomotor movements on the treadmill are recorded on videotape with a synchronised SMPTE time code and other user's data such as dates and the speed of the treadmill. Digitisation with a Peak Performance system of the light-reflecting spots glued to the various joint axes permit the reconstruction of the kinematics and the synchronisation of these mechanical events to the EMG activity with a resolution of one video field (16.7 ms). Usually a single bolus of 100 µl of the drug solution is injected through the inlet and is pushed out of the cannula with another bolus of 100 µl of sterile saline to fill the dead space of the cannula. Note that only the hindlimbs are on the treadmill, the forelimbs remain on a fixed platform above the belt.

this situation to train cats to walk daily on a treadmill (75-110 mins/day) with a daily i.p. injection of clonidine (Barbeau, Chau & Rossignol, 1993). Following such intense training, some cats could walk, with their hindlimbs on the treadmill, without drugs after 7-11 days and maintain this performance for several weeks.

The need to inject drugs daily and the variable importance of side effects of clonidine i.p. in cats (sleepiness and nausea) led us to use a chronic intrathecal (i.t.) cannula which is illustrated in Figure 1 with other essential points of methodology. The following examples illustrate three distinct situations where different aspects of these drugs can be considered: initiation of locomotion, modulation of an existing locomotor pattern and improvement of a defective locomotor pattern.

Figure 2 illustrates the initiation of locomotion after an i.t. injection of clonidine (100 μg/100μl ≈ 4mM) in a cat spinalised 3 days prior to the experiment. The stick figures and EMG records of Figure 2A show that, prior to the injection, the cat is incapable of any significant rhythmic movement on the treadmill. Thirty minutes after the injection, the cat can walk vigorously when the treadmill belt is moved. The EMGs are well organised bilaterally and the stick diagrams of the swing and stance of one step illustrate the amplitude of the overall limb movement. In some cases we could demonstrate that within only 3-5 minutes of the i.t. injection of clonidine, the cats' hindlimbs which had been completely paralysed were walking on the treadmill. Since the intrathecal route opens up the possibility of using chemicals that do not cross the blood brain barrier, we have started to evaluate in more detail the effects of various other α_2 noradrenergic agonists such as tizanidine and oxymetazoline. Whereas tizanidine has about the same time course as clonidine (about 6

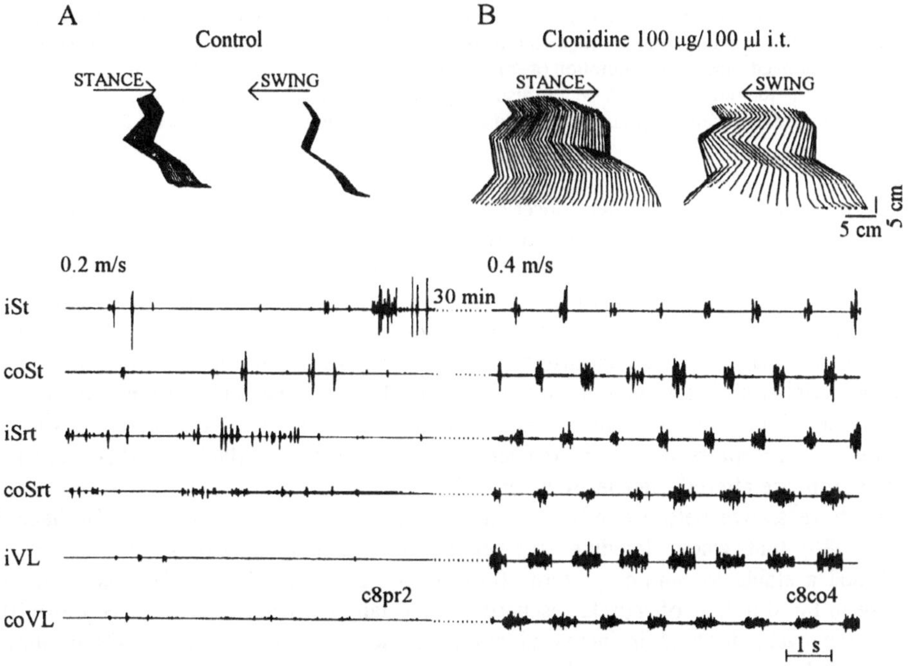

Figure 2. Initiation of locomotion in a 3-day post-spinal cat with an intrathecal injection of clonidine. A. Stick figures of the hindlimb representing the swing and stance phase of 1 step cycle and the raw electromyogram (EMG) of hindlimb muscles of a spinal cat during treadmill locomotion before clonidine injection. B. At 30 minutes after 100μg of clonidine (i.t. 100μg/100 μl, ≈ 4mM), with perineal stimulation, a good locomotor pattern is triggered. Abbreviations of the muscles are St: Semitendinosus; Srt: Sartorious; VL: Vastus Lateralis.i: ipsilateral; co: contralateral.

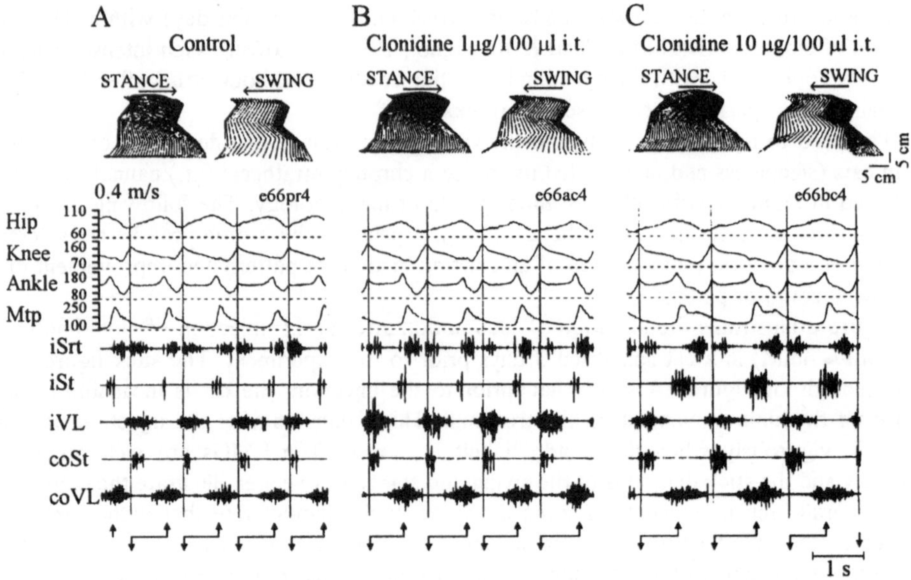

Figure 3. Effects of increasing doses of intrathecal clonidine on the locomotor pattern in a cat 275 days post-spinalisation. A. Stick figures showing joint angle displacement synchronised with the raw EMG of various hindlimb muscles, before clonidine injection (mean duration of 8 cycles = 1.07s ± S.D. 0.030s). B. At 22 minutes after 1 μg clonidine (i.t. 1μg/100μl), a slight increase in cycle duration is seen (mean duration of 7 cycles = 1.17s ±S.D. 0.026s). C. At 23 minutes after 10 μg clonidine (i.t 10μg/100μl, injected at 31 minutes following the previous 1 μg clonidine injection), the cycle duration, particularly the swing phase, was prolonged (mean cycle duration (n=8) increased to 1.462 s (± S.D. 0.035s). A marked increase in toe drag during the initial swing phase was also observed. The vertical lines delineate each step cycle, horizontal bars indicate stance; upward arrows indicate foot lift; downward arrows indicate foot contact.

hours), oxymetazoline has a slower onset of action but can last for more than 48 hours. This means that during that effective period, the cat can readily walk when its hindlimbs are placed over the moving treadmill belt. It should be emphasised that although rhythmic movements are easily elicited during that period by various stimuli when the cat is lying down in its cage or on a treadmill, spontaneous rhythmic movements are rare. Therefore, it should not be imagined that such drugs elicit a perpetual stepping pattern; the cats indeed need the peripheral afferent inputs normally provided by the treadmill or other stimuli such as perineal stimulation to express the locomotor pattern. The activation of the α_2 noradrenergic receptors prime the relevant locomotor circuits which are then activated when appropriate afferent signals are given.

We have shown before (Barbeau, Julien & Rossignol, 1987, Rossignol, Barbeau & Julien, 1986) that when clonidine is given i.p. to a late-spinal cat which has already established a stable locomotor pattern (Barbeau & Rossignol, 1987), there is a marked increase in the duration of muscle discharges, especially of flexor muscles. Therefore, the whole step cycle duration is increased although the amplitude of the EMGs is almost unchanged. Figure 3 shows how clonidine can modulate the expression of the locomotor pattern in a cat which has reached a good stable locomotor performance (Figure 3A) after being trained on a treadmill for several weeks. A very small dose of clonidine (1 μg/100μl i.t.) is sufficient to induce some changes in the step cycle duration. Note how the foot is placed at a longer distance in front of the hip joint and also how the ankle is slightly more extended. With higher doses, a deterioration of the walking pattern can actually be observed. Indeed, even before any drug, the cat can typically have a short period of foot drag. After clonidine this foot drag can be exaggerated (see in particular the metatarso-phalangeal (Mtp) joint in Figure 3C). Cycle duration is much increased due principally to a prolongation of the flexor bursts, especially semitendinosus muscles on both sides.

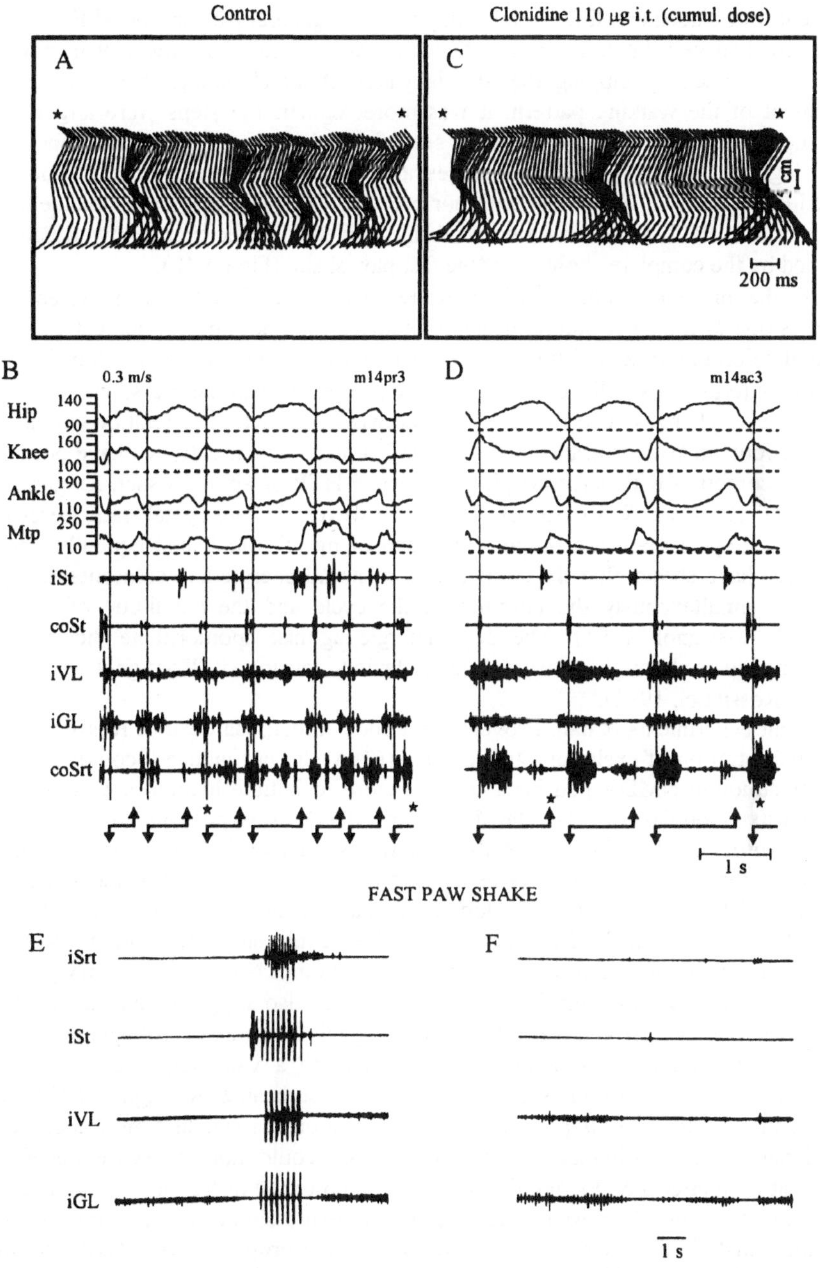

Figure 4. Improvement of the locomotor pattern in a spinal cat. A. Stick figures representing a continuous sequence (about 3.3s) of irregular stepping of the hindlimb on the treadmill. B. The corresponding sequence after clonidine. Stars indicate the starting and ending point of the kinematic analysis in A. C. Kinematics following three successive doses of 10μg, 25μg, and 75μg of clonidine injected within an hour. D. Corresponding angular and EMG data. E. Fast paw shake response which was abolished after clonidine (F).

TOWARDS A LOCOMOTOR PHARMACOTHERAPY?

Figure 4 illustrates an example where clonidine improved the characteristics of a walking pattern which had deteriorated and had become somewhat irregular due to a lack of training. In the control situation (Figure 4A & B), the cat had irregular steps with inconsistent foot placement, little weight support and a generally disorganised EMG pattern. This cat also had very brisk extensor reflexes as can be seen from the violent paw shake (Figure 4E) induced by dipping the foot in water. After clonidine, there was a general improvement of the walking pattern: it was more regular, the steps were larger and the weight support was increased as can be seen from the larger joint excursions and the "longer" legs in Figure 4C. These changes are, of course, reflected by a much better organised EMG pattern. Consistent with our previous findings using i.p. clonidine injection (Barbeau et al., 1987), there was a decrease in cutaneous reflex excitability, which is best exemplified by the complete abolition of the fast paw shake (Figure 4F).

Thus, the present results clearly indicate that the activation of noradrenergic α_2 receptors is one of the most potent means of inducing locomotion in spinal cats, that it can modulate the duration of an existing locomotor pattern and improve the rhythmicity of poor locomotor patterns. It is believed that the effects of α_2 noradrenergic stimulation are the manifestations of changes occurring in reflex pathways and rhythm generation circuits.

What about other neurotransmitter systems? In early-spinal cats, we have been unable to initiate locomotion with serotonergic precursor 5-HTP or agonists such as quipazine or 5-MeO-DMT (Barbeau & Rossignol, 1990). However, in late-spinal cats, serotonergic agonists markedly increase the output amplitude of muscles and, consequently, the cycle duration. We have shown that the combination of noradrenergic and serotonergic agonists can increase simultaneously the duration of the cycle and the amplitude of the EMGs (Barbeau & Rossignol, 1991). The dopaminergic agonist apomorphine did not induce locomotion either, although a marked hyperflexion developed (Rossignol et al., 1986, Barbeau & Rossignol, 1991).

In recent experiments (Chau, Provencher, Lebel, Jordan, Barbeau & Rossignol, 1994), the i.t. administration of excitatory amino acids (EAA) did not induce locomotion in early-spinal cats although NMDA had been shown before to initiate locomotion in decerebrate paralysed cats (Douglas, Nogas, Dai & Jordan, 1994). It is possible that activation of NMDA receptors in this situation leads to a widespread excitation of many conflicting pathways and that disorganised movements generated by NMDA such as paw-shake might have prevented the expression of the locomotor pattern. However, in walking late-spinal cats AP5 (15-20 mM), an NMDA receptor blocker, could stop locomotion which could, however, be partially reinstated with a further i.t. injection of NMDA (10-15 mM).

Recent clinical work in spinal cord injured patients also suggests that the combination of a noradrenergic agonist and a serotonergic antagonist such as cyproheptadine, together with locomotor training on treadmill, might constitute a valid approach to locomotor pharmacotherapy (see Rossignol & Barbeau, 1993 and Barbeau & Rossignol, 1994 for brief reviews), and even allow some patients to benefit from other rehabilitation procedures such as functional electrical stimulation from which they could not otherwise benefit. It is possible that such effects on locomotion may be related to the decrease in reflex excitability (spasticity) that would otherwise prevent the smooth expression of locomotion, the priming of relevant spinal circuits that could be entrained by the proper afferent feedback and the ensuing beneficial effect of locomotor training. More research is needed to understand the relative importance of all these interlinked aspects.

ACKNOWLEDGEMENTS

This work has been supported by The Network of Centres of Excellence (NCE) in Neural Regeneration and Functional recovery and a Group Grant from the Medical Research Council of Canada. C.C. is an NCE trainee partly funded by the FCAR Group; H.B. is an FRSQ chercheur-boursier. All our thanks to Janyne Provencher and France Lebel for their active participation in all experiments and G. Filosi for his graphics skills.

REFERENCES

BAEV, K. V. (1977). Rhythmic discharges in hindlimb motor nerves of the decerebrate, immobolized cat induced by intravenous injection of DOPA. *Neurophysiol.* **9**, 165-167.

BARBEAU, H., JULIEN, C. & ROSSIGNOL, S. (1987). The effects of clonidine and yohimbine on locomotion and cutaneous reflexes in the adult chronic spinal cat. *Brain Res.* **437**, 83-96.

BARBEAU, H., CHAU, C. & ROSSIGNOL, S. (1993). Noradrenergic agonists and locomotor training affect locomotor recovery after cord transection in adult cats.. *Brain Res. Bull.* **30**, 387-393.

BARBEAU, H. & ROSSIGNOL, S. (1987). Recovery of locomotion after chronic spinalization in the adult cat. *Brain Res.* **412**, 84-95.

BARBEAU, H. & ROSSIGNOL, S. (1990). The effects of serotonergic drugs on the locomotor pattern and on cutaneous reflexes of the adult chronic spinal cat. *Brain Res.* **514**, 55-67.

BARBEAU, H. & ROSSIGNOL, S. (1991). Initiation and modulation of the locomotor pattern in the adult chronic spinal cat by noradrenergic, serotonergic and dopaminergic drugs. *Brain Res.* **546**, 250-260.

BARBEAU, H. & ROSSIGNOL, S. (1994) Spinal cord injury: enhancement of locomotor recovery. *Current Opinion in Neurology.* 7, 517-524.

CHAU, C., PROVENCHER, J., LEBEL, F., JORDAN, L., BARBEAU, H. & ROSSIGNOL, S. (1994). Effects of intrathecal injection of NMDA receptor agonist and antagonist on locomotion of adult chronic spinal cats. *Soc. Neurosci. Abstr.* **20**, 573.

DOUGLAS, J. R., NOGAS, B. R., DAI, X. & JORDAN, L. M. (1993). The effects of intrathecal administration of excitatory amino acid agonists and antagonists on the initiation of locomotion in the adult cat. *J. Neurosci.* **13**, 990-1000.

FLESHMAN, J. W., LEV-TOV, A. & BURKE, R. E. (1984). Peripheral and central control of flexor digitorium longus and flexor hallucis longus motoneurons: the synaptic basis of functional diversity. *Exp. Brain Res.* **54**, 133-149.

FORSSBERG, H. & GRILLNER, S. (1973). The locomotion of the acute spinal cat injected with clonidine i.v.. *Brain Res.* **50**, 184-186.

GRILLNER, S. & ZANGGER, P. (1975). How detailed is the central pattern generation for locomotion?. *Brain Res.* **88**, 367-371.

GRILLNER, S. & ZANGGER, P. (1979). On the central generation of locomotion in the low spinal cat. *Exp. Brain Res.* **34**, 241-261.

HINSEY, J. C., RANSON, S. W. & ZEISS, F. R. (1931). Observations on reflex activity and tonicity in acute decapitate preparations with and without ephedrine. *J. Comp. Neurol.* **53**, 401-407.

JANKOWSKA, E., JUKES, M. G., LUND, S. & LUNDBERG, A. (1967a). The effect of DOPA on the spinal cord. 5. Reciprocal organization of pathways transmitting excitatory action to alpha motoneurones of flexors and extensors. *Acta physiol. scand.* **70**, 369-388.

JANKOWSKA, E., JUKES, M. G., LUND, S. & LUNDBERG, A. (1967b). The effects of DOPA on the spinal cord. 6. Half centre organization of interneurones transmitting effects from the flexor reflex afferents. *Acta physiol. scand.* **70**, 389-402.

PEARSON, K. G. & ROSSIGNOL, S. (1991). Fictive motor patterns in chronic spinal cats. *J. Neurophysiol.* **66**, 1874-1887.

ROSSIGNOL, S., BARBEAU, H. & JULIEN, C. (1986). Locomotion of the adult chronic spinal cat and its modification by monoaminergic agonists and antagonists. In *Development and plasticity of the mammalian spinal cord*, eds. GOLDBERGER, M., GORIO, A. & MURRAY, M., pp. 323-345. Liviana Press, Padova.

ROSSIGNOL, S. & BARBEAU, H. (1993). Pharmacology of locomotion: an account of studies in spinal cats and spinal cord injured subjects. *J. Am. Paraplegia Soc.* **16**, 190-196.

CRITIQUE OF PAPERS DEALING WITH PHARMACOLOGY OF CENTRAL CONTROL

Arthur W. Duggan

Department of Preclinical Veterinary Sciences
Royal (Dick) School of Veterinary Studies
Summerhall, University of Edinburgh
Edinburgh, EH9 1QH

Pharmacological studies of descending controls have two main aims: firstly to use drugs (both agonists and antagonists) as aids to an understanding of the transmitters mediating such controls and secondly to define actions of drugs which might be used therapeutically. As with all pharmacological studies, the methods of drug administration and the events measured have been diverse.

The papers of this session dealt with control of spinal events by release of neuroactive compounds from the spinal terminations of brain stem derived fibres. In all cases the primary synaptic event produced by activity in these fibres was not studied, but rather how such activity ultimately affected a polysynaptic spinal event. This can create problems in determining where drugs act when given systemically or topically but this does not detract from the overall importance of the work. The papers will be discussed according to the compound being released from descending fibres.

5-HYDROXYTRYPTAMINE (5HT)

5-HT receptors are multiple. They can be summarised as follows:

5-HT1	subtypes A, B, D, E, F.
5-HT2	subtypes A, B, C.
5-HT3	
5-HT4	
5-HT5	subtypes A, B.
5-HT6	
5-HT7	

The paper of Aggelopoulis, Clarke and Edgeley proposed a role for 5-HT1A receptors in a brain stem - derived tonic facilitation of inhibition of extensor motoneurones by impulses in contralateral group II muscle afferents. This short latency inhibition was converted to a longer latency excitation by sectioning of one dorsolateral spinal funiculus. In this state the inhibition was restored by systemic administration of the 5-HT1A receptor agonist 8-OH-DPAT and the effect of this compound was antagonised by systemic WAY 100135.

The 5-HT1A receptor is regarded as inhibitory in terms of (i) reducing 5-HT release from 5-HT releasing cells and (ii) activating a K^+ conductance in target neurones, producing a hyperpolarization. Thus the conclusion of Aggelopoulis et al that 5-HT was tonically facilitating a crossed inhibitory pathway suggests that 5-HT was inhibiting a further tonically active inhibitory pathway. Disinhibition is a well known mechanism of facilitation and it would be of considerable interest to know the location of the relevant inhibitory interneurones.

ADRENALINE AND NORADRENALINE

The papers of Pompeiano and of Rossignol et al. both dealt with release of noradrenaline from descending fibres. In the cat and rat the relevant fibres are nearly all derived from midbrain nuclei including the locus coeruleus, nucleus subcoeruleus and nucleus Kölliker-Füse.

Receptors for noradrenaline can be summarised as :

1. subtypes A, B, C and D; all are positively coupled to inositol triphosphate/diacylglycerol (IP3/DAG) and hence are likely to excite central neurones.
2. subtypes A, B, C; all reduce cAMP production and are likely to inhibit central neurones.
3. subtypes 1, 2, 3; all increase cAMP production and hence likely result in excitation.

The paper of Rossignol, Chau and Barbeau summarised studies of the effects of adrenomimetics on spinal reflexes and locomotion in the cat. It is an extraordinary result that systemic or intra-thecal administration of an α_2 adrenomimetic such as clonidine can induce rhythmic locomotion in the paralysed limbs of a chronic spinal cat positioned on a treadmill. The authors emphasise that these movements are not spontaneous but require an afferent input from the hindlimbs and/or perineum for initiation.

Systemic α_2 adrenoceptor agonists are used in veterinary medicine for sedation and analgesia which is accompanied by significant hypotonia. At first sight this seems at variance with the findings of Rossignol et al, but in their work the rhythmic movements were produced by an action restricted to α-adrenoceptors in the spinal cord (disconnected from the brain stem).

The paper of Pompeiano dealt with the overall physiology of the descending noradrenergic system. Thus the hypotonia produced by α_2 agonists given systemically to intact animals probably results from inhibition of the firing of neurones of the locus coeruleus or a partial shut-down of the noradrenergic system acting on all adrenoceptors. Pompeiano presented considerable evidence on the enhanced extensor tone associated with activity of the descending noradrenergic system and the role of this system in posture and its control by vestibulo-spinal reflexes.

It is not possible to reconcile fully these two papers dealing with differing functional aspects of the noradrenergic system but it should be emphasised that one is dealing with the whole system in anaesthetised animals while the other considers the consequence of simultaneous activation of one receptor type only in the spinal cord.

OPIOIDS

The paper of Steffens and Fronhöfer dealt with the consequences of μ opioid receptor activation in the spinal cord. All three opioid receptors are negatively linked to cAMP production and hence result in neuronal inhibition. A major difficulty in understanding this system is the plethora of potential ligands and the finding that no opioid peptide has a very high affinity only for one opioid receptor type. Thus using synthetic ligands with selectivity for one receptor will probably only reveal part of the physiological situation since it is unlikely that an endogenous opioid acts on one receptor only. The paper of Steffens and Fronhöfer is particularly interesting since it places a primary site of action of μ opiates on the spontaneous activity of FRA interneurones. They show this by a combination of:

(1) Extracellular recording from these cells and showing a powerful inhibition of their firing by i.v. DAGO and its reversal by naloxone.
(2) Demonstrating that interactions between the descending (rubrospinal) influence and the local FRA pathway were modified by DAGO and naloxone.

Their FRA interneurones were 1 - 1.5 mm from the dorsal surface and this places these cells in the dorsal horn down to lower lamina IV. Work from this laboratory has convincingly shown that opioids have potent effects on many spinal events unrelated to nociception. It is still unknown however, whether the primary trigger to the release of opioids in experiments of this type is the extensive surgery needed to prepare the animals for recording.

GLYCINE

The paper of Fyffe, Alvarez, Harrington & Dewey is an early example of what will probably be an explosion in immunocytochemistry - the localisation of neurotransmitter receptors by antibodies to defined sequences of these proteins. In this case the receptor itself was not studied but rather a receptor linked protein - gephyrin. This is a tubulin binding protein associated with the glycine receptor. The glycine system is relatively simple in that there is but one ligand and one receptor and the latter is a ligand-gated ion channel producing inhibition. Fyffe et al. found that glycine receptors occur almost exclusively in sub-synaptic locations and occupy a larger area as distance from the soma increases. They propose this increase as a means for ensuring that distally located inhibitory synapses have a significant effect on soma excitability despite their distance from the soma.

Glycine receptors on motoneurones are important in recurrent inhibition from Renshaw cells. Anatomists have upset some of the conclusions from neuropharmacological studies of motoneurones inhibition by finding co-existence of glycine and GABA in nerve terminals. Given that antibodies to some of the subunits of GABA receptors are available, this work from Robert Fyffe's laboratory can be extended in several directions including the question as to whether glycine and GABA receptors co-exist at single synapses on motoneurones. A decade ago glycine-mediated and GABA-mediated inhibition were regarded as separate and distinct entities but these recent anatomical findings cast doubt on the validity of this complete separation.

REFERENCES

BJORKLUND, A. & SKAGERBERG, G. (1982) Descending monoaminergic projections to the spinal cord. In *Brain Stem Control of Spinal Mechanisms.* eds. SJOLUND, B. & BJORKLUND, A. pp. 55-88. Elsevier Biomedical, Amsterdam.

BOHLHALTER, S., MOHLER, H. & FRITSCHY, J.-M. (1994) Inhibitory neruotransmission in rat spinal cord: Co-localization of glycine- and $GABA_A$-receptors at GABAergic synaptic contacts demonstrated by triple immunofluorescence staining. *Brain Res.* **642**, 56-69.

BOWKER, R.M., STEINBUSCH, H.W.M. & COULTER, J.D. (1981) Serotonergic and peptidergic projections to the spinal cord demonstrated by combined retrograde HRP histochemical and immunocytochemical staining method. *Brain Res.* **211**, 412-417.

CLARKE, R.W., FORD, T.W. & TAYLOR, J.S. (1988) Adrenergic and opioidergic modulation of a spinal reflex in the decerebrated rabbit. *J. Physiol.* **404**, 407-417.

CURTIS, D.R. (1978) Gabaergic transmission in the mammalian central nervous system. In *GABA-Neurotransmitters - Alfred Benzon Symposium.* eds. KROGSGAARD-LARSEN, P., SCHEEL-KRUGER, J. & KOFOD, H. pp. 17-27. Munksgaard, Copenhagen.

HERZ, A. (1993) *Opioids 1.* Springer-Verlag, Berlin.

HUMPHRIES, P.P.A. (1993) A proposed new nomenclature for 5-HT receptors. *TIPS.* **14**, 233-236.

KOBILKA, B.K. (1992) Adrenergic receptors as models for G protein-coupled receptors. *Ann. Rev. Neurosci.* **15**, 87-114.

PROUDLOCK, F., SPIKE, R.C. & TODD, A.J. (1993) Immunocytochemical study of somatostatin, neurotensin, GABA, and glycine in the rat spinal dorsal horn. *J. Comp. Neurol.* **327**, 289-297.

PART 9

CLINICAL IMPLICATIONS

THE CLINICAL VIEWPOINT

L.S. Illis

Wessex Neurological Centre
Southampton University Hospitals
Southampton, Hants., UK

Studies on inputs, discharges, and connectivity from cortex to muscle, comprise the major part of this session. My task is to give some clinical viewpoint to this important work. An impossible task, unless I take one small area and so I propose to single out spinal injury as an example of clinical disturbance where tremendous changes are occurring in the experimental field. The regeneration of CNS axons, particularly the study of growth inhibitory molecules and glia-neuronal relationships is one of the most exciting aspects of neuroscience. Sooner or later these results will be applied clinically and this is where I see the essential role of the neurophysiologist. The work on regeneration, like so much work in the central nervous system, which concentrates on the lesion, ignores the way that the rest of the central nervous system reacts to the lesion. This can be summarised as follows.

DISTANT EFFECTS

The nerve cell surface is not simply the site of termination of a few hundred boutons termineaux. On a large nerve cell such as the anterior horn cell in the cat spinal cord, the motor neurone surface has up to 30,000 synapses occupying up to 70% of cell and dendrite surface, clothing the cell surface like a mosaic, the constituent parts of which are separated by bare areas and by glial cells and their processes. Within the area encompassed by one glial cell and its processes there are inputs from many different sources. This synaptic zone, if disturbed by a partial denervation, will not only show degeneration of those synapses which had their afferent fibres cut, but also neighbouring synapses, whose afferent fibres were intact, will show a temporary reversible change. This disorganisation has been linked to the phenomena of spinal shock and to von Monakow's diaschisis (Illis, 1963; 1967).

SPROUTING

Collateral sprouting of new connections following partial damage to the nervous system is recognised as a widespread phenomenon and has been demonstrated in peripheral, central and autonomic nervous systems. Sprouting may occur in a variety of ways: as growth from undamaged fibres near denervated tissue or as growth from one branch of the axonal tree when another branch is severed. The demonstration of sprouting has had a marked effect on the way that clinicians and experimental biologists have come to look at the adult nervous system and the problem of recovery (Illis, 1994; Devor, 1994). One of the effects, of course, of sprouting is that the intact nervous system has altered structurally and will now, therefore, react in a different way.

UNMASKING

P.D. Wall and his colleagues in a series of experiments have demonstrated (Wall, 1989) that when cells are deafferented, and therefore lose their normal input, they begin to respond to new inputs. Namely, inputs which could produce no response in the intact animal. Again, a partially damaged nervous system, instead of losing it's ability to react, is reacting in a changed fashion. For example the occurrence of reflex activity unmasked by a chronic deafferentation. The lesion has altered dominant systems of connections by sprouting and by the unmasking of less dominant, but pre-existing, anatomical systems.

ABNORMAL SENSITIVITY

Alteration in synaptic transmission or effectiveness following axonal damage, is well documented. Alteration of signal traffic may change synaptic transmission and effectiveness. The best known example of this is post-tetanic potentiation. In spinal injury, why do small stimuli produce such massive responses such as, for example, spasms and spasticity? Is it simply loss of inhibition from above? In experiments with tetanus toxin (Illis & Mitchell, 1970) it is clear that inhibitory synapses are affected and anatomically these abnormal synapses are localised to dendrites. Clinically, patients with tetanus show widespread loss of inhibition but no long-tract signs.

In separate experiments, with repetitive electrical stimulation (Illis, 1969), appropriate synapses are enlarged and resemble giant synapses which are found normally in Clarke's column. In the synaptic zones of Clarke's column there is normally a system by which synaptic transmission is much more readily accomplished than at the anterior horn cells. So we have the possibility that in the altered synaptic zones in the distal segment, the removal of major pathways means that perhaps new pathways are subject to more traffic and may respond more readily to stimuli so that the altered physiology in the distal segment may not simply be "loss of inhibition" from above but an intrinsic increase in excitability.

In dealing with the partially denervated CNS we are not dealing with a linear system in which the whole is equal to the sum of its parts. In fact, what we have is a non-linear web in which the slightest change in one component causes changes elsewhere. Moreover, the hundreds of thousands of synapses will acquire a property which none of them have alone; the property of complexity. According to complexity theory, there is an incessant drive to organise and to form ever more complex structures in the face of the forces of dissolution To some extent this is exemplified in Le Chatelier's principle: if a system is in equilibrium and one of the conditions of this system is altered, the system will adjust itself in such a way as to neutralise partially the change of condition. This principle, well known in physical and chemical science was first applied to biological systems, specifically recovery (Illis, 1967).

In terms of complexity theory, in spinal injury we have, as it were, a catastrophic event which results in the apparent collapse of the subnetwork which depended on the intact nervous system. This results in a new network being built up and the more this new network fills gaps made by the original catastrophe, the more difficult it will be to change, and this network will be stagnant. What this means, of course, is that the clinical syndrome we see after a CNS lesion is partly due to the damage but is also partly due to the reaction of the nervous system to that damage in terms of failure of dominant pathways, unmasking of new pathways, alteration of synaptic effectiveness and disturbance of the balance of inhibition and excitation. In most instances the lesion is untreatable at present, but the reaction to the lesion is theoretically or potentially treatable.

In conventional neurology this new state after a lesion, produces a fixed neurological deficit, and conventional teaching is that a fixed neurological deficit cannot be altered. Can this state in fact be altered by external stimuli? If it cannot be altered then not only is the possibility of functional regeneration rather gloomy, but the whole of rehabilitation is reduced to prevention and treatment of complications, and most of the theories of recovery would be false. In fact, stimulation may alter inhibition and produce functional changes.

Which target is going to be contacted when axonal regeneration becomes a possibility? How is this going to affect connectivity and, therefore, outcome in terms of, for example, an increase in spasticity etc. Failure to appreciate changes in the distal segment may condemn excellent research in terms of regrowth of axons to a fascinating intellectual exercise with no practical application.

The changes in the intact nervous system can easily be investigated in Man. We can record segmental reflex activity, long tract sensory and motor responses, peripheral and central blood flow, and urodynamics. And we can, therefore, give a picture of the physiological state of the intact central nervous system and everything which depends upon it. I feel that this model of neurophysiological assessment in Man is not only essential but is also becoming urgent as more techniques will be available for the treatment of neurological disorder. Unless a model of neurophysiological assessment is established then any improvement seen clinically is unlikely to be taken seriously and a good example of that is the early work on transplantation in Parkinson's disease.

Finally, the initial effect of applying therapy in some conditions, for example, to a spinal injured person, may well be an increase in spasticity or an increase in pain or potentially disastrous effects as regards cardiovascular reflexes and problems relating to abnormalities of bone and muscle. This makes neurophysiological assessment even more important.

REFERENCES

DEVOR, M. (1994) Plasticity in the neonatal and adulat nervous system. In *Neurological Rehabilitation*. ed. ILLIS, L.S. pp. 59-81. Blackwell Scientific Publications, Oxford.

ILLIS, L.S. (1963) Changes in spinal cord synapses and a possible exlanation for spinal shock. *Exp. Neurol.* **8**, 328-335.

ILLIS, L.S. (1967) The motoneurone surface and spinal shock. In *Modern Trends in Neurology*. ed. WILLIAMS, D. Butterworths, London.

ILLIS, L.S. (1969) Enlargement of spinal cord synapses after repetitive stimulation. *Nature*. **223**, 76-77.

ILLIS, L.S. (1994) Rehabilitation theory. In *Neyrological Rehabilitation*. ed. ILLIS, L.S. pp. 33-44. Blackwell Scientific Publications, Oxford.

ILLIS, L.S. & MITCHELL, J.(1970) The effect of tetanus toxin on boutons termineaux. *Brain Res.* **18**, 283-295.

Wall, P.D. (1989) Recruitment of ineffective synapses after injury. In *Functional Recovery in Neurological Disease*. ed. WAXMAN, S.G. pp. 387-400. Raven Press, New York.

MECHANISMS UNDERLYING MUSCLE SYNERGY STUDIED IN MAN

J. Gibbs, L.M. Harrison, M.J. Mayston
and J.A. Stephens

Department of Physiology, UCL
Gower Street, London WC1E 6BT, UK

One simple mechanism that could account for the co-contraction of synergistic muscles is that it is brought about by the sharing of common excitatory drive. To test this hypothesis we have performed cross-correlation analysis of motor unit discharges recorded during steady voluntary contractions in man (Bremner, Datta & Stephens, 1989; Bremner, Baker & Stephens, 1991b; Carr, Harrison & Stephens, 1994). The presence of a shared input is indicated by a narrow central peak in the cross-correlogram constructed from the times of occurrence of motor unit spikes in the two muscles. An example is shown in Figure 1A for gastrocnemius and soleus (2 - 3 units in each record). Correlogram peaks have also been found for bilateral homologous muscles such as masseter, diaphragm and rectus abdominis which are normally active together to produce movements symmetrical about the mid-line. Correlogram peaks have not been found when recording from muscle pairs that do not share a common action about a common joint or axis (Gibbs, Topham, Mackenzie, Harrison & Stephens, 1994b). From these experiments a simple new generalisation has emerged for the organisation of synaptic drive to motoneurones which we call the *sharing principle* - motoneurones innervating muscles that share a common mechanical action share a common presynaptic input. The converse is also true - motoneurones innervating co-contracting muscles that do not share a common action do not share a common input (Gibbs, Harrison & Stephens, 1995). This extends to different motoneurone pools the generalisation envisaged by Henneman to govern the organisation of inputs to a single motoneurone pool - motoneurones innervating a single muscle share common reflex inputs (reviewed by Henneman, 1980). This is suported by the results of cross-correlation analysis which show that motoneurones innervating the same muscle share common presynaptic input (Bremner, Baker & Stephens, 1991a); a mechanism that ensures that during individual muscle contractions motor units act synergistically.

In recent experiments we have investigated the possibility that a counterpart inhibitory mechanism may exist that would bring about the reciprocal action of antagonistic muscles. Central cross-correlogram troughs have been found in cross-correlograms constructed from the times of occurrence of motor unit spikes recorded during voluntary co-contraction of tibialis anterior and soleus, biceps and triceps, and index flexor digitorum sublimis and index extensor digitorum communis muscles using monopolar concentric needle elctrodes. (Gibbs, Harrison, Mayston & Stephens, 1994a). An example is shown in Figure 1B for tibialis anterior and soleus, plotting the probability of occurrence of soleus motor unit spikes before and after each tibialis anterior motor unit spike (2-3 units in each record).

One mechanism that could account for such cross-correlogram troughs would be that some last order input fibres that excite motoneurones innervating one muscle branch to excite inhibitory interneurones that innervate motoneurones supplying the other muscle (Moore, Segundo, Perkel & Levithan, 1970). To test this idea we have extended the theoretical framework of Kirkwood in Kirkwood & Sears (1978) to include this double reciprocal inhibitory connection. The solid line in Figure 1B plots out the expected time course for the cross-correlogram with EPSP and IPSP parameters 10-90% rise time 1.97 ms, half width 7.85 ms and primary correlation kernal derivative/non derivative operator ratio 0.8. For a given set of EPSP/IPSP and primary correlation kernal operators we have calculated the expected time course for the reciprocal connection in each direction and have then performed a multiple linear correlation between the correlogram as the dependent variable and the two time courses to make a least squares estimate of their relative strength.

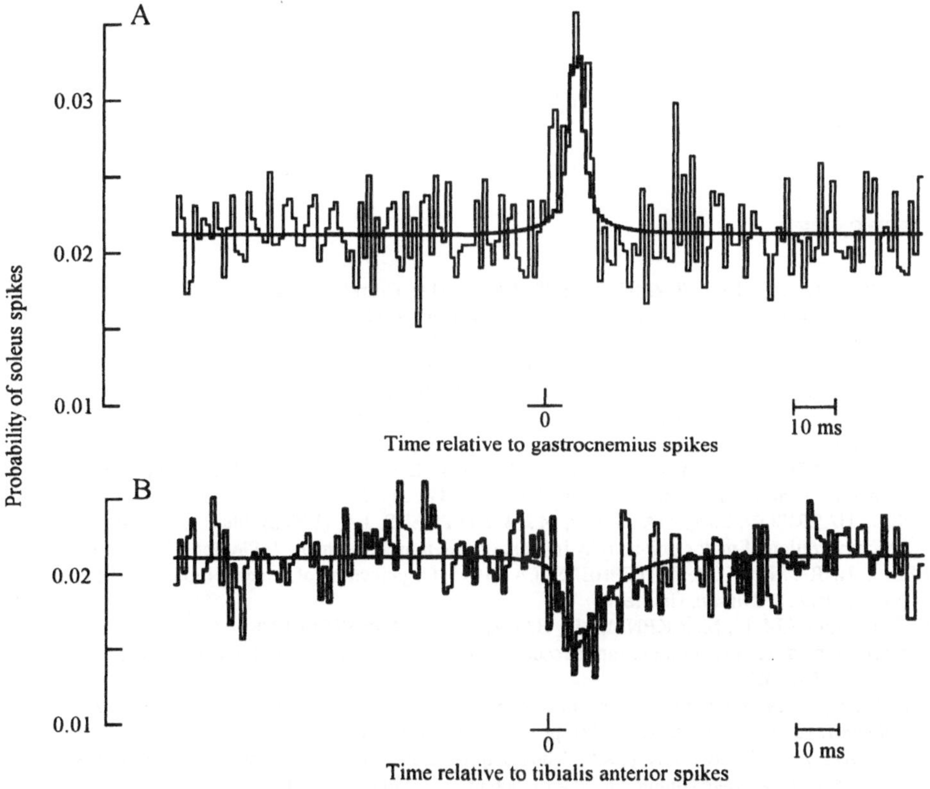

Figure 1. A. Probability of occurrence of motor unit spikes recorded from soleus before and after the occurrence of motor unit spikes recorded from gastrocnemius during steady voluntary plantar flexion of the ankle. B. Probability of occurrence of motor unit spikes in soleus before and after the occurrence of spikes recorded from tibialis anterior during voluntary co-contraction of the two muscles. Same subject in A & B. Continuous line in A & B are best fit values for the time course of correlation calculated using equations developed from the theoretical model by Kirkwood in Kirkwood & Sears (1978). Same parameters in each case (see text for details). Bin width 1 ms. 5000 spikes.

For the example in Figure 1B the reciprocal connection from soleus to tibialis anterior was not found to make a significant contribution. This is reflected in the asymmetrical time course of the original correlogram trough which begins steeply and then returns to the baseline more slowly towards the right. Had the reciprocal connection been in the opposite direction then the correlogram trough would have begun sloping steeply in the opposite direction and then returned towards baseline more slowly to the left. If the double reciprocal inhibitory connection had been equally powerful in both directions then the expected correlogram would have been symmetrical on each side.

The size of the cross-correlogram trough for the antagonist muscle pair tibialis anterior and soleus in Figure 1 is similar to that of the cross-correlogram peak for the synergists gastocnemius and soleus (cf Figure 1A & B). Not only does the counterpart inhibitory mechanism producing common drive for synergistic motoneurones exist for the organisation of last order input innervating motoneurones supplying antagonistic muscles but in this case it had a similar strength. On the basis of similarly sized cross-correlogram peaks, the strength of shared common excitatory drive between motoneurones has been estimated to be about 20% of the total input to each (Bremner et al., 1989; 1991b). A similar figure can thus be estimated for this reciprocal inhibitory drive.

ACKNOWLEDGEMENTS

J. Gibbs was supported by the Scholl Trust. M.J. Mayston was in receipt of an MRC studentship.

REFERENCES

BREMNER, F.D., BAKER, J.R. & STEPHENS, J.A. (1991a). Correlation between the discharges of motor units recorded from the same and from different finger muscles in man. *J. Physiol.* **432**, 355-380.

BREMNER, F.D., BAKER, J.R. & STEPHENS, J.A. (1991b). Variation in the degree of synchronization exhibited by motor units lying in different finger muscles in man. *J. Physiol.* **432**, 381-399.

BREMNER, F.D., DATTA, A.K. & STEPHENS, J.A. (1989). A Mechanism for muscle synergy. In *Perspectives in Motor Control.* eds. HENATSCH, H.D., WINDHORST, U., LAOURIS, Y. & MEYER-LOHMANN, J. pp 44-50. AIM Verlag, Gottingen.

CARR, L.J., HARRISON, L.M & STEPHENS, J.A. (1994). Evidence for bilateral innervation of certain homologous motoneurone pools in man. *J. Physiol.* **475** 217-227.

GIBBS, J., HARRISON, L.M., MAYSTON, M.J. & STEPHENS, J.A. (1994a). Short-term anti-synchronization of motor unit activity in antagonistic muscles in man. *J. Physiol.* **476**, 20-21P.

GIBBS, J., HARRISON, L.M. & STEPHENS, J.A. (1995) Organization of synaptic inputs to motoneurone pools in man. *J. Physiol.* (in press).

GIBBS, J., TOPHAM, L., MACKENZIE, L., HARRISON, L.M. & STEPHENS, J.A. (1994b) Correlation between the discharges of motor units recorded from different muscles in the lower limb in man. *J. Physiol.* **459**, 457P.

HENNEMAN, E. (1980) Organization of the motoneurone pool: the size principle. In *Medical Physiology, Volume 1.* ed. MOUNTCASTLE, V.B. pp. 718-741. Mosby, St. Louis.

KIRKWOOD,P.A. & SEARS, T.A. (1989). The synaptic connexions to intercostal motoneurones as revealed by the average common excitation potential. *J. Physiol.* **275** 103-134.

MOORE, G.P., SEGUNDO, J.P., PERKEL, D.H. & LEVITAN, H. (1970). Statistical signs of synaptic interaction in neurons. *Biophys. J.* **10** 876-900.

CHANGES IN MOTONEURONE CONNECTIVITY ASSESSED FROM NEURONAL SYNCHRONIZATION ANALYSIS

A. Schmied[1], J.-P. Vedel[1], J. Pouget[2],
R. Forget[3], Y. Lamarre[3] and J. Paillard[1]

[1]"Physiologie et Physiopathologie
Neuromusculaire Humaine"
CNRS-NBM, 13402 Marseille Cedex 20, France
[2]Centre Hospitalier Universitaire-La Timone
Marseille, France
[3]Université de Montréal, Montréal, PQ, Canada

In the first large scale analysis of synchronization of motor unit activity, Buchtal & Madsen (1950) reported that the incidence of synchronization was much higher than expected by chance in the hand muscles of normal subjects and was significantly increased in other muscles in cases of muscular atrophy of central origin. The first theoretical and experimental evidence concerning the synaptic origin of synchronization of discharges of mammalian motoneurones was established by Sears & Stagg (1976). Further evidence has been provided since and thoroughly reviewed (Kirkwood, 1979; Kirkwood & Sears, 1991).

SHORT-TERM AND BROAD PEAK SYNCHRONIZATION IN ANIMALS

Whatever their origin, single afferents are likely to make connections with numerous motoneurones in a given muscle pool, virtually all of them in the case of Ia afferents from muscle spindles (Mendell & Henneman, 1971). In response to simultaneous excitatory post-synaptic potentials (EPSPs) generated at terminals of branched axons, target motoneurones may fire synchronously, mostly during the rising phase of EPSPs. Given its expected brief time-course, this synchronization process is usually referred to as short-term synchronization (Sears & Stagg, 1976) and in principal occurs independently of input synchrony.

If afferents to motoneurones are presynaptically synchronized by their own common inputs, they will generate near-simultaneous EPSPs in the target motoneurones which may also result in synchronization. The resulting coupling is expected to be looser over a longer time scale than in the case of short-term synchronization.

In cross-correlograms which describe the probability of firing of one motoneurone with respect to the firing of another, synchronous action potentials are grouped in central peaks, the time course of which depends on the synchronizing synaptic processes involved. In the case of respiratory motoneurones recorded in anaesthetised cats, the narrowest peaks (half-width: 1.5 - 6 ms) were interpreted initially in terms of short-term synchronization (Sears & Stagg, 1976; Kirkwood & Sears, 1978; Kirkwood, Sears, Tuck & Westgaard, 1982). Broader peaks (up to 20 ms) that were too long to result from EPSPs generated by branched axons were considered to reflect the action of inputs (common or not) that were synchronized at a premotoneuronal level (Kirkwood et al., 1982). This synchronization process was subsequently referred to as broad-peak synchronization (Kirkwood et al., 1982). There is no clear-cut boundary, however, between the maximum and minimum durations that can be expected in the case of short-term or broad peak synchronization.

MOTOR UNIT SYNCHRONIZATION IN HUMANS

In healthy subjects, motor unit synchronization has been reported to occur quite extensively in all the muscles tested. The most striking observation was the large range of peak durations (3 - 30 ms), with mean values around 10 ms (Adams, Datta & Gutz, 1989; Powers, Vanden Noven & Rymer, 1989; Davey, Ellaway, Friedland & Short, 1990; Datta & Stephens, 1990; Bremner, Baker & Stephens, 1991; Schmied, Ivarsson & Fetz, 1993). Although this was clearly broader than the values observed for short-term synchronony in the animal studies referred to above, the correlogram peaks observed in human limb muscles have been generally interpreted as also reflecting the action of shared inputs. Differences in conduction time of the common input terminal branches, differences in the processing time of EPSPs at the motoneurone membrane and differences in conduction times both of the motor axons themselves and of their terminal arborizations in the muscle may all contribute to some dispersion of motor unit synchronization.

On this basis, the amplitude of the synchronization peaks was used as an index to assess the strength of the supposed common inputs to motoneurones. Short-term synchronization was reported to be greater when motor units were recorded in the same muscle rather than in synergist or unrelated muscles (Powers et al., 1989; Bremner et al., 1991), greater in distal than in proximal muscles (Bremner et al., 1991) and greater when motor units had similar recruitment thresholds or were slowly contracting (Datta & Stephens, 1990; Schmied et al., 1993). This suggested the existence of a gradient in the synaptic efficacy of common inputs among the different motoneurone pools and among the different motoneurone types. The picture, however, was complicated by the observation that the synchronization peaks tended to be broadest in the case of high threshold and fast contracting motor units (Datta & Stephens, 1990; Schmied et al., 1993).

MOTOR UNIT SYNCHRONIZATION IN RELATION TO MUSCLE USE

Using a global method to assess the degree of motor unit synchronization, Milner-Brown, Stein & Lee (1975) observed that synchronization was most prominent in the hand muscles of weight-lifters or manual labourers. They also showed that synchronization could be enhanced after 6 weeks of intensive muscle exercise and proposed that "supraspinal connections from the motor cortex directly to motoneurones may be enhanced as a result of training". Recently, however, the surface EMG index of synchronization used in this study has been shown to be biased by the level of muscle activity (Yue, Fugelvand, Nordstrom & Enoka, 1992).

In order to reinvestigate the possible influence of muscle use on motoneurone connectivity as revealed by motor unit synchronization, we compared the characteristics of synchronization peaks detected in the cross-correlation histograms computed with the discharges of pairs of motor units recorded in the left and right extensor carpi radialis muscles (ECR) of left-handed and right-handed male subjects (Schmied, Vedel & Pagni, 1994).

In the case of the left-handers as well as the right-handers, the incidence of significant synchronization peaks was higher in the preferred arm (Figure 1A). Upon pooling the data, the incidence of synchronization was clearly higher in the preferred arm, irrespective of handedness. Arm preference was associated with marked changes in the duration and amplitude of synchronization peaks: the peaks were much wider and larger in the preferred arm, as shown in Figure 1B, C & D. Examination of the relationships between synchronization and motor unit discharge characteristics confirmed that the stronger motor unit synchronization observed in the preferred arm could not be related to coincident changes in frequency and variability of motor unit discharges (Schmied et al., 1994). The distribution of the widths of synchronization peaks in the preferred and the non-preferred arms is shown in the Figure 2A. In the preferred arm (hatched histogram), a population of rather narrow peaks (1.5 to 6 ms) could be distinguished from broader peaks ranging from 6 to 25 ms. In the non-preferred arm, narrow peaks were predominant. The increase in amplitude found in the whole population of synchronization peaks in the preferred arm (Figure 2B, left) was still present when the comparison was restricted to narrow peaks (Figure 2B, right).

These results suggest that the preferential use of one arm is associated with change in the synaptic processes revealed by the synchronization of motor unit discharges. The

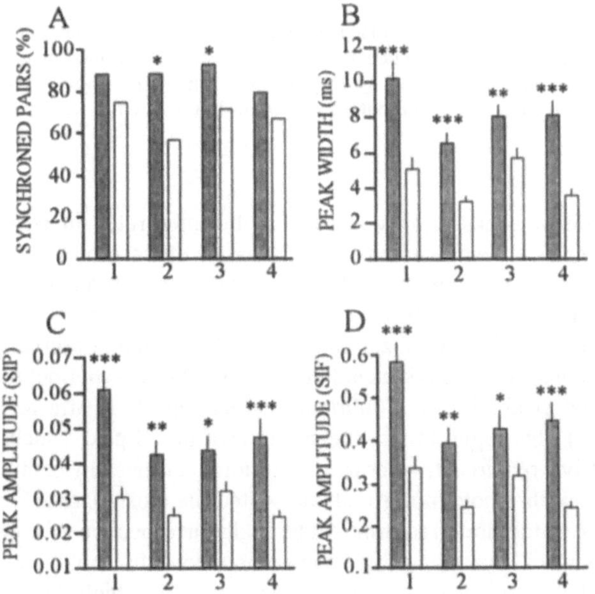

Figure 1. Motor unit synchronization in relation to muscle use in two left-handed (1 & 2) and two right-handed (3 & 4) subjects. A. Rate of occurence. B. Mean duration. C. Mean area of the synchronization peaks normalised with respect to the number triggers. D. As in C but normalised with respect to the recording duration. Data for motor unit pairs in the preferred (hatched bars) and non-preferred arm (open bars) of each subject. Levels of signficance. *: 0.01<p<0.05; **: 0.001<p<0.01; ***: p<0.001; ****: p<0.0001.

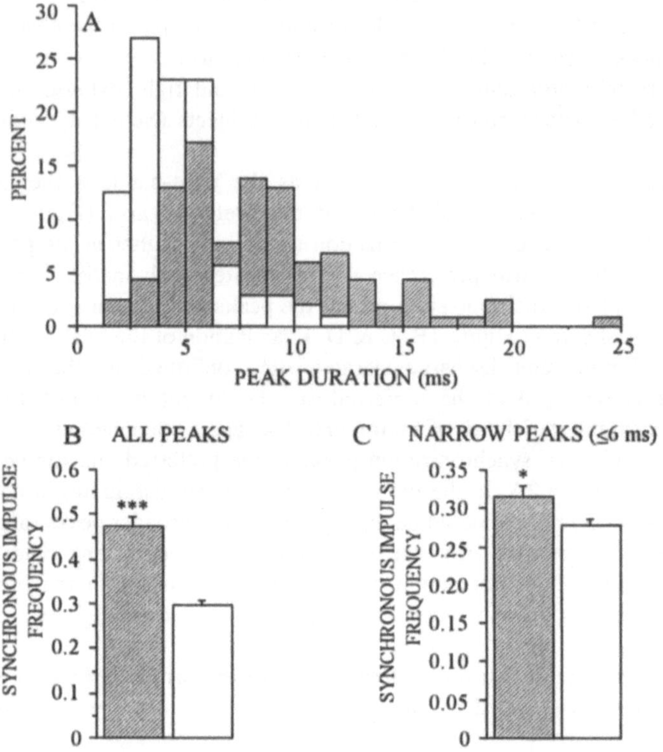

Figure 2. Changes in synchronization time course in relation to muscle use for motor unit pairs in preferred arm (hatched bars) and non-preferred arm (open bars). A. Distribution of the duration of synchronization peaks. B. Strength of synchronization assessed by the peak area normalised with respect to the record duration for the whole population. C. As in B but only for peaks lasting < 6 ms. Levels of signficance as for Figure 1.

increase in amplitude of peaks narrow enough to be interpreted in terms of short-term synchronization suggests an enhancement of the synaptic efficacy of the common inputs to motoneurones. The same mechanism could also explain the broadening of the peaks in terms of a stronger presynaptic synchronization of inputs to motoneurones.

The higher incidence of synchronization and peak broadening observed in the preferred arm affected predominantly the pairs including one or two fast-contracting high-threshold motor units (Schmied et al., 1994). Confirming earlier reports (Datta & Stephens, 1990; Schmied et al., 1993), this suggests that the strength of broad-peak synchronization might be related to the drive required to recruit these motoneurones. In contrast, the pairs of slowly-contracting low-threshold motor units presented the most marked increases in short-term synchronization in the preferred arm. Additionally, irrespective of arm preference, the narrow peaks produced by the slowly-contracting low-threshold motor units were significantly larger than those produced by the fast-contracting high-threshold motor units. Assuming that these peaks reflect mainly the action of the inputs shared by the different motoneurone types, this would indicate that, during voluntary contraction, motoneurones are controlled by inputs distributed with a gradient similar to what has been reported in the case of the muscle spindle primary afferents (Burke, Rymer & Walsh, 1976). These data strongly advise to take into account handedness in any clinical study of motor unit synchronization.

CONTRIBUTION OF PRIMARY SENSORY INPUTS TO MOTOR UNIT SYNCHRONIZATION

The widely branched muscle spindle primary afferents could contribute to short-term synchronization by providing excitatory common inputs to most of the motoneurones supplying homonymous muscles (Mendell & Henneman, 1971). Inhibitory and excitatory interneurones which relay to motoneurones the inputs generated by the afferents from tendons, skin, joints and muscles might also contribute, depending on their terminal branching throughout the motoneurone pools.

In decerebrate cats, a high degree of synchronization was observed between the discharges of alpha motoneurones reflexly activated by cutaneous stimulation (Connell, Davey & Ellaway, 1986). In anaesthetised cats, however, dorsal root section did not suppress the short-term synchronization of respiratory alpha motoneurones (Kirkwood et al., 1982) since it depended on the central drive as well as on peripheral inputs. A similar observation was made in a patient who had lost all the large myelinated afferents below the neck following a post-infective neuropathy. The 3 pairs of motor units tested in the first dorsal interosseus muscle of this patient presented synchronization peaks which were not found to differ from those observed in the same muscle of control subjects (Baker, Bremner, Cole & Stephens, 1988).

Recently, we have had the opportunity to reinvestigate the possible changes in

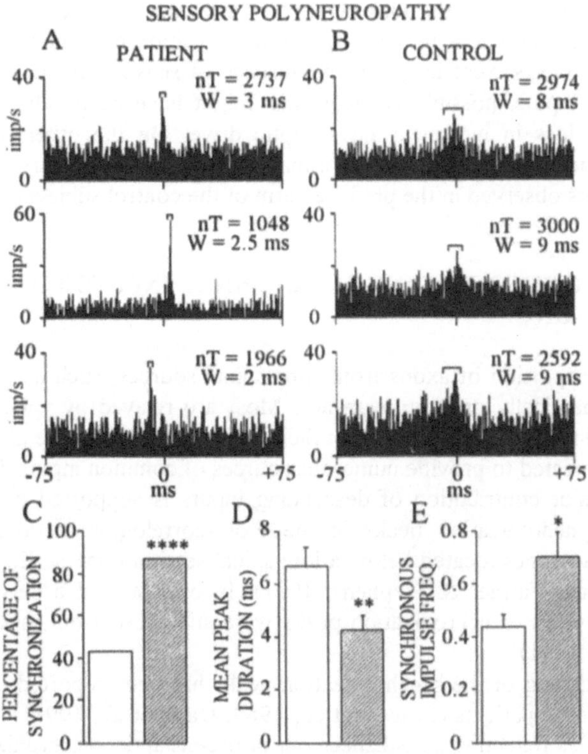

Figure 3. Changes in synchronization associated with sensory polyneuropathy. Examples of cross-correlograms computed in the case of three motor unit pairs in the right arm of a right-handed patient (A) and in right-handed control (B). Significant differences are shown in the comparison of the rate of occurrence (C), the duration (D) and the amplitude (E) of the synchronization peaks of the patient (hatched bars) and in four controls (open bars). Levels of signficance as for Figure 1.

synchronization in a patient showing a total loss (up to the forehead) of large myelinated sensory afferents as a consequence of a Guillain-Barré episode which had occurred 15 years earlier. No sign of neuromuscular dysfunction could be found in a recent clinical examination. Synchronization was tested by cross-correlation analysis of the discharges of motor unit pairs recorded in the right extensor carpi radialis muscles (e.g. in the preferred arm) during a mild clenching of the hand for a duration of 1 to 4 min. A total of 29 pairs were tested in the patient (female) and 67 pairs in four age-matched female controls.

Examples of the cross-correlograms observed in the patient (left) and in one control (right) are shown in the Figure 3A. In the patient, the peaks were apparently narrower and higher than in the control. This was confirmed by the quantitative analysis of the data shown in the Figure 3B. In the patient, the synchronization peaks were significantly more frequent, predominantly short (4 out 25 were longer than 6 ms as compared to 13 out 28 in the controls) and consistently larger. Examination of the motor unit discharge characteristics confirmed that the difference in the strength of synchronization could not be accounted for by differences in the frequency and the variability of motor unit activity. The increase in the strength of synchronization observed for the whole population of peaks detected in the patient was still present when the population of peaks lasting less than 6 ms were compared, that is, those which can be the most confidently ascribed to common inputs to motoneurones.

Although it should be noted that an unknown amount of synaptic reorganization may have occurred in the 15 years following the degenerative episode, the marked changes in synchronization associated with the loss of primary afferents suggest that these inputs do not necessarily play a major role in short-term synchonization of motor unit activity. The fact that the narrow peaks were larger in the absence of sensory inputs suggests that the unaffected common inputs, possibly corticospinal, might have been enhanced in order to compensate for the loss in motoneurone synaptic drive. On the other hand, the lower incidence of broad peaks suggests that the primary afferents might contribute to the broad synchronization peaks observed in the preferred arm of the control subjects.

CONTRIBUTION OF DESCENDING SUPRASPINAL INPUTS TO MOTOR UNIT SYNCHRONIZATION

Only a small proportion of axons from supraspinal sources, such as the corticospinal axons, act monosynaptically on motoneurones. Most are relayed by spinal interneurones and/or propriospinal neurones. Depending on their final divergence, these direct and indirect pathways may be expected to provide numerous sources of common inputs during voluntary contraction. The major contribution of descending inputs is supported by the systematic disappearance of synchronization peaks in the cross-correlograms computed from the discharges of motoneurones located below a low spinal section (Davey, Ellaway, Friedland & Short, 1990; Datta, Farmer & Stephens, 1991). In one case of a cervical lesion, the effects were less obvious: synchronization peaks were still present but much broader than normal (Datta et al., 1991).

A similar broadening of the synchronization peaks has been reported in various cases of strokes and in paretic patients (Powers et al., 1989; Datta et al., 1991). This broadening has been considered as reflecting an enhancement of pre-synaptic synchronization involving spinal interneurones, after the release of some supraspinal inhibitory control.

A much lower incidence of synchronization has been observed, however, in paretic patients as compared to controls. This suggests a more direct contribution of supraspinal inputs in the synchronization of motoneurone activity (Powers et al., 1989). This was confirmed in a recent study of motor unit synchronization in 4 cases of amyotrophic lateral sclerosis (ALS) and one case of primary lateral sclerosis (PLS). ALS is a degenerative disease which affects motoneurones and their supraspinal inputs, particularly the corticospinal ones. In PLS, the loss is mainly restricted to the corticospinal pathway.

Motor unit pairs were recorded in the right extensor carpi radialis muscles of five right-handed patients (3 females, 2 males) and of five age-matched right-handed controls (3 males, 2 females), during a wrist isometric extension for a duration of 1 to 4 minutes.

In both groups of patients, the incidence of synchronization was consistently lower than in the controls (Figure 4C). Typical examples of cross-correlograms observed in one ALS patient (left) and in one control (right) are shown in Figure 4A & B. In this patient, only 1 out of 22 motor unit pairs presented a significant synchronization peak as compared to 29 out of 34 pairs in the case of the control. The synchronization peaks observed in the ALS patients and in the controls did not differ significantly in duration and amplitude, as shown in Figure 4D & E. The same observation was made in the PLS patient.

The lower incidence and, in some cases, the total disappearance of motor unit synchronization observed in these patients suggests that the corticospinal pathway, which is disrupted in both diseases, may make a major contribution to the synchronization of motoneurone activity during voluntary contraction.

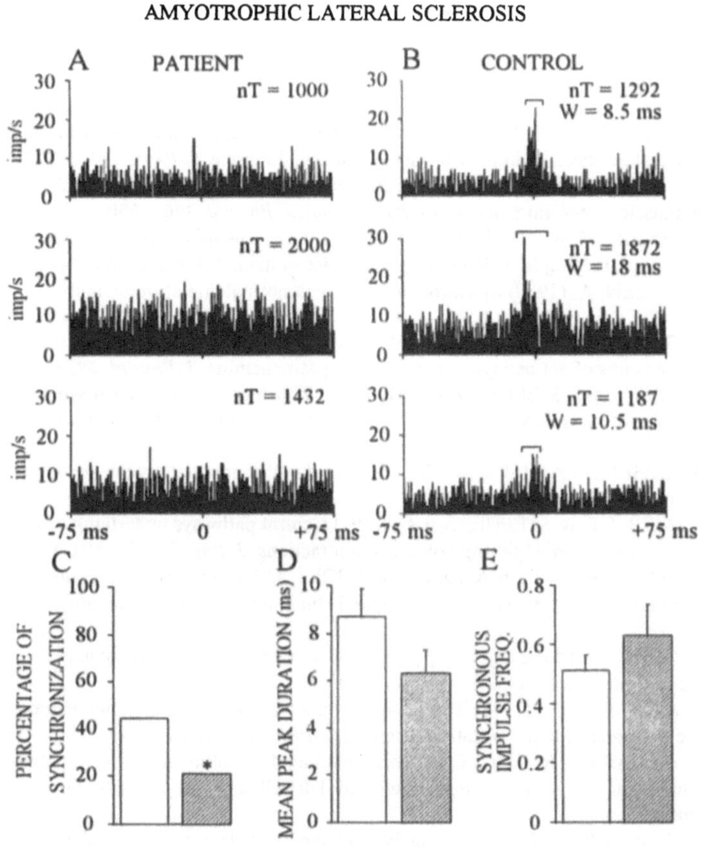

Figure 4. Changes in synchronization associated with amyotrophic lateral sclerosis. Examples of cross-correlograms computed in the case of three motor unit pairs in the right arm of a right-handed patient (A) and in right-handed control (B). Significant differences are shown in the comparison of the rate of occurrence (C), the duration (D) and the amplitude (E) of the synchronization peaks observed in the patient (hatched bars) and in five controls (open bars). Levels of signficance as for Figure 1.

CONCLUDING REMARKS

Marked changes in motoneurone synchronous activity were observed in relation to muscle use (depending on the subject's handedness). These changes may reflect a structural spinal asymmetry or an activity-dependent modulation of the synaptic efficacy of motoneurone inputs. The marked differences in synchronization observed between the preferred and the non-preferred arm must be taken into account when comparing data obtained in patients and healthy subjects. The marked changes in motoneurone synchronization observed in relation to peripheral and partial central deafferentation suggest that sensory and supraspinal inputs may contribute through distinct processes to the synchronization of motoneurone activity.

ACKNOWLEDGEMENTS

This work was supported by grants from the Direction des Recherches, Etudes et Techniques du Ministère de la Défense (DRET n° 91/199) and from the Association Française contre les Myopathies.

REFERENCES

ADAMS, L., DATTA, A.K. & GUZ, A. (1989) Synchronization of motor unit firing during different respiratory and postural tasks in human sternocleidomastoid muscle. *J. Physiol.* **413**, 213-231.

BAKER, J.R., BREMNER, F.D., COLE, J.D. & STEPHENS, J.A. (1988) Short-term synchronization of intrinsic hand muscle motor units in a 'deafferented' man. *J. Physiol.* **396**, 155P.

BREMNER, F.D., BAKER, J.R. & STEPHENS, J.A. (1991) Variations in the degree of synchronization exhibited by motor units lying in different finger muscles in man. *J. Physiol.* **432**, 381-399.

BUCHTAL, F. & MADSEN, A. (1950) Synchronous activity in normal and atrophic muscle. *Electroenceph. Clin. Neurophysiol.* **2**, 425-444.

BURKE, R.E., RYMER, W.Z., WALSH, J.V. (1976) Relative strength of synaptic inputs from short-latency pathways to motor units of defined type in cat medial gastrocnemius. *J. Physiol.* **39**, 447-458.

CONNELL, L.A., DAVEY, N.J. & ELLAWAY, P.H. (1986) The degree of short-term synchrony between alpha and gamma motoenurones coactivated during the flexion reflex in the cat. *J. Physiol.* **376**, 47-61.

DATTA, A.K. & STEPHENS, J.A. (1990) Synchronization of motor unit activity during voluntary contraction in man. *J. Physiol.* **422**, 397-411.

DATTA, A.K., FARMER, S.F. & STEPHENS, J.A. (1991) Central pathways underlying synchronization of human motor unit firing studied during voluntary contractions. *J. Physiol.* **432**, 401-425.

DAVEY, N., ELLAWAY, P.H., FRIEDLAND, C.L. & SHORT, D.J. (1990) Motor unit discharge characteristics and short-term synchrony in paraplegic humans. *J. Neurol. Neurosurg. Psych.* **53**, 764-769.

KIRKWOOD, P.A. (1979) On the use and interpretation of cross-correlation measurement in the mammalian central nervous system. *J. Neurosci. Meth.* **1**, 107-133.

KIRKWOOD, P.A. & SEARS, T.A. (1978) The synaptic connexions to intercostal motoneurones as revealed by the average common excitation potential. *J. Physiol.* **275**, 103-134.

KIRKWOOD, P.A. & SEARS, T.A. (1991) Cross-correlation analyses of motoneurone inputs in a coordinated motor act. In *Neuronal Cooperativity*, ed. KRÜGER, J., pp. 225-248. Springer Verlag, Berlin-Heidelberg.

KIRKWOOD, P.A., SEARS, T.A., TUCK, D.L. & WESTGAARD, R.H. (1982) Variations in the time course of the synchronization of intercostal motoneurones in cat. *J. Physiol.* **327**, 105-135.

MENDELL , L.M. & HENNEMAN, E. (1971) Terminals of single Ia fibres : location, density and distribution within a pool of 300 homonymous motoneurones. *J. Neurophysiol.* **34**, 171-187.

MILNER-BROWN, H.S., STEIN, R.B. & LEE, R.G. (1975) Synchronization of human motor units: possible roles of exercise and supraspinal reflex. *Electroencephal. Clin. Neurophysiol.* **38**, 245-254.

POWERS, R.K., VANDEN NOVEN, S. & RYMER, W.Z. (1989) Evidence of shared, direct input to

motoneurons supplying synergist muscles in humans. *Neurosci. Lett.* **102**, 76-81.
SCHMIED, A., IVARSSON, C. & FETZ, E.E. (1993) Short-term synchronization of motor units in human extensor digitorum communis muscle: relation to contractile properties and voluntary control. *Exp. Brain Res.* **97**, 159-172.
SCHMIED, A. VEDEL, J-P. & PAGNI, S. (1994) Human spinal lateralization assessed from motoneurone synchronization: dependence on handedness and motor unit type. *J. Physiol.* **480**, 369-387.
SEARS, T.A. & STAGG, D. (1976) Short term synchronization of intercostal motoneurone activity. *J. Physiol.* **263**, 357-381.
YUE, G., FUGLEVAND, M.A., NORDSTROM, M. & ENOKA, R.M. (1992). Effects of muscle activity on surface-EMG estimates of motor unit synchronization. *Annual Meeting Neuroscience Society.* Abstr. 590-12.

DISTURBANCES OF MOTONEURONE DISCHARGE IN MUSCLES OF SUBJECTS WITH HEMIPARETIC STROKE

W.Z. Rymer & J.J. Gemperline

Sensory Motor Performance Program
Rehabilitation Institute of Chicago
345 east Superior Avenue
Chicago, Illinois, 60611, USA

INTRODUCTION

Following injury to the cerebrum or spinal cord, human subjects experience a cluster of clinical signs widely described as the "upper motoneurone syndrome". This syndrome consists of "spasticity", (which is an increase in muscle tone), muscular weakness, and disturbances in movement coordination. The mechanisms of muscular weakness have received relatively limited scrutiny. Hughlings Jackson (see Lassek, 1970) proposed that they resulted from a loss of net descending excitatory drive from cortical centres, and (in our view), this is quite likely to be a factor. However, it is also possible that changes in the pattern of motoneurone activation may occur following lesions to cortex or descending pathways, and that these changes may contribute to the disturbances in voluntary force generation, which include muscle weakness, increased effort and an exaggerated sense of fatigue.

There have been a number of studies that have evaluated changes in motoneuronal discharge patterns in subjects with cerebral or spinal cord lesions. These studies, which include those of Rosenfalck & Andreassen (1980), Freund, Dietz, Wita & Kapp (1973), Wiegner, Wierzbicka, Davies & Young (1993), and other authors report systematic changes in firing rate of motor units in upper motor neurone lesions, and in some instances, alterations in higher order statistics of interspike interval, including changes in interval serial correlation statistics and in joint interval distributions. While these abnormalities in unitary discharge have been described in a variety of CNS disorders (such as stroke, spinal cord injury and Parkinsons Disease), there is no current assessment of their overall incidence in specific neurologic lesions, nor is the magnitude of their effect on physical performance clear. Finally, the potential mechanism(s) of such changes have not been established.

In the present study, we report on comparisons made between paretic muscles of hemiparetic stroke subjects and contralateral non-paretic muscles during voluntary isometric contraction. Specifically, we compared surface electromyographic recordings for biceps brachii muscles in both the paretic and contralateral limbs, at several joint torque levels, in six adult hemiparetic subjects. In addition, we obtained intramuscular recordings of single motor unit discharge from both paretic and contralateral muscles, and derived estimates of mean motor unit firing rate over a range of static forces and (in some instances) during ramp isometric force changes as well.

Our findings were that in 3 of 6 subjects, there were systematic reductions in mean firing rates in paretic muscles for matched joint torques. We also observed alterations in the patterns of rate change during ramp increases in voluntary force, and compression of motor unit recruitment into a narrower band of isometric forces than was evident on the contralateral side.

METHODS

We studied six hemiparetic subjects, in whom the paresis followed a stroke (in 5) or traumatic brain injury (1). Subjects were chosen on the basis of a unilateral lesion producing significant impairment of voluntary force in the contralateral upper limb. Subjects were able to comprehend instructions and to participate voluntarily in this study. (Each subject gave informed consent in writing). Evidence of prior stroke, of concurrent neurologic deficit, of cognitive deficit, of significant receptive aphasia, or of cardiovascular impairment sufficient to degrade the subject's capacity to participate in our study provided grounds for exclusion from the study.

Subjects were seated with the shoulder abducted to 60° and the elbow flexed to 90°. The forearm was attached to a supporting frame, and the arm was mounted above a load cell, which was placed at the centre of rotation of the elbow. This displayed the torque generated at the elbow. We recorded surface EMG from the biceps brachii, triceps brachii, and brachioradialis, using disc electrodes placed over the muscles in a consistent location, referenced to bony landmarks. EMG signals were amplified and then band-pass filtered (20 Hz to 600 Hz). In addition, to guarantee constancy of recording conditions, we recorded elbow and shoulder joint angles with a goniometer.

Surface EMG was sampled at 2.5 kHz by A/D converters, via a National Instruments Data Acquisition system, running under LABVIEW software within a MacIntosh computer. Joint torque was filtered at 40 Hz, and also sampled at 2.5 kHz. Surface EMG was processed by calculating the mean root mean square (RMS) values for steady torque intervals of 0.5-1 second.

Single motor units were recorded using 25 μm wire intramuscular teflon-covered stainless steel wires. 3 to 5 fine wires were inserted into the muscle using a 27 gauge needle. Recordings were made after preamplification (x 300), and initial bandpass filtering (100 Hz-10 kHz). Unitary data were digitized at 12.5 kHz. Wires were retracted progressively until single motor units were visible. After digital acquisition, data was stored immediately on a computer hard disk.

After data compression to remove baseline records, and additional high-pass digital filtering, single motor units were identified by an interactive program which relied on templates to separate out individual single motor units. Each new motor unit action potential (MUAP) was initially assigned a different template, and each subsequent spike was then compared against the catalog of templates. If the MUAP fell within a designated least squares error limit, the MUAP was assigned to that particular motor unit train. All the units in a train were stored separately. Superpositions were resolved by developing linear combinations of existing templates. These identification procedures produced spike trains of identified single motor units, typically containing 50 or more intervals, from which recruitment force, initial firing rate, rate modulation sensitivity and other interval statistics could be derived.

RESULTS

We compared surface EMG values and single motor unit discharge characteristics for biceps brachii muscles in paretic and contralateral limbs of hemiparetic subjects over a range of matched elbow joint torques. Our findings were that force-EMG slopes were statistically greater in three of the six subjects, with differences in slope ranging from 355 % to 121% in these three. The other three subjects showed relative reductions in EMG on the paretic side per unit force (80%, 80%, 87%). There were no correlations between the severity of paresis and the presence of or magnitude of shifts in slope of the force-EMG relations.

Recordings of single motor unit discharge were taken from biceps brachii and (in one instance) from brachioradialis muscles as well. Our findings were that in three of the six subjects there were systematic and statistically significant reductions in mean firing rate from these muscles (compared at matched joint torques) across a broad range of joint torques. These differences ranged from a mean decrement of 6.08 imp/s (across all units), to the smallest difference of 2.8 imp/s. Results from 2 of these subjects are shown in Figure 1, which illustrates a broad and stastically significant reduction in firing rates across a substantial torque span (5 and 8 n-m). There were no systematic differences in the statistics of interspike interval trains. Serial correlation coefficients were not routinely different, nor were joint interval (t_i, t_{i+1}) plots.

We estimated the force at which motor units were first recruited by recording the the lowest joint torque at which sustained motor unit discharge was present. Our findings here were that there was a systematic downward shift in recruitment force, (so that more motor units were recruited at lower forces), leading to a compression of recruitment force in all six subjects examined. This compression was manifested regardless of alterations in torque-EMG relations, and of significant reductions in motoneurone firing rates. Figure 2 shows two examples, drawn from different subjects, illustrating the shift of recruitment to lower forces. The recruitment compression is substantial, and there is almost no overlap in either subject.

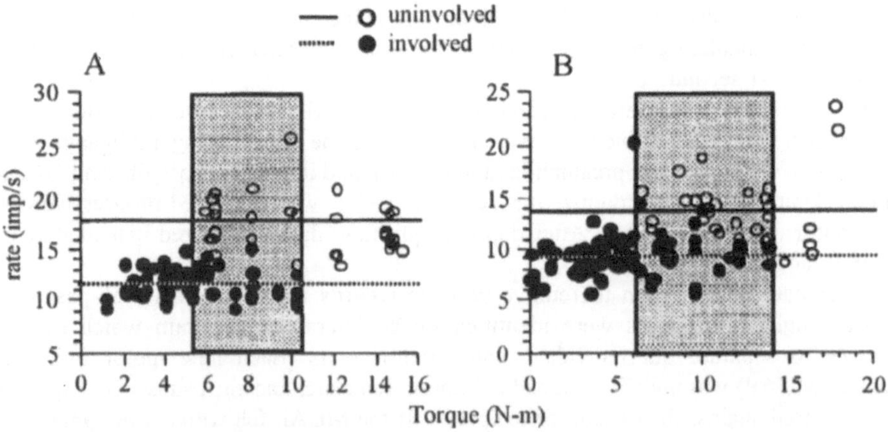

Figure 1. Motor unit rates as a function of steady-state isometric torque for 2 hemiparetic subjects. Stippled areas indicate the region over which motor unit rate comparisons were made. Horizontal lines indicate the mean rate observed in these regions. All motor units recorded from biceps muscle.

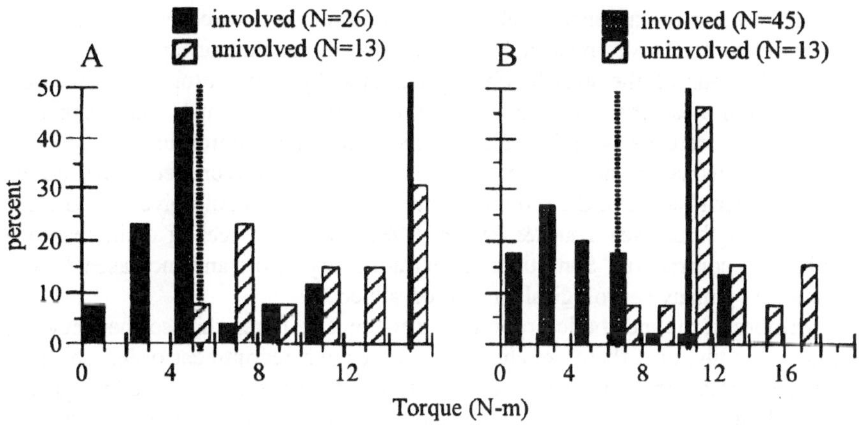

Figure 2. Histogram of lowest steady-state torques at which motor units were recorded for 2 subjects. Solid vertical line in each figure is 25% max. torque for uninvolved side. Dashed vertical line is 25% max. for involved side.

DISCUSSION

Our findings are that 3 of our 6 hemiparetic subjects demonstrated systematic alterations in both the slope of the force-EMG relations and in motor unit discharge rate in the biceps brachii muscles in the paretic limb (relative to the responses of the contralateral biceps muscle). All 6 subjects showed systematic alterations in motor unit recruitment force, displaying a compression of the torque range over which motor units were recruited.

There are two different aspects to these finding that warrant discussion. The first is the potential importance of *alterations in motor unit discharge in mediating the clinical symptoms of weakness and enhanced fatigability*. The second concerns the potential *mechanisms by which these motoneurone changes develop*. Before considering these issues, it is important to determine whether our use of the contralateral limb as a control limb is entirely appropriate.

Comparison between Impaired and Contralateral limb

We chose to draw comparisons between the paretic and contralateral limbs of the same subject (rather than against matched controls) because of the inherent similarity of the muscular arrangement, limb segment lengths, and muscle architecture on the two sides of the same subject. We are fully aware that the contralateral limb of the hemiparetic subject is not entirely normal (from the standpoint of neurologic function). There are often sensory disturbances discernible, and also abnormalities of coordination and reflex function. None the less, to the extent that such abnormalities existed (and they were not detectable on standard clinical examination), they would have probably reduced the differences between the two sides of the same subject, *diminishing* our ability to detect significant differences between the two sides. Our findings in paretic muscles may well have been more dramatic if compared against normal muscles.

Clinical Implications

The alterations in motor unit discharge rate were sometimes substantial, amounting to a 20-30% reduction in mean firing rates (relative to the contralateral limb). If such rate changes take place in the absence of alterations in motor unit twitch properties, then they would give rise to a relative "mismatch" between the motor unit firing properties and the mechanical behaviour of the muscle fibres innervated by that motor axon (see Kernell, 1979). This would mean that motor unit output becomes sub optimal, and a rather bumpy and partially fused tetanus would result. To the extent that motor unit force output is degraded by this mismatch, more motor units would need to be recruited at any designated force level in order to achieve the required motor output. This could give rise to increased EMG activity/unit force (see Tang & Rymer, 1981), and to a feeling of increased effort during voluntary contraction. Sensations of extraordinary effort and increased fatigue are widely reported by many neurologically impaired subjects.

Furthermore, if the metabolic properties of the muscle fibres shift so that they become fatigued more readily, or if there are changes in motor unit recruitment order, so that more of the fatigable motor units are recruited prematurely, then the disturbance in regulation of the muscle will be further exaggerated. At present, the relative magnitude of these effects is not clear, primarily because we have not yet characterized the contractile properties of the activated motor units. This information is vital for any assessment of the functional consequences of motor unit rate alterations.

Pathophysiology of motor unit discharge abnormalities

The mechanisms of these changes in motoneurone rate and recruitment order are presently unknown. There is likely to be a prolonged alteration in limb usage in chronically disabled subjects, which could give rise to a relative or even absolute disuse of the relevant limb muscles. This disuse would promote muscle fibre atrophy, and changes in unit contractile properties (Pierotti, Roy, Bodine-Fowler, Hodgson & Edgerton, 1991; Duchateau & Hainut, 1990) and the concurrent spasticity might also induce unhelpful mechanical changes in muscle properties (such as fibre loss and contracture of aponeuroses and muscle sheaths). While it is conceivable that these rate and recruitment changes are due solely to altered usage patterns, it is quite likely that other central neurological disturbances may also contribute. For example, reorganization of segmental spinal circuits may occur because of loss of descending synaptic input, and there may be altered patterns of interneuronal excitability that may also contribute. While all of the latter factors is important, there is a strong likelihood that at least some component of the reductions may result from *loss of neuromodulator input to the spinal cord*. For example, it is now established that monoaminergic (serotonin and norepinephrine) input to motoneurones and interneurones can influence their input-output properties. Such changes have been actively studied using intracellular recordings from unanesthetised decerebrate preparations (Hounsgard, Hultborn, Jespersen & Kiehn, 1988), or in slice preparations (Hounsgaard & Kiehn, 1985).

At one extreme, systemic administration of a serotonin re-uptake blocker (such as ketanserin) to a decerebrate cat preparation induces high ambient levels of serotonin, which give rise to motoneuronal plateau potentials and to "bistable" behaviour (Carp & Rymer, 1986). Here, transient excitatory synaptic inputs give rise to sustained depolarization, and to sustained motoneuronal discharge. On the other hand, reduction or even elimination of monoaminergic inputs (arising because of direct damage to the raphe or to locus ceruleus gives rise to a loss of neuromodulator mediated spinal facilitation, altering the relations between synaptic current and motoneuronal discharge in potentially adverse ways.

CONCLUSIONS

In many hemiparetic subjects there appear to be alterations in the patterns of motor unit activity in paretic limbs. In some instances, these changes are accompanied by alterations in the relationship between surface EMG and force (or torque), but in others these changes are not evident. Nevertheless, to the extent that abnormal discharge patterns are present, they would certainly promote the sense of exaggerated effort, enhanced fatigability, and weakness that is experienced by many of these subjects.

While the origins of these changes are likely to be complex, because of changes in usage patterns of muscles and intrinsic circuitry in accord, it is quite likely that loss of descending neuromodulator input may contribute significantly to their occurrence.

REFERENCES

CARP, J.S & RYMER, W.Z. (1986) Enhancement by serotonin of tonic vibration and stretch reflexes in the decerebrate cat. *Exp. Brain Res.* **62**, 111-122.

DUCHATEAU, J. & HAINUT, K. (1990) Effects of immobilization on contractile properties, recruitment and firing rates of human motor units. *J. Physiol.* **422**, 55-65.

FREUND, H.J., DIETZ, V., WITA, C.W. & KAPP, H. (1973) Discharge characteristics of single motor units in normal subjects and patients with supraspinal motor disturbances. In *New Developments in Electromyography and Clinical Neurophysiology*. ed. DESMEDT, J.E. pp. 242-250. Basel, Karger.

HOUNSGAARD, J., HULTBORN, H., JESPERSEN, B. & KIEHN, O. (1988) Bistability of α-motoneurones in the decerbrate cat and in the acute spinal cat after intravenous 5-hydroxytryptophan. *J. Physiol.* **405**, 345-367.

HOUNSGAARD, J. & KIEHN, O. (1985) Ca^{++} dependent bistability induced by serotonin in spinal motoneurons. *Exp. Brain Res.* **57**, 422-425.

KERNELL, D. (1979) Rhythmic properties of motoneurons innervating muscle fibers of different speed in m. gastrocnemius medialis of the cat. *Brain Res.* **160**, 159-162.

LASSEK, A.M. (1970) *The Unique Legacy of Doctor Hughlings Jackson*. Charles C. Thomas, Springfield.

PIEROTTI, D.J., ROY, R.R., BODINE-FOWLER, S.C., HODGSON, J.A. & EDGERTON, V.R. (1991) Mechanical and morphological properties of chronically inactive cat tibialis anterior motor units. *J. Physiol.* **444**, 175-192.

ROSENFALCK, A. & ANDREASSEN, S. (1980) Impaired regulation of force and firing patterns of single motor units in patients with spasticity. *J. Neurol. Neurosurg. Psychiat.* **43**, 907-916.

TANG, A. & RYMER, W.Z. (1981) Abnormal force-EMG relations in paretic limbs of hemiplegic human subjects. *J. Neurol. Neurosurg. Psychiat.* **8**, 690-698.

WIEGNER, A.W., WIERZBICKA, M.M., DAVIES, L. & YOUNG, R.R. (1993) Discharge properties of single motor units in patients with spinal cord injuries. *Muscle Nerve* **16**, 661-671.

DO ALTERED POTASSIUM CHANNELS CAUSE THE FASCICULATIONS AND CELL DEATH IN MOTOR NEURONE DISEASE?

H. Bostock[1], M.K. Sharief[2],
G. Reid[1] and N.M.F. Murray[2]

[1]Sobell Department of Neurophysiology
Institute of Neurology, Queen Square
London WC1N 3BG, UK
[2]Department of Clinical Neurophysiology
National Hospital for Neurology and Neurosurgery
Queen Square, London WC1N 3BG, UK

Fasciculations due to spontaneous discharges of motor units, are a characteristic feature of motor neurone disease (amyotrophic lateral sclerosis, ALS). They are multifocal in origin, arising from distal or proximal axon or soma in different motor neurones. To explore the nature of the altered membrane properties responsible for these discharges, we have measured the responses of the axons to subthreshold polarizing currents by the method of 'threshold electrotonus' (Bostock & Baker, 1988).

The study was carried out with the approval of the Joint Medical Ethical Committee. Excitability was tested with 1 ms current pulses applied over the ulnar nerve at the wrist, adjusted in amplitude to keep the compound motor action potential of abductor digiti minimi constant at ca. 30% of maximal. The nerve was stimulated at 1 Hz and five stimulus conditions tested in turn: 1 ms test stimulus alone (control), and test stimulus superimposed on 100 ms conditioning stimuli, set to 20%, -20%, 40% and -40% of the control stimulus. The starting time of the conditioning stimuli was stepped from 2 ms after the test stimulus to 198 ms before it, over a period of 15-20 min. Recordings were made from 15 normal controls, 11 patients with definite ALS (i.e. with loss of both upper and lower motor neurones), 6 with fasciculations but no evidence of motor neurone pathology, and 19 with denervation due to various lower motor neurone disorders.

The responses in ALS patients to the 40% depolarizing currents were of two types: some were of normal shape, but with a greater slow increase in excitability than normal (Type 1), while others were more unusual, with abrupt or profound decreases in excitability, suggesting transitions to a less excitable, depolarized state (Type 2). The threshold reduction occurring 10-15 msec after the start of the 40% depolarizing current clearly separated Type 1 ($75.0 \pm 3.1\%$, mean $\pm$ s.e., $n = 7$) and Type 2 ($23.9 \pm 12.3\%$, $n = 4$) ALS responses from normal controls (64.1 ± 1.1%, n = 15), and from patients with benign fasciculations (65.2 ± 2.3%, n = 6) or lower motor neurone disorders (64.6 ± 0.9%, n = 19).

We have found that the Type 1 and the Type 2 ALS responses can be reproduced in rat axons *in vitro* by successive action of the potassium channel blockers 4-aminopyridine and tetra-ethylammonium, and in a model human motor axon (Bostock, Baker & Reid, 1991) by successive reduction of fast and slow potassium conductances. When the potassium currents became insufficient to counteract the steady state inward current generated by internodal sodium channels, the internodes depolarized regeneratively. Regenerative depolarization by this mechanism may underlie not only the Type 2 threshold electrotonus responses, but also the fasciculations and cell death in ALS.

REFERENCES

BOSTOCK, H. & BAKER, M. (1988) Evidence for two types of potassium channel in human axons in vivo. *Brain Res.* **462**, 354-358.

BOSTOCK, H., BAKER, M. & REID, G. (1991) Changes in excitability of human motor axons underlying post-ischaemic fasciculations: evidence for two stable states. *J. Physiol.* **441**, 537-557.

Editorial Note: The above paper was presented in full at the Symposium but is here given in abstract form only at the authors' wishes.

PROPRIOCEPTION IN HUMAN BASAL GANGLIA DISORDERS

Frederick W.J.Cody[1] and Chris Rickards[2]

[1]School of Biological Sciences
University of Manchester
[2]Department of Neurology
Manchester Royal Infirmary
Manchester, UK

INTRODUCTION

Several lines of evidence indicate that both reflex and voluntary motor responses to proprioceptive input are abnormal in human basal ganglia disorders. Long-latency stretch reflexes are pathologically enhanced in Parkinson's disease (PD; Tatton & Lee, 1975; Cody, MacDermott, Matthews & Richardson, 1986) and may be reduced or absent in Huntington's disease (HD; Noth, Podoll & Friedemann, 1985). Additionally, the finding of Moore (1987) that Parkinsonian patients with asymmetrical disease overestimate the trajectory of the more bradykinetic limb when attempting to match slow, active movements of the two arms suggests a disturbance of proprioceptive guidance in basal ganglia dysfunction.

In the present experiments we have tested this possibility by applying muscle vibration to interfere with the natural patterns of proprioceptive discharge generated during voluntary wrist movements and have compared the vibration-induced trajectory errors of PD and HD patients with those of healthy subjects.

METHODS

In the main series of experiments twenty-nine patients with idiopathic PD and twenty-three age-matched control subjects were studied. Four HD patients were also investigated. All subjects participated with their informed consent and Local Ethical Committee approval.

Subjects grasped the handle of a manipulandum. They were initially trained, using a visual feedback monitor, to make reproducible extension movements of the wrist at a steady target velocity of 8.9°/s against a load of approximately 15% of their individual maximum. A visual "go" cue signalled the required onset time of movement. After training, visual feedback was withdrawn and subjects were denied sight of the moving arm. Vibration

(100Hz; 0.7mm peak-to-peak) was applied transcutaneously to the tendon of flexor carpi radialis (see Cody et al., 1986) throughout 50% of the experimental trials to stimulate antagonist muscle spindle afferents whilst movement trajectories and wrist extensor and flexor surface EMGs were recorded; vibrated (V) and non-vibrated (NV) trials were interspersed pseudorandomly. The amplitude of angular movement was measured, at time 1.65s after the "go" cue, from averaged (typically 12 trials) trajectories.

RESULTS

Vibration-induced movement errors in PD

Figure 1A plots the amplitudes of wrist extension movements made by healthy subjects and PD patients in the absence and presence of flexor (antagonist) vibration. In cases of asymmetrical parkinsonism data refer to the wrist of the more affected side. NV movement amplitudes did not differ significantly between the PD group and control subjects ($P > 0.8$, unpaired t-test).

Antagonist vibration produced a significant reduction in the amplitude of extension movements in both PD (mean V and NV amplitudes in °, respectively, 17.6 ± 6.3 (SD) and 20.4 ± 6.6; $P < 0.002$, paired t-test) and control (13.8 ± 6.3 and 20.1 ± 6.6; $P < 0.001$) groups. However, Parkinsonian patients showed a far smaller extent of vibration-induced undershooting than did healthy subjects. Figure 1B plots the mean ratios of V/NV movement amplitudes of the patient and control groups which were, respectively, 0.88 and 0.66 and were significantly larger for the PD group ($P < 0.001$, Mann-Whitney U test).

In both patients and control subjects undershooting was associated with an appreciable, sustained reduction in wrist extensor EMG; small vibration reflexes were observed in the flexors of both PD and control subjects but did not differ systematically between the groups.

In a subgroup of fourteen PD patients with clearly asymmetrical signs a comparison was made of the effects of antagonist vibration on movement performance on the two sides. Figure 2 presents records of averaged extension movements, in the absence and presence of flexor tendon vibration, made by the less and more affected wrist of a representative patient with asymmetrical parkinsonism. A far more pronounced vibration-induced undershooting

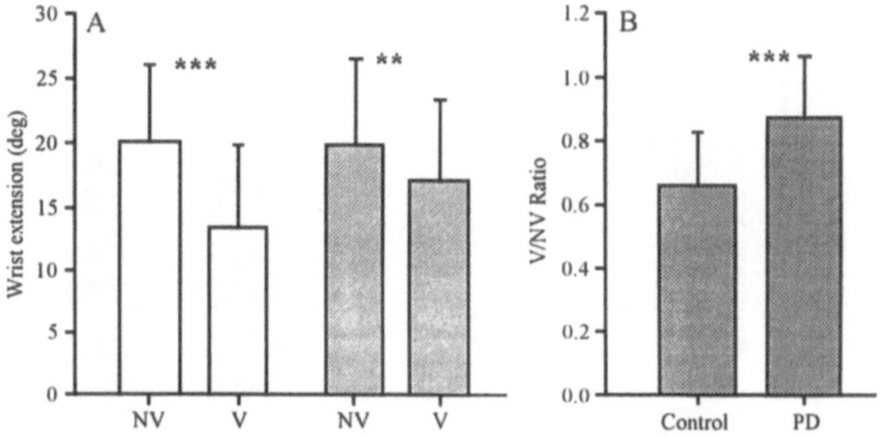

Figure 1. A, mean (+SD) amplitudes of non-vibrated (NV) and vibrated (V) wrist extension movements made by control and PD groups. *** $P < 0.001$, ** $P < 0.002$ (paired t-test). B, corresponding V/NV amplitude ratios. *** $P < 0.001$ (Mann Whitney U test).

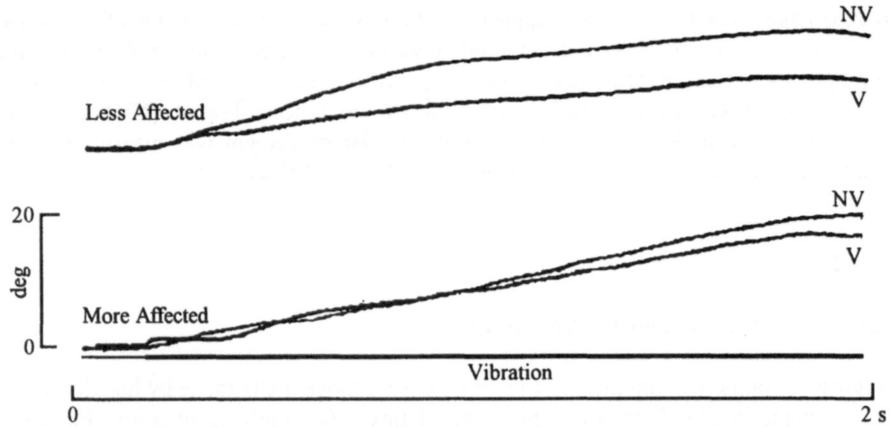

Figure 2. Averaged (12 trials) wrist extension trajectories of the less and more affected limb of a representative patient with asymmetrical Parkinsonism made in the absence (NV) and presence (V) of flexor tendon vibration.

of extension trajectories is evident on the less affected side than on the more affected side.

Data from the fourteen patients with asymmetrical PD are presented in Figure 3. Figure 3A plots the mean amplitudes of NV and V movements of the more and less affected wrists. The amplitudes of NV movements did not differ significantly between the two sides ($p = 0.78$, paired t-test). On the less affected side vibration produced a significant reduction in movement amplitude ($p < 0.001$, paired t-test). By contrast, the amplitudes of V and NV wrist movements did not differ significantly on the more affected side. In Figure 3B the corresponding V/NV ratios are presented. Pair-wise analysis indicated that V/NV ratios were significantly greater on the more affected side ($p < 0.05$, Wilcoxon).

Vibration-induced movement errors in HD

The effects of antagonist vibration varied considerably within the small ($n = 4$) group of

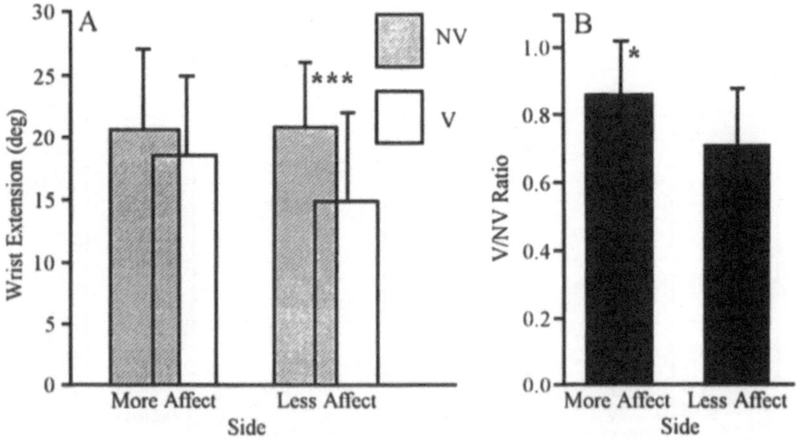

Figure 3. A, mean (+SD) amplitudes of non-vibrated (NV) and vibrated (V) extension movements made by the more and less affected wrists of patients with asymmetrical Parkinsonism. *** $P < 0.001$ (paired t-test). B, V/NV amplitude ratios for the more and less affected wrists. * $P < 0.05$ (Wilcoxon).

HD patients studied. In only one did vibration elicit definite undershooting of wrist extension movements (V/NV ratio = 0.48). In the remainder, vibration produced either virtually no alteration in movement amplitude or a tendency to overshooting (V/NV ratios 1.0 - 1.2).

DISCUSSION

Previous studies in healthy subjects have shown that muscle vibration can produce kinaesthetic illusions (Goodwin, McCloskey & Matthews, 1972) and alter the trajectories of learned voluntary movements (Capaday & Cooke, 1981; Appenteng & Prochazka, 1983, Cody, Schwartz & Smit, 1990). The latter effect is usually attributed to the CNS misinterpreting vibration-induced, artificial patterns of proprioceptive (particularly spindle Ia) afferent discharge as representing erroneous rates of change of muscle length and joint angle; it suggests that movements are normally under continuous proprioceptive guidance. The present findings that flexor (antagonist) vibration produces undershooting of voluntary wrist extension movements in parkinsonism, as was also observed in control subjects, suggest that proprioceptive guidance operates in a qualitatively normal manner in PD. However, our observation that abnormally reduced vibration-induced errors were smaller in PD implies that a quantitative impairment of proprioceptive guidance exists in this basal ganglia disorder. The situation in HD is unclear since there was considerable inter-individual variation in vibration effects in the small sample studied.

One possible type of explanation of our results is that vibration is a relatively ineffective stimulant of muscle spindle receptors in PD. However, the available evidence from microneurographic recordings in Parkinsonians indicates that vibration elicits an essentially normal pattern of spindle discharge (Burke, Hagbarth & Wallin 1977; Mano, Yamazaki & Takagi, 1979). Alternatively, it might be suggested that a disturbance of vibration reflexes in PD contributed. However, this would require that any braking effect of vibration reflexes evoked in the stimulated (antagonist) muscle were less than normal in PD. In fact, appreciable reflexes were not observed in either control or Parkinsonian subjects, in the vibrated muscle, which was essentially quiescent during the task. Furthermore, consistent abnormalities of neither phasic (Cody et al., 1986) nor tonic (Burke, Andrews & Lance, 1972) vibration reflexes have been noted in PD. Thus, the reduced vibration-induced movement errors shown by our PD patients probably reflect a derangement of higher-level integration of proprioceptive input and corollary discharge.

REFERENCES

APPENTENG, K. & PROCHAZKA, A. (1983) Feedback-controlled vibration used to improve motor performance in normal humans. *J. Physiol.* **339**, 11P.

BURKE, D., ANDREWS, C.J. & LANCE, J.W. (1972) Tonic vibration reflex in spasticity, Parkinson's disease and normal subjects. *J. Neurol. Neurosurg. Psychiat.* **35**, 477-486.

BURKE, D., HAGBARTH, K-E. & WALLIN, B.G. (1977) Reflex mechanisms in Parkinsonian rigidity. *Scand. J. Rehab. Med.* **9**, 15-23.

CAPADAY, C. & COOKE, J.D. (1981) The effects of muscle vibration on the attainment of intended final limb position during voluntary human arm movements. *Exp. Brain Res.* **42**, 228-230.

CODY, F.W.J., MACDERMOTT, N., MATTHEWS, P.B.C. & RICHARDSON, H.C. (1986) Observations on the genesis of the stretch reflex in Parkinson's disease. *Brain* **109**, 229-249.

CODY, F.W.J., SCHWARTZ, M.P. & SMIT, G.P. (1990) Proprioceptive guidance of human voluntary wrist movements studied using muscle vibration. *J. Physiol.* **427**, 455-470.

GOODWIN, G.M., McCLOSKEY, D.I & MATTHEWS, P.B.C. (1972) The contribution of muscle afferents to kinaesthesia shown by vibration-induced illusions of movement and by the effects of

paralyzing joint afferents. *Brain* **95**, 705-748.
MANO, T., YAMAZAKI, Y. & TAKAGI, S. (1979) Muscle spindle activity in parkinsonian rigidity. *Acta neurol. scand.* **60**, 176.
MOORE, A.P. (1987) Impaired sensorimotor integration in Parkinsonism and dyskinesia: a role for corollary discharges. *J. Neurol. Neurosurg. Psychiat.* **50**, 544-552.
NOTH, J., PODOLL, K. & FRIEDEMANN, H.H. (1985) Long-loop reflexes in small hand muscles studied in normal subjects and in patients with Huntington's disease. *Brain* **108**, 65-80.
TATTON, W.G. & LEE, R.G. (1975) Evidence for abnormal long-loop reflexes in rigid Parkinsonian patients. *Brain Res.* **100**, 671-676.

CORTICOSPINAL INPUTS TO SINGLE HUMAN SPINAL MOTONEURONES: WHAT CAN WE LEARN FROM PATIENTS WITH NEUROLOGICAL DISEASE

K.R. Mills

Clinical Neurophysiology Unit
University of Oxford
The Radcliffe Infirmary
Oxford, UK

INTRODUCTION

The human motor system can now be studied by brain stimulation but the interpretation of muscle responses relies heavily on analogy with earlier primate work (see Porter & Lemon, 1993). Currently available magnetic and electrical stimulators are able to activate the human upper motor tracts through the scalp, and although the mechanisms and sites of action of stimuli are still poorly defined, useful information on the human corticospinal projection has been obtained. Stimuli at appropriate strength are capable of exciting single motor units (Hess, Mills & Murray, 1987); the lowest threshold motor units recruited by voluntary action are also the ones excited by scalp stimuli. Indeed it has been shown that the same principles of size and frequency coding of motor unit recruitment which obtain during voluntary action also operate during scalp stimulation (Bawa & Lemon, 1993). The current work has examined the modulation by magnetic stimuli of the firing of low threshold, tonically active, human motor units in the first dorsal interosseous muscle of healthy subjects, and of patients with multiple sclerosis (MS), hereditary motor and sensory neuropathy type I (HMSN I) and motor neurone disease (MND). By studying such patients, the aim was not only to describe more accurately the motor deficits from which they suffer, but also to gain more information on the functioning of normal corticospinal connections.

METHODS

The times of discharge of single motor units, recorded with fine needle electrodes, were logged with respect to the times of magnetic stimuli applied with a circular coil centred over the vertex (Mills, 1991). Counts of discharges in an epoch of ± 250 ms were accumulated in 0.2 or 1 ms bins over 100 - 500 trials and displayed as a peristimulus time histogram (Figure 1A & C) and its cusum (Figure 1D). In many experiments a range of stimulus intensities was explored. The timings of periods of changed firing probability and their magnitudes were defined with respect to the mean baseline firing probability in the prestimulus period; the timing of more subtle changes in firing probability were determined by eye from the cusum (Ellaway, 1978).

NORMAL RESPONSES TO MAGNETIC STIMULI

In normal subjects, just suprathreshold stimuli cause a fall in firing probability (Boniface, Schubert & Mills, 1994), often detectable only in the cusum, at an onset latency

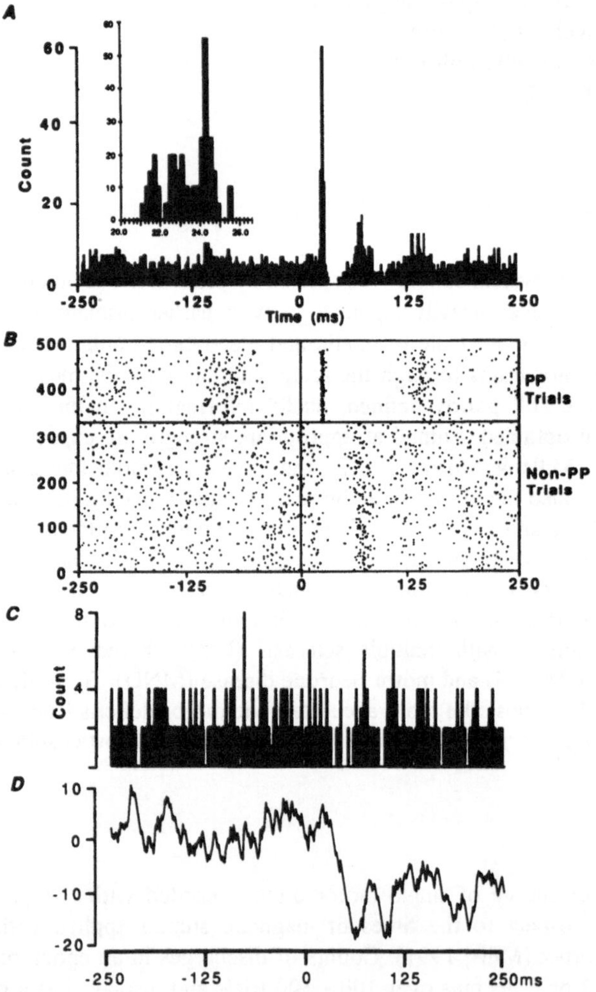

Figure 1. A. Peristimulus time histogram (1 ms binwidth, stimulus at time zero) of normal motor unit discharge in response to 490 magnetic stimuli. The primary peak at 25 ms is seen (inset, 0.2 ms binwidth) to consist of three subpeaks. The secondary peak has a modal latency of 72 ms. B. Raster sorted according to discharge in the primary peak (PP trials) showing the time history in each trial. The secondary peak is not due to a resumption of firing after the primary peak. C. Peristimulus time histogram and CUSUM (D) of normal motor unit driven by stimuli below threshold for the primary peak. Inhibition commences at about 30 ms.

of some 30 ms (Figure 1D). This is usually followed by the expected rise of firing following a period of inhibition. A small increase of stimulus intensity produces a brief period of high firing probability, referred to as the primary peak (Figure 1A); its latency varies from 24 - 32 ms in different units (Day, Dressler, Maertens de Noordhout, Marsden, Nakashima, Rothwell & Thompson, 1989; Boniface, Mills & Schubert, 1991; Palmer & Ashby, 1992). The size and complexity of the primary peak increases with stimulus intensity. When displayed with higher time resolution (Figure 1A, inset), the primary peak is seen to consist of a number of distinct subpeaks with a mean inter-modal interval of 1.8 ms. The primary peak is thought to represent the consequences of monosynaptic activation of the motoneurone by a train of impulses set up in the corticospinal tract, the individual subpeaks reflecting single composite excitatory postsynaptic potentials corresponding to each descending volley (Mills, 1991). The primary peak is followed by an obligatory period of zero firing probability, the initial part at least of which must be due to refractoriness of the spinal motoneurone. In some 50% of normal motor units, a secondary peak is also seen at an onset latency of 50 - 70 ms; the secondary peak is less synchronized than the primary peak (Mills, Boniface & Schubert, 1991). That the secondary peak does not merely represent a synchronized resumption of firing is seen by considering the sorted raster display (Figure 1B), where it can be seen that a normal motor unit never fires in both primary and secondary peaks but is constrained by its prior firing history. The origin of the secondary peak is still uncertain but may represent the operation of a long loop reflex.

RESPONSES IN NEUROLOGICAL DISORDERS

MS is a condition in which the primary pathological process is demyelination of central nerve fibres. Plaques of disease can occur throughout the central nervous system and produce a wide variety of physical impairments. If the central motor fibres subserving the upper limbs are affected, the patient may have weakness, spasticity, hyper-reflexia or impaired fine finger movements. Responses of single motor units in a patient with MS are seen in Figure 2C, D & E. The primary peak is often delayed and dispersed, and subpeaks within the primary peak may not be discernible. These phenomena presumably reflect the slowing of central conduction and the desynchronization of descending volleys which might be expected from differential demyelination of fibres. During natural voluntary action such time dispersion may reduce opportunities for temporal summation of post-synaptic potentials at the motoneurone, leading to weakness. A second phenomenon which has been observed in MS is seen in Figure 2E where subpeaks of normal duration are separated by an interval, some twice as long as that found in normals, suggesting that a single subpeak is missing. The well known increased refractoriness of demyelinated nerve fibres might be causing a frequency-dependent conduction block, allowing the first of a train of descending impulses to pass, but blocking the next before recovering sufficiently for the third to pass. This again might be a cause of weakness especially in situations in which central fibres are called upon to fire at high rates, such as might be envisaged in ballistic movements.

Patients with HMSN I have been studied to give a clue as to the origin of the secondary peak. These patients have demyelinated peripheral nerves with conduction velocities often of half normal, but the central conduction is normal. The spinal reflex loop time is therefore increased and if the secondary peak is mediated over a peripheral nerve pathway, it would be expected to be delayed in HMSN I; conversely if it is mediated over a purely central pathway, its latency would be expected to be at the normal interval after the primary response. Figure 2B shows that the secondary peak is delayed in HMSN I, implicating a peripheral nerve component. Similar considerations of the secondary peak latency in patients with MS in whom peripheral conduction is normal, but central conduction slowed, however, gives evidence of an additional central component in the mediation of the secondary peak. Possibilities therefore include a long-loop reflex, or the activation of gamma motoneurones by cortical stimuli and subsequent operation of the spinal reflex loop.

MND is a condition in which both cerebral motor cells and spinal motoneurones degenerate. The balance between cerebral and cord degeneration determines whether the patient exhibits primarily upper or lower motor neurone signs, but often both are evident. In some patients with MND, it is not possible with current magnetic stimulators to excite the motor cortex. Even with maximal strength stimuli, motor unit discharge is not modulated. In most patients with MND, threshold intensity for any effect to be produced is raised compared with normal. As in healthy subjects, the effects produced by just suprathreshold stimuli are inhibitory (Figure 2F). Increasing stimulus intensity leads to a primary response which may be delayed up to 10 ms compared with normal motor unit responses; this does not necessarily mean slowed central conduction, but could be caused by a failure of temporal summation at the motoneurone. Indeed, the subpeaks in MND motor unit responses are often not only separated by greater intervals than normal but have longer durations implying that the underlying excitatory postsynaptic potentials have slower rise times; this suggests a defect in the motoneurone membrane properties. Interestingly, motor units with delayed primary responses and wide inter-subpeak intervals tend to be found in MND patients with impaired fine finger movements but not other types of motor deficit (Mills, unpublished observations).

In normal motor units, the range of stimulus intensity over which inhibition proceeds to primary excitation is narrow (less than 5% of maximal stimulator output). Once an excitatory response is present, it becomes impossible to decide if the following period of

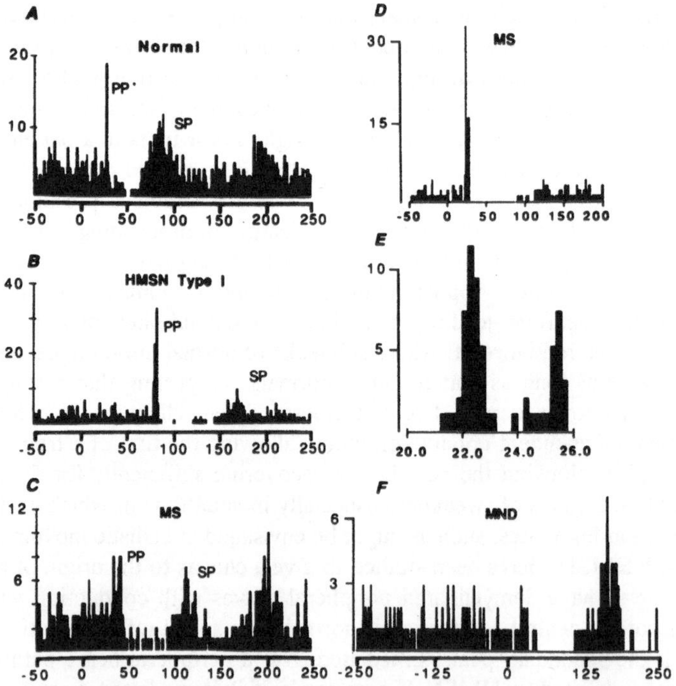

Figure 2. Peristimulus time histograms (1ms binwidth stimulus at time zero) of motor units in a normal subject (A), a patient HMSN I (B), patients with MS (C & D) and patient with MND (F). Primary peaks (PP) and secondary peaks (SP) are marked. The PP-SP interval is prolonged in HMSN I and to a lesser extent in MS. The primary peak in the histogram in D is expanded in E (0.2 ms binwidth) showing subpeaks with a greater than normal inter-subpeak interval. In F, a prolonged period of inhibition is seen.

reduced firing is due to an inhibitory process or is merely part of the obligatory pause in firing after the primary peak. By contrast, in MND, stimulus intensity can be raised considerably before an excitatory response is obtained and this has allowed the relationship between stimulus intensity and the length of inhibition to be explored. In fact the length of inhibition is linearly related to stimulus intensity, suggesting that a progressively deeper inhibition is being induced either at a cortical or spinal level by increasing stimulus intensities.

REFERENCES

BAWA, P. & LEMON, R. N. (1993). Recruitment of motor units in response to transcranial magnetic stimulation in man. *J. Physiol.* **471**, 445-464.

BONIFACE, S. J., MILLS, K. R. & SCHUBERT, M. (1991). Responses of single spinal motoneurons to magnetic brain stimulation in healthy subjects and patients with multiple sclerosis. *Brain.* **114**, 643-662.

BONIFACE, S. J., SCHUBERT, M. & MILLS, K. R. (1994). Suppression and long latency excitation of single spinal motoneurones by transcranial magnetic stimulation in health, multiple sclerosis and stroke. *Muscle Nerve.* **17**, 642-646.

DAY, B. L., DRESSLER, D., MAERTENS DE NOORDHOUT, A., MARSDEN, C.D., NAKASHIMA, K., ROTHWELL, J. C., THOMPSON P. D. (1989). Electric and magnetic stimulation of human motor cortex: surface EMG and single motor unit responses. *J. Physiol.* **412**, 449-473.

ELLAWAY, P. H. (1978). Cumulative sum technique and its application to the analysis of peristimulus time histograms. *EEG & Clin. Neurophysiol.* **45**, 302-304.

HESS, C. W., MILLS, K. R. & MURRAY, N. M. F. (1987). Responses in small hand muscles from magnetic stimulation of the human brain. *J. Physiol.* **388**, 397-419.

MILLS, K. R. (1991). Magnetic brain stimulation: a tool to explore the action of the motor cortex on single human spinal motoneurones. *TINS.* **14,** 401-405.

MILLS, K. R., BONIFACE, S. J. & SCHUBERT, M. (1991). Origin of the secondary increase in firing probability of human motor neurons following transcranial magnetic stimulation. *Brain.* **114**, 2451-2463.

PALMER, E. & ASHBY, P. (1992). Corticospinal projections to upper limb motoneurones in humans. *J. Physiol.* **448**, 397-412.

PORTER, R. & LEMON, R. (1993). *Corticospinal function and voluntary movement.* Oxford University Press, Oxford.

CRITIQUE OF THE PAPERS IN THE SESSION ON CLINICAL IMPLICATIONS

Hans-Joachim Freund

Neurologischen Universitatklinik
Moorenstrasse 5, D-4000 Dusseldorf
Germany

In the context of this session on clinical implications there are four contributions on single motor unit studies. The first two of them use the technique of cross-correlation analysis between motor unit (MU) discharges recorded during steady voluntary contractions in human forearm muscles.

Gibbs, Harrison, Mayston & Stephens examined the existence of common inputs in MU pairs from either agonist or antagonist muscles. They found peaks in the correlograms indicating common excitatory input between MUs of muscle pairs acting upon a common joint, and troughs in the correlograms of MUs from antagonist muscles. Model calculations could provide additional evidence that these inhibitory interactions may be asymmetric in some cases. This approach opens interesting possibilities for the examination of connections. From the clinical viewpoint it would be of particular significance, if alterations of these patterns could be observed in the chronic stage after central motor disturbances, such as hemiparesis or dystonia. Before this can be examined, however, there are some aspects, particularly in the forearm muscles, which should be clarified. The pattern of synergism which exists in the forearm and hand muscles is not fixed. For example, the hand and finger extensors are synergists when moving the hand and stretching the fingers dorsally, whereas the equivalent hand and finger flexors are synergists during volar movements. However, this synergy is altered when the subject is asked to make a fist. In this condition the hand extensor and finger flexors are synergists, whereas during hand opening the hand flexor and finger extensors are synergists. This example illustrates that the synergistic action of muscles acting on opposite sides of the hand is not a rare condition, but a frequent natural movement. Consequently, the question arises as to how the cross-correlation analysis would appear in such altered patterns of synergy. Would we still expect to see a trough in the cross-correlogram when synergism changes or would there be different peaks? In other words, the method could be used to explore different synergetic patterns and their changes in pathological conditions.

In the next contribution by Schmied, Vedel, Pouget, Forget & Lamarre, cross-correlation analysis was performed between MU pairs from the same muscle. Short- and long-term synchronisations were observed without a clear separation between the two distributions. The patients with sensory neuropathy or primary lateral sclerosis were chosen specifically to show the influence of sensory input or supraspinal descending pathways on the formation of synchronisation peaks. What is missing is the recognition that tremor is a major variable influencing the amount of long-term synchronisation (Dietz, Bischofsberger, Wita & Freund, 1976). It is well known that tremor increases with usage of the arm and shows a hand preference. It would therefore be necessary to show that this major variable for long-term synchronisation is controlled and compared with the data. Only then could other influences be sorted out.

In the contribution by Rymer & Gemperline the discharge rates of single MUs are related to the force exerted by the arm. The examination of six hemiparetic subjects showed that three of them had discharge rates clearly below the normal range for the corresponding forces. Since it was shown that the recruitment of MUs occurs at lower force levels than on the unaffected side, recruitment is obviously taken as a compensation for the impairment of firing rate modulation. What is unclear to me is how the attainment of a similar force level as on the other side is possible on the basis of the reduced firing rate modulation. At the torque levels chosen one can take it for granted that the high threshold units are already recruited, so that in order to compensate for the decrease in firing rates the only possibilities would be an improvement in the fusion properties of the muscle fibres, or an altered pattern of muscle synergies. The changes in muscle mechanics required would be an increase in twitch-contraction forces of the single motor units, or an increase in the duration of the twitches, which has been shown to occur during fatigue (Freund, 1983).

In the next study by Mills, the effects of transcranial magnetic stimulation on the discharge probability of MUs were described for normal subjects as having a typical pattern, with the primary and the secondary peaks separated by a strong intervening inhibition, though not in the same trials. Patients with hereditary motor sensory neuropathy, MS and motor neurone disorders were examined and single examples presented. There were interesting observations on various alterations of facilitory and inhibitory influences in different diseases. The main questions are how far the altered patterns are specific, and hence useful for diagnostic purposes, and whether they are sufficiently consistent to be used to evaluate the course of the disease. For this a systematic study would be necessary in order to distinguish effects attributable specifically to the disease from the pathophysiology of individual MUs.

Taken together, these studies used methods currently available to obtain information about disturbances of motor control at the level of the microcircuit. They showed that these methods have the potential to elucidate relevant aspects of pathophysiology, and may even be able to disclose relevant aspects of pathogenetic mechanisms. Certainly this was the case in the contribution by Bostock, Sharief, Reid & Murray who demonstrated that changes in the responses to subthreshold polarizing currents (using the method of threshold electrotonus) can not only monitor alterations secondary to damage, but also can disclose important steps in the pathogenesis of a disorder. This can be considered a break-through that also opens new possibilities for therapeutical trials to control the degenerative processes with agents specifically directed to act on the suspected channel deficiency or by the action of cytokines such as leukotriene B4, which is known to selectively block potassium channels (Köller, Siebler, Pekel & Müller, 1993).

The examination of vibration-induced trajectory errors of patients with Parkinson's or Huntingdon's disease by Cody and Rickards provided evidence for a quantitative impairment of proprioceptive guidance in the former. The atrribution of this effect to the suggested derangement of higher level integration remains speculative. The major problem here is the gross alteration of muscle mechanics in this condition. Consequently, the different transduction of the vibratory effects on the proprioceptive sensors must be taken into account.

REFERENCES

DIETZ, V., BISCHOFSBERGER, E., WITA, C. & FREUND, H.-J. (1976) Correlation between discharges of two simultaneously recorded motor units and physiological tremor. *EEG & Clin. Neurophysiol.* **40**, 97-105.

FREUND, H.-J. (1983) Motor unit and muscle activity in voluntary motor control. *Physiol. Rev.* **63**, 387-436.

KÖLLER, H., SIEBLER, M., PEKEL, M. & MÜLLER, H.W. (1993) Depolarization of cultured astrocytes by leukotriene B4. Evidence for the induction of a K+ conductance inhibitor. *Brain Res.* **612**, 28-34.

MOTOR CONTROL OF JAW VS. NECK MUSCLES USING A TRACKBALL FOR COMMUNICATION APPLICATIONS

Reinhilde Jacobs[1], Mieke Verheust[2],
Elke Hendrickx[1], Arthur Spaepen[2],
Kirstie Edwards[2] and Daniel van Steenberghe[1]

[1]Laboratory of Oral Physiology
Department of Periodontology
and [2]Department of Ergonomy
Catholic University of Leuven
Leuven, Belgium

INTRODUCTION

Many changes are underway in communications world-wide. Technical advances may create new possibilities for people with special needs, although their use may present certain problems to such a user group. The aim of the present research was to evaluate whether a trackball can be used for communication by non-vocal physically challenged people by means of isolated jaw movements or rather free head movements. As these patients often have to cope with uncontrollable motor responses, the use of isolated mandibular movements for trackball-operations instead of free head movements was considered.

METHODS

Eighteen healthy students in physical education aged 19 to 21 (13 females), free of overt symptoms of craniomandibular dysfunction, gave informed consent to participate in the present experiments. A tasks consisted of activating a trackball with the chin to typewrite a standardised text with which subjects were very familiar. Subjects were seated in a wheelchair with the trackball positioned under the chin. There were two experimental situations: 9 subjects were instructed to perform the typewriting task with the head fixed to a headrest by means of a tight strap while 9 others could perform this task with free head movements. An alphabetical keyboard on a computer screen allowed character selection and confirmation by means of two dimensional trackball movements. A standardised instruction form was read by the observer before the actual experiments were carried out. It was stated that they had to perform the tasks as quickly and accurately as possible. Prior to the presently reported task, subjects carried out two similar tasks to get familiarised with the experimental set-up. For statistical analysis, a one-way ANOVA was used to detect differences in speed and accuracy between subjects using isolated jaw movements or combined jaw and neck movements to perform the typewriting task. A level of significance of 5 % was chosen ($F(1,16) = 4.5$).

RESULTS

Subjects with their head free to move, typed the text significantly faster than subjects, who had to perform the task by means of mandibular movements only ($F = 4.8$). The mean time to type one character was also significantly shorter for the former subjects (Figure 1; $F = 6.5$). The mean (s.d.) time to type one character was 2.4 (0.3) s when using free head movements, while it was 2.7 (0.3) s for typewriting by means of mandibular movements only. With regard to the accuracy of the typewritten text, such a significant difference was not present ($F = 1.1$).

DISCUSSION

The lower speed for typewriting the text with mandibular movements only, could be explained by the more limited freedom of movement of the mandible. It could also be assumed that the worse performance when typewriting with mandibular movements only (a non-familiar task) is related to the fact that jaw motor control does not benefit from the presence of visual feedback. The latter hypothesis is in accordance with previous studies demonstrating the limited impact of visual feedback on jaw positioning or tracking and the predominant role of visual information during head or finger tracking (Jacobs, van

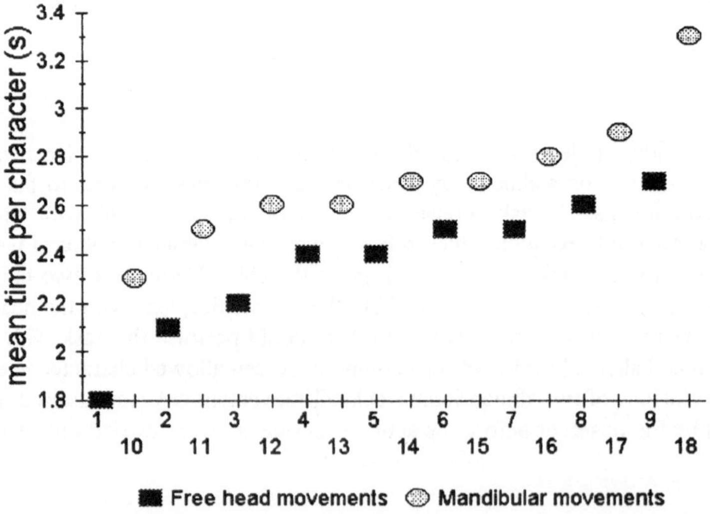

Figure 1. For each of the subjects presented in the X-axis, the mean typewriting time per character is indicated. Subjects 1 to 9 performed the typing task with free head movements, while subjects 10 to 18 used mandibular movements only.

Steenberghe & Schotte, 1992; Jacobs & van Steenberghe, 1993; Jacobs & van Steenberghe, in press). Further experiments are actually carried out to evaluate whether through learning, a chin-operated trackball could be used for communication applications by non-vocal physically challenged people.

ACKNOWLEDGEMENTS

This research was supported by the Belgian National Fund for Scientific Research (N.F.W.O. & F.G.W.O.). R. Jacobs is a post-doctoral researcher of the N.F.W.O.

REFERENCES

JACOBS, R., VAN STEENBERGHE, D. & SCHOTTE, A. (1992) The importance of visual feedback on the accuracy of jaw and finger positioning in Man. *Arch. Oral Biol.* **37**, 677-683.

JACOBS, R. & VAN STEENBERGHE, D. (1993) Jaw, head and finger tracking behaviour with delayed visual feedback. *J. Electromyography Kinesiol.* **3**, 103-111.

JACOBS, R. & VAN STEENBERGHE, D. Effects of delayed visual feedback on jaw, finger and toe positioning in man. *J. Motor Behav.* (in press).

COORDINATED HUMAN JAW AND HEAD-NECK MOVEMENTS DURING NATURAL JAW OPENING-CLOSING: REPRODUCIBLE MOVEMENT PATTERNS INDICATE LINKED MOTOR CONTROL

H. Zafar[1], P.-O. Eriksson[1],
E. Nordh[2] & N. Al-Falahe[1]

Departments of [1]Clinical Oral Physiology
and [2]Clinical Neurophysiology
Umeå University, S-901 87 Umeå, Sweden

INTRODUCTION

The complexity of the movements and muscle attachments between head, neck and mandible in man implies the existence of close functional relationships between the corresponding motor systems (see Figure 1A). Recent kinesiographic studies by our group have revealed concomitant head-neck movements during different voluntary jaw movement tasks (Nordh, Eriksson, Zafar & Al-Falahe, 1993). In the present report, we analyse the short and long term reproducibility of the temporal and spatial patterns of associated head-neck movements during voluntary jaw opening-closing tasks in human subjects.

METHODS

Ten healthy young adults volunteered for the study and were unaware of the underlying aim. They were seated comfortably in an upright position, with firm back support up to the mid-scapular level only, and without head support. They were instructed to perform one "fast" and one "slow" maximal jaw opening-closing movement within a period of 12 seconds, with no feedback or detailed instruction (see Figure 1B). Head and jaw movements were separately monitored with a precise opto-electronic motion analysis system (MacReflex®). Triplets of retro-reflective markers were fixed to the upper and lower front teeth. The location of each marker's geometrical centroid was three-dimensionally (3-D) sampled at a frequency of 50 Hz with a spatial resolution of 0.02 mm. The marker triplets determined two arbitrarily defined planes, each with fixed relation to the lower jaw and head, respectively, which allowed computation of 3-D jaw movements by deduction of superimposed head movements.

RESULTS

All subjects exhibited individual and reproducible temporal and spatial patterns of concomitant head-neck movements during fast as well as slow jaw opening-closing tasks, examples of which are shown in Figure 1B (lower panel). These movement patterns were reproducible not only when repeated during the same test session, but also when repeated after several months. There were clear differences in head-neck trajectories both between jaw opening and closing phases, and between head-neck movements at fast and slow speeds. Figure 1C shows examples of short and long term reproducibility of 3-D spatial head-neck movement trajectories in one subject. The figure also shows differences in trajectories related to phase of the movement as well as to speed.

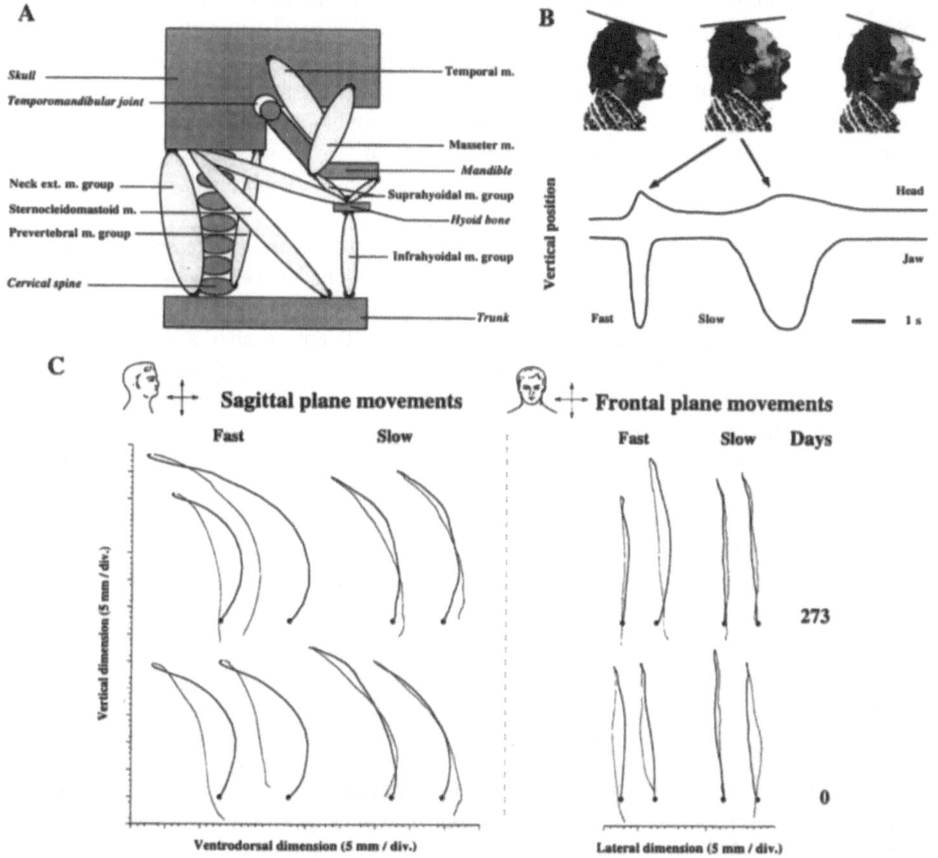

Figure 1. A. Anatomical and mechanical relationships of the human temporomandibular and the craniocervical regions. B. Upper panel: Head-neck positions at rest and at maximal jaw opening, (lines above the head indicate change in head-neck position). Lower panel: Examples of the temporal sequences of jaw and head-neck movements in vertical dimension during fast (left) and slow (right) jaw opening-closing tasks. C. Examples of the head-neck movements of a subject during fast and slow jaw opening-closing tasks. The spatial movement patterns of 3-D recorded head-neck movements during two fast and two slow jaw opening-closing tasks, and from two recording sessions, 273 days apart, are shown in the sagittal plane (left) as well as in the frontal plane (right). The starting points of the head-neck movements are marked by solid circles. The bold and the thin lines represent movements during jaw opening and jaw closing phases, respectively.

DISCUSSION

The present results support and extend our previous findings of concomitant head-neck movements during voluntary jaw opening-closing tasks (Nordh et al., 1993), indicating close connections between the human temporomandibular and craniocervical motor systems. The observations of reproducible 3-D head-neck movement patterns in short term as well as long term perspectives seem to reflect the existence of highly reproducible neuromuscular synergies (Macpherson, 1991) for coordinated head-neck movements in relation to mandibular motor tasks. It should also be noted that the head-neck movements were involuntary in the sense, that they were elicited in conjunction with jaw opening-closing activities.

The relative constancy of the movement patterns are notable, with respect to the multitude of possible movements permitted by the intricate architecture and biomechanics of the temporomandibular and craniocervical regions. The recorded trajectories indicate complex joint movements at the craniocervical junction as well as in the cervical spine. These findings imply that the head-neck movements are not just plain extensions-flexions, but that they are compound intersegmental movements related to specific tasks. Furthermore, the clear difference in the head-neck movement trajectories between jaw opening and closing phases, and between fast and slow speed, suggest that the neuromuscular synergies in the jaw and head-neck regions are flexible and related to motor context.

In the light of the present results, it seems justified to regard the temporomandibular and craniocervical motor systems as functionally united. The neural mechanisms involved are as yet unknown, however it is of interest that the human masticatory muscles contain many complex spindles (Eriksson & Thornell, 1987). There is also a report of direct trigeminal afferent projection to the cerebellum (Campos-Torres, Strazielle, Mahler & Jacquart, 1993). Finally, the inherent possibilities of the present methodological approach for studies of neuromuscular control should be noted, as well as the implications of the findings in clinical management of functional disorders of the jaw and head-neck regions.

ACKNOWLEDGEMENTS

Supported by the Swedish Medical Research Council (6874), the Swedish Dental Society and the Faculty of Odontology, Umeå University, Sweden.

REFERENCES

CAMPOS-TORRES, A., STRAZIELLE, C., MAHLER, P. & JACQUART, G. (1993) Organization of direct primary trigemino cerebellar afferences in the rat: structural study for a functional approach. *Eur. J. Neurosci. Supp.* **6**, A1034.

ERIKSSON, P.-O. & THORNELL, L.-E. (1987) Relation to extrafusal fibre-type composition in muscle spindle structure and location in the human masseter muscle. *Arch. Oral Biol.* **32**, 483-491.

MacPHERSON, J. M. (1991) How flexible are muscles synergies? In *Motor control: concepts and issues.* eds. HUMHPREY, D. R. & FREUND, H.-J., pp. 33-47. John Wiley & Sons, Chichester.

NORDH, E., ERIKSSON, P.-O., ZAFAR, H. & AL-FALAHE, N. (1993) Concomitant mandibular and head-neck movements during natural jaw opening-closing indicates parallel neuromuscular activation of jaw and head-neck systems. *Eur. J. Neurosci. Supp.* **6**, A1083.

REFLEX CONTROL OF HUMAN JAW MUSCLES BY THE MECHANORECEPTORS IN THE PERIODONTIUM OF TEETH

P. Brodin[1] and K.S. Türker[2]

[1]University of Oslo, Oslo, Norway
[2]University of Adelaide, Adelaide, Australia

INTRODUCTION

There is controversy in the literature regarding the nature of feedback from the periodontal mechanoreceptors to the jaw closer motoneurones. Most of these studies suggest that this feedback is negative and that it prevents large forces from developing between the teeth. Therefore, it is argued that, stimulation of the mechanoreceptors in the periodontium may evoke a protective reflex similar in essence to the flexor withdrawal reflex of limbs. However, other equally well designed studies suggest a positive feedback system that originates from these receptors and which may help increase forces between the teeth (reviewed in Linden, 1990).

It is possible that both lines of evidence are, in fact, correct and that there may be both excitatory and inhibitory connections between the periodontal mechanoreceptors and the jaw closer motoneurones. It is also plausible that under differing experimental conditions and stimulation algorithms, either one of these connections can become the dominant response.

The purpose of this study, therefore, was to study this controversial connection using different mechanical stimulus rates to determine whether it is possible to elicit separate excitatory and inhibitory responses.

METHODS

Reflex responses of the masseter muscle of consenting adult volunteers to mechanical stimulation of the upper central tooth were recorded using surface EMG and single motor unit potentials. In each experiment, a fine-wire electrode was inserted into the muscle to record the action potentials from single motor units. One clearly identified unit was selected as the feedback unit. With the help of audio and visual feedback the subject was able to fire the unit regularly, and hence keep the level of excitation of the motoneurone at a constant

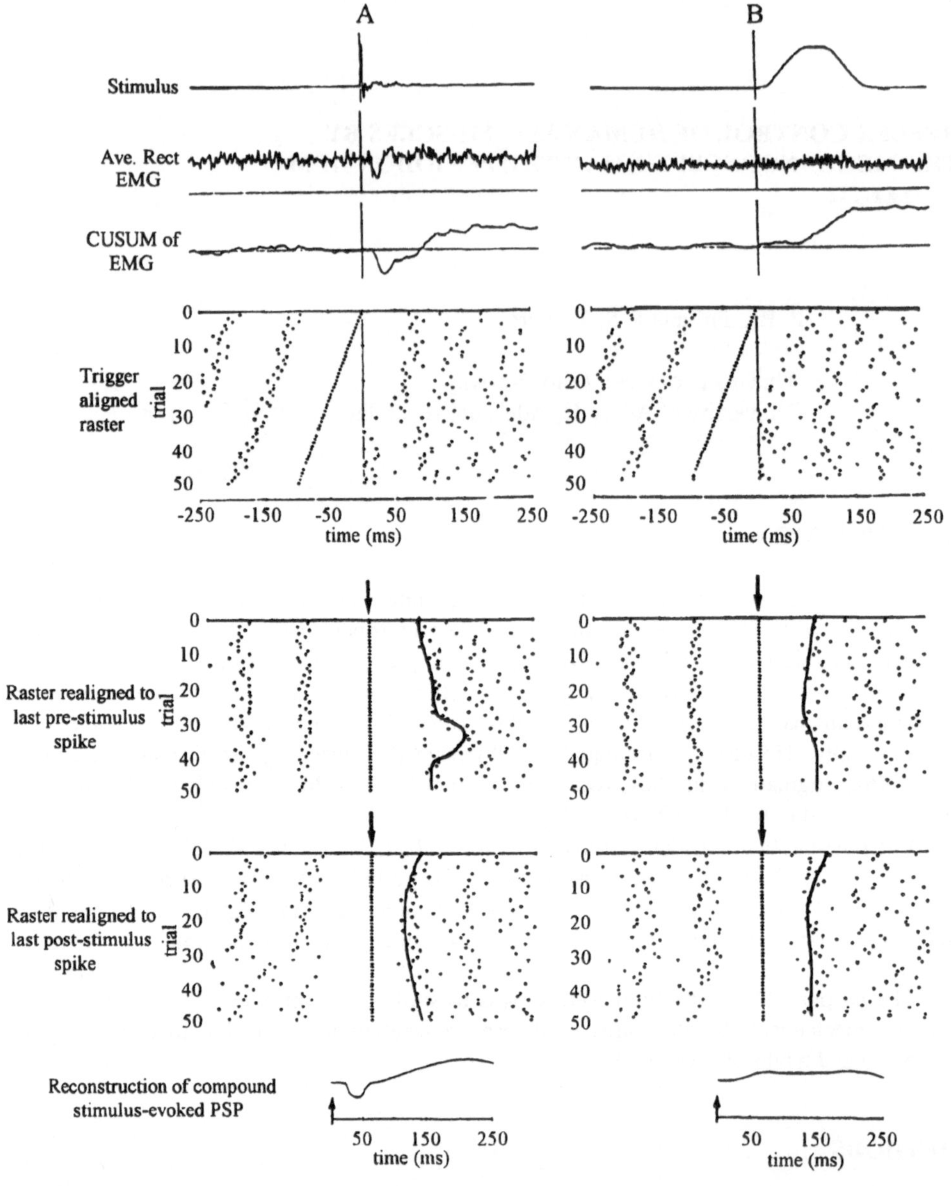

Figure 1. Reflex responses of human masseter muscle to mechanical stimulation of a tooth. The response to 2 N taps (column A) and to 2 N pushes (column B) are shown on surface EMG and dot raster recordings from single motor units. (Modified from Türker et al., 1994).

the unit regularly, and hence keep the level of excitation of the motoneurone at a constant level (for details see Brodin, Türker & Miles, 1993a and Türker, Brodin & Miles, 1994). Rectified averaged surface EMG records and the peri-stimulus time histograms (PSTH) of the motor unit potentials were used in order to determine the nature and the strength of the periodontal reflex. Two types of stimuli were delivered to the tooth using an electromechanical vibrator: brisk tap stimulus (2 ms rise time) and slow push stimulus (50 ms rise time). The amplitude of the stimulus was 0.5 - 3 N. The timing of the stimulus was controlled by a program that triggered the stimulator only when the unit was firing steadily.

RESULTS

Surface EMG records showed that the taps elicited a short-latency inhibition, often followed by an excitatory peak. The inhibitory reflex response increased as the taps became stronger. Slow pushes, however, induced a long-latency excitatory reflex response (Brodin et al, 1993a). The motor unit records illustrated basically the same pattern of response (see Figure 1).

The reflex inhibition following a 2 N tap in 16 of the 20 units had a latency of 13 ms and duration of 37 ms. This inhibition was followed by a reflex excitation in 11 of the 20 units at a latency of 71 ms and duration of 29 ms. The short-latency response to slow pushes was mixed (inhibition in 4 units and excitation in 1 unit out of 20). However, the slow pushes evoked a long-latency excitatory reflex response in 12 out of 20 units at latency of 77 ms and lasting 40 ms (Türker et al, 1994). All the reflex responses occurred at shorter latencies than the quickest reaction time of the subject to the same stimulus. The reaction time for the taps was about 80 ms and for the pushes about 140 ms (Brodin, Miles & Türker, 1993b).

CONCLUSION

It is concluded that stimulation of periodontal mechanoreceptors usually activates an excitatory reflex response to the jaw closer motoneurones. This may be a significant factor in helping to grip and crush the food bolus between the teeth during chewing. However, when the rate of force application is large enough, a short-latency inhibitory reflex response is evoked which may over-ride the subsequent excitatory reflex response. Inhibition of the jaw-closers tends to protect the teeth and soft tissues when one bites unexpectedly on a hard object while chewing.

ACKNOWLEDGMENTS

This study is supported by the NH and MRC of Australia. We thank Dr. T.S. Miles for his constructive assistance during the course of these experiments.

REFERENCES

BRODIN, P., TÜRKER, K.S. & MILES, T.S. (1993a) Mechanoreceptors around the tooth evoke inhibitory and excitatory reflexes in the human masseter muscle. *J. Physiol.* **464**. 711-723.

BRODIN, P., MILES, T.S. & TÜRKER, K.S. (1993b) Simple reaction time responses in human masseter muscle. *Arch. Oral Biol.* **38**, 221-226.

LINDEN, R.W.A. (1990) Periodontal receptors and their functions. In *Neurophysiology of the Jaws and Teeth.* ed. TAYLOR, A., pp. 52-95. Macmillan Press, London.

TÜRKER, K.S., BRODIN, P. & MILES, T.S. (1994) Reflex responses of motor units in human masseter to mechanical stimulation of a tooth. *Exp. Brain Res.* (in press).

ARE MUSCLE SPINDLES RESPONSIBLE FOR POST-CONTRACTION AND POST-VIBRATION MOTOR EFFECTS IN HUMANS?

E. Ribot-Ciscar, J.P. Roll, J.C. Gilhodes,
M.F. Tardy-Gervet and J.L. Demaria

Laboratoire de Neurobiologie Humaine
URA-CNRS 372, Faculté de Sciences
et Techniques de St. Jèrôme
Avenue Escadrille Normandie Niemen
13397 Marseille Cedex 20, France

INTRODUCTION

Many people have carried out the experiment which consists of pressing the arm strongly against a wall for 15 to 30 s, then moving aside, and perceiving the arm rising alone in the frontal plane. This uncontrolled movement, which is often accompanied by a strange sensation of lightness, reflects an involuntary muscular activity which occurs subsequent to a strong isometric contraction of the same muscle. These after-effects are called the "post-contraction response" or "Kohnstamm phenomenon". The neurophysiological mechanisms underlying these after-effects as well as their functional role in motor control are still largely unknown. It was recently suggested that the phenomena may reflect the existence of central oscillatory mechanisms subserving voluntary motor behaviour in humans (Craske & Craske, 1986). In the present study, psychcphysiological and neurophysiological methods are used to study the sensory origin of these post-contractions, first by describing their numerous analogies with post-vibration motor effects and secondly by recording the unitary muscle spindle activity occurring in response to a contraction or a vibration of the parent muscle.

METHODS

The psychophysiological experiment was conducted with 14 human volunteers. They were seated comfortably in an armchair with their forearms resting on horizontal supports. First, they were asked to strongly contract the biceps or triceps brachii for 30s, with the support fixed at 95°. When the isometric contraction was released, a powerful, long-lasting involuntary contraction developed in the same muscle. Secondly, involuntary muscular activities were also found to occur after a 30-s vibration (70 Hz, 0.5 mm, duration 30 s) of the biceps or triceps brachii distal tendons in subjects at rest (see Figure 1).

RESULTS

These post-contraction and post-vibration motor responses were found to be similar in many respects : (1) they developed in the same muscle : the contracted or vibrated muscle ; (2) their amplitudes and time courses (slow development, long duration) were analogous; (3) neither type could be elicited by co-contracting at the same level or co-vibrating at the same frequency two antagonist muscles; (4) visual stimulation could lead both types of motor responses to switch from one muscle to its antagonist (for more details see Gilhodes, Gurfinkel & Roll, 1992).

A neurophysiological experiment was planned in order to assess whether or not these motor post-effects may be attributable to post-contraction or post-vibration sensory discharges. For this purpose, we recorded the unitary activities of muscle spindle primary endings from the lateral peroneal nerve using the microneurographic method. The results showed that only rarely did a Ia fibre increase its resting discharge after a 15-s voluntary isometric contraction (2/14), and never after the end of a 30-s muscle tendon vibration (0/12). Most of the time the muscle afferents were silent or reverted to their baseline activity prior to muscle activation when a voluntary isometric contraction was released (for more details see Ribot-Ciscar, Tardy-Gervet, Roll & Vedel, 1991) or a tendon vibratory stimulation was withdrawn (see Figure 2).

CONCLUSIONS

To conclude, the motor post-effects described do not seem to involve proprioceptive post-discharges. The close similarities found to exist between the post-contraction and post-vibration motor responses suggest however that they may have the same origin. Thanks to

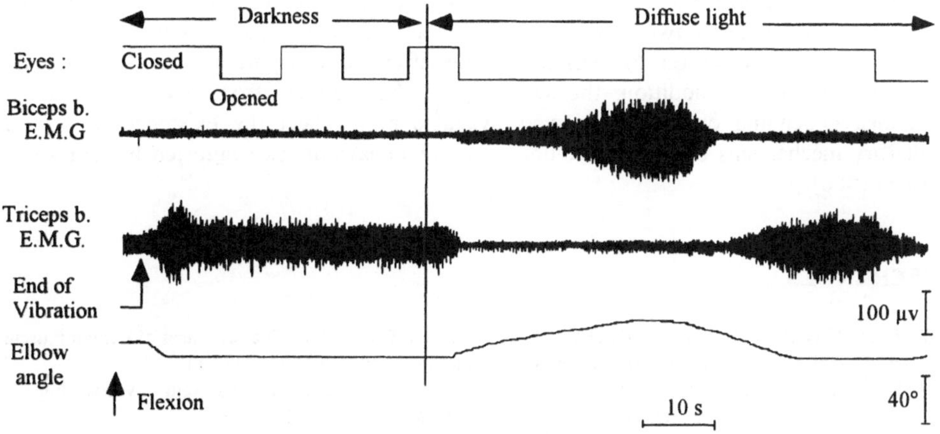

Figure 1. This figure shows the motor after-effects which occur subsequent to 30-s vibration of the triceps distal tendon : an involuntary and long lasting EMG activity develops in the muscle shortly after the offset of vibration. The upper trace gives the visual conditions : eyes open or closed, in darkness or diffuse light. This post-effect is not affected by opening and closing the eyes in darkness. The same manoeuvre performed in diffuse light causes the contraction to switch to the antagonist muscle. The bottom trace shows the accompanying forearm movements.

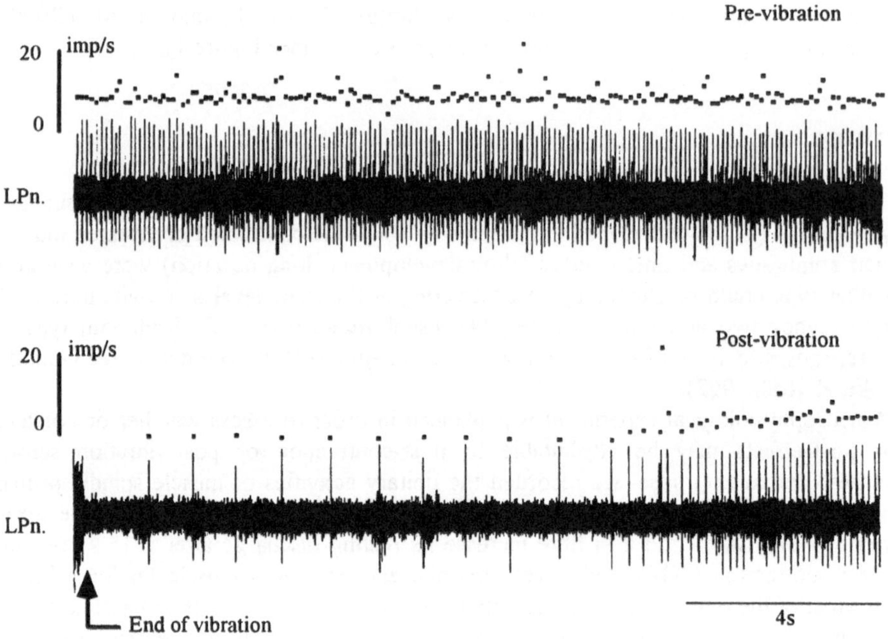

Figure 2. Example of muscle spindle primary ending discharge recorded in man from the Lateral Peroneal nerve (LPn.) using the microneurographic method. Most of the Ia afferent fibres revert to their pre-existing resting discharge, whereas a 30-s muscle tendon vibratory stimulation may lead to a short silencing of the unit (or a reduced discharge as shown in this example) before it progressively returns to the static level of discharge prior to the stimulation.

microneurographic studies, it is now well known that muscle tendon vibration as well as isometric voluntary contraction lead to an increase in the muscle spindle primary ending discharge (Burke, Hagbarth, Löfstedt & Wallin, 1976; Roll, Vedel & Ribot, 1989). Furthermore, these post-effects cannot be induced by co-contracting or co-vibrating the two antagonist muscles. These two sets of arguments seem to indicate that they may result from the asymmetrical activation of muscle spindle afferents originating from agonist and antagonist muscles. In addition, the switching of these effects from one muscle to its antagonist depending on the visual cues available may reflect the existence of central oscillatory mechanisms underlying voluntary motor behaviour, as suggested by Craske & Craske (1986).

REFERENCES

BURKE, D., HAGBARTH, K.E., LÖFSTEDT, L. & WALLIN, B.G. (1976) The responses of human muscle spindle endings to vibration during isometric contraction. *J. Physiol.* **261**, 695-711.

CRASKE, B. & CRASKE, J.D. (1986) Oscillator mechanisms in the human motor system: investigating their properties using the after contraction effect. *J. Motor Behav.* **18**, 117-145.

GILHODES, J.C., GURFINKEL, V.S. & ROLL, J.P. (1992) Role of Ia muscle spindle afferents in post-contraction and post-vibration motor effect genesis. *Neurosci. Lett.* **135**, 247-251.

RIBOT-CISCAR, E., TARDY-GERVET, M.F., ROLL, J.P. & VEDEL, J.P. (1991) Post-contraction changes in human muscle spindles resting discharge and stretch sensitivity. *Exp. Brain Res.* **86**, 673-678.

ROLL, J.P., VEDEL, J.P. & RIBOT, E. (1989) Alteration of proprioceptive messages induced by tendon vibration in man: a microneurographic study. *Exp. Brain Res.* **76**, 213-222.

PROPRIOCEPTIVE PERFORMANCE IN NORMAL AND ABNORMAL HUMAN KNEES

M.G. Hall[1,2], R.H. Baxendale[1],
W.R. Ferrell[1] and D.L. Hamblen[2]

Departments of Physiology[1]
and Orthopaedic Surgery[2]
The University, Glasgow G12 8QQ, UK

INTRODUCTION

The contribution to proprioceptive performance made by muscle, joint and skin afferents has been discussed for many years. The role of joint afferents has been questioned because of the persistence of proprioception after joint replacement. We have investigated the proprioceptive performance in normal and abnormal knees by using a threshold detection paradigm to test proprioceptive acuity to the onset of passive movement.

METHODS

Subjects lay on the contralateral side to the knee being tested on an examination couch. A motor and pulley system imposed flexion or extension movements to the knee (which was fixed in space) at a constant angular velocity of 0.4°/s. The foot and ankle were immobilised by a rigid air splint, whilst the thigh was fixed to a support bar. The direction of movement and the variable delay before its onset were computer controlled. 5 flexion and extension movements along with 3 dummy trials were performed in a pseudorandom order. Dummy trials consisting of the motor stepping backwards and forwards repeatedly were used to discourage guessing.

RESULTS

Proprioceptive acuity was tested at a range of starting knee flexion angles on two separate test days in a group of 10 young normal subjects (age = 23.8 ± 4.1). On the first day knee flexion angles of 15°, 30° and 45° were tested, whilst on the second day, angles of 5°, 30° and 60° were considered. Pooled results for the detection angles for all the angles T tested are seen in Figure 1.

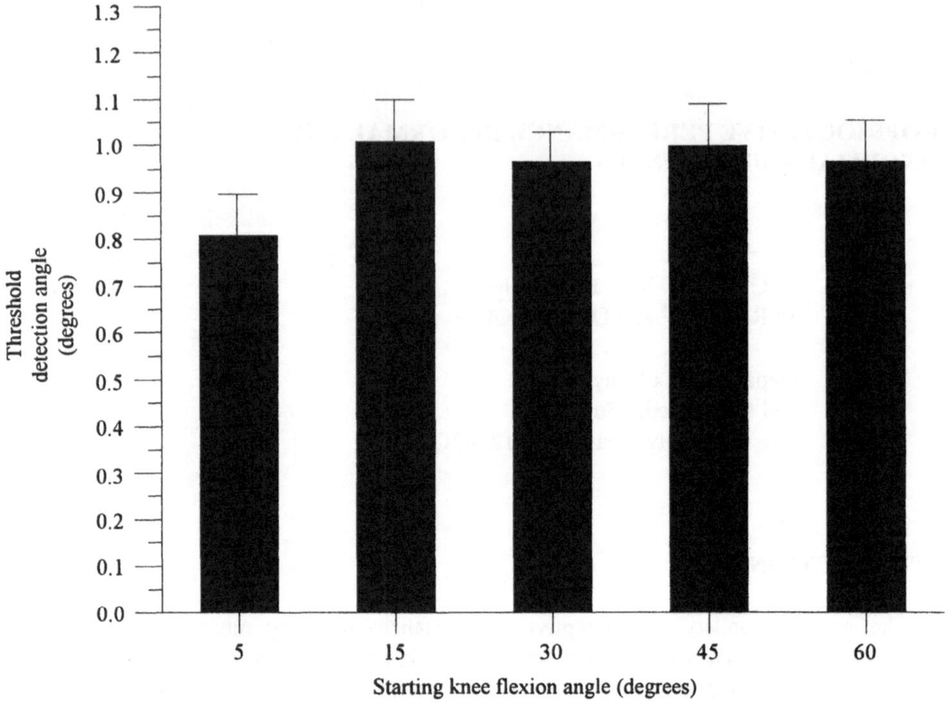

Figure 1. Mean detection angles with 95% confidence intervals for the different starting knee flexion angles.

Results of the testing at the 30° starting angle revealed no significant differences between days, indicating the threshold detection was consistent between days. Acuity significantly increased towards the end of the range of movement, at 5° starting angle (anova, $p=0.008$).

10 patients (age = 30.3 ± 6.1) with Hypermobility Syndrome (HMS), i.e. with generalised joint laxity (Kirk, Ansell & Bywaters, 1967), were tested at 5° and 30° starting knee flexion angles. HMS subjects showed a significantly poorer acuity compared to a group of age and sex matched controls, at both the mid (t-test, $p<0.001$) and end (t-test, $p<0.001$) of range of movement (see Table 1). The increased acuity at the end of range seen with the normal knee was absent. HMS knees show an increased range of motion due to laxity in the collateral and posterior cruciate ligaments along with the posterior capsule. The observed proprioceptive deficit is likely to be related to these structures, again indicating that ligamentous and capsular afferents have an important contribution to this sensation.

Proprioceptive acuity was also compared in 3 other groups: osteoarthritic (OA) patients tested immediately prior to total knee replacement, patients 1-3 years after knee

Table 1. Mean detection angles (±SD) for HMS and controls.

Group	5° starting angle	30° starting angle
HMS	1.477° ± 0.478	1.523° ± 0.419
Controls	0.964° ± 0.294	1.133° ± 0.335

Table 2. Mean detection angles (±SD) and age for the three groups.

Group	Detection angle	Mean age and range (yrs)
Control	1.855° ± 0.429	65.9 (53-73)
OA	2.296° ± 0.930	69.7 (51-84)
TKR	3.111° ± 1.834	67.9 (54-74)

replacement (TKR) and age matched controls. Degeneration of the joint capsule and ligaments are associated with OA, whilst TKR involves sub-total excision of the joint. Each group consisted of 21 subjects, with testing performed at a starting knee flexion angle of 30°. Summary results are seen in Table 2. Significant differences were found between the three groups (anova, p=0.005). Controls showed significantly greater acuity than the TKR (t-test, p=0.0058). The OA group tended to have a lower acuity compared to the controls (t-test, p=0.058), but a higher acuity when compared to the TKR (t-test, p=0.08), but these tendencies failed to reach significance. These results tend to suggest that progressive loss of the capsule and ligaments of the knee causes significant detriment to the proprioceptive sensation.

Joint afferent discharge has been shown to be greatest at the end of range in animal experiments (Burgess & Clark, 1969). Proprioceptive acuity in the normal knee is seen to be greatest where joint afferent discharge is greatest. In HMS patients ligamentous and capsular laxity appears to be associated with decreased proprioceptive performance. Osteoarthritic and replaced knees, with their associated reduced ligamentous and capsular components, both tend to show diminished proprioceptive performance. It is difficult to explain these deficits by changes in muscle afferent behaviour. We conclude that joint afferents must contribute to normal proprioception.

The cause of degenerative joint diseases remains unclear, but unfavourable biomechanical loading is one probable cause. Poor proprioceptive feedback is a possible mechanism by which this situation could arise (Barrett et al, 1991). It is interesting to note that HMS patients are a group who tend to suffer from premature OA (Bridges et al, 1992), show decreased proprioceptive acuity compared to controls. This may indicate a link between the two. A longitudinal study is under way to consider changes in proprioception with OA subjects 1 year pre- and post-TKR.

ACKNOWLEDGEMENTS

This research was supported by the MacFeat Bequest, University of Glasgow.

REFERENCES

BARRETT, D.S., COBB, A.G. & BENTLEY, G. (1991) Joint proprioception in normal, osteoarthritic and replaced knees *J Bone Joint Surgery.* **73**, 53-56.

BRIDGES, A.J., SMITH, E. & REID, J. (1992) Joint hypermobility in adults referred to rheumatology clinics *Annals Rheumatic Diseases.* **51**, 793-796.

BURGESS, P.R. & CLARK FJ (1969) Characteristics of knee joint receptors of the cat *J. Physiol.* **203**, 317-335.

KIRK, J.A., ANSELL, B.M. & BYWATERS, E.G.L. (1967) The hypermobility syndrome *Annals Rheumatic Diseases.* **26**, 419-425.

DEPRESSION OF RECURRENT INHIBITION OF THE SOLEUS AND WRIST FLEXOR MOTOR NUCLEI BY MAGNETIC BRAIN STIMULATION IN HUMANS

R. Mazzocchio[1], J.C. Rothwell[2] and A. Rossi[1]

[1]Unita di Neurofisiologia dell'Istituto
di Scienze Neurologiche
dell'Universita di Siena
Viale Bracci, 53100 Siena
Italy
[2]MRC Human Movement and Balance Unit
The Institute of Neurology
Queen Square
London WC1N 3BG, UK

INTRODUCTION

Recurrent inhibition of alpha-motoneurones via motor axon collaterals and Renshaw cells is regarded as a prominent organisational feature of the motor nuclei to the limb muscles. In the cat, various supraspinal structures can influence the activity of Renshaw cells, including the pyramidal tract the activation of which depresses recurrent inhibition (see Baldissera, Hultborn & Illert, 1981). The question is whether such cortico-spinal depression of Renshaw cells is also present in humans and whether it represents a general strategy of cortical commands.

METHODS

In nine normal subjects, homonymous recurrent inhibition of soleus (Sol) and flexor carpi radialis (FCR) motoneurones (see Rossi & Mazzocchio 1991; Katz, Mazzocchio, Penicaud & Rossi, 1993) was investigated after evoking cortico-spinal volleys through magnetic stimulation of the motor cortex of intensity insufficient for eliciting motor responses in the respective muscles. Recurrent inhibition brought about by an H1 reflex discharge was estimated by a test H' reflex. Modifications of recurrent inhibition (test H' reflex) and motoneuronal excitability (reference H reflex) following conditioning cortical stimulation were evaluated in parallel.

RESULTS

In the Sol, a conditioning cortical shock produced inhibition of the reference H at minimum conditioning-test intervals of -2 ms (H reflex stimulus before magnetic). The test H' reflex showed an opposite behaviour showing a long-lasting phase of facilitation (up to 20 ms) starting at minimum conditioning-test intervals of +1 ms. The threshold intensity for cortical facilitation of the test H' reflex was significantly lower than that necessary to obtain cortical inhibition of the reference H reflex. In the FCR, both reference and test reflexes were facilitated. However, by decreasing the intensity of cortical stimulation, facilitation of the test H' reflex alone was obtained. This started at conditioning-test intervals of +1 ms and lasted for about 10 ms.

DISCUSSION

To summarise: 1) Subthreshold cortical stimulation produces long-lasting depression of recurrent inhibition in both Sol and FCR motor nuclei; 2) The threshold intensity for cortical depression of Sol and FCR recurrent inhibition is lower than that for producing modifications in the relative motoneurones 3) In both Sol and FCR motor nuclei, depression of recurrent inhibition begins 3 - 4 ms later than the onset of motoneuronal changes. Since (1) magnetic stimulation of the brain appears predominantly to activate large, fast conducting cortico-motoneuronal axons, which are known to display phasic properties (see Mazzocchio, Rothwell, Day & Thompson, 1994) and (2) depression of Renshaw cell activity from the motor cortex appears to be a general strategy given that the same changes were observed in two functionally different muscles such as the Sol and FCR, it could be proposed that recurrent inhibition is little involved in the dynamic aspect of limb movements. As a corollary, in view of the abundant evidence indicating that recurrent inhibition mainly contributes to the adaptation of postural responses to voluntary movements (see Rossi, Decchi & Vecchione, 1992), it should not be surprising that recurrent inhibition is lacking in muscles not engaged in postural activity such as the intrinsic hand muscles (Katz et al., 1993).

REFERENCES

BALDISSERA, F., HULTBORN, H. & ILLERT, M. (1981) Integration in spinal neuronal systems. In *Handbook of Physiology, Section 1, The Nervous System, Volume 2, Motor Control.* ed. BROOKS, V.B. pp. 509-595. American Physiological Society, Bethesda.

KATZ, R., MAZZOCCHIO, R., PENICAUD, A. & ROSSI, A. (1993) Distribution of homonymous and heteronymous recurrent inhibition in the human upper limb. *Acta physiol. scand.* **149**, 183-198.

MAZZOCCHIO, R., ROTHWELL, J.C., DAY, B.L. & THOMPSON, P.D. (1994) Effect of tonic voluntary activity on the excitability of human motor cortex. *J. Physiol.* **474**, 261-267.

ROSSI, A., DECCHI, B. & VECCHIONE, V. (1992) Supraspinal influences on recurrent inhibition in humans. Paralysis of descending control of Renshaw cells in patients with mental retardation. *EEG & Clin. Neurophysiol.* **85**, 419-424.

ROSSI, A. & MAZZOCCHIO, R. (1992) Renshaw recurrent inhibition to motoneurones innervating proximal and distal muscles of the human upper and lower limbs. In *Muscle Afferents and Spinal Control of Movement.* eds. JAMI, L., PIERROT-DESEILLIGNY, E. & ZYTNICKI, D. pp. 313-319, Pergamon Press, Oxford.

SUPPRESSION OF SINGLE SPINAL MOTONEURONES BY TRANSCRANIAL MAGNETIC STIMULATION: STUDIES IN HEALTHY SUBJECTS, MULTIPLE SCLEROSIS AND STROKE

S.J. Boniface[1], K.R. Mills[1] and M. Schubert[2]

[1]Department of Clinical Neurophysiology
Radcliffe Infirmary, Oxford, UK
[2]Neurologische Universitaetsklinik
Bonn, Germany

INTRODUCTION

Transcranial magnetic stimulation produces at least two peaks in firing probability in single human spinal motoneurones (Boniface, Mills & Schubert, 1991, Mills, Boniface & Schubert, 1992). These have latencies varying from 20 - 31 ms and 64 - 100 ms in the first dorsal interosseus muscle and have been termed the primary and secondary peaks, respectively. These events are also detectable in patients with motoneurone disease, multiple sclerosis and stroke (Mills, Boniface & Schubert, 1991, Schubert, Mills, Boniface, Konstanzer & Dengler, 1991), but often requiring higher stimulus intensities. The aim of this study was to determine whether or not periods of reduced firing probability of motor units were produced at stimulus intensities sub-threshold for an early excitatory response (primary peak) in health and disease. Findings have also been reported in full elsewhere (Boniface, Mills & Schubert, 1994).

METHODS

A total of 14 tonically active low threshold single motor units were studied in the first dorsal interosseous muscle of 4 healthy subjects, 5 patients with multiple sclerosis and 1 patient with stroke. All of the patients had upper motor neurone signs. Motor units were tonically activated under voluntary control with the aid of audio visual feedback at a firing rate in the region of 10 imp/s. Stimuli were delivered at random with respect to the spike train with a 9 cm coil centred at the vertex with the initial current flowing in the direction optimal for an excitatory response (Magstim). Stimulus intensities ranged from 36% of maximum output up to 70% for some of the patients. Stimulus intensity was set below that required for the production of primary peaks in the PSTH of index motor units and was less than threshold for a surface recorded response by 7 - 30%. For healthy subjects, this was usually within 5% of the minimal intensity required for the production of discharges at the primary peak latency.

RESULTS

Significant periods of reduced firing probability were present in the peri-stimulus time histogram (PSTH) and corresponding cumulative sum in 12 of the 14 motor units (Figure 1). Onset ranged from 18 - 59 ms after the stimulus and the duration ranged from 27 - 133 ms. The PSTHs and cusums of 7 motor units contained late rises in firing probability at 65-140 ms. In 5 of these, there was a second period of reduced firing probability which lasted until 120 - 237 ms.

DISCUSSION

It is argued that the mechanism for early suppression is more likely to be a transient withdrawal of excitatory drive than an inhibitory postsynaptic potential (IPSP) at the motoneurone. There are several factors which make an IPSP unlikely: the large amplitude and duration required for effective reduction of firing probability at a firing rate of 10 imp/s, the ineffectiveness of IPSPs induced in the initial phase of the interspike interval, and the length biased sampling of interspike intervals which make it more likely for postsynaptic potentials to occur during the course of long interspike intervals which have the least effect on poststimulus firing probability when expressed as a proportion of the pre-stimulus level. The late rise in firing probability caused by the decay of a presumed IPSP was also absent in some motor units. It is suggested, therefore, that transcranial magnetic stimulation can

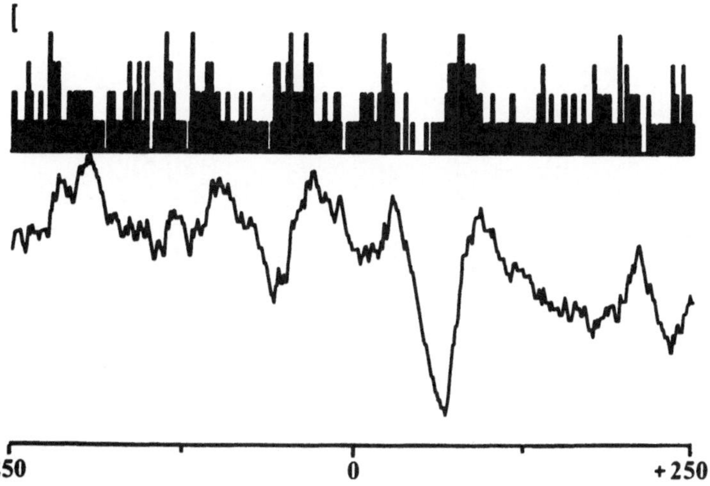

Figure 1. Peristimulus time histogram of the discharge of a single motor unit in the first dorsal interosseus muscle of a healthy subject, with the corresponding cumulative sum. Firing probability is suppressed between 29 and 66 ms after the stimulus. Vertical bar: 1 count; horizontal axis: time (ms); stimuli delivered at time zero.

suppress single motoneurones by a process of transient partial withdrawal of excitatory drive. The challenge is to test this central hypothesis further and to exploit the existence of suppression in patients with upper motoneurone lesions in the design of experiments aimed at increasing our understanding of their disorder at the level of the single spinal motoneurone and its central connections.

REFERENCES

BONIFACE, S.J., MILLS, K.R. & SCHUBERT, M. (1991) Responses of single spinal motoneurones to magnetic brain stimulation in healthy subjects and patients with multiple sclerosis. *Brain* **114**, 643-662.

BONIFACE S.J., MILLS K.R. & SCHUBERT, M. (1994). Suppression and long latency excitation of single spinal motoneurons by transcranial magnetic stimulation in health, multiple sclerosis, and stroke. *Muscle Nerve* **17**, 642-646.

MILLS, K.R., BONIFACE, S.J. & SCHUBERT, M. (1991) The firing probability of single motor units following transcranial magnetic stimulation in healthy subjects and patients with neurological disease. In: *Magnetic Motor Stimulation. Basic Principles and Practice.* eds. LEVY, W.J., CRACCO, R.Q., BARKER, A.T. & ROTHWELL, J.C.

MILLS, K.R., BONIFACE, S.J. & SCHUBERT, M. (1992) Origin of the secondary increase in firing probability of human motoneurones following transcranial magnetic stimulation. Studies in healthy subjects, type I hereditary motor and sensory neuropathy and multiple sclerosis. *Brain* **114**, 2451-2463.

SCHUBERT, M., MILLS, K.R., BONIFACE, S.J., KONSTANZER, A. & DENGLER, R. (1991) Changes in the response pattern of single motor units to transcranial magnetic stimulation in patients with multiple sclerosis and stroke. *Zeitschrift fur Elektroenzephalographie Elektromyographie und Verwandte Gebiete* **22**, 28-36.

CHARACTERISTICS OF THE FLEXION REFLEX IN THE LOWER LIMBS OF CHRONIC SPINAL MAN

R.H. Baxendale[1], D.J. Nicol[2] and M.H. Granat[3]

[1]Institute of Physiology, The University
[2]Department of Biological Sciences
Glasgow Caledonian University
[3]Bioengineering Unit, Strathclyde University
Glasgow G12 8QQ, UK

INTRODUCTION

Our attempts to reconstruct 'walking' in humans who have suffered mid to low thoracic spinal lesions employ a lightweight mechanical orthosis and 4 channels of transcutaneous electrical stimulation (Andrews, Barnett, Phillips, Yamazaki, Baxendale, Paul & Freeman, 1988). The trunk is supported during the stance phase by the action of the orthosis and stimulation of quadriceps. This is interrupted by stimulation of the peroneal nerve at the knee to elicit a flexion reflex which is generates a 'swing phase' for that limb. Variations of this technique have been used in many laboratories since it was first described by Kralj (reviewed by Kralj & Bajd, 1989). The principal problem encountered by users is that of obtaining reliable hip and knee flexion to allow for ground clearance by the toes as the limb swings through to a more advanced position. Rapid habituation of the response brings walking to a halt. We have attempted to characterise the flexor movements in the lower limbs of chronic spinal humans in the hope of improving the gait by appropriate conditioning. These studies have provided considerable experience of the behaviour of flexor reflex responses in chronically spinalised humans.

METHODS

Observation were performed on adult volunteers from the Queen Elizabeth National Spinal Injuries Unit, Southern General Hospital, Glasgow. The experimental protocols have ethical committee approval. The volunteers are young persons aged 20-40 years who have suffered traumatic cord injuries in the thoracic region at least 2 years before experimental tests begin. They are supported upright by in a rigid frame with one leg braced in extension and load bearing. The other leg is free to move and its position is recorded by electrogoniometers. Flexor reflexes are elicited by electrodes approx. 2cm in diameter attached to skin over nerve trunks. Trains of monophasic rectangular pulses of 100-300 microseconds duration lasting 0.5 seconds are applied at 25-50 Hz. The intensity of stimulation is increased progressively until reflex responses are evoked generally at about 5-10 times the current necessary to stimulate the motor axons in the nerve.

RESULTS

Responses vary greatly in magnitude, latency and the rate at which they habituate between subjects, though within any one subject they may be relatively consistent. The causes of these variations are not easily explained. It is not due to muscle fatigue since direct stimulation continues to cause movements long after reflex effects are extinguished.

The rate of decline of reflex magnitude on successive stimulation is a particular problem in gait reconstruction since this slows gait by shortening stride length and ultimately stops any further progress. We have attempted to maintain the magnitude of the reflex movement by 'conditioning' the pathway with additional period of stimulation to the saphenous and sural nerves, changes in stimulation frequency or introducing very intense shocks into the train. Contrary to the expectations raised by the scientific literature describing transmission in flexor pathways in acute spinal animal preparations it was difficult to modulate flexor reflexes in chronic spinal humans. In particular, the pattern of progressive decrement of reflex size seemed very resistant to change. Single volleys in cutaneous C fibres are known to potentiate flexor reflexes for periods in excess of 10 minutes in spinal rats (Wall & Woolf, 1986). Our attempts to reproduce this effect consistently in spinal man were unsuccessful.

DISCUSSION

Several other observations suggest that the observed phenomena may be the product of a spinal 'stepping' or 'swing phase generator', rather than a simple reflex arc. Repeated stimulation with bursts lasting 0.5 seconds at intervals shorter than 3 seconds often evokes responses on alternate presentations of the stimulus with almost complete absence of movement in between. Longer cycles of stimulation could avoid this alternating pattern. This may suggest that the natural period of a 'generator' is at least 3 seconds. It is interesting to note that the normal stride duration is 2 to 3 seconds. In addition, near synchronous stimulation applied to both limbs fails to elicit reflex responses in either limb which might suggest a measure of interlimb co-ordination rather than a simple reflex pathway.

It might be concluded that there are at least two mechanisms contributing to the observed pattern of flexor movements; the well known segmental flexor reflex pathway and a pattern generator capable of being turned on or entrained by the intermittent stimulation of quadriceps or the peroneal nerve. The former could explain the observed habituation of the movement whilst the latter could explain the consistent responses, particularly of the alternating type seen in some volunteers.

ACKNOWLEDGEMENTS

This project was supported by the Medical Research Council.

REFERENCES

ANDREWS, B.J., BARNETT, R., PHILLIPS, G.F., YAMAZAKI, T., BAXENDALE, R.H., PAUL, J.P. & FREEMAN, P. A. (1988) A hybrid orthosis incorporating closed loop control and sensory feedback. *J. Biomed. Eng.* **10**, 189-195.

KRALJ, A. & BAJD, T. (1989) Functional electrical stimulation: standing and walking after spinal cord injury. CRC Press, Boca Raton.

WALL, P. D. & WOOLF, C. J. (1986) Relative effectiveness of C primary afferent fibres of different origins in evoking a prolonged facilitation of the flexion reflex in the rat. *J. Neurosci.* **6**, 1433-1422.

TASK-RELATED CHANGES IN THE TENDON TAP MONOSYNAPTIC RESPONSE OF THE EXTENSOR CARPI RADIALIS MOTOR UNITS IN HUMANS

S. Calvin, A. Schmied, C. Rossi-Durand,
J.-P. Vedel and S. Pagni

"Physiologie et Physiopathologie
Neuromusculaire Humaine", CNRS-NBM
13402 Marseille Cédex 20, France

INTRODUCTION

It is known, in humans, that during isometric contraction Ia fibre discharge frequency increases as the contraction force and EMG activity increase (Burke, Hagbarth & Skuse, 1978). Such an observation suggests that, at high level of EMG activity, the Ia presynaptic inhibition mediated by Ia homonymous afferents might become stronger resulting in a reduction of the tendon tap reflex activation. If this is the case, some differences in the tendon tap response might be seen according to the task in which the muscle is involved.

METHODS

The tendon tap monosynaptic responses of 87 motor units (MU) recorded in the extensor carpi radialis muscles (ECR) of the right forearm of four right-handed subjects were compared during wrist extension and hand clenching around a manipulandum.

In both motor tasks the subjects were asked to maintain the MUs firing at similar frequency using visual and auditory biofeedback. Tendon taps were delivered using an electromagnetic hammer triggered by the MU impulses after a delay of 20 ms. An non-triggering period of 300 ms following each tendon tap allowed to kept the tendon stimulation frequency to around 3 Hz. The Ia afferent burst induced by the tendon tap triggered 20 ms after the MU action potential reached the motoneurone in the first 35 - 50 ms of the after-hyperpolarisation. This was intended, first, to induce submaximal reflex activation allowing a modulation of the reflex processes to be detected, and secondly, to avoid interference between reflexly induced impulses and spontaneous impulses (mean interspike interval ± SEM : 93.9 ± 0.9 ms).

Single MU monosynaptic responses to tendon taps were analysed in peri-stimulus time histograms (analysis period : 160 ms before and after tendon tap occurrence, bin duration : 0.5 ms) where they appeared in the form of a narrow peak (duration : 6.01 ± 0.11 ms; latency : 26.9 ± 0.5 ms, Figure 1A). The monosynaptic MU activation was expressed in term of response probability per trigger (number of impulses in the peak/1000 tendon taps). The recruitment threshold of each MU was measured during a stereotyped ramp contraction. The integrated EMG activity (Figure 1A) and the resulting force were averaged over each tendon stimulation sequence.

RESULTS

Overall data analysis showed that the MU response probability was generally lower during wrist isometric extension that during hand clenching (Figure 1A). The change in the MU reflex activation was closely related to the level of the EMG activity. At a low level of EMG activity (< 0.4 mV), the response probability of the MUs (mean recruitment threshold:

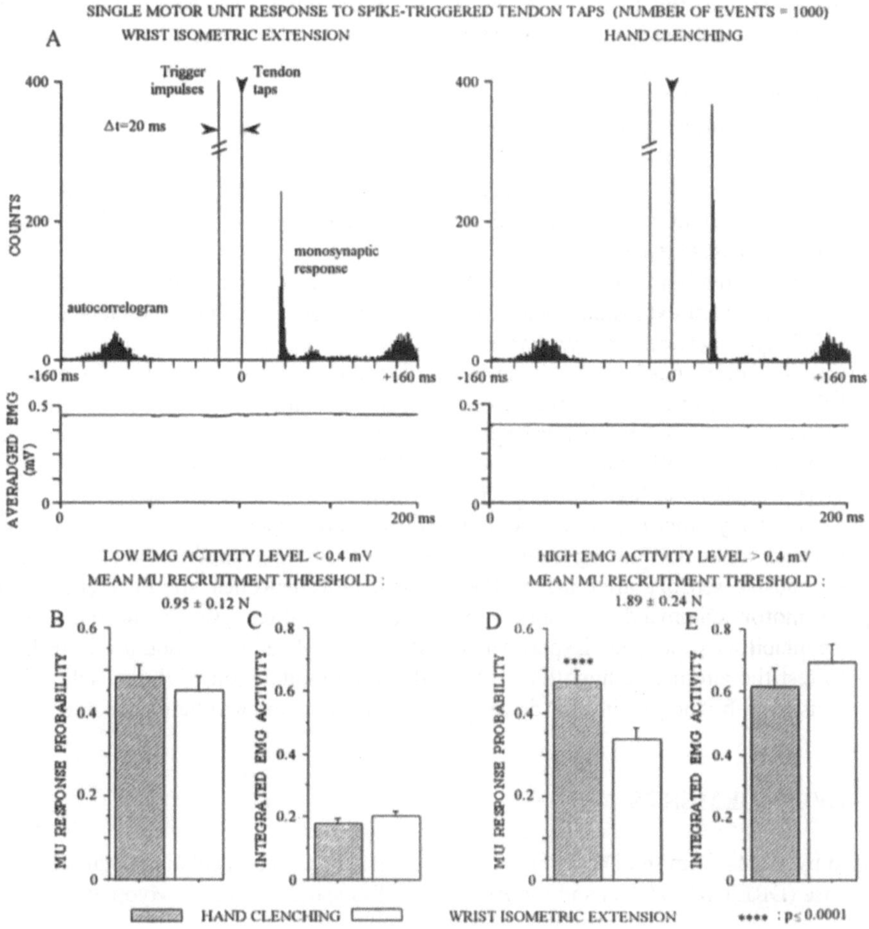

Figure 1. A. Analysis in peri-stimulus time histogram of single motor unit response to spike-triggered tendon taps in the human carpus extensor muscles during wrist extension and hand clenching. B,C, D & E. Comparative analysis of the motor unit (n = 87) response probability per trigger (B, D) during low (C) and high (E) EMG activity levels.

0.95 ± 0.12 N) was quite similar during hand clenching (0.48 ± 0.03) and wrist extension (0.45 ± 0.03, Figure 1B). At a high level of EMG activity (> 0.4 mV), however, the response probability of the MUs (mean recruitment threshold: 1.89 ± 0.24 N) was significantly smaller ($p \leq 0.0001$) during wrist extension (0.33 ± 0.02) as compared to hand clenching (0.47 ± 0.02, Figure 1D). This decrease was found to be unrelated to MU discharge frequency. At low as well as at high levels of contraction, the mean integrated EMG activities remained roughly similar in both motor tasks (Figure 1C & E).

DISCUSSION

The apparent decrease of the motoneuronal excitability at a high level of EMG activity during wrist extension as compared to hand clenching is tentatively interpreted in terms of Ia fibre presynaptic inhibition on the basis of previous data and observations in the present work.

In keeping with this hypothesis, we observed that the earliest component of the tendon tap MU response, i.e. the first 0.5 ms bin of the peak which can be considered to be uncontaminated by polysynaptic effects (Hultborn, Meunier, Morin & Pierrot-Deseilligny, 1987), was also significantly smaller during wrist isometric extension as compared to hand clenching, at a high level of EMG activity.

Moreover, whatever the EMG activity level, the latency of the tendon tap monosynaptic response was shown to be slightly but significantly ($p < 0.0001$) shorter during hand clenching (26.7 ± 0.5 ms) than during wrist extension (27.4 ± 0.5 ms). This would also be consistent with the hypothesis of a reduced efficiency of Ia postsynaptic potentials during wrist extension.

If a presynapic inhibition modulates the MU response probability at high level of muscle contraction during wrist extension, then, why would it disappear during hand clenching? An important difference between the two motor tasks is the strong activation of a large number of cutaneous receptors particularly dense in the finger tips and in the palm (Johanson & Vallbo, 1979) by the pressure exerted around the manipulandum during hand clenching. Such a strong cutaneous mechanoreceptor activation might inhibit the interneurones mediating the Ia presynaptic inhibition (Rudomin, 1990) and so, prevent a reduction of the MU response probability at high level muscle contraction during hand clenching.

Obviously, many other hypothesis could be advanced to explain the difference observed in the MU reflex activation during both motor tasks, considering notably the effects of the wrist flexor muscle sensory afferents on the extensor motoneurones, the difference in the supraspinal motor commands in both motor tasks and their effects on the various presynaptic inhibitory processes. Experiments based on the H reflex method are in progress in order to test the alternative hypothesis of a differential modulation of the muscle spindle dynamic sensitivity by their fusimotor drive during wrist extension and hand clenching.

ACKNOWLEDGEMENTS

Supported by grants from the Direction des Recherches, Etudes et Techniques du Ministère de la défense (DRET n° 91/199) and the Association Française contre les Myopathies.

REFERENCES

BURKE, D., HAGBARTH, K-E. & SKUSE, N.F. (1978) Recruitment order of human spindle endings in

isometric voluntary contractions. *J. Physiol.* **285**, 101-112.
JOHANSSON, R.S. & VALLBO, Å.B. (1979) Tactile sensibility in the human hand: relative and absolute densities of four types of mechanoreceptive units in glabrous skin. *J. Physiol.* **286**, 283-300.
HULTBORN H., MEUNIER, S., MORIN C. & PIERROT-DESEILLIGNY, E. (1987) Assessing changes in presynaptic inhibition of Ia fibres: a study in man and the cat. *J. Physiol.* **389**, 729-756.
RUDOMIN, P. (1990) Presynaptic control of synaptic effectiveness of muscle spindle and tendon organ afferents in the mammalian spinal cord. In *The Segmental Motor System.* eds. BINDER, M.D. & MENDELL, L.M. pp. 349-380. Oxford University Press, Oxford.

THE EFFECT OF INCREASING THE NUMBER OF STIMULI AND INTENSITY OF MUSCLE CONTRACTION ON THE CUTANEOMUSCULAR REFLEXES IN TIBIALIS ANTERIOR IN MAN

H. Bagheri and R.H. Baxendale

Division of Neuroscience
The University
Glasgow G12 8QQ, UK

INTRODUCTION

Research on cutaneomuscular reflexes of the leg originated from studies of the Babinski sign. Cutaneomuscular reflexes (CMRs) with polyphasic excitatory and inhibitory components can be elicited in normal adults following innocuous electrical stimulation of the toes. The general pattern is polyphasic with an excitation followed by an inhibition and then a second excitation (Gibbs, Shiers, Mallik, Harrison & Stephens, 1993; Bagheri & Baxendale, 1993, 1994). The earliest components of CMRs have latencies consistent with segmental spinal pathways but the later components are slow enough to allow for supraspinal pathways. We have examined the effect of repetitive stimuli and the magnitude of the background muscle contraction on the amplitude of CMRs.

METHODS

With local ethical committee approval, experiments were performed in 10 neurologically normal subjects, aged between 21 and 36 years. The subjects were seated in a relaxed position with knee and ankle at 90° of flexion. The cathode was placed on the skin covering the plantar surface of hallux and the anode on the dorsum of the foot. CMRs were elicited by rectangular monophasic pulses of 100 μsecs duration, with an intensity up to 3 times perceptual threshold at a frequency of about 2Hz. EMG electrodes were attached over tibialis anterior. The subject was instructed to maintain a constant force and received visual feedback of his force output from a transducer attached to the foot plate. Peristimulus time averages of 500 stimulus repetitions were prepared and CMRs were identified by comparison the averaged rectified EMG in the 100 msec before and 200 msec after stimulation.

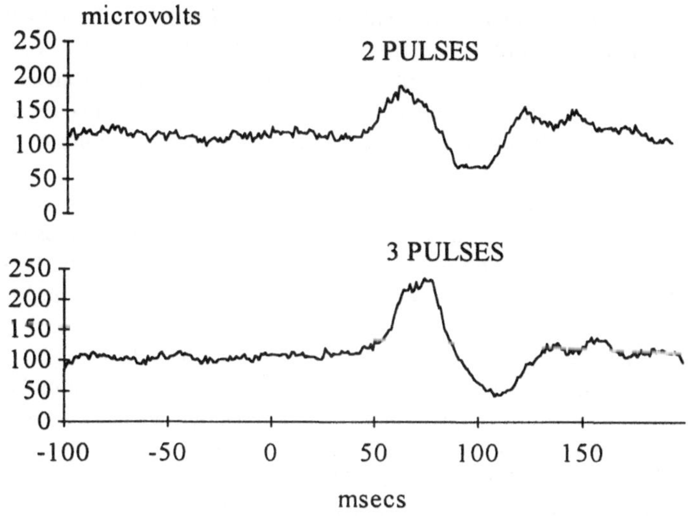

Figure 1. The effect of increasing the number of stimuli on the cutaneomuscular reflexes in tibialis anterior. The first stimulus is at time zero.

RESULTS

In the first part of our experiments CMRs were elicited by single shocks while voluntary contractions were made at 5%, 10% & 20% of maximal voluntary contraction (MVC). In the second part, CMRs were elicited by 2 or 3 shocks at 5 msec intervals. This stimulation did not evoke painful sensations. However, the addition of a fourth pulse changed the character of the sensation in many subjects.

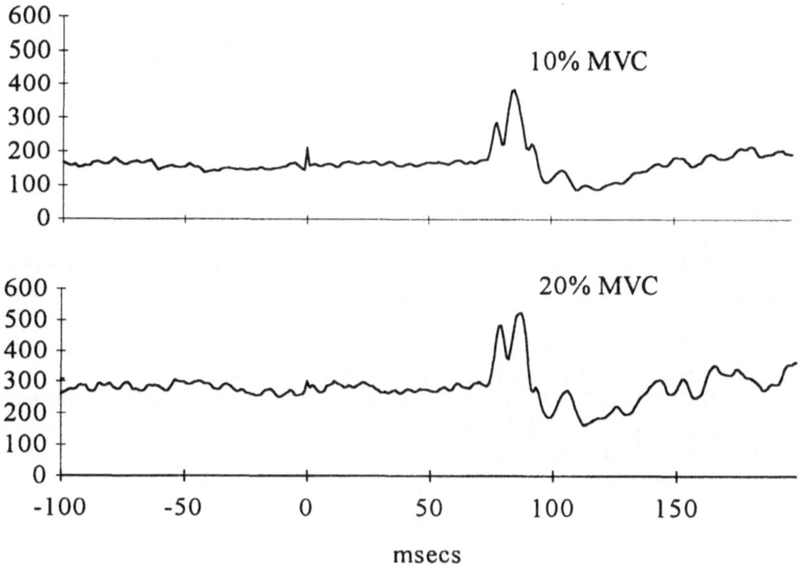

Figure 2. The effect of increasing the intensity of contraction on the cutaneomuscular reflexes in tibialis anterior. The first stimulus is at time zero.

CMR were easily elicited in the anterior tibial muscle of all subjects. Concurrent responses were detected on other muscles of the limb but these are not reported here. The mean latency for the first component of the CMR was 64 ± 9 ms (mean ± SD). The mean delays to the second and third compenents were 72 and 105 ms respectively. The latencies of the various components did not change significantly when a second or third pulse was added to the stimulus.

However, the amplitude of the CMR, expressed as percentage modulation of background EMG, increased significantly when a second or third stimulus pulse was added to the train (Figure 1). The addition of a second pulse 5 msec after the first increased the amplitude of the first excitatory component by 18% (± 9% SD) and a third pulse caused an increase by 34% (± 17% SD).

The size of the background contraction against which the CMR is elicited had a strong effect on the magnitude of the response (Figure 2). Innocuous skin stimulation was never powerful enough to recruit any EMG in an initially inactive muscle. With backgrounds of about 5% of MVC the mean excitation of the E1 was 19% (± 16%, SD) and increasing the background to 10% of MVC increased the E1 amplitude to 29% (± 18%, SD). CMRs were usually at their greatest at this contraction level and further force increases to 20% of MVC caused a slight reduction in E1 amplitudes to 21% (± 17%, SD). The reasons for this are not clearly established. Occlusion between competing excitatory drives on the motoneurone pool is the most likely cause but the effects of fatigue during sustained higher force contractions cannot be excluded.

DISCUSSION

These results show innocuous stimulation of hallux at three times perceptual threshold elicit cutaneomuscular reflexes in tibialis anterior. The earliest CMR have latencies consistent with a spinal reflex pathways. There is evidence that CMRs can be influenced by summation processes and that they are likely to be most effective when muscles are operating in a low force range at about 20% or less of their maximum force. these results are in general agreement with observations made in the upper limb (Durbaba, Cynk & Davies, 1992; Evans, Harries & Harrison, 1992)

REFERENCES

DURBABA, R., CYNK, M. & DAVIES, T.W. (1992) Force production in the human cutaneomuscular reflex. *J. Physiol.* **446**, 603P.

EVANS, P., HARRIES, D. & HARRISON, P.J. (1992) Correlation of the components of the reflex modulation of force with the underlying EMG in the human cutaneomuscular reflex. *J. Physiol.* **452**, 207P.

GIBBS, J., SHIERS, P., MALLIK, A.K., HARRISON, L.M. & STEPHENS, J.A. (1993) Cutaneomuscular reflexes recorded from leg and trunk muscles in man. *J. Physiol.* **467**, 197P.

BAGHERI, H. & BAXENDALE, R.H. (1993) Non-noxious cutaneomuscular reflexes in the human lower limb. *Proc XXXII Cong. I.U.P.S.* 170.39/P.

BAGHERI, H. & BAXENDALE, R.H. (1994) Does posture alter cutaneomuscular reflexes in the Human lower limb? *J. Physiol.* (in press).

DO SECONDARY SPINDLE AFFERENT FIBRES PLAY A ROLE IN THE LATE RESPONSE TO STRETCH OF LEG MUSCLES IN HUMANS?

Marco Schieppati[1], Antonio Nardone[2]
and Stefano Corna[2]

[1]Institute of Human Physiology
University of Genoa, Viale Benedetto XV 3
I-16132 Genoa, Italy
[2]Posture and Movement Laboratory
Medical Centre of Rehabilitation
'Clinica del Lavoro' Foundation
IRCCS, I-28010 Veruno (NO), Italy

INTRODUCTION

Upright stance perturbations induced by a movable platform elicit reflex responses in lower limb muscles (Figure 1). Foot dorsiflexion evokes a short- (SLR) and a medium-latency response (MLR) in both soleus (Sol) and flexor digitorum brevis (FDB) muscles. Foot plantarflexion evokes a MLR in the tibialis anterior (TA). While the SLR is considered to be the counterpart of the stretch reflex (Siliotto, Grasso, Nardone & Schieppati, 1995), no general agreement exists about the origin of the MLR. It has been suggested to be either a spinal response transmitted through group II afferents (Dietz, 1992) or a long-loop reflex initiated by Ia afferents (Fellows, Dömges, Töpper, Thilmann & Noth, 1993). Inspired by the hypothesis that SLR and MLR are transmitted through different neural circuits, we studied a) the effect of height and age on the latency of the responses, b) the modulation of response size induced by changes in the postural "set", c) the peripheral conduction velocity of the responsible fibres and the central delay of the responses, d) the effect of a substance known to selectively depress the transmission through group II-recipient interneurones in the cat.

METHODS

Dorsiflexing (upward tilt) and plantarflexing (downward tilt) perturbations (3 deg rotation) were performed at a velocity of 50 deg/s. Subjects were perturbed under both free stance and stabilised conditions (holding a stable frame placed in front of them). Surface EMG was recorded from the soleus (Sol), flexor digitorum brevis (FDB) and tibialis anterior (TA). The other methods employed in the different experimental situations will be reported under the relevant headings.

RESULTS AND CONCLUSIONS

Bilateral occurrence of medium-latency responses (5 subjects)

Stance perturbations induced by the movable platform consistently evoked in all subjects the response pattern described in the Introduction. When subjects maintained one foot only on the platform and the other on firm ground, no SLRs were evoked in the Sol or FDB muscles of the unperturbed limb, while a clear MLR was still observed in both muscles. The bilateral occurrence of MLR is in keeping with the hypothesis of a different neural circuit mediating the SLR and MLR.

Effect of age (73 subjects)

Age had a significant effect on the latency of Sol SLR and TA MLR. The equations of the regression line fitted through the data points were: Sol SLR, $y = 0.09x + 40.7$, $P<0.001$, $R^2 = 0.16$; TA MLR, $y = 0.35x + 77.4$, $P<0.001$, $R^2 = 0.23$. The slope of the regression lines was significantly ($P<0.01$) different (Nardone, Siliotto, Grasso & Schiepatti, 1995). This finding may be explained by a greater slowing by age of impulse conduction along the small diameter group II fibres responsible for the MLR.

Effect of height (52 subjects)

Height influenced Sol SLR and TA MLR to a different extent. The equations of the regression line fitted through the data points were: Sol SLR, $y = 0.21x + 6.99$, $P<0.001$, R^2

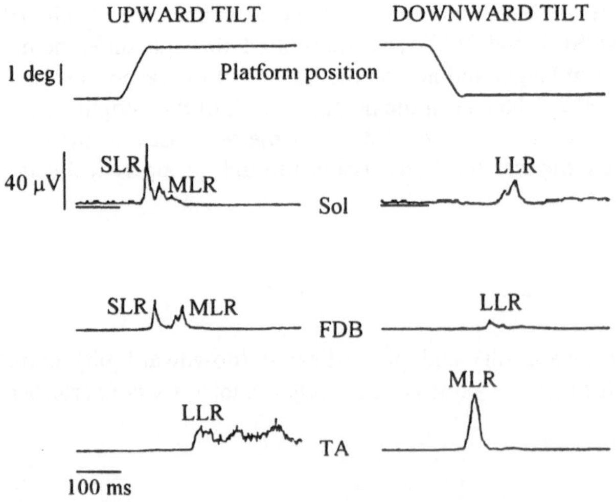

Figure 1. Rectified and averaged (n = 30 trials) EMG of Sol, FDB and TA during upward tilt (left column) and downward tilt (right column) induced during upright stance.

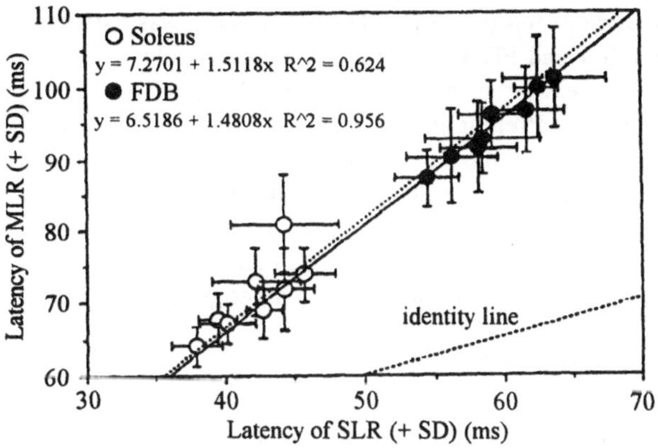

Figure 2. Relationship between MLR- and SLR-latency in Sol (open circles) and FDB (filled circles) muscles during upward tilt. The slopes of the two best-fit lines differ from the identity (dashed line), thus pointing to a transmission of the MLRs through afferents slower than Ia fibres. Each symbol is the mean value (+ S.E.) of the data obtained from eight subjects.

= 0.28; TA MLR, y = 0.42x + 15.2, P<0.02, R^2 = 0.11. The slope of the TA MLR was double that of the Sol SLR. This suggests that the MLR is transmitted through slower-conducting fibres than those mediating the SLR.

Effect of postural set (75 subjects)

SLR and MLR were affected to a different extent by changes in the postural set. When the subjects were standing and holding a stable frame, the MLRs of TA and FDB were strongly reduced in amplitude, whilst the Sol SLR was unaffected. This suggests that the spinal pathways mediating the SLR and MLR receive a different descending control from the supraspinal centres.

Test of the group II hypothesis (8 subjects)

If the MLRs were mediated by group II afferents, the response latencies would appropriately change with muscle distance from the spinal cord. Both the SLRs and MLRs were delayed in FDB as compared to Sol (Figure 2), but the latency difference between the

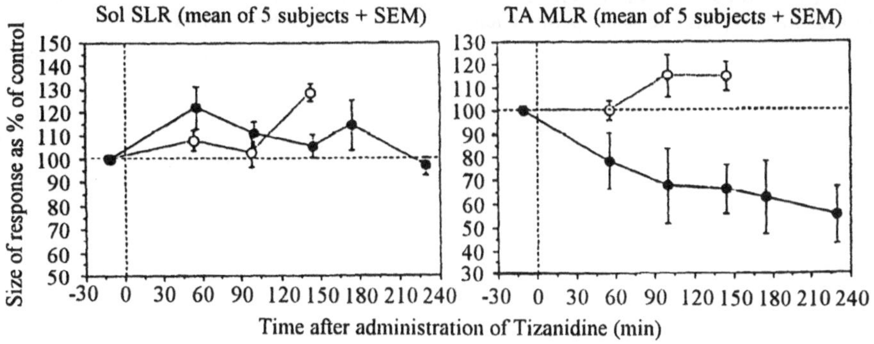

Figure 3. Time-course of the changes in the area of the Sol-SLR (left graph) and TA-MLR (right graph) after administration of tizanidine (filled circle) and placebo (open circle).

MLRs was larger than between the SLRs. The extra delay was calculated in 5 subjects and the distance between Sol and FDB recording sites was divided by this value to obtain the conduction velocity of the afferent fibres responsible for the MLR of FDB. This proved to be 27 m/s, on average. The afferent and efferent conduction velocity of the fibres mediating the H reflex in FDB was obtained by stimulating the tibial nerve in two sites, and the distance from the motor pool was calculated by using the reflex latency and appropriate correction factors. The central delay for MLR proved to be about 13 ms, while that for the FDB SLR was instead 2 ms. These figures suggest that MLRs are mediated by the spindle secondaries, relayed to the motoneurones through an oligosynaptic pathway.

Effect of tizanidine (5 subjects)

The effect of oral intake of a substance having an α_2 agonist action on the size of SLR and MLR was also tested. Tizanidine was orally administered (single-blinded) in a single dose of 150 μg/kg of body weight. Perturbations were continuously delivered prior to intake until four hours afterwards. No changes in response latencies were seen. On average, a decrease in size of TA-MLR and FDB-MLR occurred after 1 hour of drug administration, whilst Sol-SLR was unaffected (Figure 3). The degree of response depression was almost superimposable on that induced by holding. The same paradigm was repeated after a few days with intake of placebo, and no change in the amplitude of responses occurred. These findings confirm the hypothesis that different pathways subserve the two responses, since tizanidine selectively depresses transmission through group II-recipient interneurones (Jankowska, 1992). On the other hand, they show that these postural responses are under powerful control from brain stem monoaminergic centres.

REFERENCES

DIETZ, V. (1992). Human neuronal control of automatic functional movements: interaction between central programs and afferent input. *Physiol. Revs.* **72**, 33-69.

FELLOWS, S.J., DÖMGES, F., TÖPPER, R., THILMANN, A.F. & NOTH, J. (1993). Changes in the short- and long-latency stretch reflex components of the triceps surae muscle during ischaemia in man. *J. Physiol.* **472**, 737-748.

JANKOWSKA, E. (1992). Interneuronal relay in spinal pathways from proprioceptors. *Prog. Neurobiol.* **38**, 335-378.

SILIOTTO, R., GRASSO, M., NARDONE, A. & SCHIEPPATI, M. (1995). Contribution of foot muscle responses to perturbation of stance. In *Motor Control VII.* eds. STUART, D.G., GANTCHEV, G.N., GURKINKEL, V.S. & WIESENDANGER, M. (in press).

NARDONE, A., SILIOTTO, R., GRASSO, M. & SCHIEPATTI, M. (1995). Influence of aging on leg muscle reflex responses to stance perturbation. *Arch. Phys. Med. Rehabil.* (in press).

MODULATION OF RECIPROCAL Ia INHIBITION FROM PRETIBIAL FLEXORS TO EXTENSORS DURING TONIC PRETIBIAL CONTRACTION IN MAN: ANALYSIS WITH SINGLE MOTOR UNIT RECORDING

Masaomi Shindo

Department of Medicine (Neurology)
Shinshu University School of Medicine
Asahi 3-1-1, Matsumoto 390, Japan

INTRODUCTION

Change in reciprocal Ia inhibition during tonic voluntary contraction of antagonists has been controversial. Early studies reported that inhibition increased in extent during contraction (Tanaka, 1974; Shindo, Harayama, Kondo, Yanagisawa & Tanaka, 1984), whereas later one showed no evidence for the increase (Iles, 1986; Crone, Hultborn, Jespersen & Nielsen, 1987). The discrepancy may possibly be caused by physiological characteristics of the motoneuronal pool such as a nonlinear input-output relationship of the pool (Kernell & Hultborn, 1990) and dependence of the conditioning effect on the test reflex size (Crone, Hultborn, Mazières, Morin, Nielsen, & Pierrot-Deseilligny, 1990). To know whether the inhibition actually increases or not during tonic antagonist contraction, we studied this issue on single motor units with a new method, avoiding the methodological problems which derive from the motoneuronal pool.

METHODS

The experiments were performed on 6 normal subjects. The examined muscle was the soleus, and the compound H-reflex and a single motor unit were recorded, stimulating the tibial nerve. Conditioning stimuli were given to the common peroneal nerve. The experiments were done at rest and during weak tonic voluntary contraction of the pretibial flexors, and these two situations were alternated. Since the principle of the new method using a single motor unit is described in detail elsewhere (Shindo, Yanagawa, Morita & Hashimoto, in press), it is here outlined briefly. The excitability of a single soleus motoneurone was assessed at a test stimulus intensity which produced a firing probability of 50% in a particular motor unit. Initially, the test stimulus was manually set to a certain intensity around the firing threshold of a motor unit. If this stimulus fired the motor unit, the next stimulus was weakened automatically by a single step with a computer, and vice versa. By repeating such procedure, the firing probability of this motor unit converges to 50%. This intensity naturally depends on the excitability of the single motoneurone. The threshold intensity for the Ia fibres was further obtained with the method of PSTH during tonic contraction of the examined soleus muscle, and it was subtracted from the intensity with the firing probability of 50%. The resulting intensity we called the 'critical firing stimulus (CFS)' which was used as an indicator of the motoneuronal excitability. The effect of the conditioning stimulus to the common peroneal nerve was assessed by the change in the CFS, and its extent was expressed as a percentage of the unconditioned control CFS. The firing threshold of a motor unit was determined in terms of 'critical firing level (CFL)' after Henneman, Clamann, Gillies & Skinner (1974), the compound H-reflex being substituted for the monosynaptic reflex in the cat. The CFLs of the motor units examined in this study distributed from 0.3 to 20 % of the maximal M-response.

RESULTS

Figure 1 shows a time course of inhibition of a soleus motor unit produced by the common peroneal nerve stimulation in the resting state and during tonic voluntary contraction of the pretibial muscles. In the resting state inhibition appeared at a conditioning-test stimulus interval of 2.0 ms, reached a peak at 4.0 ms, and the CFS returned to the control level at 7.0 ms. This inhibition appeared at 0.75 times the motor threshold (not shown). The short latency and low threshold inhibition is compatible with

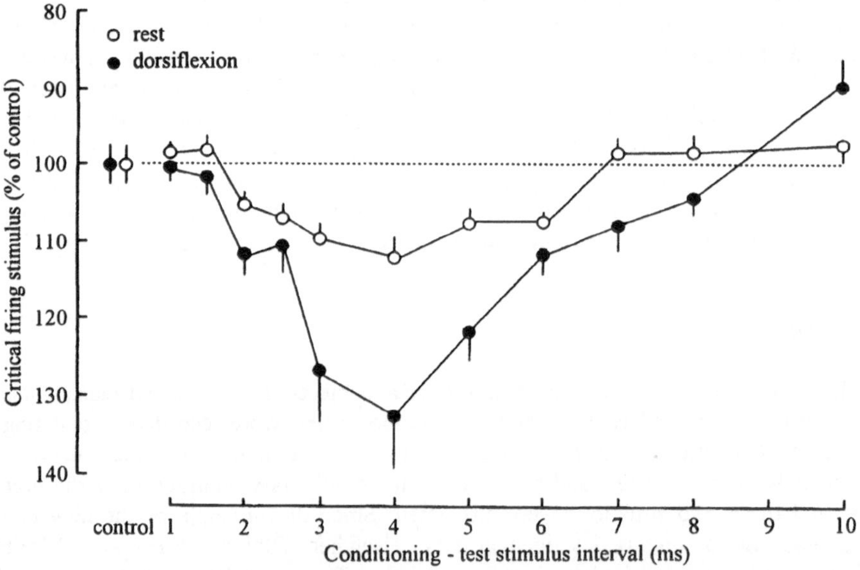

Figure 1. Time courses of inhibition of a soleus motor unit produced by conditioning stimulation to the common peroneal nerve. The critical firing stimuli (CFS, ordinate) were plotted against the conditioning-test stimulus intervals (abscissa). In the resting state (open circles), inhibition appeared at the interstimulus interval of 2.0 ms, reaching the peak at 4.0 ms. During tonic weak contraction (2 % of the maximum), the extent of inhibition increased throughout its whole time course, including the very early part of the inhibition. Note that the larger CFS, representing lower excitability of the corresponding motoneurone, was represented downward in order to have a similar image with IPSPs. Each symbol and bar are the means and S.D.s from 50 measurements.

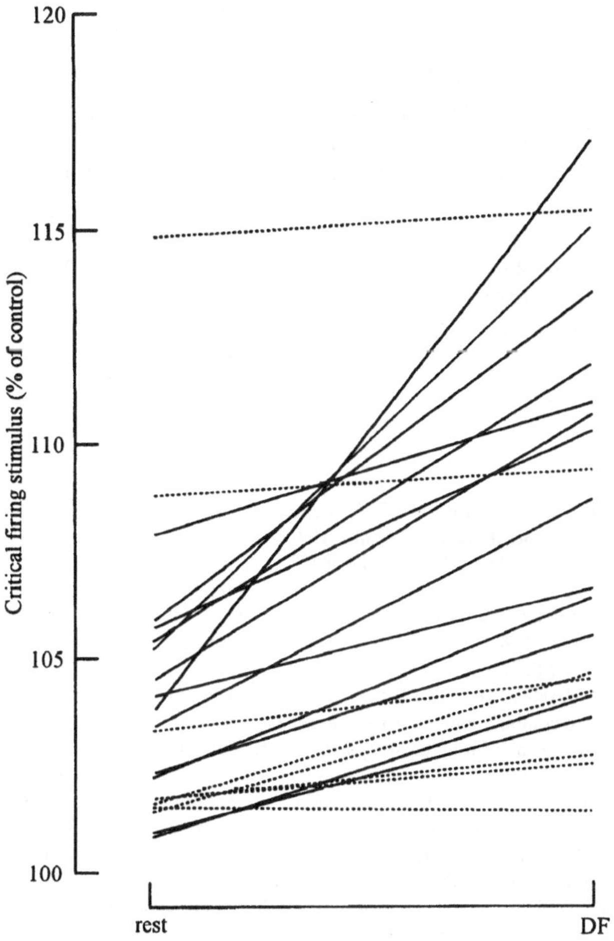

Figure 2. Change in the extent of reciprocal Ia inhibition of all the soleus motor units by tonic voluntary dorsiflexion (DF). The conditioning-test stimulus intervals were within 0.7 ms after the onset of inhibition in each time course, and the conditioning intensity was set at 0.80 - 0.95 times the motor threshold so that only small inhibition was recognised in the resting state. The strength of contraction was 1 - 2 % of the maximum. The conditioned CFS (ordinate) was expressed as a percentage of the unconditioned CFS, and larger values are represented upwards. Solid lines, significant increase in inhibition; broken lines, non-significant change. In none of the motor units did the inhibition decrease in extent during voluntary dorsiflexion.

reciprocal Ia inhibition. During weak tonic pretibial contraction (2 % of the maximum), the inhibition increased in extent through its whole time course including the earliest part. The changes in Ia inhibition during tonic dorsiflexion are summarised in Figure 2 where the conditioning-test stimulus intervals were chosen within 0.7 ms after the onset of inhibition in each time course, and the conditioning intensity was set at 0.80-0.95 times the motor threshold so that even a small inhibition was recognised in the resting state. The strength of contraction was very weak (1-2 % of the maximum) in all the subjects. Out of 21 motor units, 14 had inhibition at rest. Among these 14, 11 units showed increase in inhibition but in 3 motor units the extent of inhibition did not change. In the remaining 7 motor units that had no inhibition at rest, significant inhibition appeared in 2. In none of the motor units did the inhibition decrease in extent during voluntary dorsiflexion.

CONCLUSION

Seventy-six percent (16/21) of the soleus motor units received Ia inhibitory input from the common peroneal nerve. The inhibition was explored in single motoneurones, thus avoiding any problems related to the motoneuronal pool properties. Since the relationship between the test stimulus intensity and the size of the test EPSP in motoneurones is likely to be linear (Eccles, Eccles & Lundberg, 1958), we conclude that activity of Ia inhibitory interneurones increases with weak tonic contraction of the antagonists. The new method using a single motor unit appears to be superior to the usual H-reflex method due to the fact that the physiological characteristics of the motoneuronal pool can be neglected in interpreting the results. Such combined utilisation of the new and conventional reflex methods may be expected to yield a better understanding of the spinal circuitry.

REFERENCES

CRONE, C., HULTBORN, H., JESPERSEN, B. & NIELSEN, J. (1987) Reciprocal Ia inhibition between ankle flexors and extensors in man. *J. Physiol.* **389**, 163-185.

CRONE, C., HULTBORN, H., MAZIÈRES, L., MORIN, C., NIELSEN, J. & PIERROT-DESEILLIGNY, E. (1990) Sensitivity of monosynaptic test reflexes to facilitation and inhibition as a function of the test reflex size: a study in man and the cat. *Exp. Brain Res.* **81**, 35-45.

ECCLES, J.C., ECCLES, R.M. & LUNDBERG, A. (1958) Synaptic actions on motoneurones in relation to the two components of the group I muscle afferent volley. *J. Physiol.* **136**, 527-546.

HENNEMAN, E., CLAMANN, H.P., GILLIES, J.D. & SKINNER, R.D. (1974) Rank order of motoneurons within a pool: law of combination. *J. Neurophysiol.* **37**, 1338-1349.

ILES, J.F. (1986) Reciprocal inhibition during agonist and antagonist contraction. *Exp. Brain Res.* **62**, 212-214.

KERNELL, D. & HULTBORN, H. (1990) Synaptic effects on recruitment gain: a mechanism of importance for the input-output relations of motoneurone pools. *Brain Res.* **507**, 176-179.

SHINDO, M., HARAYAMA, H., KONDO, K., YANAGISAWA, N. & TANAKA, R. (1984) Changes in reciprocal Ia inhibition during voluntary contraction in man. *Exp. Brain Res.* **53**, 400-408.

SHINDO, M., YANAGAWA, S., MORITA, H. & HASHIMOTO, T. Conditioning effect in single human motoneurones: a new method using the unitary H reflex. *J. Physiol.* (in press).

TANAKA, R. (1974) Reciprocal Ia inhibition during voluntary movements in man. *Exp. Brain Res.* **21**, 529-540.

STUDIES OF PRESYNAPTIC INHIBITION IN THE STRETCH REFLEX PATHWAY OF THE HUMAN AND THE CAT

Charles Capaday and Brigitte A. Lavoie

Centre de Recherche en Neurobiologie
Université Laval, Québec, G1J 1Z4, Canada

INTRODUCTION

Presynaptic inhibition of muscle afferent terminals in the spinal cord, including perhaps those of Ib afferents, is a potential and direct mechanism for control of the stretch reflex parameters. Eccles (1964), in reviewing the extensive studies of his group on the cellular basis and network organisation of presynaptic inhibition in the spinal cord, expressed the view that this form of inhibition was in fact more potent than postsynaptic inhibition. Despite this, the functional role of presynaptic inhibition in motor control has, only recently, begun to be studied. Matthews (1972) commented that presynaptic inhibitory mechanisms have not, in general, been incorporated into ideas of how the spinal cord controls movements because of a lack of quantitative measures of its effects. In this chapter we describe the results of two experimental studies bearing on the functional role of presynaptic inhibition during walking in humans and on the biomechanical consequences of presynaptic inhibition of the stretch reflex (s.r.).

In the first experiment, we tested in freely moving human subjects the hypothesis (Morin, Katz, Mazieres & Pierrot-Deseilligny, 1982; Stein & Capaday, 1988) that the decrease of the soleus H-reflex which occurs in going from standing to walking is due, at least in part, to an increase of presynaptic inhibition of the intraspinal terminals of the group Ia muscle afferents. The results of the second experiment address the issue of how presynaptic inhibition of muscle spindle afferents affects the mechanical parameters (stiffness and threshold) of the s.r.

EXPERIMENTS IN HUMANS

It is well known from experiments in the cat that stimulation of flexor nerves at group I strength can induce presynaptic inhibition of the intraspinal terminals of extensor group Ia afferents (reviewed by Rudomin, 1990). Indeed this also seems to be the case in humans, since we found that in normal subjects a single electrical stimulus to the common peroneal nerve (1 - 1.5 x MT) delivered during quiet standing produces a strong inhibition of the soleus H-reflex, by up to 50%. (Figure 1). The inhibition becomes evident at conditioning - testing (C-T) intervals of about 50ms, and reaches its maximum at C-T intervals of 100 - 120 ms. The conditioning stimulus, when it is given alone, produces little or no change in the background EMG measured at the time of the test stimulus. In fact, based on the relation between the H-reflex amplitude and the mean rectified background EMG during standing, the inhibition of the H-reflex was always greater than could be accounted for by changes in the background EMG. Furthermore, the conditioning stimulus has little effect on the soleus evoked motor response to stimulation of the ankle area of the motor cortex. Taken together, these observations strongly suggest that the inhibition of the H-reflex is due to an increase of presynaptic inhibition of the intraspinal terminals of Ia muscle spindle afferents. When the conditioning stimulus is given in the early part of the stance phase of walking, at a time when the H-reflex is smaller than during standing at similar levels of background EMG (Morin et al., 1982; Stein & Capaday, 1988), it has little if any effect on the H-reflex (average inhibition 45.8% vs. 11.6%, n = 14 subjects). The soleus background EMG, and the soleus and tibialis anterior M-waves were essentially the same in the two tasks. This result is surprising since we had expected that presynaptic inhibition would be greater during walking, because our hypothesis predicts that the presynaptic inhibitory neurons should be more excitable in this phase of the step cycle. There are at least two explanations of this striking, task dependent, modulation of the effects of a conditioning input from a flexor nerve. First, it is possible that the conditioning peripheral input may be itself inhibited during walking, either at the presynaptic or the interneuronal level. Secondly, during walking, the presynaptic inhibitory network may be saturated as a result of central or afferent activity and therefore unresponsive to an added conditioning peripheral input. Thus, while the present experiments did not provide a direct demonstration of an increase of presynaptic inhibition of the soleus Ia-afferents during walking, the results may, nonetheless, be compatible with that idea; and also suggest that in this task the presynaptic inhibitory network is possibly under central control. An important methodological issue is also raised by these results. Neurophysiological studies using C-T paradigms depend on maintaining a

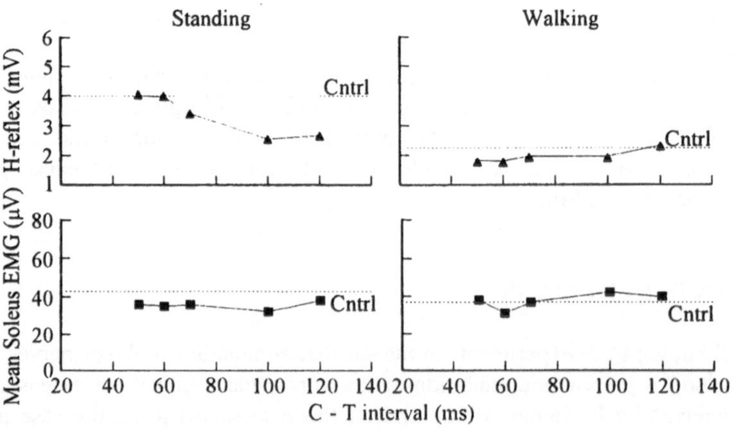

Figure 1. During walking the same conditioning stimulus (1.5 x MT) which produced an inhibition of the soleus H-reflex during standing, at C-T intervals of 100 - 120 ms, did not produce any inhibition during walking. The results were obtained in the same subject during the same experimental session and are shown side by side for each task. Note also that the H-reflex inhibition during standing occurs without a significant change of the background EMG. The natural inhibition of the H-reflex in going from standing to walking is readily apparent, as can be seen from the amplitude of the control H-reflex response in each task.

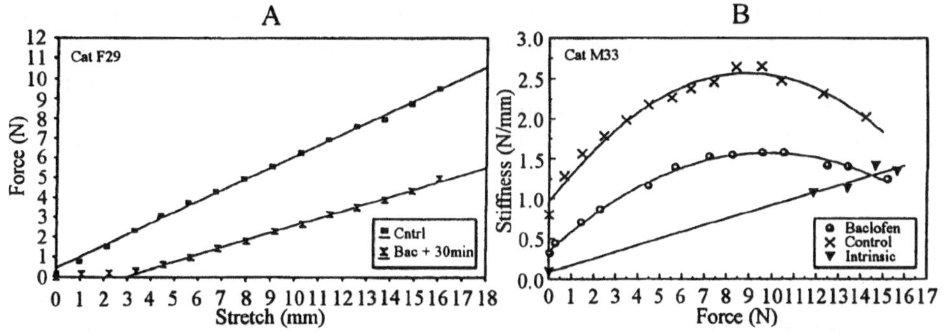

Figure 2. Examples of the effects of Baclofen on the length-tension (A) and stiffness-force (B) relations of the cat stretch reflex. There are two main points summarised in these figures. Firstly, the decrease of the stiffness of the tonic stretch reflex (A) following the injection of Baclofen is always accompanied by an increase of its threshold. Secondly, Baclofen can decrease the stiffness of the stretch reflex, measured "dynamically", at all background forces (B).

constant conditioning stimulus. This criterion can usually be met when dealing, for example, with anaesthetised animal preparations. However, when this paradigm is used during natural motor tasks, as in the present study, it is always tacitly assumed that the task itself would not affect the conditioning stimulus. This is clearly not a tenable assumption without direct corroborating experimental evidence.

EXPERIMENTS IN CATS

These experiments were done in cats decerebrated at the precollicular postmammillary level to determine how a tonic increase of presynaptic inhibition of the intraspinal terminals of muscle spindle afferents changes the mechanical properties of the soleus s.r. Baclofen, a specific $GABA_B$ receptor agonist, was injected i.v. (1 mg/kg) so as to induce a tonic increase of presynaptic inhibition (Edwards, Harrison, Jack & Kullman, 1989). The effects of Baclofen on the stiffness and threshold of the s.r. were determined from plots of stiffness vs. background force and length vs. force, respectively. Baclofen, at these doses, had no effect on the excitation-contraction coupling properties of muscle, nor on the intrinsic stiffness-force relation. Changes of the soleus background force, required to obtain the stiffness vs. force plots, were produced by stimulation of the contralateral common peroneal nerve or the posterior tibial nerve. The stiffness of the s.r. as a function of the background force level was determined by stretching the muscle with a square pulse of 1 - 2 mm amplitude and 200 - 300 ms duration. The stiffness at each force level was calculated by dividing the change in force by the change in length, at a point where the force trace had stabilised. The length-tension relation of the soleus stretch reflex was determined by stretching the muscle 14 - 17 mm at a constant rate of 1 - 2 mm/sec. Baclofen produced a significant decrease (25 - 40% or more) of the s.r. stiffness, at all force levels, within 10 - 15 min. of injection as determined from the stiffness-force plots (Figure 2B). The length-tension plots revealed that the decrease of s.r. stiffness was always accompanied by an increase of the s.r. threshold, typically 2 - 3mm (Figure 2A). Since the fusimotor system, reciprocal inhibition, and activation of descending motor pathways, mainly change the s.r. threshold with little effect on its stiffness (see Matthews, 1959; 1972), we suggest that the effects of Baclofen on the stretch reflex stiffness occur as a result of an increase of presynaptic inhibition of the muscle afferent terminals. Thus, it is concluded that it may be possible for the CNS to adaptively modify the s.r. stiffness via presynaptic inhibition of the of the intraspinal terminals of muscle afferents, and thereby modify the mechanical impedance at a single joint. However, any such change of s.r. stiffness will be accompanied by a change of the s.r. threshold. Therefore, the s.r. threshold is not an independent variable, it depends on the level of presynaptic inhibition of the muscle spindle afferent terminals.

CONCLUSION

Presynaptic inhibition of muscle spindle afferent terminals in the spinal cord is potentially a direct and independent neural mechanism specific for the control of stretch reflex parameters. The findings in the cat describe what would be expected if a change of presynaptic inhibition were to occur during a motor task. The challenge will be to identify clearly such tasks, which can be fraught by unexpected complications as described above, and to correlate the neurophysiological changes with the biomechanical changes.

ACKNOWLEDGMENTS

This work was supported by the MRC of Canada. Charles Capaday is a research scholar of the Fond de la Recherche en Santé du Québec (FRSQ). The expert and dedicated technical assistance of Louise Bertrand and François Comeau is gratefully acknowledged.

REFERENCES

ECCLES, J.C. (1964). *The Physiology of Synapses.* Academic Press Inc., New York.

EDWARDS, F.R., HARRISON P.J., JACK J.J.B. & KULLMANN, D.M. (1989). Reduction by baclofen of monosynaptic EPSPs in lumbosacral motoneurones of the anaesthetized cat. *J. Physiol.* **416**, 539-556.

MATTHEWS, P.B.C. (1959). A study of certain factors influencing the stretch reflex of the decerebrate cat. *J. Physiol.* **147**, 547-564.

MATTHEWS, P.B.C. (1972). *Mammalian Muscle Receptors and their Central Actions.* Arnold, London.

MORIN, C., KATZ, R., MAZIERES, L. & PIERROT-DESEILLIGNY, E. (1982). Comparison of Soleus H-reflex facilitation at the onset of Soleus contractions produced voluntarily and during the stance phase of human gait. *Neurosci. Lett.* **33**, 47-53.

RUDOMIN, P. (1990). Presynaptic inhibition of muscle spindle and tendon organ afferents in the mammalian spinal cord. *TINS.* **13**, 499-505.

STEIN, R.B. & CAPADAY, C. (1988). The modulation of human reflexes during functional motor tasks. *TINS.* **11**, 328-332.

THE EFFECTS OF FATIGUING VOLUNTARY ACTIVITY ON H REFLEX EXCITABILITY IN MAN

G.R. Olyaei and R.H. Baxendale

Division of Neuroscience
The University
Glasgow G12 8QQ, UK

INTRODUCTION

It is well known that the EMG activity associated with maximal voluntary contraction is reduced during muscle fatigue and it has been suggested that one reason for this decline might lie in reflex inhibition of motoneurone pools by muscle afferents sensitised by the fatiguing contraction. Garland & McComas (1990) demonstrated that fatigue of soleus causes a depression of its H reflex excitability. Soleus is difficult to fatigue and so we have sought to extend this work by examining the reflex effects of fatiguing contractions in other muscles. We have examined the effects of fatiguing the anterior tibial muscles on their own H reflex excitability and on the H reflex of their antagonist muscle, soleus.

METHODS

All experimental protocols were approved by the local ethics committee. Experiments were performed on 17 neurologically normal subjects. Their ages ranged from 17 to 42 years. Subjects were seated in a semi-reclined position with their knee and ankle supported at 110° and 90° respectively. Maximal voluntary dorsiflexions were recorded at intervals through the experiment. EMG was recorded with surface electrodes placed over the anterior tibial muscles or over soleus. H reflexes of about half maximal amplitude were elicited by stimulation of the tibial nerve on the popliteal fossa or common peroneal nerve on the head of fibula with single pulses, 0.1 ms duration, at intervals of 5 sec. In most experiments it was necessary for subjects to make a weak voluntary contraction of the anterior tibial muscles before an H reflex could be elicited. However, consistent H reflexes were recorded in relaxed soleus in all subjects. The mean of 10 successive tests were compared before and after anterior tibial muscle activity. At least 1 minute elapsed after the end of contraction before the first test was made. In addition, a wider range of stimulus intensities were employed to identify maximal M and H waves (Figure 1).

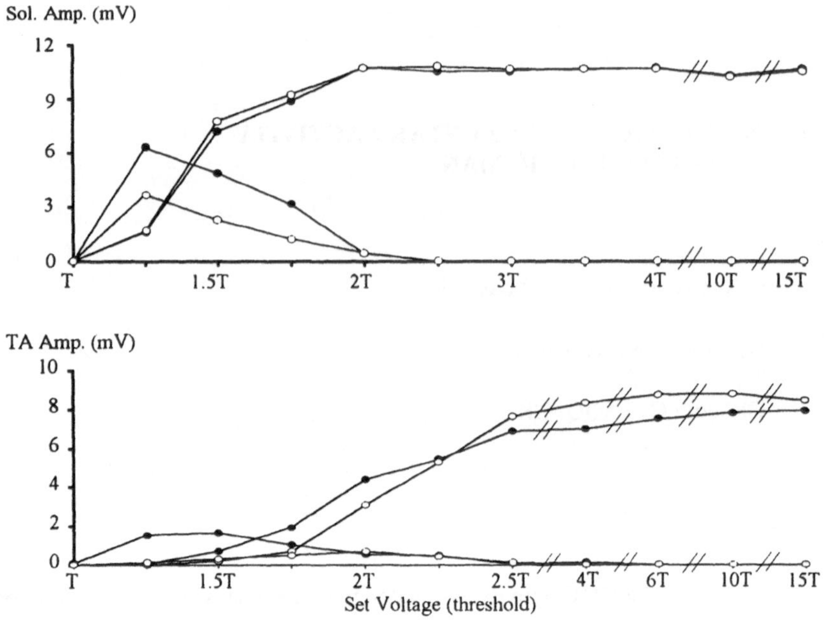

Figure 1. A specimen of soleus and anterior tibial M waves and H reflexes recruitment curves from one subject. Filled circles show data obtained before activity in anterior tibial muscle and open circles in the minute following activity.

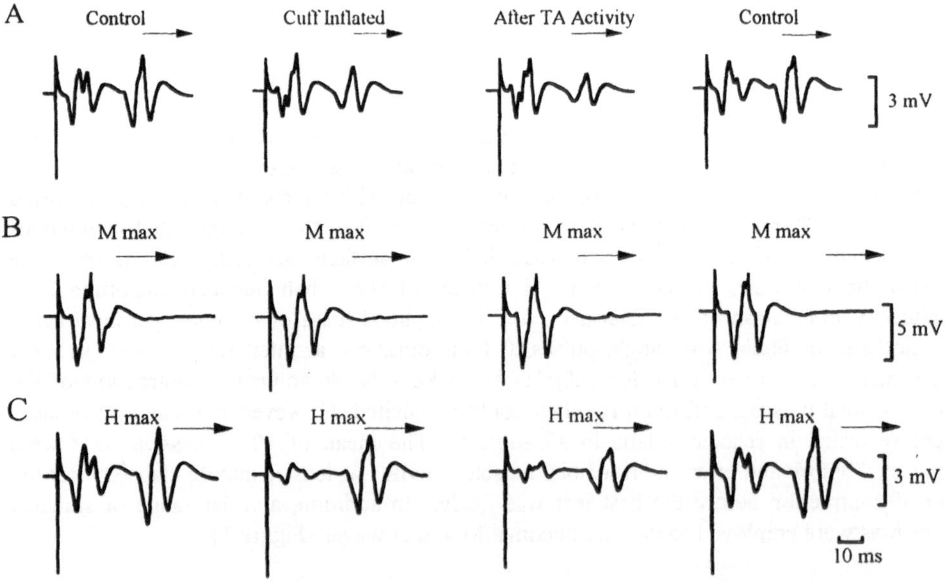

Figure 2. M_{max} waves and H_{max} reflexes in soleus in control conditions with normal blood flow, cuff inflated, after 3 minutes activity in TA (with blood flow occluded), and after 10 minutes recovery period. Sequence A- averaged submax soleus H reflexes obtained in control conditions, after inflation of BP cuff, after TA activity and in control conditions. Sequence B- in the same subject, maximal M waves in the same sequence. Sequence C- maximal H waves.

RESULTS

Anterior tibial muscles were fatigued by intermittent voluntary isometric contractions at 30% of MVC, 7 sec on, 3 sec off, sustained for up to 9 minutes. These periods of muscle activity were performed with and without an arterial occlusion cuff round the mid thigh inflated to 150 mmHg to obstruct the circulation. Alternatively, the anterior tibial muscles were fatigued by transcutaneous electrical stimulation of the muscle belly at 20 Hz at the highest intensity the volunteer found tolerable. Stimulation was delivered for 7 seconds and then turned off for 3 seconds. This stimulation cycle was repeated until the tetanic force had fallen to about half its original value.

Approximately equivalent fatigues, as judged by reductions in MVC and slowing of relaxation rates, were produced by electrical and voluntary exercise.

No significant difference was observed between the reflex changes accompanying voluntary and electrically induced muscle activity. The mean post activity MVC was 86 ± 7% (mean ± SD, range 71-93%) with normal blood flow. The amplitude of anterior tibial muscles M waves was reduced (mean 90 ± 10%, range 76-107%), but the anterior tibial muscles H wave amplitude fell rather more to 60 ± 28%, (range 10-86%). The anterior tibial muscles H wave reduction was significantly greater than the anterior tibial muscles M wave reduction ($P<0.02$, student's t test). The ratio of anterior tibial muscles H_{max}/M_{max} showed a statistically significant reduction after activity, falling to 58 ± 22% of preactivity values (range 34-85%, $P<0.005$).

After 3 minutes exercise when blood flow had been occluded the mean MVC was reduced to 79% ± 5% (mean ± SD, range 72-88%). The amplitude of soleus M and H waves were not significantly different from pre-exercise values. Maximal soleus M waves were not affected by anterior tibial muscles activity but the maximal soleus H reflexes were significantly reduced. The ratio of Soleus H_{max}/M_{max} showed a significant reduction after activity, falling to 75 ± 22% control values (range 45-100%, $P<0.008$). These results are shown in Figure 2. MVCs, M and H waves were restored to control values after 10 minutes recovery.

DISCUSSION

These results show that H reflexes in anterior tibial muscles are reduced after fatiguing voluntary contractions even with normal blood flow. However only maximal H reflexes in soleus can be reduced after severe fatiguing voluntary contractions in its antagonist anterior tibial muscles, when blood pressure cuff was inflated. The relative stability of the M and M_{max} waves whilst anterior tibial muscles and soleus H waves are reduced suggests that this is a reflex phenomenon rather than a change in excitability of neuromuscular junctions.

REFERENCES

GARLAND, S. J. & MCCOMAS, A. J. (1990) Reflex inhibition of human soleus muscle during fatigue. *J. Physiol.* **429**, 17-27.

"SPATIAL ACTIVATION MAPS" OF HUMAN ALPHA-MOTONEURONES IN RELATION TO THREE-DIMENSIONALLY RECORDED MOVEMENTS

E. Nordh[1], N. Al-Falahe[2],
H. Zafar[2] and P.-O. Eriksson[2]

Departments of Clinical Neurophysiology[1]
and Clinical Oral Physiology[2]
Umeå University, S-901 87 Umeå, Sweden

INTRODUCTION

Analyses of movement-related nerve discharge characteristics are usually performed on data obtained from experiments involving spatially restricted motor tasks. In reduced preparations, the muscular action often comprises only one or two dimensions, and in behavioural studies of animals or humans, the motor task is often stereotyped and spatially limited, compared to the natural movement range. These approaches allow analyses of nerve cell activation in relation to temporal aspects of movements, but relations to the spatial aspects of the movements are easily lost. Furthermore, relating neuronal discharge patterns to multi-dimensional movements is conceptually demanding. The investigator has to comprehend simultaneously the neuronal discharge rate and details of the movements, as for example extent, velocity or acceleration. Additional methods for presentation of nerve cell activity in relation to the three-dimensional (3-D) movement range of a body segment, which would facilitate the analysis of neuromuscular activation patterns during complex movements, would thus be of value. This report describes a technique by which electrophysiological data from unitary recordings can be qualitatively related to the movements of a body segment in 3-D space.

METHODS

Single nerve fibre activity was recorded together with natural finger movements in man. The subject was seated, with his left arm firmly supported by an individually shaped vacuum-cast, and with the hand strapped to a mechanical support. Activity from efferent axons to the finger extensor muscles in the forearm was recorded from the left radial nerve with the microneurographic technique of Vallbo & Hagbarth (1968). Classification of the fibre as an efferent was based on (i) isolated nerve discharges during contractions forearm extensors, (ii) absence of response to manipulation of the skin and the deep tissues within the innervation area of the radial nerve, and (iii) absence of response to direct electrical muscle stimulation or passive stretch of the extensors. Finger movements were performed with the left index finger, restricted to the metacarpo-phalangeal joint by splinting the interphalangeal joints. Optionally, the finger movements were guided and restricted in different degrees of adduction-abduction, by a low friction sliding plane. The spatial position of the finger was digitally recorded with a wireless, high-resolution 3-D movement recording system (MacReflex®). A light-weight retro-reflective marker, positioned on the finger tip, was stroboscopically illuminated with infra-red light, and its geometrical centroid was computed on-line. The 3-D coordinates for this point were sampled at a frequency of 50 Hz, and with a spatial resolution of 0.02 mm. A digital device (Biopac®) was used for multi-channel sampling and A/D conversion of electrophysiological signals. Standard software packages (Wingz®; MacSpin®; Spyglass®) were used for off-line data analysis.

RESULTS

Figure 1A shows a recording of the discharge of a presumed α-motoneurone to the extensor indicis muscle during free finger movements. Although this record is informative,

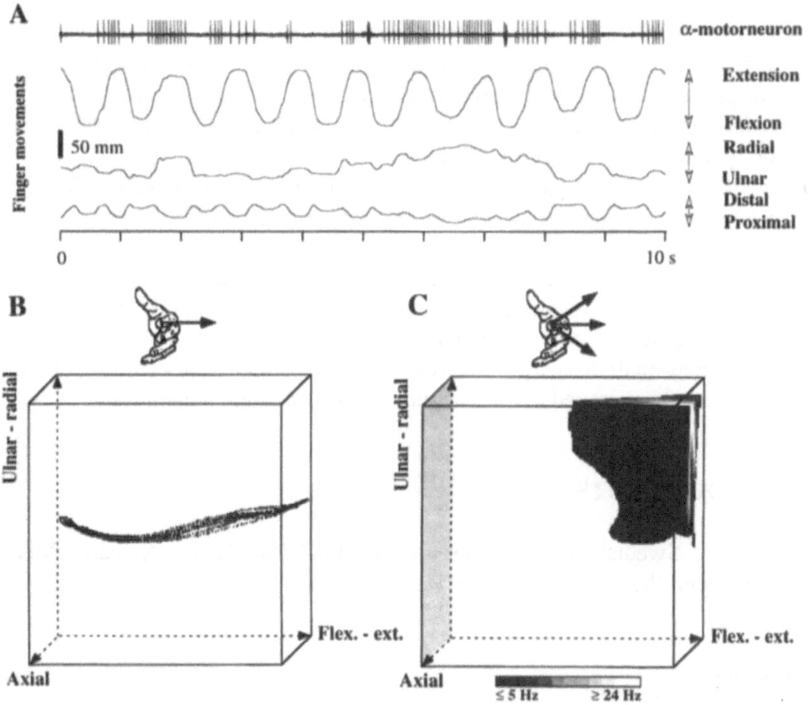

Figure 1. Firing of a human single α-motoneurone of extensor indicis muscle (threshold ≤15% of the maximum voluntary force), and the position of the index finger tip in three dimensions, displayed as; (A) a 10 s excerpt from a recording of nerve activity and position signals *vs.* time during free finger extension-flexion movements, (B) a 'spatial plot' (oblique top view) of the finger tip movements in a simulated 3-D space during 10 guided (see text) index finger extensions, mathematically selected from a 10 s recording with extension-flexion movements, and (C) a 'spatial map' of the discharge rate within the 3-D movement range of the finger tip during free finger extensions selected from a 90 s recording of natural finger movements (regions associated with discharge rates below 5 Hz are transparent).

it is less useful in estimating the neural activation in relation to the spatial range of the movements. As the motor task was freely and "non-orthogonally" performed, most movements were not optimally projected in any given dimension. Consequently, in order to comprehend fully the 3-D spatial aspect of the movements from the 2-D display, an observer has to conceptually relate the three position signals to the bioelectric signal.

Figure 1B is a temporal record of a finger movement in three dimensions displayed as a simulated 3-D *'spatial plot'*. In this plot, the discharge rate of the studied neurone can be introduced by colouring the data spots. When several superimposed natural movements are displayed, the spatial plot becomes complex, and the richness in details may disguise general features of the neuronal discharge patterns. A more comprehensive qualitative description of the bioelectric activity during such conditions can be obtained from a *'spatial activity map'* (Figure 1C). In this display, the bioelectrical isoactivity levels within the movement range were created, by smoothing and interpolation of the data. This gives a more continuous display of nerve activity levels at the different body segment locations, which were encompassed by the movement. This volumetric display was further refined by arbitrary 'cut-off limits', set at freely chosen discharge rates, below which the computed data are transparent in the display.

CONCLUSIONS

This report describes a conceptually new way to relate qualitatively nerve discharge data to 3-D recorded body segment movements. In a previous report, the rationale has also been used for presentation of electromyographic activity from intramuscular recordings (Al-Falahe, Nordh, Eriksson & Zafar, 1993). The technique complements existing statistical methods for quantitative evaluation of multidimensional relationships, and simplifies the visualisation of movement related discharge characteristics. It is intended for analysis of complex movements, which is of increasing interest in motor control research (cf. Humphrey & Freund, 1991). The suggested spatial plot (cf. Figure 1B), which has the character of a four-dimensional spatial scattergram, simplifies the understanding of the neuronal activity in relation to movements, while still retaining the original character of the recorded data. The spatial activation map (Figure 1C) is of value to visualise general features of the neuronal discharge recorded during complex movements, as this 'map' delineates the spatially averaged activity within the recorded movement range.

ACKNOWLEDGEMENTS

Supported by the Swedish Medical Research Council (6874), the Swedish Medical and Dental Societies and the Umeå University Research Fund.

REFERENCES

AL-FALAHE, N., NORDH, E., ERIKSSON, P.-O. & ZAFAR, H. (1993) Mapping of natural electromyographic activation patterns to three dimensionally recorded limb movement territories. *Eur. J. Neurosci. Suppl.* **6**, A1082.

HUMHPREY, D. R. & FREUND, H.-J. (1991) *Motor control: concepts and issues.* John Wiley & Sons, Chichester.

VALLBO, Å. B. & HAGBARTH, K.-E. (1968) Activity in skin mechanoreceptors recorded percutaneously in awake human subjects. *Exp. Neurol.* **26**, 270 - 289.

CORRELATION BETWEEN Ia AFFERENT DISCHARGES, EMG AND TORQUE DURING STEADY ISOMETRIC CONTRACTION OF HUMAN FINGER MUSCLES

D.M. Halliday[2], N. Kakuda[1], J. Wessberg[1], Å.B. Vallbo[1], B.A. Conway[3] & J.R. Rosenberg[2]

[1]Department of Physiology, Göteborg University
Göteborg, S-413 90, Sweden
[2]Department of Physiology, Glasgow University
Glasgow G12 8QQ, UK
[3]Bioengineering Unit, Strathclyde University
Glasgow G4 0NW, UK

INTRODUCTION

Muscle spindle discharge in the cat is sensitive to the mechanical activity of anatomically close motor units (e.g. Binder & Stuart, 1980; Conway, Halliday & Rosenberg, 1993). In man similar have been demonstrated by electrical stimulation of skeletomotor axons (McKeon & Burke, 1983). However, studies of voluntary contractions in man have failed to reveal effects of this nature when single afferent firing was correlated with gross surface EMG in leg muscles (Gandevia, Burke & McKeon, 1986). Here we examine this problem in the case of finger extensor muscles during low force voluntary isometric contractions.

METHODS

Unitary afferent discharges were recorded by microneurography (Vallbo & Hagbarth, 1968). Afferents were classified as Ia by a series of tests (Edin & Vallbo, 1990). Surface EMG of finger extensor muscles was also recorded, along with torque measured at the metacarpophalangeal joint. Subjects maintained an isometric contraction at fixed torque levels (0.025 and 0.1 Nm). Recording periods of 20 s were used to avoid fatigue. EMG, torque and the time of Ia spikes were recorded at a sampling rate of 1600/s. Consecutive records, judged to be stationary at a given torque level, were joined to form single data sets for analysis, totalling 1 to 5 mins. Twelve data sets were obtained from 6 different afferents.

Rectified EMG and torque data were treated as stationary time series, x(t), and Ia spike trains as stochastic point processes, $N_1(t)$. All analyses were performed via the frequency domain using the methods of Rosenberg, Amjad, Breeze, Brillinger & Halliday (1989), with the extension proposed by Jenkins (1963) for dealing with hybrid point process/time series data. In the frequency domain, coherence functions, $|R_{x1}(f)|^2$, were estimated, and in the time domain hybrid cumulant densities, $q_{x1}(u)$, were estimated to assess the dependence of the afferent discharge on both EMG and torque. Hybrid cumulant densities estimate the statistical dependence between a spike train and a time series as a function of lag u, and have an interpretation similar to a spike triggered average.

RESULTS

Figure 1 shows examples of estimates of $q_{x1}(u)$ and $|R_{x1}(f)|^2$ between the discharge of a single afferent and the corresponding rectified EMG and torque records, typical of all data sets.

The time domain results suggest an organisation of muscle spindle Ia discharge with respect to the EMG and torque signals. In general a small increase in EMG activity followed by a marked depression in activity could be identified prior to Ia discharge. The increased EMG activity was apparent at negative lags ranging from 25 - 80 ms (mean = 58 ms), whilst the depression was centred between negative lags of 10 - 25 ms (mean = 15 ms) with a width ranging from 14 - 31 ms (mean = 22 ms; Figure 1A). Spindle discharge is seen in 5 out of the 6 afferents to occur as the torque begins to fall (Figure 1C). For these 5 examples peak torque occured at negative lags of 15 - 30 ms (mean = 24ms). The sixth had a peak at zero lag indicating synchronisation between peak torque and the Ia discharge.

In the frequency domain, the relationship between EMG and Ia afferent discharge is centred about a narrow band (Figure 1B) of 7 - 13 Hz (mean = 9.5 Hz), whilst the relationship between torque and afferent firing (Figure 1D) is centred about a similar band of 5 - 11 Hz (mean = 9 Hz). These frequencies do not simply reflect the mean Ia discharge

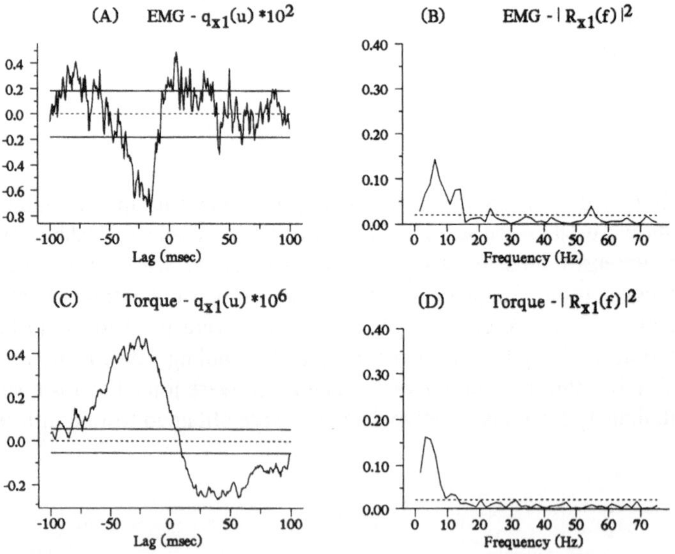

Figure 1. Hybrid cumulant estimates between (A) EMG and afferent, and (C) Torque and afferent. Horizontal lines represent the expected value (dashed), and the upper and lower 95% confidence limits (solid) for the case of independent processes. (B) and (D) are the corresponding coherence estimate. The horizontal dashed line represents the upper 95% confidence limit for the case of independent processes.

rate, but indicate a modulating effect on the afferent discharge. For the afferent data used to calculate the examples in Figure 1B & D the mean rate of discharge was 9.8 imp/s (range of mean rates for all the Ia afferents studied was 4.5 - 20 imps/s).

DISCUSSION

These results indicate that human muscle spindle Ia afferents are sensitive to the fluctuations in torque associated with the maintenance of a constant low force isometric contraction. The interactions described are all peripheral, reflecting mechanical coupling within the muscle and most likely local length changes in the vicinity of the receptors. No consistent evidence of a central interaction was detected (Figure 1A, u > 0). In all but one case the correlated spindle discharge was associated with a falling torque indicating that the spindle discharge tends to occur during the relaxation phase. During the rising phase of a twitch the Ia afferent shows a decreased likelihood of activity. In the one case where the peak torque occurred synchronously with the Ia afferent discharge (lag u = 0) it is suggested that this result may reflect either correlated extrafusal and intrafusal motor activity or may simply be a consequence of receptor location with respect to the active motor-units.

This study illustrates the sensitivity of human muscle spindles to voluntary extrafusal activity. This is emphasised by the frequency bands at which significant coherence occurs between the Ia afferent discharge and EMG and torque signals. As the frequencies common to the EMG/torque signals and the Ia afferent discharges do not simply correspond to the mean rate of afferent firing it seems that the coherence illustrates a modulating effect of extrafusal activity on the Ia discharge. Spindles may have a role in coding intramuscular events in finger muscles beyond those which become manifest in joint movements.

ACKNOWLEDGEMENTS

Supported by the Swedish Medical Research Council (project No. 3548), the Bank of Sweden Tercentenary Foundation (project 94-0040), The Wellcome Trust, and a Swedish-British Academic Cooperation Programme travel grant.

REFERENCES

BINDER, M.D. & STUART, D.G. (1980) Responses of Ia and spindle group II afferents to single motor unit contractions. *J. Neurophysiol.* **43**, 621-629.

CONWAY, B.A., HALLIDAY, D.M. & ROSENBERG, J.R. (1993). Detection of weak synaptic interactions between single Ia-afferents and motor-unit spike trains in the decerebrate cat. *J. Physiol.* **471**, 379-409.

EDIN, B.E. & VALLBO, Å.B. (1990). Classification of human muscle stretch receptor afferents: A bayesian approach. *J. Neurophysiol.* **63**, 1314-1322.

GANDEVIA, S.C., BURKE, D. & MCKEON, B. (1986). Coupling between human muscle spindle endings and motor units assessed using spike-triggered averaging. *Neurosci. Lett.* **71**, 181-186.

JENKINS, G.M. (1963). Discussion in: Bartlett, M.S. (1963). The spectral analysis of point processes. *J. R. Stat. Soc. B* **25**, 264-280.

MCKEON, B.& BURKE, D (1983). Muscle spindle discharge in response to contraction of single motor units. *J. Neurophysiol.* **49**, 291-302.

ROSENBERG, J.R., AMJAD, A.M., BREEZE, P., BRILLINGER, D.R. & HALLIDAY, D.M. (1989). The Fourier approach to the identification of synaptic coupling between neuronal spike trains. *Prog. Biophys. & Mol. Biol.* **53**, 1-31.

VALLBO, Å.B. & HAGBARTH, K-E. (1968). Activity from skin mechanoreceptors recorded percutaneously in awake human subjects. *Exp. Neurol.* **21**, 270-289.

VISUAL TARGET VELOCITY CODING THROUGH OCULAR MUSCLE PROPRIOCEPTION

Gabriel M. Gauthier, Jean-Louis Vercher
and Jean Blouin

Laboratoire de Contrôles Sensorimoteurs
URA CNRS 1166, Université de Provence
Avenue Normandie-Niémen
F-13397 Marseille cedex 20, France

INTRODUCTION

Considering the eyes as contributing only to visual function is extremely restrictive in that the eyes are also involved, with the head as their carrier, in target position and velocity coding, a vital function for our daily activities. Indeed, to follow a visual object with the hand, the brain has first to code both the object position and velocity with respect to the body and second to program and control the appropriate commands to the arm and hand muscles. As compared to visual function *per se* and to eye movement control function (saccade, smooth pursuit), the target position and velocity coding with respect to the body is still not elucidated and some basic observations are still controversial. The position of the object with respect to the body has to be computed from the distance of the object image to the fovea and the angular position of the eye in the orbit (with respect to a head-centric reference). Moreover, since the head carries the eyes in space, determination of the target position with respect to the body frame of reference also requires computation of the head angular position with respect to the body. In binocular viewing, both eyes must be aligned on the target and both eye positions must be properly sensed to provide accurate target position coding. We shall limit our analysis of the overall problem of eye-hand tracking of visual target to the coding of a visual target velocity with respect to the body in terms of what is commonly referred to as retinal and extra-retinal coding.

Anatomical studies describing the presence of receptors in all extraocular muscles and their central projection sites have suggested the involvement of ocular muscle proprioception in eye movement control and related sensory functions. However, most of these functions remain speculative and tentative demonstrations of proprioceptive participation in target position and velocity coding have yielded contradictory results. For instance, the role of ocular muscle proprioception in eye alignment and visual target localization has been demonstrated in strabismic patients by Steinbach & Smith (1981) and

Gauthier, Bérard, Deransard, Semmlow & Vercher (1985), as well as in normals (Gauthier, Nommay & Vercher, 1990; Bridgeman & Stark, 1991). Conversely, Bock & Kommerell (1986), using a similar approach, have provided data that rather supported the involvement of the copy of the eye muscle command.

As mentioned earlier, the question as to how the brain codes the velocity of a visual target with respect to the body has been documented less than that related to the coding of the visual target position. Obviously, position and velocity coding mechanisms share common features. A simplistic approach consists in assuming that the brain extracts target velocity from a compound signal that codes target position with respect to the head (if the head is stationary). This signal is composed of at least two sensori-motor components (efferent copy and proprioception from eye muscles) and a retinal error if the target is not directly projected on the fovea. Retinal slip is the direct counterpart of the position retinal error signal. Indeed, retinal velocity error, i.e. retinal slip, is particularly efficient for driving the eye as demonstrated by optokinetic stimulation carried out in man and animals (cf. the results from one eye stimulation experiments by Grüsser, Kulikowski, Pause & Wollensak, 1981). On the other hand, patients having one eye paralysed still have a fairly accurate cognitive sensing of visual stimulation velocity when submitted to rotatory random dot surround. Nonetheless, such experiments are not conclusive as to what signal is being cognitively sensed and what the sensed signal has in common with the optokinetic signal that drives the eyes.

A model in which the coding of target velocity directly emerges from neuro-differentiation of proprioceptive and efferent copy eye position signals is worth considering. Indeed, when retinal slip does not occur as in most daily situations where the retinal image of the visual target remains relatively stationary on the observer's retina through appropriate smooth pursuit and optokinetic responses (if the head is stationary), eye-in-the head velocity, hence target-in-space velocity, ought to emerge largely from non-retinal signal.

To test the hypothesis that ocular muscle proprioception contributes to eye velocity hence target velocity sensing, we have recently run an experiment where human subjects tracked a visual target with one eye while the other eye was mechanically (passively with respect to central nervous system command) rotated in phase or out of phase with the visual target or simply prevented from moving.

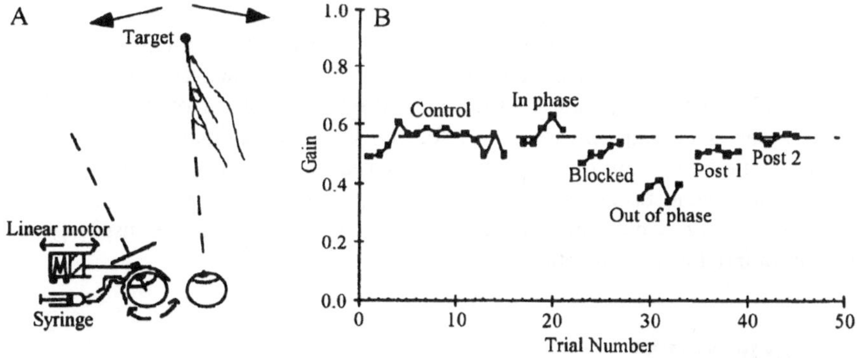

Figure 1. A. Experimental arrangement. The subject was seated in front of a screen on which a red target moved sinusoidally. Subject carried a pointer with his preferred, unseen hand and followed the perceived position of the target. A suction lens was used to either immobilized or move the non-viewing eye. A linear motor immobilised or rotated the eye around a vertical axis either in-phase or out-of-phase with the visual target motion. B. Perceived visual target motion (displayed in terms of gain i.e hand amplitude over target amplitude), as indicated by the unseen hand used as a pointer. The motion (amplitude and velocity) of the target was perceived as being significantly larger when the covered eye was moved in phase with the target, hence with the tracking eye, than when the covered eye was either immobilised or moved out of phase.

METHODS

The method we used to rotate or to immobilise the eye involved a scleral lens and was similar to that used in the eye position sense studies we carried out earlier (Gauthier et al., 1990). Here, the lens was attached to a linear motor (Links Dynamics) positioned on the observer headrest so that it rotated the eyeball around its vertical axis by applying a torque tangent to the eyeball at equatorial level (Figure 1A). The visual target (red dot, 3 mm in diameter, produced by a laser diode) was presented at eye level on a semi-circular screen positioned 57 cm from the observer. The target motion resulted from the activation of a mirror galvanometer by means of a wave generator. The latter also activated the linear motor. A phase control processor was used to insure proper in-phase or out-of-phase drive of the linear motor with the visual target. The peak to peak motion of the eye was set at 4.8 mm, resulting in a close to 16° rotation. Subjects, with the head immobilised, tracked the target with the viewing eye and concomitantly pursued with the unseen arm and hand, the perceived position and velocity of the visual target. A light rotating pointer was used to code the arm (and hand) motion.

Results from two subjects tested four times each definitely show that when the covered eye is moved out of phase with the visual target, hence with the tracking eye, the perceived target motion amplitude and velocity is significantly smaller than when the covered eye is either moved in phase with the target or is immobilized in its orbit. Figure 1B shows the average data pooled for right and left eye testing and the two subjects. Since the eye manipulation used here results essentially in an alteration of ocular muscle proprioception without changing the muscle command and its efferent copy, these observations support the idea that ocular muscle proprioception intervenes in the coding of eye-in-head motion, hence in the sensing of target-in-space velocity. Our data are also in agreement with those obtained recently by Donaldson & Knox (1993), showing that passive eye movement imposed to one eye during vestibular stimulation, affects the motion of the other eye. These authors' interpretation was that the brain monitors eye movement (velocity) through peripheral mechanisms i.e. ocular muscle proprioception to achieve such control. A role for proprioception through its velocity component has also recently been demonstrated in adaptive control of ocular motor functions in the rat by Gauthier, De'Sperati, Tempia, Marchetti & Strata (in press) who showed that alteration of ocular muscle proprioception results in changes of VOR gain adaptive ability to visuo-vestibular conflicts.

Further work is necessary to model the process by which the brain codes visual target position and velocity. What appears evident from our results however (see Figure 1B), is that extraretinal signals, when isolated from normal retinal feedback of the environment in which the target is moving, provide cues for the brain to compute target velocity. A comprehensive model of how the brain monitors target position and velocity is of interest not only for neuroscientists. The present data and related literature dealing with visual target position and velocity coding in man and animals may prove to be inspiring for the development of intelligent automata.

ACKNOWLEDGEMENTS

This work was supported by CNRS UA 372, INSERM 896007 and Human Frontier Science Program grants. J. Blouin was a postdoctoral fellow supported by HFSF grant and by Natural Sciences and Engineering Research Council of Canada Fellowships.

REFERENCES

BOCK, O. & KOMMERELL, G. (1986) Visual localization after strabismus surgery is compatible with the "Outflow" theory. *Vis. Res.* **26**, 1825-1829.

BRIDGEMAN, B. & STARK, L. (1991) Ocular proprioception and efference copy in registering visual direction. *Vis. Res.* **31**, 1903-1913.

DONALDSON, I.M.L. & KNOX, P.C. (1993) Evidence for corrective effects of signals from the extraocular muscles on single units in the pigeon vestibulo-ocular system. *Exp. Brain Res.* **95**, 240-250.

GAUTHIER G.M., NOMMAY, D. & VERCHER, J.-L. (1990) The role of ocular muscle proprioception in visual localization of targets. *Science.* **249**, 58-61.

GAUTHIER G.M., BÉRARD, P. V., DERANSARD, J., SEMMLOW, J. L. & VERCHER, J-L. (1985) Adaptation processes resulting from surgical correction of strabismus. In *Adaptive Processes in Visual and Oculomotor Systems*, eds. KELLER, E. L. & ZEE, D.S. pp 185-189, Pergamon Press, Oxford.

GAUTHIER, G.M., DE'SPERATI, C., TEMPIA, F., MARCHETTI, E. & STRATA, P. (1995) Influence of eye motion on adaptive modifications of the vestibulo-ocular reflex in the rat. *Exp. Brain Res.* (in press).

GRÜSSER, O.J., KULIKOWSKI, J., PAUSE, M. & WOLLENSAK, M. (1981) Optokinetic nystagmus, sigmaoptokinetic nystagmus and eye pursuit movements elicited by stimulation of an immobilized human eye. *J. Physiol.* **320**, 21-22.

STEINBACH, M.J. & SMITH, D.R. (1981) Spatial localization after strabismus surgery: evidence for inflow. *Science.* **213**, 1407-1409.

PART 10

NATURAL MOTOR PATTERNS - 1

OPTICAL RECORDING AND LESIONING OF SPINAL NEURONES DURING RHYTHMIC ACTIVITY IN THE CHICK EMBRYO SPINAL CORD

Michael J. O'Donovan and Amy Ritter

Section on Developmental Neurobiology
Laboratory of Neural Control
NINDS, NIH, Bethesda, MD 20892, USA

INTRODUCTION

The development of *in vitro* preparations of the spinal cord has catalysed our understanding of locomotor network organisation in simple vertebrates (Grillner & Matsushima, 1991; Roberts & Tunstall, 1990). This is because such preparations are stable enough to allow long term single or dual intracellular penetrations of spinal neurones to establish their firing behaviour, synaptic drive and network connections. Although *in vitro* spinal preparations of higher vertebrates have been available for many years (Otsuka & Konishi, 1974) it is only relatively recently that they have become widely used for studies of locomotion and respiration (Landmesser & O'Donovan, 1984; Smith & Feldman 1987; Kudo & Yamada, 1987). Application of the classical electrophysiological methods will be more challenging in the these preparations because of their greater complexity and large cell complement. Furthermore, single cell methods are a laborious way of documenting the activity of neuronal populations and cannot readily provide information about the complex spatiotemporal patterns of network activation. Another limitation of this approach is that it is purely descriptive. It provides no information about the causal role of particular cell classes in the genesis of the behaviour. In invertebrate systems the functional role of a neurone in a network is often inferred after lesioning the cell or hyperpolarising its membrane potential during network behaviour. While such a strategy has limitations it can be successfully applied to identify the role of certain classes of cell in rhythm generation. One good example of this approach is the lesioning of efferent neurones in the mollusc *Clione limacina* to show that they are not required for rhythmogenesis (Arshavsky, Orlovsky & Panchin, 1985).

It seems likely that progress in understanding network function will require the application of methods for recording and manipulating neuronal populations as an adjunct to the classical electrophysiological methods. In this paper we describe our progress in the development of these methods and their application to rhythmogenesis in the developing spinal cord of the chick embryo.

MATERIALS AND METHODS

Isolation of the spinal cord and muscle nerve recording

All experiments were performed on isolated spinal cord preparations from white Leghorn chicken embryos aged E9 - E13. The embryos were rapidly decapitated and eviscerated while continuously superfused with cooled (12 - 15°C) Tyrode's solution oxygenated with $95\%O_2/5\%CO_2$. The spinal cord was isolated from the vertebral column together with muscle nerves innervating the thigh and transferred to a perfusion chamber. The perfusion solution was heated to 27 - 29°C to facilitate the expression of spontaneous rhythmic activity (Landmesser & O'Donovan, 1984). Episodes of rhythmic activity could also be evoked by a single stimulus to the surface of the spinal cord. Electrical activity was recorded from muscle nerves or the roots using suction electrodes. After the signals were amplified (x10,000) and filtered (DC to 2 - 10 kHz) they were digitised and stored on videotape or recorded on one of the audio channels of the videotape player used for recording the optical data (see below).

Optical recording and dye-loading

Spinal neurones were loaded with calcium dyes to monitor their population activity using video-imaging methods (O'Donovan, Ho, Sholomenko, Yee, 1993; O'Donovan, Ho & Yee, 1994). Neurones were loaded by bath application of the membrane-permeant calcium dye fura-2am to the cut transverse or horizontal surface of the lumbosacral cord. This loading method labels cell 1 - 2 layers deep in the cut surface (O'Donovan et al., 1994). In a second method interneurones were retrogradely labelled with calcium green coupled to dextran (O'Donovan et al., 1993). The dye was applied to the cut axonal tract running in the ventrolateral white matter and was allowed to diffuse for several hours back to the interneuronal cell bodies before recordings were made. In most experiments the spinal cord was cut transversely between LS3 and LS4 and mounted in a perfusion chamber on the stage of a microscope.

Using epifluorescence optics, fura-loaded cells were generally excited at 380 nm whereas calcium green-loaded cells were excited at 450 - 490 nm. Activity-dependent changes in fluorescence above 510 - 520 nm were detected using an intensified CCD camera. In some experiments, calcium green-loaded cells were viewed with a video-rate, slit scanning confocal microscope (Meridian Insight plus). Under these conditions the dye was excited using the 488 nm line of the laser. Video signals were stored on video tape for later analysis. The output of the camera was also fed to a frame grabber for further processing.

The spread of optical activity at the onset of rhythmic bursting (see Figure 1) was quantified in the following manner. All of the fura-2 images were first inverted because the fluorescence at 380 nm decreases with activity (O'Donovan et al., 1994). Initially a control image was acquired before the onset of rhythmic activity. This image was the average of 30 frames. Then each video frame, starting just before the onset of activity, was subtracted from the control image to produce a difference image reflecting the change in fluorescence occurring in that frame. This difference image was smoothed with a 15x15 kernel and thresholded and smoothed again. The resulting image displayed the spatial extent of the fluorescence change in that video frame. These regions were color coded in successive frames to illustrate the spread of fluorescent activity (see Figure 1).

Optical lesioning of motoneurones

Motoneurones were filled retrogradely with a 25 - 50 mM solution of dextran-fluorescein applied to the cut end of the crural plexus originating from the first three lumbosacral segments (LS1 - LS3). After the motoneurones were labelled a section of cord (LS1 - LS3) was isolated, hemisected and placed in a perfusion chamber on the stage of a microscope. The cord section was then exposed to continuous or intermittent high intensity illumination at 480 nm to induce phototoxic damage of the labelled neurones. Continuous illumination was used in initial experiments until it was determined that such illumination can depress rhythmic activity. The effect of lesioning motoneurones on the rhythmic motor output was assessed by recording the electrical activity from the ventral roots or interneuronal axons.

RESULTS

Optical recording of neuronal population activity

Motor activity in the isolated spinal cord consists of recurring episodes of rhythmic activity that can be recorded as cyclical discharges from muscle or motor nerves (Landmesser & O'Donovan, 1984). To document the location and dynamics of active neurones we imaged the fluorescence of fura2-labeled motoneurones and interneurones in the transverse face of the spinal cord during episodes of rhythmic activity (O'Donovan et al., 1994). Both cell classes exhibited rhythmic fluorescence transients that accompanied episodes of motor activity. The fluorescent transients in motoneurones were synchronised with each other and with the electrical activity recorded from a muscle nerve. The neurones outside the lateral motor column (LMC), presumed to be interneurones, displayed more varied responses than the motoneurones. Some behaved similarly to motoneurones showing rhythmic transients synchronised with motoneurones. Others exhibited tonic changes in fluorescence that lasted for the duration of the episode. The majority of the active interneurones were located in the lateral 2/3 of the grey matter.

At the beginning of each cycle of rhythmic activity the earliest optical activity was observed in and around the lateral motor column. From here the fluorescence change spread ventrodorsally and ventromedially to encompass the whole transverse face of the spinal cord (O'Donovan et al., 1994; see also Figure 1). This surprising result raised two possibilities. First, that the early activation of motoneurones and the ventrodorsal spread represented the sequence of synaptic activation of neurones in the cut face. The axons running in the ventrolateral white matter tracts were a good source for these inputs because they were located adjacent to the LMC and they had been shown previously by lesion experiments to mediate some of the synaptic excitation of motoneurones (Ho & O'Donovan, 1993). A second possibility was that the motoneurones might be the source of the rhythm. Motoneurones have not generally been considered to be part of the rhythmic generator in other species. However, in the chick embryo rhythmic activity can still be recorded from ventral or lateral strips of the spinal cord providing that the LMC is preserved (Ho & O'Donovan, 1993).

If the ventrolateral tract is responsible for some of the rhythmic excitation of neurones in the cut face this implies that neurones whose axons project into the tract would be rhythmically active. To test this possibility we retrogradely filled the neurones of the ventrolateral tract with calcium green dextran (O'Donovan et al., 1993) and imaged the activity of the labelled cells. Filling the tract on one side of the cord labeled two populations of cells: an ipsilateral group dorsomedial to the lateral motor column and a contralaterally projecting group near the central canal. Video-rate imaging using conventional (Ho & O'Donovan, 1992) and confocal microscopy (Ritter & O'Donovan, 1994) showed that some, but not all, of the cells in both groups were rhythmically active, suggesting that only a subset of the labeled interneurones is involved in rhythmic behaviour.

While this observation and the results of cutting the tract were consistent with a role for these interneurones in the synaptic activation of motoneurones they could not eliminate the possibility that motoneurones themselves were somehow involved in rhythmogenesis. To address this possibility we developed a method for lesioning motoneurones optically to assess their role in rhythmic behaviour.

The effects of optical lesions of motoneurones on rhythmic interneurone activity

To establish if motoneurones were required for the rhythmic activation of interneurones we developed a method for killing the majority of motoneurones in a spinal cord section. This was done by loading the motoneurones with dextran-fluorescein in a 3 segment (LS1 - LS3) hemisection of the lumbosacral cord. Three segments are sufficient to

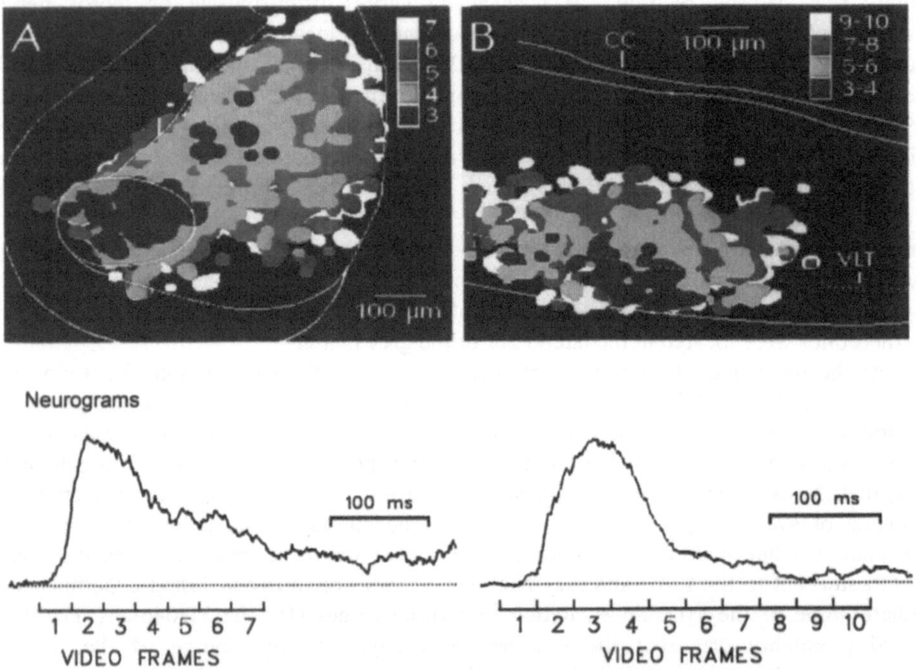

Figure 1. Optical recordings of the spread of fluorescent activity at the onset of an episode of activity in neurones loaded with the calcium-sensitive dye fura-2am. Each color indicates the spread of fluorescence in successive frames (A) or pairs of frames (B). See methods for details of processing. A. Spread of fluorescence at the onset of a spontaneous episode of rhythmic activity in a transverse section of an E11 spinal cord stained with fura2-am. B. Spread of fluorescence accompanying the onset of a single cycle of evoked activity in a horizontal section of the lumbosacral cord viewed from the dorsal aspect. The dorsal part of the spinal cord has been removed with a vibrating needle down to the level of the central canal. The cut horizontal surface was labelled with the calcium-sensitive dye fura2-am and imaged after a single electrical stimulus to the spinal cord. Although this preparation only produced a single cycle the optical activity was observed to begin laterally and then to spread medially, as it does in the transverse plane.

produce a robust rhythm while allowing almost all of the motoneurones in the hemisection to be loaded. Control experiments were performed which showed that continuous illumination, in the absence of labeling, reduced and could block the rhythmic activation of motoneurones and interneurones. Therefore, we devised an intermittent mode of illumination that did not compromise motor function in the unstained cord. Application of this illumination protocol to loaded motoneurones resulted in a progressive decline in the amplitude of rhythmic ventral root potentials. We confirmed that the motoneurones were killed by staining sections of the cord with Trypan Blue which is excluded from living cells and selectively stained motoneurones after this type of lesion. In most experiments the amplitude of the electrotonic potentials recorded from the ventrolateral tract (VLT) declined substantially and the duration of each episode was reduced. Despite these changes in motor output some cords remained spontaneously active, sometimes at a higher rate than before the lesion. These findings suggested that motoneurones were probably not crucial for rhythmic interneuronal activity but it also raised the possibility that they might be involved somehow in the activation of interneurones. We were concerned, however, that the depression of interneuronal activity might be due to a secondary effect of the motoneurone lesion rather than to a loss of motoneurones. Such secondary effects could include release of some noxious material from the damaged motoneurones or, alternatively, glutamate toxicity of interneurones driven by injury discharges or potassium release from lesioned motoneurones. For this reason we repeated the motoneurone lesions in a solution containing low calcium and high manganese to block synaptic transmission. The motoneurones were exposed to the same duration and intensity of light that we used when performing the experiments in normal Tyrode's solution. After the lesion the low calcium solution was replaced with normal Tyrode's solution to assess the effects on motoneuronal and interneuronal activity. Our preliminary data indicate that optical lesions performed under these conditions did not produce as marked a change in interneuronal output as lesions performed in normal Tyrode's solution. At present the mechanism responsible for the protective effect of low calcium solutions is not clear because these solutions also reduce the severity of the motoneurone lesion. As a consequence, the improvement of interneuronal output in low calcium solutions might be because fewer motoneurones were lesioned.

DISCUSSION

We have demonstrated that optical methods can be used both to record the activity of neuronal populations and also to lesion cell classes to investigate their causal role in behaviour. Calcium imaging allowed us to map the location of some of the cells activated during episodes of rhythmic motor activity. These interneurones were distributed throughout the gray matter although many were concentrated in the region dorsal to the lateral motor column where the motoneurone dendrites are situated. Video-rate measurements of the fluorescent transients revealed that most cells were synchronized with each other and with the electrical activity recorded from the muscle nerve. The synchrony in calcium transients originates in part from the synchronized rhythmic synaptic drive that these interneurones receive during motor activity (Ritter & O'Donovan, 1994). Such synchrony is probably mediated by recurrent synaptic excitation rather than electrical connections because interneurones at this age are not dye-coupled when injected with neurobiotin (Ritter & O'Donovan, 1994, O'Donovan & Ritter, 1994).

At the onset of each cycle of activity the earliest activity begins in the vicinity of the lateral motor column and spreads from there to encompass the rest of the cord. This spatiotemporal pattern of activation has not been identified in the cord before and would be difficult to detect using classical electrophysiological methods. However, since calcium imaging is an indirect measure of neuronal activity it will be important to verify this pattern of activation using more direct measures of neuronal activity including voltage sensitive dyes and possibly multi-electrode arrays. At least two mechanisms could contribute to the observed pattern of activation. First, the spread of optical activity could reflect the sequence of synaptic activation of the neurones in the cut face. This hypothesis is plausible because the ventrolateral axonal tracts have been shown previously to provide some of the rhythmic excitatory drive to motoneurones (Ho & O'Donovan, 1993). Furthermore, as shown in this paper, we observed a flow of excitation from lateral to medial in sections of the spinal cord in which the dorsal 2/3 had been removed (Figure 1B). In such preparations the only remaining white matter was ventrolateral.

Another possibility to account for the optical data is that the motoneurones are the source of the rhythm or contribute in some way to rhythmogenesis. To test this idea we photo-lesioned the lateral motor column and determined the effects on the activity of interneurones. We established that it is possible to lesion a substantial number of the motoneurones in a three segment section of the cord. This was demonstrated by showing that illumination virtually abolished the ventral root potentials recorded during rhythmic activity and by the specific labelling of motoneurones with the a dye (Trypan Blue) that is excluded from living cells. The persistence of rhythmic potentials recorded from the VLT axons in motoneurone-lesioned cords suggests that the motoneurones are not essential for the rhythmic activity of these interneurones. One caveat with this conclusion is that we were never able to accomplish a total lesion of the motoneurones, so that it remains possible that survival of a few motoneurones could maintain rhythmic interneuronal activity.

Participation of motoneurones in the activity of interneurones was suggested by the lesion-induced decline in both the amplitude of interneuronal potentials and in the number of cycles generated in each episode. It is not yet clear, however, to what extent these changes in interneuronal function depend on the loss of motoneurones or on interneuronal damage secondary to the motoneurone lesion. We did observe occasional Trypan blue labelled cells outside the lateral motor column, but their number appeared to be too low to account for the changes in the output of the interneurones. It is possible, however, that interneuronal function could be compromised as a result of the motoneurone lesion and yet not be manifest as Trypan blue staining. For instance, potassium released from damaged motoneurones could directly excite adjacent interneurones which might lead to glutamate toxicity. Alternatively, the release of some noxious substance from damaged motoneurones might interfere with interneuronal function.

The decline in interneuronal activity was reduced when the motoneurone lesion was performed in a low calcium/high manganese solution. This procedure appeared also to offer some protection to the motoneurones because they were less effectively lesioned in the presence of low calcium than in normal Tyrode's solution. As a consequence, we cannot yet resolve if low calcium solutions protect interneuronal activity because of reduced motoneurone injury or, alternatively, because they minimize some type of secondary damage to the interneurones. In future experiments we plan to use excitatory amino acid antagonists during the lesion to establish if synaptic transmission is required for the alterations in interneuronal activity that accompany a motoneurone lesion.

REFERENCES

ARSHAVSKY, YU. I., ORLOVSKY, G.N. & PANCHIN, YU.V. (1985) Control of locomotion in marine mollusc - *Clione limacina*. V. Photoinactivation of efferent neurons. *Exp. Brain Res.* **59**, 203-205.

GRILLNER, S & MATSUSHIMA, T. (1991) The neural network underlying locomotion in Lamprey-synaptic and cellular mechanisms. *Neuron.* **7**, 1-15.

HO, S. & O'DONOVAN, M.J. (1992) Real-time imaging of identified propriospinal activity following bath application of fura-2am and microinjection of fura-2 dextran into fiber tracts. *Neurosci. Abstr.* **17**, 120.
HO, S. & O'DONOVAN, M.J. (1993) Regionalization and inter-segmental coordination of rhythm generating networks in the spinal cord of the chick embryo. *J. Neurosci.* **13**, 1345-1371.
KUDO, N. & YAMADA, T. (1987) *N*-methyl-D,L-aspartate induced locomotor activity in a spinal cord-hindlimb preparation of the newborn rat studied in vitro. *Neurosci. Lett.* **75**, 43-48.
LANDMESSER, L.T. & O'DONOVAN, M.J. (1984) Activation patterns of embryonic chick hindlimb muscles recorded *in-ovo* and in an isolated spinal cord preparation. *J. Physiol.* **347**, 189-204.
O'DONOVAN, M.J., HO, S., SHOLOMENKO, G. & YEE, W. (1993) Real-time imaging of neurons retrogradely and anterogradely labelled with calcium sensitive dyes. *J. Neurosci. Meth.* **46**, 91-106.
O'DONOVAN, M.J. & RITTER, A. (1994) Rhythmic activity patterns of motoneurons and interneurones in the embryonic chick spinal cord. In *Neural Control of Movement.* eds. FERRELL, W.R. & PROSKE, U. Plenum Press, New York. (in press).
O'DONOVAN, M.J., HO, S. & YEE, W. (1994) Calcium imaging of rhythmic network activity in the developing spinal cord of the chick embryo. *J. Neurosci.* **14**, 6354-6369.
OTSUKA, M & KONISHI, S. (1974) Electrophysiology of mammalian spinal cord in vitro. *Nature* **252**, 733-734.
RITTER, A. & O'DONOVAN, M.J. (1994) Video-rate, slit-scanning, confocal calcium imaging reveals patterns of interneuron activity in the embryonic chick spinal cord. *Neurosci. Abstracts.* **20.**
ROBERTS, A. & TUNSTALL, M.J. (1990) Mutual re-excitation with post-inhibitory rebound: a simulation study on the mechanisms for locomotor rhythm generation in the spinal cord of Xenopus embryos. *Eur. J. Neurosci.* **2**, 11-23.
SMITH, J.C. & FELDMAN, J. (1987) In vitro brainstem-spinal cord preparations for study of motor systems for mammalian respiration and locomotion. *J. Neurosci. Meth.* **21**, 321-333.

MODES OF RHYTHMIC MOTOR PATTERNS GENERATED BY THE SPINAL CORD IN THE CAT

E.D. Schomburg

Institute of Physiology, University of Göttingen
D-37073 Göttingen, Germany

INTRODUCTION

The spinal cord can generate rhythmic motor activity even when deprived of supraspinal control by spinalisation and of peripheral feedback by paralysis, as was first shown by Jankowska, Dukes, Lund & Lundberg (1967). The rhythmic motor activity which can be induced chemically (nialamide plus DOPA, 5-HT, clonidine) in the spinal cord resembles the main characteristics of normal locomotor activity. These preparations not only have a co-ordinated rhythmic flexor-extensor alternation but also a co-ordination between forelimb and hind limb muscles (for review see Grillner, 1985), and between limb and trunk muscles (Koehler, Schomburg & Steffens, 1984). The phenomenon of a locomotor-like spinal activity in a paralysed animal ("fictive locomotion") has been widely used for the investigation of the mechanism of spinal locomotor generation. Therefore, the main experimental interest has concentrated on rhythmic spinal patterns reproducing locomotor patterns which are as authentic and stable as possible. However, the patterns of rhythmic activity observed in the fixed spinal cat may vary considerably, not only under different experimental conditions, but also in comparable conditions. This paper will concentrate on these variations, which reflect the marked capacity of the spinal cord for generating complex movements of considerable versatility in addition to locomotor movements.

METHODS

The results were assembled from some 70 anaemically decapitated adult cats spinalised at C1 (for technical details see Kniffki, Schomburg & Steffens, 1981) from several different experimental series. Due to the technique used, any supraspinal interference could be excluded. All cats were paralysed ("Pancuronium"/"Pavulon", Organon) and received 30 - 100 mg/kg nialamide and 30 - 100 mg/kg L-DOPA. In some cats, recordings were also performed before application of nialamide and L-DOPA, before and after paralysis. Neurograms were recorded from the following dissected nerves in different combinations, some bilaterally (left, l. and right, r.). Abbreviations used are given in parenthesis: quadriceps (Q), posterior biceps semitendinosus (PBSt), gastrocnemius-soleus (GS), medial gastrocnemius (MG), lateral gastrocnemius-soleus (LGS), deep peroneal (DP), longissimus dorsi (LoD), obliquus abdominis externus (OAE), superficial peroneal, cutaneous branch (SPC). In non-paralysed spinal cats, electromyograms were taken from corresponding muscles. In most cats, the end-tidal CO_2 level was raised in order to support rhythm generation or else some cats received 4-amino-pyridine (0.25-1 mg/kg) or naloxone (0.2-1.0 mg/kg, Pearson, Jiang & Ramirez, 1992; Schomburg & Steffens, in press).

RESULTS

Three main variants were seen in addition to the characteristic rhythmic locomotor patterns: (1) rhythms with fundamentally different basic frequencies, (2) changed or changing phase relations between the activity of different motoneurone pools, and (3) different rhythms of different motoneurone pools.

Rhythms with different basic frequencies

The rhythmic motor activity induced in the spinal cat by injection of nialamide and DOPA characteristically has a very low frequency. During periods of "spontaneous" regular rhythmic activity the main range was 0.2 - 0.4 Hz (Figure 1C). This would correspond to

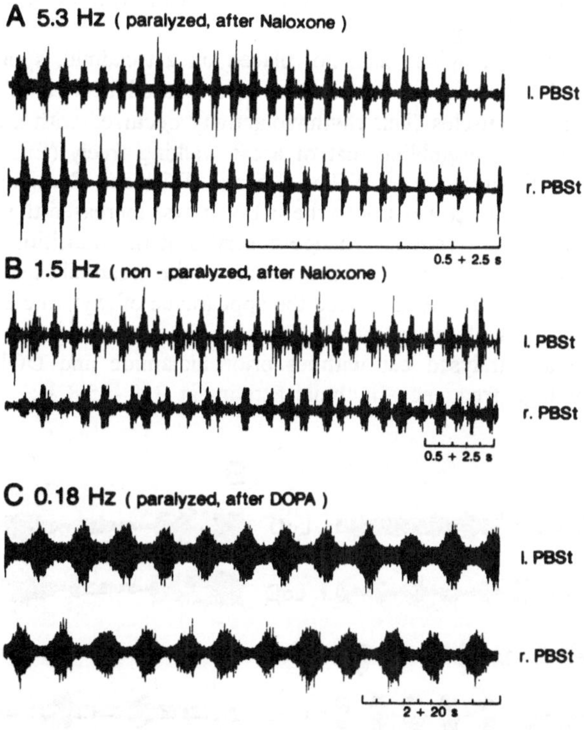

Figure 1. Different basic frequencies of rhythmic motor activity generated by the spinal cord. Acute high spinal cats. (A) paralysed; neurograms after injection of naloxone 0.5 mg/kg i.v. (without nialamide and DOPA). (B) non-paralysed, electromyograms after injection of naloxone 0.5 mg/kg i.v. (C) the same experiment as in B, after paralysis and injection of nialamide and DOPA. Note the different time scales (modified from Schomburg & Steffens, 1995).

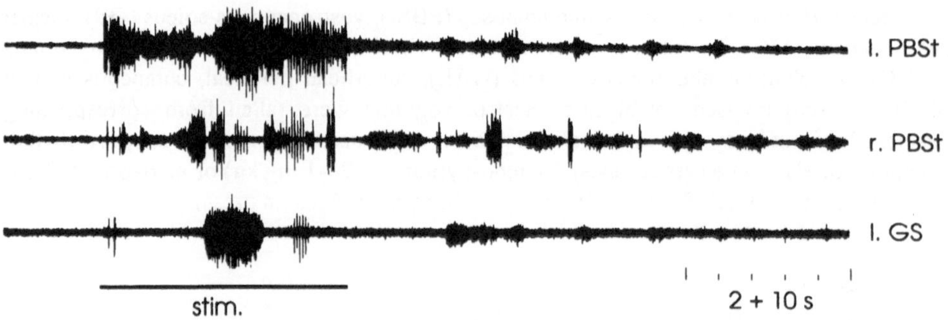

Figure 2. Coexistent occurrence of low and high frequency motor rhythms. Electroneurograms of a paralysed spinal cat after injection of naloxone (0.5 mg/kg i.v.), during suffusion of the lumbar spinal cord with a NMDA solution (10^{-3} M). Stimulation of the sural nerve (50 Hz, 2 times threshold) is marked below the recordings (stim.).

the step frequency of a cat walking very slowly (about 1 Hz corresponds to a cat walking on a treadmill with a speed of 0.3 m/s, Wisleder et al. 1990). A similar low frequency of rhythmic motor activity (slower than 1 Hz) is also induced by NMDA (cf. Figure 2), clonidine and D-amphetamine (Cazalets, Grillner, Menard, Cremieux & Clarac, 1990; Forssberg & Grillner, 1973; Viala & Buser, 1971) in spinal paralysed animals or *in vitro* spinal cord preparations, or in paralysed decorticate animals with mesencephalic stimulation (Gossard, Cabelguen & Rossignol, 1989).

In non-paralysed acute spinal cats, not placed on a treadmill, spontaneous rhythmic motor activity could be observed, which was co-ordinated between the four limbs and between limb and trunk muscles. This rhythmic activity occurred with a frequency of 1.2 - 1.9 Hz (Figure 1B), thus resembling that of a cat walking freely (Engberg & Lundberg, 1969) or on a treadmill with a speed of 0.6 - 1.5 m/s (Wisleder, Zernicke & Smith, 1990). Additional mechanical stimuli (pressure to the tail or paws) increased the frequency to 1.9 - 2.9 Hz. This resembles the frequency change observed at the transition from walk to trot (Engberg & Lundberg, 1969) or with an increase of the treadmill speed to about 2 m/s (Wisleder et al., 1990). Naloxone facilitated the appearance of rhythmic activity in the acute spinal non-paralysed cat without essentially changing the basic frequency. If naloxone was injected in the spinal paralysed cat without prior nialamide and DOPA application, it induced periods of high frequency rhythmic activity at 3.4 - 5.8 Hz, mainly with a co-

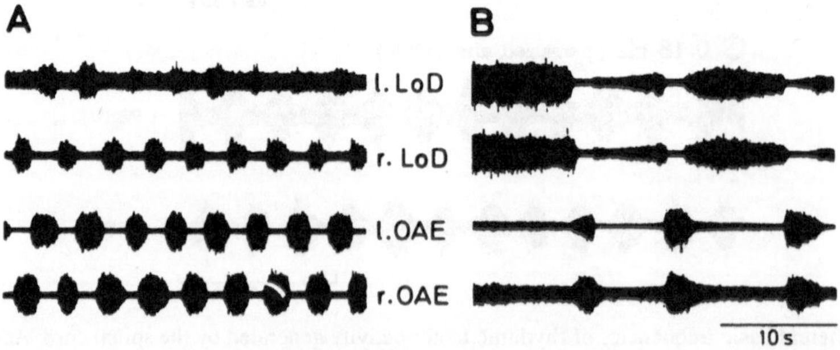

Figure 3. Different phase relations between the activity of dorsal (LoD) and ventral (OAE) trunk muscles of both sides. Neurograms in a paralysed high spinal cat after injection of nialamide and DOPA, (B) during a period of hypoxia (modified from Koehler et al., 1984).

activation of left and right PBSt (Figure 1A). This frequency is comparable to that observed during fast scratching movements (Deliagina, Feldman, Gelfand & Orlovsky, 1975) or fast gallop (Engberg & Lundberg, 1969).

The fast naloxone-induced rhythm and the slow DOPA type of rhythm seem to be based on two different generating mechanisms. There was no transition between the two frequencies, but both types of rhythm may occur in parallel, as is shown in Figure 2. The neurograms of PBSt and GS were recorded in a spinal paralysed cat after injection of naloxone (0.5 mg/kg i.v.) during superfusion of the spinal cord with NMDA solution. During and after stimulation of the sural nerve some slow rhythmic activity with long-lasting bursts was observed and interspersed with some high frequency rhythmic activity with very short-lasting bursts. These bursts could occur in longer trains, but also in short groups or even as isolated events.

Variable phase relations between the activity of different motoneurone pools

The phase relation between the burst activity of different motoneurone pools could be altered in two different ways. Either the normal pattern was replaced by another pattern with a changed phase relation, or the phase relation changed gradually over a period. Figures 3 & 4 show two different examples of the first type with changed phase relation. Figure 3A shows the normal pattern of co-activation in nerves to LoD and OAE trunk muscles on one side, alternating with the activity of the other side. In Figure 3B the pattern is reversed. Co-activation in the nerves to LoD on both sides alternate with that of OAE of both sides. The latter pattern was more rarely observed, and could be induced particularly by hypoxia. While the pattern in Figure 3A resembles the pattern observed during walking, the one in B fits the pattern during gallop. Figure 4 demonstrates the appearance of different phase relations with muscles which may be bifunctional. In Figure 4B, PBSt is synchronously active with the flexor DP, both alternating with the active phases in the nerves to the ipsilateral extensors MG and LGS and to Q. This is the predominant pattern observed during DOPA-induced fictive locomotion. In Figure 4A, Q displayed a double burst activity, and PBSt is active during one of these phases, but in this case alternating with the activity in DP and synchronously active with MG and LGS, i.e. with the extensors.

As shown in Figure 5 smoothly changing phase relations from in-phase to alternating

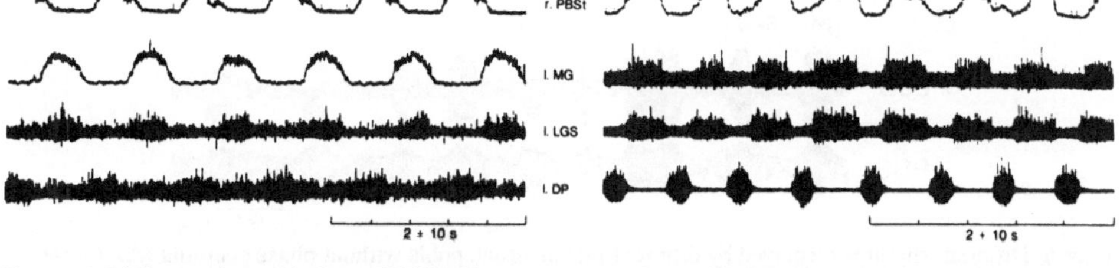

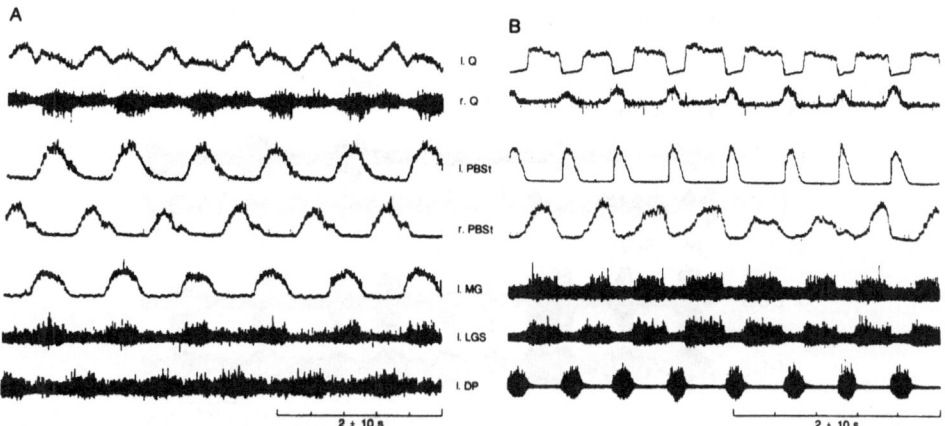

Figure 4. Different phase relations of PBSt. Neurograms, partly rectified and integrated; paralysed spinal cats after injection of nialamide, DOPA and 4-AP. (A) co-activation of PBSt with extensors (MG, LGS and second active phase of Q), alternating to flexor activity (DP). (B) co-activation of PBSt with a flexor (DP), alternating to extensor activity (Q, MG, LGS). (Schomburg, Petersen, Barajon & Hultborn, unpublished).

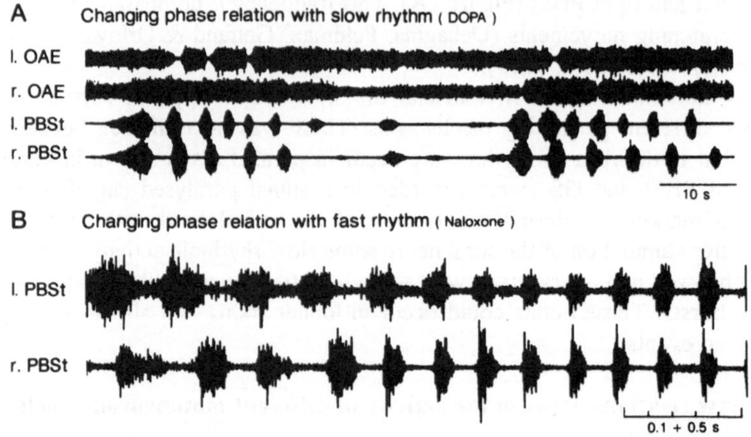

Figure 5. Changing phase relation between the activity of PBSt of both sides during low (A) and high (B) frequency rhythmic activity. Neurograms in paralysed spinal cats. (A) after injection of nialamide and DOPA, (B) after injection of naloxone (05 mg/kg i.v.) without nialamide and DOPA (modified from Koehler et al., 1984 and Schomburg & Steffens, 1995).

activity can occur in context of slow DOPA type of rhythms (A) as well as with fast rhythmic activity induced by naloxone. In the experiment of Figure 5A periods were observed repeatedly which started with a co-activation of PBSt on both sides, but then a more alternating activity gradually developed. In this case, the more tonic activity of the OAE developed a modulation which was only partly coupled. This behaviour would fit partly into the pattern described in the next paragraph, with different rhythms of different motoneurone pools. In Figure 5B at the beginning of a period of high frequency rhythmic activity there was a clear alternation between the burst activity in the left and right PBSt. Subsequently, the pattern changed to synchronous activity in both nerves.

Different rhythms of different motoneurone pools

As shown already in Figure 5A, the rhythmic modulation of the activity of different motoneurone pools is not always strictly coupled. Basically two alternate ways of achieving

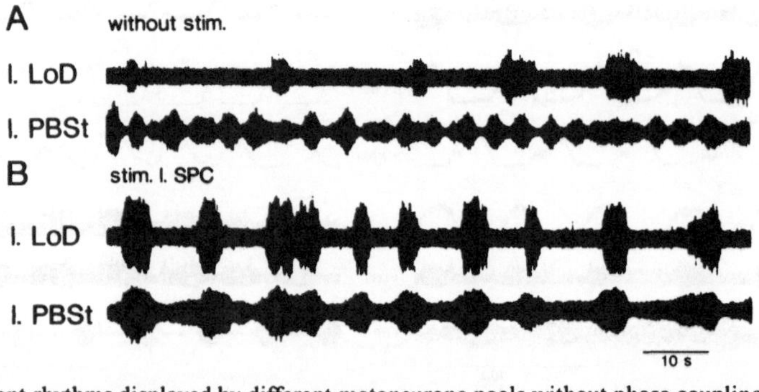

Figure 6. Different rhythms displayed by different motoneurone pools without phase coupling (A). Phase coupling is induced by cutaneous nerve stimulation (B, stimulation of SPc with 50 Hz throughout the whole recording). Neurograms of LoD and PBSt in a paralysed spinal cat after injection of nialamide and DOPA (modified from Koehler et al., 1984).

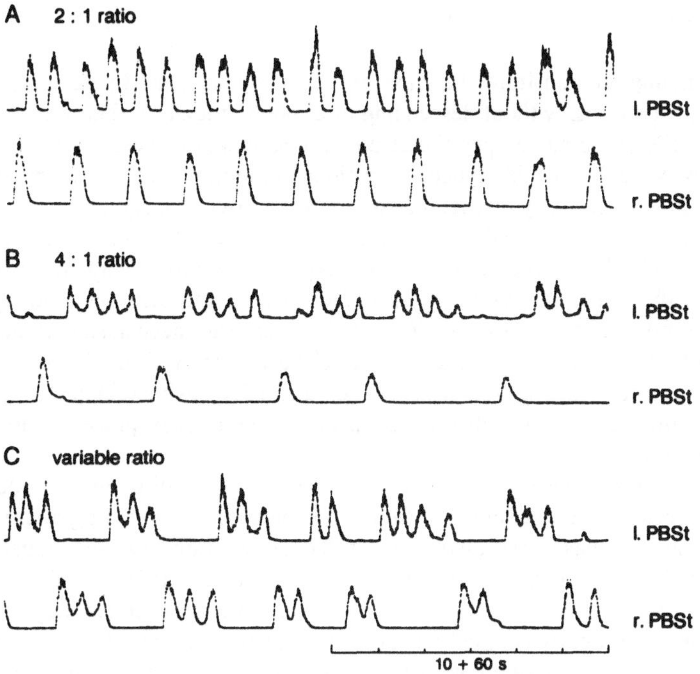

Figure 7. Different rhythms displayed by different motoneurone pools with strict phase coupling. The ratio of active phases between alternately active left (l.) and right (r.) PBSt could be quite stable (2:1 in A, 4:1 in B) or relative variable (C). Paralysed spinal cat, rectified and integrated neurograms after injection of nialamide, DOPA and 4-AP. (Schomburg, Petersen, Barajon & Hultborn, unpublished).

different rhythmic modulation of different motoneurone pools could be discerned: (1) without any phase coupling between the rhythmic burst activity in different muscles, or (2) with a phase coupling together with a fixed or variable ratio with respect to the number of bursts.

Figure 6A illustrates the first type, in which there is independent rhythmic activity of different muscle groups. There was no strict phase relation between the slow burst activity to the LoD and the faster rhythmic activity of the knee flexor PBSt. Additional stimulation of a cutaneous nerve (SPc, throughout the whole recording shown in B) accelerated the rhythm to LoD and induced a strict phase coupling between the activity of both motoneurone pools (C).

The second type of different rhythmic modulation in which there is clear phase coupling of the activity of different muscles is shown in Figure 7. There were periods with a relatively fixed ratio between the number of bursts during an active phase of the PBSt of one side and that of the other side (2:1 in A and 4:1 in B), but periods also occured with a variable number of bursts in the alternating active phases of both motoneurone pools.

Beside the type just described in which there were alternating phases of activity in different muscles, with a superimposed rhythm of higher frequency, another pattern of coupling could occur. This pattern was characterised by a stable rhythmic burst activity of one motoneurone pool, whilst the activity of another motoneurone pool started synchronously with one of the bursts of the first motoneurone pool, but then continued until the end of the second to the fourth bursts of the first motoneurone pool. This complex pattern could last for 2 to 4 periods. It is reminiscent of a pattern based on a bistable mechanism in the motoneurones of the second pool (cf. Hounsgaard, Hultborn, Jespersen & Kiehn, 1988).

DISCUSSION

These findings demonstrate the capacity of the spinal cord to generate a variety of rhythmic motor patterns, without descending or afferent influences. The versatility would be even greater if supraspinal or peripheral afferent stimulation were superimposed on this rhythmic activity, as this would induce alternations of the timing of the rhythmic phases, and change the pattern of activity distributed to different muscles (Edgerton, Grillner, Sjostrom & Zangger, 1976; Grillner, 1985; Kniffki et al., 1981; Koehler et al., 1984; Pearson & Rossignol, 1991). The generation of two types of rhythm with distinctly different frequencies in the spinal paralysed cat (slow DOPA type and fast naloxone type) suggests that there are probably two basically different rhythmogenic mechanisms in the spinal cord. Rhythms with different frequencies have also been observed in *in vitro* spinal cord preparations of the neonatal rat (Cazalets et al., 1990) which indicates that the two rhythmogenic mechanisms are intrinsic spinal properties already present before the motor system has maturated.

The versatility of phase relations between the activity of different motoneurone pools (from strictly fixed phase relations in different combinations, via shifting phase relations, to completely missing phase relations) support the assumption that individual "locomotor centres" exist for individual muscle groups (Edgerton et al., 1976; cf. also Kniffki et al., 1981; Koehler et al., 1984). These centres may be coupled in different ways depending on the motor task, from tightly coupled for the limb and trunk muscles during locomotion, to a less secure coupling in a more complex pattern of movement. The mode of coupling seems to depend partly on the afferent input. Thus, the flexibility of the spinal locomotor organisation not only enables a purely spinal adaptation of locomotion to different peripheral conditions but also higher centres may engage these spinal facilities during complex goal-directed movements (see Koehler et al., 1984; Grillner, 1981).

ACKNOWLEDGEMENTS

Supported by the Deutsche Forschungsgemeinschaft (Scho 37-3/3)

REFERENCES

CAZALETS, J.R., GRILLNER, P., MENARD, I., CREMIEUX, J. & CLARAC, F. (1990) Two types of motor rhythm induced by NMDA and amines in an in vitro spinal cord preparation of neonatal rat. *Neurosci. Lett.* **111**, 116-121.

DELIAGINA, T.G., FELDMAN, A.G., GELFAND, I.M. & ORLOVSKY, G.N. (1975) On the role of central program and afferent inflow in the control of scratching movements in the cat. *Brain Res.* **100**, 297-313.

EDGERTON, V.R., GRILLNER, S., SJÖSTRÖM, A. & ZANGGER, P. (1976) Central generation of locomotion in vertebrates. In *Neural Control of Locomotion.* eds. HERMAN, R.M., GRILLNER, S., STEIN, P.S.G. & STUART, D.G. pp. 439-469. Plenum Press, New York.

ENGBERG, I. & LUNDBERG, A. (1969) An electromyographic analysis of muscular activity in the hind limb of the cat during unrestrained locomotion. *Acta physiol. scand.* **75**, 614-630.

FORSSBERG, H. & GRILLNER, S. (1973) The locomotion of the acute spinal cat injected with clonidine i.v. *Brain Res.* **50**, 184-186.

GOSSARD, J.P. CABELGUEN, J.M. & ROSSIGNOL, S. (1989) Intra-axonal recordings of primary afferents during fictive locomotion in the cat. *J. Neurophysiol.* **62**, 1177-1188.

GRILLNER, S. (1981) Control of locomotion in bipeds, tetrapods and fish. In *Handbook of Physiology, The Nervous System*, vol. II, *Motor Control.* ed. BROOKS, V.B. pp. 1179-1236. American Physiological Society, Bethesda.

GRILLNER, S. (1985) Neural control of vertebrate locomotion - central mechanisms and reflex interaction with special reference to the cat. In *Feedback and Motor Control in Invertebrates and Vertebrates*. ed. BARNES, W.J.P. & GLADDEN, M.H. pp. 35-56. Croom Helm, London.

HOUNSGAARD, J., HULTBORN, H., JESPERSEN, B. & KIEHN, O. (1988) Bistability of α-motoneurones in the decerebrate cat and in the acute spinal cat after intravenous 5-hydroxytryptophan. *J. Physiol.* **405**, 345-367.

JANKOWSKA, E., JUKES, M.G.M., LUND, S. & LUNDBERG, A. (1967) The effect of DOPA on the spinal cord. 6. Half-centre organisation of interneurones transmitting effects from the flexor reflex afferents. *Acta physiol. scand.* **70**, 389-402.

KNIFFKI, K.-D., SCHOMBURG, E.D. & STEFFENS, H. (1981) Effects from fine muscle and cutaneous afferents on spinal locomotion in cats. *J. Physiol.* **319**, 543-554.

KOEHLER, W.J., SCHOMBURG, E.D. & STEFFENS, H. (1984) Phasic modulation of trunk muscle efferents during fictive spinal locomotion in cats. *J. Physiol.* **353**, 187-197.

PEARSON, K.G., JIANG, W. & RAMIREZ, J.M. (1992) The use of naloxone to facilitate the generation of the locomotor rhythm in spinal cats. *J. Neurosci. Meth.* **42**, 75-81.

PEARSON, K.G. & ROSSIGNOL, S. (1991) Fictive motor patterns in chronic spinal cats. *J. Neurophysiol.* **66**, 1874-1887.

SCHOMBURG, E.D. & STEFFENS, H. (1995) Influence of opioids and naloxone on rhythmic spinal motor activity in cats. *Exp. Brain Res.* (in press).

VIALA, D. & BUSER, P. (1971) Modalités d' obtention de rhythmes locomoteurs chez le lapin spinal par traitements pharmacologiques (DOPA, 5-HTP, D-amphétamine). *Brain Res.* **35**, 151-165.

WISLEDER, D., ZERNICKE, R.F. & SMITH, J.L. (1990) Speed-related changes in hindlimb intersegmental dynamics during the swing phase of cat locomotion. *Exp. Brain Res.* **73**, 651-660.

COORDINATION OF TWO MOTOR PATTERNS IN A SIMPLE VERTEBRATE

S.R. Soffe

School of Biological Sciences
University of Bristol, Woodland Road
Bristol BS8 1UG, UK

THE *XENOPUS* EMBRYO AS A SIMPLE VERTEBRATE SPINAL CORD MODEL

A number of non-mammalian vertebrates are in current use as simple models for general aspects of vertebrate motor control (including lamprey: Grillner, Wallen, Brodin & Lansner, 1991; mudpuppy: Wheatley, Edamura, Stein, 1992; turtle: Mortin & Stein, 1989; chick: O'Donovan, Sernagor, Sholomenko, Ho, Antal & Yee, 1992). This is supported by the idea that many fundamental features of neural design have been conserved during vertebrate evolution, so that features seen in lower groups can tell us something of the organisation within higher groups. Of course, any model system is to some extent a compromise. If one is interested in mammalian, particularly human function, the loss is that some details may be specific to the lower vertebrate, and therefore irrelevant to the situation in man, and also the complexity in the higher vertebrate may itself add other dimensions absent in the simpler preparation. The gain is that model systems can be chosen to be experimentally far more tractable. Amphibian embryos such as that of *Xenopus* (Figure 1), combine evolutionary simplicity with developmental simplicity. They provide what have turned out to be very profitable model systems in which to explore questions of the design of spinal cord locomotor control. It is not within the scope of this chapter to address in detail the question of how motor patterns are generated within the spinal cord. The intention here is to examine differences in the ways that the same motoneurones are driven during two different behaviours in order to produce appropriate patterns of output to the myotomes.

The trunk muscles on either side of the *Xenopus* embryo, the segmental myotomes, represent a very rudimentary system of muscular antagonists with their controlling circuitry lying within the spinal cord. At the most basic level, we can explore simply the patterns of alternating activity between the two sides. However, as we shall see, there is also a longitudinal component whose phase relationship marks a major contrast between the two main rhythmic behaviours generated by this motor system. Early work was directed at understanding generation of a single motor pattern: swimming. This is now being extended to the study of a second pattern, termed struggling, and the use of the embryo as a preparation for exploring the more general question of switching between motor patterns.

The neuroanatomy of the *Xenopus* embryo spinal cord

A very important feature to make clear from the start is the contrast in anatomical complexity between the spinal cord of the amphibian embryo and that, for example, of an adult mammal. The *Xenopus* embryo spinal cord is about 100 μm in diameter. There is no obvious segmentation actually within the cord, and sensory and motor axons do not enter and exit the cord segmentally by discrete dorsal and ventral roots. However, the cord can conveniently be divided into segments according to the pattern of adjacent myotomes and within each such segment there are only around 200 neurones. This contrasts with perhaps 40,000 neurones or more in the ventral horn alone of a segment of cat spinal cord. More importantly, a combination of anatomical, physiological and immunocytochemical studies suggest that these comprise only eight classes of neurone (Roberts & Clarke, 1982; Roberts, 1989). Undoubtedly each class shows a degree of heterogeneity, but their characterisation has been a major factor in explaining the functional organisation within the cord. Broadly, the neurones can be divided into a single class of sensory neurones (the Rohon-Beard neurones), two classes of sensory interneurones, three classes of premotor interneurones and a single class of motoneurone. There is some evidence that the motoneurones are a mixture of primary and secondary motoneurones (Clarke, Holder, Soffe, & Strom-Mathisen, 1991). There is also a class of ciliated ependymal cell (the Kolmer-Agdhur cells). Since the major behaviours can be seen in spinalised embryos (Soffe, 1991), these neuronal classes are sufficient to provide the basis for motor pattern co-ordination.

"Natural" motor patterns generated by the spinal cord

In addition to simple, non-rhythmic body flexions occasionally seen in response to a weak stimulus, *Xenopus* embryos show two main rhythmic behaviours. The most commonly occurring and best understood of these is swimming, which is self-sustaining following a brief triggering sensory stimulus. A sustained stimulus, acting through the same sensory neurones, can produce a quite distinct pattern termed struggling. Struggling differs from swimming in being slower and stronger, and in having a different longitudinal pattern (Kahn & Roberts, 1982). Body movement spreads rostro-caudally on each cycle during swimming and caudo-rostrally during struggling. These features of cycle period, strength and phase relationship are the main ones characterising any rhythmic motor pattern. By differing in all these respects, swimming and struggling are as different as any two rhythmic behaviours could be in such a simple animal. Embryos immobilised with neuromuscular blockers can generate appropriate motor responses to drive behavioural swimming and struggling, termed "fictive" swimming and struggling (Figure 1; cf. Soffe, 1991). A major advantage of the *Xenopus* embryo is that, rather unusually for a vertebrate, the motor patterns generated in the paralysed preparation can be evoked by the same sensory stimuli that evoke real behaviour, without the need for pharmacological excitants or brain stimulation. Also, there appears to be no movement-related sensory modification of the basic pattern (there are, for example, no muscle receptors). This means that the fictive swimming and struggling motor patterns represent, as closely as possible, "natural" patterns underlying real behaviour. As with the behaviour, fictive swimming and struggling (from now on referred to simply as swimming and struggling) differ in all important features. Cycle period and motor burst duration are relatively short in swimming and discharge propagates rostro-caudally on each cycle. Cycle period and burst duration are relatively long in struggling and discharge propagates caudo-rostrally. The two motor patterns are therefore both qualitatively and quantitatively distinct.

Swimming and struggling can both be evoked by stimulating one set of sensory neurones: the Rohon-Beard neurones (Soffe, 1991). Brief stimulation of the trunk skin, which contains the peripheral unmyelinated neurites of Rohon-Beard neurones, triggers long-lasting swimming episodes. Repetitive stimulation of the same neurites can evoke struggling during the period of stimulation. Selection of one or other motor pattern can therefore be achieved simply by changing the pattern of discharge in the same sensory neurones.

The synaptic drive to motoneurones during swimming and struggling

In earlier descriptions of the synaptic drive to motoneurones during swimming, three major components were distinguished, an on-cycle phasic excitation, a mid-cycle phasic inhibition corresponding to antagonistic activity on the opposite side of the cord, and a sustained background excitation (Soffe & Roberts, 1982). The mid-cycle inhibition is glycinergic and is produced by characterised inhibitory commissural interneurones, lying on the opposite side of the spinal cord (Dale, 1985). Excitatory interneurones on the same side of the spinal cord produce both quick on-cycle excitation and sustained background excitation by release of an excitatory amino acid acting at non-NMDA and NMDA receptors respectively (Dale & Roberts, 1985). However, it has recently been shown that excitatory amino acid excitation only accounts for about 30% of the on-cycle excitation to motoneurones during swimming (Perrins, 1995). There are in addition two further components, both of which appear to derive from other synergistic motoneurones, many of

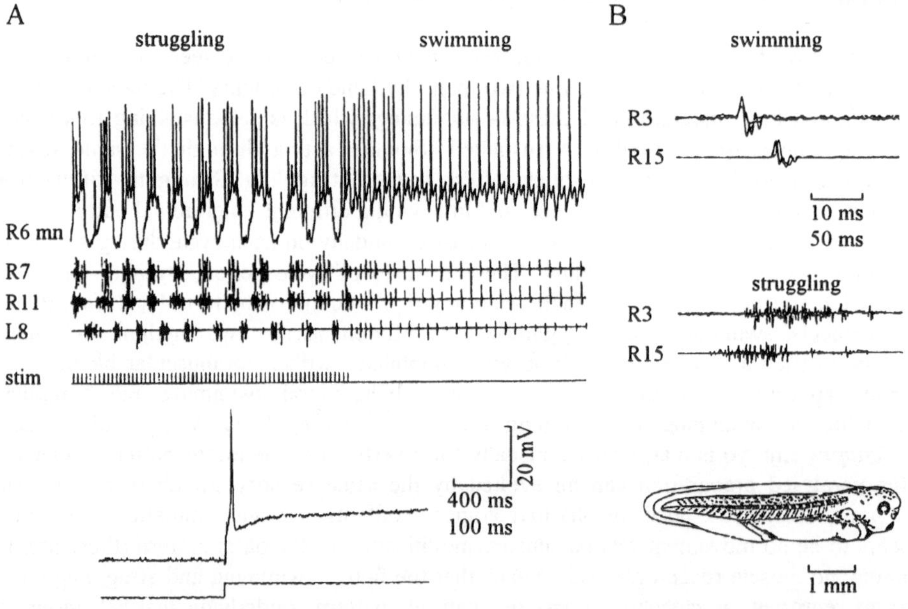

Figure 1. Swimming and struggling motor patterns in *Xenopus* embryos. Embryo illustrated lower right. A. Intracellular recording from a motoneurone (R6; mn), from rostral (R7) and caudal (R11) motor roots on the same side and rostrally on the opposite side (L8). Repetitive stimulation (stim) evokes struggling which then switches to swimming at the end of stimulation. This motoneurone only fired once to a depolarising (120 pA) current step (lower record). B. Rostral (R3) and caudal (R15) motor root recordings in a second embryo. Brief bursts propagate rostro-caudally during swimming; long bursts propagate caudo-rostrally during struggling.

which have descending central axons and can make output connections to other motoneurones (Perrins & Roberts, 1995). The first component is cholinergic, acting at central nicotinic receptors, and accounts for about 20% of fast, on-cycle excitation. The second is an electrotonic component from neighbouring motoneurones, which accounts for some 50% of the on-cycle excitation. The synaptic drive to motoneurones during swimming therefore consists of an alternating pattern of excitation and inhibition, where the exitation has an excitatory amino acid, cholinergic and electrotonic component, and the inhibition is glycinergic, all superimposed on a sustained background of excitatory amino acid excitation.

The pattern of synaptic drive to motoneurones during struggling is somewhat similar. Excitation alternates with mid-cycle inhibition (Soffe, 1993a). Once more the inhibition is glycinergic and the on-cycle excitation contains an excitatory amino acid component. This pattern reflects the fact that the same excitatory and inhibitory premotor interneurones that fire during swimming also fire during struggling, and there appears to be no recruitment of further neurone classes (Soffe, 1993a). Recruitment of more neurones within existing classes parallels the apparent increase in strength of synaptic drive during struggling. The demonstration of both a cholinergic and electrotonic component during swimming made it probable that these components would also be present during struggling. The same motoneurones are, after all, active during both patterns and the presence of cholinergic and electrotonic components would help explain the relatively weak effect of excitatory amino acid receptor blockers (Soffe, 1993a). Preliminary experiments have confirmed that these extra components are indeed present. Overall, these findings suggest that the two motor patterns for swimming and struggling, despite being quite distinct, are driven by essentially the same premotor circuitry, producing qualitatively similar patterns of synaptic drive.

WHAT CONTROLS SWITCHING BETWEEN THE SWIMMING AND STRUGGLING MOTOR PATTERNS?

We have seen that selection of either the swimming or struggling motor pattern can be produced by the particular pattern of sensory input, changing the output of essentially the same premotor circuitry. Is the role of the sensory neurones crucial, for example by providing a source of neuromodulation? Although a strength of the embryo preparation is that motor behaviour can be evoked by natural stimulation, both patterns can also be evoked in the spinal embryo by bath application of excitants. Broadly: low concentrations of excitatory amino acid excitants can evoke swimming; brief applications of higher concentrations (though not apparently NMDA) can evoke struggling (Soffe, 1993b). Neither produces firing in the sensory neurones. The implication of these findings is that the sensory neurones are not crucial and that selection of one or other pattern depends simply on the level of excitation in the spinal cord motor and premotor circuitry, with maintained sensory discharge and an associated recruitment of excitatory premotor interneurones being simply one way excitation could be raised.

For a switch from swimming to struggling, an increase in the level of excitation must produce an increase in cycle period, an increase in burst duration and a reversal in longitudinal phase delay. How this might operate is still not clear, but we can look at one of the components in more detail.

WHAT DETERMINES MOTOR BURST DURATION DURING SWIMMING AND STRUGGLING?

One striking difference between the patterns of motor root discharge during swimming and struggling (and indeed many other adult patterns of motor discharge) is that of burst duration. During swimming, firing is highly synchronous, with motoneurones (and the premotor interneurones that drive them) firing only a single spike per cycle. The result is a very short motor root burst. During struggling, the motor root burst is relatively long. Also, unlike during swimming, burst length scales with cycle period.

Many *Xenopus* embryo motoneurones fire only a single spike to a depolarising current step (Figure 1; Soffe, 1990). Firing in these motoneurones seems to be limited by membrane accommodation resulting possibly from Na^+ current inactivation, or the presence of a relatively slowly inactivating K^+ conductance. It seemed likely that this would in turn limit motor discharge to a single spike per cycle during swimming. The presence of prolonged motor bursts during struggling then made it reasonable to suggest that neuromodulation of this cellular property of neurones, allowing them to fire repetitively, might play an integral role in switching to the struggling motor pattern (Soffe, 1993a). However, this oversimplifies the situation. In fact some motoneurones (about 30%) show less strong membrane accommodation and can fire short bursts of impulses when depolarised. These neurones still only fire a single spike per cycle during swimming. Also, motoneurones can be made to fire trains of impulses to a depolarising step by injecting intracellular Cs^+ (Soffe, 1990), presumably weakening a K^+ conductance. During swimming, such motoneurones similarly only fire a single spike per cycle.

The question is then not simply how do neurones that normally fire a single spike fire bursts during struggling. Instead, we must also ask how neurones that can fire repetitively are limited to a single spike per cycle during swimming while those with strong membrane accommodation can fire bursts during struggling. It appears that, at the level of individual motoneurones, firing during swimming is limited by the pattern of synaptic drive as much as by intrinsic cellular properties. During struggling, motoneurones that normally fire only once could contribute to extension of motor bursts if their spikes become desynchronised. However, many individual motoneurones clearly themselves fire bursts of impulses. If this does not occur through neuromodulation of membrane accommodation, it must presumably result again from the particular pattern of synaptic drive. It is important to remember that since motoneurones supply a significant component of the excitatory input to other motoneurones (Perrins, 1995), changes in firing pattern of some motoneurones may reinforce the same pattern in other motoneurones.

ACKNOWLEDGEMENTS

This work was supported by the M.R.C.

REFERENCES

CLARKE, J.D.W., HOLDER, N., SOFFE, S.R. & STORM-MATHISEN J. (1991) Neuroanatomical and functional analysis of neural tube formation in notochordless *Xenopus* embryos; laterality of the ventral spinal cord is lost. *Development* **112**, 499-516.

DALE, N. (1985) Reciprocal inhibitory interneurones in the *Xenopus* embryo spinal cord. *J. Physiol.* **363**, 61-70.

DALE, N. & ROBERTS, A. (1985) Dual-component amino-acid-mediated synaptic potentials: excitatory drive for swimming in *Xenopus* embryos. *J. Physiol.* **363**, 35-59.

GRILLNER, S., WALLEN, P., BRODIN, L. & LANSNER, A. (1991) Neuronal network generating locomotor behaviour in lamprey: circuitry, transmitters, membrane properties, and simulation. *Ann. Rev. Neurosci.* **14**, 169-199.

KAHN, J.A. & ROBERTS, A. (1982b) The neuromuscular basis of rhythmic struggling movements in embryos of *Xenopus* laevis. *J. Exp. Biol.* **99**, 197-203.

MORTIN, L.I. & STEIN, P.S.G. (1989) Spinal cord segments containing key elements of the central pattern generators for three forms of scratch reflex in the turtle. *J. Neurosci.* **9**, 2285-2296.

O'DONOVAN, M., SERNAGOR, E., SHOLOMENKO, G., HO, S., ANTAL, M. & YEE, W. (1992) Development of spinal motor networks in the chick embryo. *J. Exp. Zool.* **261**, 261-273.

PERRINS, R. (1995) The roles of central cholinergic and electrical synapses made by spinal motoneurones in *Xenopus* embryos. This Volume.

PERRINS, R. & ROBERTS, A. (1995) Cholinergic and electrical synapses between synergistic spinal motoneurones in the *Xenopus* laevis embryo. *J. Physiol.* (in press).

ROBERTS, A. (1989) The neurons that control axial movements in a frog embryo. *Am. Zool.* **29**, 53-63.

ROBERTS, A. & CLARKE, J.D.W. (1982) The neuroanatomy of an amphibian embryo spinal cord. *Phil. Trans. R. Soc. B* **296**, 195-212.

SOFFE, S.R. (1990) Active and passive membrane properties of spinal cord neurons which are rhythmically active during swimming in *Xenopus* embryos. *Eur. J. Neurosci.* **2**, 1-10.

SOFFE, S.R. (1991) Triggering and gating of motor responses by sensory stimulation: behavioural selection in *Xenopus* embryos. *Proc. R. Soc. B* **246**, 197-203.

SOFFE, S.R. (1993a) Two distinct rhythmic motor patterns are driven by common premotor and motor neurons in a simple vertebrate spinal cord. *J. Neurosci.* **13**, 4456-4469.

SOFFE, S.R. (1993b) Two different rhythmic motor responses to excitatory amino acid agonists in the *Xenopus* embryo spinal cord. *J. Physiol.* **473**, 190P.

SOFFE, S.R. & ROBERTS, A. (1982) Tonic and phasic synaptic input to spinal cord motoneurons during fictive locomotion in frog embryos. *J. Neurophysiol.* **48**, 1279-1288.

WHEATLEY, M., EDAMURA, M. & STEIN, R.B. (1992) A comparison of intact and in vitro locomotion in an adult amphibian. *Exp. Brain Res.* **88**, 609-614.

CRITIQUE OF PAPERS BY O'DONOVAN & RITTER, SCHOMBURG AND SOFFE

Serge Rossignol

Department of Physiology, CNRS
Faculty of Medicine, Université de Montréal
Montréal, Canada

These three papers raise some issues which should perhaps be integrated in a more general context schematised in Figure 1. Over the last 30 years the field of motor control dealing with rhythmic behaviours such as locomotion, swimming, scratching, paw shaking and allied movements (e.g. hatching, struggling) has made enormous progress which has been reviewed several times, and has been the subject of many international meetings. The interest has been largely in defining which structures of the nervous system are essential for the generation of these rhythmic behaviours. A great deal of work was performed on supraspinal structures which can initiate and control these motor patterns (Armstrong, 1986), as well as on central spinal mechanisms that can generate these patterns in some great detail (Grillner, 1981), and on their modulation by afferent inputs (Rossignol, Lund & Drew, 1988). In the last 15 years there has been an important effort devoted to establishing the membrane properties and synaptic connectivity, as well as network properties involved in producing these motor patterns, and major advances have been made by various groups working on lower and higher vertebrates (Rossignol & Dubuc, 1994).

All the work presented in this session deals with aspects of properties of cells involved in these rhythmic processes, and this critique will address some of these aspects. One clear message in the last few years has been that studies of the nervous system under dynamic conditions reveal properties in cell membranes and circuits or reflex pathways that could hardly be suspected from the work in static conditions. However, because of the experimental approaches needed to isolate elements of these circuits and identify their properties, a certain amount of uncertainty exists (dotted arrow in Figure 1) as to the relationship between the rhythmic phenomena observed and real behaviours.

LOCALISATION AND IDENTIFICATION OF CELLS RELATED TO RHYTHMIC ACTIVITY

The paper presented by O'Donovan is a very nice attempt at visualising the spatiotemporal pattern of activation of cells in the spinal cord of chick embryo after loading the spinal cord with the ester form of fura-2 which permeates all membranes over a certain distance. Some groups of cells were also loaded more specifically by injecting ventrolateral tract white matter through which course the axons of some interneurones, or by loading motoneurones specifically in a hemisected piece of spinal cord. It is of great interest to see how the wave of calcium travels from the ventrolateral cord to the mediodorsal part of the cord, and the video sequences presented at the symposium were certainly convincing. Several questions can be raised about these observations.

Does this represent a ventrodorsal activation of neurones or could glial cells participate?

Glial cells appear to have much more important roles than might have been thought before it became known that they have voltage operated-channels, receptor-operated ion channels (GABA, Glutamate), and receptor-operated second messenger systems, including calcium, that can be activated by most neurotransmitters. Astrocytes in culture, loaded with fluo-3, have been shown to generate slow rhythmic calcium waves (with periods of 10 s) when activated by glutamate (Cornell-Bell, Finkbeiner, Cooper, Smith, 1990; Cornell-Bell & Finkbeiner, 1991). These waves, appearing in groups of astrocytes in one spot of the culture, travel at a slow rate of 10 µm/s (1000 times slower than the rate reported here by O'Donovan), and invade the whole culture before starting all over again. Calcium waves can also be generated by glutamate in astrocytes incorporated within the hippocampal network in organotypically cultured slices of rat hippocampus (Dani, Chernjavsky & Smith, 1992). Are these waves, and those reported in the spinal cord embryo of a completely different nature? Could glial cells be part of these processes, and could the rhythm and conduction velocity be much faster in the chick embryo which is closer to real networks than are cells in culture? Could glial cells participate in the slow rhythms seen after DOPA in acute spinal cats? In Schomburg's paper, confirming many other previous observations, some rhythmic cycles may be quite long (even as slow as 1 per 10 s) which is probably not incompatible with such waves. Recent work in the *in vitro* neonatal rat preparation has shown that, among the sulphur-containing excitatory amino acids capable of inducing rhythmic activity, homocysteic is the most potent and it is strictly localized within glial cells (Sqalli-Houssaini, Cazalets, Martini & Clarac, 1993).

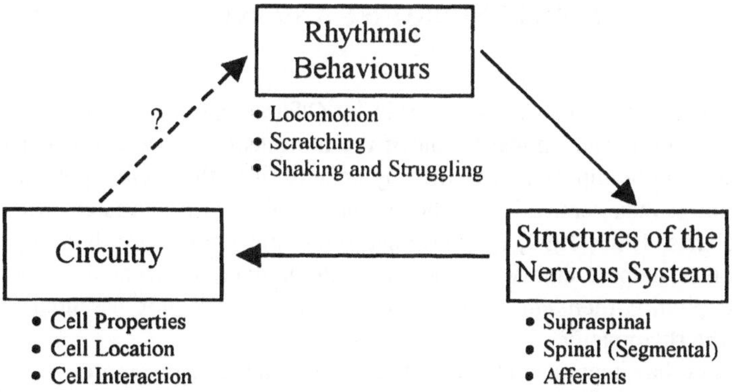

Figure 1. General context and emphasis of research on rhythmic behaviours in the last 30 years. After a period of description of locomotion, scratching etc., interest was centred on the various structures (afferent and descending input) of the nervous system need to express these behaviours. The recent trend is not to try to localise and indentify the properties of cells responsible for generating rhythmic behaviours, with the unavoidable uncertainty in relating aspects of circuitry to the behaviours which they control.

Does this represent the invasion of dendrites of motoneurones or interneurones?

The figures presented here cannot resolve that issue. Fortunately, the authors are now using a laser confocal microscope that allows them to specify which structures of the neuropile participate in such waves. Golgi stains of Ramon y Cajal show the dendrites of motoneurones invading not only the dorsal horn but also the contralateral side. The same applies to interneurones filled with HRP by Jankowska. Could the invasion of these dendrites be contribituing to the calcium waves? In a recent paper (Alford, Frenguelli, Schofield & Collingridge, 1993), it was shown, using a combination of confocal microscopy and patch clamp recordings, that the elevation of calcium within Ca1 pyramidal neurons activated by Schaeffer collaterals was complex and could involve different NMDA mechanisms at the soma and the dendrites. Of particular interest are local oscillations of calcium concentration in dendrites which have a period of about 1.5 Hz, quite close to the oscillations seen by O'Donovan in the chick preparation after a shock to the spinal cord.

Localisation of interneurones

The involvement of certain classes of interneurones in rhythmic processes has been studied either by direct recording of known interneurones (Ia, Renshaw, group II interneurones), by chemical inactivation of interneurones with known neurotransmitters or by recording from interneurones with unknown projections in different spinal laminae. Other attempts at using activity markers are now suggesting that cells around the central canal may be important for rhythm generation. Thus, using deoxyglucose (Viala, Buisseret-Demas & Portal, 1988) and c-fos (Dai, Douglas, Nagy, Noga & Jordan, 1990), labelled cells were found around the central canal in rabbits and cats. Similarly, recent experiments in neonatal rats using Sulforhodamine (Kjaerulff, Barajon & Kiehn, in press) as an activity marker point to a region of the central canal. Recording of cells in this region support their ability to generate rhythmic activity (Hochman, Jordan & MacDonald, 1994). These findings corroborate those of O'Donovan showing interneurones with crossed projection to the ventrolateral white matter in the chick embryo. The approach of O'Donovan not only can point to the participation of a certain group of cells in a given region but can also show what is the temporal sequence of its activation in relation to other groups.

NEW PROPERTIES OF MOTONEURONES AND INTERNEURONAL PATHWAYS

The path of the calcium waves reported by O'Donovan shows an early activation of motoneurones in the cycle; optical lesion of the motoneurones suggests that they are not essential but contribute importantly to this rhythmicity since the rhythm generated is of very short duration after their removal from the circuitry. This is also consistent with the findings (Perrins & Roberts, 1993; 1994) that motoneurones contribute more than 50 % of the on-cycle excitation in tadpole embryos either through cholinergic mechanisms or electrotonic coupling. The picture then emerges that motoneurones are not essential but are very important for the rhythmicity.

The evidence has been accumulating that new membrane properties of motoneurones are revealed during such rhythmic processes in turtles and cats (Hounsgaard & Kiehn, 1985; 1989; Conway, Hultborn, Kiehn & Mintz, 1988; Hounsgaard, Hultborn, Jespersen & Kiehn, 1988; Eken, Hultborn & Keihn, 1989; Brownstone, Gossard & Hultborn, in press). Drugs capable of activating locomotor circuits such as DOPA and clonidine as well as 5-HT may activate mechanisms of bistability in motoneuronal membranes. These plateau potentials are thus a contribution by the motoneurones to their own rhythmicity. Similarly, recent evidence in the neonatal rat suggest that NMDA can induce oscillatory properties in lumbar motoneurones (Sqalli-Houssaini, Cazalets & Clarac, 1993) which are TTX-resistant (Hochman, Jordan & Schmidt, 1994). Whether these phenomena, coupled to similar properties triggered in interneurones, may contribute to maintaining the rhythm will have to be elucidated. In that context, one could ask whether the struggling reaction described in the tadpole by Soffe is mainly an amplification of the swimming pattern or whether some specific afferent pathways could trigger certain properties in interneurones that will lead to the longer and more vigorous rhythmic pattern observed.

How motoneurones could feed back to interneurones generating the rhythm itself is a difficult question. However, as mentioned earlier, new properties and new pathways are brought to light during such dynamic processes. One good example of this is the remarkable phase-dependent modulation of reflexes during locomotion (Rossignol et al., 1988). The same stimulation can evoke different responses in different parts of the cycle, thus revealing the dynamic opening and closing of pathways. Even a classical pathway, such as the Ib inhibitory pathway, is now seen to generate an autogenic excitation during locomotion (Pearson & Collins, 1993; Gossard, Brownstone, Barajon & Hultborn, 1994). It is probable that such alternative pathways will be revealed in similar rhythmic conditions and perhaps some of these (other than the well known Renshaw-Ia interneurone pathway) can provide feedback from motoneurones to interneurones.

RELATIONS TO BEHAVIOUR

Each of the three papers presented raise the question - although to different degrees - whether there is any relation between the experimental observations and natural behaviour. In O'Donovan's paper, the behaviour has the characteristics of embryonic motility which is itself not easy to define (see Bekoff, 1992 for recent review). In Schomburg's paper, some of the many patterns described do not necessarily relate to any known behaviours.

A clear relationship to natural behaviour is sometimes evident. In Soffe's paper, the two behaviours are clear and simple. In cats, some of the characteristics of locomotion are preserved, in the spinal state and fictive conditions, such as the cycle time, the pattern of discharge (double burst) of certain muscles such as the knee flexor semitendinosus, the relative duration of knee and hip flexors and finally the symmetrical alternation on both sides (Pearson & Rossignol, 1991; Rossignol, Saltiel, Perreault, Drew, Pearson & Bélanger, 1992). Thus it is possible to be confident that the fictive rhythm observed is very close indeed to the locomotor pattern.

All rhythms in the cord are not necessarily locomotor as can be shown in behaving chronic spinal cats as well as in fictive preparations: real and fictive scratching (Deliagina, Feldman, Gelfand & Orlovsky, 1975), paw shake and struggling (Pearson & Rossignol, 1991) can all be evoked by specific sensory triggers. Indeed, in fictive conditions, pinching the paw generates a pattern where all flexors, mainly ipsilaterally, are strongly activated as if to release the paw from a pinch. Squirting water on the central pad can generate a fast paw shake with a typical 8-12 Hz cycle and the particular synergy described by Judy Smith (Smith, Hoy, Koshland, Phillips & Zernicke, 1985): TA and VL, which are antagonists during walking, discharge at the same time during fast paw shake.

We should try our best to stick to patterns with characteristics resembling known behaviours. Although the fictive situation is useful, the mismatch between central rhythms and the afferent feedback may, at times, generate patterns of coupling between muscles which are difficult to relate to behaviour. The point made by Schomburg is however well taken that the spinal cord is capable of generating a variety of patterns as already shown

(Grillner & Zangger, 1979) in similar preparations. In addition Schomburg showed that different patterns can also be generated by different drugs. It was reported recently that glutamate can yield episodes of fictive paw shake in decerebrate paralysed cats (Douglas, Noga, Dai & Jordan, 1993). Similarly, glutamate injected intrathecally can generate fast paw shake in early chronic spinal cats (Chau, Provencher, Lebel, Jordan, Barbeau & Rossignol, 1994).

CONCLUSIONS

These three papers raise important issues which will be central in the coming years. We must try very hard to identify the rhythmic behaviour we are dealing with in fictive conditions in higher vertebrates. We should try to identify where the critical cells are, using activity markers as well as calcium imaging with confocal microscopy, and look for properties of cells and circuits which may be quite different from what could be expected from a quasi-quiescent nervous system.

REFERENCES

ALFORD, S., FRENGUELLI, B.G., SCHOFIELD, J.G. & COLLINGRIDGE, G.L. (1993). Characterization of Ca^{2+} signals induced in hippocampal CA1 neurones by the synaptic activation of NMDA receptors. *J. Physiol.* **469**, 693-716.

ARMSTRONG, D.M. (1986). Supraspinal contributions to the initiation and control of locomotion in the cat. *Prog. Neurobiol.* **26**, 273-361.

BEKOFF, A. (1992). Neuroethological approches to the study of motor development in chicks: achievements and challenges. *J. Neurobiol.* **23**, 1486-1505.

BROWNSTONE, R.M., GOSSARD, J-P. & HULTBORN, H. (1995). Voltage-dependent excitation of motoneurones from spinal locomotor centres in the cat. *Exp. Brain Res.* in press.

CHAU, C., PROVENCHER, J., LEBEL, F., JORDAN, L., BARBEAU. H. & ROSSIGNOL, S. (1994). Effects of intrathecal injection of NMDA receptor agonist and antagonist on locomotion of adult chronic spinal cats. *Soc. Neurosci. Abstr.* **20**. 573.

CONWAY, B.A., HULTBORN, H., KIEHN, O. & MINTZ, I. (1988). Plateau potentials in alpha-motoneurones induced by intravenous injection of L-Dopa and clonidine in the spinal cat. *Prog. Brain Res.* **405**, 369-384.

CORNELL-BELL, A.H., FINKBEINER, S.M., COOPER, M.S. & SMITH, S.J. (1990). Glutamate induces calcium waves in cultured astrocytes: long-range glial signaling. *Science* **247**, 470-473.

CORNELL-BELL, A.H. & FINKBEINER, S.M. (1991). Ca^{2+} waves in astrocytes. *Cell Calcium* **12**, 185-204.

DAI, X., DOUGLAS, J.R., NAGY, J.I., NOGA, B.R. & JORDAN, L.M. (1990). Localization of spinal neurons activated during treadmill locomotion using the c-fos immunohistochemical method. *Soc. Neurosci. Abstr.* **16**, 368.4.

DANI, J.W., CHERNJAVSKY, A. & SMITH, S.J. (1992). Neuronal activity triggers calcium waves in hippocampal astrocytes networks. *Neuron* **8**, 429-440.

DELIAGINA, T.G., FELDMAN, A.G., GELFAND, I.M. & ORLOVSKY, G.N. (1975). On the role of central program and afferent inflow in the control of scratching movements in the cat. *Brain Res.* **100**, 297-313.

DOUGLAS, J.R., NOGA, B.R., DAI, X. & JORDAN, L.M. (1993). The effects of intrathecal administration of excitatory amino acid agonists and antagonists on the initiation of locomotion in the adult cat. *J. Neurosci.* **13**, 990-1000.

EKEN, T., HULTBORN, H. & KIEHN, O. (1989). Possible functions of transmitter-controlled plateau potentials in alpha-motoneurones. *Prog. Brain Res.* **80**, 257-267.

GOSSARD, J.-P., BROWNSTONE, R.M., BARAJON, I. & HULTBORN, H. (1994). Transmission in a locomotor-related group Ib pathway from hindlimb extensor muscles in the cat. *Exp. Brain Res.* **98**, 213-228.

GRILLNER, S. (1981). Control of locomotion in bipeds, tetrapods, and fish. In *Handbook of physiology.*

The nervous system II., eds. BROOKHART, J. M. & MOUNTCASTLE, V. B., pp. 1179-1236. American Physiological Society, Bethesda.

GRILLNER, S. & ZANGGER, P. (1979). On the central generation of locomotion in the low spinal cat. *Exp. Brain Res.* **34**, 241-261.

HOCHMAN, S., JORDAN, L.M. & MACDONALD, J.F. (1994). N-methyl-D-aspartate receptor-mediated voltage oscillations in neurons surrounding the central canal in slices of rat spinal cord. *J. Neurophysiol.* **72**, 565-577.

HOCHMAN, S., JORDAN, L.M. & SCHMIDT, B.J. (1994) TTX-Resistant NMDA receptor-mediated voltage oscillations in mammalian lumbar motoneurons. *J. Neurophysiol.* **72**, 1-4.

HOUNSGAARD, J., HULTBORN, H., JESPERSEN, J. & KIEHN, O. (1988). Bistability of alpha-motoneurones in the decerebrate cat and in the acute spinal cat after intravenous 5-hydroxytryptophan. *Prog Brain Res.* **405**, 345-367.

HOUNSGAARD, J. & KIEHN, O. (1985). Ca^{++} dependent bistability induced by serotonin in spinal motoneurons. *Exp. Brain Res.* **57**, 422-425.

HOUNSGAARD, J. & KIEHN, O. (1989). Serotonin-induced bistability of turtle motoneurones caused by a nifedipine-sensitive calcium plateau potential. *Prog. Brain Res.* **414**, 265-282.

KJAERULFF, O., BARAJON, I. & KIEHN, O. (1995) Distribution of sulforhodamine-labelled cells in the neonatal rat spinal cord following induced locomotor activity in vitro. *J. Physiol.* in press.

PEARSON, K.G. & COLLINS, D.F. (1993). Reversal of the influence of group Ib afferents from plantaris on activity in medial gastrocnemius muscle during locomotor activity. *J. Neurophysiol.* **70**, 1009-1017.

PEARSON, K.G. & ROSSIGNOL, S. (1991). Fictive motor patterns in chronic spinal cats. *J. Neurophysiol.* **66**, 1874-1887.

PERRINS, R. & ROBERTS, A. (1993). Electrical interactions between spinal motoneurones in the *Xenopus laevis* embryo. *J. Physiol.* **473**, 194P.

PERRINS, R. & ROBERTS, A. (1994). Motoneurones contribute cholinergic and elctrical excitation to each other during fictive swimming in *Xenopus* embryos. *J.Physiol.* **476**, 25P.

ROSSIGNOL, S. & DUBUC, R. (1994) Spinal pattern generation. *Current Opinion in Neurobiology.* **4**, 894-902.

ROSSIGNOL, S., LUND, J.P. & DREW, T. (1988). The role of sensory inputs in regulating patterns of rhythmical movements in higher vertebrates. A comparison between locomotion, respiration and mastication. In *Neural control of rhythmic movements in vertebrates*, eds. COHEN, A., ROSSIGNOL, S. & GRILLNER, S., pp. 201-283. Wiley and Sons, New York.

ROSSIGNOL, S., SALTIEL, P., PERREAULT, M.-C., DREW, T., PEARSON, K. & BÉLANGER, M. (1992). Intralimb and interlimb coordination in the cat during real and fictive rhythmic motor programs. *Seminars in the neurosciences* **5**, 67-75.

SMITH, J.L., HOY, M.G., KOSHLAND, G.F., PHILLIPS, D.M. & ZERNICKE, R.F. (1985). Intralimb coordination of the paw-shake response: a novel mixed synergy. *J. Neurophysiol.* **54**, 1271-1281.

SQALLI-HOUSSAINI, Y., CAZALETS, J.R. & CLARAC, F. (1993) Oscillatory properties of the central pattern generator for locomotion in neonatal rats. *J. Neurophysiol.* **70**, 803-813.

SQALLI-HOUSSAINI, Y., CAZALETS, J.R., MARTINI, F. & CLARAC, F. (1993). Induction of fictive locomotion by sulphur-containing amino acids in an in vitro newborn preparation. *Eur. J. Neurosci.* **5**, 1226-1232.

VIALA, D., BUISSERET-DELMAS, C. & PORTAL, J.J. (1988). An attempt to localize the lumbar locomotor generator in the rabbit using 2-deoxy-[14C]glucose autoradiography. *Neurosci. Lett.* **86**, 139-143.

MODULATION OF SPINDLE SENSITIVITY AND REFLEXES IN DECEREBRATE AND SPINAL WALKING CATS

D.J. Bennett, S.J. De Serres and R.B. Stein

Department of Physiology
University of Alberta
Edmonton, Canada T6G 2S2

INTRODUCTION

Recordings from awake behaving cats indicate that muscle spindle sensitivity is lower during walking than during postural tasks (Prochazka, 1989), due to a predominant static fusimotor drive. Thus, during walking the effectiveness of the stretch reflex should be relatively reduced by this static fusimotor drive. We investigated this possibility in acute locomoting spinal cats, where we could directly measure spindle sensitivity and reflex responsiveness. For comparison, measurements were also taken in the quiescent state (resting) and during tonic contractions. As tonic contractions are unusual in the spinal cat, we elicited them in the decerebrate cat prior to spinalisation.

METHODS

Soleus or triceps surae muscles were held on a muscle puller during tonic crossed extensor reflex contractions in the decerebrate cat and during walking induced by Clonidine and Naloxone in the spinal cat (Pearson & Collins, 1993). Small sinusoidal stretches (0.5 mm peak-to-peak, 4 - 5 Hz) were applied during the contractions, allowing estimation of reflex responsiveness from EMG recordings and spindle sensitivity from Ia afferent recordings (18 units). Responses to stretch were averaged at 8 distinct force levels and the amplitude of a sine wave fit to the averages (linear least squares) was used as either the estimate of reflex responsiveness or of spindle sensitivity. As some units appeared to unload during contractions (decrease firing), as though they had little gamma drive, five units were studied after severing the ventral roots. To mimic the unloading and relengthening effects produced by alpha activity, slow ramp stretches (7 mm, 0.25 Hz) were applied and sensitivity to a superimposed sinusoidal stretch was measured as the units fired at different mean rates. A quadratic curve fit to the sensitivity vs. rate scatter plots of these de-efferented units gave a reference to test whether intact units had significant fusimotor drive.

RESULTS

The reflex responses to stretch were markedly reduced during tonic contractions and during walking, as compared to the resting state. These reflex results are reported more extensively in Bennett et al. (1994, in preparation), and are not discussed further here due to a lack of space. During walking the spindle sensitivity was usually lowered in comparison to the sensitivity of the same units in the resting state prior to walking. Results from 18 units (6 cats) were analyzed during both resting and walking conditions (swing and stance phases analyzed separately). The high firing rates observed in many units suggest the presence of a strong gamma bias. This bias was probably of gamma static origin, due to the low sensitivity in comparison to the de-efferented sensitivity at the same firing rates. The alpha-gamma linkage of this gamma bias was highly variable. Some units fired as if they were driven with static fusimotor drive during both stance (with EMG) and swing (without EMG). Others only fired during the stance phase. Finally, about a third of the units had low mean firing rates, and decreased their firing rates and sensitivities during a contraction due to unloading of the spindle. Thus, these units appeared to have little or no gamma drive in spite of alpha activity, a type of alpha-gamma dissociation which has rarely been reported.

The results during tonic contractions were more consistent: units always had a strong alpha-linked gamma bias (about 100 imp/s) and a decreased sensitivity, suggesting tonic gamma static drive. The drops in sensitivity during tonic contractions were greater than during walking, probably due to 1) greater dynamic fusimotor bias prior to contractions and 2) greater static fusimotor drive during contractions. In spite of the greater drop in sensitivity during tonic contractions, the decrease in reflex responsiveness was less than the drop observed during walking, suggesting the presence of a contribution from other sources during walking such as presynaptic inhibition.

REFERENCES

PEARSON, K. G. & COLLINS D. F. (1993) Reversal of the influence of group Ib afferents from plantaris on activity in medial gastrocnemius muscle during locomotor activity. *J. Neurophysiol.* **70**, 1009-1017.

PROCHAZKA, A. (1989) Sensorimotor gain control: A basic strategy of motor systems? *Prog. Neurobiol.* **33**, 281-307.

FORELIMB PROPRIOCEPTORS RECORDED DURING VOLUNTARY MOVEMENTS IN CATS

D.F. Collins, M.A. Gorassini
and A. Prochazka

Division of Neurosciences
University of Alberta
Edmonton, Alberta
Canada, T6G 2S2

Most of our knowledge regarding proprioceptive activity in freely moving cats comes from recordings from hindlimb afferents. Currently, little is known about this activity in the forelimbs. One might speculate that this could be extrapolated from existing hindlimb data. Indeed, assuming similarities in receptor morphology and sensitivity, this may be true for receptors not under efferent control. However, there are several reasons to suspect that this extrapolation is less secure for the muscle spindle receptor; 1) The forelimb performs reaching and manipulative tasks requiring more supraspinal control than hindlimb movements (Pettersson, 1990), 2) Supraspinal sites branch more extensively to cervical regions (Kuypers & Martin, 1982), 3) Presumed spindle afferents in arm and hand muscles of monkeys generally had more complex firing characteristics than cat hindlimb spindles (Schieber & Thach, 1985), 4) There may be differences in spindle receptor morphology, as demonstrated between hindlimb and neck musculature (Dutia, 1991). A knowledge of spindle receptor firing patterns from the cat forelimb may also shed some light on the differences in firing rates in humans (peak approx. 85 imp/s) compared to cats (peak approx. 600 imp/s). We have therefore been developing a technique to obtain forelimb afferent recordings in awake cats.

Initially, we intended to implant electrodes into the cervical dorsal root ganglia, along similar lines to the lumbar hindlimb recordings (Prochazka, Westerman & Ziccone, 1977). However, the lateral location of the ganglia and the large range of motion between the cervical vertebrae thwarted our attempts to obtain stable afferent recordings. Therefore, we modified the technique so as to record from the primary afferent fibres as they ascend the dorsal columns.

Prior to implantation a flexible microwire loop was fabricated and attached to a cable of Cooner wires (AS632). During the surgical procedure the dorsal aspect of one side of vertebrae C2-C3 was exposed. The exposed C3 lamina was removed to provide access to the dorsal columns. Initial dissections revealed that the movement between the vertebrae and the spinal cord was the least at C2-C3. The Cooner wire-microwire loop junction was anchored to the caudal aspect of C2 using dental acrylic and cyano-acrylate (Figure 1). The cable was led subcutaneously to a dental acrylic headpiece. The 25 mm loop was led down, and sutured to, the exposed dura at the rostral end of C3. Its flexibility allowed movement between vertebrae and spinal cord without dislodging the 6 microwire electrodes. The electrodes (8-12 mm long from point of suture to deinsulated, bevelled 17 μm dia. tip) were implanted into the fasciculus cuneatus region through a slit cut in the dura. Cats recovered over a 24 hour period. Length data were obtained via an external mercury-in-rubber length gauge. EMG data were obtained via needle electrodes inserted into the receptor bearing muscle through a patch of skin anaesthetised with lidocaine cream.

Data have been obtained from forelimb and neck afferents in 3 cats. These recordings depend on a balance between a migration of the electrode tip to favourable positions near afferent axons and the maintenance of those positions long enough for data collection and unit identification. So far we have obtained periodic, stable recordings for up to 6 weeks. In one case the same slowly-adapting skin receptor was recorded for over 3 weeks. It was located over the dorsal ridge of the scapula and was highly sensitive to skin stretch along a medial-lateral axis. Peak firing rates reached 500 imp/s with dynamic indices up to 200 imp/s. Over 3-4 consecutive stretches the dynamic indices dropped to about 50 imp/s, indicating a decline of the phasic component of response.

Data has also been obtained from a Golgi tendon organ in the long head of triceps brachii, identified by suxamethonium and twitch tests. Peak firing rates during imposed ramp and hold stretches reached 150 imp/s. An average of 15 step cycles revealed that the receptor was active during the stance phase of locomotion with maximal discharge of 125 imp/s around foot contact, declining steadily to near zero around lift off. This pattern was similar to that of tendon organs of the hindlimb, though more recordings are needed before firm conclusions can be drawn.

Some data have also been obtained from two units presumed to be of muscle spindle origin. One recording, from a receptor located in a flexor/adductor of the lateral toe and suspected to be a Ia afferent, had peak firing rates of 250 imp/s and a dynamic index of 150

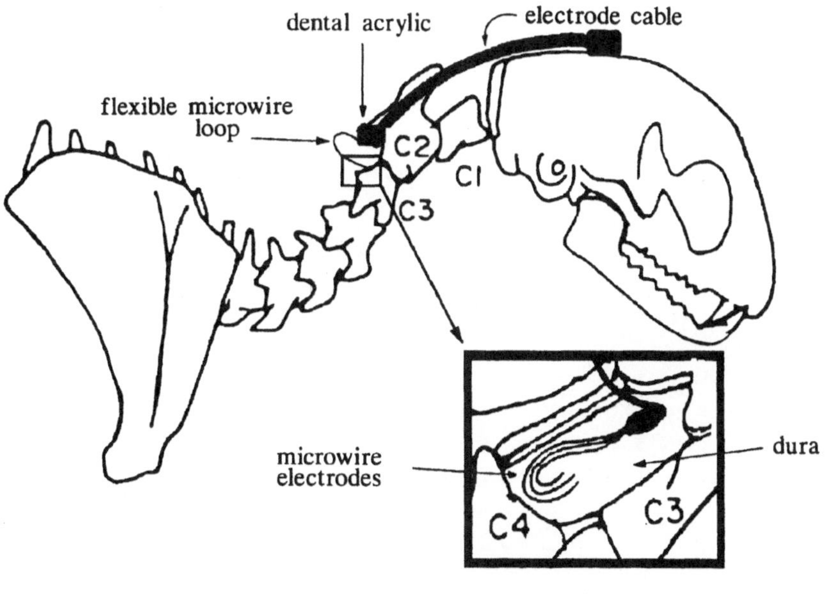

Figure 1. Diagram of cat skeleton showing neurogram implant method. (adapted from Dutia, 1991).

imp/s during imposed ramp and hold stretches. The second, from a receptor in the neck musculature which we suspect was a b_2c or a group II afferent, was very length-sensitive and had relatively small dynamic components of response. This unit fired at rates up to 140 imp/s during sinusoidal imposed movements.

CONCLUSION

We have developed a viable technique to obtain stable afferent recordings from cat forelimb via microelectrodes implanted into the cervical dorsal columns. Preliminary data reveal receptor firing rates similar to those in the cat hindlimb. In the future we intend to investigate the fusimotor control of the muscle spindle during movements such as reaching and manipulative tasks. We also intend to contribute to a further characterisation of the activity of neck proprioceptors.

REFERENCES

DUTIA, M.B. (1991). The muscles and joints of the neck: their specialisation and role in head movement. *Prog. Neurobiol.* **37**, 165-178.

KUYPERS, H.G.J.M. & MARTIN, G.F. (1982). Anatomy of descending pathways to the spinal cord. *Prog. Brain Res.* **57**, 329-360.

PETTERSSON, L.-G. (1990). Forelimb movements in the cat; kinetic features and neuronal control. *Acta physiol. scand.* **140**, 1-60.

PROCHAZKA, A., WESTERMAN, R.A. & ZICCONE, S. (1977). Discharges of single hindlimb afferents in the freely moving cat. *J. Neurophysiol.* **39**, 1090-1104.

SCHIEBER, M.H. & THACH, W.T. (1985). Trained slow tracking. II Bidirectional discharge patterns of cerebellar nuclear, motor cortex, and spindle afferent neurons. *J. Neurophysiol.* **54**, 1228-1270.

ON THE FUNCTIONAL SIGNIFICANCE OF LONG MONOSYNAPTIC DESCENDING PATHWAYS TO SPINAL MOTONEURONES

P.A. Kirkwood[1] and J.D. Road[2]

[1]Sobell Department of Neurophysiology
Institute of Neurology, Queen Square
London WClN 3BG, UK
[2]Department of Medicine
University of British Columbia
2211 Wesbrook Mall
Vancouver, BC, Canada V6T 2B5

Studies in behaving animals show that the activities of neurones with direct long descending connections to motoneurones, such as corticomotoneuronal cells to hand muscles in the primate, often do not co-vary with the activities of the target muscles. Explanations for this include the operation of populations of neurones involving many input cells with different patterns affecting many motoneurones of different muscles, estimated via network analysis (Fetz, 1992), or the effects of other pathways setting recruitment patterns at motoneuronal level (Bennett & Lemon, 1994). The pathways that include these long monosynaptic connections are accorded a great deal of importance in controlling the movements concerned and it seems to be most often assumed that it is the excitation derived via this direct link that is the most important part of the control. In contrast we present here results from another system which shows an extreme case of lack of co-variance, and we argue from these that rather more importance should be given to events at spinal level in producing the patterns of motoneurone output.

The experiments consisted of measuring monosynaptic connections between expiratory bulbospinal neurones and the expiratory motoneurones of different expiratory muscles in the barbiturate anaesthetized cat. Cross-correlations were performed between the extracellularly recorded discharges of single bulbospinal neurones in the caudal medulla (antidromically identified from the upper lumbar spinal cord) and the efferent discharges in intercostal and abdominal muscle nerves at both T8 and L1 levels. Addition of $C0_2$ to the inspired gases was used in order to promote a strong respiratory output.

A typical cross-correlation histogram produced from the nerve we are most concerned with here, that innervating external abdominal oblique at T8, is shown in Figure 1. This shows a clear narrow peak which indicates a monosynaptic connection (Davies, Kirkwood

& Sears, 1985a). Nearly all of the neurones (12/15) gave such a peak and the overall excitation derived from these connections, as calculated below, was high. Following the logic and assumptions of Davies, Kirkwood & Sears (1985b), we started by assessing the factor k, the ratio of peak count to baseline counts. The mean value (for contralaterally located neurones) (flat histograms taken as k = 1.0) was 1.174, or 1.130 if bulbospinal units with axons terminating between T9 and L1-L3 were included. Even this latter figure was about twice as strong as for any of the other expiratory nerves, including the internal intercostal branch at T8 and, incidently, more than twice as strong as the similar connection between inspiratory bulbospinal neurones and phrenic motoneurones (Davies et al., 1985b), which is generally regarded as a significant connection.

We assume conservatively that there are 400 expiratory bulbospinal neurones on each side (cf. Merrill & Lipski, 1987, who estimated the number as 700). The mean firing rate of the neurones giving connections was 94.5 imp/s, leading to a total impulse rate of 400 x 94.5 imp/s = 37,800 imp/s. We also assume that a value of k = 1.5 is equivalent to a 100μV EPSP, which, for k = 1.130, gives an average EPSP amplitude of 26μV. We further assume that EPSPs add linearly and that a mean rate of 14,000 EPSPs per second (100μV amplitude) will bring a motoneurone from its resting potential into its firing range. Here, then, the contralateral input is equivalent to 37.8/14 x 26% of this, i.e., about 70%. Other experiments in this laboratory show that ipsilateral bulbospinal neurones, via collaterals crossing the cord, give a connection 36% of the strength of that from contralateral ones (Kirkwood, 1995), leading to a final figure of 70% x 1.36 = 95%. Thus this single monosynaptic input alone under these circumstances (a strong respiratory drive under anaesthesia) is sufficient to provide more or less all the motoneurone excitation.

This result would appear to make this connection of prime importance in the control of these motoneurones in respiration and, indeed, would appear to give a special role to the bulbospinal neurones in other movements in which they are active. However, observations from other studies severely limit this view. Merrill & Lipski (1987) sampled expiratory motoneurones at T8 intracellularly. For most cells the depolarization during expiration was subliminal and the monosynaptic input from the bulbospinal neurones appeared much weaker than here (2 EPSPs, 55 and 62μV in amplitude, in 57 trials). If both these observations and our present ones are correct then the direct connections must be concentrated on only a minority of motoneurones, possibly as few as 10%. These

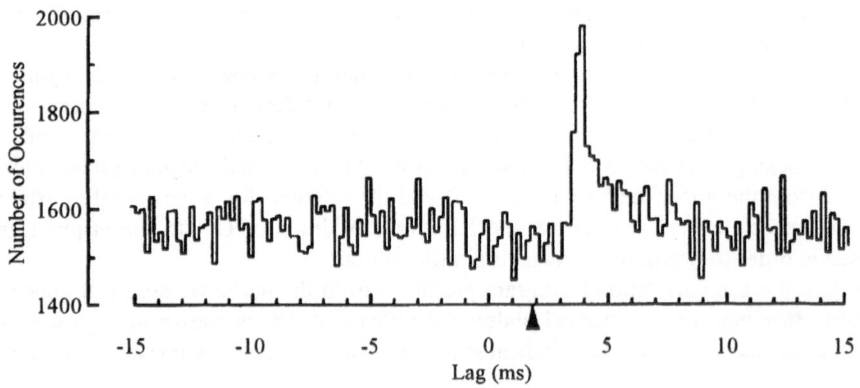

Figure 1. Typical cross-correlation histogram from the discharges of an expiratory bulbospinal neurone (reference spikes) and efferent discharges in the nerve to external abdominal oblique at T8. Both the duration of the peak (half-width 0.6ms) and its timing (arrowhead shows time of the impulse in the stem axon at T8) indicate that this peak was derived from a monosynaptic connection. Bin-width 0.192ms. k estimated as 1980/1570 = 1.26.

connections are probably the determining factor in the recruitment of these 10% or so of the motoneurones under the conditions of our experiments, but in the circumstances when any of the other 90% are recruited one must assume that their activity will mostly be determined by other inputs, the most obvious ones being those from spinal interneurones.

These calculations, which explicitly suggest that motoneurones are controlled from different sources in different behaviours, also provide a ready explanation for results from the corticomotoneuronal system. The apparently variable strength of post-spike facilitation with different forces or phases of a movement in Bennett & Lemon (1994) could be explained entirely by the involvement, as here, of only a minority of motoneurones in the connection.

Results from the study of Miller, Tan & Suzuki (1987), who recorded the behaviour of the expiratory bulbospinal neurones during fictive vomiting, also emphasise spinal mechanisms. In the rhythmic retching movements of vomiting, the abdominal muscles are active in the opposite phase to the internal (expiratory) intercostal muscles and in-phase with the diaphram, instead of in-phase with internal intercostal and opposite to the diaphgram, as in respiration. Miller et al. (1987) found that a majority of the bulbospinal neurones (17/27) fired in-phase with the internal intercostals, only 9/27 in-phase with the abdominal muscles. Thus during retching a large proportion of those neurones which we have shown here to have strong monosynaptic connections to the motoneurones of the external abdominal oblique at T8 must be active in the opposite phase to that when the activation of this muscle is required.

Although, by calculations similar to above, this input will not neccessarily provide enough excitation to depolarize the motoneurones to threshold, fewer bulbospinal neurones being involved and these firing at relatively low rates (Miller et al., 1987), the observations do suggest a particular requirement for inhibition during the "off" phase of the retching cycle, as is known to be present for expiratory motoneurones during inspiration (Sears, 1964). The observations also raise the question of why these bulbospinal neurones should be recruited at all during this action, when their most obvious, direct connection to the abdominal motoneurones is not useful. One answer may lie in their connections to the intercostal motoneurones, but given that these connections are much weaker, this explanation is not attractive. Rather, the observation suggests that these neurones are recruited for their effects via some other, unknown connection, again most likely via spinal interneurones. The monosynaptic connections would not then be regarded as being of special importance for the control of the motoneurone pattern, but instead as providing more general excitation to the expiratory motoneurones during either respiration or vomiting, the pattern being created largely by spinal interneurones. For respiratory movements of the thorax this explanation has already been independently suggested by the observation that the most rhythmically firing thoracic interneurones seem to be inhibitory, whereas respiratory-modulated excitatory interneurones most often had a tonic background activity (Kirkwood, Schmid & Sears, 1993), implying that inhibition at spinal cord level is of prime importance in producing the rhythmic motor output pattern, even when the descending drive itself has a clear phasic pattern.

Extrapolating once more to the corticomotoneuronal system, such a view would not be inconsistent with Bennett & Lemon (1994; see also this volume), but instead of regarding monosynaptic cortical control as being imposed on motoneurones whose recruitment patterns had been additionally set by segmental or propriospinal circuits, it would place those spinal circuits central to the control from the cortex; i.e., the cortex would be seen as operating via these circuits, the monosynaptic link being just one of many excitatory inputs to the motoneurones. This view would also be quite consistent with Fetz's (1992) approach, except that in his analysis the "hidden units" would have to include a large proportion of spinal interneurones. A similar view has recently been put forward by Lundberg (1992) who pointed out certain specific interneurone circuits which should be considered. Note that a prime role for the corticomotoneuronal cells in the control of hand movements is not denied, any more than is the prime role of the bulbospinal neurones in transmitting the respiratory drive. The problem lies in determining the importance of the monoynaptic link itself within that role. We suggest that among the various approaches needed to sort this out, some estimation of the amplitudes and rates of EPSPs from each potential source would be very valuable.

ACKNOWLEDGEMENTS

Support from the Canadian MRC and the Jeanne Anderson Fund (Institute of Neurology) is acknowledged.

REFERENCES

BENNETT, K.M.B. & LEMON, R.N. (1994) The influence of single monkey cortico-motoneuronal cells at different levels of activity in target muscles. *J. Physiol.* **477**, 291-307.

DAVIES, J.G.McF., KIRKWOOD, P.A. & SEARS, T.A. (1985a) The detection of monosynaptic connexions from inspiratory bulbospinal neurones to inspiratory motoneurones in the cat. *J. Physiol.* **368**, 33-62.

DAVIES, J.G.McF., KIRKWOOD, P.A. & SEARS, T.A. (1985b) The distribution of monosynaptic connexions from inspiratory bulbospinal neurones to inspiratory motoneurones in the cat. *J. Physiol.* **368**, 63-87.

FETZ, E.E. (1992) Are movement parameters recognizably coded in the activity of single neurons? *Behav. Brain Sci.* **15**, 679-690.

KIRKWOOD, P.A. (1995) Synaptic excitation in the thoracic spinal cord from expiratory bulbospinal neurones in the cat. *J. Physiol.* **484**, 201-225.

KIRKWOOD, P.A., SCHMID, K. & SEARS, T.A. (1993) Functional identities of thoracic respiratory interneurones in the cat. *J. Physiol.* **461**, 667-687.

MERRILL, E.G. & LIPSKI, J. (1987) Inputs to intercostal motoneurons from ventrolateral medullary respiratory neurons in the cat. *J. Neurophysiol.* **57**, 1837-1853.

LUNDBERG, A. (1992) To what extent are brain commands for movements mediated by spinal interneurones? *Behav. Brain Sci.* **15**, 775-776.

MILLER, A.D., TAN, L.K. & SUZUKI, I. (1987) Control of abdominal and expiratory intercostal muscle activity during vomiting: role of ventral respiratory group expiratory neurons. *J. Neurophysiol.* **57**, 1854-1866.

SEARS, T.A. (1964). The slow potentials of thoracic respiratory motoneurones and their relation to breathing. *J. Physiol.* **175**, 404-424.

THE EFFECTS OF PLANTAR NERVE STIMULATION ON LONG LATENCY FLEXION REFLEXES IN THE ACUTE SPINAL CAT

B.A. Conway[1], D.T. Scott[2] and J.S. Riddell[2]

[1]The Bioengineering Unit, University of Strathclyde
[2]Institute of Physiology, University of Glasgow
Glasgow, UK

INTRODUCTION

Recently, it has been shown that knee and ankle extensor Ib afferents can reset or entrain locomotor activity by promoting extensor burst activity whilst suppressing flexor burst generation (Conway, Hultborn & Kiehn, 1987; Pearson, Ramirez & Jiang, 1992). This indicates that limb loading during stance is important in the reflex regulation of stepping and we have now investigated whether other afferent systems, which might signal limb loading, show similar reflex effects. In the acute spinal cat treated with L-Dopa, stimulation of high threshold afferents results in the generation of late and long lasting flexion reflexes (LFRs), which are believed to reflect the organisation of a spinal locomotor generator (Lundberg, 1979). In this study, we have examined whether electrical stimulation of low threshold afferents from the plantar aspect of the hind foot (including cutaneous and muscle afferents) modify the transmission of LFRs.

METHODS

In cats anaesthetised with halothane, a decerebration, lumbar laminectomy (L4-S1) and spinal cord section (T12) were performed. Both hindlimbs were extensively denervated and selected nerves dissected for stimulation/recording purposes. These nerves included medial, deep and lateral branches of the ipsilateral plantar nerve (PN). Following surgery, anaesthesia was discontinued and the cats paralysed and artificially ventilated. Blood pressure, end tidal PCO_2 and temperature were maintained within physiological limits. Following IV administration of nialamide (30-50 mg/kg) and L-Dopa (50-100 mg/kg), stimulation of high threshold afferents generated LFRs, which were recorded as electroneurograms (ENGs) or as DC intracellular recordings from identified motoneurones.

The effect of low intensity stimulation of the PNs on transmission of these reflexes was tested.

RESULTS

Figure 1A shows an example of an LFR recorded from a flexor muscle nerve. Figure 1B shows that transmission of this LFR was completely suppressed during continous stimulation (< 2T) of the plantaris muscle nerve (cf. Conway et al., 1987). the remaining traces in Figure 1 show that transmission of LFRs could also be modified during low intensity stimulation of each of the main branches of PN. As can be seen from Figure 1C & D stimulation of deep and lateral PN respectively, like stimulation of plantaris, completely suppressed the LFR. In contrast, stimulation of medial PN resulted in potentiation of the LFR. This is seen in Figure 1E as an increased amplitude and a reduced latency of the LFR. In some experiments, lateral PN stimulation was seen to promote flexor activity (see below).

Short bursts of low intensity stimulation (< 2T) applied to deep PN during ongoing LFR activity resulted in a rapid and abrupt termination of flexor ENG. Intracellular recordings from flexor motoneurones reveal that this was due to a disfacilitation of the motoneurones rather than a result of direct inhibition. The suppression of flexor activity was

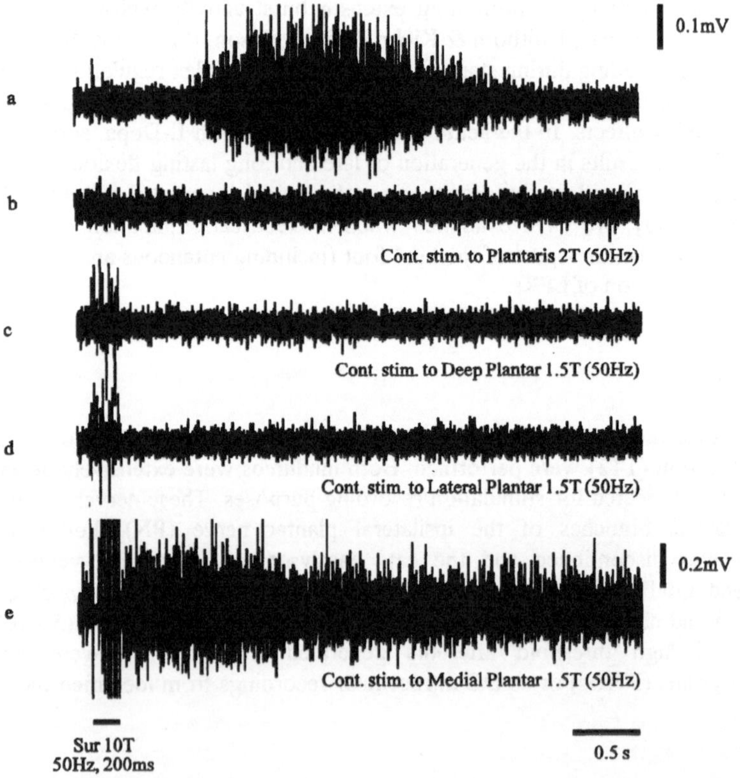

Figure 1. Effect of continuous low threshold afferent stimulation on the transmission of an LFR recorded as ENGs from PbSt. A. Control reflex seen following stimulation of the sural nerve (time of stimulation indicated by bar, 10T, 50Hz, 200ms). B to E. Effect on reflex of continual stimulation of plantaris (B; 2T, 50Hz), deep plantar (C; 1.5T, 50Hz), lateral plantar (D; 1.5T, 50Hz) and medial plantar (E; 1.5T, 50Hz). Note that voltage calibration for E is twice that for A-D.

accompanied by a maintained increase in extensor activity, seen both in intracellular recordings and ENG's. If a similar stimulus was delivered to medial PN the ongoing LFR was potentiated and no increase in extensor activity was observed. In the case of lateral PN stimulation the results were mixed. In most cases, we have observed a consistent suppression of the LFR similar to that observed during deep PN stimulation. However, in other experiments the effects were equivalent to results associated with stimulation of medial PN. This variability of action seen in different experiments may result from differences in the positioning of the stimulating electrode relative to the small muscle and cutaneous branhces of lateral PN (i.e. proportion of cutaneous and muscle afferents activated).

DISCUSSION

The present results show that a population of low threshold afferents from the plantar aspect of the hind foot can influence the behaviour of lumbar interneurones associated with the generation of long lasting flexor and extensor motor output. Stimulation of low threshold afferents in deep PN during the expression of LFRs results in the disfacilitation of flexor motoneurones and the excitation of extensor motoneurones. This behaviour is identical to the reflex actions associated with stimulation of extensor Ib afferents during LFRs and locomotion (Conway et al, 1987; Pearson et al., 1992). Given that the majority of low threshold deep PN afferents innervate muscle, it is likely that the actions described in this report are also associated with stimulation of Ib afferents. The less consistent actions produced by low intensity stimulation of lateral PN may reflect variations in the recruitment of muscle and cutaneous afferents in different experiments. This view is supported by the observation that low intensity stimulation of the predominately cutaneous medial PN consistently resulted in facilitation of LFRs. It is worth noting, however, that in spontaneously walking premammillary cats Duysens (1977) reported that stimulation of presumed cutaneous afferents from the foot results in a phase-dependent extensor and flexor burst prolongation. Accordingly, further work, including studies on fictive locomotion, is required to clarify the role of different populations of low threshold muscle and cutaneous foot afferents in the reflex regulation of locomotion. Nevertheless, the present results coupled with previous descriptions of the reflex actions of knee and ankle extensor Ib afferents during stepping suggests that a widely distributed system of proprioceptors associated with the detection of limb loading exists and aids in the reflex regulation of stepping.

ACKNOWLEDGEMENTS

This study was supported by a grant from the MRC.

REFERENCES

CONWAY, B.A., HULTBORN, H. & KIEHN, O. (1987) Proprioceptive input resets central locomotor rhythm in the spinal cat. *Exp. Brain Res.* **68**, 643-656.

DUYSENS, J. (1977) Reflex control of locomotion as revealed by stimulation of cutaneous afferents in spontaneously walking premammillary cats. *J. Neurophysiol.* **40**, 737-751.

LUNDBERG, A. (1979) Multisensory control of spinal reflex pathways. *Prog. Brain Res.* **50**, 11-28.

PEARSON, K.G., RAMIREZ, J.M. & JIANG, W. (1992) Entrainment of the locomotor rhythm by group Ib afferents from ankle extensors in spinal cats. *Exp. Brain Res.* **90**, 557-566.

ON THE RELATION BETWEEN MOTOR-UNIT DISCHARGE AND PHYSIOLOGICAL TREMOR

B.A. Conway[1], S.F. Farmer[2],
D.M. Halliday[3] and J.R. Rosenberg[3]

[1]Bioengineering Unit, University of Strathclyde
Glasgow, UK
[2]The National Hospital for Neurology
and Neurosurgery, Queen Square, London, UK
[3]Department of Physiology, University of Glasgow
Glasgow, UK

INTRODUCTION

Physiological tremor has two main spectral components. One is a consequence of mechanical resonance and its frequency changes with inertial loading (Stiles & Randall, 1967). The other is load independent (referred to as neurogenic) and contributes to two distinct frequency bands of the spectrum (Amjad, Conway, Farmer, O'Leary, Halliday & Rosenberg, 1994). Here we describe the relation between the neurogenic components of tremor and the discharges of single motor-units. While Elble & Randall (1976) studied extensor digitorum communis with a single load of between 200 - 250g or an equivalent voluntarily generated force, we have been concerned with loads not exceeding 25g, which is less than a 10% maximum voluntary contraction.

METHODS

Recordings were made in 24 normal adult subjects with local Ethics Committee approval. The tremor signal was derived from an accelerometer fixed to the distal phalanx of the middle finger, while the other fingers, wrist and forearm were immobilised. Subjects were asked to extend and hold the middle finger approximately horizontal. Tremor signals were sampled at 1 ms intervals for 60 s with the finger unloaded and during loading in 5g increments up to 25g. Pairs of single motor-units were recorded with separate concentric needle electrodes. Firing times of single motor-units were measured with a 1ms resolution.

The tremor signal, $x(t)$, was assumed to be a stationary zero-mean time series and spike trains, N_1 and N_2, were assumed to be stationary point processes that are orderly and satisfy a mixing condition in which events widely separated in time become independent (Rosenberg, Amjad, Breeze, Brillinger & Halliday, 1989). Frequency domain analyses (according to Rosenberg et al. (1989) and Brillinger (1983)) yielded the auto-spectrum of the tremor signal, $G_{xx}(f)$, the coherence between spike trains, $|R_{12}(f)|^2$, and the coherence between a spike train and the tremor signal, $|R_{x1}(f)|^2$. In the time domain the point-process cumulant density between spike trains at lag u, $q_{12}(u)$, and the hybrid cumulant density between a spike train and the tremor signal, $q_{x1}(u)$, were estimated. The coherence provides a normalised measure of the strength of association between two processes on a scale from zero to one, with zero occurring in the case of independence. The function $q_{12}(u)$ has an interpretation similar to a cross-correlation histogram, whereas $q_{x1}(u)$ has an interpretation similar to a spike-triggered average.

RESULTS

Figure 1A & B shows that a 20g load resulted in a shift in the maximum of the tremor

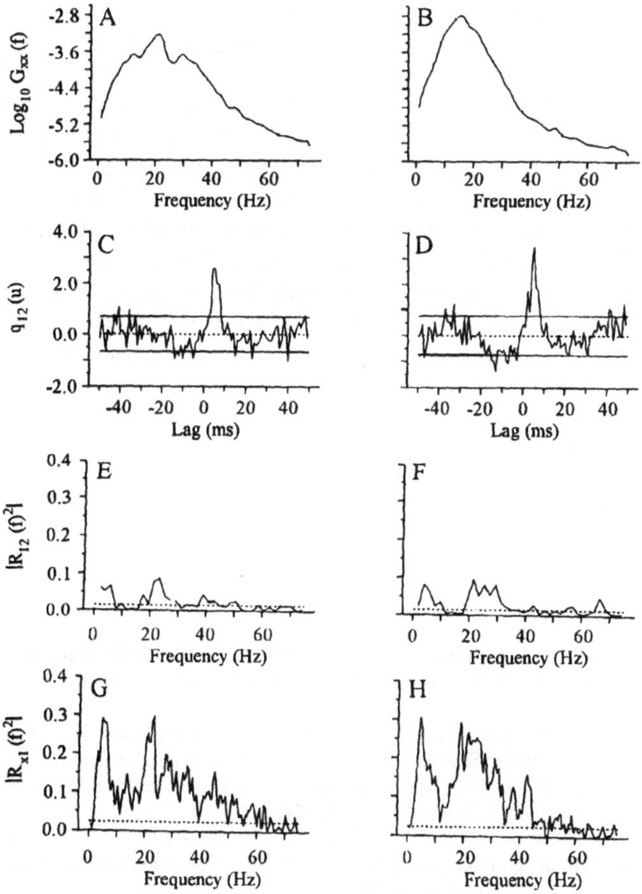

Figure 1. Auto-spectra of finger tremor at 0g (A) and 20g (B) load. Cumulant density between motor-units at 0g (C) and 20g (D) load. Coherence between motor-units at 0g (E) and 20g (F) load. Coherence between motor-unit and tremor at 0g (G) and 20g (H) load. The solid horizontal lines in C & D represent the approximate 95% confidence interval under the assumption of independence. The dashed lines in E-H represent the upper level of the 95% confidence interval under the assumption of independence.

spectrum from about 22 Hz to 16 Hz. There was also an overall increase in the magnitude of the tremor at all frequencies which obscured the neurogenic components observed as small peaks on either side of the dominant load-dependent peak. The cumulant density (Figure 1C & D) showed that the two motor-units remained correlated as the load was increased. The corresponding coherences between the motor-units (Figure 1E & F) had two small, but significant peaks at about 5 Hz and between 20 - 30 Hz. Increasing the load did not change the position or magnitude of the significant peaks in the coherence as shown by a statistical test (Rosenberg et al., 1989). In this example the mean firing rates of the two motor-units changed from 12.6 to 13.1 imp/s and from 9.9 to 11.9 imp/s with increased load. The coherence between a single motor-unit and the tremor signal showed dominant peaks centred about the same frequencies over which the two motor-units were correlated (Figure 1G & H). Increasing the load did not change the contribution that a single motor-unit made to the tremor signal.

DISCUSSION

The above results show that within the range of loads used in our experiment the strength of coupling between motor-units does not change with increasing load or EMG activity, and suggests that correlation of motor-units may be a controlled variable. Also, once recruited, the contribution that a single motor-unit makes to the tremor remains fixed over the range of loading. The coherence between a single motor-unit and the tremor has significant peaks over the same range of frequencies as the coherence between motor-units. Moreover, the frequency corresponding to the mean rate of discharge of either motor-unit does not contribute to the coherence between the motor-unit and the tremor. Farmer, Bremner, Halliday, Rosenberg & Stephens (1993) showed that the 16 - 32Hz component of the coherence between motor-units was most likely a consequence of a central descending effect on the motor-unit pair. Consequently, the spectrum of physiological tremor may contain a contribution which reflects the frequency content of a descending signal.

ACKNOWLEDGEMENTS

Supported by the Wellcome Trust.

REFERENCES

AMJAD, A.M., CONWAY, B.A., FARMER, S.F., O'LEARY, C., HALLIDAY, D.M. & ROSENBERG, J.R. (1994). A load-independent 30-40Hz component of physiological tremor in man. *J. Physiol.* **476**, 21P.

BRILLINGER, D.R. (1983). The finite Fourier transform of a stationary process. In *Handbook of Statistics*. ed. BRILLINGER, D.R. & KRISHNAIAH, P.R., pp. 21-37. Elsevier, Amsterdam.

ELBLE, R.J. & RANDALL, J.E. (1976). Motor-unit activity responsible for 8- to 12-Hz component of human physiological tremor. *J. Neurophysiol.* **39**, 370-383.

FARMER, S.F., BREMNER, F.D., HALLIDAY, D.M., ROSENBERG, J.R. & STEPHENS, J.A. (1993). The frequency content of common synaptic inputs to motoneurones studied during voluntary isometric contraction in man. *J. Physiol.* **470**, 127-155.

ROSENBERG, J.R., AMJAD, A., BREEZE, P., BRILLINGER, D.R. & HALLIDAY, D.M. (1989). The Fourier approach to the identification of synaptic coupling between neuronal spike trains. *Prog. Biophys. & Mol. Biol.* **53**, 1-31.

STILES, R.N. & RANDALL, J.E. (1967). Mechanical factors in human tremor frequency. *J. Appl. Physiol.* **23**, 324-330.

PART 11

NATURAL MOTOR PATTERNS - 2

ARE SLOW FINGER MOVEMENTS PRODUCED BY A PULSATILE MOTOR COMMAND?

Å.B. Vallbo and J. Wessberg

Department of Physiology
Medicinaregatan 11, Göteborg University
S-413 90 Göteborg, Sweden

INTRODUCTION

Tremor during position holding has been analysed in many studies. In recent years evidence has accumulated that the spinal stretch reflex loop is essential for the generation of physiological tremor as well as enhanced physiological tremor when the subject is asked to keep a finger in an outstretched position (Bigland & Lippold, 1954; Halliday & Redfearn, 1956; Lippold, 1970; Hagbarth & Young, 1979; Young & Hagbarth, 1980; Marsden, 1984). Positional tremor at the finger has a frequency of 8 - 10 Hz which roughly fits with the loop time of the spinal stretch reflex. In some tremor studies it is stated in passing that tremor with this frequency increases during voluntary movements (Young & Hagbarth, 1980; Marsden, 1984). However, finger tremor during voluntary movements has not been described except in one study by Marshall & Walsh (1956) who reported, on the basis of accelerometer recordings, tremor at 7 - 12 Hz in a number of different joints during voluntary movements.

In the present study the kinematics of finger movements were studied in relation to proprioceptive afferent activity. It was found that these movements are characterised by discontinuities recurring at 8 - 10 Hz. Evidence will be presented that these discontinuities are not accounted for by the spinal stretch reflex loop alone. It is suggested that the descending motor command for finger movements might be pulsatile rather than continuous.

METHODS

The subject was instructed to move one finger at its metacarpophalangeal joint. The joint angle and speed of movement were recorded with transducers and acceleration derived off-line from the velocity record. EMG from the finger extensor muscles was recorded with surface electrodes while single afferent activity from this muscle was recorded by microneurography (Vallbo, Hagbarth, Torebjörk & Wallin, 1979).

RESULTS

Kinematic structure of self-generated finger movements

When human subjects perform finger movements which involve a single joint at low or moderate speed these movements were not smooth but were characterised by recurring speed variations as illustrated in Figure 1 (Vallbo & Wessberg, 1993). Further analysis revealed a dominance of discontinuities with a period length of 100 - 120 ms, which were particularly obvious in the traces of speed and acceleration.

It was found that the amount of speed variations at 8 - 10 Hz varied between subjects whereas it was relatively constant within the same subject as illustrated in Figure 2 showing recordings from 2 different subjects. The uniformity persisted between experimental sessions separated by several months. Hence the individual kinematic pattern appears to be a kind of finger print of the subject's control system for finger movements.

The 8 - 10 Hz discontinuities were present under a variety of conditions, e.g. with different size of loads, with multi-joint movements as well as single joint movements, with precision movements as well as routine type of movements. Moreover, they were present within a range of overall velocity of movement, as illustrated in Figure 3, where it may be appreciated that the faster movements were implemented by a series of large steps whereas slower movements were implemented by smaller steps repeated at the same rate of 8 - 10 Hz.

Mechanisms accounting for 8 - 10 Hz discontinuities during finger movements

Among the mechanisms that may be involved in the generation of the 8 - 10 Hz discontinuities, it is obvious that mechanical resonance can be ruled out because the resonance frequency of the finger system (20 - 25 Hz) falls way outside the range of the 8 - 10 Hz discontinuities (Halliday & Redfearn, 1956; Stiles & Randall, 1967).

Moreover, it was confirmed by EMG recordings that the 8 - 10 Hz were of muscular origin, as these recordings demonstrated that the muscular activity was modulated in phase with the speed variations (Figure 4). The modulation involved primarily the shortening muscle, i.e. the muscle which drives the movement in the desired direction. The antagonist

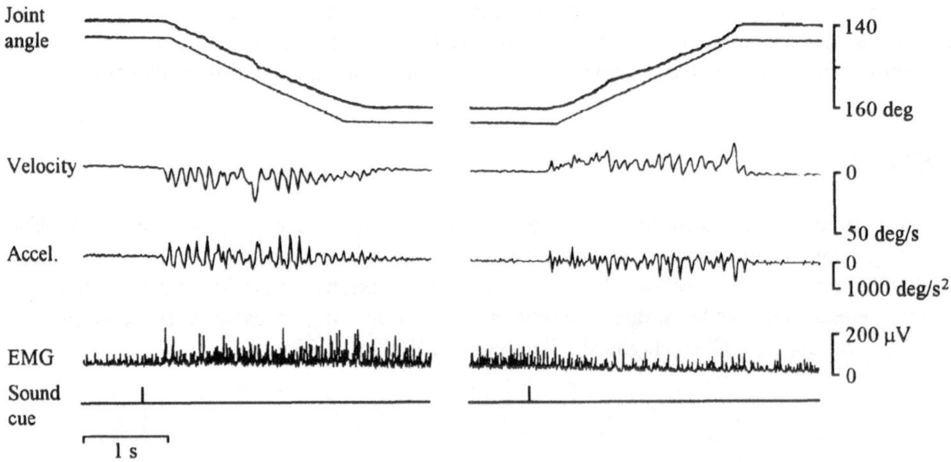

Figure 1. Kinematics of slow finger movements. The subject performed an extension and flexion movement at a single metacarpophalangeal joint in a visual tracking paradigm (Vallbo & Wessberg, 1993).

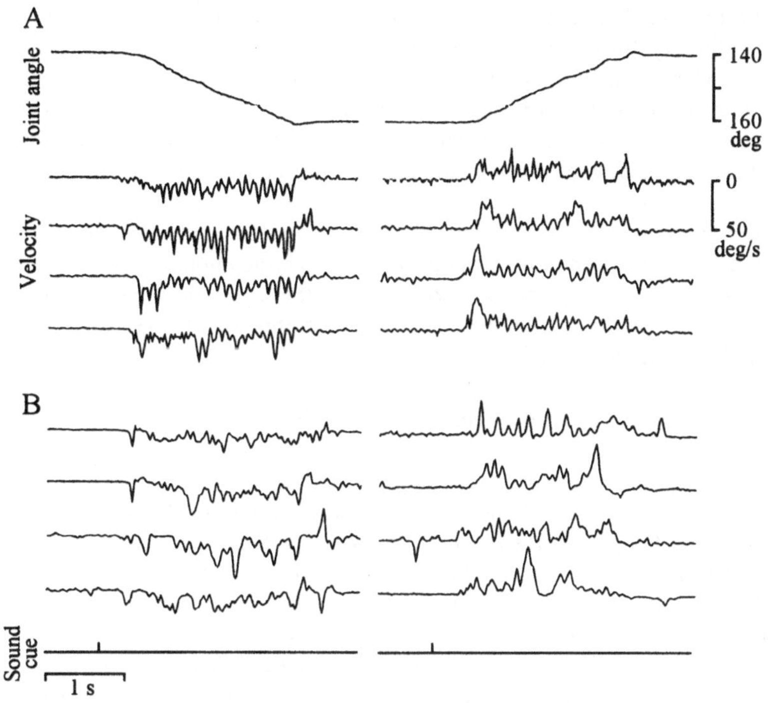

Figure 2. Intraindividual uniformity of finger movement kinematics. Upper and lower four velocity traces from two different subjects. Same type of movement as in Figure 1 (Vallbo & Wessberg, 1993).

was sometimes silent, sometimes active in these movements, but when it was active its activity was modulated at 8 - 10 Hz as well.

Hence it was obvious that the 8 - 10 Hz discontinuities are caused by a series of motor pulses, often a series of double pulses, i.e one driving pulse produced by the agonist and, a few tens of ms later, a braking pulse produced by the antagonist.

The stretch reflex and muscle spindle response

Considering the loop time of the spinal stretch reflex which is in the order of 50 ms, including the electromechanical delay, it seems reasonable to ask if the 8 - 10 Hz discontinuities are manifestations of oscillations in this feedback loop. Simple principles say that oscillations in a feedback loop may have a period time which is roughly twice the reflex loop time. The stretch reflex hypothesis would then imply that 8 - 10 Hz discontinuities are added to a smoothly changing central command. On the other hand, it should be noted that the simple relation between period time and loop time does not hold under all circumstances because it may be modified by a number of factors. In any case, an important issue in relation to the hypothesis that the spinal stretch reflex accounts for the 8 - 10 Hz discontinuities is to what extent muscle spindles respond to these discontinuities.

It is well known from animal work that muscle spindle primary afferents may be very sensitive to minute speed variations (Matthews & Stein, 1969; Hasan & Houk, 1975; Cussons, Hulliger & Matthews, 1977; Hulliger, Matthews & Noth, 1977; Houk, Rymer & Crago, 1981). On the other hand, these findings do not allow a prediction of spindle response from human finger muscles during voluntary movements, because the sensitivity to speed variations is not an invariant characteristic, but it is dependent on a number of factors, e.g. range of movement, amount of pre-stretch of the muscle, and fusimotor activity.

When voluntary finger movements were analysed it was found that spindles could respond very clearly to the discontinuities. An example with a Ia afferent during a shortening movement is shown in Figure 5. It may be seen that the impulses occurred in strict relation to the discontinuities and at the moments when the speed attained its minimum. It seems likely that this was the result of a balance between the fusimotor drive, which tends to maintain the firing and the shortening movement which tends to silence the afferent. During lengthening movements the correlation of spindle firing to the 8 - 10 Hz discontinutites could be even more pronounced although the impulses then tended to appear at the phase of maximal lengthening velocity. Such findings would be consistent with the interpretation that spindles code the occurrence of the 8 - 10 Hz discontinuities and therefore might contribute to their genesis.

The individual cycle may exhibit complex kinematics

On the other hand, there are several indications that the spinal stretch reflex is not the main factor in the generation of the 8 - 10 Hz discontinuties. One is that the individual discontinuity had a fairly complex kinematic structure, which is not what one might expect with oscillations in a simple feedback loop. This was particularly clear for slow movements (Figure 6A). In each cycle of discontinuity two phases could be discriminated from the velocity record: stand still and movement. The acceleration records showed three phases: acceleration, deceleration, and stand still. Moreover, there was a close relationship between the size of the peak acceleration and the peak deceleration, the latter being regularly higher than the former. This feature is also illustrated in Figure 1 during both lengthening and shortening movements. For faster movements these kinematic details were not as clear; sometimes merely a notch was seen in the acceleration record (Figure 6B).

Although, the complex kinematic structure of the discontinuities at first sight appears as a strong argument against oscillations in a feedback loop because one would expect more

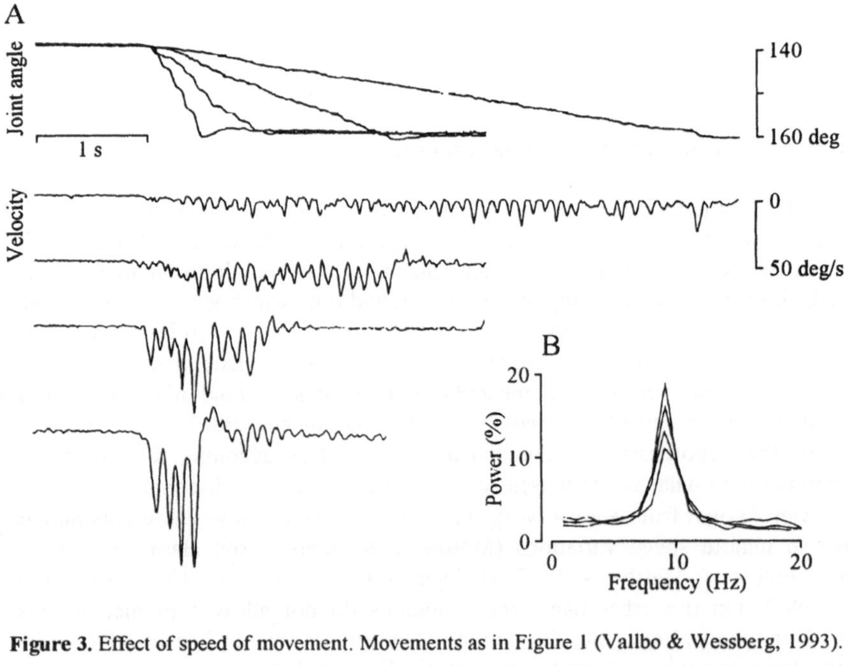

Figure 3. Effect of speed of movement. Movements as in Figure 1 (Vallbo & Wessberg, 1993).

sinusoidal movements, it is probably not a tight argument because it is feasible that non-linear characteristics might account for complex kinematics in a feed-back loop.

The spinal stretch reflex seems to lack to power to produce 8 - 10 Hz discontinuities during movements

An additional argument against the spinal stretch reflex is that it is not very strong. The lack of adequate power of the spinal stretch reflex could be demonstrated in actual experiments when perturbations were injected during the voluntary movements. It was then found that the spinal stretch reflex gives very small responses to large perturbations. It seemed therefore likely that the heavy modulations of the EMG as seen in the voluntary movements can not be accounted for by the spinal stretch reflex.

DISCUSSION

Central pulse generator

By exclusion, it seemed reasonable to propose that the 8 - 10 Hz discontinuities reflect

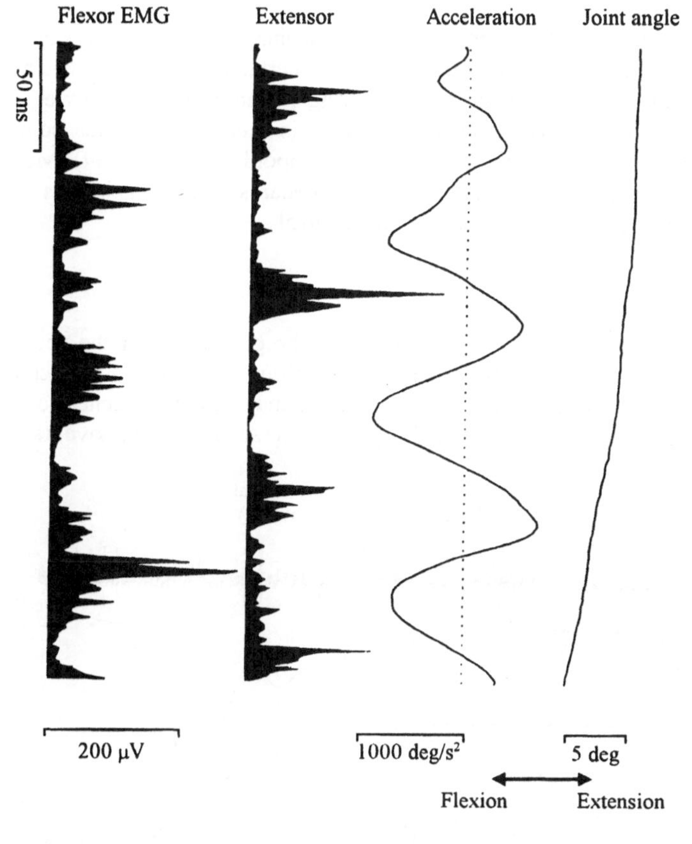

Figure 4. EMG activity in relation to 8 - 10 Hz discontinuities during slow finger movements at a single metacarpophalangeal joint. EMG recorded with surface electrodes from the finger extensor and finger flexor muscles (Vallbo & Wessberg, 1993).

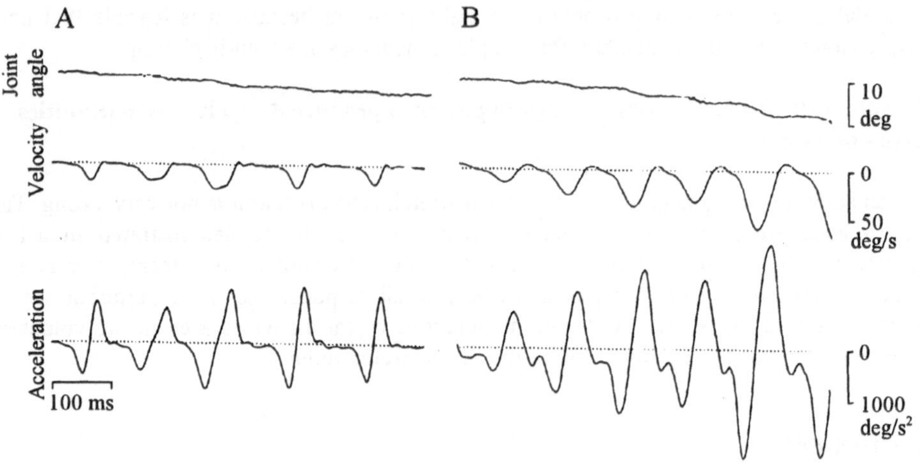

Figure 5. Muscle spindle firing in relation to kinematics of a slow shortening finger movement. The unit which originated from the finger extensor muscle was classified as a muscle spindle primary afferent on the basis of 8 functional characteristics.

a pulsatile descending motor command from supraspinal centres. The idea of a central pulse generator and an intermittency in motor control is certainly not new. It has been discussed in several studies on physiological tremor during position holding. However, supraspinal mechanisms have largely been rejected as the main mechanism in this context (Lippold, Redfearn & Vuco, 1957; Lippold, 1970; Burne, Lippold & Pryor, 1984). Moreover, Llinás (1991) has recently proposed that the olivocerebellar system constitute a functional unit which may account for intermittency in motor control.

Model

As a framework for further discussion it may be proposed that the control system for slow finger movements realises three separate functions. One would be a clock to account for a rate of 8 - 10 pulses per second. Second, there may be a pulse generator producing the motor command pulses to the segmental organisation. As indicated above, the intermittency

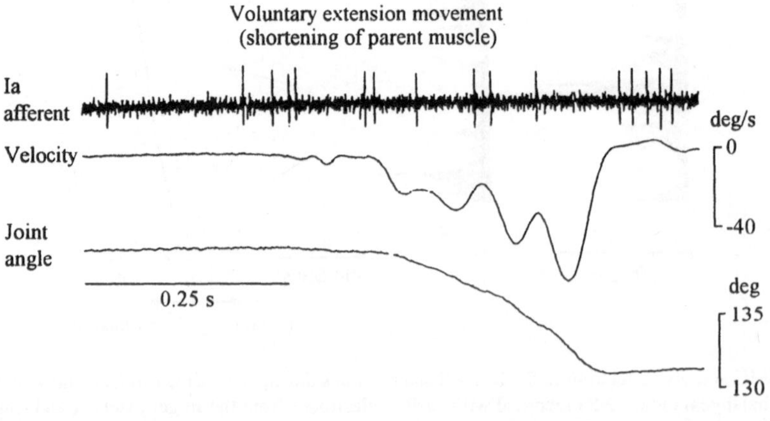

Figure 6. Detailed kinematics of individual movement cycles during slow finger movements (Vallbo & Wessberg, 1993) .

is not limited to simple modulations of the agonist muscle. Rather, it seems to involve double pulses, one to the agonist which drives the movement in the desired direction, followed by a braking pulse to the antagonist. Thus, the pattern of motor activity is akin to the triphasic pattern produced during fast movements (Hallet, Shahani & Young, 1975; Ghez & Gordon, 1987; Cooke & Brown, 1990). Finally, one might conceive a functional unit which sets the size of the pulses and hence the sizes of the individual step in order to accomplish the desired overall speed of movement (cf. Figure 3).

Proprioceptive implications

If finger movements are implemented by an intermittent controller rather than a continuous one, it is of interest to consider whether the signal from the intramuscular proprioceptors has a role in relation to such a controller. Since the proprioceptor population seems to code the discontinuities one might ask whether this input has a conditional role for the running of the clock. There are also indications that the proprioceptive input in addition might code the size of the individual discontinuity and the characteristics of the separate kinematic phases, although this remains to be explored further.

Reservations

It should be emphasised that the model proposed above is speculative because it is reasonable that the motor performance described in slow finger movements are implemented by other circuits. Moreover, it is obvious that other control systems than a strict 8 - 10 Hz pulse generator are involved as well, because the 8 - 10 Hz discontinuities are often mixed with a number of other frequencies, suggesting that a final summation point for motor commands is downstream of a postulated 8 - 10 Hz generator.

However, the present analysis is consistent with the interpretation that the descending motor command contains a modulation at 8 - 10 Hz which involves the agonist as well as the antagonist. In this respect the findings suggest a pulsatile control for the co-ordination between agonist and antagonist activity in slow finger movements. Moreover, the findings suggest that muscle afferents may code the occurrence of 8 - 10 Hz discontinuities as part of their high sensitivity to the velocity of muscle shortening and lengthening. This finding offers additional significance for the high dynamic sensitivity of spindle endings, i.e to code the effects of minute force pulses which may constitute 'quantum' outputs of the motor systems in their role of implementing slow finger movements. Moreover, the findings invite speculations that the response of muscle spindles to small steps of joint movement may be a reflection of their role in coding primarily intramuscular events rather than joint position or joint movements. It may be surmised that information on intramuscular events may be more pertinent for the control and coordination of muscles in self-generated finger movements than information on joint kinematics.

ACKNOWLEDGEMENTS

This study was supported by the Swedish Medical Research Council (Grant 14X-3548), Magn Bergvalls Stiftelse, and Knut och Alice Wallenbergs Stiftelse.

REFERENCES

BIGLAND, B. & LIPPOLD, O.C.J. (1954). The relation between force, velocity and integrated electrical activity in human muscles. *J. Physiol.* **123**, 214-224.

BURNE, J.A., LIPPOLD, O.C.J. & PRYOR, M. (1984). Proprioceptors and normal tremor. *J. Physiol.* **384**, 559-572.

COOKE, J.D. & & BROWN, S.H. (1990). Movement-related phasic muscle activation. II. Generation and functional role of the triphasic pattern. *J. Neurophysiol.* **63**, 465-472.

CUSSONS, P.D., HULLIGER, M. & MATTHEWS, P.B.C. (1977). Effects of fusimotor stimulation on the response of the secondary ending of the muscle spindle to sinusoidal stretching. *J. Physiol.* **270**, 835-850.

GHEZ, C. & GORDON, J. (1987). Trajectory control in targeted force impulses. Role of opposing muscle. *Exp. Brain Res.* **67**, 225-240.

HAGBARTH, K.-E. & YOUNG, R.R. (1979). Participation of the stretch reflex in human physiological tremor. *Brain* **102**, 509-526.

HALLET, M., SHAHANI, B.T. & YOUNG, R.R. (1975). EMG analysis of stereotyped voluntary movements in man. *J. Neurol. Neurosurg. Psychiat.* **38**, 1154-1162.

HALLIDAY, A.M. & REDFEARN, J.W.T. (1956). An analysis of the frequency of finger tremor in healthy subjects. *J. Physiol.* **134**, 600-611.

HASAN, Z. & HOUK, J.C. (1975). Analysis of response properties of deefferented mammalian spindle receptors based on frequency response. *J. Neurophysiol.* **38**, 663-672.

HOUK, J.C., RYMER, W.Z. & CRAGO, P.E. (1981). Dependence of dynamic response of spindle receptors on muscle length and velocity. *J. Neurophysiol.* **46**, 143-166.

HULLIGER, M., MATTHEWS, P.B.C. & NOTH, J. (1977). Static and dynamic fusimotor action on the response of Ia fibres to low frequency sinusoidal stretching of widely ranging amplitude. *J. Physiol.* **267**, 811-838.

LLINÁS, R.R. (1991) The noncontinuous nature of movement execution. In *Motor control: Concepts and Issues.* eds. HUMPHREY, D. R. & FREUND, H-J. pp 223-242. Wiley, New York.

LIPPOLD, O.C.J. (1970). Oscillation in the stretch reflex arc and the origin of the rhythmical, 8-12 c/s component of physiological tremor. *J. Physiol.* **206**, 359-382.

LIPPOLD, O.C.J., REDFEARN, J.W.T. & VUCO, J. (1957). The rhythmical activity of groups of motor units in the voluntary contraction of muscle. *J. Physiol.* **137**, 473-487.

MARSDEN, C.D. (1984) Origins of normal and pathological tremor. In *Movement Disorders: Tremor.* eds. FINDLEY, L. J. & CAPILDEO, R. pp 37-84. Macmillan, London.

MARSHALL, J. & WALSH, E.G. (1956). Physiological tremor. *J. Neurol. Neurosurg. Psychiat.* **19**, 260-267.

MATTHEWS, P.B.C. & STEIN, R.B. (1969). The sensitivity of muscle spindle afferents to small sinusoidal changes of length. *J. Physiol.* **200**, 723-743.

STILES, R.N. & RANDALL, J.E. (1967). Mechanical factors in human tremor frequency. *J. App. Physiol.* **23**, 324-330.

VALLBO, Å.B., HAGBARTH, K.E., TOREBJÖRK, H.E. & WALLIN, B.G. (1979). Somatosensory, proprioceptive, and sympathetic activity in human peripheral nerves. *Physiol. Rev.* **59**, 919-57.

VALLBO, Å.B. & WESSBERG, J. (1993). Organization of motor output in slow finger movements. *J. Physiol.* **469**, 673-691.

YOUNG, R.R. & HAGBARTH, K.-E. (1980). Physiological tremor enhanced by manoeuvres affecting the segmental stretch reflex. *J. Neurol. Neurosurg. Psychiat.* **43**, 248-256.

CUTANEOUS INHIBITION OF DESCENDING COMMANDS VIA THE PROPRIOSPINAL RELAY MAY HELP TERMINATE HUMAN ARM MOVEMENTS

E. Pierrot-Deseilligny, D. Mazevet
and S. Meunier

Rééducation, Hôpital de la Salpétrière
47 bd de l'Hôpital, 75651 - Paris cedex 13
France

INTRODUCTION

In a unique series of experiments in the cat, involving both electrophysiological analysis and behavioural studies on the effects of selective spinal cord lesions, it has been established that the descending command for visually guided target-reaching is mediated through a system of propriospinal neurones (PNs) located at the C3-C4 level (for review, see Alstermark & Lundberg, 1992). PNs receive strong inhibitory feedback (● in Figure 1) from cutaneous and muscle afferents in the moving limb, and behavioural experiments have shown that this feedback inhibition plays a decisive role for controlling execution of target-reaching and termination of the movement.

There is a similarly organized system in man (Pierrot-Deseilligny & Mazevet, 1993). The relevant neurones, which are referred to as propriospinal neurones (PNs) in this paper, are located above the cervical enlargement and project to motoneurones of all muscles acting at the shoulder, elbow or wrist which have been examined (Gracies, Meunier, Pierrot-Deseilligny & Simonetta, 1991). These PNs receive both excitation and inhibition from peripheral afferents (Malmgren & Pierrot-Deseilligny, 1988a, b). In addition to the monosynaptic connection between motor cortex and motoneurones supplying upper limb muscles, there is good evidence for a parallel disynaptic pathway involving a cortical projection via these PNs (Gracies, Meunier & Pierrot-Deseilligny, 1994), and it has been shown that this pathway transmits a significant part of the command for voluntary movement (Burke, Gracies, Mazevet, Meunier & Pierrot-Deseilligny, 1994). During movement, the descending facilitation of PNs has a characteristic pattern in that the subsets selected by higher centres are those receiving the afferent feedback from the contracting muscle (Mazevet & Pierrot-Deseilligny, 1994), and each subset receives specific cutaneous inhibition from the skin field which will meet the target at the end of the movement (Nielsen & Pierrot-Deseilligny, 1991).

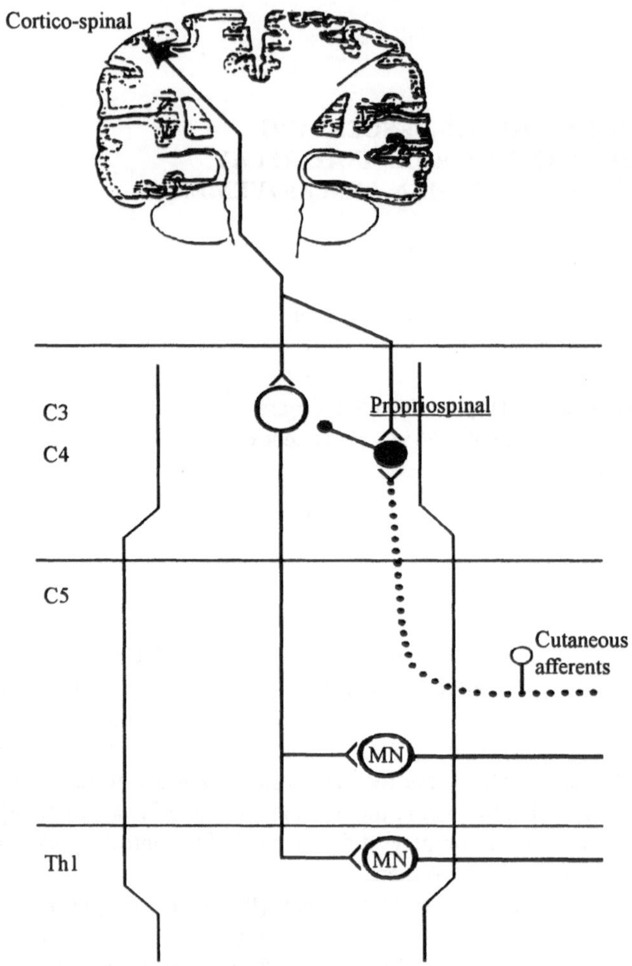

Figure 1. schematic representation of the C3-C4 propriospinal system. A. Propriospinal neurone (large open circle) is shown with its projections to motoneurones (MN), its excitatory input from the corticospinal tract and its inhibition (through an inhibitory interneurone: black circle) from upper limb cutaneous afferents. Note that the inhibitory interneurone receives facilitation from the corticospinal tract (see Alstermark & Lundberg, 1992).

The particular pattern of cutaneous inhibition of PNs found in human experiments favours the view that this inhibition could also contribute in man to termination of movement : the exteroceptive activity evoked by contact with the target would inhibit the descending command passing through the propriospinal relay. However, animal experiments have shown that the transmission in this inhibitory pathway is controlled (facilitated, Figure 1) from the sensori-motor cortex (Alstermark, Lundberg & Sasaki, 1984). If the functional hypothesis concerning the role of this pathway in the termination of the movement is correct, one would expect the cortical facilitation of these inhibitory interneurones to be stronger at the end of the movement. In the present experiments the amount of cutaneous inhibition of PNs was therefore compared at the onset and at the offset of movement.

In a recent investigation (Burke et al., 1994), we have shown that stimulation of the cutaneous superficial radial nerve at 3 x perceptual threshold depresses the *tonic* ongoing EMG of extensor carpi radialis (ECR) and the discharge of single ECR motor units, both with a central delay of about 4 ms. Because this stimulation also depresses the response to transcranial magnetic stimulation of the motor cortex but has little effect on the H reflex it has been argued that the cutaneous-induced depression reflects a disfacilitation, namely an inhibition of the descending excitation to ECR motoneurones at a premotoneuronal level. The finding that the central latency of this depression increases with the rostro-caudal location of the explored motoneurone implies that the site of disfacilitation is located above the cervical enlargement and favours the view of an inhibition of PNs. At the onset of phasic voluntary wrist extension, the same cutaneous stimulation also inhibits the descending excitation passing through the PN relay, since the ongoing EMG activity (Pierrot-Deseilligny & Mazevet, 1993) and the cortically evoked response are clearly depressed, whereas the H reflex is little reduced. However, there was no evidence that this disfacilitation was significantly larger at the end of such a selective voluntary wrist extension (Burke, Meunier & Pierrot-Deseilligny, unpublished results). This negative result might be related to the fact that the contractions studied were a simple fragment of a movement, without real functional significance. For these reasons, in the present experiments cutaneous inhibition of PNs was compared at the onset and offset of a coordinated reaching movement. The main component of this movement was a visually guided tracking elbow extension and the effects of cutaneous stimulation were assessed on ongoing EMG activity of the triceps brachii.

METHODS

Experiments were performed on 7 normal subjects with Ethical Committee approval. Surface EMG recordings were taken from the triceps brachii during tonic contraction and during visual tracking movements of elbow extension lasting 1 s. The EMG activity was rectified and averaged against the cutaneous stimulation which was either a single shock (3-4 times perceptual threshold) or a train of 3 shocks (300 Hz) applied to the superficial radial nerve (SR). Trials with and without (control) stimulation were randomly alternated. Cutaneous stimulation was given during tonic contraction and either during the first (onset) or the last (offset) 100 ms of voluntary elbow extension, the amount of control EMG activity being about the same in the three situations. The motor cortex was stimulated much as described by Gracies et al. (1994), using a Magstim 200 with a 9 cm coil at the optimal position for a twitch in elbow extensors. Because the H reflex is usually too small in the triceps, the excitability of its motoneurone pool was tested by eliciting the tendon reflex. The tap was applied to the distal tendon of the muscle by an electromagnetic hammer (Brüel et Kjaer, model 4809). Both cortically evoked and reflex responses were measured from peak to peak.

RESULTS

The first step has been to establish that superficial radial nerve stimulation inhibits the transmission of the descending excitation to triceps motoneurones at a premotoneuronal level, thus producing disfacilitation of these motoneurones. To that end, the effects induced by cutaneous stimulation on ongoing EMG activity (○), cortical-evoked responses (Δ) and tendon reflexes (◆) were investigated during tonic contraction and are illustrated in Figure 2, where the responses conditioned by cutaneous stimulation are expressed as a percentage of their unconditioned value. The latency of the cutaneous-induced changes in EMG activity was expressed as the central transmission ("central delay") calculated from the expected time of arrival of the cutaneous volley at the same segmental level as the motoneurone pool (zero of the abscissa). The calculations involved measuring the latency of the H reflex and adding to this value the additional peripheral conduction time for cutaneous afferents to the stimulus site for the H reflex (about 5.5 ms). The zero of the abscissa for cortically evoked and tendon responses corresponds to the simultaneous arrival of the cutaneous and test (either descending or peripheral Ia respectively) volleys at the same segmental level as the motoneurone pool. As described in full previously (Burke et al., 1994), the times of arrival of the corticospinal and the tap-evoked volleys at the triceps motoneurone pool were determined from the timing of the corresponding EMG potentials during contraction.

Open circles in Figure 2 show that SR stimulation (single shock) evoked clear inhibition in two periods (6 - 16 and 17 - 35 ms), the EMG being reduced to 60% of its control value at the 10 ms interval. SR stimulation reduced the cortically evoked response (Δ) to the same extent but hardly depressed the tendon reflex (◆). As previously argued (Burke et al., 1994), this differential effect must reflect an inhibition of the descending excitation at a premotoneuronal level. The delay of the first EMG depression implies a spinal pathway, compatible with a C3-C4 propriospinal mechanism, but, as already argued (Burke et al., 1994), supraspinal mechanisms could contribute to the second period of the depression.

Figure 2 shows that the central delay of the SR-induced depression of the cortically evoked response (Δ) was only 2ms, ie 4ms shorter than that of the ongoing EMG depression. This may be due to the multiple (I-waves) volley set up by magnetic stimulation of the motor cortex (see Day, Dressler, Hess, Maertens de Noordhout, Marsden, Nakashima, Rothwell & Thompson, 1989). As in the case of the monosynaptic reflex (see Araki, Eccles & Ito, 1960), it may be assumed that the shortest conditioning-test interval with a depression of the cortical-evoked response corresponds to an inhibition of PNs when the latest corticospinal volley (I3) is passing through the PN relay. Accordingly, the inhibition of the first (I1) cortical volley should require a longer inter-stimulus interval.

In Figure 3, a comparison is drawn between the amount of cutaneous-induced inhibition of ongoing triceps EMG activity occurring at the onset (○) and at the offset (●) of a reaching movement involving elbow extension. In each case, 300 trials (conditioned and

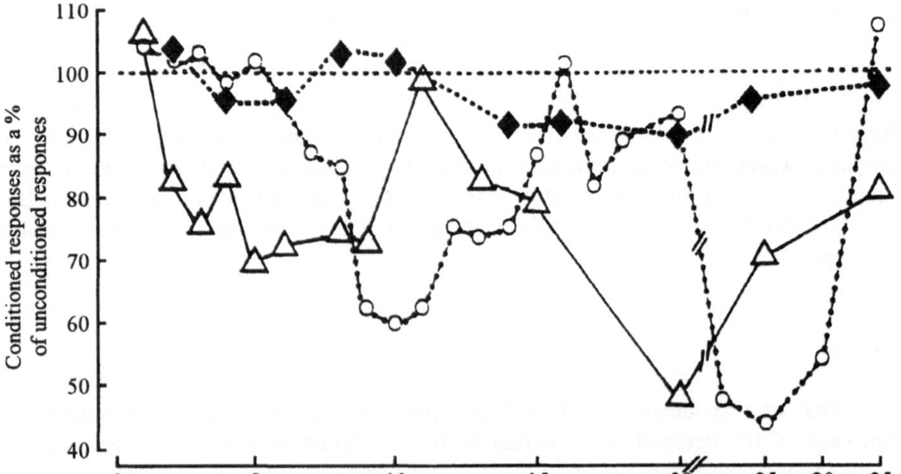

Figure 2. Changes in the ongoing (rectified) EMG activity (○), the cortically evoked response (Δ) and the tendon reflex (◆) of the triceps brachii evoked by stimulation of the superficial radial nerve during triceps tonic contraction. Responses are expressed as a percentage of their unconditioned value. each point represents the mean of 100 (○) or 10 (Δ,◆) responses.

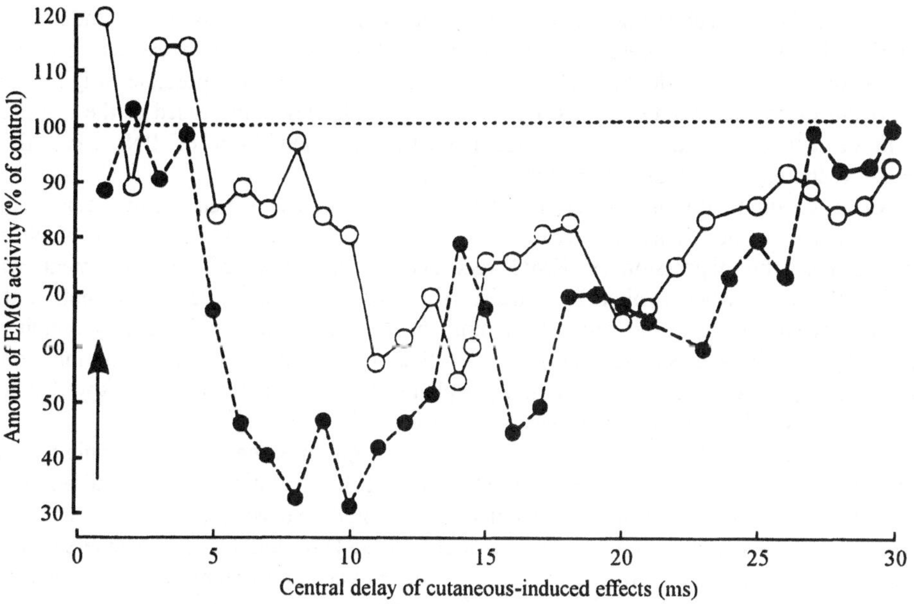

Figure 3. Cutaneous-induced changes in the voluntary (rectified) EMG of the triceps recorded during the first (onset, ○), or the last (offset, ●) 100 ms of a visual-tracking movement of elbow extension. Each point represents the average of 300 trials, and the EMG conditioned by SR stimulation is expressed as a percentage of its unconditioned value.

unconditioned) were averaged, and results obtained with cutaneous stimulation were expressed as a percentage of the control average EMG activity. At the onset, the depression occurred with a central delay of 5 ms, was moderate during its first 6 ms, reached its maximum at the 11 ms delay where the EMG was depressed to 55% of its control value, and was over at 30 ms. At the offset, the depression had the same central latency and the same duration but was both significantly larger and much more abrupt, reaching a maximum at 8 ms where the EMG was depressed to 32% of its control value. By contrast with this initial part of the depression, which is supposed to reflect a propriospinal mechanism, the second part of the depression, which is probably explained by cutaneous inhibition of motor cortex excitability, as described by Maertens de Noordhout, Rothwell, Day, Dressler, Nakashima, Thompson & Marsden (1992), was the same magnitude at the onset and at the offset of the movement. A similar increase in the initial part of the depression at the offset, with respect to its value at the onset, was found in all 7 subjects. In two, this increase appears to be particularly dramatic since there was no significant depression at the onset.

CONCLUSION

For obvious methodological reasons it was impossible to study the SR-induced changes in the cortically evoked response and tendon reflex during such phasic movements of elbow extension. However, it was confirmed that, as during tonic contraction, at the onset and at the offset of a phasic isometric contraction SR stimulation had little effect on the tendon reflex but depressed both the ongoing EMG activity and cortically evoked response, and it was found that this depression was more pronounced at the offset of the contraction. As argued above, this strongly suggests that the depression of ongoing EMG activity reflects an inhibition of the cortical excitation of triceps motoneurones at the PN level. Under these conditions the increased depression (of both ongoing EMG and cortically evoked response) seen at the offset of voluntary elbow extension may have two possible origins: the component of cortical excitation passing through the PN relay is increased at the offset of the movement and/or the interneurones mediating the inhibition of PNs have their excitability increased, because of an increased cortical drive.

An argument favouring the latter can be drawn from experiments using trains of 3 shocks applied to the SR nerve. Because of temporal summation at the level of inhibitory interneurones, the depression of EMG recorded at the onset of the movement was significantly larger when evoked by a train. However, at the offset of movement, if the inhibition evoked by a single shock was more pronounced, the supplementary inhibition evoked by a train was smaller. This could reflect occlusion, at the level of inhibitory interneurones, between large cortical and peripheral inputs, and support the view that the cortical drive to inhibitory interneurones is increased at the end of movement.

The finding that descending excitation of triceps motoneurones is inhibited by cutaneous afferents when passing through PNs and that this inhibition is more marked at the offset of the movement suggests that this mechanism could be used to curtail a movement of target-reaching. As emphasised by Alstermark et al. (1984) it would be a reasonable control strategy to delegate part of the termination of the movement to spinal cord mechanisms, since termination must be one of the most difficult parameters of a movement for the brain to calculate. This could be particularly important if the movement meets an unexpected obstacle.

This provides a good example of modulation of the descending command by peripheral afferents. It can be speculated that it is advantageous that this integration takes place at a premotoneuronal level. An inhibition exerted at the motoneuronal level could of course have the same inhibitory effect. However, because of the inertia related to the duration of the IPSPs in motoneurones, the movement could not start again before these IPSPs had vanished. In the case of an inhibition at the premotoneuronal level the movement can start again, if necessary, immediately after activation of another excitatory pathway, since the motoneurones themselves are not inhibited.

ACKNOWLEGDMENTS

This work was supported by grants from DRED, AP-HP, INSERM (92 08 11) and IRME.

REFERENCES

ALSTERMARK, B. & LUNDBERG, A. (1992) The C3-C4 propriospinal system: target-reaching and food-taking. In *Muscle Afferents and Spinal Control of Movement.* eds. JAMI, L., PIERROT-DESEILLIGNY, E. & ZYTNICKI, D. pp. 327-354, Pergamon Press, London.

ALSTERMARK, B., LUNDBERG, A. & SASAKI, S. (1984) Integration in descending motor pathways controlling the forelimb in the cat. 12. Interneurones which may mediate descending feed-forward inhibition and feed-back inhibition from the forelimb to C3-C4 propriospinal neurones. *Exp. Brain Res.* **56**, 308-322.

ARAKI, T., ECCLES, J.C. & ITO, M. (1960) Correlation of the inhibitory post-synaptic potential of motoneurones with the latency and time course of inhibition of monosynaptic reflexes. *J. Physiol.* **154**, 354-377.

BURKE, D., GRACIES, J.M., MAZEVET, D., MEUNIER, S. & PIERROT-DESEILLIGNY, E. (1994) Non monosynaptic transmission of the cortical command for voluntary movement in man. *J. Physiol.* **480**, 191-207.

DAY, B.L., DRESSLER, D., HESS, C.W., MAERTENS DE NOORDHOUT, A., MARSDEN, C.D., NAKASHIMA, K., ROTHWELL, J.C. & THOMPSON, P.D. (1989) Electric and magnetic stimulation of human motor cortex: surface EMG and single motor unit responses. *J. Physiol.* **412**, 449-473.

GRACIES, J.M., MEUNIER, S. & PIERROT-DESEILLIGNY, E. (1994) Evidence for corticospinal excitation of human propriospinal-like neurones. *J. Physiol.* **475**, 509-518.

GRACIES, J.M., MEUNIER, S., PIERROT-DESEILLIGNY, E. & SIMONETTA, M. (1991) Pattern of propriospinal-like excitation to different species of human upper limb motoneurones. *J. Physiol.* **434**, 151-167.

MAERTENS DE NOORDHOUT, A., ROTHWELL, J.C., DAY, B.L., DRESSLER, D., NAKASHIMA, K., THOMPSON, P.D. & MARSDEN, C.D. (1992) Effect of digital nerve stimuli on responses to electrical or magnetic stimulation of the human brain. *J. Physiol.* **447**, 535-548.

MALMGREN, K. & PIERROT-DESEILLIGNY, E. (1988a) Evidence for non-monosynaptic Ia excitation of wrist flexor motoneurones, possibly via propriospinal neurones. *J. Physiol.* **405**, 747-764.

MALMGREN, K. & PIERROT-DESEILLIGNY, E. (1988b). Inhibition of neurones transmitting non-monosynaptic Ia excitation to human wrist flexor motoneurones. *J. Physiol.* **405**, 765-783.

MAZEVET, D. & PIERROT-DESEILLIGNY, E. (1994) Pattern of descending excitation of presumed propriospinal neurones at the onset of voluntary movement in man. *Acta physiol. scand.* **150,** 27-38.

NIELSEN, J. & PIERROT-DESEILLIGNY, E. (1991) Pattern of cutaneous inhibition of the propriospinal-like excitation to human upper limb motoneurones. *J. Physiol.* **434**, 169-182.

PIERROT-DESEILLIGNY, E. & MAZEVET, D. (1993) Propriospinal transmission of voluntary movement in humans. In *Spasticity: Mechanisms and Management.* eds. THILMANN, A.F., BURKE, D.J. & RYMER, W.Z. pp. 40-56, Springer-Verlag. Berlin-Heidelberg.

PROPRIOCEPTIVE INPUT AND CONDITIONAL LOGIC IN THE CONTROL OF VOLUNTARY MOVEMENT

Arthur Prochazka and Monica Gorassini

Division of Neuroscience
University of Alberta
Edmonton, Alberta
Canada

INTRODUCTION

In the 14 years since the last symposium on mammalian muscle receptors in London (Taylor & Prochazka, 1981), considerable progress has been made in understanding how muscle afferents fire during normal motor behaviour and how sensory input is used by the CNS to control movement. The first part of this chapter provides an overview of how ensembles of cat muscle receptors fire during locomotion and other types of movement. The fusimotor action which might underlie muscle spindle firing in different situations is discussed. We then review some recent work on the sensorimotor interactions which occur during cat locomotion. Certain aspects of these interactions indicate that conditional rules govern the control of phase transitions in the step cycle. We discuss some theoretical approaches to conditional control, which may become useful in dealing with the multisensory conditional logic underlying movement sequences.

MUSCLE AFFERENT FIRING DURING NORMAL MOVEMENTS

At the time of the 1980 symposium, the data on the activity of single afferents recorded in awake animals was new and somewhat anecdotal. Recordings from muscle spindle afferents obtained in cats and monkeys showed three major differences in firing from that of human spindle afferents recorded by neurography and from that expected on the basis of strong α-linked γ action. First, changes in spindle firing attributable to γ action depended not so much on phasic contractions of the receptor-bearing muscles as on the overall task and its context. Second, the modulation of firing rate of spindle afferents in the animals was often most closely related to changes in muscle length. Third, the peak firing rates of the animal muscle afferents were four or five times those in humans.

In the intervening time, many more afferent recordings have been obtained in cats, monkeys and humans. In some cases repeated observations of the firing of afferents of a given modality in a given muscle during stereotyped movements such as stepping have allowed ensemble response profiles to be collated. In Figure 1, firing profiles of Ia, II and Ib afferents in the locomotor step cycle recorded in awake cats are combined in small ensembles. From these data, one can see that to a first approximation spindle Ia and II afferents signal muscle length changes in the step cycle, albeit with some distortion due to dynamic response properties. Ensembles of 4 - 5 Ib afferents provide smooth profiles which match the EMG generated by the muscle. This is true even when staircase increments of firing are present in single cycles contributing to the ensemble profiles (Prochazka, 1995). The Ib profiles are remarkably similar to those obtained during cat locomotion from sensors implanted on tendons (Walmsley, Hodgson & Burke, 1978; Herzog, Leonard & Guimaraes, 1993) and so the ensemble firing of several tendon organs is probably closely related to whole muscle force, again with some distortion due to dynamic responsiveness and non-linear behaviour.

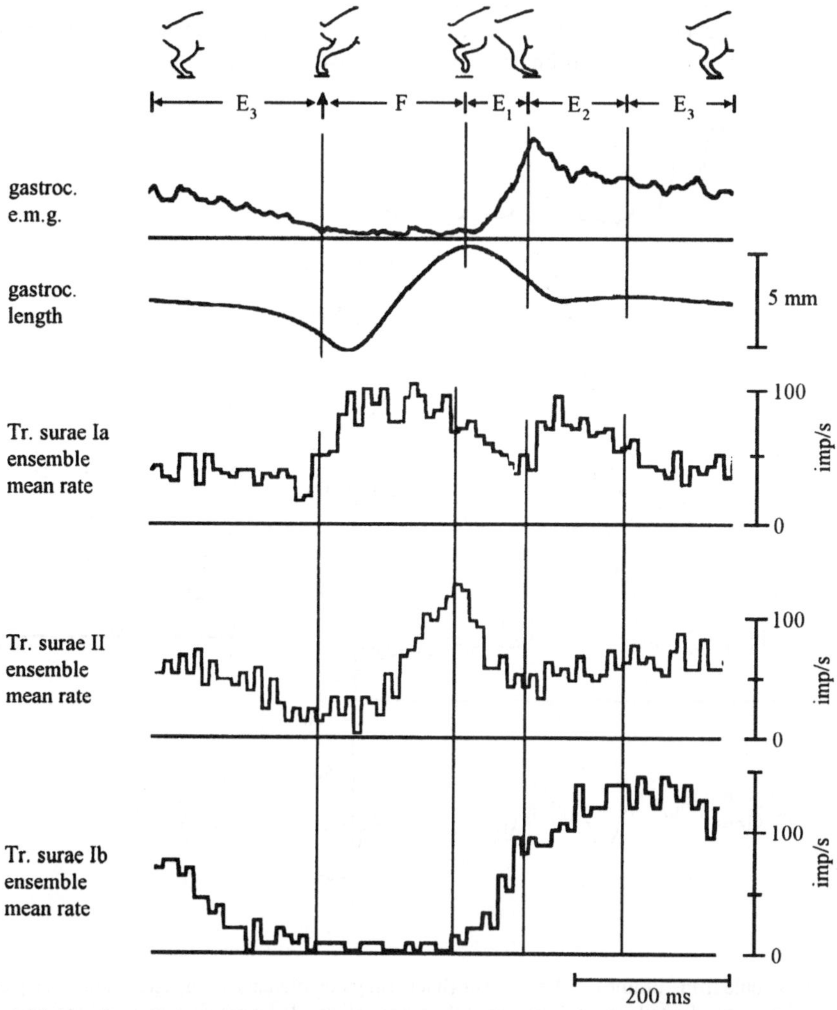

Figure 1. Firing of small ensembles of Ia, II and Ib afferents in the step cycle of normal cats. Top: stick figures showing phases of step cycle. Note the correspondence between that the Ia and II spindle afferent profiles and muscle length. Note too the high mean firing rate of Ib tendon organ afferents. Reproduced from Elek et al., 1990.

Qualitatively, we have found that once 4 or 5 afferents have contributed to an ensemble profile, the addition of data from more afferents does not change the profile substantially. Thus one may infer that the convergence of 4 or 5 muscle spindle afferents on an interneurone in the CNS is sufficient to provide that interneurone with high-resolution information on muscle length changes. Second, when linear models of Ia transduction are used to filter the averaged length signals the resulting profiles match the corresponding ensemble firing rates rather well (see below). Third, deviations from the modelled profiles suggest α-linked γ action and mechanical transients in the step cycle. Finally, if the firing of all the spindles in a typical cat hindlimb muscle is summed, the peak Ia input to the CNS in the step cycle is 20 - 40 kilo-impulses/sec (Prochazka, Trend, Hulliger & Vincent, 1989).

LINEAR MODELLING

In the 1970s, several groups did detailed linear harmonic analyses of spindle responses to muscle length changes (Matthews & Stein, 1969; Hulliger, Matthews & Noth, 1977; Chen & Poppele, 1978). Non-linearity was acknowledged at the outset and further

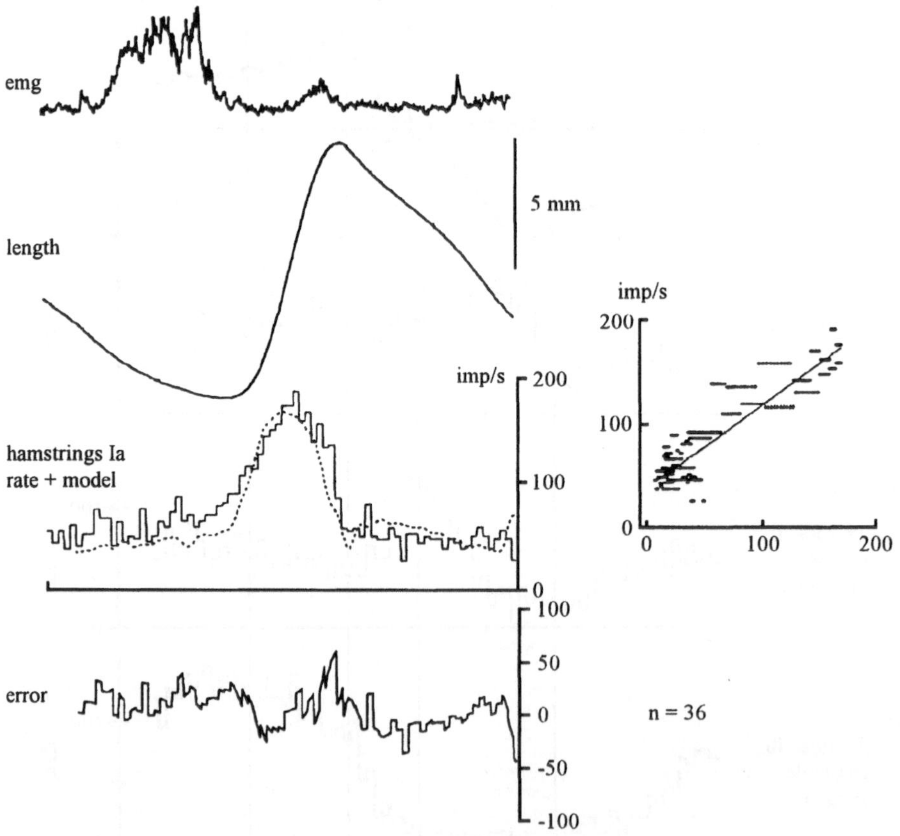

Figure 2. Ensemble firing profile of 9 knee flexor (hamstrings) Ia afferents, each contributing 4 step cycles. The dashed line is the digitally filtered version of the averaged muscle length signal, using the Chen & Poppele (1978) large-signal transfer function for Ia responses. The bottom trace shows the error between the model and the actual Ia profile. The plot on the right is of actual firing rate (y-axis) versus model.

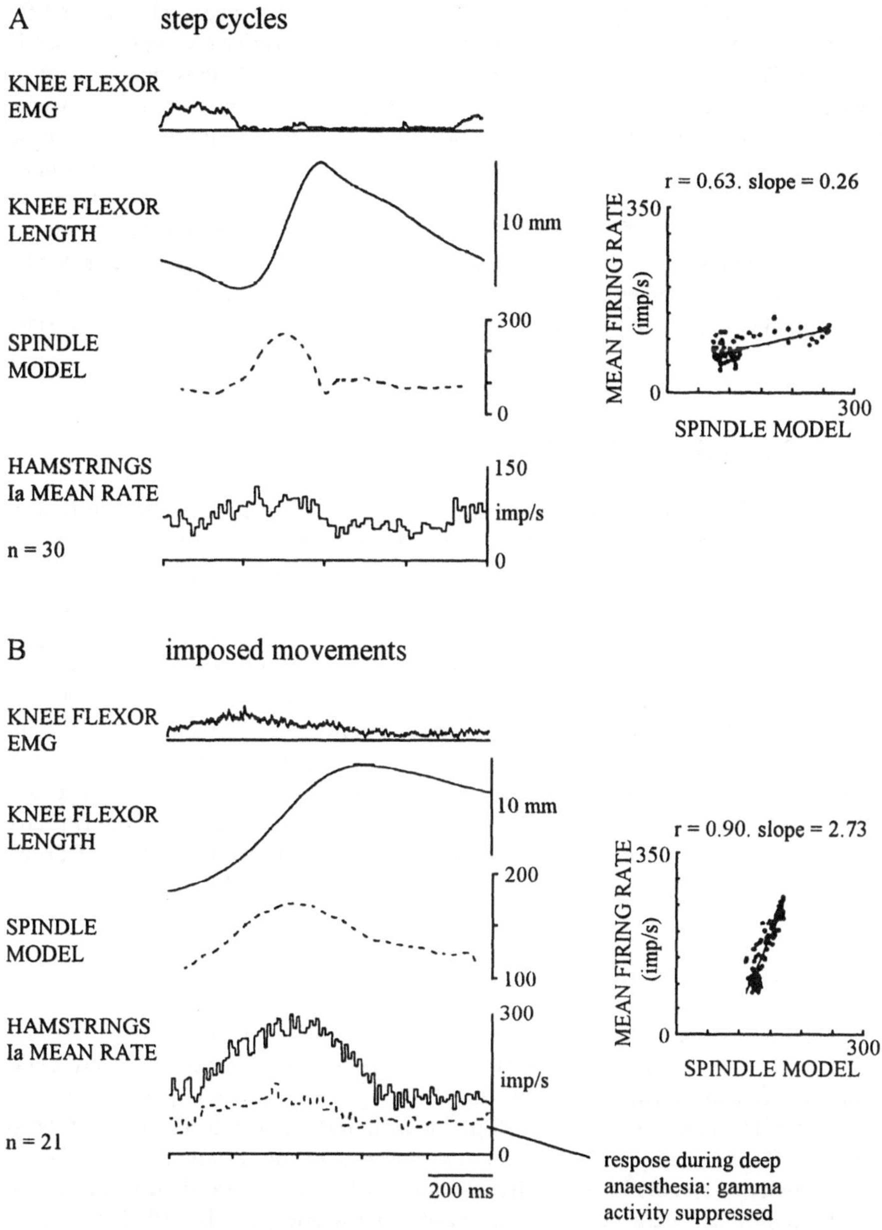

Figure 3. As in Figure 2, showing the difference in Ia length-sensitivity during stepping (A) compared to when slow movements were imposed on the cats' limbs (B). The slope of the regression line in the plot in B is much higher than that in A, indicating that in the imposed movements, there was substantial γ_d drive. The averaged responses of all the units to similar imposed stretches during deep anaesthesia, when γ action is suppressed, is shown as the bottom trace in B.

emphasized in the 1980s (e.g. Houk, Rymer & Crago, 1981) and this led to the formulation of non-linear models (Hasan, 1983; Schaafsma, Otten & van Willigen, 1991). However, these are not easy to implement in laboratories such as ours, so as a first approximation we used the large-signal transfer function for Ia afferents developed by Chen & Poppele (1978).

We digitally filtered the averaged length profiles in our ensemble step cycle data and found that the resultant profiles fitted the Ia ensemble firing data surprisingly well (e.g. see knee flexor data in Figures 2 and 3). Similar fits were obtained from Ia ensemble data of other muscle groups (Prochazka & Gorassini, in preparation). The difference or error between the filtered length signal and the afferent firing profile is also seen in Figure 2. There is some evidence in this error signal of inadequacies of the model, and deviations due to α-linked γ action, tendon stretch, muscle "bounce" and muscle slackening (Rack & Westbury, 1984; Hoffer, Caputi, Pose & Griffiths, 1989; Murphy, Stein & Taylor, 1984). Nonetheless, the errors are small enough that we are now applying the inverse of the Chen & Poppele (1978) transfer function to Ia profiles to obtain estimates of length changes of muscles inaccessible to length gauges such as extensor digitorum longus (Prochazka & Gorassini, in preparation).

NOVEL TASKS: FUSIMOTOR SET

Tasks involving novelty, difficulty or arousal tend to be associated with high Ia stretch-sensitivity in cats (Prochazka, Hulliger, Zangger & Appenteng, 1985). This implies task- and context-dependent γ_d action as a manifestation of preparatory set. The most reliable way to evoke the phenomenon in the conscious animal is to impose stretch on the receptor-bearing muscle. This presumably evokes increased alertness or wariness. The Ia firing rates can be quite remarkable: > 600 ips in some cases. In Figure 3B, the ensemble mean Ia firing rate during slow imposed movements reached 300 ips. When plotted against the signal obtained from the length profile using the Chen & Poppele (1978) model, the slope in the imposed movements was much higher than that in the step cycles (Figure 3A), indicating increased γ_d action in the imposed movements. This illustrates in the ensemble data a general finding obtained previously from single afferent data by linear modelling (Prochazka & Wand, 1981) and verified in acute "reconstruction" experiments (Hulliger, Dürmüller, Prochazka & Trend, 1989).

CEREBELLAR CONTROL OF FUSIMOTOR SYSTEM (?)

Recently we tested the hypothesis that fusimotor set is controlled from the cerebellum (Granit & Kaada, 1952, Gilman, 1969; Brooks & Thach, 1984). Recordings were obtained from spindle afferents in awake cats before and during a 5 - 10 minute period of cerebellar ataxia resulting from an injection of 2 - 6 μl lidocaine into the cerebellar nuclei (Gorassini, Prochazka & Taylor, 1993). We did not find any statistically significant changes in the stretch sensitivity of spindle Ia or II afferents during the ataxia. Task-dependent changes in inferred fusimotor set were not abnormal during the ataxic periods. All three cerebellar nuclei have now been tested, with similar negative results. Thus we conclude that the cerebellum is not primarily responsible for fusimotor control and that the ataxia we evoked was not attributable to aberrant fusimotor action.

CONDITIONAL REFLEX CONTROL

Linear control theory is useful in studying simple single-joint movements, particularly when combined with data on intrinsic muscle properties (Bennett, Gorassini & Prochazka, 1994). However, in natural motor sequences such as stepping, it has been known for some time that sensorimotor transmission changes in different phases of the locomotor cycle (e.g.

phase-dependent stumble reaction: Forssberg, Grillner & Rossignol, 1975). Pearson (1976) stated a rule by which the CNS locomotor pattern generator might switch control from stance phase to swing: "two conditions seem to be necessary for the swing phase to be initiated. First, the hip joint must be extended; second, the extensor muscles must be unloaded" (Rossignol, Grillner & Forssberg, 1975; Pearson & Duysens, 1976; Conway, Hultborn & Kiehn, 1987). In the last two years, our own "foot-in-hole" experiments in

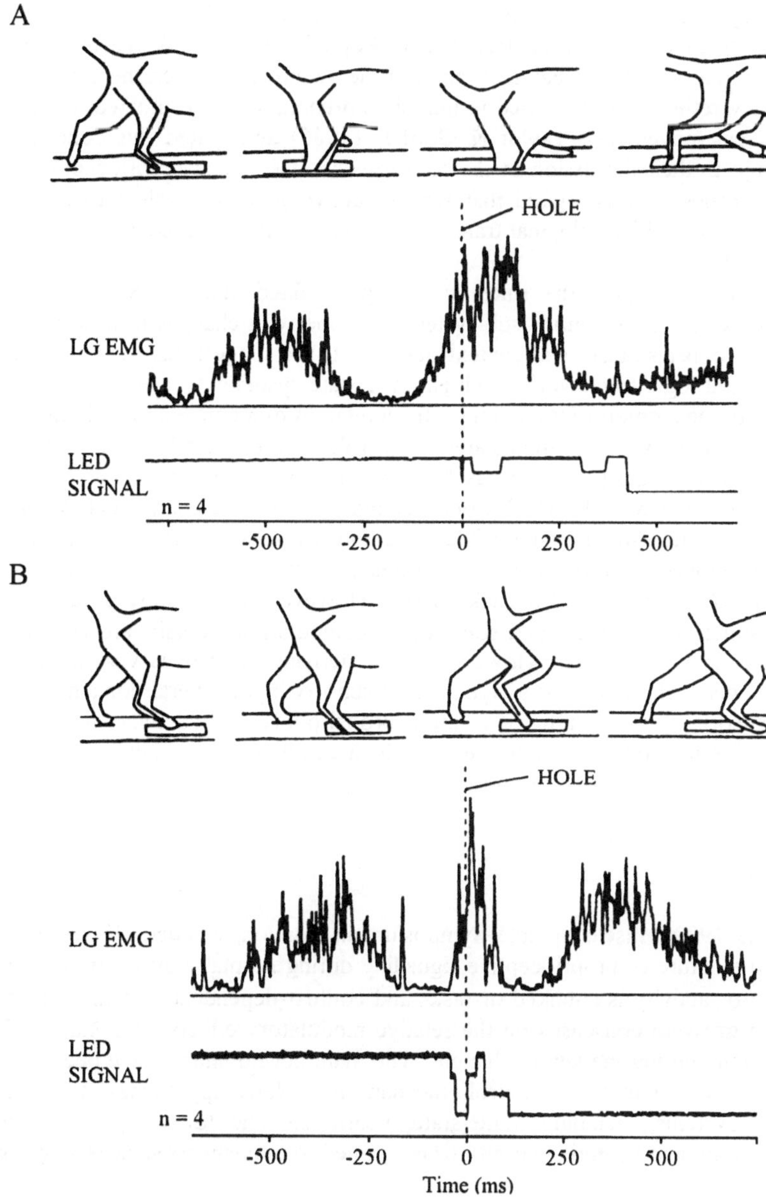

Figure 4. Averaged foot-in-hole trials in normal cats A) the contralateral paw had left the ground at the moment of foot entry into the hole. The ipsilateral leg then extends deep into the hole, with prolonged extensor activity. B) the contralateral leg is still in contact with the ground, and the ipsilateral foot is rapidly withdrawn from the hole. This reveals state-dependent conditional logic at work in the cat step cycle. Reproduced from Gorassini et al. (1994).

normal and spinal locomotor cats have revealed some additional rules related to speed of locomotion, the load-bearing state of the contralateral limb and descending adaptive control (Gorassini, Prochazka, Hiebert & Gauthier, 1994; Hiebert, Gorassini, Jiang, Prochazka & Pearson, 1994). For example, Figure 4 shows that in trials in which the contralateral paw has left the ground (A), the foot entering the hole stays in extension for much longer than if the contralateral leg still has ground support (B). Thus a more complete rule for slow gait might be: IF gait is slow AND contralateral limb is loaded AND extensor force is low AND hip is extending THEN initiate SWING, ELSE prolong stance. The use of IF-THEN rules, which derives from finite state control theory (Popovic, Tomovic & Tepavac, 1991), is rather different than the traditional approach to reflex control, in which the combined effect of short-latency reflex actions studied in immobile preparations is estimated, often without reference to real movements. Taylor & Gottlieb (1985) anticipated this conceptual shift some years ago in their discussion of the convergence of skin and proprioceptive input on spinal interneurones. It was argued that sensory convergence equivalent to AND or OR functions would allow kinesiological transitions to be identified and control to be switched from one variable to another.

We should not expect the rules to be rigidly fixed. The CNS clearly modifies, modulates or overrides its control sub-systems to cope with changes in task and context. Fusimotor control is also task- and context-dependent. We suggest that the time is now ripe to use conditional logic frameworks such as finite state systems, to make sense of reflexes and higher-level sensorimotor interactions. One limitation of a conventional finite-state rule-base should be acknowledged immediately. The rules are essentially all-or-none, and are triggered by precise transitions such as "soleus force has dropped below 5 N". This is unlikely to reflect the way the CNS works. Rather, we need a conditional system which allows for graded transitions and flexible weighting of sensory inputs. Fuzzy logic was specifically developed to provide these features and may serve as a useful framework initially (Kosko & Isaka, 1993; Prochazka, 1995). However, the structure of fuzzy logic is a little cumbersome at this stage, and some of its computational details are not necessarily comparable to neural processes. However, the underlying concept is very interesting for motor control neurophysiologists: assign to each sensory input several domains of context-dependent sensorimotor weighting (e.g. interneuronal pools) and combine the output of all of these domains to produce a motor result, which can be either continuous or threshold-based (Driankov, Hellendoorn & Reinfrank, 1993).

CONCLUSION

Since the 1980 Muscle Spindle Symposium in London, a much clearer picture has emerged of the nature of proprioceptive signalling during normal movement. The extent to which fusimotor activity is α-linked or task- and context-dependent remains controversial, but there is a growing consensus on the relative modulatory effects on spindle endings of fusimotor action and muscle length changes. The main developments have been in the area of analysing and identifying control mechanisms. New approaches to control in technological systems, including finite state, neural net and fuzzy logic structures are proving interesting and potentially useful when applied to movements such as locomotion.

ACKNOWLEDGEMENTS

We thank Janet Taylor, Keir Pearson, Gordon Hiebert and Michel Gauthier for their help. Funded by the Canadian MRC and the Alberta Heritage Foundation for Medical Research.

REFERENCES

BENNETT, D.J., GORASSINI, M. & PROCHAZKA, A. (1994) Catching a ball: contributions of intrinsic muscle stiffness, reflexes and higher-order responses. *Can. J. Physiol. Pharmacol.* in press.

BROOKS, V.B. & THACH, W.T. (1984) Cerebellar control of posture and movement. In: *Handbook of Physiol., The Nervous System II*, ed. BROOKS, V.B., pp. 877-946. American Physiological Society, Bethseda.

CHEN, W.J. & POPPELE, R.E. (1978) Small-signal analysis of response of mammalian muscle spindles with fusimotor stimulation and a comparison with large-signal properties. *J. Neurophysiol.* **41**, 15-27

CONWAY, B.A., HULTBORN, H. & KIEHN, O. (1987) Proprioceptive input resets central locomotor rhythm in the spinal cat. *Exp. Brain Res.* **68**, 643-656.

DRIANKOV, D., HELLENDOORN, H. & REINFRANK, M. (1993) *An Introduction to Fuzzy Control.* Springer, New York.

ELEK, J., PROCHAZKA, A., HULLIGER, M. & VINCENT, S. (1990) In-series compliance of gastrocnemius muscle in cat step cycle: do spindles signal origin-to-insertation length? *J. Physiol.* **429**, 237-258.

FORSSBERG, H. GRILLNER, S. & ROSSIGNOL, S. (1975) Phase dependent reflex reversal during walking in chronic spinal cats. *Brain Res.* **85**, 103-107.

GILMAN, S. (1969) The mechanism of cerebellar hypotonia. An experimental study in the monkey. *Brain* **92**, 621-638.

GORASSINI, M., PROCHAZKA, A. HIEBERT, G.W. & GAUTHIER, M. (1994) Adaptive responses to loss of ground support during walking. I. Intact cats. *J. Neurophysiol.* **71**, 603-610.

GORASSINI, M., PROCHAZKA, A. & TAYLOR, J. (1993) Cerebellar ataxia and muscle spindle sensitivity. *J. Neurophysiol.* **70**, 1853-1862.

GRANIT, R. & KAADA, B.R. (1952) Influence of stimulation of central nervous structures on muscle spindles in cat. *Acta physiol. scand.* **27**, 130-160.

HASAN, Z. (1983) A model of spindle afferent response to muscle stretch. *J. Neurophysiol.* **49**, 989-1006.

HERZOG,W., LEONARD,T.R. & GUIMARAES, A.C. (1993) Forces in gastrocnemius, soleus and plantaris tendons of the freely moving cat. *J. Biomech.* **26**, 945-953.

HIEBERT, G., GORASSINI, M., JIANG, W., PROCHAZKA, A. & PEARSON, K.G. (1994) Adaptive responses to loss of ground support during walking. II. Comparison of intact and chronic spinal cats. *J. Neurophysiol.* **71**, 611-622.

HOFFER, J.A., CAPUTI, A.A., POSE, I.E. & GRIFFITHS, R.I. (1989) Roles of muscle activity and load on the relationship between muscle spindle length and whole muscle length in the freely walking cat. *Prog. Brain Res.* **80**, 75-86.

HOUK, J.C. RYMER, W.Z. & CRAGO, P.E. (1981) Dependence of dynamic response of spindle receptors on muscle length and velocity. *J. Neurophysiol.* **46**, 143-166.

HULLIGER, M., DÜRMÜLLER, N., PROCHAZKA, A. & TREND, P. (1989) Flexible fusimotor control of muscle spindle feedback during a variety of natural movements. *Prog. Brain Res.* **80**, 87-102.

HULLIGER, M., MATTHEWS, P.B.C. & NOTH, J. (1977) Static and dynamic fusimotor stimulation on the response of Ia fibres to low frequency sinusoidal stretching of widely ranging amplitudes. *J. Physiol.* **267**, 811-838.

KOSKO, B. & ISAKA, S. (1993) Fuzzy Logic. *Sci. Amer.* **269**, 76-81.

MATTHEWS, P.B.C. & STEIN, R.B. (1969) The sensitivity of muscle spindle afferents to small sinusoidal changes of length. *J. Physiol.* **200**, 723-743.

MURPHY, P.R. STEIN, R.B. & TAYLOR, J. (1984) Phasic and tonic modulation of impulse rates in - motoneurons during locomotion in premammillary cats. *J. Neurophysiol.* **52**, 228-243.

PEARSON, K.G. (1976) The control of walking. *Sci. Amer.* **235**, 72-86.

PEARSON, K.G. & DUYSENS, J. (1976) Function of segmental reflexes in the control of stepping in cockroaches and cats. In *Neural Control of Locomotion*, eds. HERMAN, R.M., GRILLNER, S., STEIN, P.S.G. & STUART, D.G. pp 519-537. Plenum Press, New York.

POPOVIC, D., TOMOVIC R. & TEPAVAC, D. (1991) Control aspects of active above-knee prosthesis," Int. *J. Man-machine Studies.* **35**, 751-767.

PROCHAZKA, A. (1995) Proprioceptive feedback and movement regulation. In *American Handbook of Physiology. Section A. Neural Control of Movement.* ed. SMITH, J.L. In revision.

PROCHAZKA, A., HULLIGER, M., ZANGGER, P. & APPENTENG, K. (1985) "Fusimotor set": new evidence for α-independent control of γ-motoneurones during movement in the awake cat. *Brain Res.* **339**, 136-140.

PROCHAZKA, A., TREND, P., HULLIGER, M. & VINCENT, S. (1989) Ensemble proprioceptive activity in the cat step cycle: towards a representative look-up chart. *Prog. Brain Res.* **80**, 61-74.

PROCHAZKA, A. & WAND, P. (1981) Independence of fusimotor and skeletomotor systems during voluntary movement. In *Muscle Receptors and Movement.* eds. TAYLOR, A. & PROCHAZKA, A., pp. 229-243. Macmillan, London.

RACK, P.M.H. & WESTBURY, D. (1984) Elastic properties of the cat soleus tendon and their functional importance. *J. Physiol.* **347**, 479-495.

ROSSIGNOL, S. GRILLNER, S. & FORSSBERG, H. (1975) Factors of importance for the initiation of flexion during walking. *Soc. Neurosci. Abstr.* **5**, 181.

SCHAAFSMA, A. OTTEN, E. & VAN WILLIGEN, J.D. (1991) A muscle spindle model for primary afferent firing based on a simulation of intrafusal mechanical events. *J. Neurophysiol.* **65**, 1297-1312.

TAYLOR, A. & GOTTLIEB, S. (1985) Convergence of several sensory modalities in motor control. In *Feedback and Motor Control in Invertebrates and Vertebrates.* eds. BARNES, W.J.P. & GLADDEN, M.H. pp77-91. Croom Helm, London.

TAYLOR, A. & PROCHAZKA, A. (1981) *Muscle Receptors and Movement.* Macmillan, London.

WALMSLEY, B. HODGSON, J.A. & BURKE, R.E. (1978) Forces produced by medial gastrocnemius and soleus muscles during locomotion in freely moving cats. *J. Neurophysiol.* **41**, 1203-1216.

CRITIQUE: RULES FOR MOTOR CONTROL IN NATURAL MOVEMENTS

S.C. Gandevia

Prince of Wales Medical Research Institute
and University of New South Wales
Sydney 2031, Australia

The three papers in the final session deal differently with how the motor comand system delivers the output to motoneurones. They highlight the limitations that still remain in understanding the control of even simple movements. Vallbo & Wessberg describe the motoneuronal output during simple flexion and extension movements at the metacarpophalangeal joint - is the output continuous or does it contain 8-10 Hz steps? Pierrot-Deseilligny and colleagues ask how the propriospinal-like system acts during a simple elbow movement. Prochazka & Gorassini seek the global rules which govern the motor repertoire. What is the principle, "fuzzy" logic or otherwise, which produces simple movements as well as complex multijoint tasks such as locomotion and grasping?

Vallbo & Wessberg bring new data to an old problem. Finger movements at the metacarpophalangeal joint contain small irregularities which are just visible in the position records but clearly evident in the velocity and acceleration records. While the deviations from target position look small they are effectively encoded in the discharge of presumed primary and secondary muscle spindle endings. The arguments are those of exclusion: the phenomenon does not reflect the resonance of the arm, and the micro-movements are accompanied by appropriate bursts of EMG. However, I am not sure that the coherence between the movement and the level of agonist EMG is especially high. This coherence will give a clue as to the extent to which limb (and measurements) mechanics contribute. The authors suggest that the discontinuities are produced by "a series of motor pulses ... often double pulses with a braking one delivered to the antagonist". While this may be true, further analysis would be required to establish it.

The next step is to argue that while spindle afferents respond to the changes in velocity (discharging preferentially as the speed slows) the strengths of the proprioceptive reflexes are too weak for the spinal stretch reflex, or too slow for the long-latency reflex, to produce the irregularities. Several ways exist to pursue this further. First, if, as suggested, these irregularities appear with movements of various speeds and against different loads, then what happens when the movements occur at a shortening velocity which unloads the spindles (Al-Falahe, Nagaoka & Vallbo, 1990), or when intrafusal thixotropic effects are used to diminish the agonist spindle input (see Proske, Morgan & Gregory, 1993). This would help to rule out agonist spindle afferents. A further argument against them may be the temporal dispersion arising from variable intramuscular location of the receptors and the range of their conduction velocities. Secondly, are these motor pulses evident during contractions under isometric conditions? Tremor can develop prior to fatigue during tracking under isometric conditions, but it has a broader bandwidth (5-10 Hz) than that observed under non-isometric conditions (Gillies, 1994). If so, this argues for a peripheral component. Indirect evidence based on motoneurone firing rates during the acute abolition of spindle afferent facilitation suggests that muscle afferents provide a fair degree of motoneuronal support during isometric contractions (e.g. Macefield, Gandevia, Bigland-Ritchie, Gorman & Burke, 1993). Finally, the phenomenon could be sought in muscles from different parts of the body where both the biomechanics and reflex-loop times vary. In summary, while the case for the discontinuities is strong, the extent to which this is a subtle afferent modulation of a continuous output or a precisely timed intermittent controller remains to be dissected.

Turning to the propriospinal-like system. It is always prudent not to weigh the favourable evidence too much and the contrary evidence too little. However, there is now impressive evidence that a system exists in the upper cervical cord in humans which delivers some of the motor command to motoneurones. This pathway was proposed as a parallel to the C3 - C4 propriospinal system elucidated in the cat (Alstermark & Lundberg, 1992). It mediates non-monosynaptic muscle afferent excitation to various species of motoneurones (including wrist flexors and extensors, biceps, triceps and deltoid). Its location in the spinal cord has been inferred from the latency of effects in motoneuronal pools at different spinal levels. This system is facilitated at the onset of voluntary movement (Baldissera & Pierrot-Deseilligny, 1989) and more recently evidence has accumulated that a significant part of the command reaching motoneurones travels via this disynaptic route (Gracies, Meunier & Pierrot-Deseilligny, 1994; Burke, Gracies, Mazevet, Meunier & Pierrot-Deseilligny, 1994). The pattern of cutaneous inhibition of this system is topographically organized with inputs from the front and back of the hand producing differential effects (Nielsen & Pierrot-Deseilligny, 1991; Burke, Gracies, Mazevet, Meunier & Pierrot-Deseilligny, 1992). This pattern has led to the studies here which aimed to support the contention that the system is organized to curtail voluntary movement. The authors feel that this function should be left to the spinal cord since "it must be one of the most difficult parameters for the brain to calculate". Although hardly sound in logic, this argument must please those working on purely spinal mechanisms! One could ask facetiously whether the initiation of movement should be left to the spinal cord as well!

What specific issues arise from these results? First, the peripheral input from the whole arm must differ when it is flexed compared with extended. However, we are expected to accept that the increased depression of the cutaneous inhibition at the end of the movement purely results only from decreased excitation of the propriospinal neurones (or increased excitation of the interneurones mediating the inhibition of the propriospinal neurones). The extent of this problem is not so simple to exclude because of the difficulty of checking for post-synaptic events at the motoneurone *during* movement. Secondly, transcranial magnetic stimulation is used to help assess the timing of the propriospinal inhibition. The difficulty here is the complex multi-peaked descending volley dispersed over several milliseconds produced by suprathreshold stimulation (e.g. Burke, Hicks, Gandevia, Stephen, Woodforth & Crawford, 1993). Thirdly, the proposal that this system is organized to curtail movement following a cutaneous cue, must take into account that tasks may require a continuation of movement despite the cue. At least for the hand the input from the glabrous skin may favour a grasp (e.g. Gandevia & McCloskey, 1976; Johansson & Westling, 1984). Even skin in the superficial radial territory can facilitate low-threshold motoneurones involved in flexion of the thumb (Wiranski, Kilbreath & Gandevia, 1994). Finally, the motor unit firing frequency - force relation shows hysteresis, so that firing rates will probably be lower in the falling rather than the rising phase of the force during the movement. If so, the same inhibitory input might produce an apparently larger effect in the ongoing EMG.

Overall, while the parallels with the system of propriospinal neurones in the cat are strong and attractive, the human arm makes not only reaching movements but also relatively independent movements of the digits even with the extrinsic muscles (such as flexor digitorum profundus, Kilbreath & Gandevia, 1994). Future studies may well consider the operation of comparable interneuronal organizations for muscles moving the digits.

How can we begin to make sense of the supposed logic systems that underlie voluntary movement? Prochazka & Gorassini point out several critical constraints. First, if it is accepted that something like locomotion can be broken down into fragments to look for the "finite state rules" governing the system, then certain things follow inevitably. The "*if* this happens, *then* produce that response" rule is another way of stating that the system behaves in a task-dependent way. Locomotor responses vary with the speed of movement and load-bearing, but there are presumably only a finite set of basic responses which can be modified according to circumstances in which, for example, the cat finds its "foot in a hole". An element of short-term adaptation and prediction, and ultimately longer term memory is crucial here. The most appropriate behaviour is not to put the foot in the hole in the first place!

Since, the Muscle Spindle Symposium in London in 1980, the term 'task-dependent' has become widely used and abused. It encapsulates the fullness of the motor repertoire but should not be allowed to overshadow what is actually going on centrally. Thus, the "set" of supraspinal structures, propriospinal paths and the spinal machinery will differ, but the peripheral input will also differ. Although difficult, the role of these four (at least) levels must be teased out. In this regard, the results of Prochazka and colleagues argue convincingly against the view that the cerebellum produces the full repetoire of fusimotor outputs.

One intriguing aspect of the debates on the roles of muscle afferents in movement control is the possibility of differences between cats and humans. Flexibility in alpha-gamma linkage is present in human movement, although it is probably difficult to show its full colours within the limits imposed by microneurographic recordings. There is now evidence that the dissociation can occur in humans during standing (Aniss, Diener, Hore, Burke & Gandevia, 1990) and in relaxed upper limb muscles (Gandevia, Wilson, Cordo & Burke, 1994). However, it is not clear why the discharge frequencies of human spindle endings are so low: the averages for slow movements may be well under 10 Hz. The role of tendon and muscle compliance and effectively "in series" connections may be critical here. It is also notable that Golgi tendon organs discharge at lower rates in human recordings.

The papers in this section all push out the boundaries of motor control and all rely on argument by exclusion: Vallbo & Wessberg argue for discontinuous motor outputs because they cannot be explained by an afferent mechanism. Prochazka & Gorassini argue for the need to develop alternative logic approaches to complex movements, because the approach derived from the single joint behaviour is artificial and insufficient. Pierrot-Deseilligny and colleagues point out the limitations of direct cortical control of motoneurones and thus force us to examine the detailed properties of indirect disynaptic systems controlling motoneurones. Intuition is notoriously deceptive but I suspect that these three approaches will still receive acclamation at the next Muscle Spindle Symposium. If confusion still remains, then at least it will be on a higher plane.

REFERENCES

AL-FALAHE, N.A., NAGAOKA, M. & VALLBO, Å.B. (1990) Response profiles of human muscle afferents during active finger movements. *Brain* **113**, 325-346.

ALSTERMARK, B & LUNDERBERG, A.. (1992) The C3-4 propriospinal system: target-reaching and food-taking. In *Muscle Afferents and Spinal Control of Movement.* eds. JAMI, L., PIERROT-DESEILLIGNY, E. & ZYTNICKI, D. pp. 327-354. Pergamon. London.

ANISS, A.M., DIENER, H.-C., HORE, J., BURKE, D. & GANDEVIA, S.C. (1990) Reflex activation of muscle spindles in human pretibial muscles during standing. *J. Neurophysiol.* **64**, 671-679.

BALDISSERA, F. & PIERROT-DESEILLIGNY, E. (1989) Facilitation of transmission in the pathway of non-monosynaptic Ia excitation to wrist motoneurones at the onset of voluntary oveent in man. *Exp. Brain Res.* **56**, 308-322.

BURKE, D., GRACIES, J.M., MAZEVET, D., MEUNIER, S. & PIERROT-DESEILLIGNY, E. (1992) Convergnece of descending and various peripheral inputs onto common propriospinal-like neurones in man. *J. Physiol.* **449**, 655-671.

BURKE, D., GRACIES, J.M., MAZEVET, D., MEUNIER, S. & PIERROT-DESEILLIGNY, E. (1994) Non-monosynaptic transmission of the cortical command for voluntary movement in man. *J. Physiol.* **480**, 191-202

BURKE, D., HICKS, R., GANDEVIA, S.C., STEPHEN, J., WOODFORTH, I. & CRAWFORD, M.. (1993) Direct comparison of corticospinal volleys in human subjects to transcranial magnetic and electrical stimulation. *J. Physiol.* **470**, 383-393.

GANDEVIA, S.C. & McCLOSKEY, D.I. (1976) Perceived heaviness of lifted objects and effects of sensory inputs from related non-lifting parts. *Brain Res.* **109**, 399-401.

GANDEVIA, S.C., WILSON, L.R., CORDO, P.J.& BURKE, D. (1994). Fusimotor reflexes in relaxed forearm muscles produced by cutaneous afferents from the human hand. *J. Physiol.* **479**, 499-508.

GILLIES, J.D. (1994). Measurement of physiological and essential tremor. In *Science and Practice of Clinical Neurology*. eds. GANDEVIA, S.C., BURKE, D. & ANTHONY, M. pp. 191-201. Cambridge University Press. Cambridge.

GRACIES, J.M., MEUNIER, S. & PIERROT-DESEILLIGNY, E. (1994) Evidence for corticospinal excitation of human propriospinal-like neurones. *J. Physiol.* **475**, 509-518.

HULLIGER, M. (1984) The mammalian muscle spindle and its central control. *Rev. Physiol. Biochem. Pharmacol.* **101**, 1-111.

JOHANSSON, R.S. & WESTLING, G. (1984) Roles of glabrous skin receptors and sensorimotor memory in automatic control of precision grip when lifting rougher or more slippery objects. *Exp. Brain Res.* **56**, 550-564.

KILBREATH, S.L. & GANDEVIA, S.C. (1994). Limited independent flexion of the thumb and fingers in human subjects. *J. Physiol.* **479**, 487-497

MACEFIELD, V.G., GANDEVIA, S.C., BIGLAND-RITCHIE, B., GORMAN, B.R. & BURKE, D. (1993). The firing rates of human motoneurones voluntarily activated in the absence of muscle afferent feedback. *J. Physiol.* **471**, 429-443.

NEILSON, J. & PIERROT-DESEILLIGNY, E. (1991) Pattern of cutaneous inhibition of the propriospinal-like excitation to human upper limb motoneurones. *J. Physiol.* **434**, 169-182.

PROSKE, U., MORGAN, D.L. & GREGORY, J.E. (1993). Thixotropy in skeletal muscle and in muscle spindles: a review. *Prog Neurobiol.* **41**, 705-721.

WIRANSKI, A., KILBREATH, S.L. & GANDEVIA, S.C. (1994). Is accuracy dependent on cutaneous afferents and/or co-contraction when weights are lifted with flexor pollicis longus? *Proc. Aust. Neurosci. Soc.* **5**, 223.

SOME EVIDENCE FOR A MID-THORACIC NUCLEUS IN THE HUMAN MOTOR PATHWAY

B.A. Taylor[1], M.C. Ridding[2]
and J.C. Rothwell[2]

[1]Royal National Orthopaedic Hospital
Brockley Hill, Stanmore
Middlesex HA7 4LP, UK
[2]MRC Human Movement and Balance Unit
Institute of Neurology, Queen Square
London WC1N 3BG, UK

During surgery for both spinal cord injury and spinal deformity, evoked potential techniques have been developed to reduce the risk of iatrogenic spinal cord damage. Initial techniques used somatosensory recording. Motor pathway recording has been more difficult and only recently developed. The technique in this paper involves stimulation of the upper thoracic cord from the posterior epidural space with a bipolar catheter type electrode. Recordings were made from lower limb muscles, commonly gastrocnemius, using surface or concentric neddle electrodes. Two key areas were found to contribute to the success of recording during surgery, namely indentification of appropriate anaesthetic agents, and the method of spinal cord stimulation.

This study describes systematic intra-operative observations in ten patients undergoing surgery for spinal deformity. The most important observations were:

1. Lower limb EMG responses could not be evoked by stimulating the spinal cord above vertebral level T8 with a single stimulus of any intensity.
2. Stimulation of the spinal cord below vertebral level T8 with a single stimulus evoked large lower limb EMG responses.
3. Stimulation above T8 with paired pulses of lower total electrical energy and of appropriate inter-stimulus interval evoked large EMG responses from the lower limbs.
4. Stimulation at successive vertebral segmental levels between T3 and L2, maintaining stimulus and recording parameters (see Figure 1), revealed a step change in pathway characteristics at T8.
5. Lower limb evoked reponses to paired pulse stimulation in the upper thoracic cord were sensitive to the anaesthetic agents nitrous oxide and isofluorane, and virtually insensitive to the effects of Propafol.

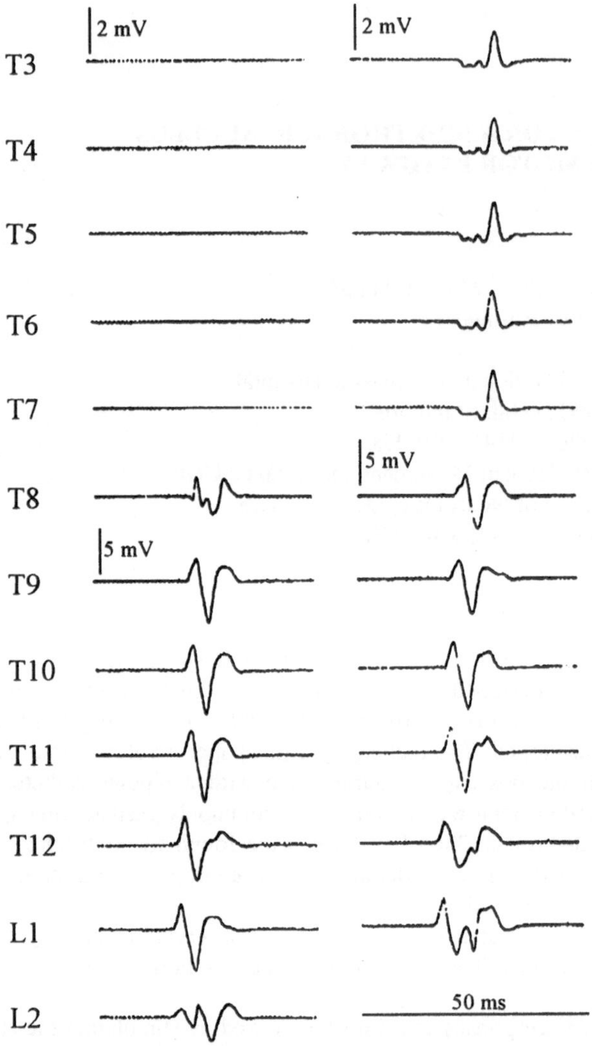

Figure 1. Spinal motor evoked potentials recorded with a concentric needle electrode from the medial head of gastrocnemius in response to single pulse (left traces) or double pulse (right traces) bipolar stimulation of the spinal cord at each vertebral level from T3 to L2. Stimulus intensity 50V, pulse width 0.5 ms, interstimulus interval for double pulse stimulation 2 ms. Each record is a single sweep, negative up. Intra-operative recordings during posterior spinal fusion for adolescent idiopathic scoliosis. No response to single stimuli above T8, maximal responses below T8. Maximal responses above T8 with dual stimulus with a different form below T8.

DISCUSSION

These observations are consistent with the presence of a spinal pathway which projects from the T8 level to motoneurones of the gastrocnemius muscle. This pathway may be either interposed in the projection from more proximal spinal segments or separate from it. If it were interposed, a single proximal cord stimulus might be incapable of discharging motoneurones through two synaptic relays (at T8 and at the motoneurones), whereas the temporal summation provided by paired volleys would be sufficient. Single stimuli at T8 or below might readily produce motoneuronal activity via only a single synapse (at the motoneurone).

If the pathway at T8 were separate from the proximal descending routes, then it may be that a single volley in the long spinal pathway would be insufficient to discharge the motoneurone pool, whereas a pair of volleys might summate at the motoneurone synapse. Stimulation at T8 or below might recruit a sufficiently large descending volley because of the additional activity in a local link from T8.

Alternatively, it is possible that antidromic conduction of Ia afferents, activation of the monosynaptic reflex and consequent motoneurone firing is responsible for the success of single stimuli at and below T8. An interesting observation has been that the apparent localisation to the T8 level of the change in conduction characteristics is shared by several target muscles studied, namely quadriceps, tibialis anterior and gastrocnemius.

It is interesting to speculate on the nature of the local pathway and the concentration of function at the T8 vertebral level. It is possible that this is similar to the interneuronal mechanism described by Jankowska (1989) in the cat, and that it may in some way be important for integration of gait patterns in man.

REFERENCE

JANKOWSKA, E. (1989) A neuronal system of movement control via muscle spindle secondaries. *Prog. Brain Res.* **90**, 299-303.

AUTHOR INDEX

* Symposium Attendee

SUBJECT INDEX